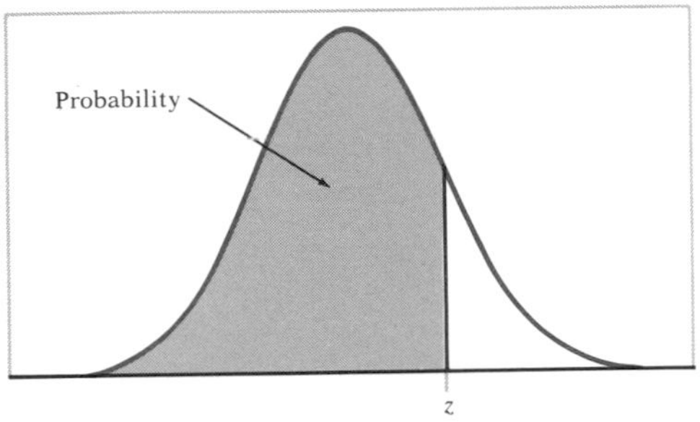

Table entry is probability at or below z.

Table A *(Continued)*

z	.00	.01	.02	.03	.04	.05	.06	.07	.08	.09
0.0	.5000	.5040	.5080	.5120	.5160	.5199	.5239	.5279	.5319	.5359
0.1	.5398	.5438	.5478	.5517	.5557	.5596	.5636	.5675	.5714	.5753
0.2	.5793	.5832	.5871	.5910	.5948	.5987	.6026	.6064	.6103	.6141
0.3	.6179	.6217	.6255	.6293	.6331	.6368	.6406	.6443	.6480	.6517
0.4	.6554	.6591	.6628	.6664	.6700	.6736	.6772	.6808	.6844	.6879
0.5	.6915	.6950	.6985	.7019	.7054	.7088	.7123	.7157	.7190	.7224
0.6	.7257	.7291	.7324	.7357	.7389	.7422	.7454	.7486	.7517	.7549
0.7	.7580	.7611	.7642	.7673	.7704	.7734	.7764	.7794	.7823	.7852
0.8	.7881	.7910	.7939	.7967	.7995	.8023	.8051	.8078	.8106	.8133
0.9	.8159	.8186	.8212	.8238	.8264	.8289	.8315	.8340	.8365	.8389
1.0	.8413	.8438	.8461	.8485	.8508	.8531	.8554	.8577	.8599	.8621
1.1	.8643	.8665	.8686	.8708	.8729	.8749	.8770	.8790	.8810	.8830
1.2	.8849	.8869	.8888	.8907	.8925	.8944	.8962	.8980	.8997	.9015
1.3	.9032	.9049	.9066	.9082	.9099	.9115	.9131	.9147	.9162	.9177
1.4	.9192	.9207	.9222	.9236	.9251	.9265	.9279	.9292	.9306	.9319
1.5	.9332	.9345	.9357	.9370	.9382	.9394	.9406	.9418	.9429	.9441
1.6	.9452	.9463	.9474	.9484	.9495	.9505	.9515	.9525	.9535	.9545
1.7	.9554	.9564	.9573	.9582	.9591	.9599	.9608	.9616	.9625	.9633
1.8	.9641	.9649	.9656	.9664	.9671	.9678	.9686	.9693	.9699	.9706
1.9	.9713	.9719	.9726	.9732	.9738	.9744	.9750	.9756	.9761	.9767
2.0	.9772	.9778	.9783	.9788	.9793	.9798	.9803	.9808	.9812	.9817
2.1	.9821	.9826	.9830	.9834	.9838	.9842	.9846	.9850	.9854	.9857
2.2	.9861	.9864	.9868	.9871	.9875	.9878	.9881	.9884	.9887	.9890
2.3	.9893	.9896	.9898	.9901	.9904	.9906	.9909	.9911	.9913	.9916
2.4	.9918	.9920	.9922	.9925	.9927	.9929	.9931	.9932	.9934	.9936
2.5	.9938	.9940	.9941	.9943	.9945	.9946	.9948	.9949	.9951	.9952
2.6	.9953	.9955	.9956	.9957	.9959	.9960	.9961	.9962	.9963	.9964
2.7	.9965	.9966	.9967	.9968	.9969	.9970	.9971	.9972	.9973	.9974
2.8	.9974	.9975	.9976	.9977	.9977	.9978	.9979	.9979	.9980	.9981
2.9	.9981	.9982	.9982	.9983	.9984	.9984	.9985	.9985	.9986	.9986
3.0	.9987	.9987	.9987	.9988	.9988	.9989	.9989	.9989	.9990	.9990
3.1	.9990	.9991	.9991	.9991	.9992	.9992	.9992	.9992	.9993	.9993
3.2	.9993	.9993	.9994	.9994	.9994	.9994	.9994	.9995	.9995	.9995
3.3	.9995	.9995	.9995	.9996	.9996	.9996	.9996	.9996	.9996	.9997
3.4	.9997	.9997	.9997	.9997	.9997	.9997	.9997	.9997	.9997	.9998

Introduction to the Practice of Statistics

Introduction to the Practice of Statistics

• • •

SECOND EDITION

David S. Moore
George P. McCabe
PURDUE UNIVERSITY

W. H. FREEMAN AND COMPANY
NEW YORK

Cover illustration by Salem Krieger

Library of Congress Cataloging in Publication Data

Moore, David S.
Introduction to the practice of statistics / David S. Moore, George P. McCabe—2nd ed.
p. cm.
Includes bibliographical references.
ISBN 0-7167-2250-X
1. Mathematical statistics. I. McCabe, George P. II. Title.
QA276. 12.M65 1993
519.5—dc20 92-22880
CIP

Printed in the United States of America

2 3 4 5 6 7 8 9 0 **MB** 9 9 8 7 6 5 4 3

Contents

Starred sections are optional.

Preface

We are pleased that so many students and teachers have been receptive to a text that attempts to focus on data and on statistical reasoning. Our second edition has nonetheless been revised more thoroughly than many new editions, though without altering the substance and style of the text. In this preface we will first describe our overall philosophy and then discuss the changes that appear in the new edition.

Introduction to the Practice of Statistics is an elementary but serious introduction to modern statistics for general college audiences. This book is elementary in the level of mathematics required and in the statistical procedures presented. It is serious because our aim is to help readers think about data and use statistical methods with understanding. Students need only a working knowledge of algebra; that is, they must be able to read and use formulas without a detailed explanation of each step.

Statistics is interesting and useful because it is a means of using data to gain insight into real problems. As the continuing revolution in computing relieves the burden of calculating and graphing, an emphasis on statistical concepts and on insight from data becomes both more important and more practical. We have seen many statistical mistakes, but few that involved simply getting a calculation wrong. We therefore ask students to think about the background of the data, the design of the study that produced the data, the possible effect of outlying observations on their conclusions, and the reasoning that lies behind standard methods of inference. Users of statistics who form these habits from the beginning are well prepared to learn and use more advanced methods.

The title of the book expresses our intent to introduce readers to statistics as it is used in practice. Statistics in practice is concerned with gaining understanding from data; it is focused on problem solving rather than on methods that may be useful in specific settings. A text cannot fully imitate practice, because it must teach specific methods in a logical order and must use data that are not the reader's own. Nonetheless, our interest and experience in applying statistics have influenced the nature of this book in several ways.

Focus on Statistical Reasoning and Data We share the emerging consensus among statisticians that statistical education should focus on data and on statistical reasoning rather than on either the presentation of as many methods as possible or the mathematical theory of inference. Understanding statistical reasoning should be the most important objective of any reader. We attempt to present statistics as a coherent discipline with important modes of thought that recur in many specific settings. The first two chapters, for example, concern the art of organizing and exploring data. We hope that readers will draw from these chapters some strategies for understanding data and not just a kit of useful tools. Later chapters similarly attempt to make clear the fundamental modes of thought in designs for data production and in the probabilistic reasoning of formal inference.

An emphasis on data accompanies the emphasis on conceptual understanding. Not all numbers are data. The number 10.3 alone is meaningless; it acquires meaning when we are told that it is the birth weight of a child in pounds or the percent of teenagers who are unemployed. Because context makes numbers meaningful, our examples and exercises are presented in the context of real-world problems. We often comment on issues of statistical practice raised by particular examples. The data presented in examples and in exercises are mostly real and, even when not, are based on real problems. Many of these problems are drawn from those brought to the statistical consulting service at Purdue University by students and faculty from many disciplines. We hope that the presence of background information, even in exercises intended for routine drill, will encourage readers to consider the meaning of their calculations as well as the calculations themselves.

Computers and Statistical Calculations Statistical calculations are in practice performed by software packages on a computer. Many instructors make use of statistical software in the first course. We use either Minitab or SAS in our own teaching, the choice depending on the needs of the students; many other packages are equally satisfactory. We have included some topics that reflect the dominance of software in practice, such as the interpretation of normal quantile plots and an explanation of the two-sample t statistic with approximate degrees of freedom. Some exercises require graphs and calculations that are tedious without a computer, and others present computer output as a basis for further work. But we have been careful to make the book easily usable by students without access to computing facilities. A scientific calculator with statistical functions is helpful, but any calculator that will give the mean and standard deviation from keyed-in data is adequate.

Judgment in Statistics Statistics in practice requires judgment. It is easy to list the mathematical assumptions that justify use of a particular procedure, but not always easy to decide when the procedure can be used in practice. Because judgment is developed by experience, an introductory course should present firm guidelines and not make unreasonable demands on the judgment of students. We have given guidelines—for example on using the t procedures for means and avoiding the F procedures for variances—that we follow ourselves. Although we have cited the literature on which our recommendations rest, we recognize that other statisticians may follow different guidelines. We prefer to give imperfect rules rather than avoid the issue of when procedures are in fact useful in analyzing real problems. Similarly, some exercises require the use of judgment in addition to right-or-wrong calculations and conclusions. Both students and teachers should recognize that not every part of every exercise has a single correct answer. We enjoy and encourage classroom discussion of questions of interpretation. But, as is appropriate in a first course, most exercises are straightforward even at the cost of some oversimplification.

Teaching Experiences In writing this book, we drew on our experience with two groups of students. In teaching general undergraduates from a variety of disciplines, we cover the first nine chapters omitting all but a very few of the starred sections. This results in a modern introduction to basic statistics that is quite standard in content, though with some reordering of material and with a strong emphasis on data and reasoning. The Annenberg/Corporation for Public Broadcasting telecourse *Against All Odds: Inside Statistics* uses the text in this way. Our second group of students are advanced undergraduates and beginning graduate students in such fields as education, social sciences, and health sciences. These students are more scientifically sophisticated than general undergraduates, though not more mathematically advanced. In teaching this second group, we cover nearly the entire text except for Sections 4.4 and 10.2. The starred sections therefore help distinguish between two somewhat different courses that can be taught from this book. Of course, any starred section can be omitted without impeding the reading of later chapters.

The Second Edition

The focus of the changes in this Second Edition is to make the book more accessible to students and more teachable for instructors. In undertaking the revision, we have been guided by our own teaching experience and by comments from many students and teachers. There are numerous small improvements in the writing, simpler notation where possible, and several interesting new real data sets. More important are several major changes:

- Readers will find a substantial reorganization early in the book. Chapter 2 of the first edition, which dealt with data on a variable changing over time, has vanished. Because most teachers were treating these topics only lightly, they are now integrated with other material, resulting in a clearer path for students. The most important improvement is a single unified treatment of fitting lines to data early in the new Chapter 2. Plots against time now appear in Chapter 1 as part of data analysis for a single variable; exponential growth is an optional topic in Chapter 2, as an example of transforming data to obtain linearity; and statistical control is discussed in Chapter 5 along with $\bar{x}$ control charts. As a result, Chapters 1 and 2 present a more straightforward introduction to the tools and tactics of data analysis.
- We have modified the presentation of probability in Chapter 4 at the two points at which students encountered the most difficulty: The introduction of random variables is now delayed until the second section to allow better digestion of probability basics, and the exposition of the rules for means and variances of random variables has been slowed for easier comprehension. In addition, Bayes's rule appears in an optional section for those who need it. In Chapter 5, we make a clearer distinction between sample counts and sample proportions in discussing sampling distributions.
- We have completely rewritten the presentations of the chi-square test for two-way tables in Section 8.3 and of inference in the simple linear regression setting in Section 9.1. Students will now find a more direct path to the main results, with refinements placed in later optional subsections. Instructors and students will enjoy new data sets such as the lean of the Tower of Pisa over time and Florida manatees killed by power boats against boat registrations.
- We tried to help teachers who (as we recommend) use software. A data disk available in several formats contains all large and many moderate-size data sets. Exercises that are feasible only if students use software are separated out as "computer exercises" at the end of each chapter. Many of these exercises are new. A completely new Minitab Guide is available.

The practice of even elementary statistics is more intellectually challenging than is suggested by the limited mathematics that is required.

We present real or realistic data and at least hint at the complexities of statistics in practice; we regularly ask students to draw a substantial conclusion rather than simply report the result of a calculation, and we try to explain the reasoning (but not the mathematics) behind recipes. Most students and teachers report that the added effort is rewarded by added understanding. In this new edition we have done our best to remove unnecessary challenges. There is, to paraphrase Euclid's advice to King Ptolemy of Egypt long ago, no royal road to learning. We hope that this road to learning is now a bit smoother.

Supplements

Several supplements are available free to adopters of the Second Edition of *Introduction to the Practice of Statistics:*

- The *Instructor's Guide* contains the following: overviews and teaching suggestions for each chapter; suggested excerpts from the telecourse *Against All Odds* and tips for using these videos in a nontelecourse environment; sample examinations with solutions; and worked-out solutions to all exercises.
- A *data disk,* available in several formats, enables instructors to easily enter all our larger data sets and many moderate-size data sets into their local computer system.
- A *test bank,* available both in computerized and printed versions, for generating quizzes and exams.
- *Transparency masters* of many of the text's figures and tables.

In addition, students may purchase a *Minitab Guide,* written by Betsy Greenberg and Mark Serva of the University of Texas, Austin, that accompanies the text and gives detailed instructions in using the Minitab statistical software.

We have not developed software specifically to accompany this text because we strongly encourage the use of professionally written software, which is both more capable and more transportable to other settings. W. H. Freeman and Company offers a special discount on the student version of *Data Desk, Learning Data Analysis with Data Desk,* by Paul F. Velleman, to users of this text.

The Telecourse *Against All Odds*

Against All Odds: Inside Statistics is a 26-program telecourse on statistics and its applications sponsored by the Annenberg/Corporation for Public Broadcasting Project. This telecourse, developed by David S.

Moore, offers a visual introduction to modern statistics that closely parallels *Introduction to the Practice of Statistics*. In addition to broadcast of the series by public television stations, the programs are inexpensively available on videotape to individuals and institutions. Instructors who do not follow the telecourse format may find excerpts from this unique video series valuable as supplements to classroom instruction. More information and demonstration videotapes can be obtained by telephoning 1-800-LEARNER.

Two supplements are available that link the television programs to this text.

- The *Telecourse Faculty Guide* contains an overview of the material in each video program and provides detailed advice on implementing a telecourse.

- The *Telecourse Study Guide*, available for sale to students, guides students in their study of the course material. Each unit in the study guide corresponds to a video program, and provides an overview of the content, learning objectives, assigned reading and exercises from this text, and self-test questions with fully worked solutions. The Study Guide also contains sample examinations with worked solutions. Instructors who are not using the video material may find the large number of additional exercises in the Study Guide useful.

Acknowledgments

We are grateful to colleagues who commented on the manuscript and to students who studied from it. In particular, we would like to thank the following colleagues who offered specific comments on the second edition:

Richard Berk, UCLA
Trudy Ann Cameron, UCLA
Philip B. Ender, UCLA
Eugene A. Enneking, Portland State University
James Finch, University of San Francisco
Evan Fisher, Lafayette College
Chris Franklin, University of Georgia
Chris Freiling, UCLA
Gavin G. Gregory, University of Texas at El Paso
Pete Herron, Suffolk County Community College
Piet de Jong, University of British Columbia
Ita G. G. Kreft, UCLA

N. T. Longford, Educational Testing Service
J. David Mason, University of Utah
David Meyers, Bakersfield College
Cheryl McKeeman, Vancouver Community College
Mary R. Parker, Austin Community College
Clifford O. Pope, Atlantic Union College
Thomas P. Ryan, Texas A & M University
W. Robert Stephenson, Iowa State University
Frances P. Stewart, University of South Carolina
George Sturm, Grand Valley State University
Nola D. Tracy, McNeese State University
Roy H. Williams, Memphis State University
Bob Woodle, Jamestown College
Mary Rising, University of Louisville

We would also like to acknowledge the ongoing influence of reviewers of the first edition:

Rudolf J. Freund, Texas A & M University
Bernard Harris, University of Wisconsin, Madison
David Hildebrand, University of Pennsylvania
Lambert Koopmans, University of New Mexico
David R. Lund, University of Wisconsin, Eau Claire
Donald E. Ramirez, University of Virginia
Charles W. Sinclair, Portland State University
Jessica Utts, University of California, Davis
Robert Wardrop, University of Wisconsin, Madison
Anne Watkins, Los Angeles Pierce College
Jeffrey Witmer, Oberlin College
Mary Sue Younger, University of Tennessee, Knoxville
Douglas Zahn, Florida State University

Most of all, we are grateful to the many people in varied disciplines and occupations with whom we have worked to gain understanding from data. They provided both material for this book and the experience that enabled us to write it. Perhaps even more important, working with people from many fields has constantly reminded us of the importance of statistical fundamentals in an age when computer routines and professional advice quickly handle detailed questions. If the publisher would allow it, we would call this book "What you should know before you talk to a statistician." We hope that users and potential users of statistics will find it helpful.

David S. Moore

George P. McCabe

Introduction: What Is Statistics?

Statistics is the science of collecting, organizing, and interpreting numerical facts, which we call *data*. We are bombarded by data in our everyday life. Most of us associate "statistics" with the bits of data that appear in news reports: baseball batting averages, imported car sales, the latest poll of the president's popularity, and the average high temperature for today's date. Advertisements often claim that data show the superiority of the advertiser's product. All sides in public debates about economics, education, and social policy argue from data. Yet the usefulness of statistics goes far beyond these everyday examples.

The study and collection of data are important in the work of many professions, so that training in the science of statistics is valuable preparation for a variety of careers. Each month, for example, government statistical offices release the latest numerical information on unemployment and inflation. Economists and financial advisors as well as policy makers in government and business study these data in order to make informed decisions. Doctors must understand the origin and trustworthiness of the data that appear in medical journals if they are to offer their patients the most effective treatment. Politicians rely on data from polls of public opinion. Business decisions are based on market research data that reveal consumer tastes. Farmers study data from field trials of new crop varieties. Engineers gather data on the quality and reliability of manufactured products. Most areas of academic study make use of numbers, and therefore also make use of the methods of statistics.

We can no more escape data than we can avoid the use of words. Just as words on a page are meaningless to the illiterate or confusing to the partially educated, so data do not interpret themselves but must be read with understanding. Just as a writer can arrange words into convincing arguments or incoherent nonsense, so data can be compelling, misleading, or simply irrelevant. Numerical literacy, the ability to follow and understand numerical arguments, is important for everyone. The ability to express yourself numerically, to be an author rather than just a reader, is a vital skill in many professions and areas of study. The study of statistics is therefore essential to a sound education. We must learn how to read data, critically and with comprehension; we must learn how to produce data that provide clear answers to important questions; and we must learn sound methods for drawing trustworthy conclusions based on data.

Historically, the ideas and methods of statistics developed gradually as society grew interested in collecting and using data for a variety of applications. The earliest origins of statistics lie in the desire of rulers to count the number of inhabitants or measure the value of taxable land in their domains. As the physical sciences developed in the seventeenth and eighteenth centuries, the importance of careful measurements of weights, distances, and other physical quantities grew. Astronomers and surveyors striving for exactness had to deal with variation in their measurements. Many measurements should be better than a single measurement, even though they vary among themselves. How can we best combine many varying observations? Statistical methods that are still important were invented in order to analyze scientific measurements.

By the nineteenth century, the agricultural, life, and behavioral sciences also began to rely on data to answer fundamental questions. How are the heights of parents and children related? Does a new variety of wheat produce higher yields than the old, and under what conditions of rainfall and fertilizer? Can a person's mental ability and behavior be measured just as we measure height and reaction time? Effective methods for dealing with such questions developed slowly and with much debate.[1]

As methods for producing and understanding data grew in number and sophistication, the new discipline of statistics took shape in the twentieth century. Ideas and techniques that originated in the collection of government data, in the study of astronomical or biological measurements, and in the attempt to understand heredity or intelligence came together to form a unified "science of data." That science of data—statistics—is the topic of this text.

The first two chapters deal with statistical methods for organizing and describing data. These chapters progress from simpler to more complex data. Chapter 1 examines data on a single variable, Chapter 2 is devoted to relationships among two or more variables. You will learn both how to examine data produced by others and how to organize and summarize your own data. These summaries will be first graphical, then numerical, then when appropriate in the form of a mathematical model that gives a compact description of the overall pattern of the data. Chapter 3 outlines arrangements (called designs) for producing data that answer specific questions. The principles presented in this chapter will help you to design proper samples and experiments, and to evaluate such investigations in your field of study.

The remaining seven chapters discuss statistical inference—formal methods for drawing conclusions from properly produced data. Statistical inference uses the language of probability to describe how reliable its conclusions are, so some basic facts about probability are needed to understand inference. Probability is the subject of Chapters 4 and 5. Chapter 6, perhaps the most important chapter in the text, introduces the reasoning of statistical inference. We emphasize that effective inference is based on good procedures for producing data (Chapter 3), careful examination of the data (Chapters 1 and 2), and an understanding of the nature of statistical inference as discussed in Chapter 6. Chapters 7 through 10 describe some of the most common specific methods of inference: for drawing conclusions about means and proportions from one and two samples, about relations in categorical data, regression and correlation, and analysis of variance.

The practice of statistics involves the use of many recipes for numerical calculation, some quite simple and some very complex. As you learn how to use these recipes, remember that the goal of statistics is not calculation for its own sake, but gaining understanding from numbers. Many of the calculations can be automated by a calculator or computer, but you must supply the understanding. Chapters 7 to 10 present only a few of the many specific procedures for inference. The more complex procedures are always carried out by computers using specialized software. A thorough grasp of the principles of statistics will enable you to quickly learn more advanced methods as needed. On the other hand, a fancy computer analysis carried out without attention to basic principles will often produce elaborate nonsense. As you read, seek to understand the principles as well as the necessary details of methods and recipes.

NOTE

1. The rise of statistics from the physical, life, and behavioral sciences is described in detail by S. M. Stigler, *The History of Statistics: The Measurement of Uncertainty Before 1900,* Harvard-Belknap, Cambridge, Mass. 1986. Much of the information in the brief historical notes appearing throughout the text is drawn from this book.

Introduction to the Practice of Statistics

CHAPTER

1

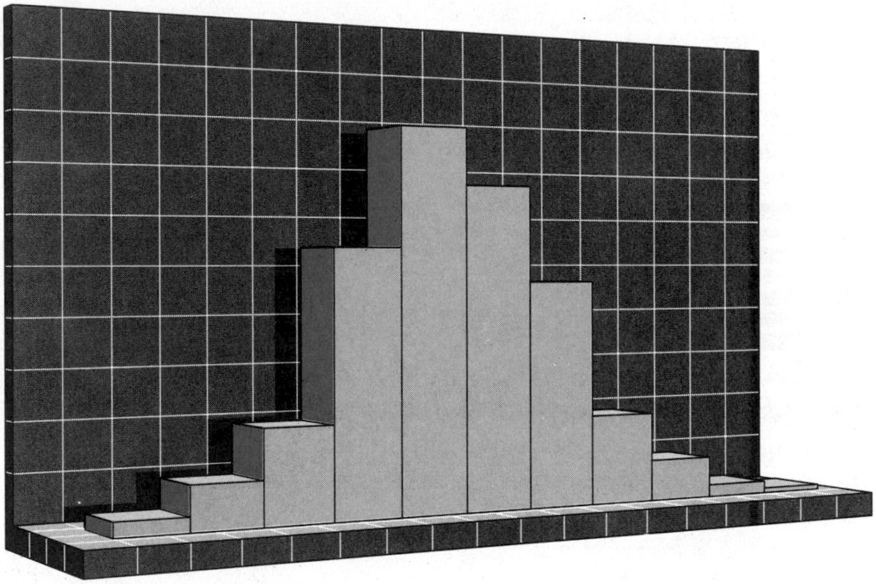

Looking at Data—Distributions

PRELUDE

The goal of statistics is to gain information from data. The first step is to display the data in a graph so that our eyes can take in the overall pattern and spot unusual observations. Next, we often summarize specific aspects of the data, such as the average value of a variable, by numerical measures. As we study graphs and numerical summaries, we keep firmly in mind where the data come from and what we hope to learn from them. Graphs and numbers are not ends in themselves, but aids to understanding.

In this chapter we begin the study of statistical tools and ideas for describing data with an examination of data that consist of observations on only a single variable, such as the heights of a group of people, the list prices of new cars, or the daily numbers of customers at a theme park. Several types of graphs are available to picture such data, and several types of numerical measures summarize them. Good data analysis requires not only that we be able to make the graphs and calculate the numerical summaries, but that we know which graphs and numerical summaries to choose and how to interpret in plain language the results of our work. Here are some questions we will meet, and answer from data, in this chapter:

- Government regulations group hot dogs into three types: beef, meat, and poultry. Are there systematic differences in how many calories hot dogs of these types contain?
- Is there a regular pattern in the changing price of fresh oranges over time? What about the changing value of stock-market indexes?
- The mean (average) price of new houses is much higher than the median (typical) price. Why is this gap not surprising?
- Eleanor scores 680 on the mathematics Scholastic Aptitude Test. Gerald scores 27 on the American College Testing examination. How can we compare their scores?

• • •

variable

value

Any characteristic of a person or thing that can be expressed as a number is called a *variable*. A *value* of the variable is the actual number that describes a particular person or thing. Height, sex, and annual income are variables that describe people; your specific height, sex, and income are the values of these variables that apply to you. An important initial step in organizing the way you think about data is to observe *how many variables* are present. This chapter looks at data that consist of values of a single variable, such as heights of people, incomes of families, or scores of schoolchildren on a reading test. Chapter 2 examines relationships between two variables and the complexities of relationships among several variables. In this way we will advance from simpler to more complex data, gaining experience and learning new principles at each stage.

In addition to the number of variables, you should note what *types of variables* are present in a set of data. We will work with two types of variables: quantitative variables and categorical variables.

QUANTITATIVE VARIABLE, CATEGORICAL VARIABLE

A quantitative variable takes numerical values for which arithmetic operations such as differences and averages make sense. A categorical variable simply records into which of several categories a person or thing falls.

Variables measured as numbers in a scale of equal units, such as height in centimeters or income in dollars, are quantitative variables. So are counts of individuals and percents or fractions based on counts. The most common statistical methods are appropriate for quantitative variables. Examples of categorical variables include the sex or political party affiliation of a person, the type of insecticide applied to a field, and the make of a car. The values of categorical variables are sometimes expressed as numbers, such as

$$\text{Female} = 1$$

$$\text{Male} = 0$$

However, we cannot meaningfully compute the "average sex" from a set of 0s and 1s. In general, it does not make sense to do arithmetic with a categorical variable. Furthermore, it is usually more convenient to use labels such as F for female and M for male rather than numerical codes. This chapter deals primarily with quantitative variables. The distinction between quantitative and categorical variables becomes important in Chapter 2.

1.1 DISPLAYING DISTRIBUTIONS

How do we begin to examine intelligently a set of values of a single quantitative variable? As a case study in the preliminary examination of data, we will look at the results of an important scientific study.

EXAMPLE 1.1

Light travels fast, but it is not transmitted instantaneously. Light takes over a second to reach us from the moon and over 10 billion years to reach us from the most distant objects yet observed in the expanding universe. Because radio and radar also travel at the speed of light, an accurate value for that speed is important in communicating with astronauts and orbiting satellites. An accurate value for the speed of light is also important to computer designers because electrical signals travel only at light speed.

The first reasonably accurate measurements of the speed of light were made a little over 100 years ago by A. A. Michelson and Simon Newcomb. Table 1.1 contains 66 measurements made by Newcomb between July and September 1882.[1] ●

Measurement

A set of numbers such as those in Table 1.1 is meaningless without some background information. We must ask several preliminary questions about any set of data. First, *What variable is being measured*? Newcomb measured how long light took to travel from his laboratory on the Po-

TABLE 1.1 Newcomb's measurements of the passage time of light

28	22	36	26	28	28
26	24	32	30	27	24
33	21	36	32	31	25
24	25	28	36	27	32
34	30	25	26	26	25
−44	23	21	30	33	29
27	29	28	22	26	27
16	31	29	36	32	28
40	19	37	23	32	29
−2	24	25	27	24	16
29	20	28	27	39	23

tomac River to a mirror at the base of the Washington Monument and back, a total distance of about 7400 meters. Just as you can compute the speed of a car from the time required to drive a mile, Newcomb computed the speed of light from the travel time.

Answering the question "What variable is being measured?" requires a description of the *instrument* used to make the measurement. Then we must judge whether the variable measured is appropriate for our purpose. This judgment often requires expert knowledge of the particular field of study. For example, Newcomb invented a novel and complicated apparatus to measure the passage time of light. We accept the judgment of physicists that this instrument is appropriate for its intended task and more accurate than earlier instruments.

Newcomb's study of the speed of light measured a clearly defined and easily understood variable, the time light takes to travel a fixed distance. Questions about measurement are often harder to answer in the social and behavioral sciences than in the physical sciences. We can agree that a tape measure is an appropriate way to measure a person's height. But how shall we measure her intelligence? Questionnaires and interview forms as well as tape measures are measuring instruments. A psychologist wishing to measure "general intelligence" might use the Wechsler Adult Intelligence Scale (WAIS). The WAIS is a standard "IQ test" that asks subjects to solve a large number of problems. The appropriateness of this instrument as a measure of intelligence is not accepted by all psychologists, many of whom disagree about what "intelligence" is and how to measure it. Because questions about measurement usually require knowledge of the particular field of study, we will say little about them.

Users of data should nonetheless be aware that taking numbers at face value, without thinking about the variable measured and the process used to measure it, can produce misleading results. The following examples illustrate the dangers.

EXAMPLE 1.2

A teenager argues that young people are safer drivers than the elderly. He cites government data showing that in 1989, 5426 drivers 65 years of age and over were involved in fatal accidents. In contrast, only 2900 drivers aged 16 and 17 had fatal accidents. So the older drivers had more fatal accidents. The teenager's argument is incorrect because the variable (the count of drivers who had fatal accidents) is not appropriate. There are more than five times as many licensed drivers over age 65 as there are drivers aged 16 and 17. The larger group of drivers naturally had a larger number of accidents. A more appropriate variable is the *rate* of fatal accidents, that is, the fraction of drivers who were involved in fatal accidents in 1989. In fact, 70 of every 100,000 teenage drivers but only 26 of every 100,000 elderly drivers had fatal accidents in 1988. Fatal accidents are much more common among younger drivers. The use of *counts* where *rates* are more appropriate is a common misuse of statistics. ●

rate

EXAMPLE 1.3

One important measure of a nation's economic success is its unemployment rate, the percent of the active labor force that cannot find jobs. The unemployment rate in Japan has long been much lower than that in the United States. In 1991, for example, the U.S. unemployment rate was about 7%, whereas Japan's was about 2%. But examination shows that the measurement systems are not the same in the two countries. In the United States, any person without a job who is actively seeking work is counted as unemployed. In Japan, school-leavers seeking their first job are not considered unemployed, and self-employed workers are also omitted from the data. Because of these and other differences, the Japanese unemployment rate would almost double if measured by the American method. Conclusions about the two economies drawn without regard for such measurement differences may be incorrect.[2] ●

The two remaining questions you should ask about any set of data are more straightforward: *What are the units of measurement?* and *How are the data recorded?* Newcomb's first measurement of the passage time of light was 0.000024828 second. So his unit of measurement was seconds. But the entries in Table 1.1 do not look at all like 0.000024828. Such numbers are awkward to write and to do arithmetic with. We therefore move the decimal point nine places to the right, giving 24828, and then record only the deviation from 24800. The table entry 28 is short for the original 0.000024828, and the entry −2 stands for 24798, or 0.000024798. This procedure is called *coding* the data. You should code whenever the actual data values contain many digits and only a few digits vary from observation to observation. The data in coded form are easier to read. In addition, when you are using a calculator or a computer, reducing many digits to a few helps keep the results of arithmetic within the fixed number of digits that the machine can work with and store.

coding

We now have some understanding of what the numbers in Table 1.1 mean and where they came from. It is easier to understand data in your own area of study, but you should always ask: What variables were measured? What are the units of measurement? Are the data coded? Now we can begin to look more closely at the data themselves.

Variation

The first feature of the entries in Table 1.1 is that they vary. Newcomb's observations of the passage time of light are not all the same. Why should this be? After all, each observation records the travel time of light over the same path, measured by a skilled observer using the same apparatus each time. Newcomb knew that careful measurements almost always vary. The environment of every measurement is slightly different. The apparatus changes a bit with the temperature, the density of the atmosphere changes from day to day, and so on. Newcomb did his best to eliminate the sources of variation that he could anticipate, and his experimental skill gave measurements with less variation than those taken by less able scientists. But even the best experiments produce variable results. That is why Newcomb took many measurements rather than just one. The average of 66 passage times should be less variable than the result of a single measurement, because the average does not depend on the temperature and atmospheric density at a single time. (Section 1.2 shows how to compute several "averages," and Chapter 5 shows how to express numerically the advantage of averaging 66 observations.)

Putting ourselves in Newcomb's place, we are tempted to compute the average passage time, convert this time to a new and better estimate of the speed of light, and rush off to make our reputations by publishing the result. The temptation to do a routine calculation and announce an answer arises whenever data have been collected to answer a specific question. Because computers are very good at routine calculations, succumbing to temptation is the easy road. The first step toward statistical sophistication is to resist the temptation to calculate without thinking. Since variation is surely present in any set of data, we must first examine the nature of the variation. We may be very surprised at what we see.

Variation is even more obviously a fact of life when different people or things are measured. The reading ability of seventh graders varies from child to child, so many children must be tested to get a picture of seventh-grade reading skills. Likewise, because the concentration of a drug may vary from lot to lot, the manufacturer measures many lots to keep the concentration within the desired range. In such cases, the nature and magnitude of the variation may be as important as the average outcome.

> **DISTRIBUTION**
>
> The pattern of variation of a variable is called its distribution. The distribution records the numerical values of the variable and how often each value occurs.

The distribution of a variable is best displayed graphically. Some graphical tools for displaying distributions are introduced in the next few pages. With a better-stocked tool kit, we will then return to Newcomb's measurements.

Stemplots

A *stemplot* (also called a stem-and-leaf plot) offers a quick way to picture the shape of a distribution while including the actual numerical values in the graph. Stemplots work best for small numbers of observations that are all greater than 0.

EXAMPLE 1.4

Here are the number of home runs that Babe Ruth hit in each of his 15 years with the New York Yankees, 1920 to 1934:[3]

54 59 35 41 46 25 47 60 54 46 49 46 41 34 22

To make a stemplot for these data:

stem

leaf

1. Separate each observation into a *stem* (the first digit in this case) and a *leaf* (the second digit). Thus 54 produces 5 as the stem and 4 as the leaf. Stems may have as many digits as needed, but each leaf contains only a single digit.
2. List the stems vertically in increasing order from top to bottom, draw a vertical line to the right of the stems, and add the leaves to the right of the line.
3. Arrange the leaves in increasing order from left to right.

Here is the resulting stemplot.

2	25
3	45
4	1166679
5	449
6	0

Once a distribution has been displayed by a stemplot, we can see its important features as follows:

median

1. *Locate the center of the distribution*, either by eye or by counting in from either end of the stemplot until half the observations are counted. Because Ruth hit 46 homers three times, less than 46 six times, and more than 46 six times, the distribution of his home runs is centered at 46. (This simple measure of the center of a distribution is called the *median*. We will use the term "median" informally until we study measures of center in detail in Section 1.2.)

symmetric

skewed

2. *Examine the overall shape of the distribution*. Turn the stemplot on its side so that the larger observations fall to the right. Does the distribution have one peak or several peaks? Is it approximately *symmetric* or is it *skewed* in one direction? A distribution is symmetric if the portions above and below its center are mirror images of each other. It is skewed to the right if the right tail (higher values) is much longer than the left tail (lower values). We do not insist on exact symmetry in giving a general description of the shape of a distribution. Babe Ruth's home runs have an approximately symmetric distribution.

3. *Finally, look for marked deviations from the overall shape.* These may be *gaps* in the distribution, or they may be *outliers*, individual observations that fall well outside the overall pattern of the data. There are no gaps or outliers in the distribution of Ruth's home runs. In particular, his famous 60 home runs in 1927 was not unusual in the context of his career distribution of home runs per year.

gaps

outliers

Ruth's record for a single year was broken by another Yankee, Roger Maris, who hit 61 home runs in 1961. Here is a stemplot of Maris's home run production during his 10 years in the American League:

0	8
1	346
2	368
3	39
4	
5	
6	1

Maris's record year is an outlier, an individual value that stands apart from his usual pattern. You should search for an explanation for any outlier. Sometimes outliers point to errors made in recording the data. In other cases, the outlying observation may be caused by equipment failure or other unusual circumstances. The outlier in Roger Maris's home run data marks an exceptional individual achievement.

When you wish to compare two related distributions, a *back-to-back stemplot* with common stems is useful. The leaves on each side are ordered out from the common stem. Here is a back-to-back stemplot comparing the home run data for Ruth and Maris. Ruth's superiority as a home run hitter is clearly evident.

back-to-back stemplot

Ruth		Maris
	0	8
	1	346
52	2	368
54	3	39
9766611	4	
944	5	
0	6	1

Stemplots do not work well for large data sets, where each stem must hold a large number of leaves. Fortunately, there are several modifications of the basic stemplot that are helpful when plotting the distribution of a moderate number of observations. You can increase the number of stems in a plot by *splitting each stem* into two, one with leaves 0 to 4 and the other with leaves 5 through 9. When the observed values have many digits, it is

splitting stems

truncate

often best to *truncate* the numbers by dropping all but a few digits before making a stemplot. You must use your judgment in deciding whether to split stems or truncate. Remember that the purpose of a stemplot is to display the shape of a distribution. If a stemplot has fewer than about five stems, you should usually split the stems unless there are few observations. If many stems have no leaves or only one leaf, see if truncation will help. Here is an example making use of both of these modifications.

EXAMPLE 1.5

A marketing consultant observed 50 consecutive shoppers at a grocery store. One variable of interest was how much each shopper spent in the store. Here are the data (in dollars), arranged in increasing order.

2.32	6.61	6.90	8.04	9.45
10.26	11.34	11.63	12.66	12.95
13.67	13.72	14.35	14.52	14.55
15.01	15.33	16.55	17.15	18.22
18.30	18.71	19.54	19.55	20.58
20.89	20.91	21.13	23.85	26.04
27.07	28.76	29.15	30.54	31.99
32.82	33.26	33.80	34.76	36.22
37.52	39.28	40.80	43.97	45.58
52.36	61.57	63.85	64.30	69.49

To make a stemplot of this distribution, we truncate the purchases by dropping the cents so that the stems are tens of dollars and the leaves are dollars. This gives us the single-digit leaves that a stemplot requires. Then we split each stem. For example, we truncate 34.76 to 34 and write it as a leaf of 4 on the first 3 stem.

0	2
0	6689
1	0112233444
1	556788899
2	00013
2	6789
3	012334
3	679
4	03
4	5
5	2
5	
6	134
6	9

Turning the stemplot on its side so that the larger values fall to the right, we see that the distribution of amount spent is strongly skewed to the right. The center of the distribution is at \$20 (found by counting 25 observations up from the lowest value). But 10% of the customers spent over \$50. The shape revealed by the stemplot will interest the store management. If the big spenders can be studied in more detail, steps to attract them may be profitable. ●

These examples illustrate the varied shapes of distributions. They also illustrate how simple calculations (of center, for example) can be helpful in clarifying the nature of a distribution. Making a graphical display is the first step toward understanding data; performing well-chosen calculations is the second. Section 1.2 will give more systematic suggestions for doing such calculations.

Most statistical software packages will produce a stemplot for you. If the observations have been entered into the Minitab package as column C1, for example, then the command

```
MTB > STEM C1
```

will make a stemplot. The program will make a decision about splitting stems. You can change that decision by a second command if you wish.

Histograms

A virtue of stemplots is that they display the actual values of the observations for detailed examination. But this feature makes stemplots awkward for large data sets. Moreover, the picture presented by a stemplot divides the observations into groups (stems) determined by the number system rather than by judgment. Histograms do not have these limitations. A *histogram* breaks the range of values of a variable into intervals and displays only the count (or percent) of the observations that fall into each interval. Making a histogram is much simpler if you choose intervals of equal width. You can choose any convenient number of intervals. Histograms are slower to construct by hand than stemplots and do not display the actual values observed. For these reasons we prefer stemplots for small data sets. The construction of a histogram is best shown by example.

histogram

EXAMPLE 1.6

A study of the teaching of reading in the public schools of Gary, Indiana, began by examining the performance of seventh graders on the reading portion of the Iowa Test of Basic Skills.[4] In all, 947 Gary seventh graders took this test. Their vocabulary scores, expressed as an equivalent grade level, ranged from 2.0 (that is, equivalent to a beginning second grader) to 12.1. The scores of the first few of the 947 students are

5.4 6.8 2.0 7.6 6.6 7.8 8.1 5.6 6.0 7.9 ...

To make a *histogram* of the distribution of scores, proceed as follows:

1. *Divide the range of the data into classes of equal width.* In this case, it is natural to use grade levels, so the classes are

$$2.0 \leq \text{score} < 3.0$$

$$3.0 \leq \text{score} < 4.0$$

$$\vdots$$

$$12.0 \leq \text{score} < 13.0$$

We have specified the classes precisely so that each score falls in exactly one class. A score of 2.9 falls in the first class, for example, and a score of 3.0 falls in the second.

frequency

frequency table

2. *Count the number of observations in each class*. These counts are called *frequencies*, and a table of frequencies for all classes is a *frequency table*. Table 1.2 is a frequency table of the Gary vocabulary scores. A frequency table is a convenient summary of a large set of data.
3. *Draw the histogram*. In the histogram in Figure 1.1, the vocabulary score scale is horizontal, and the frequency scale is vertical. Each bar represents a class. The base of the bar covers the class, and the bar height is the class frequency. The graph is drawn with no horizontal space between the bars (unless a class is empty, so that its bar has 0 height). ●

The distribution in Figure 1.1 is centered in the 6.0–6.9 class and is roughly symmetric in shape. The choice of classes can slightly affect the appearance of a histogram, as can the automatic choice of classes forced by a stemplot. This is another reason why exact symmetry is not necessary in order to describe a distribution as "symmetric." The distribution in Figure 1.1 is quite regular. The histogram shows no large gaps or obvious outliers, and both tails fall off quite smoothly from a single center peak.

TABLE 1.2 Reading achievement scores for seventh graders

Class	Number of students	Percent
2.0–2.9	9	.95
3.0–3.9	28	2.96
4.0–4.9	59	6.23
5.0–5.9	165	17.42
6.0–6.9	244	25.77
7.0–7.9	206	21.75
8.0–8.9	146	15.42
9.0–9.9	60	6.34
10.0–10.9	24	2.53
11.0–11.9	5	.53
12.0–12.9	1	.11
Total	947	100.01

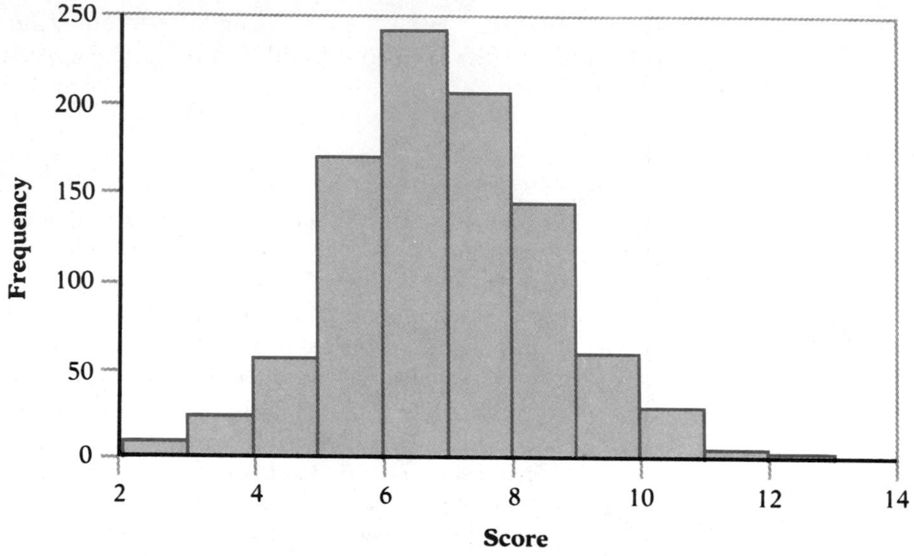

FIGURE 1.1 Histogram of the vocabulary scores of 947 seventh graders in Gary, Indiana.

This is a bit surprising. Since adolescent girls as a group read better than boys of the same age, we might have expected a "double-peaked" distribution with this general shape:

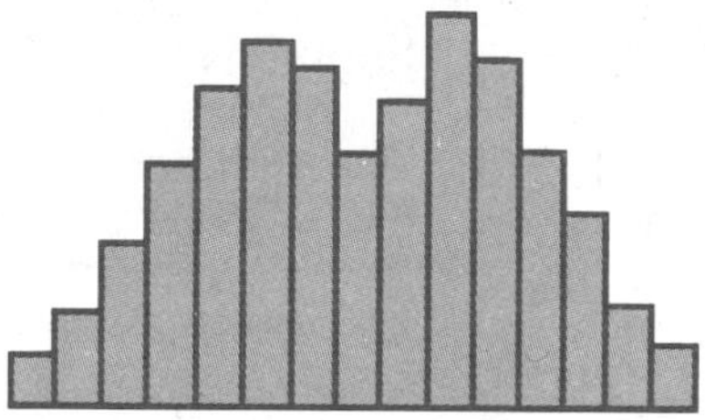

More elaborate calculations show that there is indeed an effect based on sex in the Gary reading scores, but the effect is not large enough to change the shape of the histogram. We are still at the first step of examining data, looking for major patterns and clear deviations from them. In time we will learn how to discern finer detail.

Large sets of data are almost always reported in the form of frequency tables, since it is not practical to publish the individual observations. In addition to the frequency (count) for each class, the fraction or percentage of the observations that fall in each class can be reported. These fractions are sometimes called *relative frequencies*. Table 1.2 gives both frequencies and relative frequencies for the Gary reading scores. When you are reporting statistical results, however, it is clearer to use the nontechnical terms

relative frequency

"number" and "percent." Expressed as percents, the relative frequencies of all of the classes should add to 100%. The percents in Table 1.2 total 100.01% because of *roundoff errors*. Each percent is rounded to the nearest 0.01%, so that the rounded numbers do not have a sum exactly equal to 100%.

roundoff error

A histogram of relative frequencies has the same appearance as a frequency histogram such as Figure 1.1. Simply relabel the vertical scale to read in percents. Histograms of relative frequencies are preferable for comparing several distributions with different numbers of observations. The bars of a frequency histogram for reading scores in Los Angeles would tower over a similar histogram from Gary, because there are many more seventh graders in Los Angeles. Both relative frequency histograms have the same vertical scale (0% to 100%), so that we can more easily compare the shapes of the two distributions.

Because the purpose of a histogram is to display the shape of a distribution, we must be attentive to the visual aspects of the display. Our eyes respond to the *area* of the bars, so the area of each bar should be proportional to the frequency of its class. When the classes are of equal width, we need only make the bar heights proportional to the frequencies. We recommend equal class widths whenever the raw data are available to you. However, at times you may need to draw a histogram with unequal class widths, particularly when examining data published in the form of a frequency table with unequal classes. Many government economic and social data, for example, are published in this form. In such cases, you must vary the bar heights in such a way that the areas (height × width) are proportional to the frequencies. See Exercise 1.30 for guidance.

Before drawing a histogram you must decide how many classes to use. This is a matter of judgment. Too few classes will lump most observations together, and too many will place only a few observations in each class. Neither extreme will show clearly the shape of the distribution. Narrow classes preserve more detail, but the heights of the bars often vary irregularly. Wide classes often give a more regular picture of the overall shape, but they lose detail by lumping a wider interval of values into each class. Use natural classes, such as the grade levels in Example 1.6, whenever possible. Fortunately, a broad range of choices will usually give similar impressions of the distribution.

Statistical software on a computer can ease the task of drawing a histogram. The program will choose the number of classes and produce a histogram for you to inspect. You can then instruct the program to change the choice of classes if you wish. If the data have been entered into the Minitab statistical package as column C1, for example, then the command

```
MTB > HISTOGRAM C1
```

will make a histogram. The artistic quality of a computer-drawn histogram depends on the output device you use.

Looking at data

Some principles have emerged from our initial experiences with data that will remain valid as we advance. These are guidelines based on experience rather than hard and fast rules that must always be followed.

1. To interpret data, you must first learn something of their context: What exactly was measured? What are the units? How was the measurement carried out? Are the data intended to answer specific questions? Are they appropriate for that purpose?
2. Always examine your data. An informative picture comes first, usually supplemented by some numerical calculations.
3. Look first for an overall pattern and then for deviations from that pattern, such as outliers.

What of Simon Newcomb and the speed of light? A laboratory scientist knows that careful measurements will vary but still hopes for a symmetric single-peaked distribution like that of Figure 1.1. Then the best estimate of the true value of the measured quantity is the center of the distribution. (This is not true if the measurement process is biased so as to produce systematically high or low readings—that is a fatal flaw that statistics cannot remedy.) The histogram in Figure 1.2 shows that Newcomb was not so fortunate; his data are afflicted with outliers. What can be done?

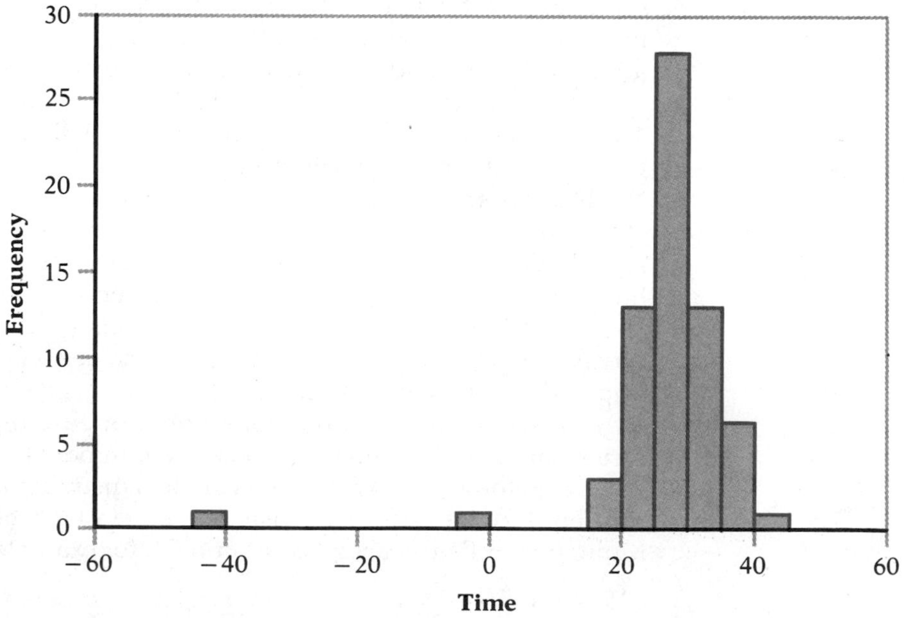

FIGURE 1.2 Histogram of Simon Newcomb's 66 measurements of the passage time of light, from Table 1.1.

The handling of outliers is another matter for judgment. Sometimes outliers are of special interest as evidence of an extraordinary event. An outlier from the background distribution of brightness seen from a surveillance satellite may represent a missile launch. An outlier from the background measurement of electric activity in a detector used by high-energy physicists may be evidence of a new elementary particle. In such cases the overall distribution simply provides the background against which extraordinary events stand out. But Newcomb hoped for a well-behaved distribution with a clear center, and the two outliers (−44 and −2 in Table 1.1) were disturbing.

When outliers are surprising and unwanted, you should first search for a clear cause for each outlier, such as equipment failure during the experiment or an error in typing the data. Almost all large data sets have errors, often caused by mistakes in entering the data into a computer. Outliers sometimes betray these errors and allow us to correct them by looking at the original data records. If equipment failure or some other abnormal condition caused the outlier, we can delete it from the data with a clear conscience. When no cause is found, the decision is difficult. The outlier may be evidence of an extraordinary occurrence or of unexpected variability in the data. Newcomb finally dropped the worst outlier (−44) but retained the other. He based his estimate of the speed of light on the average (the mean, in the language of Section 1.2) of his observations. The mean of all 66 observations is 26.21; the mean of the 65 retained observations is 27.29. The strong effect of the single value −44 on the mean is one motivation for discarding it when our interest is in the center of the distribution as a whole.

Adjustments based on judgment alone, as Newcomb's were, are risky. As the following example illustrates, it is always best to find the reason for an outlier, if possible.

EXAMPLE 1.7

In 1985 British scientists reported a hole in the ozone layer of the earth's atmosphere over the South Pole. This news is disturbing, because ozone protects us from cancer-causing ultraviolet radiation. The British report was at first disregarded, because it was based on ground instruments looking up. More comprehensive observations from satellite instruments looking down had shown nothing unusual. Then, examination of the satellite data revealed that the South Pole ozone readings were so low that the computer software used to analyze the data had automatically set these values aside as suspicious outliers. Readings dating back to 1979 were reanalyzed and showed a large and growing hole in the ozone layer that is unexplained and considered dangerous.[5] Computers analyzing large volumes of data are often programmed to suppress outliers as protection against errors in the data. As the example of the hole in the ozone illustrates, suppressing an outlier without investigating it can conceal valuable information. ●

Time plots

We can gain more insight into Newcomb's data from a different kind of graph. When data represent similar observations made over time, it is wise to plot them against either time or against the order in which the observations were taken. Figure 1.3 plots Newcomb's passage times for light in the order that they were recorded. There is some suggestion in this plot that the variability (the vertical spread in the plot) is decreasing over time. In particular, both outlying observations were made early on. Perhaps Newcomb became more adept at using his equipment as he gained experience. Learning effects like this are quite common; the experienced data analyst will check for them routinely. If we allow Newcomb 20 observations for learning, the mean of the remaining 46 measurements is 28.15. The best modern measurements suggest that the "true value" for the passage time in Newcomb's experiment is 33.02. Allowing for learning does move the average result closer to the true value.

Whenever data are collected over time, it is a good idea to plot the observations in time order. Always put time on the horizontal scale of your plot and the variable you are following on the vertical scale. Connecting the data points by lines helps emphasize the change over time. Patterns in a time plot can help us to better understand the data. Summaries of

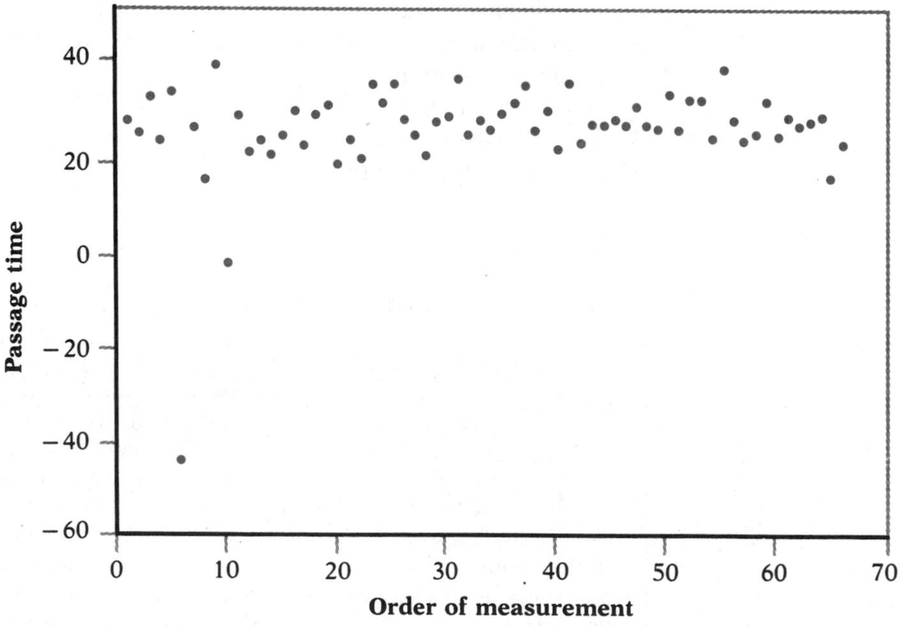

FIGURE 1.3 Plot of Newcomb's measurements against the time order of the observations.

the distribution of a variable that ignore time order, like stemplots and histograms, can be misleading when there is systematic change over time.

time series

Many interesting data sets are *time series*, measurements of a variable taken at regular intervals over time. The government often publishes economic and social data as time series, such as the monthly unemployment rate and the quarterly gross domestic product. Weather records, the demand for electricity, and measurements on the items produced by a manufacturing process are other examples of time series.

Plots against time can reveal the main features of a time series. As in our examination of distributions, we look first for overall patterns and then for striking deviations from those patterns. Some examples will illustrate common patterns in time series.

EXAMPLE 1.8

index number

Figure 1.4 is a graph of the average price of fresh oranges over the decade 1981 to 1990.[6] This information is collected each month by the Census Bureau as part of the government's reporting of retail prices. The monthly Consumer Price Index is the most publicized product of this effort. The price is presented as an *index number*. That is, the price scale gives the price as a percent of the average price of oranges in 1967. This is indicated by the legend "1967 = 100." The first data point is 272.9 for January 1981, so at that time oranges cost about 273% of their 1967 price. The index number is based on the retail price of oranges at many stores in all parts of the country. The price of oranges at a single store will drop when the manager decides to advertise a sale on oranges, and it will rise when the sale ends. The index number is a kind of nationwide average price that is less variable than the price at one store. ●

trend

seasonal variation

Figure 1.4 shows several features that are common in time series. There is a *trend*, a long-term change in the level of the variable. In this case there is a trend of increasing price. Superimposed on this trend is a strong *seasonal variation*, a regular rise and fall that recurs each year. Orange prices are usually highest in August or September, when the supply is lowest. Prices then fall in anticipation of the harvest and are lowest In January or February, when the harvest is complete and oranges are plentiful. But perhaps the most notable features of this time series are the dramatic increases in price in 1982 and again in 1984. The seasonal low in the winter of 1984–1985 is higher than any previous seasonal high except those of 1982 and 1984. These *irregular fluctuations* are price shocks due to unusual events—a freeze in Florida and a failure of the orange crop in Brazil.

seasonally adjusted

Because many economic time series show strong seasonal variation, government agencies often adjust for this variation before releasing economic data. The data are then said to be *seasonally adjusted*. Seasonal adjustment helps avoid misinterpretation. A rise in the unemployment rate from December to January, for example, does not mean that the economy

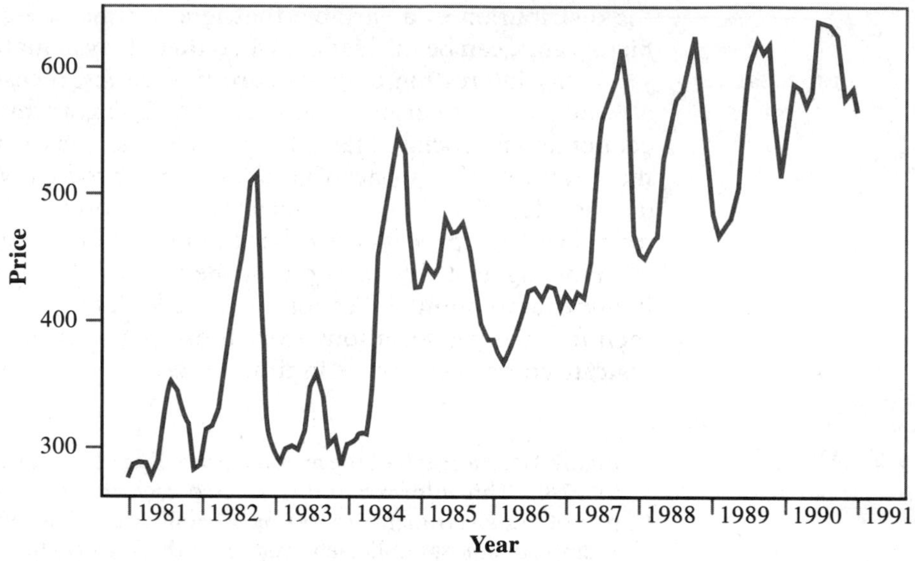

FIGURE 1.4 Index number (1967 = 100) for monthly prices of fresh oranges, 1981 to mid-1991.

is slipping. Unemployment almost always rises in January as temporary Christmas help is laid off and outdoor employment in the north drops because of bad weather. The seasonally adjusted unemployment rate reports an increase only if unemployment rises more than normal from December to January.

In addition to trends, seasonal variation, and irregular fluctuations, time series sometimes show *cycles*, distinct up and down movements that are less regular than seasonal variation and are not explained by seasonal effects. Figure 1.5 is a time series of great interest to many people—common stock prices. It may surprise you to learn that some people suspect regular cycles in stock prices.

cycles

EXAMPLE 1.9

The variable graphed in Figure 1.5 is the Standard & Poor's 500 Composite Stock Price Index, commonly known as the S&P 500. The S&P 500 is an average of the stock prices for 500 large companies, but it is not simply the mean of the prices. It is instead a weighted average in which each company's stock influences the index in proportion to its total market value (stock price × number of shares). A change in the price of IBM stock, for example, moves the index more than the same change in the price of any other stock, because IBM leads the pack in the total market value of its shares. The S&P 500 is a widely used measure of the overall level of U.S. common stock prices. It is one of the Commerce Department's official economic indicators, for example. ●

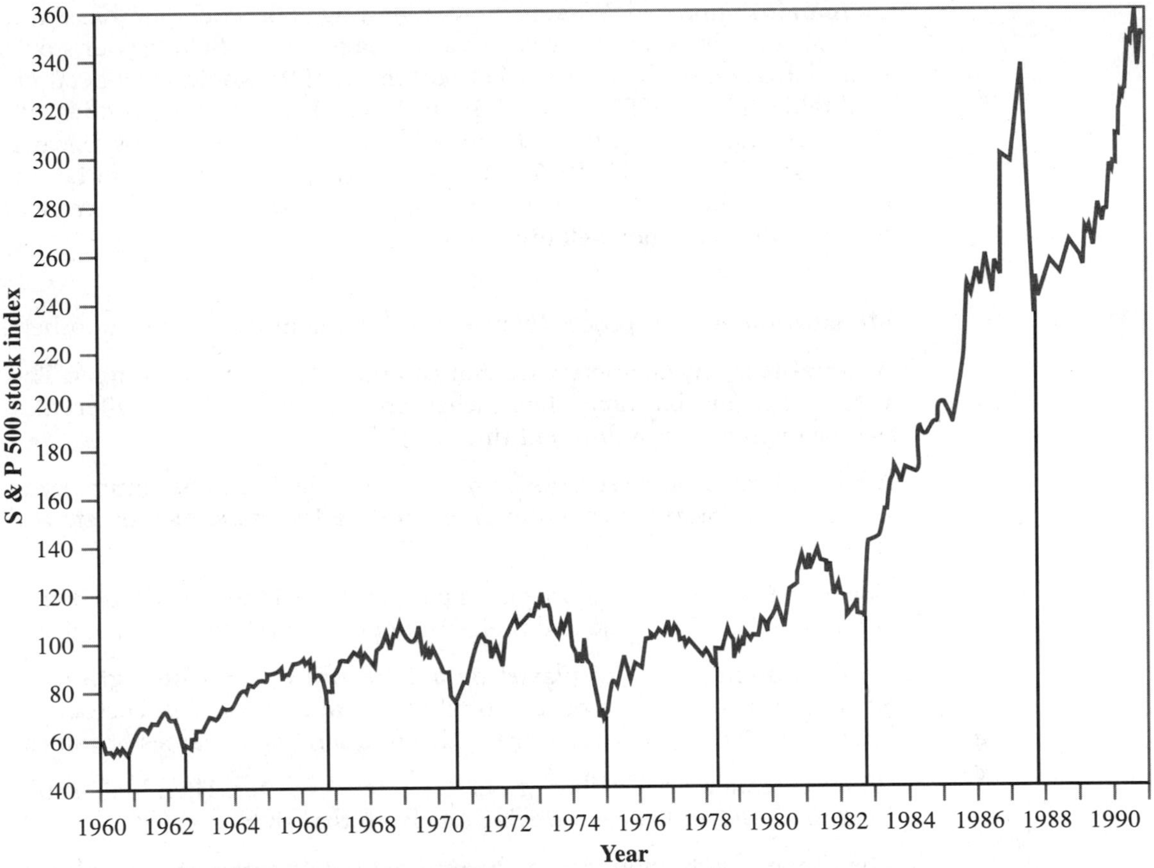

FIGURE 1.5 The S&P 500 stock index, 1960 to 1991, with market bottoms marked.

Figure 1.5 shows an increasing trend, though with a long trendless period during the 1970s. For more than 20 years there also appeared to be a 4-year cycle. Stock prices bottomed out in 1962, 1966, 1970, 1974, 1978, and 1982. The regular downturn vanished in 1986 but reappeared in 1990. The peaks are less regular but occurred roughly midway between the bottoms.

Why should stock prices show a 4-year cycle? Cynics suggest that the 4-year cycle of American politics may have something to do with it. Stock market tops occur roughly at presidential election time, with the bottoms midway between major elections. Is it possible that the party in power attempts to pump up the economy, and with it the stock market, as the election approaches? Necessary but unpleasant economic medicine could then be administered during the period between elections. This hypothesis remains unproved. What is more, planning your stock purchases and sales around the 4-year cycle is no way to make money. Selling your stocks in anticipation of the next market bottom in 1986 would have been unprofitable: The S&P 500 rose 14.6% in 1986. Buying stocks in mid-1987 in anticipation of the 1988 election would have been disastrous, because the crash of October 19, 1987, intervened. You should be wary of betting on a pattern based solely on inspection of a time series when the reasons for the pattern are not well understood.

SUMMARY

Measurement is the process of representing a characteristic by numbers.

A **variable** is any characteristic that can be expressed as a number. The values of a variable vary when measurements are made on different people or things or at different times.

A large number of observations on a single variable can be summarized in a table of **frequencies** (counts) or **relative frequencies** (percents or fractions).

The **distribution** of a variable is its pattern of variation, as described by the values of the variable and their frequencies or relative frequencies.

A distribution can be displayed by a **stemplot** or by a **histogram.** Stemplots separate each observation into a **stem** and a **leaf.** Histograms are based on the frequencies or relative frequencies of classes of values.

When examining a distribution, first locate its **center.** Then look at the **overall shape** and at clear **deviations** from that shape.

The shape of a distribution can be approximately **symmetric** (each side of the center is a mirror image of the other) or **skewed** (one tail extends farther from the center than the other). The number of peaks is another aspect of overall shape.

Deviations from the overall shape of a distribution include gaps and **outliers** (individual observations that appear not to be in accord with the other data).

Time plots show variation over time. Patterns in a time plot may include **trends, seasonal variation,** and **cycles,** often combined with large **irregular fluctuations.**

SECTION 1.1 EXERCISES

1.1 You want to compare the "size" of several statistics textbooks. Describe at least three possible numerical variables that describe the "size" of a book. In what *unit* is each variable measured? What *measuring instrument* does each require? Which of these variables is most appropriate for estimating how long it would take you to read the book? Which is most appropriate for deciding whether the book will fit easily into your book bag?

1.2 You are studying the relationship between political attitudes and length of hair among male students. You will measure political attitudes with a standard questionnaire. How will you measure length of hair? Give precise instructions that an assistant could follow. Include a statement of the *unit* and the *measuring instrument* that your assistant is to use.

1.3 Various studies have attempted to rank cities in terms of how desirable it is to live and work in each city. Describe five variables that you would measure for each city if you were designing such a study. Give reasons for each of your choices.

1.4 The number of deaths from cancer in the United States has risen steadily over time. In 1991, for example, about 514,000 people died of cancer, up from 331,000 deaths in 1970. A member of Congress says that these numbers show that no progress has been made in treating cancer. Explain how the number of people dying of cancer could increase even if treatment of the disease were improving. Then describe a variable that would be a more appropriate measure of the effectiveness of medical treatment for a potentially fatal disease.

1.5 Scientists who study human growth use different measures of the size of an individual. Weight, height, and weight divided by height are three of the most common measures. If you were interested in studying the short-term effects of a digestive illness, which of these three variables would you study? Why?

1.6 You are writing an article for a consumer magazine based on a survey of the magazine's readers on the reliability of their household appliances. Of 13,376 readers who reported owning Brand A dishwashers, 2942 required a service call during the past year. Only 192 service calls were reported by the 480 readers who owned Brand B dishwashers. Describe an appropriate variable to measure the reliability of a make of dishwasher, and compute the values of this variable for Brand A and for Brand B.

1.7 The usual method of determining heart rate is to take the pulse and count the number of beats in a given time period. The results are generally reported as beats per minute; for instance, if the time period is 15 seconds, the count is multiplied by 4. Take your pulse for two 15-second periods, two 30-second periods, and two 1-minute periods. Convert all the counts to beats per minute and report the results. Which procedure do you think gives the most accurate results? Why?

1.8 In the previous exercise, some beats occurred just before or after the end of your time interval. In what way would this problem affect your results? Does the seriousness of the problem depend upon the length of the time interval? Consider the following alternative procedure. Count the time that it takes for the heart to beat a certain number of times, say 80. Describe how you would convert this measure to beats per minute. Make two measurements on yourself and report the results. Do you think this alternative procedure is better? Give reasons.

1.9 Each year *Fortune* magazine lists the top 500 companies in the United States, ranked according to their total annual sales in dollars. Describe three other variables that could reasonably be used to measure the "size" of a company.

1.10 Cost is often an important factor in deciding which variable to measure in a study. Sometimes the cost can be determined directly in dollars; in other cases it may be expressed in terms of the time required to take the measurement. A crude but inexpensive measure is sometimes preferable to a more precise measure that is too expensive. A blood test, for example, is a faster and cheaper way to diagnose many diseases than exploratory surgery.

Foresters need quick measurements of the size of a tree, since they must measure many trees in a woodlot in order to assess its overall condition. List three variables that measure the size of a tree, and arrange them in order from most costly to least costly.

1.11 A quality engineer in an automobile engine plant measures a critical dimension on each of a sample of crankshafts at regular intervals. The dimension is supposed to be 224 millimeters (mm), but some variation will occur in production. Here are the latest measurements:

224.120	224.001	224.017	223.982	223.989	223.961
223.960	224.089	223.987	223.976	223.902	223.980
224.098	224.057	223.913	223.999		

The engineer *codes* these measurements to make them easier to work with. The coded value is the number of thousandths of a millimeter above 223 mm. (For example, 224.120 mm is coded as 1120.) Give the coded value for each of the measurements in the sample.

1.12 A quality engineer in an auto plant measures the eccentricity of a valve assembly. (Eccentricity is a measure of how off-center two supposedly concentric circles are.) Here are the data in inches:

.001084	.001131	.000887	.000639	.001216	.000903
.000977	.001088	.000940	.001069	.000667	.000536

Explain how you would code these data for easier calculation. (A good coding results in positive whole numbers.) Then give the coded values for each of the eccentricities listed.

1.13 The histograms in Figure 1.6 display the distributions of two weather-related variables. Figure 1.6(a) shows the number of days of frost (minimum temperature below freezing) in Greenwich, England, in the month of April over a 65-year period. Figure 1.6(b) shows the number of hurricanes reaching the east coast of the United States each year over a 70-year period. Give a brief description of the overall shape of each of

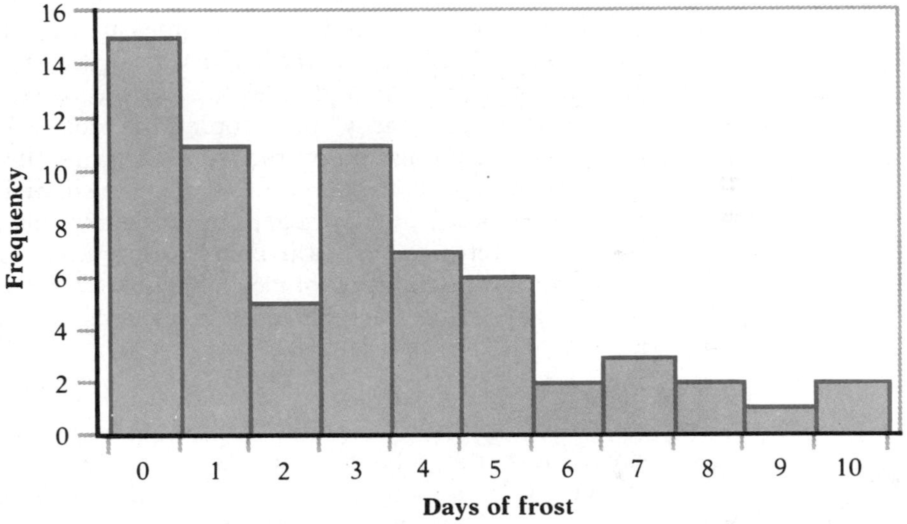

FIGURE 1.6(a) Histogram of the number of frost days during April at Greenwich, England, over a 65-year period, for Exercise 1.13.

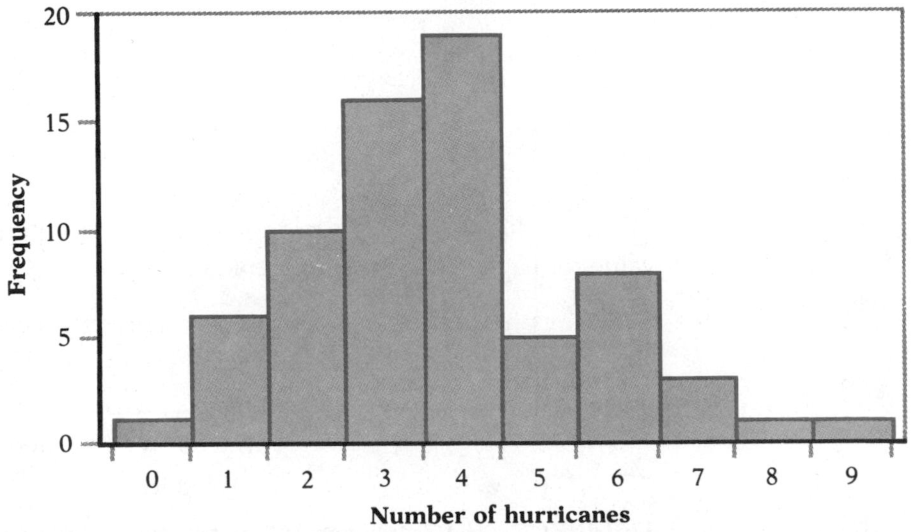

FIGURE 1.6(b) Histogram of the annual number of hurricanes on the U.S. east coast over a 70-year period, for Exercise 1.13.

these distributions. What is the most important difference between the two shapes? About where does the center of each distribution lie? (Data on frosts from C. E. Brooks and N. Carruthers, *Handbook of Statistical Methods in Meteorology*, Her Majesty's Stationery Office, London, 1953. Hurricane data from H. C. S. Thom, *Some Methods of Climatological Analysis*, World Meteorological Organization, Geneva, Switzerland, 1966.)

1.14 The number of letters in a word is a measure of its length. Figure 1.7 displays the distribution of the lengths of words used in articles in *Popular Science* magazine. These data were collected by students who opened the magazine arbitrarily and recorded the lengths of all words in the first complete paragraph on the page. Describe the main features of this distribution: Is it symmetric, right skewed, or left skewed? Single or double peaked? Are there gaps or outliers? (Statistical methods have been effectively used to distinguish between works by different authors. Although the distribution of word lengths is a rather crude measure, *Popular Science* uses notably more long words than do authors of serious fiction.)

1.15 Figure 1.8 is a histogram of the mean scores on the verbal part of the Scholastic Aptitude Test (SAT) in each of the 50 states in 1990. In some states most college-bound students take the SAT, while in others most students take the American College Testing (ACT) examination instead. In the ACT states, only students who hope to enter selective colleges take

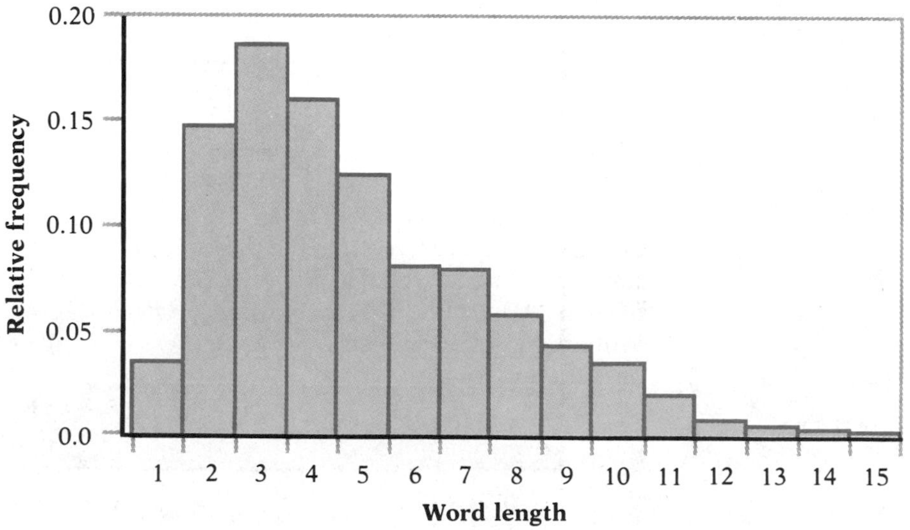

FIGURE 1.7 The distribution of the lengths of words appearing in articles in *Popular Science* magazine, for Exercise 1.14.

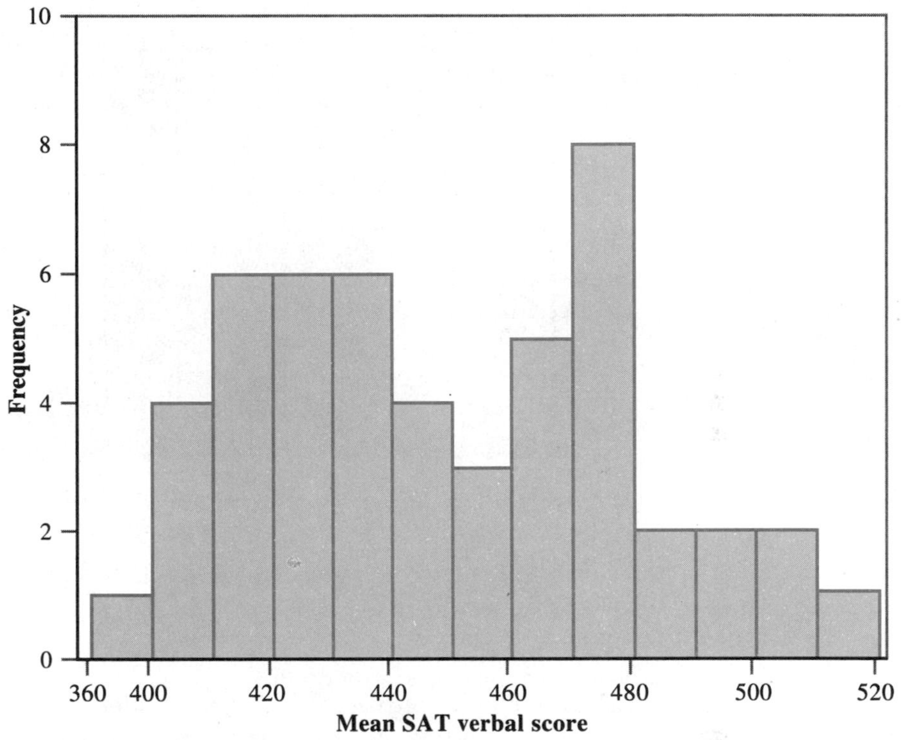

FIGURE 1.8 Histogram of the mean scores on the verbal part of the SAT examination for the 50 states in 1990, for Exercise 1.15.

the SAT. Describe the overall shape of this distribution. How are the two groups of states just described visible in the histogram?

1.16 Figure 1.9 displays the distribution of batting averages for all 167 American League baseball players who batted at least 200 times in the 1980 season. Is the overall shape (ignoring the outlier) roughly symmetric or clearly skewed? What is the approximate batting average of a typical American League regular player? (The outlier is the .390 batting average of George Brett, the highest batting average in the major leagues since Ted Williams hit .406 in 1941. See Exercise 1.80 [page 81] for a comparison of Brett and Williams.)

1.17 The Survey of Study Habits and Attitudes (SSHA) is a psychological test that evaluates college students' motivation, study habits, and attitudes toward school. A selective private college gives the SSHA to a sample of 18 of its incoming first-year college women. Their scores are

154	109	137	115	152	140	154	178	101
103	126	126	137	165	165	129	200	148

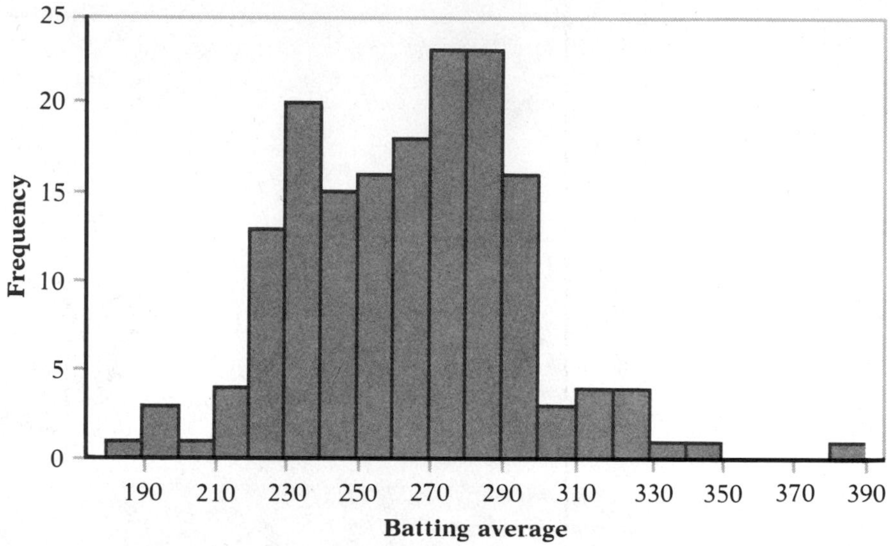

FIGURE 1.9 Batting averages of American League players in 1980, for Exercise 1.16.

Make a stemplot of these data. The overall shape of the distribution is indistinct, as often happens when only a few observations are available. Are there any outliers? What is the approximate value of the median score (the score such that half are higher and half lower)?

1.18 Here are data on the percent of residents 65 years of age and over in each of the 50 states. Make a stemplot of these data using percent as stems and tenths of a percent as leaves. Then make a second stemplot splitting each stem in two (for leaves 0 to 4 and leaves 5 to 9). Which display do you prefer? Describe the shape of the distribution. Is it roughly symmetric or distinctly skewed? Where is the center of the distribution? Are there clear outliers?

Ala.	12.5	Hawaii	10.4	Mass.	13.7	N. Mex.	10.3	S. Dak.	14.0
Alaska	3.8	Idaho	11.7	Mich.	11.7	N.Y.	13.0	Tenn.	12.5
Ariz.	12.8	Ill.	12.2	Minn.	12.5	N.C.	11.9	Tex.	9.9
Ark.	14.6	Ind.	12.2	Miss.	12.3	N.Dak.	13.5	Utah	8.4
Calif.	10.6	Iowa	14.9	Mo.	13.8	Ohio	12.6	Vt.	11.8
Colo.	9.5	Kans.	13.5	Mont.	12.8	Okla.	13.0	Va.	10.6
Conn.	13.4	Ky.	12.4	Nebr.	13.8	Oreg.	13.8	Wash.	11.8
Del.	11.6	La.	10.9	Nev.	10.7	Pa.	14.9	W.Va.	14.3
Fla.	17.8	Maine	13.4	N.H.	11.3	R.I.	14.7	Wis.	13.2
Ga.	10.0	Md.	10.8	N.J.	13.1	S.C.	10.9	Wyo.	9.4

1.19 Thomas the cat is helping researchers study the fleas that cause discomfort to him and his fellow cats. One part of the research concerns the egg production of the flea *Ctenocephalides felis*. The researchers deposit 25 female and 10 male fleas in Thomas's fur and count the number of flea eggs produced each day. The number of eggs produced by the fleas in 27 consecutive days is as follows:

436	495	575	444	754	915	945	655	782	704
590	411	547	584	550	487	585	549	475	435
523	390	425	415	450	395	405			

(Data provided by Sayed Gaafar and Michael Dryden, Purdue University School of Veterinary Medicine.)

(a) Truncate each count by dropping the last digit, and present the distribution in a stemplot. Describe the general shape and the center of the distribution. Are there any unusual values?
(b) Make a time plot of the data (the order of time is along the rows). Are there any trends or other patterns in the daily number of eggs produced by this population of fleas?

1.20 Climatologists interested in flooding gather statistics on the daily rainfall in various cities. The following data set gives the *maximum* daily rainfall (in inches) for each of the years 1941 to 1970 at South Bend, Indiana. (Successive years follow each other across the rows in the table.)

1.88	2.23	2.58	2.07	2.94	2.29	3.14	2.15	1.95	2.51
2.86	1.48	1.12	2.76	3.10	2.05	2.23	1.70	1.57	2.81
1.24	3.29	1.87	1.50	2.99	3.48	2.12	4.69	2.29	2.12

(a) Make a stemplot for these data. Describe the general shape of the distribution and any prominent deviations from the overall pattern. (You are expected to truncate and split stems as needed in making stemplots.)
(b) Make a time plot of the data. Is there any suggestion of a long-term change in maximum rainfall at South Bend?

1.21 Make a stemplot for the data of Exercise 1.11 after you have coded the original measurements. Would a stemplot of the original data have the same appearance? Describe the shape of the distribution.

1.22 The *Old Farmer's Almanac* gives the growing season for major U.S. cities as reported by the National Climatic Center. The growing season is defined as the average number of days between the last frost in the spring and the first frost in the fall. The values are

279	244	318	262	335	321	165	180	201	252
145	192	217	179	182	210	271	302	169	192
156	181	156	125	166	248	198	220	134	189
141	142	211	196	169	237	136	203	184	224
178	279	201	173	252	149	229	300	217	203
148	220	175	188	160	176	128			

(a) Make a stemplot for these data. Describe the general shape of the distribution. Are there any unusual values?
(b) The values for Los Angeles, San Diego, San Francisco, and Miami are represented by an asterisk (*) in the almanac rather than a number. There is a footnote explaining why. Given the above definition of growing season, why do you think there were difficulties with measuring this variable for these cities?

1.23 There are many ways to measure the reading ability of children. Research designed to improve reading performance requires good measures of the outcome. One frequently used test is the Degree of Reading Power (DRP). In a research study on third-grade students, the DRP was administered to 44 students. Their scores were

40	26	39	14	42	18	25	43	46	27	19
47	19	26	35	34	15	44	40	38	31	46
52	25	35	35	33	29	34	41	49	28	52
47	35	48	22	33	41	51	27	14	54	45

Make a stemplot of these data. Then make a histogram. Which display do you prefer, and why? Describe the main features of the distribution. (Data provided by Maribeth Cassidy Schmitt, from her Ph.D dissertation, "The effects of an elaborated directed reading activity on the metacomprehension skills of third graders," Purdue University, 1987.)

1.24 Make a histogram of the distribution of growing season from the data of Exercise 1.22. You must first choose classes and make a frequency table.

1.25 In 1798 the English scientist Henry Cavendish measured the density of the earth by careful work with a torsion balance. The variable recorded was the density of the earth as a multiple of the density of water. Here are Cavendish's 29 measurements.

5.50	5.61	4.88	5.07	5.26	5.55	5.36	5.29	5.58	5.65
5.57	5.53	5.62	5.29	5.44	5.34	5.79	5.10	5.27	5.39
5.42	5.47	5.63	5.34	5.46	5.30	5.75	5.68	5.85	

Present these measurements graphically by either a stemplot or a histogram and explain the reason for your choice. Then briefly discuss the main features of the distribution. In particular, what is your estimate of the density of the earth based on these measurements? (Cavendish's data are reported by S. M. Stigler in the article cited in Note 1.)

1.26 The distribution of the ages of a nation's population has a strong influence on economic and social conditions. The following table shows the age distribution of U.S. residents in 1950 and 2075, in millions of persons. The 1950 data come from that year's census, and the 2075 data are projections made by the Census Bureau.

Age group	1950	2075
Under 10 years	29.3	34.9
10–19 years	21.8	35.7
20–29 years	24.0	36.8
30–39 years	22.8	38.1
40–49 years	19.3	37.8
50–59 years	15.5	37.5
60–69 years	11.0	34.5
70–79 years	5.5	27.2
80–89 years	1.6	18.8
90–99 years	0.1	7.7
100–109 years	—	1.7
Total	151.1	310.6

(a) Make a histogram of the 1950 age distribution, and describe the main features of the distribution. In particular, look at the number of children relative to the rest of the population.

(b) Make a histogram of the projected age distribution for the year 2075. What are the most important changes in the U.S. age distribution projected for the 125-year period between 1950 and 2075?

1.27 There is some evidence that increasing the amount of calcium in the diet can lower blood pressure. In a medical experiment one group of men was given a daily calcium supplement, while a control group received a placebo (a dummy pill). The seated systolic blood pressure of each man was measured before the treatments began and again after 12 weeks. The blood pressure distributions in the two groups should have been similar at the beginning of the experiment. Here are the initial blood pressure readings for the two groups.

Calcium group

107 110 123 129 112 111 107 112 136 102

Placebo group

123 109 112 102 98 114 119 112 110 117 130

Make a back-to-back stemplot of these data. Does your plot show any major differences in the two groups before the treatments began? In particular, are the centers of the two blood pressure distributions close together?

1.28 Plant scientists have developed varieties of corn that have increased amounts of the essential amino acid lysine. In a test of the protein quality of this corn, an experimental group of 20 one-day-old male chicks was fed a ration containing the new corn. A control group of another 20 chicks was fed a ration that was identical except that it contained normal corn. Here are the weight gains (in grams) after 21 days:

Control				Experimental			
380	321	366	356	361	447	401	375
283	349	402	462	434	403	393	426
356	410	329	399	406	318	467	407
350	384	316	272	427	420	477	392
345	455	360	431	430	339	410	326

Make a back-to-back stemplot of these data. Does it appear that the chicks fed high-lysine corn grew faster? Are there any outliers or other problems? (Based on G. L. Cromwell et al., "A comparison of the nutritive value of *opaque-2, floury-2* and normal corn for the chick," *Poultry Science*, 47 (1968), pp. 840–847.)

1.29 A manufacturing company is reviewing the salaries of its full-time employees below the executive level at a large plant. The clerical staff is almost entirely female, but a majority of the production workers and technical staff are male. As a result, the distributions of salaries for male and female employees may be quite different. Table 1.3 gives the frequencies and relative frequencies for women and men. Make histograms from these data, choosing the type that is most appropriate for comparing the two distributions. Then describe the overall shape of the two salary distributions and the chief differences between them.

TABLE 1.3 Salary distributions of female and male workers in a large manufacturing plant

Salary	Women		Men	
($1000)	Number	%	Number	%
10–15	89	11.8	26	1.1
15–20	192	25.4	221	9.0
20–25	236	31.2	677	27.9
25–30	111	14.7	823	33.6
30–35	86	11.4	365	14.9
35–40	25	3.3	182	7.4
40–45	11	1.5	91	3.7
45–50	3	.4	33	1.4
50–55	2	.3	19	.8
55–60	0	0	11	.4
60–65	0	0	0	0
65–70	1	.1	3	.1
Total	756	100.1	2451	100.0

1.30 A report on the recent graduates of a large state university includes the following relative frequency table of the first-year salaries of last year's graduates. Salaries are in $1000 units, and it is understood that each class includes its left endpoint but not its right endpoint—for example, a salary of exactly $15,000 belongs in the second class.

Salary	10–15	15–20	20–25	25–30	30–35	35–45	45–55	55–75
Percent	4	10	21	26	18	13	5	3

The last three classes are wider than the others. An accurate histogram must take this into account. If the base of each bar in the histogram covers a class and the height is the percent of graduates with salaries in that class, the areas of the three rightmost bars will overstate the percent who have salaries in these classes. To make a correct histogram, the area of each bar must be proportional to the percent in that class. Most classes are $5000 wide. A class *twice* as wide ($10,000) should have a bar *half* as tall as the percent in that class. This keeps the area proportional to the percent. How should you treat the height of the bar for a class $20,000 wide? Make a correct histogram with the heights of the bars for the last three classes adjusted so that the areas of the bars reflect the percent in each class.

1.31 The National Health Survey measured the heights of 52,744 males between the ages of 18 and 79. Here is a frequency table of the data as given in a government publication:

Height (inches)	Count	Height (inches)	Count	Height (inches)	Count
Under 62	675	66	7021	71	3216
62	874	67	6249	72	2817
63	1720	68	9379	73	1103
64	3691	69	5421	74	581
65	3488	70	6239	75 and over	270

This table has *open classes* on both ends. For example, we know that 675 men were less than 62 inches tall, but we do not know the lower endpoint of this interval. In making a histogram we cannot draw bars for open classes. Instead, put a sign such as * at each end and write a note stating how many observations fell below (or above) the classes drawn. Make a histogram for the National Health Survey data, and discuss the center and shape of the distribution of heights of American men.

1.32 The years around 1970 brought unrest to many U.S. cities. Here are data on the number of civil disturbances in each 3-month period during the years 1968 to 1972:

Period	Count	Period	Count
1968, Jan.–Mar.	6	1970, July–Sept.	20
Apr.–June	46	Oct.–Dec.	6
July–Sept.	25	1971, Jan.–Mar.	12
Oct.–Dec.	3	Apr.–June	21
1969, Jan.–Mar.	5	July–Sept.	5
Apr.–June	27	Oct.–Dec.	1
July–Sept.	19	1972, Jan.–Mar.	3
Oct.–Dec.	6	Apr.–June	8
1970, Jan.–Mar.	26	July–Sept.	5
Apr.–June	24	Oct.–Dec.	5

(a) Make a time plot of these counts. Connect the points in your plot by straight line segments to make the pattern clearer.
(b) Describe the trend and seasonal variation in this time series. Can you suggest an explanation for the seasonal variation in civil disorders?

1.33 We often look at time series data to see the effect of a social change or new policy. Here are data on motor vehicle deaths in the United States. Because the *number* of deaths will tend to rise as motorists drive more miles, we look instead at the *rate* of deaths, which is the number of deaths per 100 million miles driven.

Year	Rate	Year	Rate
1960	5.1	1976	3.3
1962	5.1	1978	3.3
1964	5.4	1980	3.3
1966	5.5	1982	2.8
1968	5.2	1984	2.6
1970	4.7	1986	2.5
1972	4.3	1988	2.4
1974	3.5	1990	2.2

(a) (a) Make a time plot of these death rates. Describe the overall pattern of the data.
(b) In 1974 the national speed limit was lowered to 55 miles per hour in an attempt to conserve gasoline after the 1973 Mideast war. In the mid-1980s most states raised speed limits on interstate highways to 65 miles per hour. Some said that the lower speed limit saved lives. Is the effect of lower speed limits between 1974 and the mid-1980s visible in your plot?

1.34 Babe Ruth was a pitcher for the Boston Red Sox in the years 1914 to 1917. In 1918 and 1919 he played some games as a pitcher and some as an outfielder. From 1920 to 1934 Ruth was an outfielder for the New York

Yankees. He ended his career in 1935 with the Boston Braves. Here are the number of home runs Ruth hit in each year:

1914	1915	1916	1917	1918	1919	1920	1921	1922	1923	1924
0	4	3	2	11	29	54	59	35	41	46

1925	1926	1927	1928	1929	1930	1931	1932	1933	1934	1935
25	47	60	54	46	49	46	41	34	22	6

Earlier we examined the distribution of Ruth's home run totals during his Yankee years. Now make a time plot and describe its main features.

1.35 Here are the numbers of international airline passengers (in thousands) for each month of the years 1954 to 1956. Plot this time series. Identify the major patterns present (trend, cycles, seasonal variation). Suggest an explanation for the patterns that you see. (Part of a larger data set given by G. E. P. Box and G. M. Jenkins, *Time Series Analysis*, Holden-Day, Oakland, Calif., 1976, p. 531.)

Month	1954	1955	1956
Jan.	204	242	284
Feb.	188	233	277
Mar.	235	267	317
Apr.	227	269	313
May	234	270	318
June	264	315	374
July	302	364	413
Aug.	293	347	405
Sept.	259	312	355
Oct.	229	274	306
Nov.	203	237	271
Dec.	229	278	306

1.36 As the earth revolves around the sun, the number of daylight hours varies in a regular pattern. The following table gives the daylight hours (in hours and minutes) for Boston on the first day of the month for each month in 1987:

Month	Hours	Month	Hours	Month	Hours
Jan.	9:08	May	14:04	Sept.	13:10
Feb.	10:00	June	15:04	Oct.	11:45
Mar.	11:14	July	15:14	Nov.	10:21
Apr.	12:43	Aug.	14:28	Dec.	9:19

(a) Plot the length of daylight versus month. (Hint: First convert all numbers to minutes.) Connect the points with a smooth curve.

(b) Explain why the point for January at the beginning of the plot should have a value close to that for December at the end of the plot.
(c) Recall that the values are given for the first of each month. Using your smooth curve, give an estimate for the length of the day on December 15. The actual value is 9:06. What is the difference between the actual value and your guess?
(d) Again using your smooth curve, estimate the length of the day for April 15. The actual value is 13:22. Compute the difference between the actual value and your guess.
(e) Using your smooth curve you could give an estimate for any day of the year. From what you learned in parts (c) and (d) of this exercise, give an estimate of the worst error (difference between the actual value and your guess) that you would expect if you guessed the number of daylight hours for each day of the year in Boston using your smooth curve. Justify your answer by a careful consideration of your plot.

1.37 The following table gives the winning times (in minutes, rounded to the nearest minute) for the Boston Marathon in the years 1959 to 1991:

Year	Time	Year	Time	Year	Time
1959	143	1970	131	1981	129
1960	141	1971	139	1982	129
1961	144	1972	136	1983	129
1962	144	1973	136	1984	131
1963	139	1974	134	1985	134
1964	140	1975	130	1986	128
1965	137	1976	140	1987	132
1966	137	1977	135	1988	129
1967	136	1978	130	1989	129
1968	142	1979	129	1990	128
1969	134	1980	132	1991	131

(a) Plot the winning times against year.
(b) Give a brief description of the pattern of Boston Marathon winning times over these years.

1.38 There are seasonal patterns in the growth of children. We can examine the pattern for some children using data from the Egyptian village of Kalama. The following tables give the average weights (in kilograms) and heights (in centimeters) of Egyptian toddlers who reach the age of 24 months in each month of the year. Because there are only a few children who are exactly 24 months old in any one month, a fairly complicated statistical procedure has been used to obtain the table entries from the data for all children between 18 and 30 months of age. Changes in these standardized weights and heights indicate slower or faster average growth in the entire group of children. (Data courtesy of Linda D. McCabe and the Agency for International Development.)

Month	Weight (kg)	Month	Weight (kg)	Month	Weight (kg)
Jan.	11.26	May	10.87	Sept.	10.87
Feb.	11.39	June	10.80	Oct.	10.74
Mar.	11.37	July	10.77	Nov.	10.89
Apr.	11.10	Aug.	10.79	Dec.	11.25

Month	Height (cm)	Month	Height (cm)	Month	Height (cm)
Jan.	79.80	May	80.10	Sept.	79.94
Feb.	79.66	June	80.23	Oct.	79.80
Mar.	79.91	July	80.07	Nov.	79.39
Apr.	80.00	Aug.	79.79	Dec.	79.51

(a) Plot the weight data. Describe any seasonal patterns that appear in the plot. In particular, which months correspond to the highest weights? Which correspond to the lowest?

(b) Plot the height data. Describe any seasonal patterns that appear in the plot. Which months correspond to the largest heights? Which correspond to the smallest?

(c) Describe the similarities and differences between the weight plot and the height plot.

1.39 Refer to the previous exercise. In villages such as Kalama, the prevalence of illness may slow the growth of children. In particular, diarrhea is a serious problem and a major cause of death for small children. The following table gives the average percentage of days ill with diarrhea per month for the children whose weight and height information appears in the previous exercise:

Month	Percent	Month	Percent	Month	Percent
Jan.	2.54	May	8.42	Sept.	1.20
Feb.	1.82	June	6.87	Oct.	2.51
Mar.	2.28	July	4.51	Nov.	2.34
Apr.	6.86	Aug.	3.42	Dec.	1.27

(a) Plot the data. Describe any seasonal patterns that appear in the plot. Which months have the highest average percentage of days ill with diarrhea? Which months have the lowest?

(b) Egyptian physicians say there are two peaks in the frequency of this disease, a primary peak and a secondary, lower peak. Is there evidence in these data to support this claim?

(c) Compare your plot with those of the previous exercise. Can you formulate any theories concerning the relationship between diarrhea and the growth of these children? Note that illness in one month may not have an effect on growth until subsequent months. On the other hand, a child who is not growing well may be more susceptible to illness.

1.2 DESCRIBING DISTRIBUTIONS

"Face it. A hot dog isn't a carrot stick." So said *Consumer Reports*, commenting on the low nutritional quality of the all-American frank. Table 1.4 shows the magazine's laboratory test results for calories and milligrams of sodium (mostly due to salt) in a number of major brands of hot dogs. There are three types: all beef, "meat" (mainly pork and beef, but government regulations allow up to 15% poultry meat), and poultry. Because people concerned about health may prefer low-calorie, low-salt hot dogs, we ask: Are there any systematic differences among the three types in these two variables? We can begin to answer this question by comparing stemplots of the three distributions of calorie content, one for each type, and by making a similar comparison for sodium content. The comparisons can be made more explicit by using some numerical tools for describing distributions. This section will develop these tools.

Measuring center

We saw in Section 1.1 that a numerical measure of the center of a distribution is helpful in inspecting a stemplot or histogram. The most common measure of center is the ordinary arithmetic average, or mean.

TABLE 1.4 Calories and sodium in hot dogs by type

Beef hot dogs		Meat hot dogs		Poultry hot dogs	
Calories	Sodium	Calories	Sodium	Calories	Sodium
186	495	173	458	129	430
181	477	191	506	132	375
176	425	182	473	102	396
149	322	190	545	106	383
184	482	172	496	94	387
190	587	147	360	102	542
158	370	146	387	87	359
139	322	139	386	99	357
175	479	175	507	170	528
148	375	136	393	113	513
152	330	179	405	135	426
111	300	153	372	142	513
141	386	107	144	86	358
153	401	195	511	143	581
190	645	135	405	152	588
157	440	140	428	146	522
131	317	138	339	144	545
149	319				
135	298				
132	253				

Source: *Consumer Reports*, June 1986, pp. 366–367.

MEAN

If n observations are denoted by $x_1, x_2, \ldots, x_n$, their mean is

$$\overline{x} = \frac{1}{n}(x_1 + x_2 + \cdots + x_n)$$

or in more compact notation

$$\overline{x} = \frac{1}{n}\sum x_i \qquad (1.1)$$

The $\sum$ (capital Greek sigma) in the formula for the mean is short for "add them all up." The bar over the x indicates the mean of all the x-values. This notation is so common that writers in many fields use $\overline{x}$, $\overline{y}$, etc., without additional explanation. The subscripts on the observations x_i are just a way of keeping the n observations distinct. They do not necessarily indicate order or any other special facts about the data.

EXAMPLE 1.10

Here are the number of home runs hit by Babe Ruth in each of his seasons as a Yankee:

54 59 35 41 46 25 47 60 54 46 49 46 41 34 22

Ruth's mean number of home runs hit in a year is

$$\overline{x} = \frac{1}{15}(54 + 59 + \cdots + 22) = \frac{659}{15} = 43.9$$

Roger Maris's mean, from the data on page 8, is

$$\overline{x} = \frac{1}{10}(8 + 13 + \cdots + 61) = \frac{261}{10} = 26.1$$

Ruth's superiority is evident from these averages. ●

The distribution of Maris's yearly home run production shows a clear outlier, his record 61 homers in 1961. If we exclude this as atypical, the mean number of home runs in the other 9 years of his American League career is 22.2. The single outlier has increased the 10-year average by about four homers per year. This illustrates an important weakness of the mean as a measure of center: It is sensitive to the influence of a few extreme observations. These may be outliers, but a skewed distribution that has no outliers will also pull the mean toward its long tail. Because the mean cannot resist the influence of extreme observations, we say that it is not a *resistant measure* of center. A measure that is resistant does more than limit the influence of outliers; its value does not respond strongly to changes in a few observations, no matter how large those changes may be.

resistant measure

The mean fails this requirement, because we can make the mean as large as we wish by making a large enough increase in just one observation.

We made informal use of a resistant measure of center, the median, in the previous section. The median is the midpoint of a distribution, the point such that half of the observations fall above it and half below. We now make this concept more specific by giving a rule for calculating the median.

> **MEDIAN**
>
> To compute the median of a distribution:
>
> 1. Arrange all observations in order of size, from smallest to largest.
> 2. If the number n of observations is odd, the median M is the center observation in the ordered list. The location of the median is found by counting $(n+1)/2$ observations up from the bottom of the list.
> 3. If the number n of observations is even, the median M is the average of the two center observations in the ordered list. The location of the median is again $(n+1)/2$ from the bottom of the list.

Note that the formula $(n+1)/2$ does *not* give the median, only the location of the median in the ordered list. Medians require little arithmetic, so they are easy to find by hand for small sets of data. Arranging even a moderate number of observations in order is very tedious, however, so it is often quicker to calculate the mean with a calculator than it is to locate the median.

EXAMPLE 1.11

To find the median number of home runs hit in a year by Babe Ruth, we arrange the data in increasing order:

22 25 34 35 41 41 46 **46** 46 47 49 54 54 59 60

The median is the bold 46, the eighth observation in the ordered list. You can find the median by eye—there are seven observations to the left and seven to the right. Or you can use the recipe $(n+1)/2 = 16/2 = 8$ to locate the median in the list.

Roger Maris's median home run production is found in a similar way. His yearly totals in increasing order are

8 13 14 16 **23** **26** 28 33 39 61

Since the number of observations $n = 10$ is even, there is a center pair of numbers rather than a single center point. This pair of numbers is shown in bold in the list. The median M is therefore

$$M = \frac{23 + 26}{2} = \frac{49}{2} = 24.5$$

The recipe $(n + 1)/2 = 11/2 = 5.5$ for the position of the median in the list means that the median is at location "five and one-half," that is, halfway between the fifth and sixth observations. ●

Notice that because a stemplot arranges the observations in increasing order you can compute the median from a stemplot without rewriting the data. Notice also that the median, unlike the mean, is resistant. Maris's record 61 home runs is simply one observation falling above the median. Even if this observation were erroneously entered as 610, the median would not change.

The mean and median have the same value for symmetric distributions. But because the mean is sensitive to extreme observations, it will move away from the median toward the long tail in skewed distributions. This effect can be substantial. For example, the Insurance Information Institute once pointed out that the "average" award in product liability cases exceeded \$1 million. This, said the institute, is why liability insurance coverage is so expensive. Opponents of increases in insurance premiums replied that the median award was only \$271,000. The two measures differ dramatically because the average (the mean) allows a few huge awards to overwhelm the effects of many smaller cases. The lack of resistance of the mean should not lead you to think that the median is always preferred. The mean is computed from the actual values of the observations and contains more information than the median. Insurance companies, for example, are concerned about the total claims payments they must make. This total can be recovered from the mean award, but not from the median award.

We now have two general strategies for dealing with outliers and other irregularities in data. The first strategy requires that outliers be detected and their causes investigated. The offending observations can then be corrected, deleted for good reason, or otherwise given individual attention. The second strategy is to use resistant methods, so that outliers have little influence over our conclusions. In general, we prefer to examine data carefully and consider any irregularity in the light of the individual situation. This will sometimes lead to a decision to employ resistant methods.

In the case of Roger Maris, for example, the median is more typical of his performance than is the mean, but it obscures the great event of his career. A better summary might be: "Maris hit 61 home runs in 1961 and averaged 22.2 homers in his other 9 years in the league." Newcomb's data, on the other hand, are supposed to be repeated measurements of the same quantity, but they contain unexplained outliers. Even though the outlying values may be unexplained, they should not be given much influence over our estimate of the velocity of light. We can delete the outliers from a computation of the mean, or we can report the median. If the two outliers are omitted, the mean of the remaining 64 measurements is 27.75. The median of 66 observations has position $(66 + 1)/2 = 33.5$ in the ordered list of passage times from Table 1.1 and can be found to be 27. Either of these values is preferable to the overall mean (26.21) as a statement of Newcomb's value for the travel time of light.

Resistant measures of spread

Even the briefest summary of a distribution requires an indication of how spread out, or variable, the data are in addition to a measure of their center. Two nations with identical median personal incomes are very different if in one there are extremes of wealth and poverty, while in the other equality of incomes is the rule. A drug whose mean potency is just what the doctor ordered is defective if some lots have dangerously high potency and others are too weak to be effective.

percentile

The spread or variability of a distribution can be indicated by giving several percentiles. The *pth percentile* of the distribution is the value such that p percent of the observations fall at or below it. The median is just the 50th percentile, so the use of percentiles to report spread is particularly appropriate when the median is our measure of center. The most commonly used percentiles other than the median are the *quartiles*. The first quartile is the 25th percentile, and the third quartile is the 75th percentile. (The second quartile is the median itself.) To calculate a percentile, arrange the observations in increasing order and count up the required percent from the bottom of the list. Our definition of percentiles is a bit inexact, because there is not always a value with exactly p percent of the data at or below it. We will be content to take the nearest observation for most percentiles, but the quartiles are important enough to require an exact recipe.

QUARTILES

To calculate the quartiles, first locate the median in the ordered list of observations. The first quartile is the median of the observations below the location of the median, and the third quartile is the median of the observations above that location.

EXAMPLE 1.12

In Example 1.5 (page 9), we arranged the shopping data in order in a stemplot, after truncating to eliminate the cents. The truncated data are sufficient to give us a quick numerical summary of the distribution. Here they are, rewritten for convenience from the stemplot:

2	6	6	8	9	10	11	11	12	12	13	13	14	14	14	15	15
16	17	18	18	18	19	19	20	‖	20	20	21	23	26	27	28	29
30	31	32	33	33	34	36	37	39	40	43	45	52	61	63	64	69

There are $n = 50$ observations, so the position of the median is $(50 + 1)/2 = 25.5$, or midway between the 25th and 26th observations. This location is marked by ‖ in the list. Both center observations are \$20, so the median is \$20. To find the first quartile, consider the 25 observations falling below the location ‖ of the median. Note carefully that the lower of the two center observations is included in this group, because the location ‖ falls between two data points. The median of these 25 observations is the thirteenth in order, or \$14. This is the first quartile. You can find the third quartile similarly as the thirteenth value above ‖, or \$33. We can summarize these results in compact form as

$$Q_1 = \$14 \quad M = \$20 \quad Q_3 = \$33$$

Other percentiles are found more informally. For example, we take the 95th percentile for the 50 observations to be the 48th in the ordered list, because $0.95 \times 50 = 47.5$, which we round to 48. The 95th percentile is therefore \$63. ●

When there are an odd number of observations, the median is the unique center observation, and the rule for finding the quartiles excludes this center value. For example, the median of Ruth's 15 home run totals

22 25 34 35 41 41 46 **46** 46 47 49 54 54 59 60

is the bold 46 (the eighth entry). The first quartile is the median of the seven observations falling below this point in the list, $Q_1 = 35$. Similarly, $Q_3 = 54$. Some statistical computing systems use slightly different rules to compute the median and the quartiles, so the results given by a computer may not agree exactly with the results found by using our rules. However, any differences will be small.

The quartiles together with the median give some indication of the center, spread, and shape of a distribution. In Example 1.12, the fact that Q_3 lies farther from M than does Q_1 reflects the right skewness of the data. The distance between the quartiles is a simple measure of spread that gives the range covered by the middle half of the data. This distance is called the *interquartile range*, or in symbols

interquartile range

$$IQR = Q_3 - Q_1$$

In Example 1.12, $IQR = \$33 - \$14 = \$19$. The quartiles and the IQR are not affected by changes in either tail of the distribution. They are therefore resistant, because changes in a few data points have no effect once these points move outside the quartiles. You should be aware that no single numerical measure of spread, such as IQR, is very useful for describing skewed distributions. The spreads of the two sides of a skewed distribution are unequal, so that the two quartiles Q_1 and Q_3 (given with the median M) are more informative.

One common rule of thumb for identifying suspected outliers singles out values falling at least $1.5 \times IQR$ above the third quartile or below the first quartile. In Example 1.12, $1.5 \times IQR = 1.5 \times \$19 = \$28.5$, so any values at or below $\$14 - \$28.5 = -\$14.5$ and at or above $\$33 + \$28.5 = \$61.5$ are flagged. In the case of the strongly skewed distribution of grocery purchases, the flagged values (\$63, \$64, and \$69) do not appear to be outliers in the sense of deviating from the overall pattern of the distribution. Using the $1.5 \times IQR$ rule to spot outliers does not replace using judgment, but it is helpful when the process must be automated.

$1.5 \times IQR$

Because Q_1, M, and Q_3 contain no information about the tails, a fuller summary of the shape of a distribution can be obtained by giving the highest and lowest values as well. This is known as the five-number summary.

> **FIVE-NUMBER SUMMARY**
>
> The five-number summary of a distribution consists of the median M, the quartiles Q_1 and Q_3, and the smallest and largest individual observations, written in the order
>
> $$\text{Minimum}, \ Q_1, \ M, \ Q_3, \ \text{Minimum}$$

The five-number summary for the grocery spending data of Example 1.12 is

$$\$2 \quad \$14 \quad \$20 \quad \$33 \quad \$69$$

The IQR can be computed from the five-number summary, as can other helpful differences such as the distances of the quartiles from the median.

The five-number summary leads to another visual representation of a distribution, the *boxplot*. Figure 1.10(a) is a boxplot of the distribution of grocery store purchases from Example 1.12. In a boxplot,

boxplot

1. The ends of the box are at the quartiles, so that the box length is the IQR.
2. The median is marked by a line within the box.
3. The two lines (called whiskers) outside the box extend to the smallest and largest observations.

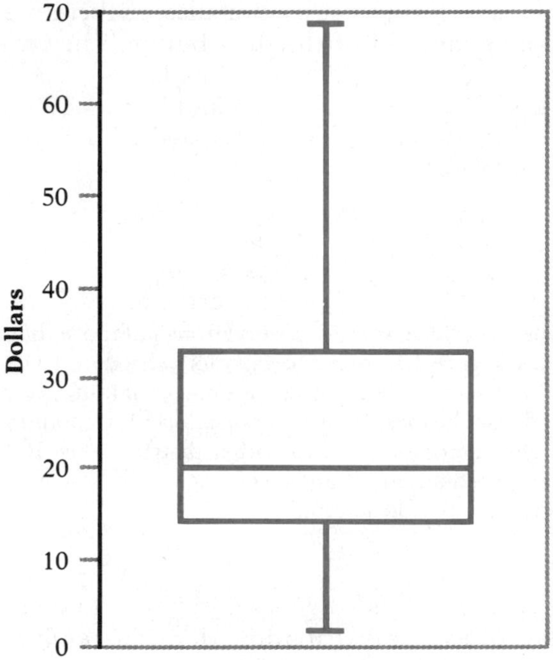

FIGURE 1.10(a) Boxplot of the amount spent by 50 shoppers at a grocery store, from Example 1.12.

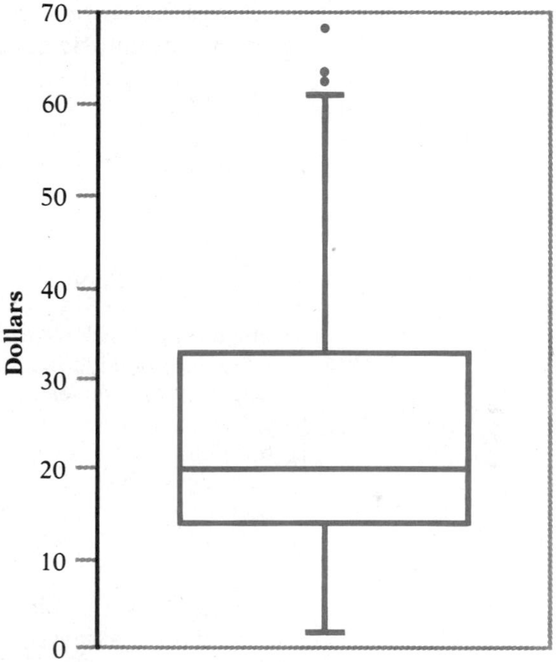

FIGURE 1.10(b) Modified boxplot of the grocery spending data.

The center, spread, and overall range of the distribution are immediately apparent from a boxplot. There are many variations on the boxplot idea. Modifications appropriate for large numbers of observations are particularly important. Because the smallest and largest observations in a large data set are often extreme observations that tell us little about the shape of the distribution, we can replace them in both the five-number summary and the boxplot by the 5th and 95th percentiles. This would be appropriate, for example, in a summary of the incomes of all 600,000 lawyers in the United States.

modified boxplot

When dealing with a moderate numbers of observations, it is worthwhile to plot potential outliers individually. To do this in a boxplot, extend the whiskers to the extreme high and low observations *only if* these values are less than $1.5 \times IQR$ beyond the quartiles. Otherwise, end the whiskers at the most extreme observations still within $1.5 \times IQR$ of the quartiles and plot the remaining cases individually. This plot is a *modified boxplot*. Figure 1.10(b) is a modified boxplot of the grocery purchase distribution. The upper whisker extends to 61, which is the largest observation within the $1.5 \times IQR$ limit, and the suspected outliers \$63, \$64, and \$69 are marked separately. We recommend modified boxplots except for very large numbers of observations.

Boxplots are most useful for comparing distributions. Let us therefore return to the calorie content of hot dogs.

EXAMPLE 1.13

The five-number summaries of the distributions of calories for the three types of hot dogs in Table 1.4 on page 36 are

Beef	111	140	152.5	178.5	190
Meat	107	139	153	179	195
Poultry	86	102	129	143	170

Calculate the *IQR* for each type, and check that no observations fall more than $1.5 \times IQR$ outside the quartiles. Figure 1.11 presents boxplots based on these calculations. We see at once that poultry hot dogs as a group contain fewer calories than beef or meat hot dogs. The median number of calories in a poultry hot dog is less than the first quartile of either of the other distributions. But each type shows quite a large spread among brands, so that simply buying a poultry frank does not guarantee a low-calorie food. ●

Figure 1.11 also illustrates why boxplots are generally inferior to stemplots and histograms as displays of a single distribution. Here is a stemplot of the calorie content of the 17 brands of meat hot dogs:

10	7
11	
12	
13	5689
14	067
15	3
16	
17	2359
18	2
19	015

There are two distinct clusters of brands and one outlier in the lower tail. The boxplot hid the clusters. The quartiles $Q_1 = 139$ and $Q_3 = 179$ are roughly at the centers of the clusters, so that much of the *IQR* is the spread between the two clusters. Because of this, the $1.5 \times IQR$ rule used in drawing the boxplot did not call attention to the outlier. Returning to the *Consumer Reports* article from which the data were taken supplies an explanation for the shape of the distribution. Not all brands of hot dogs have the same weight. It turns out that all of the brands in the high-calorie cluster weigh 2 ounces each, but 7 of the 8 in the lower cluster weigh 1.6 ounces each. The heavier franks naturally tend to have more calories. The outlier is unusual in two ways. It is the calorie content of Eat Slim Veal

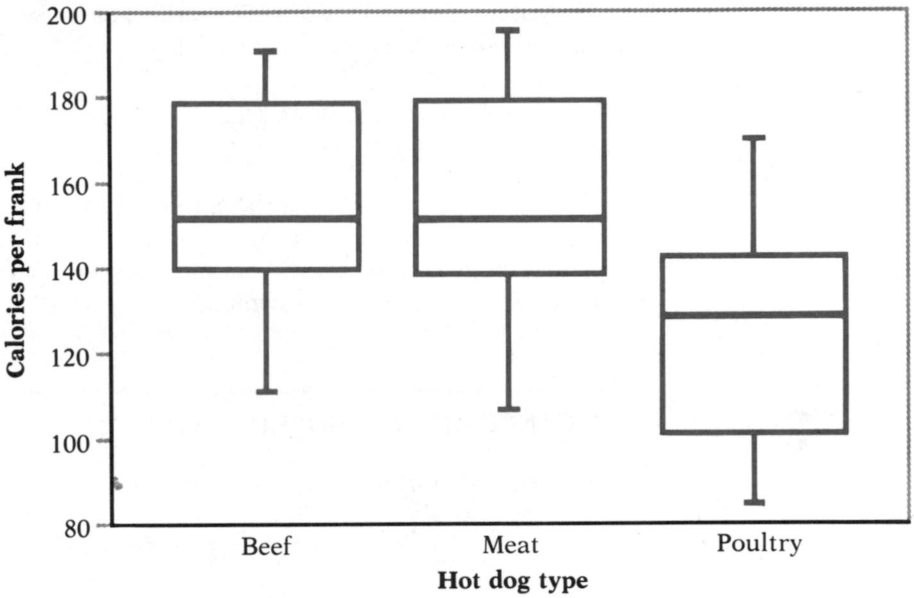

FIGURE 1.11 Boxplots of the calorie content of three types of hot dogs, from Table 1.4.

Hot Dogs, the only frank in the group that is not composed of some mix of pork, beef, and poultry. But this brand also weighs only 1.4 ounces. If *Consumer Reports* had reported calories per ounce, Eat Slim would not be an outlier.[7]

As this example illustrates, boxplots are best used for comparing several distributions. A stemplot (or, for large data sets, a histogram) provides a clearer display of a single distribution, especially when accompanied by the median and quartiles as numerical signposts.

The distribution of calorie content of meat hot dogs also illustrates the limitations of numerical summaries of distributions. Neither the mean nor the median conveys much useful information about this distribution, because a clear "center" is not present. The median falls near the gap between the two clusters and is not a typical calorie value. The main message about a double-peaked distribution is simply its shape, and shape is best portrayed visually. The routine methods of statistics compute numerical measures and draw conclusions based on their values. These methods are extremely useful, and we will study them carefully in later chapters. But they cannot be applied blindly, by feeding data to a computer program, because *statistical measures and methods based on them are generally meaningful only for distributions of sufficiently regular shape*. This principle will become clearer as we progress, but it is good to be aware at the beginning that quickly resorting to fancy calculations is the mark of a statistical amateur. The trained practitioner looks, thinks, and chooses calculations selectively.

The standard deviation

The most common measure of the spread of a distribution is not the *IQR* or other resistant measure, but the standard deviation. This is a measure of spread about the mean and should only be used when the mean is employed as your measure of center.

> **VARIANCE AND STANDARD DEVIATION**
>
> The variance of n observations $x_1, x_2, \ldots, x_n$ is
>
> $$s^2 = \frac{1}{n-1}[(x_1 - \bar{x})^2 + (x_2 - \bar{x})^2 + \cdots + (x_n - \bar{x})^2]$$
>
> or more compactly
>
> $$s^2 = \frac{1}{n-1}\sum(x_i - \bar{x})^2 \qquad (1.2)$$
>
> The standard deviation s is the square root of the variance s^2.

The idea behind the variance and the standard deviation as measures of spread is as follows: The deviations $x_i - \bar{x}$ display the spread of the values x_i about their mean $\bar{x}$. Some of these deviations will be positive and some negative because some of the observations fall on each side of the mean. So we cannot simply add the deviations to get an overall measure of spread; in fact, the sum of the deviations $x_i - \bar{x}$ will always be 0. Squaring the deviations makes them all positive, and observations far from the mean in either direction will have large positive squared deviations. The variance is roughly the average squared deviation. So s^2 will be large if the observations are widely spread about their mean and small if the observations are all close to the mean. Because the variance involves squaring the deviations, it does not have the same unit of measurement as the original observations. Lengths measured in centimeters have a variance measured in squared centimeters. Taking the square root remedies this, so that the standard deviation s measures dispersion about the mean in the original scale.

If the variance is the average of the squares of the deviations of the observations from their mean, why do we average by dividing by $n - 1$ rather than n? Because the sum of the deviations $x_i - \bar{x}$ is always 0, the last deviation can be found once we know the first $n - 1$. So we are not averaging n unrelated numbers. Only $n - 1$ of the squared deviations can vary freely, and we average by dividing the total by $n - 1$. The number $n - 1$ is called the *degrees of freedom* of the variance or standard deviation.

degrees of freedom

EXAMPLE 1.14

Babe Ruth's 15 yearly home run totals were

54 59 35 41 46 25 47 60 54 46 49 46 41 34 22

As we saw in Example 1.10, the mean of this distribution is

$$\bar{x} = \frac{1}{n}\sum x_i = \frac{659}{15} = 43.93$$

To compute the variance from Equation 1.2 we subtract 43.93 from each observation, square the resulting deviations, add, and divide by 14 as follows:

$$\begin{aligned} s^2 &= \frac{1}{14}[(54 - 43.93)^2 + (59 - 43.93)^2 + \cdots + (22 - 43.93)^2] \\ &= \frac{101.40 + 227.10 + \cdots + 480.92}{14} \\ &= \frac{1770.93}{14} = 126.495 \end{aligned}$$

The standard deviation is therefore $s = \sqrt{126.495} = 11.25$. ●

Computing the standard deviation is our first substantial piece of arithmetic, so some comments on calculation are in order. You will find the task of calculating s and other statistical measures more or less tedious depending on the resources at hand. Many calculators will compute s directly from keyed-in data. Anyone working with statistics should have such a calculator. Interactive computer software is even more desirable because its display and editing capabilities help prevent mistakes in entering the data. Statistical software will also do graphical displays and the more elaborate statistical procedures that we will meet later. Calculating descriptive measures like s step-by-step as in Example 1.14 is rapidly becoming outdated. We nonetheless recommend that you do a few examples in detail to be certain that you understand how the formula works.

You will also find that the exact numerical value of your answer will depend on how you did the calculation. In general, whenever you write down an intermediate result without saving all the digits produced by your calculator, you introduce *roundoff error* into the following steps in the calculation. In Example 1.14, the mean $\bar{x}$ is not exactly 43.93, so our calculation of s will suffer from roundoff error. Since calculators and computers can handle only a fixed number of digits, roundoff error is always with us. Usually these errors are small, and it is the business of program

roundoff error

designers to minimize them in constructing the programs for a calculator or computer system. Roundoff errors will therefore not be of great concern to us. If you must use intermediate results that will later be keyed back into your calculator, save at least one more digit than the data you started with had. Most of the calculations in this book are done on a computer, so if you check them without a computer you may get slightly different results.

Computing formula for the variance* If you do statistical calculations with a basic calculator, you will need to know alternative formulas that are designed for easier calculation of such quantities as s^2. Equation 1.2 is the *defining formula* for the variance. That is, this equation shows how s^2 measures spread about the mean. But Equation 1.2 is awkward to use because you must first subtract the mean $\bar{x}$ from each individual observation. A bit of algebra shows that an equivalent formula is

$$s^2 = \frac{1}{n-1}\left[\sum x_i^2 - \frac{1}{n}\left(\sum x_i\right)^2\right] \tag{1.3}$$

This is a *computing formula* for the variance; it obscures the meaning of s^2 but leads to shorter calculations. Equation 1.3 uses the basic quantities $\sum x_i$ and $\sum x_i^2$, which can be obtained quickly on a calculator with a memory and a square button without writing down intermediate results.

EXAMPLE 1.15

Returning to the calculation of Example 1.14, we first find the building block sums

$$\sum x_i = 54 + 59 + \cdots + 22 = 659$$

$$\sum x_i^2 = 54^2 + 59^2 + \cdots + 22^2 = 30,723$$

Because $n = 15$, the variance calculated from Equation 1.3 is

$$\begin{aligned} s^2 &= \frac{1}{14}\left(30,723 - \frac{659^2}{15}\right) \\ &= \frac{1}{14}\left(30,723 - \frac{434,281}{15}\right) \\ &= \frac{1}{14}(30,723 - 28,952.07) \\ &= \frac{1770.93}{14} = 126.495 \end{aligned}$$

It is considerably faster to do this arithmetic on your calculator than to write it down. The standard deviation is then $s = \sqrt{126.495} = 11.25$ as before. ●

*This explanation is optional if you have a calculator or computing system that will compute s directly from the data.

Properties of the standard deviation Attention to the detail of getting calculations right, whether through computing formulas as in Example 1.15 or through instruction sets for a computer system, is a necessary part of statistical work. But do not let this detail distract you from a clear view of what you are doing and why. What we are now doing is measuring spread. Here are the basic properties of the standard deviation s as a measure of spread:

- s measures spread about the mean and should be used only when the mean is chosen as the measure of center.
- $s = 0$ only when there is no spread, that is, when all observations have the same value. Otherwise $s > 0$.

The standard deviation is not a resistant measure of the spread of a distribution. In fact, the use of squared deviations renders s even more sensitive to a few extreme observations than is the mean. For example, Roger Maris's 61-home-run year raises his mean home runs per season from 22.2 for the remaining 9 years to 26.1, and it increases the standard deviation from 10.2 for 9 years to 15.6 for all 10 years. A skewed distribution with a few observations in the single long tail will have a large standard deviation, and the number s does not give much helpful information in such a case.

You may rightly feel that the importance of the standard deviation is not yet clear. If we must measure spread about the mean, why not just take the average distance of observations from the mean? Or the median distance, which is more resistant to outlying values? Squaring the distances seems to reduce the resistance for no good reason. These criticisms are well founded. The standard deviation is not a very helpful measure of spread for distributions in general. We commented earlier that the usefulness of many statistical procedures is tied to distributions of particular shapes. This is distinctly true of the standard deviation. The omnipresence of s in statistical work is due to its intimate connection with the normal distributions, a specific class of distributions that we shall meet shortly. For the preliminary work of describing a data set, measures based on percentiles are superior.

Changing the unit of measurement

The same variable can be recorded in different units of measurement. Americans commonly record distances in miles and temperatures in degrees Fahrenheit, while Europeans measure distances in kilometers and temperatures in degrees Celsius. Fortunately, it is easy to convert numerical descriptions of a distribution from one unit of measurement to another. This is true because a change in the measurement unit is a *linear transformation* of the measurements.

linear transformation

A linear transformation changes the original variable x into the new variable x^* given by an equation of the form

$$x^* = a + bx$$

Adding the constant a shifts all values of x upward or downward by the same amount. In particular, such a shift changes the origin (zero point) of the variable. Multiplying by the positive constant b changes the size of the unit of measurement.

EXAMPLE 1.16

(a) If a distance x is measured in miles, the same distance in kilometers is

$$x^* = 1.609x$$

That is, 10 miles is the same as 16.09 kilometers. Here $a = 0$ and $b = 1.609$. This transformation changes the units without changing the origin—a distance of 0 miles is the same as a distance of 0 kilometers.

(b) A temperature x measured in degrees Fahrenheit (F) must be re-expressed in degrees Celsius (C) to be easily understood by the rest of the world. The transformation is

$$x^* = \frac{5}{9}(x - 32) = -\frac{160}{9} + \frac{5}{9}x$$

Thus the high of 95° F on a hot American summer day translates into 35° C. In this case

$$a = -\frac{160}{9} \text{ and } b = \frac{5}{9}$$

This linear transformation changes both the unit size and the origin of the measurements. The origin in the Celsius scale (0° C, the temperature at which water freezes) is 32° in the Fahrenheit scale. ●

Linear transformations do not affect the shape of a distribution. If measurements on a variable x have a right-skewed distribution, any new variable x^* obtained by a linear transformation $x^* = a + bx$ (for $b > 0$) will also have a right-skewed distribution. Similarly, if the distribution of x is symmetric, the distribution of x^* remains symmetric.

Although a linear transformation preserves the basic shape of a distribution, the center and spread will change. Because linear changes of measurement scale are common, we must be aware of their effect on numerical descriptive measures of center and spread. To begin with a simple example, suppose that John and Julie both measure the length of four cockroaches. John measures in centimeters (cm). His data are

1.5 2.0 2.7 1.8

Julie measures the same cockroaches in millimeters (mm). Because there are 10 millimeters in a centimeter, her data are

15 20 27 18

John's mean length is 2.0 cm. It is no surprise that the mean of Julie's measurements is 20 mm; this is the same result expressed in her chosen units. John's standard deviation is 0.51 cm. You can check that Julie's standard deviation is 5.1 mm, again 10 times John's number and again the same result in different units. The facts below put commonsense relations like these in more formal language. Think of John's measurements as values of x and Julie's as values of the new variable x^*. The linear transformation is then $x^* = 10x$.

EFFECT OF LINEAR TRANSFORMATION

Suppose that $x_1, x_2, \ldots, x_n$ are observations with mean $\bar{x}$, median M, quartiles Q_1 and Q_3, interquartile range IQR, and standard deviation s. The linear transformation $x_i^* = a + bx_i$ with $b > 0$ produces observations $x_1^*, x_2^*, \ldots, x_n^*$ for which

- The mean, median, and quartiles are all transformed by multiplying by b and then adding a. For example, the mean of x^* is $a + b\bar{x}$.
- The interquartile range and standard deviation are multiplied by b. For example, the standard deviation of x^* is bs.

Notice that the two measures of spread, IQR and s, are not affected by adding the constant a to all of the observations. Adding a constant changes the location of the distribution but leaves the spread unaltered. Multiplying all observations by b does change the spread. The extremes of John's x-values are 1.5 and 2.7, and the extremes of Julie's $10x$ are 15 and 27, so any measure of spread should be 10 times as large.

In discussing Newcomb's measurements of the speed of light, we noticed that the entries in Table 1.1 had been *coded* to produce simpler numbers. Coding is also a linear transformation. That is, the coded value x^* is obtained from the original measurement x by $x^* = a + bx$. The facts we have given about linear transformations therefore relate the mean (for example) of the coded values to the mean of the original data.

SUMMARY

A **resistant measure** of any aspect of a distribution is relatively unaffected by changes in the numerical value of a small proportion of the total number of observations, no matter how large these changes are.

The **center** of a distribution can be measured by the **mean** $\bar{x}$ or by the **median** M. The mean is the arithmetic average, and the median is the midpoint. The median is a resistant measure of center, but the mean is not.

Resistant measures of the **spread** or **variability** of a distribution are provided by the **quartiles** Q_1 and Q_3 or by the **interquartile range** $IQR = Q_3 - Q_1$. These measures of spread are appropriate when the median is used to indicate the center of the distribution.

The **five-number summary** consisting of the median, quartiles, and high and low extremes provides a quick overall description of a distribution.

Boxplots based on the five-number summary are useful for comparing several distributions. The identification of potential outliers as observations falling at least $1.5 \times IQR$ outside the quartiles is helpful in drawing **modified boxplots** and in other contexts.

The **variance** s^2 and especially its square root, the **standard deviation** s, are common but nonresistant measures of spread about the mean as center. They are most useful for the normal distributions introduced in the next section.

Linear transformations have the form $x^* = a + bx$. A linear transformation changes the origin if $a \neq 0$; it changes the size of the unit of measurement if $b \neq 1$. Linear transformations do not alter the overall shape of a distribution. A linear transformation multiplies a measure of spread by b when $b > 0$, and changes a percentile or measure of center m into $a + bm$.

Numerical measures of particular aspects of a distribution, such as center and spread, do not report the entire shape of general distributions. In some cases, particularly distributions with multiple peaks and gaps, these measures may not be very informative.

SECTION 1.2 EXERCISES

1.40 A college rowing coach tests the 10 members of the women's varsity rowing team on a Stanford Rowing Ergometer (a kind of stationary rowing machine). The variable measured is revolutions of the ergometer's flywheel in a 1-minute session. The data are

446	552	527	504	450	583	501	545	549	506

Make a stemplot of these data by dropping the last digit and using the first digit as the stem. Then compute the mean and the median of the untruncated ergometer scores. Explain the similarity or difference in these two measures in terms of the symmetry or skewness of the distribution.

1.41 The scores of 18 first-year college women on the Survey of Study Habits and Attitudes, first presented in Exercise 1.17 (page 25), are

154	109	137	115	152	140	154	178	101
103	126	126	137	165	165	129	200	148

Find the mean score and the median score for this group. Explain the relationship between these two measures in terms of the main features of the distribution of scores, which you discovered in the earlier exercise.

1.42 Last year a small accounting firm paid each of its five clerks \$22,000, two junior accountants \$50,000 each, and the firm's owner \$270,000. What is the mean salary paid at this firm? How many of the employees earn less than the mean? What is the median salary?

1.43 Exercise 1.11 (page 22) gives measurements on 16 automobile engine crankshafts as follows:

224.120	224.001	224.017	223.982	223.989	223.961
223.960	224.089	223.987	223.976	223.902	223.980
224.098	224.057	223.913	223.999		

Compute the mean and the median of these measurements. Explain the relationship between the two measures of center in terms of the symmetry or skewness of the distribution.

1.44 This year, the firm in Exercise 1.42 gives no raises to the clerks and junior accountants, but the owner's take increases to \$430,000. How does this change affect the mean? How does it affect the median?

1.45 In Exercise 1.11 (page 22), you were asked to code the crankshaft measurements. Write your coding procedure in the form $x^* = a + bx$, where x is the original measurement and x^* is the coded value. Then use your results from Exercise 1.43 to give the mean and median for the coded data.

1.46 According to the Department of Commerce, the mean and median prices of new houses sold in the United States in 1989 were \$129,900 and \$159,000. Which of these numbers is the mean and which is the median? Explain your answer.

1.47 In computing the median income of any group, some federal agencies omit all members of the group who had no income. Give an example to show that the reported median income of a group can go down even though the group becomes economically better off. Is this also true of the mean income?

1.48 Exercise 1.41 gives the scores of 18 first-year college women on the Survey of Study Habits and Attitudes. The college also administered the test to a sample of 20 first-year college men. Their scores were

108	140	114	91	180	115	126	92	169	146
109	132	75	88	113	151	70	115	187	104

(a) Make a back-to-back stemplot of the men's and women's scores. What is the most noticeable contrast between the two distributions?
(b) Find the five-number summaries for both distributions. Are there any outliers according to the $1.5 \times IQR$ criterion?
(c) Make side-by-side boxplots of the distributions. Use modified boxplots that plot outliers separately.

1.49 Exercise 1.20 (page 27) gives these data on the maximum daily rainfall in inches at South Bend, Indiana, over a 30-year period:

1.88	2.23	2.58	2.07	2.94	2.29	3.14	2.15	1.95	2.51
2.86	1.48	1.12	2.76	3.10	2.05	2.23	1.70	1.57	2.81
1.24	3.29	1.87	1.50	2.99	3.48	2.12	4.69	2.29	2.12

(a) Compute the five-number summary for these data.
(b) Compute the *IQR* and draw a modified boxplot that plots individually any suspected outliers indicated by the $1.5 \times IQR$ criterion.
(c) Based on the shape of the distribution as found in the earlier exercise, do you expect the mean to fall distinctly above the median, close to the median, or distinctly below the median?

1.50 In the previous exercise you computed a five-number summary in inches. Now you are writing an article for a journal that requires use of the metric system. Give the five-number summary in centimeters (1 inch is 2.54 cm).

1.51 The following are the golf scores of 12 members of a women's golf team in tournament play:

89 90 87 95 86 81 102 105 83 88 91 79

(a) Display the distribution by a stemplot and describe its main features.
(b) Compute the mean, variance, and standard deviation of these golf scores.
(c) Then compute the median, the quartiles, and the *IQR*. Are there any suspected outliers by the $1.5 \times IQR$ criterion?
(d) Based on the shape of the distribution, would you report the standard deviation or the quartiles as a measure of spread?

1.52 The rowing ergometer scores in Exercise 1.40 appear to have a symmetric distribution. The researcher who administered the test described the results by giving the mean and the standard deviation. You found the mean in Exercise 1.40. Now find the standard deviation of the scores.

1.53 A health-conscious medical student takes her resting pulse each night after eating. The past week's measurements (in beats per minute) were

62 57 53 69 60 61 58

Compute the mean and the standard deviation of these data. (There are too few observations to provide reliable information about the shape of the distribution. The medical student chooses $\bar{x}$ and s because she knows that pulse rate under stable conditions tends to have a symmetric and roughly normal distribution.)

1.54 This is a standard deviation contest. You must choose four numbers from the whole numbers 0 to 10, with repeats allowed.

(a) Choose four numbers that have the smallest possible standard deviation.

(b) Choose four numbers that have the largest possible standard deviation.

(c) Is more than one choice possible in either (a) or (b)? Explain.

1.55 This exercise requires either a calculator that is preprogrammed to find the standard deviation or statistical software on a computer. The observations

10,001 10,002 10,003

have mean $\bar{x} = 10,002$ and standard deviation $s = 1$. By adding a 0 in the center of each number, the next set becomes

100,001 100,002 100,003

The standard deviation remains $s = 1$ as more 0s are added. Use your calculator or computer to calculate the standard deviation of these numbers, adding extra 0s until you get an incorrect answer. How soon did you go wrong? This demonstrates that calculators and computers cannot handle an arbitrary number of digits correctly. Coding the data is the remedy for this difficulty.

1.56 Exercise 1.25 (page 28) gives Cavendish's 29 repeated measurements of the density of the earth as follows:

5.50	5.61	4.88	5.07	5.26	5.55	5.36	5.29	5.58	5.65
5.57	5.53	5.62	5.29	5.44	5.34	5.79	5.10	5.27	5.39
5.42	5.47	5.63	5.34	5.46	5.30	5.75	5.68	5.85	

Compute the mean and standard deviation. Based on the general shape of this distribution, would you be willing to report the mean and standard deviation as helpful descriptive measures? If so, the mean would be used to estimate the actual density of the earth.

1.57 Compute the standard deviation of the crankshaft measurements in Exercise 1.11 (page 22) after coding. (You found the mean in Exercise 1.45.) Then report the standard deviation in the original units (mm). Based on the general shape of this distribution, would you be willing to report the mean and standard deviation as helpful descriptive measures?

1.58 From the data given in Exercise 1.18 (page 26), give a brief numerical summary of the distribution of percent of residents over 65 by state. Explain your choice of numerical measures.

1.59 In Exercise 1.19 (page 26) you displayed the distribution of counts of flea eggs on Thomas the cat. Based on the shape of the distribution and the presence or absence of outliers, would you prefer the median and quartiles or the mean and standard deviation as a helpful numerical summary? Compute the summary that you have chosen.

1.60 In each of the following cases, give the values of a and b for the linear transformation $x^* = a + bx$.
(a) x is measured in feet and transformed to x^* in yards.
(b) x is measured in hours and transformed to x^* in minutes.
(c) x is measured in degrees Fahrenheit and transformed to the difference x^* between x and the "normal" body temperature of 98.6°F.

1.61 Most statistical software will calculate our descriptive measures in response to simple commands, once the data are entered. Suppose, for example, that the growing season data of Exercise 1.22 (page 27) have been entered into the Minitab statistical system as a variable named GROW. We then have the following command and output:

```
MTB > DESCRIBE 'GROW'

              N     MEAN   MEDIAN   TRMEAN   STDEV   SEMEAN
GROW         57   203.65   192.00   200.92   51.87     6.87

            MIN      MAX       Q1       Q3
GROW     125.00   335.00   167.50   233.00
```

In this Minitab output, TRMEAN is the 5% trimmed mean (to be explained in Exercise 1.71), and SEMEAN is the standard error of the mean, a quantity we will meet in Chapter 7. The other quantities are familiar to us.
(a) Use the output to find the *IQR*. Are there any suspected outliers by the $1.5 \times IQR$ criterion?
(b) Draw a modified boxplot of the distribution. (In fact, Minitab can be instructed to do both (a) and (b) for us, although these are easy to carry out by hand once the computer has done the lengthier calculations.)

1.62 Assume that a suitably defined average growing season for the four cities mentioned in Exercise 1.22(b) (page 28) would be longer than 300 days even though not known exactly. Find the median and quartiles of the 61 observations that result when these cities are included, and compare your results with the values given in Exercise 1.61. Explain why you cannot compute the mean and standard deviation of the expanded data set.

1.63 Exercise 1.22 (page 27) gives the growing season x in days for selected U.S. cities.

(a) Find the linear transformation $x^* = a + bx$ that will express the growing season in weeks.

(b) Using what you have learned about the effect of a linear transformation and the Minitab results for the original data set from Exercise 1.61, calculate the mean, the standard deviation, and the median of the growing season in weeks.

1.64 Find the 20th and the 80th percentiles of the distribution of residents over age 65 from Exercise 1.18 (page 26).

1.65 Find the *quintiles* (the 20th, 40th, 60th, and 80th percentiles) of the growing season data in Exercise 1.22 (page 27). For quite large sets of data, the quintiles or the *deciles* (10th, 20th, 30th, etc., percentiles) give a more detailed summary than the quartiles.

1.66 We have already compared the distribution of calories in the three types of hot dogs represented in Table 1.4 (page 36). The following computer output gives the basic descriptive measures for the three distributions of sodium content. Using this information, draw side-by-side boxplots and comment on any differences between the distributions. (The notation of the output is that of PROC UNIVARIATE in the SAS software system, which also reports other measures that we do not need at this point.)

```
VARIABLE=BEEF

N              20                 100% MAX      645.00
MEAN       401.15                  75% Q3       478.00
STD DEV    102.43                  50% MED      380.50
                                   25% Q1       320.50
                                    0% MIN      253.00

VARIABLE=MEAT

N              17                 100% MAX      545.00
MEAN       418.53                  75% Q3       496.00
STD DEV     93.87                  50% MED      405.00
                                   25% Q1       386.00
                                    0% MIN      144.00

VARIABLE=POULTRY

N              17                 100% MAX      588.00
MEAN       459.00                  75% Q3       528.00
STD DEV     84.74                  50% MED      430.00
                                   25% Q1       383.00
                                    0% MIN      357.00
```

1.67 Compute five-number summaries of the blood pressure distributions for the calcium and placebo groups in the experiment in Exercise 1.27 (page 29). Draw side-by-side boxplots to compare the two distributions. Are any important differences between the groups apparent? (Remember that the boxplots are less informative than the back-to-back stemplot of Exercise 1.27.)

1.68 Compute five-number summaries of the weight gains in the experimental and control groups of chicks in Exercise 1.28 (page 29). Draw side-by-side boxplots to compare the distributions. Write a short statement about the observed differences, backing your claims by citing appropriate numbers. (We will see in Exercise 1.116 that both distributions are approximately normal. Therefore analysis of this experiment would in practice be based on means and standard deviations.)

1.69 Refer to the previous exercise on weight gains in grams (g) for two groups of chicks.

(a) One ounce equals 28.35 g. What is the linear transformation $x^* = a + bx$ that restates the weight gains in ounces?

(b) Using the results of the previous exercise and what you know about the effect of linear transformations, calculate the five-number summary for each group in ounces.

(c) Make side-by-side boxplots to compare the two groups using the transformed weight scale. Describe the similarities and differences between these side-by-side boxplots and the ones you made in the previous exercise.

1.70 Give an approximate five-number summary of the salary data in Exercise 1.30 (page 31) by pretending that all salaries fall at the midpoint of their class. (Replace the extremes by the 10th and 90th percentiles.)

1.71 The *trimmed mean* is a measure of center that is more resistant than the mean but uses more of the available information than the median. To compute the 10% trimmed mean, discard the highest 10% and the lowest 10% of the observations, and compute the mean of the remaining 80%. Trimming eliminates the effect of a small number of outliers. Compute the 10% trimmed mean of the maximum-rainfall data in Exercise 1.49 (page 54). Then compute the 20% trimmed mean. Compare the values of these measures with the median and the ordinary untrimmed mean.

1.3 THE NORMAL DISTRIBUTIONS

We now have command of a kit of graphical and numerical tools for describing distributions. But these tools alone do not give us a full picture of a substantial data set such as the 947 vocabulary scores for Gary, Indiana, seventh graders summarized in Table 1.2 (page 11). A stemplot is impractical, a frequency table and histogram suppress much detail and depend upon a choice of classes, and numerical measures such as the median and percentiles report only selected aspects of the data. Can we find a way to describe an entire distribution in a single expression? If we are willing to accept a description of the overall shape of the distribution, omitting outliers and other deviations from a regular pattern, the answer is yes. Such a description is provided by a *mathematical model* for the distribution.

The use of idealized mathematical descriptions for complex objects and phenomena is a common and powerful tool in science, not least in statistics. Suppose, for example, that we are watching a caterpillar crawl forward. Once each minute we measure the distance it has moved, plot the observed values against time, and see that the points fall close to a straight line. It would be natural to draw a straight line through our graph as a compact description of these data. The straight line, with its equation of the form $y = a + bt$ (where y is distance and t is time), is a mathematical model of the caterpillar's progress. It is an idealized description because the points on the graph do not fall *exactly* on the line. We want to give a similarly compact description of the distribution of a variable, to replace stemplots and histograms. Just as a straight line is one of many types of curves that can be used to describe a plot of distance traveled against time, there are many types of mathematical distributions that can be used to describe a set of single-variable data. This section will concentrate on one type, the normal distributions.

Density curves

The rationale for using mathematical models to describe distributions lies close at hand. We can approximate a histogram by a smooth curve that displays the shape of the distribution without the lumpiness of the histogram. Because a smooth curve can be described by a single equation, this gives us a compact description of the entire distribution. But because the curve smooths out some of the detail presented in the histogram, the description is idealized and not completely accurate. Figure 1.12 shows a histogram of the Gary vocabulary test scores. We used twice as many classes as in Figure 1.1 in order to display more detail. The corresponding normal curve is drawn over the histogram. You can see that the curve describes the overall shape in idealized form. The curve is actually a better description of the distribution of test scores than Figure 1.12 suggests, because some of the unevenness in the histogram is a consequence of the particular classes chosen.

The vertical scale for a frequency histogram depends on the total number of observations. The scale for a relative frequency histogram is less arbitrary, because only the percentage or fraction of data in each class is recorded. The areas of the bars in a histogram are proportional to these

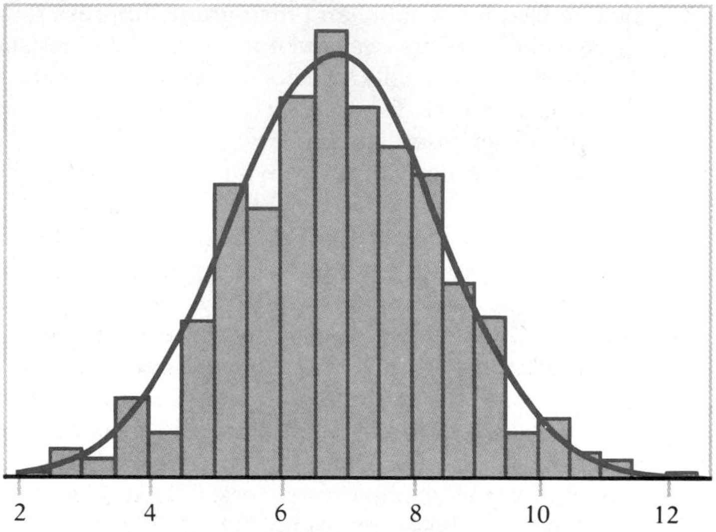

FIGURE 1.12 Histogram of the vocabulary scores from Example 1.6, showing the approximation by a normal curve.

relative frequencies. The simplest choice of a scale to describe a distribution makes area not only proportional to relative frequency but exactly equal to it. Because the relative frequency of all the observations together is 1, we require that our curve have *total area 1* underneath it. The area under the curve and above any range of values of the variable is then equal to the proportion of observations falling in this range, thought of as a fraction of the whole. The curve describes the shape of the distribution and has the advantage that *area = relative frequency*. It is called the *density curve* of the distribution.

density curve

Density curves, like distributions, come in many shapes. Figure 1.13 shows two density curves, a symmetric, normal density curve and a right-skewed curve. A density curve of the appropriate shape is often an adequate description of the overall pattern of a distribution. Outliers, which are deviations from the overall pattern, are not described by the curve.

Our measures of center and spread apply to density curves as well as to actual sets of observations. Think first about percentiles. Since a proportion p (expressed as a percent) of all observations falls below the pth percentile, *the pth percentile on a density curve is the point with p percent of the area lying to the left and the remaining* $100 - p$ *percent to the right*. In particular, the median is the "equal-area point," the point with half the total area on each side. As Figure 1.13(a) illustrates, the median of a symmetric density curve falls at the center of symmetry. On

percentile

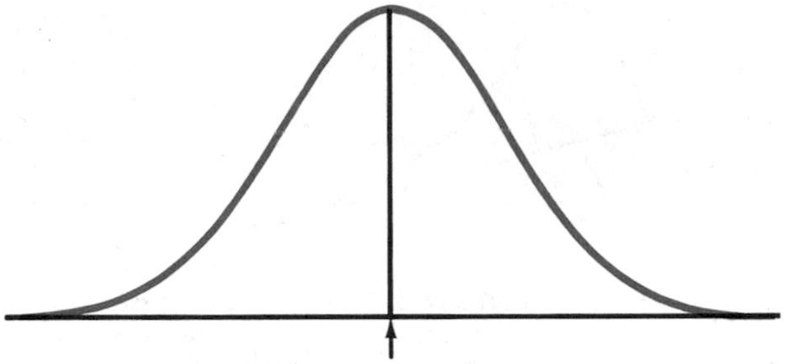

FIGURE 1.13(a) The median and mean of a symmetric density curve.

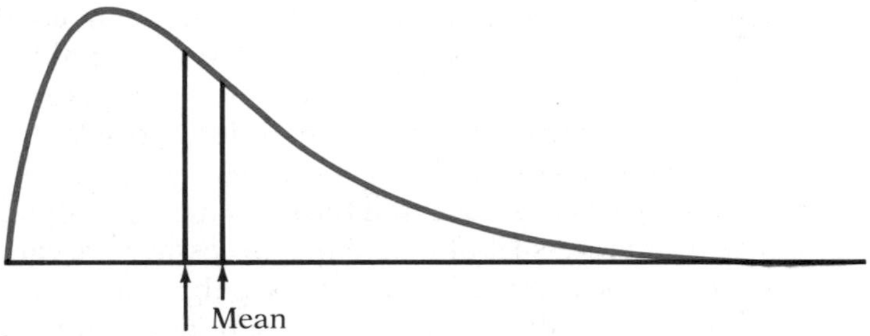

FIGURE 1.13(b) The median and mean of a right-skewed density curve.

an asymmetric curve, the equal-area point is not so easy to locate by eye. There are mathematical ways of finding the median for any density curve, which we used to mark the median on the skewed curve in Figure 1.13(b). You can judge that the areas on either side of this point are equal. You can also roughly locate the quartiles by dividing the area into quarters as accurately as possible by eye. The *IQR* is then the distance between the first and third quartiles.

What about the mean? The mean of a set of observations is just their arithmetic average. If we think of the observations as weights strung out along a thin rod, the mean is the point at which the rod would just balance on a fulcrum beneath. This interpretation extends to density curves: *The* mean *mean is the point at which the curve would balance if made of solid material.* Figure 1.14 illustrates this interpretation of the mean. The mean of a symmetric curve is therefore at the center of symmetry, and the mean and median are equal for symmetric distributions. As part of the idealization inherent in a mathematical model, symmetric density curves are exactly symmetric even though real data will rarely show perfect symmetry. The

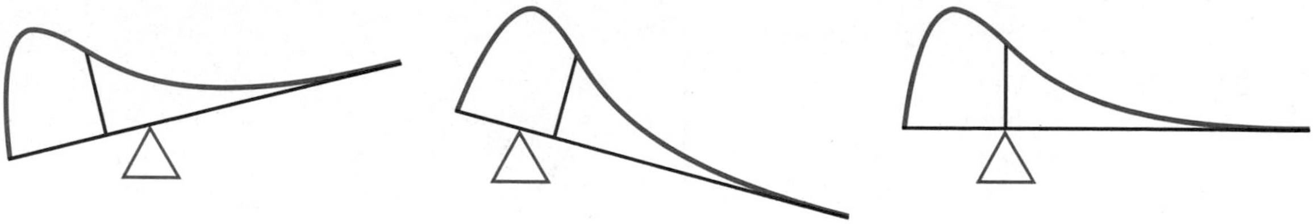

FIGURE 1.14 The mean is the balance point of a density curve.

mean (or balance point) of a skewed distribution is pulled toward the long tail of a skewed density curve more than is the median. This is because a small area (a few observations) far out in the tail can tip the curve more than the same area near the center. Locating the mean of a skewed distribution by eye is tricky. There are mathematical ways of calculating the mean for any density curve, so that we are able to mark the mean as well as the median on the skewed curve in Figure 1.13(b).

The mean and percentiles can be located at least approximately by eye on any density curve. This is not true of the standard deviation, in keeping with our earlier remark that the standard deviation is not a natural measure for most distributions. When necessary, we can call on the results of more advanced mathematics to learn the value of the standard deviation. The study of mathematical methods for performing calculations with density curves is part of theoretical statistics. Though we are concentrating on statistical practice, we often make use of the results of mathematical study. The ability to apply mathematical methods is a major advantage of using mathematical models to describe data.

Because a density curve is an idealized description of the distribution of data, we want to distinguish between the mean and standard deviation of the density curve and the numbers $\bar{x}$ and s computed from the actual observations. The usual notation for the mean of an idealized distribution is μ, the Greek letter mu. The standard deviation is denoted by σ, the Greek letter sigma.

Our usual strategy in examining data is to look first for an overall pattern and then for deviations from that pattern. The overall pattern can be presented in ways that are based entirely on the data by describing the major features of a stemplot or histogram. Sometimes no briefer description is adequate, as in the discussion of the calorie content of hot dogs that followed Example 1.13 on page 44. But statisticians often try to find a mathematical model that gives a compact description of the overall pattern of the data. When the data are observations on a single variable, the models employed are density curves for the distribution of data. As we progress, we will meet many other examples of the process of giving models for data (and then inspecting the deviations of the actual data from the model).

Normal distributions

normal distributions

One particularly important class of density curves has already appeared in Figures 1.12 and 1.13(a). The *normal distributions* are symmetric, single-peaked, bell-shaped density curves. All normal distributions have the same overall shape. The exact density curve for a particular normal distribution is described by giving its mean μ and its standard deviation σ. The mean is located at the center of the symmetric curve and is the same as the median. Changing μ without changing σ moves the normal curve to a new location without altering its spread. The standard deviation controls the spread of a normal curve. Figure 1.15 shows normal curves with the same mean but different values of σ. We can even locate σ by eye on a normal density curve. As we move out in either direction from the center μ, the curve changes from falling ever more steeply

to falling ever less steeply

The points at which this change of curvature takes place are located at distance σ on either side of the mean μ. You can feel the change as you run a pencil along a normal curve, and so find the standard deviation. Remember that μ and σ alone do not specify the shape of general distributions, and that the shape of density curves in general does not reveal σ. These are special properties of normal distributions.

There are symmetric, bell-shaped density curves that are not normal. The normal density curves are specified by a particular equation: The height of the density curve at the point x is given by

$$\frac{1}{\sigma\sqrt{2\pi}}e^{-\frac{1}{2}\left(\frac{x-\mu}{\sigma}\right)^2} \qquad (1.4)$$

We will not make direct use of this fact, although it is the basis of mathematical work with normal distributions. Note that the equation of the curve is completely specified by μ and σ.

Although there are many normal curves, as Figure 1.15 illustrates, they are tightly bound together. In fact, *all normal distributions are the same*

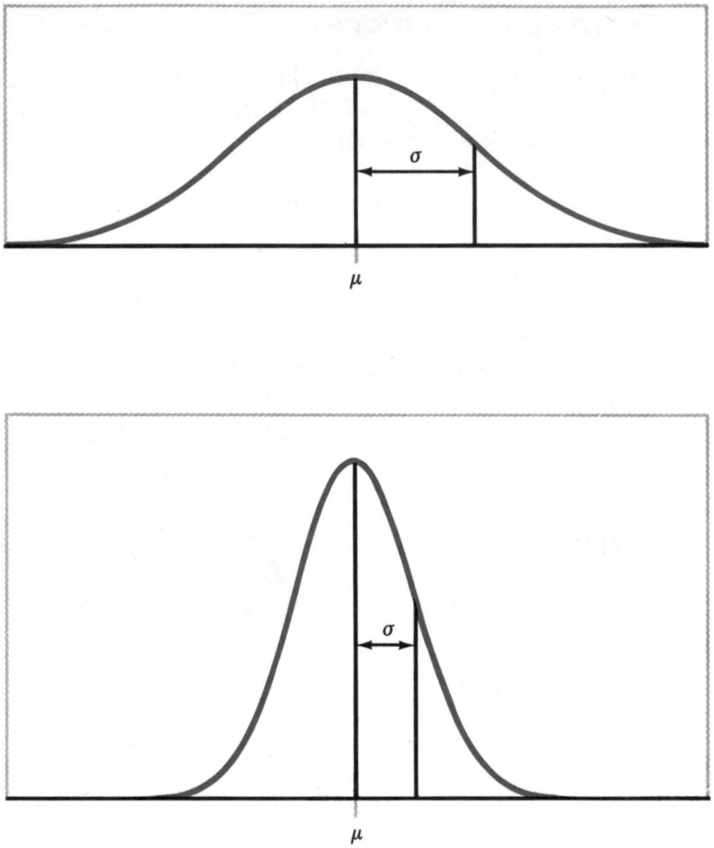

FIGURE 1.15 Two normal curves, showing the mean μ and standard deviation σ.

if measurements are made in units of σ *about the mean* μ. In particular, all normal distributions have the properties illustrated in Figure 1.16 and specified in the following rule.

THE 68–95–99.7 RULE

In any normal distribution:

- 68% of the observations fall within σ of the mean μ.
- 95% of the observations fall within 2σ of μ.
- 99.7% of the observations fall within 3σ of μ.

The 68–95–99.7 rule enables us to think about normal distributions without constantly making detailed calculations.

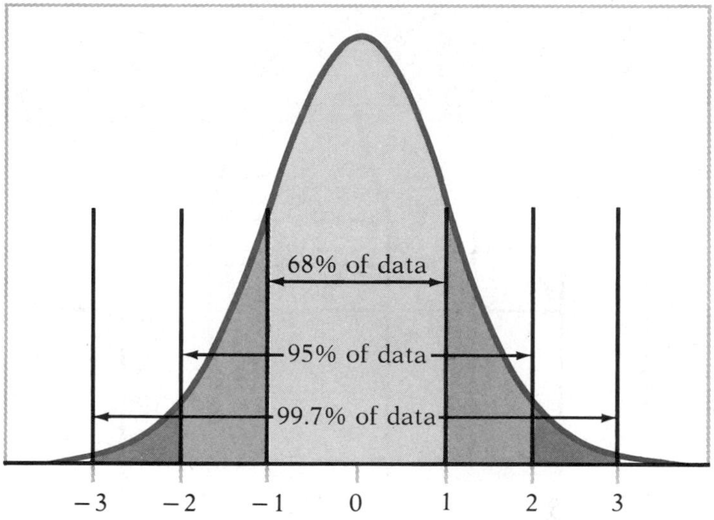

FIGURE 1.16 The 68–95–99.7 rule for normal distributions.

EXAMPLE 1.17

The distribution of heights of American women aged 18 to 24 is approximately normal with mean $\mu = 65.5$ inches and standard deviation $\sigma = 2.5$ inches. From the 68–95–99.7 rule we can estimate that the middle 95% of the distribution lies between 60.5 inches and 70.5 inches tall. Therefore, the tallest 2.5% of these women are taller than 70.5 inches. (The extreme 5% fall more than 2σ, or 5 inches, from the mean. Because the normal distributions are symmetric, half of these women are on the tall side.) Almost all young American women are between 58 inches and 73 inches in height, according to the 99.7 part of the rule. Figure 1.17 shows the application of the 68–95–99.7 rule in this example. ●

The normal distributions provide good models for some distributions of real data. Distributions that are often close to normal include scores on tests taken by a broad population (such as tests of scholastic ability and many psychological tests); repeated careful measurements of the same quantity (such as Newcomb's data in Table 1.1 without the outliers); and characteristics of homogeneous biological populations (such as lengths of cockroaches, yields of corn, and moisture loss in packaged chicken meat). The distributions of some other common variables are usually skewed and therefore distinctly nonnormal. Examples include economic variables such as personal income and gross sales of business firms, the survival times of cancer patients after treatment, and the service lifetime of mechanical or electronic components. Although experience can suggest whether or not a normal model is plausible in a particular case, it is risky to assume that a distribution is normal without actually inspecting the data.

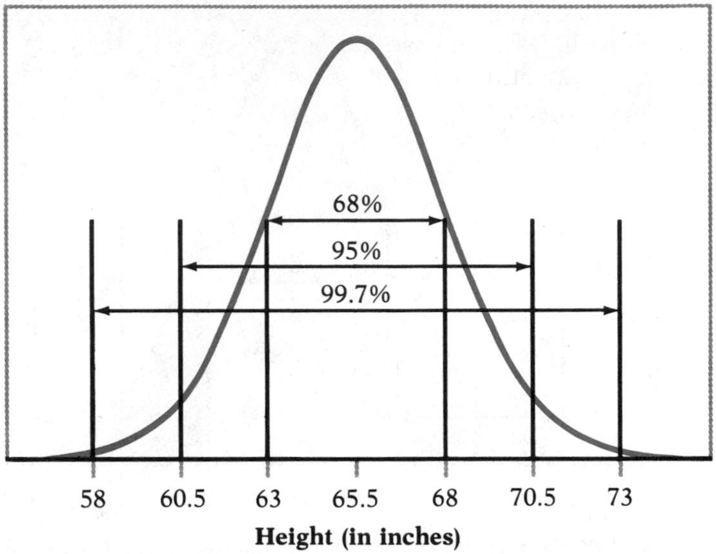

FIGURE 1.17 **The 68–95–99.7 rule applied to the heights of young women.**

EXAMPLE 1.18

We expect a large set of Iowa Test scores to be approximately normal, and in fact Figure 1.12 suggests that the normal model is adequate for the 947 Gary vocabulary scores. Because it is difficult to assess the fit of a normal curve to a histogram by eye, let us use the 68–95–99.7 rule to check more closely. Having entered the scores into a statistical computing system as the values of a variable GARY, we obtain the mean and standard deviation as follows. (We used the Minitab system; most other systems do these calculations in response to similar commands.)

```
MTB > MEAN 'GARY'
MEAN = 6.8585

MTB > STDEV 'GARY'
STDEV = 1.5962
```

Now that we know that $\bar{x} = 6.8585$ and $s = 1.5952$, we imitate the description of the normal distribution in Figure 1.16 by finding the actual counts of Gary vocabulary scores in intervals of length s about the mean $\bar{x}$. This requires more elaborate computing. Here are the counts:

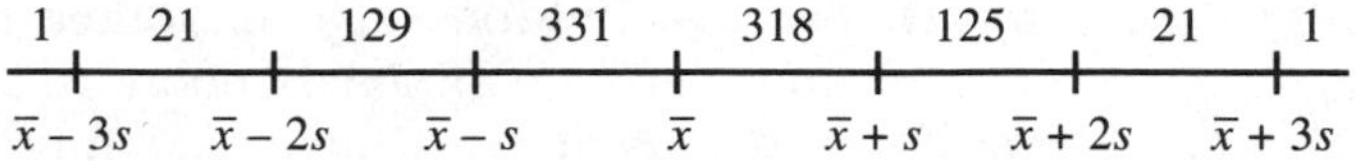

The symmetry of this distribution is evident, and its adherence to the 68–95–99.7 rule is striking: There are 649 scores (68.5%) within one standard deviation of the mean, 903 (95.4%) within two standard deviations, and 945 (99.8%) of the 947 scores within three. These counts confirm that the normal model with $\mu = 6.86$ and $\sigma = 1.595$ fits these data well. ●

Adopting a density curve as a model for a distribution should not be done lightly, because the curve simultaneously specifies every aspect of the distribution. Tests for the adequacy of particular models are therefore important. Comparing actual counts with thc 68–95–99.7 rule, as in Example 1.18, provides a quick check for normality.

What effect does changing the units of measurement or coding the data have on a normal distribution? We know that a linear transformation does not change the general shape of a distribution and that the mean and standard deviation change in a simple manner. In fact, *any variable obtained from a normal variable by a linear transformation remains normal.* Of course, the mean and standard deviation will change in the usual way, so that if x has the normal distribution with mean μ and standard deviation σ, then $x^* = a + bx$ for a positive b has the normal distribution with mean $a + b\mu$ and standard deviation $b\sigma$.

EXAMPLE 1.19

A food company bakes cakes in a large oven at a temperature of 350° F. The temperature x is not exactly constant but varies according to a normal distribution with mean 350° F and standard deviation 2° F. If instruments that measure temperature on the Celsius scale are used, what is then the distribution of the oven temperature? The temperature in the Celsius scale is

$$x^* = \frac{5}{9}(x - 32) = -\frac{160}{9} + \frac{5}{9}x$$

Therefore, the Celsius temperature x^* has the normal distribution, with mean

$$-\frac{160}{9} + \frac{5}{9}350 = 176.7^\circ$$

and standard deviation

$$\frac{5}{9}2 = 1.1^\circ$$

The change of scale did not change the normal shape of the distribution, but it did change the numerical values of the center and the spread. ●

Normal distribution calculations

The normal distributions are such commonly used models for the distributions of observed variables that a compact notation is helpful. If X is a variable having the normal distribution with mean μ and standard deviation σ, we say that X is $N(\mu, \sigma)$. We use uppercase letters, such as X, to represent variables that follow some theoretical distribution. Lowercase letters, such as x, stand for particular numerical values of the variable.

Because area under a density curve is relative frequency, any question about relative frequencies in a distribution described by a density curve can be answered by calculating areas under the curve. In the case of normal distributions, these calculations are made easier by the fact that except for changes in location and scale, there is only one normal curve. The 68–95–99.7 rule is one consequence of this fact. A more sweeping consequence is the opportunity to find relative frequencies for any normal distribution from a single table of areas.

All normal distributions are the same when measurements are made in standard deviation units about the mean as origin. Here is a more formal statement of this fact.

> **STANDARD NORMAL DISTRIBUTION**
>
> If a variable X has a normal distribution with mean μ and standard deviation σ, then the standardized variable
>
> $$Z = \frac{X - \mu}{\sigma} \tag{1.5}$$
>
> has the normal distribution $N(0, 1)$ with mean 0 and standard deviation 1. This is called the standard normal distribution.

standardized observation

When the distribution of a variable is approximately normal, observations are often *standardized* by subtracting the mean and dividing by the standard deviation. A standardized observation states how many standard deviations the original observation falls away from the mean and in which direction. Observations larger than the mean are positive when standardized, and observations smaller than the mean are negative. Standardization is a form of coding that changes the mean to 0 and the standard deviation to 1.

EXAMPLE 1.20

The heights of American young women are (approximately) normally distributed with $\mu = 65.5$ inches and $\sigma = 2.5$ inches. The standardized height

$$Z = \frac{\text{height} - 65.5}{2.5}$$

then follows the standard normal distribution. A woman's standardized height is the number of standard deviations by which her height differs from the mean height of all American young women. A woman 69 inches tall, for example, has a standardized height of

$$z = \frac{69 - 65.5}{2.5} = 1.4$$

or 1.4 standard deviations above the mean. Similarly, a woman 5 feet (60 inches) tall has a standardized height of

$$z = \frac{60 - 65.5}{2.5} = -2.2$$

or 2.2 standardized deviations less than the mean height. Note that Z represents a variable having a standard normal distribution, while $z = 1.4$ and $z = -2.2$ represent values of this variable for particular women. ●

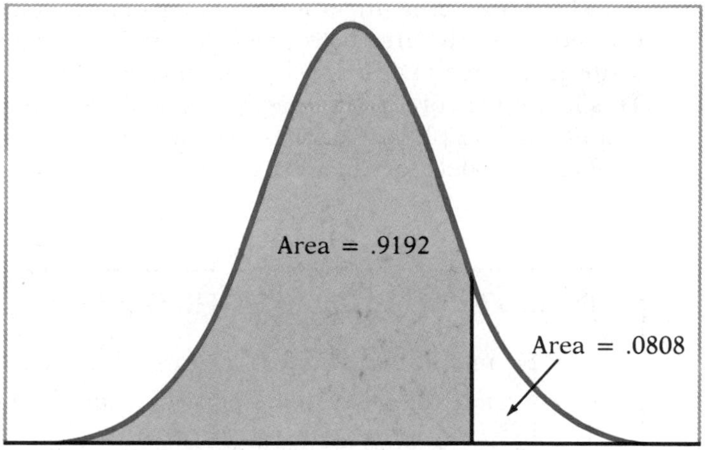

FIGURE 1.18 The area under a standard normal curve to the left of the point $z = 1.4$ is 0.9192. The total area under the curve is 1.

What fraction of all American young women are less than 69 inches tall? This relative frequency is the area under the $N(65.5, 2.5)$ curve to the left of the point 69. Because the standardized height corresponding to 69 inches is 1.4, this area is equal to the area under the *standard* normal curve below the point 1.4. *Table A in the back of the book gives areas under the standard normal curve*, calculated using Equation 1.4. This table also appears on the front end covers. The table entry for each value z is the area to the left of z.

EXAMPLE 1.21

To find the area to the left of 1.4, we enter Table A under $z = 1.40$. Locate 1.4 in the left-hand column, then the remaining digit in the top row. The entry opposite 1.4 and under 0.00 is 0.9192, which is the area we seek. Figure 1.18 illustrates the relationship between the value $z = 1.40$ and the area 0.9192. This area is the relative frequency of the event that $Z < 1.4$. Since the total area under the curve is 1, the area lying *above* 1.4 is $1 - 0.9192 = 0.0808$. Therefore, the fraction of young women who are less than 69 inches tall is 0.9192 (or about 92%) and the fraction who are taller than 69 inches is 0.0808 (or about 8%). ●

Any normal relative frequency can be found by first standardizing and then looking in Table A, as in Examples 1.20 and 1.21. Sometimes we encounter a value of z more extreme than those appearing in Table A. For example, the area to the left of $z = -4$ is not given directly in the table. But because -4 is less than -3.4, this area is smaller than the entry for $z = -3.40$, which is 0.0003.

There is *no* area under a smooth curve and exactly over the point 69. Consequently, the area below 69 (the relative frequency of $X < 69$) is the same as the area at or below this point (the relative frequency of $X \leq 69$). This is not true of actual data sets, which may contain a woman just 69 inches tall. This phenomenon is a consequence of the idealized smoothing of normal models for data.

FINDING NORMAL RELATIVE FREQUENCIES

- Formulate the problem in terms of an observed variable X.
- Restate the problem in terms of a standard normal variable Z by standardizing X, using Equation 1.5.
- Find the required area under the $N(0, 1)$ curve, making use of Table A, the symmetry of normal curves, and the fact that the total area is 1.

As the following examples illustrate, drawing a picture will help you do normal relative frequency calculations.

EXAMPLE 1.22

A video display tube for computer graphics terminals has a fine mesh screen behind the viewing surface. During assembly the mesh is stretched and welded onto a metal frame. Too little tension at this stage will cause wrinkles, and too much will tear the mesh. The tension is measured by an electrical device with output readings in millivolts (mV). At the present time, the tension readings for successive tubes follow the $N(275, 43)$ distribution. The minimum acceptable tension corresponds to a reading of 200 mV. What proportion of tubes exceed this limit?

Following the procedure above, with X denoting the tension measurement:

- We want the relative frequency that $X > 200$, where X is $N(275, 43)$.
- This is the same as the relative frequency that

$$\frac{X - 275}{43} > \frac{200 - 275}{43} = -1.74$$

or $Z > -1.74$ for a standard normal Z.

- From Table A, we see that the relative frequency that $Z < -1.74$ is 0.0409. Since the total area under the curve is 1, the area *to the right of* -1.74 is $1 - 0.0409 = 0.9591$. That is, about 96% of the tubes exceed the minimum. See Figure 1.19(a) for an illustration of the areas. ●

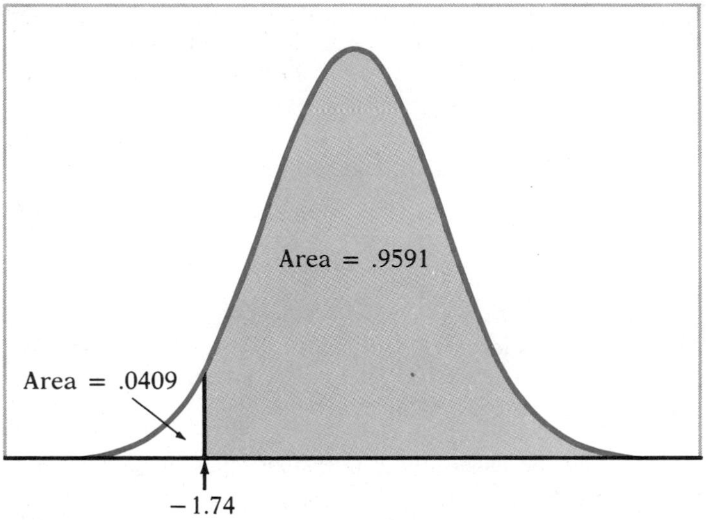

FIGURE 1.19(a) The area under the standard normal curve for Example 1.22.

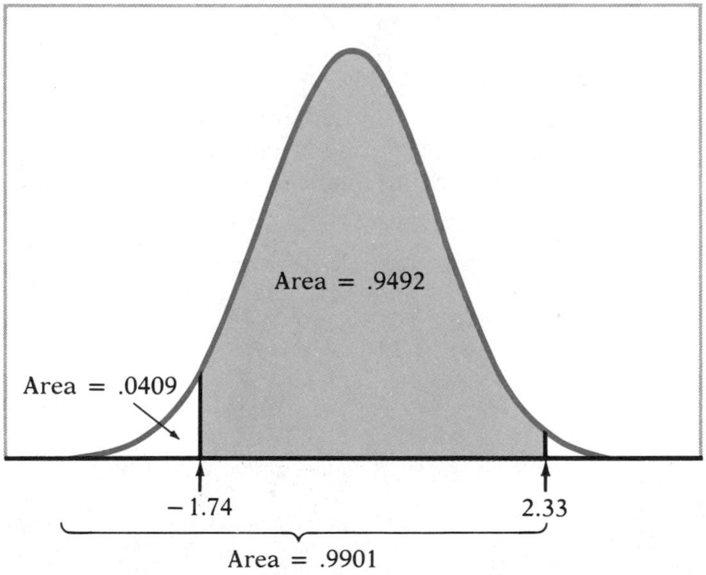

FIGURE 1.19(b) The area under the standard normal curve for Example 1.23.

Because normal curves are symmetric, the area above −1.74 is equal to the area below 1.74, so we could have found the relative frequency in Example 1.22 directly as the table entry for 1.74. There are often several correct ways of doing normal distribution calculations using Table A.

EXAMPLE 1.23

In the production setting of Example 1.22, tension above 375 mV will usually tear the mesh. The acceptable range of tension readings is therefore 200 mV to 375 mV. What proportion of tubes have acceptable tension values?

- We must find the relative frequency that $200 < X < 375$.
- Standardizing, this is the same as

$$\frac{200 - 275}{43} < \frac{X - 275}{43} < \frac{375 - 275}{43}$$

or $-1.74 < Z < 2.33$.

- The area between -1.74 and 2.33 is the area below 2.33 *minus* the area below -1.74, or from Table A, $0.9901 - 0.0409 = 0.9492$. A picture like Figure 1.19(b) helps us see the areas in question. Thus, 95% of the tubes have acceptable tension.

Can the tensioning process be improved? A moment's thought suggests that adjusting the operation so that the mean tension is exactly centered between the limits 200 mV and 375 mV will make the proportion of acceptable outcomes as high as possible for $\sigma = 43$ mV. This adjustment may be quite easy. Further improvement requires reducing the standard deviation. Because σ measures the inherent variability in the process, reducing σ will require changing the tensioning operation to make it less variable. ●

Examples 1.22 and 1.23 illustrate the use of Table A to find the relative frequency of a given event, such as "tension reading between 200 mV and 375 mV." We may instead wish to find the observed value corresponding

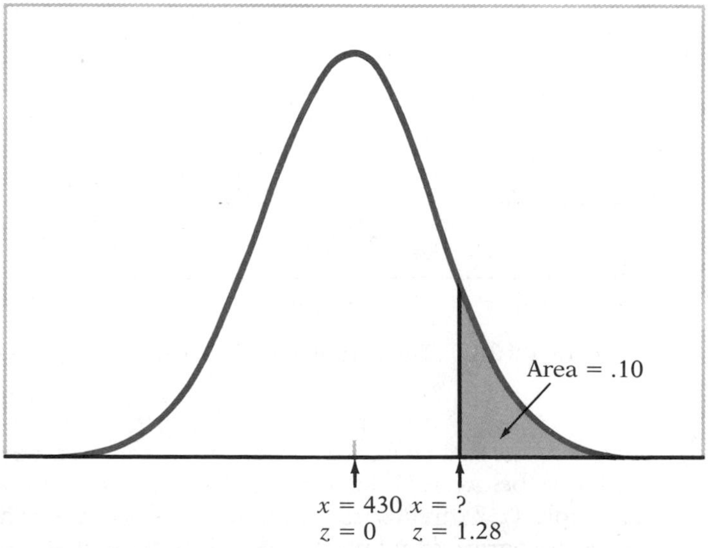

FIGURE 1.20 Locating the point on a normal curve with area 0.10 above it.

to a given relative frequency. To do this, we use Table A backward, finding a desired relative frequency in the body of the table and then reading the corresponding z from the left column and top row.

EXAMPLE 1.24

Scores on the Scholastic Aptitude Test (SAT) for verbal ability of high school seniors follow the $N(430, 100)$ distribution. (This is approximately true for current scores. The range of possible scores is 200 to 800.) How high must a student score in order to place in the top 10% of all students taking the SAT? Figure 1.20 poses this question in graphical form.

- Let X stand for a student's SAT score. We want to find the number x such that the event $X \geq x$ has relative frequency 0.1. That is, the event $X < x$ has relative frequency 0.9. (As usual, X stands for the variable and x represents a particular number.)
- In terms of standardized scores, this event is the same as

$$\frac{X - 430}{100} < \frac{x - 430}{100}$$

or

$$Z < \frac{x - 430}{100}$$

- Look in the body of Table A for the entry closest to 0.90. It is 0.8997 and is the area below 1.28. So, 1.28 is the value of a standard normal variable with relative frequency 0.1 above it. Therefore,

$$\frac{x - 430}{100} = 1.28$$

Solving this equation for x gives

$$x = 430 + (1.28)(100) = 558$$

A student must score at least 558 to place in the highest 10%. ●

Remember that these calculations are valid only for normal distributions. Nonnormal distributions, like nonnormal people, are common and interesting.

Assessing normality

The decision to describe a distribution by a normal model may determine the later steps in our analysis of the data. Both relative frequency calculations and statistical inference based on such calculations follow from the choice of a model. Such a momentous choice should be made carefully. How can we judge whether data are approximately normal?

A histogram or stemplot can reveal distinctly nonnormal features of a distribution, such as outliers (Newcomb's histogram, Figure 1.2, on page

14), pronounced skewness (the grocery purchase stemplot, Example 1.5, on page 9), or gaps and clusters (the hot dog stemplot, Example 1.13 on page 45). The effectiveness of these plots for assessing whether the distribution is normal can be improved by marking the points $\bar{x}$, $\bar{x} \pm s$, and $\bar{x} \pm 2s$ on the x axis. This gives the scale natural to normal distributions. We can then compare the count of observations in each interval with the 68–95–99.7 rule, as in Example 1.18 on page 66.

normal quantile plot

A more sensitive assessment of the adequacy of the normal model for a set of data is provided by a *normal quantile plot*. Quantiles are percentiles expressed as fractions rather than percents. For example, the median is the 50th percentile or the 0.50 quantile of a distribution. The idea of a normal quantile plot is to plot each observation x versus the corresponding quantile z of the standard normal distribution.

We can grasp the normal quantile plot idea from an example. Suppose we have 20 observations as follows:

11.5	5.1	12.1	7.8	15.9	8.2	10.7	6.8	10.7	12.9
11.7	12.4	8.1	9.4	12.9	8.2	3.8	11.4	10.3	6.1

The smallest observation is $x = 3.8$. It is the first of 20 in size, so $x = 3.8$ is the 1/20, or 0.05, quantile of the data. The 0.05 quantile of the standard normal distribution is the z with area 0.05 to its left under the $N(0, 1)$ density curve. Look for the entry closest to 0.05 in the body of Table A to see that z is about -1.65. The first point on the normal quantile plot is $(3.8, -1.65)$. The second point plots $x = 5.1$, the second smallest of the 20 observations, against $z = -1.28$, the 0.1 quantile of the standard normal distribution, and so on.

Normal quantile plots (often called normal probability plots) are the recommended tool for assessing normality. Computer software makes normal quantile plots using more sophisticated versions of the basic idea. You will not make these plots by hand, but you must interpret the computer's work.

Suppose, for example, that your observations have been entered into the Minitab statistical package as the variable X. The following commands will draw a normal quantile plot:

```
MTB > NSCORES 'X' C2
MTB > PLOT C2 'X'
```

If the distribution of the observations is close to the standard normal distribution, the points in a normal quantile plot will fall close to the line $z = x$ because the actual quantiles x of the data will be close to the standard normal quantiles z. If the observations follow *any* normal distribution $N(\mu, \sigma)$, the standardized variable $Z = (X - \mu)/\sigma$ is standard normal. There is then a straight-line relation $X = \sigma Z + \mu$ between the observed variable and a standard normal variable, so the points on a normal quantile plot will fall close to some straight line.

> **USE OF NORMAL QUANTILE PLOTS**
>
> If the points on a normal quantile plot lie close to a straight line, the plot indicates that the data are normal. Systematic deviations from a straight line indicate a nonnormal distribution. Outliers appear as points that are far away from the overall pattern of the plot.

Four normal quantile plots for data we have met earlier appear in Figure 1.21. The data x are plotted horizontally and the corresponding standard normal z's are plotted vertically. The vertical scale extends from

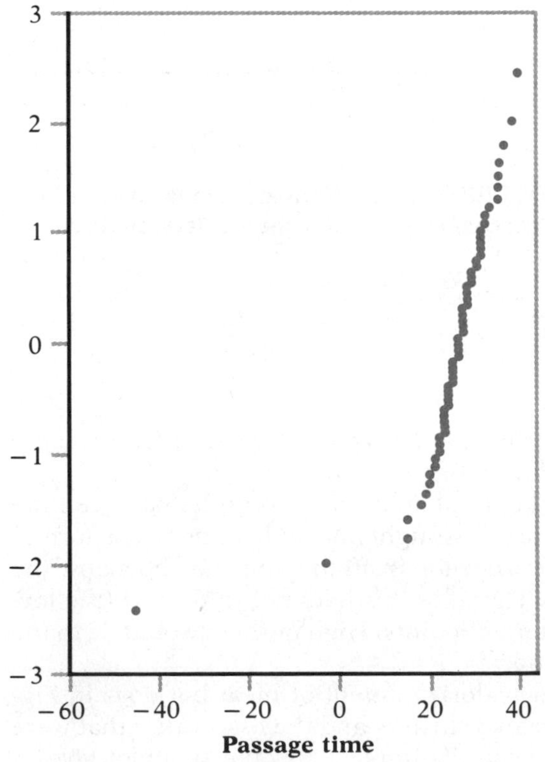

FIGURE 1.21(a) Normal quantile plot of Newcomb's passage time data, Table 1.1.

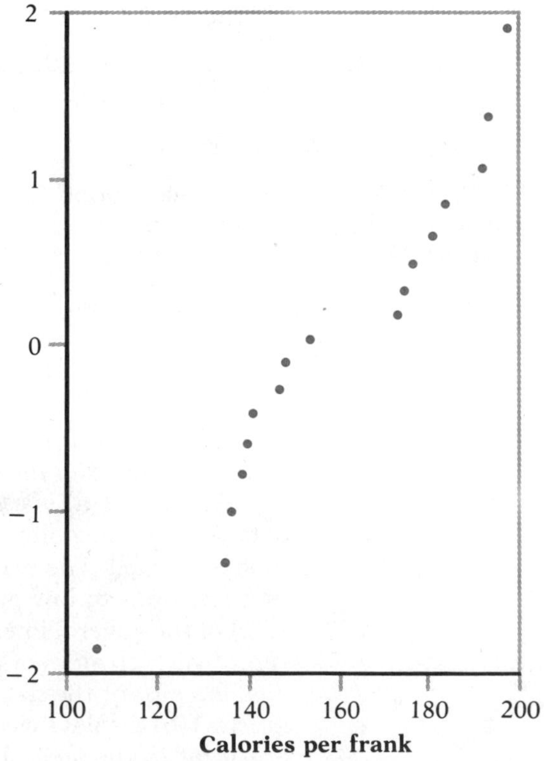

FIGURE 1.21(b) Normal quantile plot of calories in brands of meat hot dogs, Table 1.4.

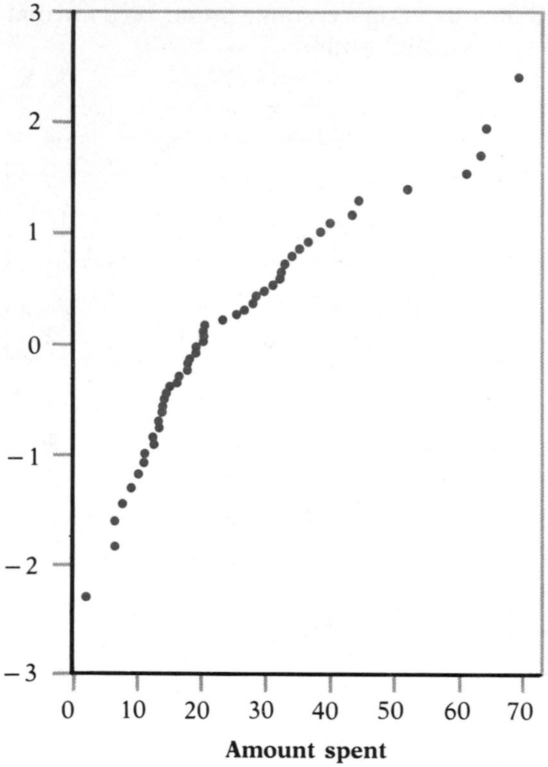

FIGURE 1.21(c) Normal quantile plot of amount spent by grocery shoppers, Example 1.12.

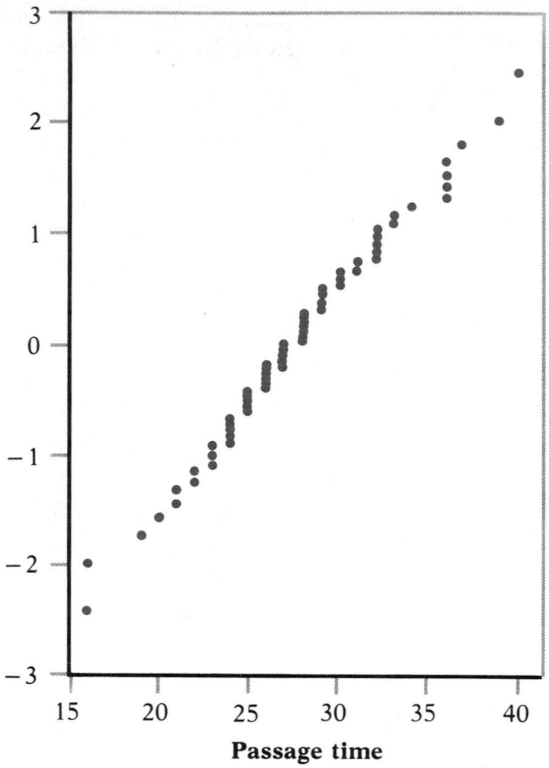

FIGURE 1.21(d) Normal quantile plot of Newcomb's data with the outliers omitted.

−3 to 3 because almost all of a standard normal curve lies between these values. Let us examine these four plots in turn.

Figure 1.21(a) is a normal quantile plot of Newcomb's passage time data. Most of the points lie close to a straight line, indicating that a normal model fits well. The two outliers deviate from the line and show how the plot responds to low outliers. These low outliers fall to the left (the low side) of the general linear pattern of points. High outliers would lie to the right (the high side) of the pattern.

The plot of the data on the calorie content of meat hot dogs in Figure 1.21(b) displays clearly the two clusters and the low outlier that were apparent in the stemplot following Example 1.13. The stemplot gives a clearer picture of this small and definitely nonnormal data set, but comparing Figure 1.21(b) with the stemplot helps us understand how normal quantile plots behave.

Figure 1.21(c) is a normal quantile plot of grocery spending by 50 shoppers, a right-skewed distribution. The right skewness is apparent if you lay a straightedge along the leftmost points, which correspond to the smaller observations. The larger observations fall systematically to the right of this line. That is, they are farther toward the right than in a normal distribution. *In a right-skewed distribution, the largest observations fall distinctly to the right of a line drawn through the main body of points.* Similarly, left skewness is evident when the smallest observations fall to the left of the line. Unlike Figure 1.21(a), there are no individual outliers.

Figure 1.21(d) is another normal quantile plot of Newcomb's measurements of the passage time of light, this time with the two outliers omitted. The effect of omitting the outliers is to magnify the plot of the remaining data. As we see from Figure 1.21(d), a normal distribution fits quite well. The only important deviation from normality is the numerous short vertical stacks of points in the plot. Each stack represents repeated observations having the same value—there are six measurements at 27 and seven at 28, for example. This phenomenon is called *granularity*, because the data come in lumps rather than being smoothly distributed as the normal model specifies. Minor granularity, due to the limited precision of the measurements, does not hinder adopting a normal distribution as a model. The data would presumably be even closer to the normal model if Newcomb had been able to measure one more decimal place, thus spreading out the stacks.

granularity

What of the Gary vocabulary scores, whose normality we have assessed by cruder means in Figure 1.12 (page 60) and Example 1.18 (page 66)? A plot of 947 individual points, many taking the same values, will wear holes in your paper. Figure 1.22 therefore presents a normal quantile plot based on every tenth score in the ordered list. It is remarkably similar to Figure 1.21(d), showing excellent fit to a normal model except for some granularity due to the reporting of scores only to the nearest tenth of a grade level. A normal distribution does provide an accurate model for this large data set.

As Figures 1.21(d) and 1.22 illustrate, real data will almost always show some departure from the theoretical normal model. It is important to confine your examination of a normal quantile plot to searching for shapes that show clear departures from normality. Don't overreact to minor wiggles in the plot. When we discuss statistical methods that are based on the normal model, we will pay attention to the sensitivity of each method to departures from normality. Many common methods work well when the data are approximately normal. It is this fact, along with the accuracy of the normal model for some common types of data, that explains the frequent use of the normal distributions in statistics.

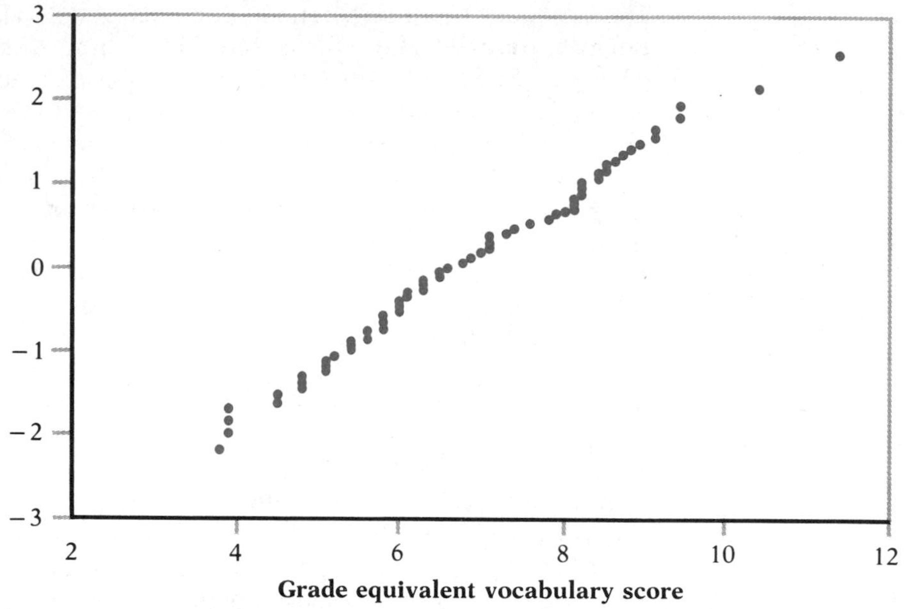

FIGURE 1.22 Normal quantile plot of the Gary vocabulary scores.

SUMMARY

The overall pattern of a distribution can often be described compactly by a **density curve.** A density curve is a curve that always remains on or above the horizontal axis and has total area 1 underneath it. Areas under a density curve give **relative frequencies** for the distribution.

The mean μ (balance point), the median (equal-area point), and other percentiles can be located on a density curve. The standard deviation σ cannot be located by eye on most density curves. The mean and median are equal for symmetric density curves, but the mean of a skewed curve is located farther toward the long tail than is the median.

The **normal distributions** are specified by bell-shaped, symmetric density curves. The mean μ and standard deviation σ completely describe the normal distribution $N(\mu, \sigma)$. The mean is the center of symmetry and σ is the distance from μ to the change-of-curvature points on either side. All normal distributions are the same when measurements are made in units of σ about the mean. These are called **standardized measurements.** In particular, all normal distributions satisfy the **68–95–99.7 rule.**

If X has the $N(\mu, \sigma)$ distribution, then the **standardized variable** $Z = (X - \mu)/\sigma$ has the **standard normal** distribution $N(0, 1)$. Relative frequencies for any normal distribution can be calculated from Table A, which gives relative frequencies for the events $Z < z$ for many values of z.

The adequacy of a normal model for a set of data is best assessed by a **normal quantile plot,** which is available on most statistical software packages. Points on such a plot that deviate substantially from a straight line indicate that the data are not normal.

SECTION 1.3 EXERCISES

1.72 Figure 1.23 displays three density curves, each with three points marked on them. At which of these points on each curve do the mean and the median fall?

1.73 Figure 1.24 displays the density curve of a *uniform* distribution. The curve takes the constant value 1 over the interval from 0 to 1, and is 0 elsewhere. This means that data described by this distribution take values that are uniformly spread between 0 and 1. Use the fact that areas under a density curve are relative frequencies to answer the following questions.
(a) What fraction of the observations lie above 0.8?
(b) What fraction of the observations lie below 0.6?
(c) What fraction of the observations lie between 0.25 and 0.75?

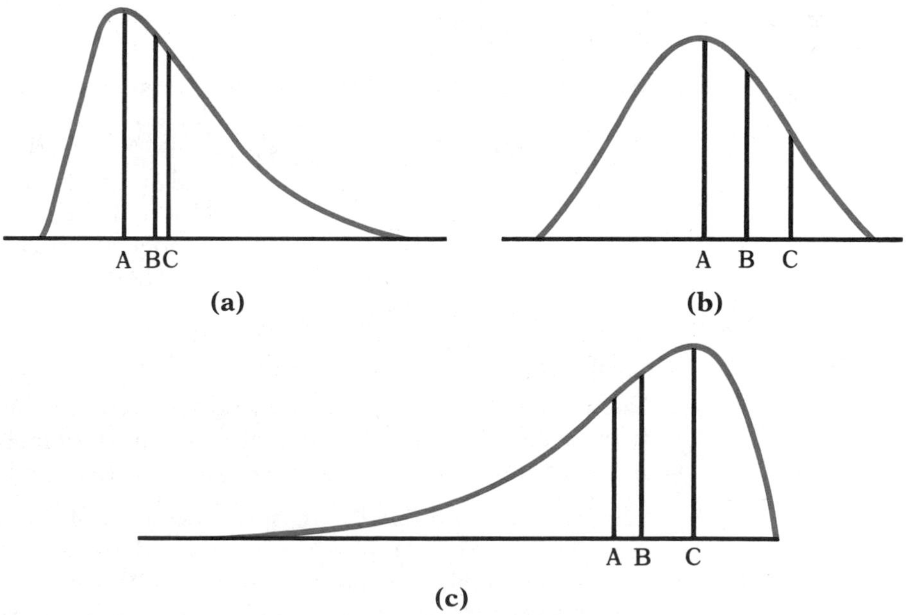

FIGURE 1.23 Density curves for Exercise 1.72

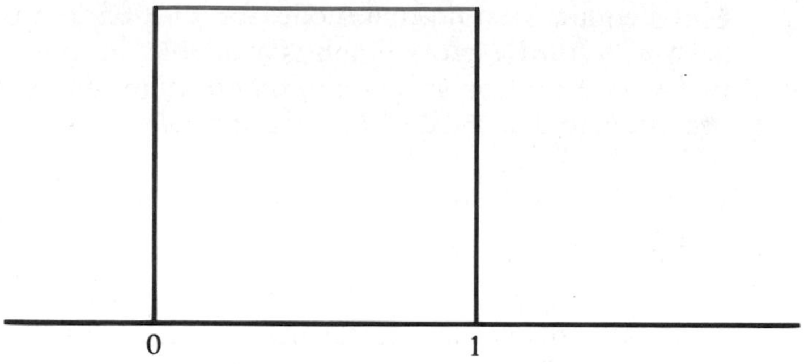

FIGURE 1.24 **A uniform density curve for Exercise 1.73.**

1.74 What are the mean and the median of the distribution in Figure 1.24? What are the quartiles?

1.75 The Environmental Protection Agency requires that the exhaust of each model of motor vehicle be tested for the level of several pollutants. The level of oxides of nitrogen (NOX) in the exhaust of one light truck model was found to vary among individual trucks according to a normal distribution with mean $\mu = 1.45$ grams per mile driven and standard deviation $\sigma = 0.40$ grams per mile. Sketch the density curve of this normal distribution, with the scale of grams per mile marked on the horizontal axis.

1.76 In a study of elite distance runners, the mean weight was reported to be 63.1 kilograms (kg), with a standard deviation of 4.8 kg. Assuming that the distribution of weights is normal, sketch the density curve of the weight distribution, with the horizontal axis marked in kilograms. (Based on M. L. Pollock et al., "Body composition of elite class distance runners," in P. Milvy (ed.), *The Marathon: Physiological, Medical, Epidemiological, and Psychological Studies*, New York Academy of Sciences, 1977.)

1.77 Give an interval that contains the middle 95% of NOX levels in the exhaust of trucks using the model described in Exercise 1.75.

1.78 Use the 68–95–99.7 rule to find intervals centered at the mean that will include 68%, 95%, and 99.7% of the weights of the elite runners described in Exercise 1.76.

1.79 Eleanor scores 680 on the mathematics part of the Scholastic Aptitude Test. The distribution of SAT scores in a reference population is normal with mean 500 and standard deviation 100. Gerald takes the American College Testing mathematics test and scores 27. ACT scores are normally distributed with mean 18 and standard deviation 6. Find the standardized scores for both students. Assuming that both tests measure the same kind of ability, who has the higher score?

1.80 Three landmarks of baseball achievement are Ty Cobb's batting average of .420 in 1911, Ted Williams's .406 in 1941, and George Brett's .390 in 1980. These batting averages cannot be compared directly because the distribution of major league batting averages has changed over the decades. The distributions are quite symmetric and (except for outliers such as Cobb, Williams, and Brett) reasonably normal. Although the mean batting average has been held roughly constant by rule changes and the balance between hitting and pitching, the standard deviation has dropped over time. Here are the facts:

Decade	Mean	Std. dev.
1910s	.266	.0371
1940s	.267	.0326
1970s	.261	.0317

Compute the standardized batting averages for Cobb, Williams, and Brett to compare how far each stood above his peers. (Data from Stephen Jay Gould, "Entropic homogeneity isn't why no one hits .400 anymore," *Discover*, August 1986, pp. 60–66. Gould does not standardize but gives a speculative discussion instead.)

1.81 Using Table A, find the proportion of observations from a standard normal distribution that satisfies each of the following statements. In each case, sketch a standard normal curve and shade the area under the curve that is the answer to the question.

(a) $Z < 2.85$
(b) $Z > 2.85$
(c) $Z > -1.66$
(d) $-1.66 < Z < 2.85$

1.82 Using Table A, find the relative frequency of each of the following events in a standard normal distribution. In each case, sketch a standard normal curve with the area representing the relative frequency shaded.

(a) $Z \leq -2.25$
(b) $Z \geq -2.25$
(c) $Z > 1.77$
(d) $-2.25 < Z < 1.77$

1.83 Use Table A to find the value z of a standard normal variable Z that satisfies each of the following conditions. (Use the value of z from Table A that comes closest to satisfying the condition.) In each case, sketch a standard normal curve with your value of z marked on the axis.

(a) The point z with 25% of the observations falling below it.
(b) The point z with 40% of the observations falling above it.

1.84 The variable Z has a standard normal distribution.

(a) Find the number z such that the event $Z < z$ has relative frequency 0.8.

(b) Find the number z such that the event $Z > z$ has relative frequency 0.35.

1.85 The height X of young American women has approximately the normal distribution with mean $\mu = 65.5$ inches and standard deviation $\sigma = 2.5$ inches. Use Table A to find the relative frequencies of each of the following events. In each case, sketch a normal curve and shade the area that represents the relative frequency.
(a) $X < 67$
(b) $64 < X < 67$

1.86 The Graduate Record Examinations (GRE) are widely used to help predict the performance of applicants to graduate schools. The range of possible scores on a GRE is 200 to 900. The psychology department at a university finds that the scores of its applicants on the quantitative GRE are approximately normal with mean $\mu = 544$ and standard deviation $\sigma = 103$. Use Table A to find the relative frequency of applicants whose score X satisfies each of the following conditions:
(a) $X > 700$
(b) $X < 500$
(c) $500 < X < 800$

1.87 A patient is said to be hypokalemic (low potassium in the blood) if the measured level of potassium is 3.5 or less. (The units for this measure are meq/l or milliequivalents per liter.) An individual's potassium level is not a constant, however, but varies from day to day. In addition, the measurement procedure itself has some variation. Suppose that the overall variation follows a normal distribution. Judy has a mean potassium level of 3.8 with a standard deviation of 0.2. If she is measured on many days, on what proportion of days will the measurement indicate hypokalemia?

1.88 The Acculturation Rating Scale for Mexican Americans (ARSMA) is a psychological test that evaluates the degree to which Mexican Americans are adapted to Mexican/Spanish versus Anglo/English culture. The range of possible scores is 1.0 to 5.0, with higher scores showing more Anglo/English acculturation. The distribution of ARSMA scores in a population used to develop the test is approximately normal with mean 3.0 and standard deviation 0.8. A researcher believes that Mexicans will have an average score near 1.7 and that first-generation Mexican Americans will average about 2.1 on the ARSMA scale. What proportion of the population used to develop the test has scores below 1.7? Between 1.7 and 2.1?

1.89 How high a score on the ARSMA test of Exercise 1.88 must a Mexican American obtain to be among the 30% of the population used to develop the test who are most Anglo/English in cultural orientation? What scores make up the 30% who are most Mexican/Spanish in their acculturation?

1.90 Refer to Exercise 1.76. If 90% of elite distance runners weigh less than Peter, what is Peter's weight?

1.91 A nutrition study of London bus drivers included measurements of daily food intake in calories. The drivers had a mean intake of 2821 calories, with a standard deviation of 436 calories. The distribution was approximately normal. Calculate the proportion of drivers whose daily consumption falls in each of the following ranges: less than 2000, 2000 to 2500, 2500 to 3000, 3000 to 3500, and above 3500.

1.92 The scores of a reference population on the Wechsler Intelligence Scale for Children (WISC) are normally distributed with $\mu = 100$ and $\sigma = 15$.

(a) What percent of this population have WISC scores below 100?
(b) Below 80?
(c) Above 140?
(d) Between 100 and 120?

1.93 The distribution of scores on the WISC is described in the previous exercise. What score must a child achieve on the WISC in order to fall in the top 5% of the population? In the top 1%?

1.94 The distribution of a critical dimension on auto engine crankshafts is approximately normal with $\mu = 224$ mm and $\sigma = 0.03$ mm. Crankshafts with dimensions between 223.92 mm and 224.08 mm are acceptable. What percent of all crankshafts produced are acceptable? A quality engineer measures a sample of 16 of these crankshafts as follows:

224.120	224.001	224.017	223.982	223.989	223.961
223.960	224.089	223.987	223.976	223.902	223.980
224.098	224.057	223.913	223.999		

What percent of these 16 crankshafts are acceptable?

1.95 What percents of the 16 crankshaft measurements in the previous exercise fall within one, two, and three standard deviations of the mean of the $N(224, 0.03)$ distribution? Compare your results with the 68–95–99.7 rule. Is there any strong indication that the data do not follow this normal distribution?

1.96 The median of any normal distribution is the same as its mean. We can use Table A to find the quartiles and related descriptive measures for normal distributions.

(a) What are the first and third quartiles of the standard normal distribution? (These numbers give the location of the quartiles for any normal distribution, in standard deviation units away from the mean.)
(b) What is the value of the IQR for the standard normal distribution?
(c) What percent of the observations in the standard normal distribution are suspected outliers according to the $1.5 \times IQR$ criterion? (This percent is the same for any normal distribution.)

1.97 Figure 1.25 is a normal quantile plot of the DRP scores from Exercise 1.23. Are these scores approximately normally distributed? Discuss any major deviations from normality that appear in the plot.

1.98 In a medical experiment, 72 guinea pigs are injected with tubercle bacilli to study their resistance to infection. The survival time (in days) after infection is measured. (The actual data appear in Exercise 7.33.) Figure 1.26 is a normal quantile plot of the survival times. In what way does the distribution of survival times depart from normality? (Based on data in T. Bjerkedal, "Acquisition of resistance in guinea pigs infected with different doses of virulent tubercle bacilli," *American Journal of Hygiene*, 72 (1960), pp. 130–148.)

1.99 Figure 1.27 is a normal quantile plot of the maximum-rainfall data from Exercise 1.20. Are there any clear deviations from normality?

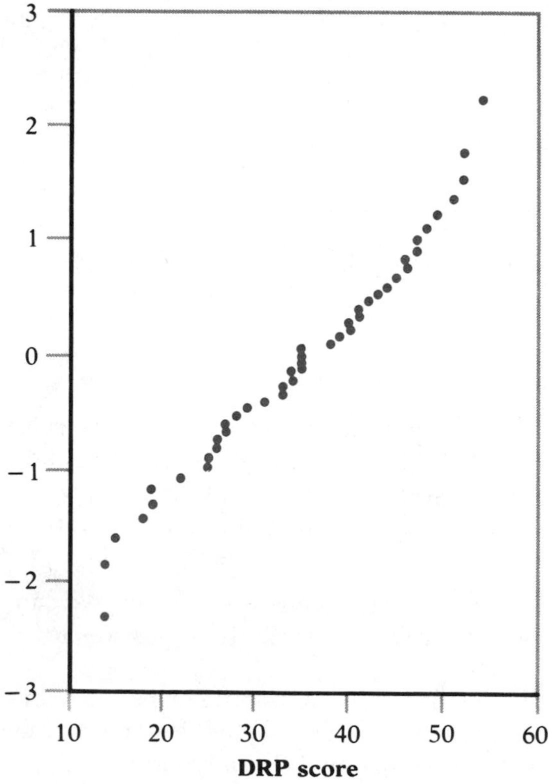

FIGURE 1.25 Normal quantile plot of the DRP scores, for Exercise 1.97.

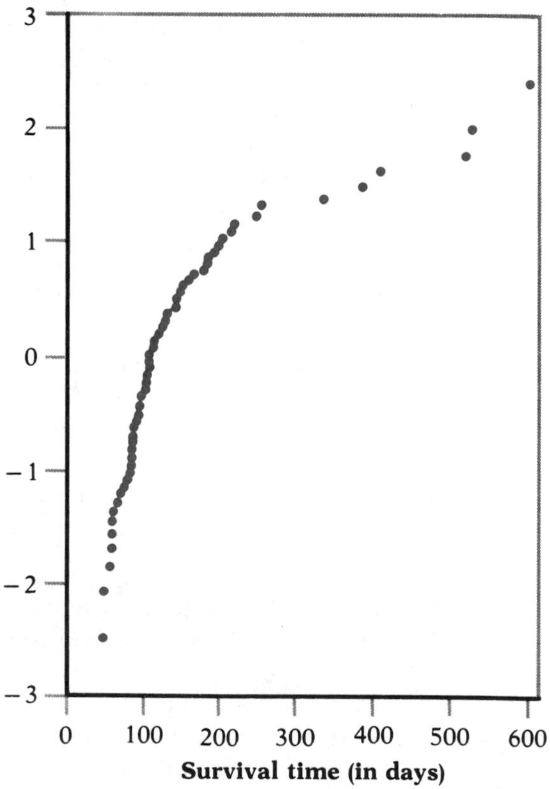

FIGURE 1.26 Normal quantile plot of guinea pig survival times, for Exercise 1.98.

1.100 The distance between two mounting holes is important to the performance of an electrical meter. The manufacturer measures this distance regularly for quality control purposes, recording the data in coded form as thousandths of an inch above 0.600 inches. For example, 0.644 is recorded as 44. Figure 1.28 is a normal quantile plot of the distances for the last 27 meters measured. Is the overall shape of the distribution approximately normal? Why do you think that many of the points appear in vertical stacks? (Data provided by Charles Hicks, Purdue University.)

1.101 Assess the normality of Cavendish's measurements in Exercise 1.25 (page 28) by finding the percent that fall within one, two, and three standard deviations of the mean. (You calculated the mean and standard deviation in Exercise 1.56 on page 55.) Is there a clear indication of nonnormality?

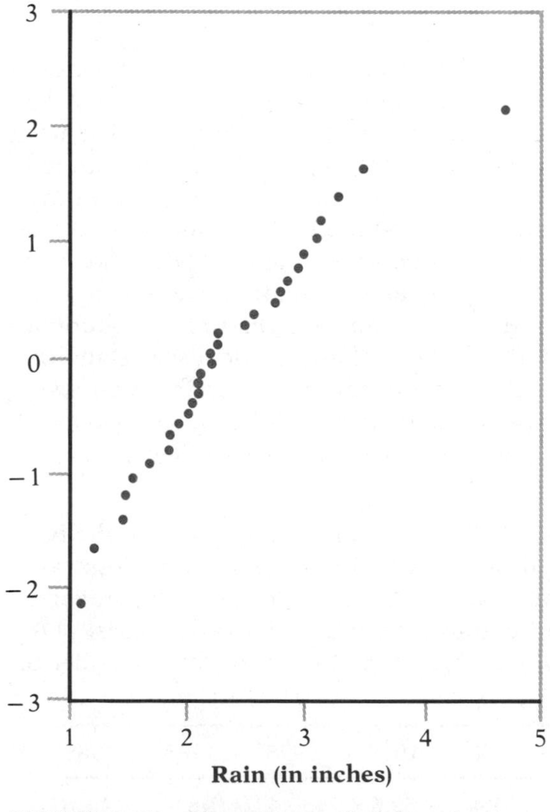

FIGURE 1.27 Normal quantile plot of maximum daily rainfall, for Exercise 1.99.

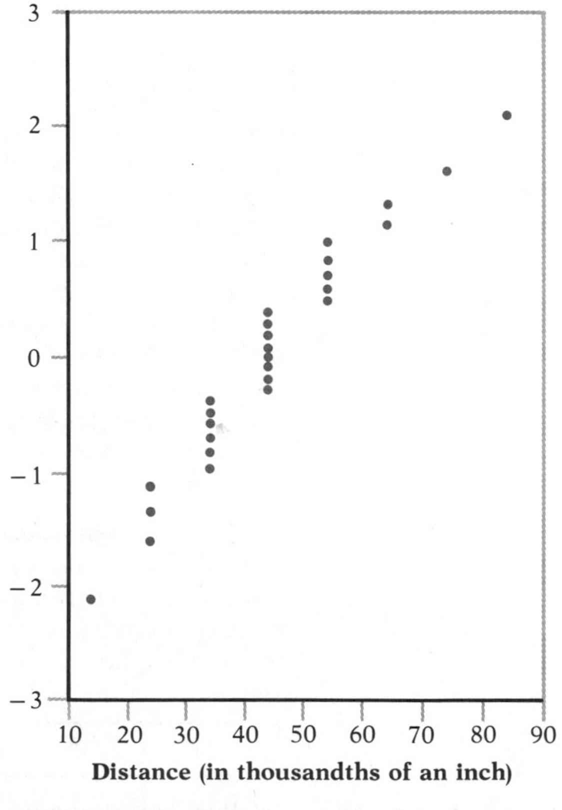

FIGURE 1.28 Normal quantile plot of distance measurements, for Exercise 1.100.

The remaining exercises for this section require the use of computer software that will make normal quantile plots.

1.102 Make a normal quantile plot for Cavendish's measurements in Exercise 1.25 (page 28). Are the data approximately normal? If not, describe any clear deviations from normality.

1.103 Make a normal quantile plot of the crankshaft measurements in Exercise 1.11 (page 22). Are the data approximately normal? If not, how do they depart from normality?

1.104 Table 1.4 (page 36) gives the calorie content of a number of brands of beef, meat, and poultry hot dogs. We have already compared these distributions to some extent in Example 1.13 (page 44) and Figure 1.11. Make normal quantile plots for the calorie content of each of the three types of hot dogs. Then briefly compare the three distributions on the basis of their normal quantile plots.

CHAPTER 1 EXERCISES

1.105 Voting patterns were affected by many social changes in the years between 1940 and 1980. Prior to the civil rights movement of the 1960s, blacks were effectively prevented from voting in some southern states. And until 1971 citizens between the ages of 18 and 21 could not vote in national elections. Figure 1.29(a) displays the distribution of the percent of the voting-age population who actually voted in the 1940 presidential election in each of the 48 states that existed at that time. Figure 1.29(b) shows the distribution of the percent of the voting-age population who voted in the 1980 presidential election in the 50 states and the District of Columbia. To allow a direct comparison, the horizontal and vertical scales and the class widths are the same for the two histograms. Describe the most important changes in the shape of the distribution between 1940 and 1980. In particular, how is the addition of black voters reflected in the shapes of the distributions?

1.106 Here are the death rates per 100,000 women age 45 to 54 from the two types of cancer that kill the most women: breast cancer and respiratory/intrathoracic (mainly lung) cancer. Plot the two time series together and describe the pattern. Now extrapolate your curves (this is risky) to predict when lung cancer will pass breast cancer as the leading killer of women.

Year	1940	1950	1960	1970	1975	1980	1985
Breast	47.5	49.9	51.4	52.6	50.4	48.9	46.7
Lung	6.2	6.7	10.1	22.2	28.0	34.8	35.9

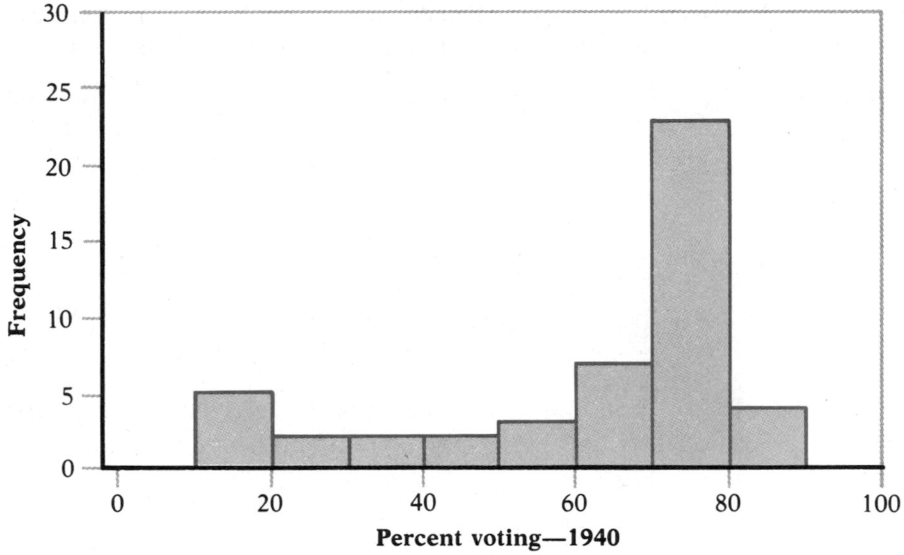

FIGURE 1.29(a) Distribution of the percent of the voting-age population who voted in each of the states in the 1940 presidential election, for Exercise 1.105.

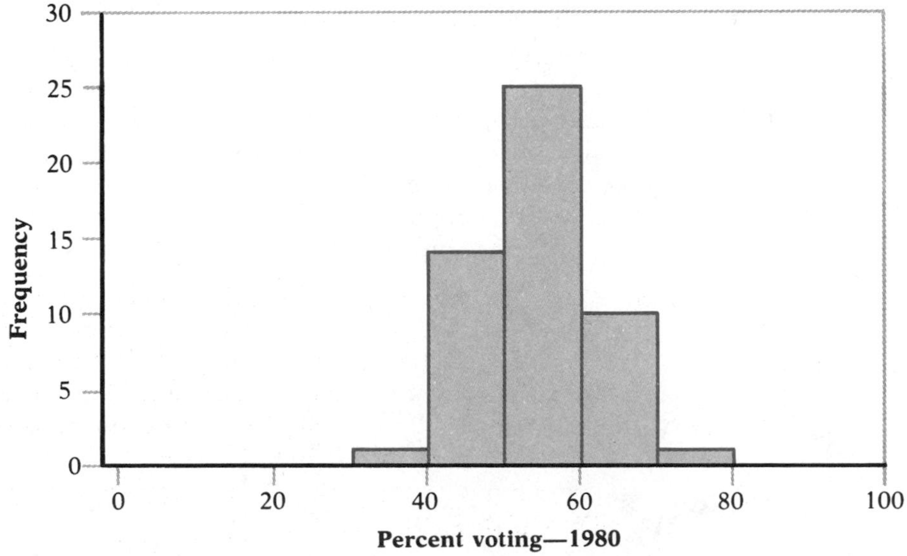

FIGURE 1.29(b) Distribution of the percent of the voting-age population who voted in each of the states in the 1980 presidential election, for Exercise 1.105.

1.107 A quality engineer in an auto plant measures the eccentricity of a valve assembly. (Eccentricity is a measure of how off-center two supposedly concentric circles are.) Here are the data in thousandths of an inch:

1.084	1.131	.887	.639	1.216	.903
.977	1.088	.940	1.069	.667	.536

Describe in detail the distribution of these measurements. Include at least one graphical display of the distribution and numerical measures that are appropriate for a distribution of this shape. Is the distribution approximately normal?

1.108 The following table gives the 1990 baseball income (in thousands of dollars) for the members of the Chicago Cubs. Examine this distribution, present it graphically and numerically in the way you think appropriate, and describe the notable features of the distribution. (Data from the *New York Times*, April 10, 1991.)

Dawson	3300	Assenmacher	1000	Dascenzo	165
Sandberg	2650	Bielecki	810	Boskie	130
Jackson	2625	Salizar	575	Villaneuva	120
Maddux	2400	Lancaster	550	Vizcaino	114
Sutcliffe	2275	Berryhill	230	McElroy	111
Bell	2100	Girardi	225	Scott	100
Dunston	2100	Smith	225	Pappas	100
Smith	1900	Harkey	220	Slocumb	100
Grace	1200	Walton	210	Walker	100

1.109 You are planning a sample survey of households in Indiana. You decide to select households separately within counties. (Chapter 3 will present more information on planning sample surveys.) To aid in the planning, here are the populations of the state's 92 counties (in thousands of persons) according to the 1990 census:

31	301	64	9	14	38	14	19	38	88
24	31	10	28	39	24	35	120	37	156
26	64	18	20	19	32	74	30	109	46
30	76	48	81	35	38	25	22	30	24
88	40	65	29	476	107	43	131	797	42
10	37	109	34	56	14	38	5	18	17
15	19	13	129	26	13	30	27	25	18
247	21	40	19	23	27	19	8	131	16
7	165	17	106	35	8	45	24	72	26
24	28								

Examine the distribution of county populations both graphically and numerically, using whatever tools are most suitable. Then write a brief description of the main features of this distribution. Suggest some groupings of counties based on population size that would be useful in conducting the sample survey.

1.110 The NASDAQ Composite Index describes the average price of common stock traded over the counter, that is, not on one of the stock exchanges. In 1991, the mean capitalization of the companies in the NASDAQ index was \$80 million and the median capitalization was \$20 million. (A company's capitalization is the total market value of its stock.) Explain why the mean capitalization is much higher than the median.

1.111 A test designed to measure reading ability of children was administered to a large group of third- and sixth-grade children. The mean score for the third graders was 75 and the standard deviation was 10. For the sixth graders the mean was 82 and the standard deviation was 11. We want to transform the scores so that a meaningful comparison can be made between the score of a third grader and the score of a sixth grader relative to the scores of other children in the same grade.

(a) Find the linear transformation, that is, the values of a and b in $x^* = a + bx$, that will transform the third-grade scores to have a mean of 100 and a standard deviation of 20. Choose the value of b to be positive so that the order of the scores is not reversed.
(b) Do the same for the sixth-grade scores.
(c) Suppose that a third-grade student scores 78 on the test. Find the transformed score. Do the same for a sixth-grade student who scores 78.
(d) Suppose that the distribution of scores in each grade is normal. Then both transformed scores have the $N(100, 20)$ distribution. What percent of third graders have scores less than 78? What percent of sixth graders have scores less than 78?

1.112 The Chapin Social Insight Test evaluates how accurately the subject appraises other people. In the reference population used to develop the test, scores are approximately normally distributed with mean 25 and standard deviation 5. The range of possible scores is 0 to 41.

(a) What proportion of the population has scores below 20 on the Chapin test?
(b) What proportion has scores below 10?
(c) How high a score must you have in order to be in the top quarter of the population in social insight?

1.113 The Chapin Social Insight Test described in the previous exercise has a mean of 25 and a standard deviation of 5. Suppose that you want to rescale the test using a linear transformation so that the mean is 100 and the standard deviation is 20. Let x denote the variable in the original scale and x^* represent the transformed variable.

(a) Find the linear transformation required, that is, find the values of a and b in the equation $x^* = a + bx$.
(b) Give the rescaled score for an individual who scores 30 in the original scale.
(c) Make a table of the deciles (10th, 20th, . . ., 80th, 90th percentiles) for the rescaled test scores.

1.114 Explain why the shape of the distribution of DRP reading scores as displayed in the normal quantile plot in Figure 1.25 (page 84) suggests that the mean and standard deviation are adequate measures of center and spread for these data. Then calculate the mean and standard deviation from the scores given in Exercise 1.23 (page 28).

1.115 Refer again to the DRP reading scores whose mean and standard deviation you found in the previous exercise. Find the linear transformation (that is, find the values of a and b in $x^* = a + bx$) that preserves the order of the scores and produces transformed scores having mean 100 and standard deviation 10. (Hint: State the relations between the means of x and x^*, and between the standard deviations of x and of x^*. Then put in the known values for x and the desired values for x^* and solve for a and b.)

1.116 Consider the weight gain data for chicks fed normal and high-lysine corn in Exercise 1.28 (page 29).
(a) Make a normal quantile plot for each group if your software allows. Both distributions are approximately normal.
(b) We can therefore summarize the distributions by computing the mean and standard deviation for each group; do this. Did the high-lysine chicks gain more weight? Is the variability in weight gains substantially different in the two groups?

CHAPTER 1 COMPUTER EXERCISES

1.117 Exercise 1.109 gives the populations of the 92 counties in Indiana.
(a) Make a normal quantile plot of these data. Explain briefly how your plot displays the shape of the distribution.
(b) Calculate both the mean and median population and explain why these values differ greatly. Can you recover the total state population from either or both of these measures of center? If so, do it.
(c) Find the counts of counties with populations within one, two, and three standard deviations of the mean. Does the 68–95–99.7 rule apply to the distribution of county populations?

1.118 The following data are the founding dates of wineries in New Zealand:

1975	1980	1973	1974	1943	1902	1947	1942
1937	1983	1934	1936	1935	1902	1937	1961
1979	1969	1976	1925	1922	1937	1973	1933
1948	1944	1904	1981	1872	1918	1860	1960
1876	1980	1967	1974	1977	1979	1981	1978

(a) Make either a stemplot or a histogram for these data. If suitable software is available, make a normal quantile plot.

(b) Compute the mean, standard deviation, five-number summary, and *IQR*. (Does your software have a single command that gives most of these measures?)

(c) Give a careful description of the main features of the distribution. How useful are the numerical measures you computed?

(d) There is another winery for which a precise founding date is not available. We do know, however, that it was founded after World War II. Assuming that the date is after 1945, recompute the median. Do you think that the median is a good descriptive measure for these data? Why or why not?

1.119 The following table gives the weights in pounds of the players on a Big Ten football team as listed in the team's preseason publicity brochure. We will investigate the shape of this distribution.

160	165	202	245	150	235	241	195	181	256
200	265	168	233	215	215	185	185	263	214
230	260	190	185	294	180	179	295	175	157
218	208	145	205	210	171	276	220	189	222
175	170	157	195	260	186	225	188	260	192
214	266	265	188	196	228	270	163	183	160
260	178	260	155	210	200	276	290	165	220
210	200	185	195	196	260	165	175	224	235
225	228	215	225	229	214	193	162	294	280
183	185	263	276	290	218	184	190	180	225
220	250	253	220	155	160	176	200	230	218
158	215	225	184	195	282	252	218	235	182
180	180	180	224	214	210	255	270	200	234
175									

(a) Make either a stemplot or a histogram, and explain your choice of graphical display. If you have suitable software, also make a normal quantile plot.

(b) Describe the main features of the distribution. Is it symmetric? Approximately normal? Does it appear that there are groups of players with different weight distributions, such as linemen and defensive backs? Are there any outliers?

(c) Give a brief numerical summary chosen in the light of your findings in (b).
(d) For publication you must restate your summary in kilograms rather than pounds; do this (1 kg equals 2.2 pounds, so 160 pounds is 72.73 kg).
(e) There is a special kind of granularity in this data set; describe it. (Hint: Look at the last digit.)

1.120 Exercise 7.33 reports the survival time (in days) of 72 guinea pigs after they were injected with disease organisms in a medical experiment. Give a brief graphical and numerical description of this distribution, using the methods you think are most appropriate. Briefly describe the main features of the distribution.

1.121 Most statistical software packages have routines for generating values of variables having specified distributions. The following Minitab commands, for example, produce 25 observations from the $N(20, 5)$ distribution and then compute several descriptive measures for these observations:

```
MTB > RANDOM 25 C1;
SUBC > NORMAL 20, 5.
MTB > DESCRIBE C1
```

Use your statistical software to generate 25 observations from the $N(20, 5)$ distribution. Compute the mean and standard deviation $\bar{x}$ and s of the 25 values you obtain. How close are $\bar{x}$ and s to the μ and σ of the distribution from which the observations were drawn?

Repeat 20 times the process of generating 25 observations from the $N(20, 5)$ distribution and recording $\bar{x}$ and s. Make a stemplot of the 20 values of $\bar{x}$ and another stemplot of the 20 values of s. Briefly describe each of these distributions. Are they symmetric or skewed? Where are their centers? (The distributions of measures like $\bar{x}$ and s when repeated sets of observations are made from the same theoretical distribution will be very important in later chapters.)

1.122 Exercise 1.108 gives the 1990 salaries of the members of the Chicago Cubs. The major league minimum salary in 1990 was \$100,000. Compute a new variable which is salary in excess of the minimum salary in thousands of dollars (In most software packages this can be done by a command similar to SALNEW = SALARY − 100.)
(a) Compute the five-number summary, the *IQR*, the mean, and the standard deviation for the salaries as given and for the the new variable. For each numerical measure explain how the value for the new variable is related to the value for salary.
(b) Is the shape of the distribution changed by the transformation? Explain why or why not.

1.123 The CHEESE data set described in the data appendix records measurements on 30 specimens of Australian cheddar cheese. Investigate the distributions of the variables H2S and LACTIC using graphical and numerical summaries of your choice. Write a short description of the notable features for each distribution.

1.124 The CSDATA data set described in the data appendix contains information on 224 computer science students. We are interested in comparing female students as a group with male students as a group. Use side-by-side boxplots to compare men and women first on SAT mathematics score and then on college grade index. Write a brief discussion of the male–female comparisons. Then, if your software allows, make normal quantile plots of grade index and SAT math scores separately for men and women. Which of the four distributions are approximately normal?

NOTES

1. Newcomb's data and the background information about his work appear in S. M. Stigler, "Do robust estimators work with real data?" *Annals of Statistics*, 5 (1977), pp. 1055–1078.
2. *The Economist*, Aug. 30, 1986.
3. Data from *The Baseball Encyclopedia*, 3rd ed., Macmillan, New York, 1976. Maris's home run data are from the same source.
4. Data from Gary Community School Corporation, courtesy of Celeste Foster, Department of Education, Purdue University.
5. James Gleick, "Hole in ozone over South Pole worries scientists," *New York Times*, July 29, 1986. Note that the outliers, although omitted from the initial analysis, were not discarded. It was therefore possible to reanalyze the data later.
6. The data on which Figure 1.4 is based appear in the monthly issues of the Bureau of Labor Statistics' *CPI Detailed Report*.
7. We thank W. Robert Stephenson of Iowa State University for pointing out the importance of weight in explaining the distribution of calories.

CHAPTER

2

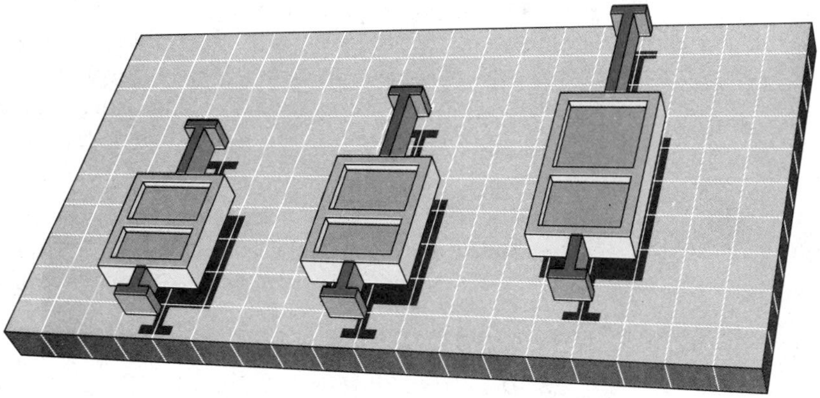

Looking at Data: Relationships

PRELUDE

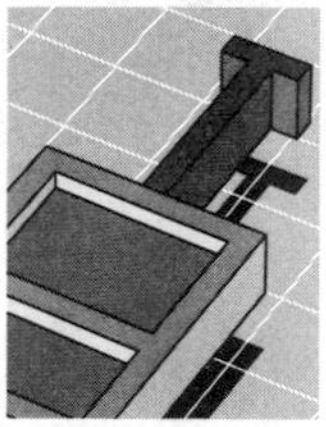

Statistical investigations only rarely focus on a single variable. We are more often interested in comparisons among several distributions, changes in a variable over time, or relationships among several variables. Chapter 1 provided graphical and numerical tools for examining the distribution of a single variable and for comparing several distributions, as well as an introduction to investigating change over time. In this chapter we will concentrate on methods for studying relations among several variables.

Once again we begin with graphs and then move to numerical summaries of specific aspects of the data. Although we will meet a variety of numerical summaries in this chapter, we emphasize those that describe straight-line relations between two variables. These methods, called regression and correlation, are among the most common statistical tools. A study of data often leads us to ask whether there is a cause-and-effect relation between two variables that are closely linked in the data. We will study data on issues such as these:

- Can we use the age at which a child begins to talk to predict later scores on a test of mental ability?
- Only a few fossil specimens of an extinct beast survive. Are all these specimens from the same species, or do they represent several different species?
- What is the relation between age and education level among American adults?
- How strong is the evidence that smoking causes lung cancer?

• • •

Many statistical questions concern relations among several variables. Here is a small portion of a large data set, as printed out by the statistical computing system used to analyze it.

OBS	ID	AGE	SEX	JOB	WT	SBP
1	1083	39	M	T	183	132
2	1381	27	F	E	116	117
3	1502	57	M	E	172	144
4	1481	26	M	T	139	110
5	1666	48	F	T	132	150

These data record medical measurements on the employees of a large company, made during the company's regular physical examinations. Each row gives data for one employee, or, in statistical language, one case.

> **CASE**
>
> A case is an individual person, animal, or thing for which values of variables are recorded.

The computing system has numbered the cases consecutively under the heading OBS. The employees are given identification numbers (ID) so that their names will not appear on the printout. Following the case number and ID, each row contains the values of five variables for one individual. The variable names appear as column headings in the printout. The variables are the employee's age, sex, job category, weight, and systolic blood pressure. Sex is coded as F (female) or M (male) and job category as E (executive) or T (technical). This general format, with each row representing a case and each column a variable, is common to most statistical software systems.

When you examine the relationship between two or more variables, you should ask the preliminary questions that are familiar from Chapter 1: What exactly are the variables? How were they measured? Are all the variables quantitative or is at least one a categorical variable? Recall that *quantitative variables* take numerical values for which descriptions such as means and standard deviations are meaningful. In the medical records, the age, weight, and systolic blood pressure of the employees are quantitative variables. *Categorical variables*, on the other hand, are merely labels that tell us into which class an individual falls. The sex and job classification of the employees are categorical variables.

quantitative variable

categorical variable

We have concentrated on quantitative variables until now. When we have data on several variables, however, categorical variables are usually present and are essential aids in organizing the data. We are interested in relations between two quantitative variables (such as a person's weight and blood pressure), between a quantitative and a categorical variable (such as blood pressure and sex), and between two categorical variables (such as sex and job category).

When you examine the relation between two variables rather than just describing a single variable, a new question becomes important. Is your purpose simply to explore the nature of the relationship, or do you hope to show that one of the variables can explain variation in the other? Suppose that a political scientist looks at the Democrats' share of the popular vote for president in each state in two consecutive presidential elections. She does not wish to explain one year's data by the other's, but rather to see a pattern that may shed light on political conditions. In other situations, the purpose of a study is to demonstrate that one variable helps explain another. Agricultural researchers, for example, carefully plant corn in different amounts per acre and record the yield because they are interested in a cause-and-effect relationship. They believe that the planting rate will affect the yield, and their purpose is to recommend the best planting rate to farmers. In this setting, we must distinguish the explanatory variable (plants per acre) from the response variable (yield of corn).

RESPONSE VARIABLE, EXPLANATORY VARIABLE

A response variable measures an outcome of a study. An explanatory variable attempts to explain the observed outcomes.

In many studies, the goal is to show that changes in one or more explanatory variables actually *cause* changes in a response variable. But not all explanatory-response relationships involve direct causation. The Scholastic Aptitude Test (SAT) scores of high school students may help predict the students' future college grades, but high SAT scores certainly don't cause high college grades. Some of the statistical techniques in this chapter require us to distinguish explanatory from response variables; others make no use of this distinction. Explanatory variables are often called *independent variables*, and response variables are often referred to as *dependent variables*. The idea behind this language is that the response variable depends on the explanatory variable. Because the words "independent" and "dependent" have other meanings in statistics that are unrelated to the explanatory-response distinction, we prefer to avoid those words here.

independent variable

dependent variable

The techniques used to study relations among variables are more complex than the methods we developed in Chapter 1 to examine the distribution of a single variable. Fortunately, statistical analysis of several-variable data builds on the tools used for examining individual variables. The principles that guide our work also remain the same:

- Combine graphical display with numerical summaries.
- Seek overall patterns and deviations from those patterns.
- Seek compact mathematical models for the data in addition to descriptive measures of specific aspects of the data.

2.1 SCATTERPLOTS

scatterplot

Relationships between two quantitative variables are best displayed graphically. The most useful graph for this purpose is a *scatterplot*. Here is an example of the effective use of scatterplots.

EXAMPLE 2.1

Ronald Reagan won election as president in 1980 and again in 1984. Reagan, who was the Republican candidate, defeated Democrats Jimmy Carter in 1980 and Walter Mondale in 1984. (In 1980 an independent candidate, John Anderson, captured 6.7% of the national vote.) Table 2.1 shows the percent of the popular vote won by the Democratic presidential candidates in the 1980 and 1984 elections.[1] We know that many states have persisting political traditions, so we expect similar behavior in two successive elections. It is possible to see this relationship in the columns of numbers in the table, but it is very difficult to assess the strength of the relationship or to see any significant changes from 1980 to 1984. A picture is needed.

Figure 2.1 is a scatterplot of the data in Table 2.1. Each point on the plot represents a single case—that is, a single state. Because there is no explanatory-response relationship in these data, we can choose to plot either the 1980 or the 1984 data horizontally. In the figure, the horizontal coordinate x is the percent who voted Democrat in a state's 1980 presidential vote. The vertical coordinate y is the percent who voted Democrat in 1984. For example, Alabama appears as the point (48.7, 38.7). Because both variables have the same units (percent), we use the same scale on both axes. The resulting plot outline is square. ●

Interpreting scatterplots

To interpret a scatterplot, look first for an overall pattern. In Figure 2.1, this pattern should reveal the direction, form, and strength of the relationship between 1980 and 1984 voting in the states. The direction is clear from Figure 2.1: States with a high percent of Democratic votes in 1980 tended to also vote Democrat in 1984. That is, the percent voting Democrat in 1980 and in 1984 are positively associated.

TABLE 2.1 Percent of presidential vote won by Democratic candidate, 1980 and 1984

State	1980 percent	1984 percent	State	1980 percent	1984 percent
Ala.	48.7	38.7	Mont.	33.3	38.7
Alaska	30.1	30.9	Nebr.	26.4	29.0
Ariz.	28.9	32.9	Nev.	27.9	32.7
Ark.	48.3	38.8	N.H.	28.6	31.1
Calif.	36.9	41.8	N.J.	39.2	39.5
Colo.	32.0	35.6	N.Mex.	37.5	39.7
Conn.	39.0	39.0	N.Y.	44.8	46.0
Del.	45.3	40.0	N.C.	47.5	38.0
Fla.	38.8	34.7	N.Dak.	26.7	34.3
Ga.	56.3	39.8	Ohio	35.5	40.5
Hawaii	45.6	44.3	Okla.	35.4	31.0
Idaho	25.7	26.7	Oreg.	40.1	43.9
Ill.	42.3	43.5	Pa.	43.1	46.3
Ind.	38.2	37.9	R.I.	48.1	48.2
Iowa	39.1	46.3	S.C.	48.6	35.9
Kans.	33.9	33.0	S.Dak.	32.1	36.7
Ky.	48.1	39.7	Tenn.	48.7	41.8
La.	46.4	38.6	Tex.	41.8	36.2
Me.	43.1	38.9	Utah	20.9	24.9
Md.	47.6	47.2	Vt.	39.3	41.3
Mass.	42.3	48.6	Va.	40.9	37.3
Mich.	43.1	40.4	Wash.	38.2	43.2
Minn.	47.7	50.1	W.Va.	50.1	44.7
Miss.	48.6	37.7	Wis.	44.0	45.4
Mo.	44.7	40.0	Wyo.	28.7	28.6

POSITIVE ASSOCIATION, NEGATIVE ASSOCIATION

Two variables are positively associated when above-average values of one tend to accompany above-average values of the other and below-average values tend similarly to occur together. Two variables are negatively associated when above-average values of one accompany below-average values of the other, and vice versa.

linear relationship

In addition to the positive association, the scatterplot in Figure 2.1 shows the form of the relationship: It is roughly *linear,* that is, has the form of a straight line, though with much scatter about the linear pattern. The large scatter indicates that the straight-line relationship is not very strong.

The most important systematic deviation from the overall linear pattern in Figure 2.1 occurs at the right of the graph. A cluster of states there

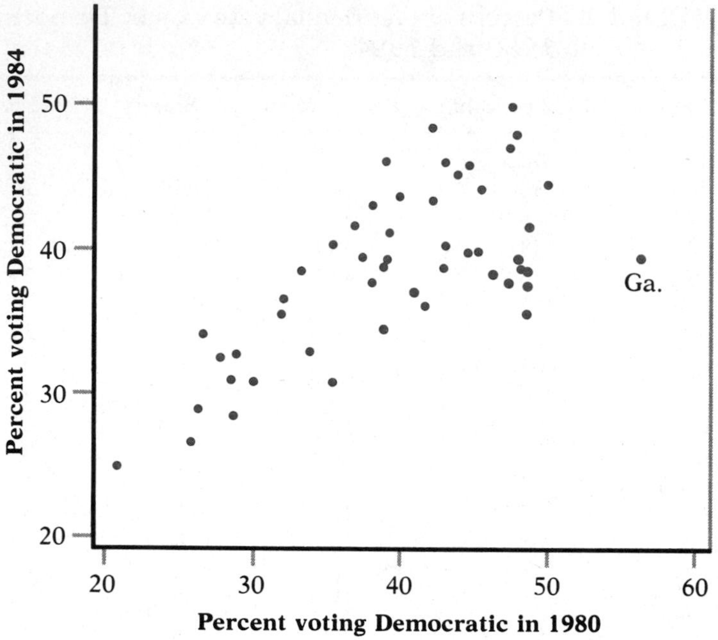

FIGURE 2.1 Percent of Democratic votes in the 1980 and 1984 presidential elections, by state (Example 2.1).

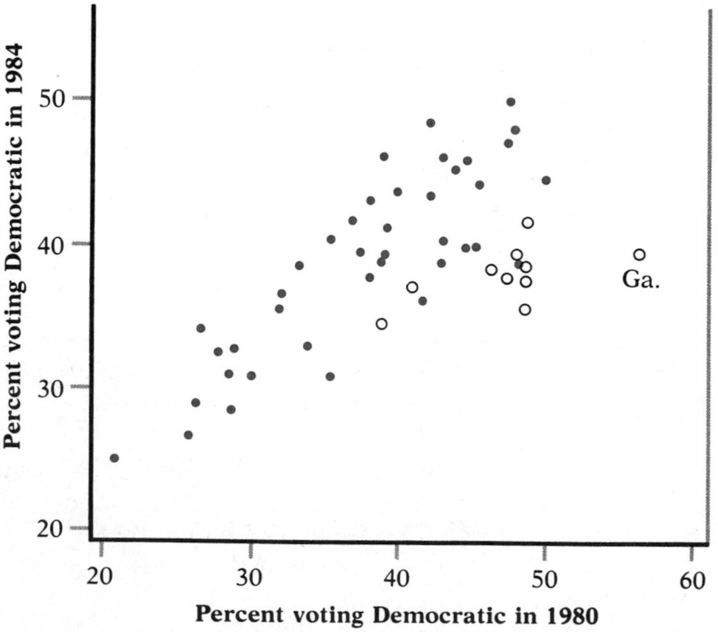

FIGURE 2.2 The 1980 and 1984 percent of Democratic votes, with the South emphasized (○).

voted most heavily for Democrats in 1980 but were markedly less favorable to the Democrats in 1984. The single outlier on the extreme right is Georgia, President Carter's home state. Following that hint, we suspect that the South as a whole was more receptive to the southerner Carter in 1980 than to the northern liberal Mondale in 1984. To show this effect on the graph, Figure 2.2 uses a different symbol (○) to represent the 10 states south of Washington, D.C., and east of the Mississippi River; Louisiana, through which the river flows, is also included.

Figure 2.2 generally sustains our surmise about the South. In fact, we can refine our crude geographical definition of "the South" by examining the political behavior shown in Figure 2.2. The solid point (•) in the middle of the southern cluster is Arkansas, which appears to be southern in voting pattern even though it lies west of the Mississippi. The two open points (○) lying in the political mainstream (to the left of the cluster) are Florida and Virginia. Neither is fully southern in its behavior.

In dividing the states into "southern" and "nonsouthern," we introduced a third variable into the scatterplot. This is a categorical variable that has only two values. The two values are displayed by the two different plotting symbols. Using different symbols to plot points is a good way to incorporate a categorical variable into scatterplot.[2]

EXAMPLE 2.2

News reports sometimes rate state educational systems by comparing the mean scores of seniors in each state on college entrance examinations. This method is misleading, because the percent of high school seniors who take any particular college entrance test varies greatly from state to state. Figure 2.3 is a scatterplot of the mean score on the SAT mathematics examination for high school seniors in each state versus the percent of graduates in each state who took the test in 1990. The percent of students who took the SAT is the explanatory variable, so it appears on the horizontal scale.

The *negative association* between these variables is evident: SAT scores tend to be lower in states where the percent of students who take the test is higher. Only 5% of the seniors in the highest-scoring state took the SAT. The overall pattern shows two *clusters* of points. At the upper left are states where only students seeking admission to colleges that require the SAT take that test. Most students in these states take a different college entrance examination, the American College Testing (ACT) examination. The students who elect to take the SAT tend to be above average academically, so the mean SAT scores for states in this cluster are high. The other cluster, at the lower right of the scatterplot, contains states in which a high percent of college-bound seniors take the SAT. The mean scores are lower here because a less selective group of students takes the test. There is little difference in mean SAT scores among these states, even though the percent of seniors who take the test varies from 40% to over 70%.

cluster

The points that are individually labeled seem to lie a bit outside the lower cluster. The mean SAT scores in these states are lower than in other states in which a similar percent of students take the test. It is possible that the test scores point to educational deficiencies in these states. ●

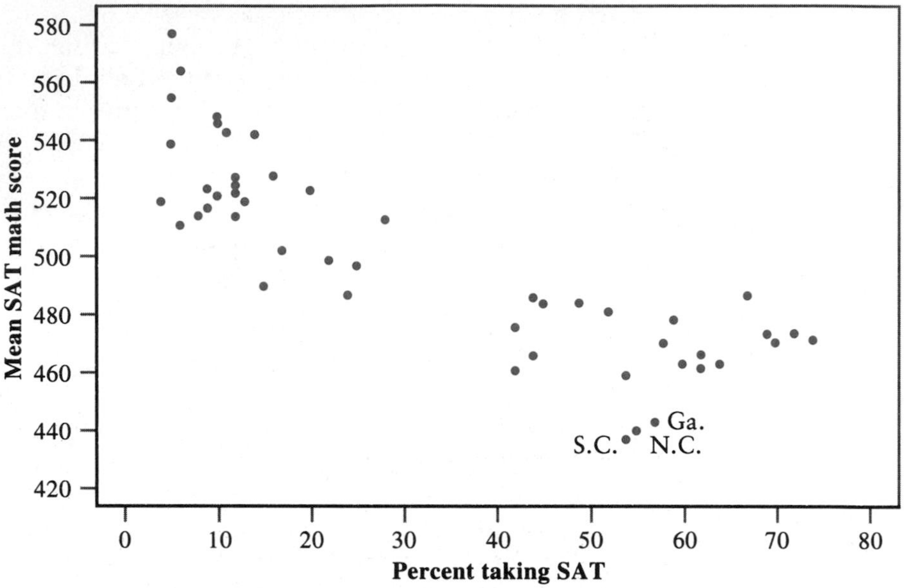

FIGURE 2.3 Mean SAT mathematics score and percent of high school seniors who took the test, by state (Example 2.2).

Some scatterplots appear quite different from the clouds a of points in Figures 2.1 and 2.3. This is particularly true in experiments in which measurements of a response variable are taken at only a few selected levels of the explanatory variable. The following example illustrates the use of scatterplots in this setting:

EXAMPLE 2.3

How much corn per acre should a farmer plant to obtain the highest yield? Too few plants will clearly result in a low yield. On the other hand, if there are too many plants they will compete with each other for moisture and nutrients, and yields will fall. Table 2.2 shows the results of several years of field experiments in Nebraska.[3] Each entry is the mean yield of four small plots planted at the indicated rate per acre. All plots were irrigated, and all were fertilized and cultivated identically. The yield of each plot should therefore depend only on the planting rate—and of course on the uncontrolled aspects of each growing season, such as temperature and wind. The experiment lasted several years in order to avoid misleading conclusions due to the peculiarities of a single growing season.

The scatterplot in Figure 2.4 displays the results of this experiment. Because planting rate is the explanatory variable, it is plotted horizontally as the x variable. The yield is the response variable y. The vertical spread of points over

TABLE 2.2 Average irrigated corn yields (bushels per acre)

Plants per acre	1956	1958	1959	1960	Mean
12,000	150.1	113.0	118.4	142.6	131.0
16,000	166.9	120.7	135.2	149.8	143.2
20,000	165.3	130.1	139.6	149.9	146.2
24,000		134.7	138.4	156.1	143.1
28,000			119.0	150.5	134.8
Mean	160.8	124.6	130.1	149.8	

each planting rate shows the year-to-year variation in yield. The overall pattern is revealed by plotting the mean yield for each planting rate (averaged over all years). These means are marked by triangles and joined by line segments. As expected, the form of the relationship is not linear. Yields first increase with the planting rate, and then decrease when too many plants are crowded in. Since there is no consistent direction, we cannot describe the association as either positive or negative. It appears that the best choice is about 20,000 plants per acre. ●

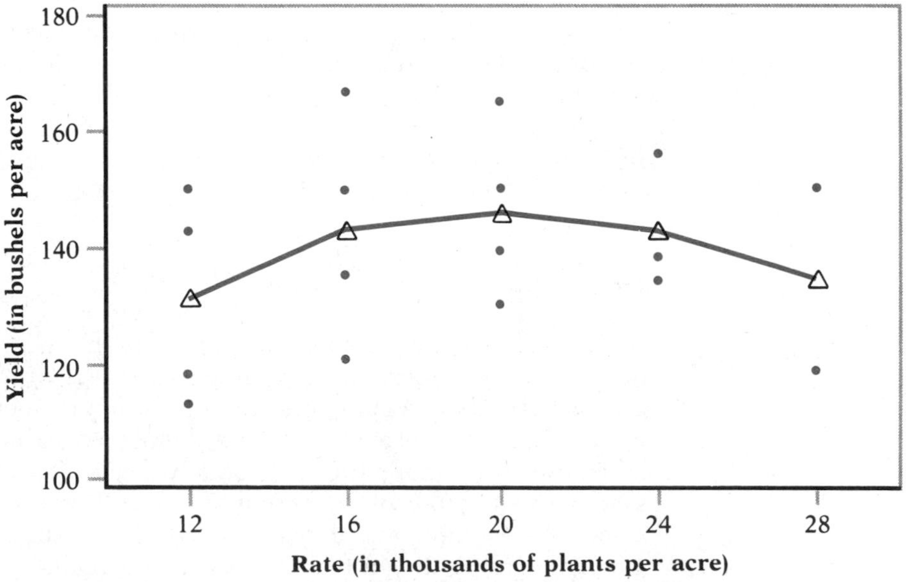

FIGURE 2.4 Yield of corn against planting rate for irrigated corn in Nebraska (Example 2.3).

Plotting the means as in Figure 2.4 is an excellent way to summarize an experiment in which observations are made at a few fixed levels of a variable, such as planting rate. The scatterplot also indicates how much variability lies behind each mean. In this case, however, we must regard the means with caution. As the gaps in Table 2.2 show, the agronomists used only the three lowest planting rates in all 4 years, adding the higher rates only as the need to study them became apparent. The means therefore cover different spans of time. In particular, 1956 was a year of very high yields, so that the means that include 1956 are biased upward relative to those that do not.

A closer look at Table 2.2 shows that 24,000 plants per acre was superior to 20,000 in 2 of the 3 years in which both rates were planted. The agronomists in fact concluded that yields continue to increase somewhat up to 24,000 plants per acre. *Incomplete data* as in Example 2.3 are common in practice. The agronomists' data would have been more convincing if all five planting rates had been used in all 4 years—but it was only after the first results were in that the researchers realized that higher planting rates might give better results. Incomplete data often complicate a statistical analysis. You should therefore be alert to missing data.

incomplete data

Examples 2.1 and 2.3, though different in most respects, share a most important feature: *The relationship between the two variables plotted cannot be fully understood without knowledge about a third variable*. Figure 2.1 plots the percent of the presidential vote won by the Democrats in 1980 versus 1984 by state, but the relationship observed is partly explained by a political and geographic grouping of the states. Figure 2.4 plots corn yield versus planting rate, but the experiment spanned several growing seasons that differed from each other. If the agronomists had not carefully controlled many other variables (moisture, fertilizer, etc.), these variables would have confused the situation completely. You should be cautious in drawing conclusions from a strong relationship appearing in a scatterplot until you understand what other variables may be lurking in the background.

Smoothing scatterplots*

A scatterplot provides a complete picture of the relationship between two quantitative variables—at least as far as that relationship is reflected in the available data. A complete picture is often too detailed for easy interpretation, so we seek to describe the plot in terms of an overall pattern and deviations from that pattern. Though we can often do this by eye, more systematic methods of extracting the overall pattern are desirable. This is called *smoothing* a scatterplot. Example 2.3 suggests how to proceed when we are plotting a response variable y against an explanatory variable x. We smoothed Figure 2.4 by averaging the y-values separately for

smoothing

*This material is not required in later parts of the text, so it can be omitted without loss of continuity.

each x-value. Though not all scatterplots have many y-values at the same value of x, as did Figure 2.4, we can smooth a scatterplot by slicing it into vertical strips and computing the mean or median of the y-values in each strip. For initial analysis, the median is preferred to the mean because it is more resistant.

MEDIAN TRACE

To construct the median trace of a scatterplot, first slice the plot into vertical strips of equal width. Compute the median of the y-values in each strip and plot this median vertically above the horizontal midpoint of the strip. Connect the medians by straight line segments to form the median trace.

The median trace displays the overall pattern of the dependence of y on x. The scatter of the observations above and below the median trace displays deviations from the pattern.

EXAMPLE 2.4

Prior to 1970, American young men were drafted into military service by local draft boards, who followed a complex system of preferences and exemptions. Congress decided that a random selection process—a draft lottery—would be fairer. The first draft lottery was held in 1970. The 366 possible birth dates were placed into identical plastic capsules, poured into a rotating drum, and picked out one by one. The first birth date drawn won draft number 1, the next 2, and so on. Eligible men were then drafted in order of their draft numbers, those with the lowest numbers first.

The outcome of the 1970 draft lottery appears in the scatterplot in Figure 2.5. Birth dates, numbered 1 to 366 beginning with January 1, are plotted horizontally. The draft number assigned each date by the lottery is plotted on the vertical scale. A properly conducted lottery should produce *no* systematic relationship between these variables. Figure 2.5 certainly shows no clear overall pattern. Yet it was charged that the lottery was biased against men born late in the year, that these men received systematically lower draft numbers than men born earlier. An investigation showed that birth dates had been inserted into capsules and poured into a box one month at a time before being placed into the drum.

With this hint that a month effect might be present, we will smooth Figure 2.5 by a median trace. Imagine that the scatterplot is sliced vertically like a loaf of bread. Each slice contains one month's birth dates. We calculate the median draft number for each month and plot it vertically above the horizontal midpoint of each slice. Figure 2.6 shows the median trace superimposed on the scatterplot. The downward trend late in the year is now apparent. The capsules for the later months were put in last and remained near the top because of inadequate mixing. In 1971, the Department of Defense reassigned the officers who had conducted the 1970 lottery and asked statisticians from the National Bureau of Standards to design a truly random selection procedure.[4] ●

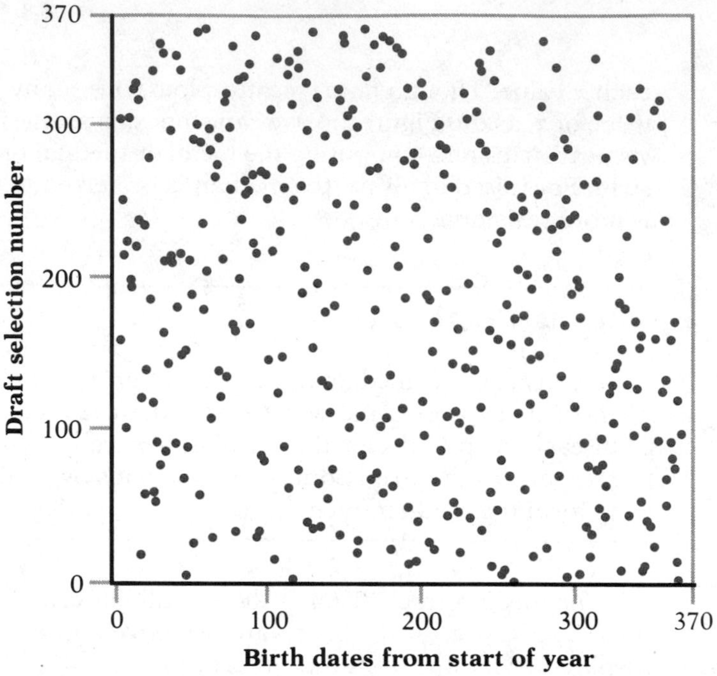

FIGURE 2.5 The 1970 draft lottery: Draft selection number against birth date (Example 2.4).

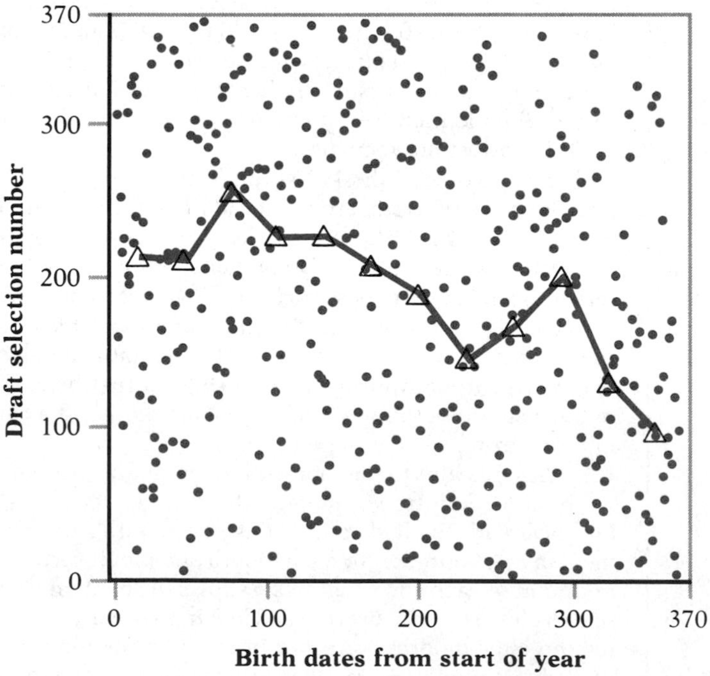

FIGURE 2.6 A median trace by month of birth for the 1970 draft lottery.

Example 2.4 demonstrates that smoothing can reveal relationships that are not obvious from a scatterplot alone. Here, there is a *negative association* between birth date and draft number: Later birth dates tend to have lower draft numbers. Once the median trace shows us what to look for, we can see that the point cloud in Figure 2.5 is a bit thin in the upper right region, indicating that men born late in the year won few high draft numbers. The combination of graphing and calculating once again proves its effectiveness.

To draw a median trace, you must first slice the scatterplot. The number of slices chosen determines the degree of smoothing provided by the median trace. Fewer slices smooth the data more; more slices allow the median trace to follow the ups and downs of y more closely. In general, the number of slices should increase with the number of cases; beyond this, choosing the slices is a matter of judgment. It is often best to take advantage of naturally occurring slices, such as the months in Example 2.4.

The 1970 draft lottery raises one final question. How can we be confident that the lower draft numbers assigned to men born later in the year were not merely the play of chance? After all, repeated random drawings of birth dates would give a different order each time. Some would appear—after the fact—to favor January, and others to favor December. So the pattern that Figure 2.6 seems to show may be an accident. We must judge whether this pattern is stronger than could reasonably arise from the play of chance in a truly random drawing. This judgment requires a calculation of probabilities. Such a calculation shows that an association between birth date and draft number as strong as that observed in 1970 would occur less than once in 1000 random drawings. There is in fact good evidence that the 1970 lottery was unfair.

Categorical explanatory variables

Variations on scatterplots are also the preferred methods for displaying relations between a categorical explanatory variable and a quantitative response. These displays are very similar to those already discussed. Suppose that the agronomists of Example 2.3 had compared the yields of five varieties of corn rather than five planting rates. The plot in Figure 2.4 remains helpful if the varieties A, B, C, D, and E are marked at equal intervals on the horizontal axis in place of the planting rate. In particular, a graph of the mean (or median) responses for each category will show the overall nature of the relationship. If there are too many observations in each category to plot individually, as we did in Figure 2.4, *side-by-side boxplots* can display the response values for each category. Figure 1.11 (page 44) is such a graph. There the categorical explanatory variable is hot dog type (beef, meat, or poultry), and the response is the number of calories in each hot dog.

side-by-side boxplots

Many categorical variables, like corn variety or type of hot dog, have no natural order from smallest to largest. In such situations we cannot speak of a positive or negative association with the response variable. If the mean responses in our plot increase as we go from right to left, we could make them decrease by writing the categories in the opposite order. The plot simply presents a side-by-side comparison of several distributions. The categorical variable labels the distributions. But some categorical variables do have a least-to-most order. We can then speak of the direction of the association between the categorical explanatory variable and the quantitative response. Here is an example.

EXAMPLE 2.5

What is the relationship between family income and the educational level of the householder? (In government records, the householder is the adult who owns or rents a dwelling.) Educational level is the explanatory variable. It appears in government data as a categorical variable with the values:

A Less than 8 years of elementary school
B 8 years of elementary school
C 1 to 3 years of high school
D 4 years of high school
E 1 to 3 years of college
F 4 or more years of college

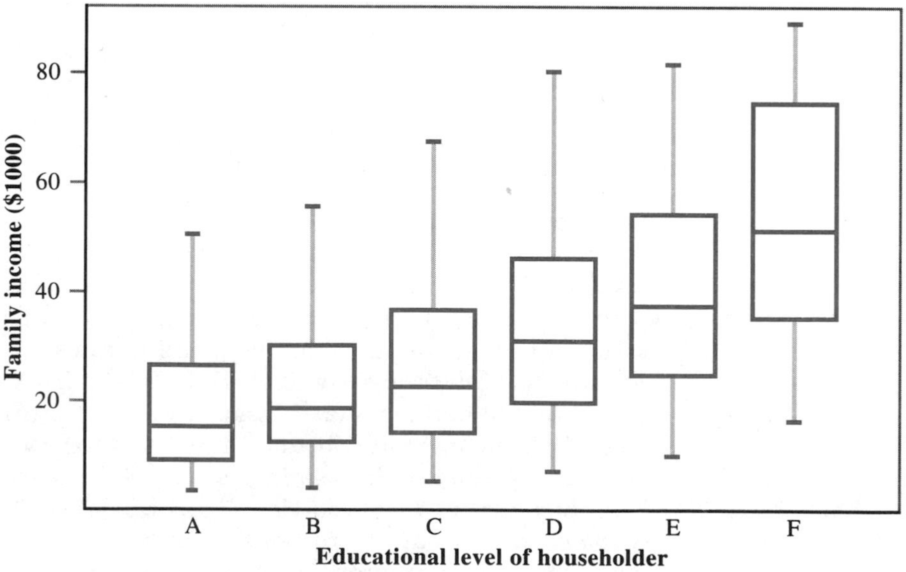

FIGURE 2.7 **Boxplots for family income in six education categories (Example 2.5).**

The categories A to F are ordered from least education to most education. The side-by-side boxplots of Figure 2.7 show the distributions of family incomes in 1987 for white families in these educational categories. The plot shows a positive association between educational level and income. That is, family income tends to increase as the educational level of the householder increases. What is more, the variability of income also increases with education. Some college-educated households have quite low incomes. Additional education offers the opportunity to earn more but does not guarantee a high income. ●

SUMMARY

A **categorical variable** records into which of two or more groups an observation falls. A **quantitative variable** takes numerical values for which arithmetic operations make sense.

When changes in a variable x are thought to explain or even cause changes in a second variable y, x is called an **explanatory variable** and y is called a **response variable.**

A **scatterplot** is a plot of observations x_i and y_i as points in the plane, where x_i and y_i are the values of quantitative variables x and y for the same **case,** that is, the same individual or object.

The explanatory variable, if any, is always plotted on the horizontal scale of a scatterplot. Plotting points with different symbols allows us to see the effect of a categorical variable in a scatterplot.

In examining a scatterplot, look for an overall pattern showing the form, direction, and strength of the relationship, and then for **outliers** or other deviations from this pattern. **Linear relationships** are an important form. If the relationship has a clear direction, we speak of either **positive association** or **negative association.** Smoothing a scatterplot by a **median trace** or other method helps reveal the nature of the dependence of y on x.

SECTION 2.1 EXERCISES

2.1 In each of the following situations, tell whether the variable is quantitative or categorical:
(a) The name of the manufacturer of a TV set
(b) The number of insects on a corn plant
(c) The score on a test of math anxiety for a student taking a statistics course
(d) The major area of study for the student in (c)
(e) The number of pages in a book
(f) Your height in inches

2.2 In each of the following situations, tell whether you would be interested simply in exploring the relationship between the two variables or whether you would want to view one of the variables as an explanatory variable and the other as a response variable. In the latter case, state which is the explanatory variable and which is the response variable.

(a) The amount of time spent studying for a statistics exam and the grade on the exam
(b) The weight and height of a person
(c) The amount of yearly rainfall and the yield of a crop
(d) A student's scores on the SAT math exam and the SAT verbal exam
(e) The occupational class of a father and that of his son

2.3 Vehicle manufacturers are required to test their vehicles for the amount of each of several pollutants in the exhaust. Even among identical vehicles the amount of pollutant varies, so several vehicles must be tested. Figure 2.8 plots the amounts of two pollutants, carbon monoxide and nitrogen oxides, for 46 identical vehicles. Both variables are measured in grams of the pollutant per mile driven. (Data from Thomas J. Lorenzen, "Determining statistical characteristics of a vehicle emissions audit procedure," *Technometrics*, 22 (1980), pp. 483–493.)

(a) Describe the nature of the relationship. Is the association positive or negative? Is it nearly linear or clearly curved? Are there any outliers?
(b) A writer on automobiles says, "When an engine is properly built and properly tuned, it emits few pollutants. If the engine is out of tune, it emits more of all important pollutants. You can find out how badly a vehicle is polluting the air by measuring any one pollutant. If that

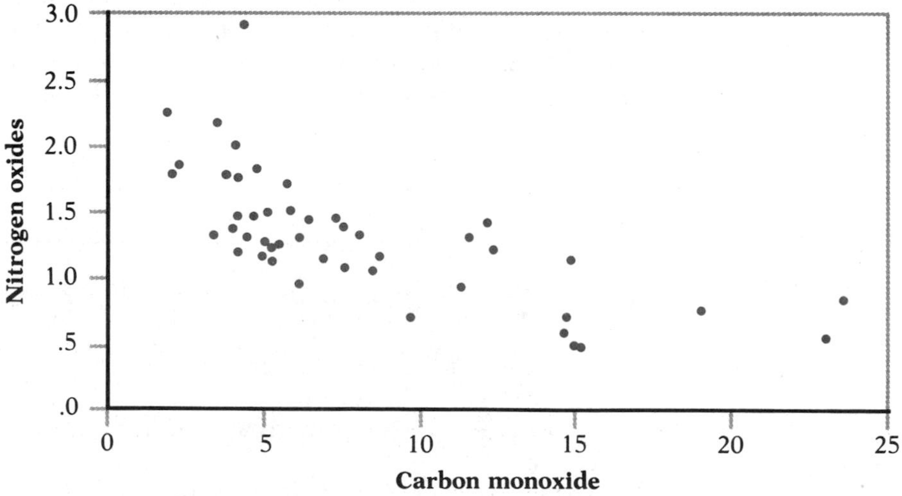

FIGURE 2.8 Nitrogen oxides versus carbon monoxide in the exhaust of 46 vehicles (Exercise 2.3).

value is acceptable, the other emissions will also be OK." Do the data in Figure 2.8 support this claim?

2.4 The following are the golf scores of 11 members of a women's golf team in two rounds of tournament play. (A golf score is the number of strokes required to complete the course, so low scores are better.)

Player	1	2	3	4	5	6	7	8	9	10	11
Round 1	89	90	87	95	86	81	105	83	88	91	79
Round 2	94	85	89	89	81	76	89	87	91	88	80

(a) Plot the round 2 scores versus the scores from round 1.
(b) Is there an association between the two scores? If so, is it positive or negative? Explain why you would expect scores in two rounds of a tournament to have an association like the one you observed.
(c) There is a generally linear pattern in the scatterplot, but one point falls clearly outside this pattern. Circle this point in your plot. A good golfer can have an unusually bad round or a weaker golfer can have an unusually good round. Can you tell from the data given whether the unusual value is produced by a good player or a poor player? What other data would you need to distinguish between the two possibilities?

2.5 When water flows across farmland, some of the soil is washed away, resulting in erosion. Researchers released water across a test bed at different flow rates and measured the amount of soil washed away. The following table gives the flow (in liters per second) and the weight (in kilograms) of eroded soil. (From G. R. Foster, W. R. Ostercamp, and L. J. Lane, "Effect of discharge rate on rill erosion," paper presented at the 1982 Winter Meeting of the American Society of Agricultural Engineers.)

Flow rate	.31	.85	1.26	2.47	3.75
Eroded soil	.82	1.95	2.18	3.01	6.07

(a) Plot the data. Which is the explanatory variable?
(b) Describe the pattern that you see. Would it be reasonable to describe the overall pattern by a straight line? Is the association positive or negative?

2.6 In 1974, the Franklin National Bank failed. Franklin was one of the 20 largest banks in the nation, and the largest ever to fail. Could Franklin's weakened condition be detected in advance by simple data analysis? The table below gives the total assets (in billions of dollars) and net income (in millions of dollars) for the 20 largest banks in 1973, the year before

Franklin failed. Franklin is case number 19. (Data from D. E. Booth, *Regression Methods and Problem Banks*, COMAP, Arlington, Mass., 1986.)

Bank	1	2	3	4	5	6	7	8	9	10
Assets	49.0	42.3	36.3	16.4	14.9	14.2	13.5	13.4	13.2	11.8
Income	218.8	265.6	170.9	85.9	88.1	63.6	96.9	60.9	144.2	53.6

Bank	11	12	13	14	15	16	17	18	19	20
Assets	11.6	9.5	9.4	7.5	7.2	6.7	6.0	4.6	3.8	3.4
Income	42.9	32.4	68.3	48.6	32.2	42.7	28.9	40.7	13.8	22.2

(a) We expect banks with more assets to earn higher income. Make a scatterplot of these data that displays the relation between assets and income. Mark Franklin (bank 19) with a separate symbol.

(b) Describe the overall pattern of your plot. Are there any banks with unusually high or low income relative to their assets? Does Franklin stand out from other banks in your plot?

2.7 Do heavier cars cost more than lighter cars? Figure 2.9 is a plot of the base price in dollars and the weight in pounds for all 1991 model four-door

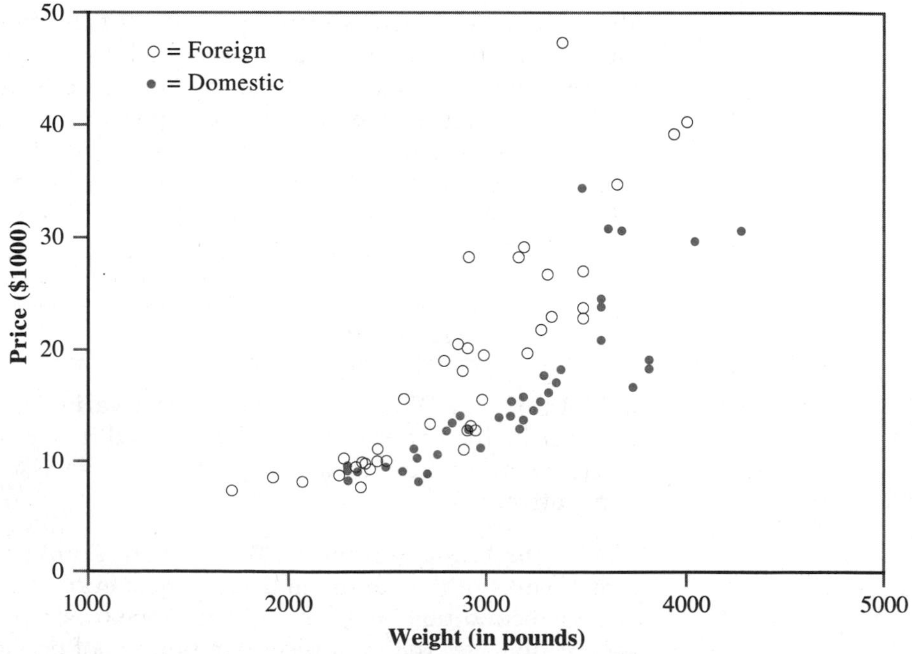

FIGURE 2.9 Base price in thousands of dollars versus weight in pounds for 1991 four-door sedans (Exercise 2.7).

sedans listed in an auto guide. Cars sold by American auto companies are plotted with filled circles and foreign brands are plotted with unfilled circles.

(a) Describe the overall relationship between the weight of a car and its price. Is the association strong (weight and price closely connected) or weak? Is it generally positive or negative?

(b) Describe the major differences between domestic and foreign cars as they appear in the plot.

2.8 The following table gives data on the lean body mass (kilograms) and resting metabolic rate for 12 women and 7 men who are subjects in a study of obesity. The researchers believe that lean body mass (that is, the subject's weight leaving out all fat) has an important influence on metabolic rate.

Subject	Sex	Mass	Rate	Subject	Sex	Mass	Rate
1	M	62.0	1792	11	F	40.3	1189
2	M	62.9	1666	12	F	33.1	913
3	F	36.1	995	13	M	51.9	1460
4	F	54.6	1425	14	F	42.4	1124
5	F	48.5	1396	15	F	34.5	1052
6	F	42.0	1418	16	F	51.1	1347
7	M	47.4	1362	17	F	41.2	1204
8	F	50.6	1502	18	M	51.9	1867
9	F	42.0	1256	19	M	46.9	1439
10	M	48.7	1614				

(a) Make a scatterplot of the data for the female subjects. Which is the explanatory variable?

(b) Is the association between these variables positive or negative? What is the overall shape of the relationship?

(c) Now add the data for the male subjects to your graph, using a different color or a different plotting symbol. Does the type of relationship that you observed in (b) hold for men also? How do the male subjects as a group differ from the female subjects as a group?

2.9 Does increasing public spending on education improve student performance? Figure 2.10 plots one measure of academic performance, the mean SAT mathematics test score in each state, against one measure of spending, the median salary paid to teachers in the state. The outlier is identified as Alaska.

(a) Is Alaska an outlier in both variables or only in one? Why do you think Alaska is an outlier?

(b) Describe the overall pattern of the relationship between teachers' salaries and students' SAT mathematics scores. (Experts in education say there is no cause-and-effect relation between spending on public

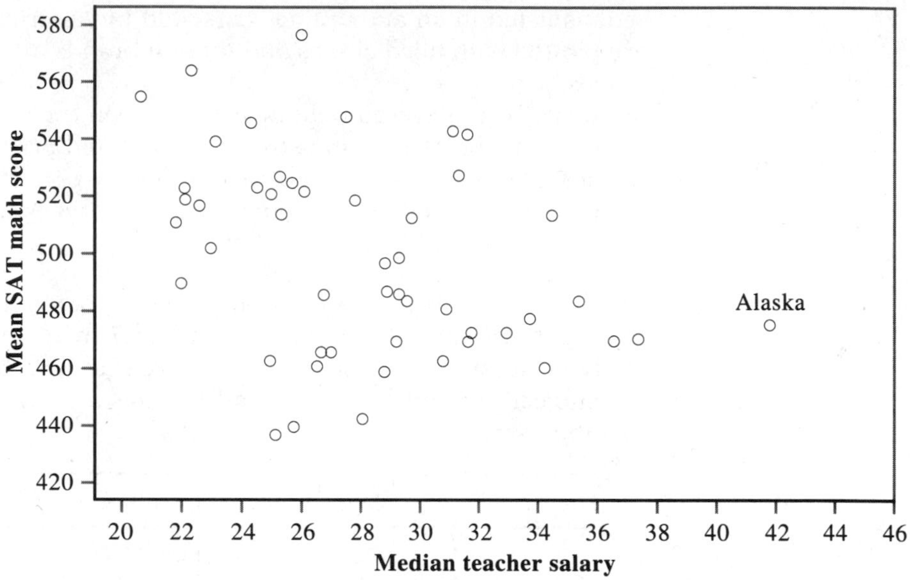

FIGURE 2.10 Mean SAT mathematics score versus median teacher salary by state (Exercise 2.9).

education and SAT scores. As Example 2.2 shows, the percent of high school students who take the SAT varies greatly from state to state, so the mean SAT score is based on different kinds of students in different states.)

2.10 The following data refer to an outbreak of botulism, a form of food poisoning that can be fatal. Each case is a person who contracted botulism in the outbreak. The variables recorded are the subject's age in years, the incubation period (the time in hours between eating the infected food and the first signs of illness), and whether the subject survived (S) or died (D). (Modified from data provided by Dana Quade, University of North Carolina.)

Case	1	2	3	4	5	6	7	8	9
Age	29	39	44	37	42	17	38	43	51
Incubation	13	46	43	34	20	20	18	72	19
Outcome	D	S	S	D	D	S	D	S	D
Case	**10**	**11**	**12**	**13**	**14**	**15**	**16**	**17**	**18**
Age	30	32	59	33	31	32	32	36	50
Incubation	36	48	44	21	32	86	48	28	16
Outcome	D	D	S	D	D	S	D	S	D

(a) Make a scatterplot of incubation period against age, using different symbols for cases that were fatal and cases where the victim survived.
(b) Is there an overall relationship between age and incubation period? If so, describe it.
(c) More important, is there a relationship between either age or incubation period and whether the victim survived? Describe any relations that seem important here.
(d) Are there any unusual cases that may require individual investigation?

2.11 We wish to predict the level of nitrogen oxide (NOX) emissions from the level of carbon monoxide (CO) emissions for a vehicle of the type considered in Exercise 2.3. Figure 2.8 displays a moderately strong relationship between the two variables. To construct a median trace, we compute the median NOX level for the vehicles in each 5-g/mile-wide slice of CO levels. Here is the result of these calculations:

CO level	Count	Median NOX
$0 \le CO < 5$	16	1.785
$5 \le CO < 10$	18	1.270
$10 \le CO < 15$	8	1.055
$15 \le CO < 20$	2	.635
$20 \le CO < 25$	2	.715

Draw the median trace on a graph with the same scales as those used in Figure 2.8. Describe the overall relation displayed by the median trace.

2.12 Thomas the cat is the subject of research on fleas. The researchers count the number of eggs produced each day by Thomas's flea population. Here are the counts for 27 consecutive days. (Data provided by Sayed Gaafar and Michael Dryden, Purdue University School of Veterinary Medicine.)

436	495	575	444	754	915	945	655	782	704
590	411	547	584	550	487	585	549	475	435
523	390	425	415	450	395	405			

A median trace will help us see the pattern of change over time. Divide the data into periods of 5 consecutive days (discard the final 2 days) and calculate the median flea count for each period. Then plot the individual counts against time (days 1 to 27) and add the median trace to your graph. Describe the pattern of change displayed by the trace.

2.13 To demonstrate the effect of nematodes (microscopic worms) on plant growth, a botanist prepares 16 identical planting pots and then introduces different numbers of nematodes into the pots. A tomato seedling is transplanted into each plot. Here are data on the increase in height of the seedlings (in centimeters) 16 days after planting. (Data provided by Matthew Moore.)

Nematodes	Seedling growth			
0	10.8	9.1	13.5	9.2
1000	11.1	11.1	8.2	11.3
5000	5.4	4.6	7.4	5.0
10,000	5.8	5.3	3.2	7.5

(a) Make a scatterplot of the response variable (growth) against the explanatory variable (nematode count). Then compute the mean growth for each group of seedlings, plot the means against the nematode counts, and connect these four points with line segments.
(b) Briefly describe the conclusions about the effects of nematodes on plant growth that these data suggest.

2.14 The presence of harmful insects in farm fields is detected by erecting boards covered with a sticky material and then examining the insects trapped on the board. Some colors are more attractive to insects than others. In an experiment aimed at determining the best color for attracting cereal leaf beetles, six boards of each of four colors were placed in a field of oats in July. The table below gives data on the number of cereal leaf beetles trapped. (Modified from M. C. Wilson and R. E. Shade, "Relative attractiveness of various luminescent colors to the cereal leaf beetle and the meadow spittlebug," *Journal of Economic Entomology*, 60 (1967), pp. 578–580.)

Board color	Insects trapped					
Lemon yellow	45	59	48	46	38	47
White	21	12	14	17	13	17
Green	37	32	15	25	39	41
Blue	16	11	20	21	14	7

(a) Make a plot of the counts of insects trapped against board color (space the four colors equally on the horizontal axis). Compute the mean count for each color, add the means to your plot, and connect the means with line segments.
(b) Based on the data, state your conclusions about the attractiveness of these colors to the beetles.
(c) Does it make sense to speak of a positive or negative association between board color and insect count?

2.15 When animals of the same species live together, they often establish a clear "pecking order." Lower-ranking individuals defer to higher-ranking animals in many ways, usually avoiding open conflict. A researcher on animal behavior wants to study the relationship between pecking order and physical characteristics such as weight. He confines four chickens in each of seven pens and observes the pecking order that emerges in each pen. Here is a table of the weights (in grams) of the chickens, arranged by pecking order. That is, the first row gives the weights of the dominant chickens in the seven pens, the second row gives the weights of the number 2 chickens in each pen, and so on. (Data collected by D. L. Cunningham, Cornell University.)

	Weight (g)						
Pecking order	Pen 1	Pen 2	Pen 3	Pen 4	Pen 5	Pen 6	Pen 7
1	1880	1300	1600	1380	1800	1000	1680
2	1920	1700	1830	1520	1780	1740	1460
3	1600	1500	1520	1520	1360	1520	1760
4	1830	1880	1820	1380	2000	2000	1800

(a) Make a plot of these data that is appropriate to study the effect of weight on pecking order. Include in your plot any means that might be helpful.
(b) We might expect that heavier chickens would tend to stand higher in the pecking order. Do these data give clear evidence for or against this expectation?

2.2 LEAST-SQUARES REGRESSION

linear relationship

regression line

Scatterplots portray the relationship between two quantitative variables. We would like to summarize the relationship more briefly, just as the five-number summary or the mean and standard deviation summarize the center and spread of a single variable. The simplest interesting relationship is *linear* (straight-line) dependence of a response variable y on an explanatory variable x. A straight line that describes the dependence of one variable on another is called a *regression line*.* How to use regression lines is our topic in this section.

*The term "regression" and the general methods for studying relationships now included under this term were introduced by Sir Francis Galton (1822–1911). Galton was engaged in the study of heredity. One of his observations was that the children of tall parents tended to be taller than average but not so tall as their parents. This "regression toward mediocrity" gave these statistical methods their name.

EXAMPLE 2.6

How do children grow? The pattern of growth varies from child to child, so we can best understand the general pattern by following the average height of a number of children. Table 2.3 presents the mean heights of a group of children in Kalama, an Egyptian village that is the site of a study of nutrition in developing countries.[5] The data were obtained by measuring the heights of a sample of 161 children from the village each month from 18 to 30 months of age.

Figure 2.11 is a scatterplot of the data in Table 2.3. Age is the explanatory variable, which we plot on the horizontal scale. Following our usual strategy for examining data, we look first for an overall pattern and then for deviations from that pattern. The overall pattern of growth is clear: The points follow a straight line. That is, the mean height of Kalama children increases by about the same fixed amount each month. The deviation from the linear pattern is the wobble that prevents the points from forming a perfectly straight line. There are no outliers—no points that fall clearly outside the straight-line pattern. The overall linear pattern describes these data very well. ●

Fitting a line to data

fitting a line to data

When a scatterplot displays a linear pattern, we can describe the overall pattern by drawing a straight line through the points. Of course, no straight line passes exactly through all of the points. *Fitting a line* to data means drawing a line that comes as close as possible to the points. The equation of a line fitted to the data gives a compact description of the dependence of the response variable y on the explanatory variable x. It is a mathematical model for the straight-line relationship.

TABLE 2.3 Mean height of Kalama children

Age in months	Height in centimeters
18	76.1
19	77.0
20	78.1
21	78.2
22	78.8
23	79.7
24	79.9
25	81.1
26	81.2
27	81.8
28	82.8
29	83.5

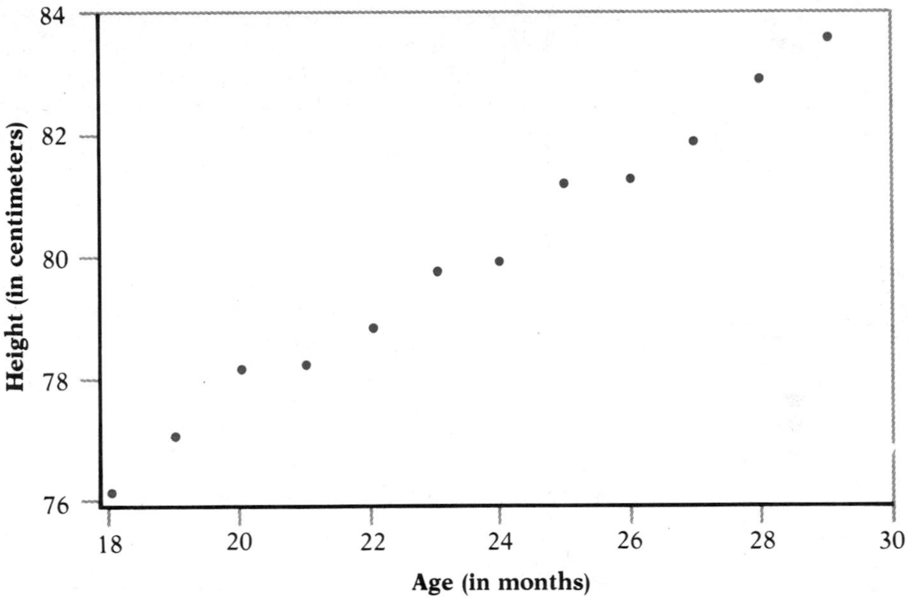

FIGURE 2.11 Mean height of children in Kalama, Egypt, plotted against age from 18 to 29 months, from Table 2.3.

> STRAIGHT LINES
>
> Suppose that y is a response variable (plotted on the vertical axis) and x is an explanatory variable (plotted on the horizontal axis). The equation of a straight line has the form
>
> $$y = a + bx$$
>
> In this equation, b is the slope, the amount by which y changes when x increases by one unit. The number a is the intercept, the value of y when $x = 0$.

Any straight line describing the Kalama data must have the form

$$\text{height} = a + (b \times \text{age})$$

We will see later that one line that fits the data well has the equation

$$\text{height} = 64.93 + (.635 \times \text{age})$$

This line appears in Figure 2.12. The slope $b = 0.635$ tells us that the height of Kalama children increases by about 0.6 centimeter (cm) for each month of age. The slope b of a line $y = a + bx$ describes the *rate of change*

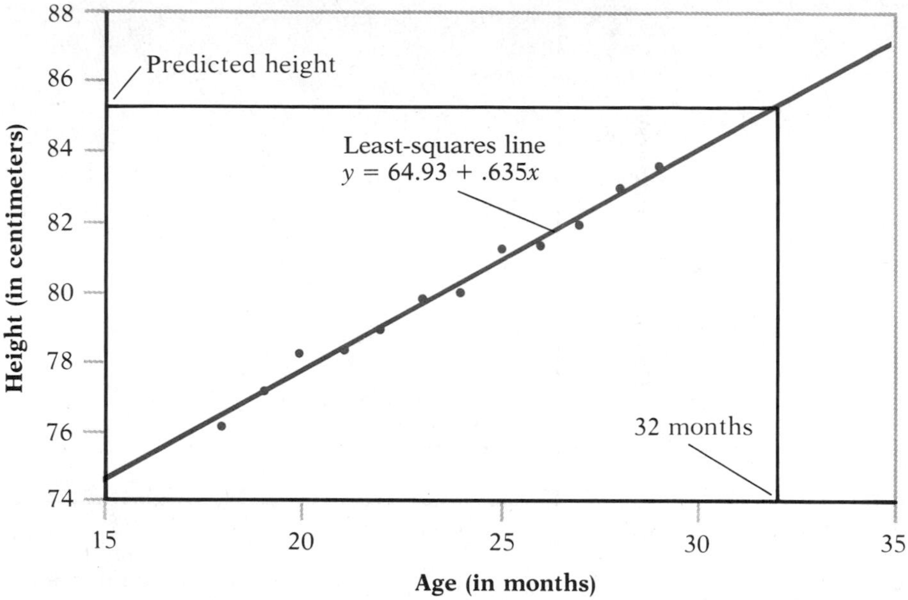

FIGURE 2.12 The least-squares line fitted to the Kalama data and used to predict height at age 32 months.

in the response y as the explanatory variable x changes. The slope of a regression line is an important numerical description of the relationship between the two variables. The intercept, $a = 64.93$ cm, would be the height at birth (age $= 0$) if the straight-line pattern of growth were true starting at birth. Children don't grow at a fixed rate from birth, so the intercept a is not important in our situation except as part of the equation of the line.

prediction

A regression line can be used to *predict* the response y for a specific value of the explanatory variable x. To predict y, just substitute the given x-value into the equation of the line.

EXAMPLE 2.7

What do we predict to be the mean height of Kalama children at 32 months of age? Substituting 32 for the age in the equation of the fitted line gives

$$\text{height} = 64.93 + (.635 \times 32) = 85.25 \text{ cm}$$

If you have fitted a line by eye on graph paper, you can predict height on the graph. Find age 32 months on the horizontal axis, then go up to the fitted line and over to the vertical axis; the point at which you hit the vertical axis is the predicted height. Figure 2.12 illustrates this prediction method. Substituting the value of x into the equation of the fitted line is just a quicker and more precise way to carry out this process. ●

The accuracy of predictions from a regression line depends on how much scatter about the line the data show. When, as in this example, the data points are all very close to the line, we are confident that our prediction is reliable. If, however, the data show a linear pattern with considerable spread (as in Figure 2.1 on page 100), we may agree that a regression line summarizes the pattern but we will put less confidence in a prediction based on the line. We will learn in Chapter 9 how to give a numerical statement of our level of confidence in predictions.

EXAMPLE 2.8

Can we predict the height of Kalama residents at age 20 years? Since 20 years is 240 months, we substitute 240 for the age. The prediction is

$$\text{height} = 64.93 + (.635 \times 240) = 217.33 \text{ cm}$$

This is more than 7 feet. Blind calculation has produced an unreasonable result. Our data covered only ages from 18 to 29 months. As people grow older, they gain height more slowly, so our fitted line is not useful at ages far removed from the data that produced it. ●

extrapolation

Making use of a regression line or other model outside the range of the data to which we actually fitted the model is called *extrapolation*. As Example 2.8 illustrates, extrapolation is dangerous. Notice that age 0 (birth) lies well outside the range of the data in Table 2.3. Taking the intercept 64.93 cm (over 25 inches) as an estimate of height at birth is therefore another example of extrapolation. The linear growth pattern does not begin at birth, and this estimate of height at birth is not accurate. In addition, we are well aware that the pattern we discovered in Figure 2.12 does not hold for all children everywhere. We cannot use data from Kalama to predict the height of children in Cairo, much less Los Angeles. Statistical methods such as fitting regression lines cannot go beyond the limitations of the data that feed them. Do not let calculation override common sense.

Least-squares regression

How shall we actually fit a line to the points in Figure 2.11? You might find a regression line by laying a transparent straightedge over the graph and moving it about until you are satisfied. This is not a bad way to draw a line on the graph, but it does not produce an equation without more work on your part. There are many mathematical ways of fitting a line to a set of data. The oldest and most common is the *method of least squares*.*

*The method of least squares was first published by the French mathematician Adrien Marie Legendre (1752–1833) in 1805. It at once became the standard method for combining observations in astronomy and in surveying, and remains one of the most-used statistical tools.

The least-squares method begins with the goal of predicting y from x. Because the response y is plotted on the vertical scale, the vertical deviations of the points in the scatterplot from a fitted line are errors in predicting y. We want to make these vertical deviations as a group as small as possible. When a line fits reasonably well, it will pass below some points and above others. That is, some of the deviations will be positive and some negative. In Figure 2.13, we magnify the center portion of Figure 2.11 and draw a line through these points to illustrate the vertical deviations. Two deviations are positive (the points are above the line) and one is negative (the point lies below the line). The *squares* of the deviations are all positive, however. The sum of the squares of the deviations measures the size of the entire set of deviations. The least-squares method fits to the data the line that makes this sum of squares smallest.

> **THE LEAST-SQUARES REGRESSION LINE**
>
> The least-squares regression line is the line that makes the sum of the squares of the deviations of the data points from the line in the vertical direction as small as possible.

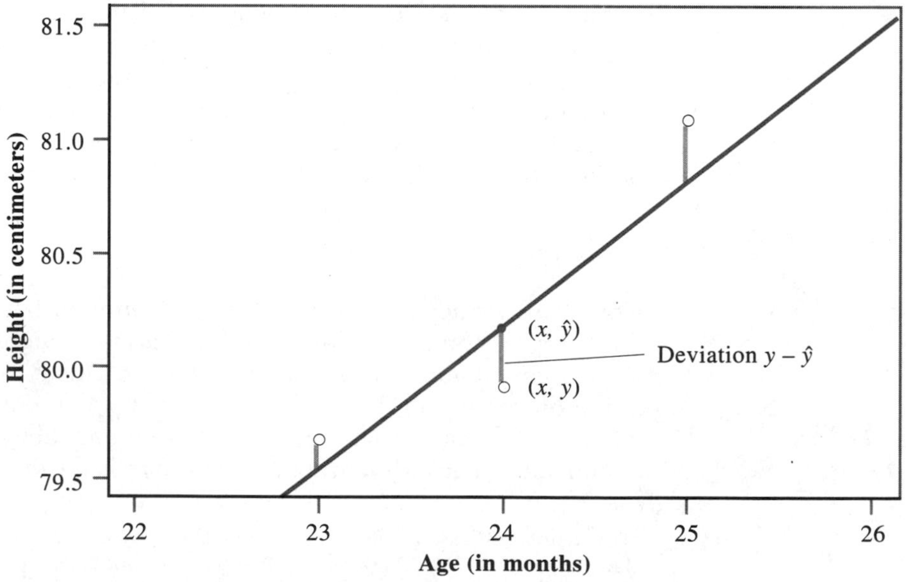

FIGURE 2.13 The least-squares method: Fitted line and deviations.

To give the equation of the least-squares line, we must use algebraic notation. We represent n observations on two variables x and y as

$$(x_1, y_1), (x_2, y_2), \ldots, (x_n, y_n)$$

If a line $y = a + bx$ is drawn through the scatterplot of these observations, the line gives the value of y corresponding to x_i as $\hat{y}_i = a + bx_i$. We use the notation $\hat{y}$ (read "y hat") to distinguish the value $\hat{y}$ predicted by the line from the actually observed y. The vertical deviation of the ith data point from the line is

$$\begin{aligned} \text{deviation} &= \text{observed } y - \text{predicted } y \\ &= y_i - \hat{y}_i \\ &= y_i - a - bx_i \end{aligned}$$

The method of least squares chooses the line that makes the sum of the squares of these deviations as small as possible. Finding this line amounts to finding the values of the intercept a and the slope b that minimize

$$\sum(\text{deviation})^2 = \sum(y_i - a - bx_i)^2$$

for the given observations x_i and y_i. In the case of the growth data in Table 2.3, we want to choose the a and b that minimize

$$(76.1 - a - 18b)^2 + (77.0 - a - 19b)^2 + \cdots + (83.5 - a - 29b)^2$$

Here is the solution to this mathematical problem.

CALCULATING THE LEAST-SQUARES REGRESSION LINE

The least-squares regression line of y on x calculated from n observations on these two variables is given by $\hat{y} = a + bx$, where

$$b = \frac{\sum xy - \frac{1}{n}(\sum x)(\sum y)}{\sum x^2 - \frac{1}{n}(\sum x)^2} \qquad (2.1)$$

$$a = \bar{y} - b\bar{x}$$

The sums in Equation 2.1 run over all of the observed values x_i or y_i; the subscripts were omitted for easier reading. There are many forms of the recipe for the slope of the least-squares line. The form in Equation 2.1 is convenient if you must use a basic calculator. Like Equation 1.3 (page 48) for the variance, this formula makes efficient use of basic sums and sums of squares. In Section 2.4 we will meet an alternative formula for the slope b that gives more insight than does Equation 2.1.

EXAMPLE 2.9

To calculate the least-squares regression line of height on age from the data in Table 2.3, we proceed as follows. First compute the building block sums starting from the values of x and y:

x	x^2	y	xy
18	324	76.1	1369.8
19	361	77.0	1463.0
20	400	78.1	1562.0
21	441	78.2	1642.2
22	484	78.8	1733.6
23	529	79.7	1833.1
24	576	79.9	1917.6
25	625	81.1	2027.5
26	676	81.2	2111.2
27	729	81.8	2208.6
28	784	82.8	2318.4
29	841	83.5	2421.5
$\sum x = 282$	$\sum x^2 = 6770$	$\sum y = 958.2$	$\sum xy = 22,608.5$

Then substitute these sums into Equation 2.1:

$$b = \frac{\sum xy - \frac{1}{n}(\sum x)(\sum y)}{\sum x^2 - \frac{1}{n}(\sum x)^2}$$

$$= \frac{22,608.5 - \frac{1}{12}(282)(958.2)}{6770 - \frac{1}{12}(282)^2}$$

$$= \frac{90.8}{143} = .635$$

$$a = \bar{y} - b\bar{x}$$

$$= \frac{958.2}{12} - .635\frac{282}{12} = 64.93$$

The resulting line is

$$\hat{y} = 64.93 + .635x$$

This is the line we plotted in Figure 2.12 and used for prediction in Example 2.7. ●

Now you know how to calculate the least-squares line. In statistical practice the arithmetic is usually automated. Many calculators and all statistical software systems will calculate the slope b and the intercept a for you after you have entered the x and y data values. In the Minitab system, for example, we find the least-squares line for the data in Table 2.3 as follows:

```
MTB> REGRESS 'HEIGHT' ON 1, 'AGE'
```

The output generated in response to this command begins as follows.

```
The regression equation is
HEIGHT = 64.9 + 0.635 AGE

PREDICTOR        COEF      STDEV     T-RATIO
CONSTANT      64.9283     0.5084      127.71
AGE           0.63496    0.02140       29.66
```

The "regression equation" in the output is the equation of the least-squares line. We will not need most of the entries in the table below the regression equation until Chapter 9. Notice that the COEF column in the table gives the intercept and the slope more accurately than the summary equation. This added accuracy is often needed in problems. Rather than print a table like this for each example, we will often abbreviate computer results by writing the regression equation with the extra decimal places obtained from the table. If you use Minitab yourself, you can duplicate our results by looking in the table that follows the regression equation.

Residuals

A mathematical model, such as a straight line, gives a compact statement of the overall pattern of a set of data. Such models have the added advantage of being completely explicit. We can use a fitted line but not a verbal description of the pattern to predict future values, for example. When a model expresses the overall pattern, deviations from that pattern can also be stated explicitly.

RESIDUAL

A residual is the difference between an observed value of the response variable and the value predicted by the model. That is,

$$\begin{aligned}\text{residual} &= \text{observed } y - \text{predicted } y \\ &= y - \hat{y}\end{aligned}$$

In examining any graph of data we look first for an overall pattern and then for striking deviations from this pattern. In the regression setting, the fitted line represents the pattern and the residuals are the deviations. In fact, adding the residual $y - \hat{y}$ to the predicted response $\hat{y}$ gives the actually observed response y. Our principle "look for pattern plus deviations" takes the form of an equation, $y = \hat{y} + (y - \hat{y})$. Or, in a form that is easy to remember,

$$\text{DATA} = \text{FIT} + \text{RESIDUAL}$$

EXAMPLE 2.10

According to Table 2.3, the observed mean height of Kalama children at 24 months was 79.9 cm. That is, the observed y is 79.9 when $x = 24$. The least-squares regression line predicts the height to be

$$\hat{y} = 64.93 + (.635 \times 24) = 80.17$$

The residual at 24 months is therefore

$$\begin{aligned} \text{residual} &= \text{observed } y - \text{ predicted } y \\ &= y - \hat{y} \\ &= 79.9 - 80.17 = -.27 \end{aligned}$$

The 12 data points used in fitting the least-squares line in Figure 2.12 produce 12 residuals. They are

−.26	.01	.47	−.06	−.10	.17
−.27	.30	−.24	−.27	.09	.16

These residuals are just the vertical deviations of the 12 data points from the least-squares line. They are the deviations that are left over after we fit the line that makes the sum of squares of the deviations as small as possible. That is why we call them "residuals." ●

Because the residuals show how far the data fall from our regression line, examining the residuals helps assess how well the line describes the data. Although residuals can be calculated from any model fitted to the data, the residuals from the least-squares line have a special property: *The mean of the residuals is always 0.* Figure 2.14 is a *residual plot* for the Kalama data. Each residual from Example 2.10 is plotted against the corresponding value of the explanatory variable x. Because the mean of the residuals is always 0, the horizontal line at 0 helps orient us. This line (residual = 0) corresponds to the fitted line in Figure 2.12. The residuals in Figure 2.14 show the irregular horizontal pattern that is typical of data that do not deviate from the model in any systematic way.

residual plot

Because there is a residual for each case, finding residuals by hand is often tedious. Most statistical software systems calculate and plot the residuals in a regression problem, either automatically or in response

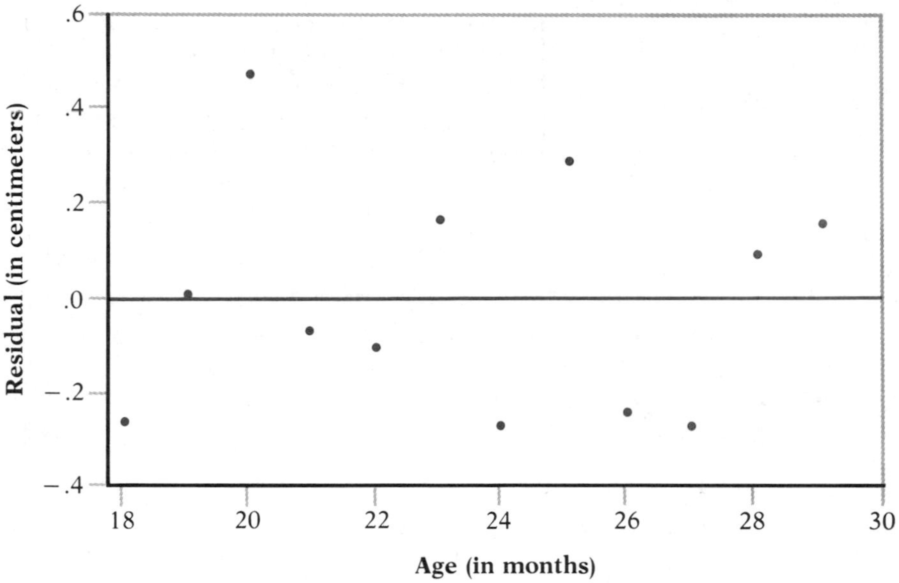

FIGURE 2.14 Residual plot for the least-squares line for the Kalama data, based on the residuals in Example 2.10.

to an additional command. Minitab, for example, gives the residuals in Example 2.10 and the plot of Figure 2.14 when you type

```
MTB> REGRESS 'HEIGHT' ON 1, 'AGE';
 SUBC> RESIDUALS C3.
MTB> PLOT C3 'AGE'
```

Figure 2.15 shows the overall pattern of some typical residual plots in idealized form. The residuals are plotted in the vertical direction against the corresponding values of the explanatory variable x in the horizontal direction. If all is well, the pattern of this plot will be an unstructured horizontal band centered at 0 (the mean of the residuals) and symmetric about 0, as in Figure 2.15(a). A curved pattern like the one in Figure 2.15(b) shows that the relation between y and x is not linear but curved. A straight line is not a good overall description of such a relation. A fan-shaped pattern like the one in Figure 2.15(c) shows that the variation of y about the line increases as x increases; predictions of y will therefore be more precise for smaller values of x, where y shows less spread about the line.

The basic plot of the residuals against x simply magnifies patterns that we could see in the original scatterplot of y versus x, although the residual plot does save us the effort of adding the fitted line to our plot. Other plots of the residuals, however, can reveal new information. They may, for example, suggest the presence of lurking variables.

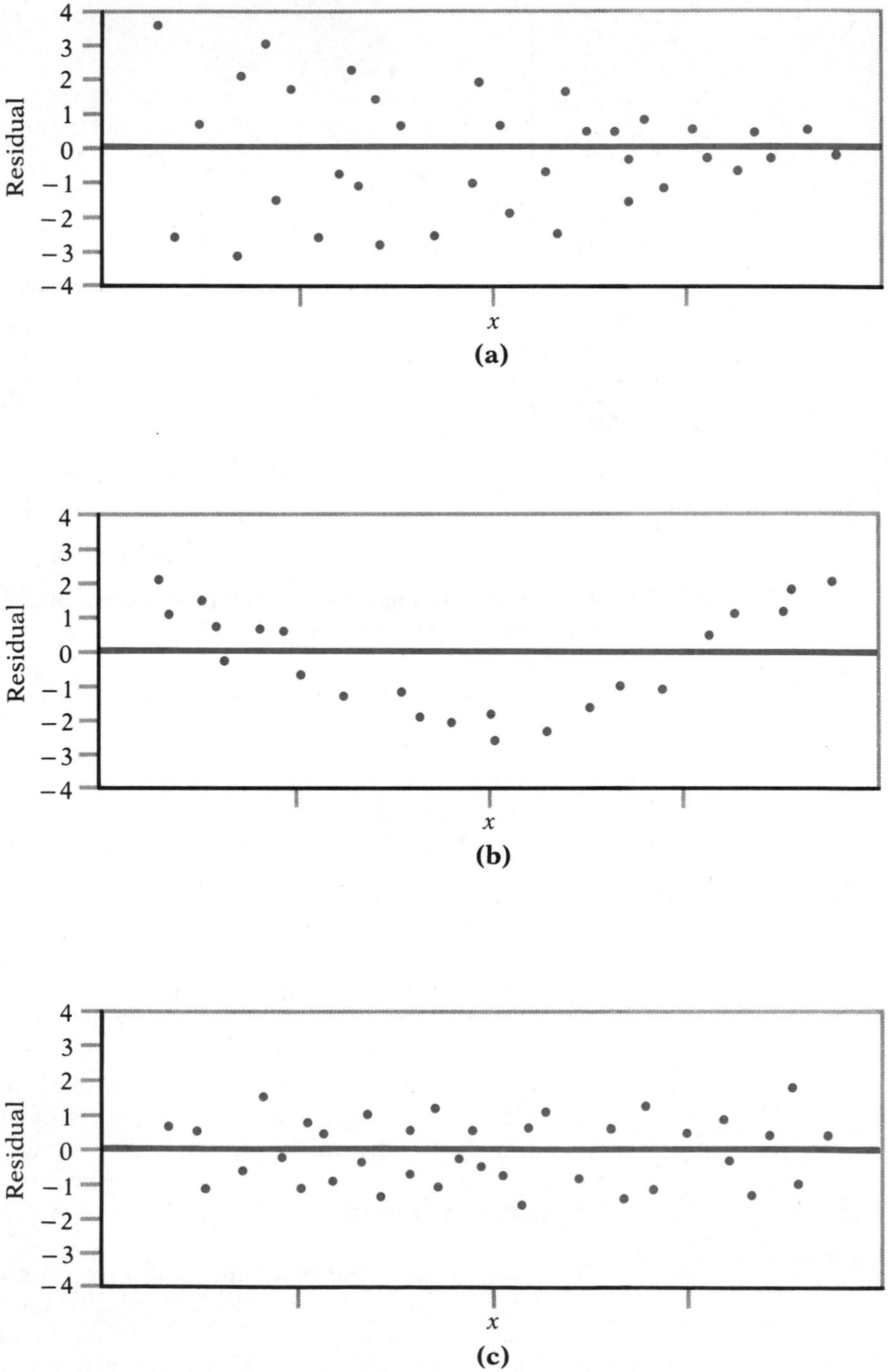

FIGURE 2.15 Idealized patterns in plots of least-squares residuals.

> **LURKING VARIABLE**
>
> A lurking variable is a variable that has an important effect on the response but is not included among the explanatory variables studied.

Lurking variables can dramatically change the conclusions of a regression study. Because lurking variables are often unrecognized and unmeasured, detecting their effect is a challenge. Many lurking variables change systematically over time. One useful method for detecting lurking variables is therefore to *plot both the response variable and the regression residuals against the time order of the observations* whenever the time order is available. An understanding of the background of the data then allows you to guess what lurking variables might be present. Here is a more elaborate example of plotting and interpreting residuals that uncovered a lurking variable.

EXAMPLE 2.11

The mathematics department of a large state university must plan in advance the number of sections and instructors required for elementary courses. The department hopes that the number of students in these courses can be predicted from the number of entering first-year students, which is known before the new students actually choose courses. The table below contains the data for recent years.[6] The explanatory variable x is the number of students in the first-year class, and the response variable y is the number of students who enroll in mathematics courses at the 100 level.

Year	1980	1981	1982	1983	1984	1985	1986	1987
x	4595	4827	4427	4258	3995	4330	4265	4351
y	7364	7547	7099	6894	6572	7156	7232	7450

Equation 2.1 could be used to compute the regression line from these data. Instead, we enter the data into the Minitab statistical computing system, calling x STUD and y MATH for easy recall. The regression command and the first part of Minitab's output are as follows:

```
MTB> REGRESS 'MATH' ON 1, 'STUD'

The regression equation is
MATH = 2492.69 + 1.0663 STUD
```

Other information, including a table of the residuals, is available at our request. We see from the output that the least-squares regression line is

$$\hat{y} = 2492.69 + 1.0663x$$

●

The scatterplot (Figure 2.16) of the data from Example 2.11 with the regression line shows a reasonably linear fit. There is a cluster of points with similar values near the center. A plot of the residuals against x (Figure 2.17) magnifies the vertical deviations of the points from the line. It is apparent from either graph that a slightly different line would fit the five lower points very well, so that the three points above the line represent a

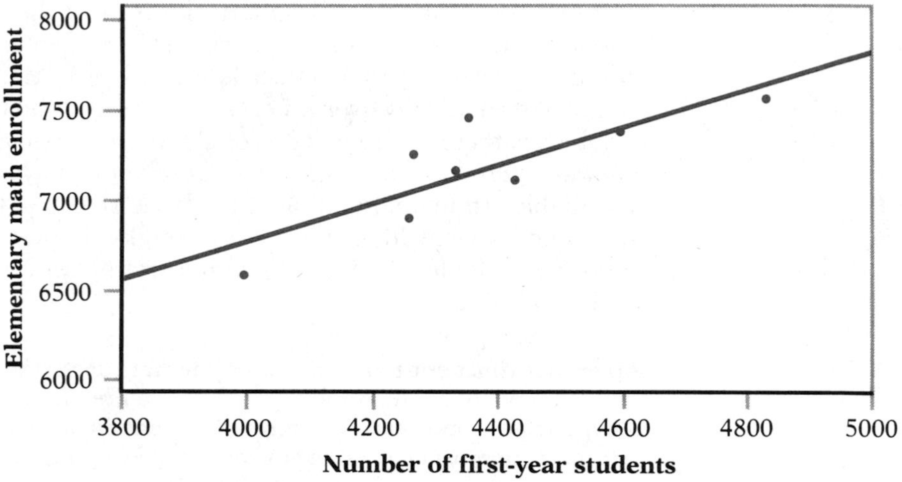

FIGURE 2.16 **The regression of enrollment in elementary mathematics on number of first-year students at a large university (Example 2.11).**

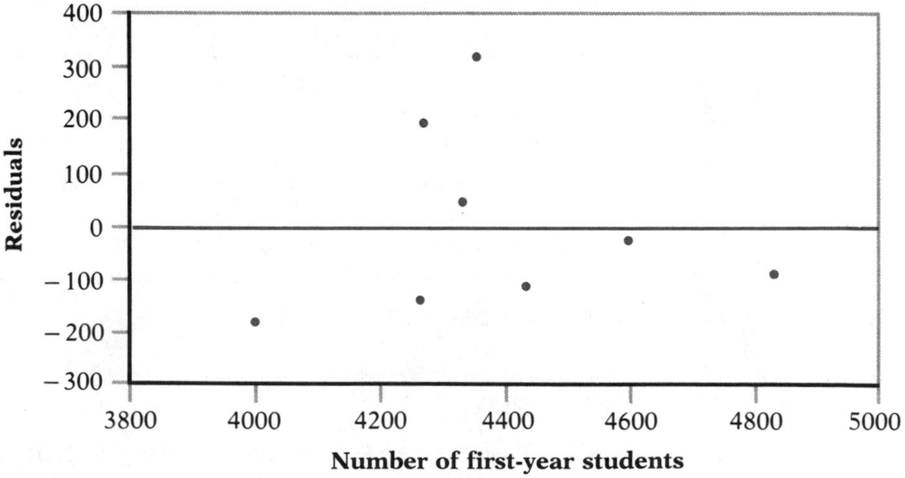

FIGURE 2.17 **Residual plot for the regression of mathematics enrollment on number of first-year students.**

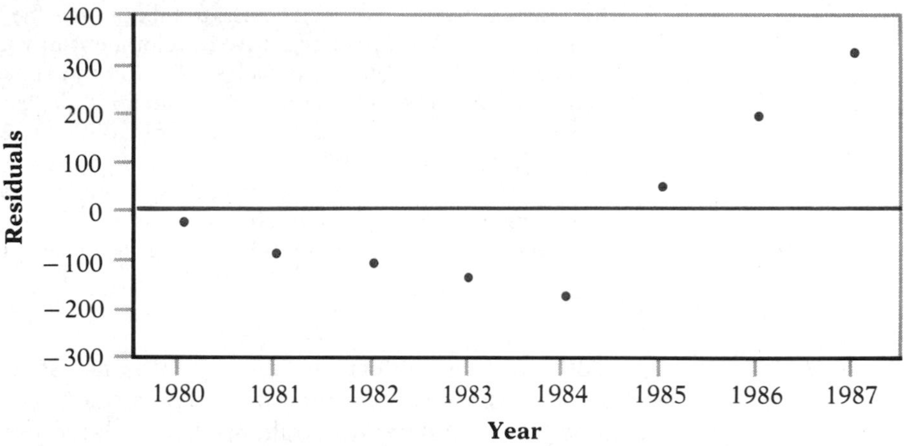

FIGURE 2.18 Plot of the residuals against time for the regression of mathematics enrollment on number of first-year students.

somewhat different relation between the number of first-year students x and mathematics enrollment y.

A second plot of the residuals clarifies the situation. Figure 2.18 is a plot of the residuals against time. We now see that the five negative residuals are from the years 1980 to 1984, and the three positive residuals represent the years 1985 to 1987. This pattern suggests that a change took place between 1984 and 1985 causing a higher proportion of students to take mathematics courses beginning in 1985. In fact, one of the schools in the university changed its program to require that entering students take another mathematics course. This change is the lurking variable that explains the pattern we observed.

The contrast in Figure 2.18 between the years to 1984 and the following years is another reminder that an observed relationship may not be trustworthy when changes occur in the underlying situation. Because the least-squares regression line from Example 2.11 makes use of data from 1984 and earlier, the mathematics department should not use it in future years to predict elementary mathematics enrollments from the count of first-year students.

Outliers and influential observations

We have to this point concentrated on the overall pattern of a scatterplot or a residual plot. Individual points that lie outside the pattern can be even more important to intelligent use of regression. Here is an example of data that contain some unusual cases.

EXAMPLE 2.12

Does the age at which a child begins to talk predict the later score on a test of mental ability? A study of cognitive development in young children recorded the age (in months) at which each of 21 children spoke their first word and their Gesell Adaptive Score, the result of an aptitude test taken much later. The data appear in Table 2.4 and in the scatterplot of Figure 2.19.[7] Notice that cases 3 and 13 and cases 16 and 21 have identical values for both variables. When drawing the scatterplot, we used a different symbol to show that one point stands for two cases. The plot shows a negative association; that is, children who begin to speak later tend to have lower test scores than early talkers. ●

Although the pattern in Figure 2.19 is not strongly linear, we want to use regression for prediction. The purpose of the study was to learn whether age at first word x could predict the later test score y. The colored line in Figure 2.20 is the regression line of y on x that we would use for prediction. The plot also highlights two unusual cases, the children numbered 18 and 19 in Table 2.4. Both points lie away from the other points in the scatterplot. There is an important distinction between these two points: Case 19 lies far from the overall linear pattern indicated by the regression line, whereas case 18 lies quite close to the line. Case 18 is unusual only because this child was slow to begin speaking. The residual plot in Figure 2.21 confirms that case 19 has a large residual, whereas the residual for case 18 is small. To emphasize the distinction between points that lie outside the general straight-line pattern and points that do not, we will refine the idea of an outlier.

TABLE 2.4 Age at first word and Gesell adaptive score

Case	Age	Score	Case	Age	Score
1	15	95	11	7	113
2	26	71	12	9	96
3	10	83	13	10	83
4	9	91	14	11	84
5	15	102	15	11	102
6	20	87	16	10	100
7	18	93	17	12	105
8	11	100	18	42	57
9	8	104	19	17	121
10	20	94	20	11	86
			21	10	100

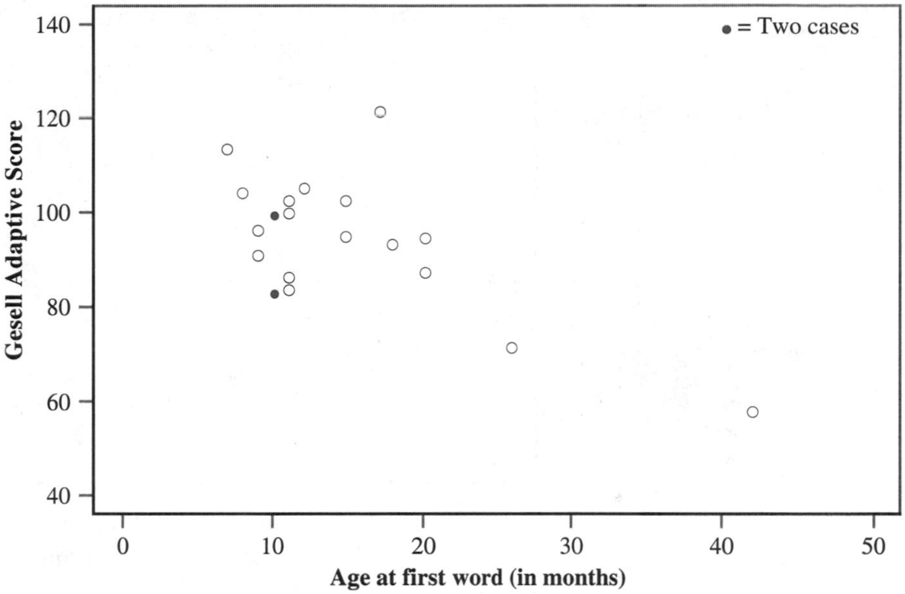

FIGURE 2.19 Gesell Adaptive Score versus the age at first word for 21 children, from Table 2.4.

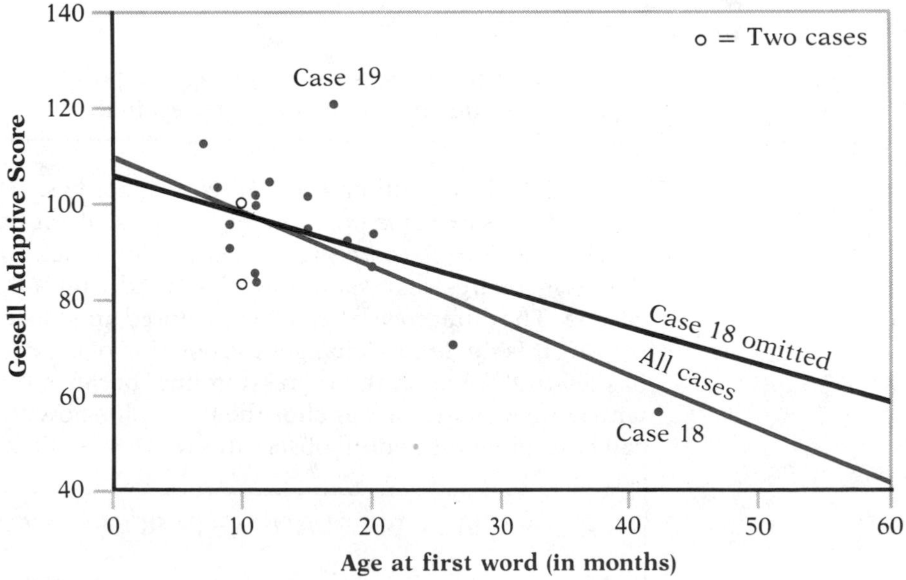

FIGURE 2.20 Least-squares regression of Gesell score on age at first word, with and without an influential case.

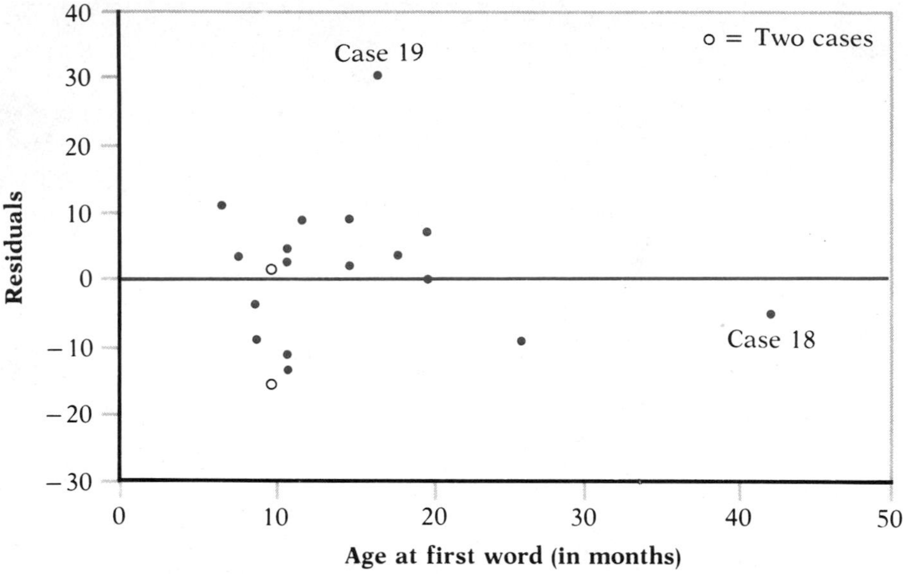

FIGURE 2.21 Residual plot for the regression of Gesell score on age at first word; case 19 is an outlier and case 18 is an influential case that does not have a large residual.

> **OUTLIERS IN REGRESSION**
>
> An outlier in the regression setting is a point that lies far from the fitted line and so produces a large residual.

Case 19 is an outlier. Case 18 is not. Yet case 18 is very important. Because of its extreme position on the age scale, it has a strong influence on the position of the regression line. The black line in Figure 2.20 is the least-squares regression line computed from all 20 cases other than case 18. The difference between the colored and black lines shows how removing this single case changes the position of the regression line. Case 19 has less influence on the regression line, because the many other points with similar values of x anchor the line well below the outlying point. We call case 18 an influential observation.

> **INFLUENTIAL OBSERVATIONS IN REGRESSION**
>
> An observation is influential if removing it would markedly change the position of the regression line. Points that are separated in the x direction from the other observations are often influential.

You must be on the alert for influential observations that are not outliers. Least-squares regression is not resistant. The position of the least-squares regression line is heavily influenced by observations that are extreme in x. Because influential observations draw the regression line toward themselves, we cannot always spot them by looking for large residuals. Many statistical computing systems produce tables and plots of regression residuals. If you rely on these alone, you may miss the most influential cases. The basic scatterplot of y versus x will alert you to observations that are extreme in x and may therefore be influential. The surest way to verify that a point is influential is to find the regression line both with and without the suspect point, as in Figure 2.20. If the line moves more than a small amount when the point is deleted, the point is influential.

An influential observation should be investigated to ensure that it is correct. Even if no error is found, you should ask whether this observation belongs to the population you are studying. The child development researcher in Example 2.12 must decide whether the child of case 18 was so slow to speak that this case should not be allowed to influence the analysis. If case 18 is excluded, much of the evidence for a connection between the age at which a child begins to talk and later aptitude scores vanishes. If case 18 is retained, we need data on other children who were also slow to begin talking, so that the analysis is no longer so heavily dependent on a single child.

SUMMARY

The simplest relation between an explanatory variable x and a response variable y is **linear dependence.** A straight line fitted to the scatterplot of y versus x then describes the overall pattern of the data.

The most common method of fitting a line to a scatterplot is least squares. The **least-squares regression line** is the straight line $\hat{y} = a + bx$ that minimizes the sum of the squares of the vertical deviations of the observed y-values from the line.

A fitted line can be used to predict the value of y for any value of x, but **extrapolation** beyond the range of x-values spanned by the data is risky.

The fit of a regression line is examined by plotting the **residuals,** or differences between the observed and predicted values of y. Be on the lookout for **outliers,** which are points with unusually large residuals, and also for nonlinear patterns and uneven variation about the line.

Evidence of the effects of **lurking variables** on y may be provided by plots of y and of the residuals against the time order of the observations.

Influential observations, individual points that substantially change the regression line, must also be detected and examined. Influential observations are often outliers in the x variable, but they need not have large residuals.

SECTION 2.2 EXERCISES

2.16 (Review of straight lines) Fred keeps his savings in his mattress. He began with \$500 from his mother and adds \$100 each year. His total savings y after x years are then given by the linear equation

$$y = 500 + 100x$$

(a) Draw a graph of this equation. (Hint: Choose two values of x, such as 0 and 10. Compute the corresponding values of y from the equation. Plot these two points on graph paper and draw the straight line joining them.)
(b) After 20 years, how much will Fred have in his mattress?
(c) If Fred had added \$200 instead of \$100 each year to his initial \$500, what is the equation that describes his savings after x years?

2.17 (Review of straight lines) Sound travels at a speed of 1500 meters per second in sea water. You dive into the sea from your yacht. Give an equation for the distance y at which a shark can hear your splash in terms of the number of seconds x since you hit the water.

2.18 (Review of straight lines) During the period after birth, a male white rat gains exactly 40 grams (g) per week. (This rat is unusually regular in his growth, but 40 g per week is a realistic rate.)
(a) If the rat weighed 100 g at birth, give an equation for his weight after x weeks. What is the slope of this line?
(b) Draw a graph of this line between birth and 10 weeks of age.
(c) Would you be willing to use this line to predict the rat's weight at age 2 years? Do the prediction and think about the reasonableness of the result. (There are 454 g in a pound. To help you assess the result, note that a large cat weighs about 10 pounds.)

2.19 (Review of straight lines) Cellular mobile telephone service costs \$30 per month plus \$33 for each hour of air time. Give an equation for the amount y of the monthly bill in terms of the number x of hours of use that month. You use about 15 hours of time each month. What will your monthly bill be? Another mobile telephone company offers air time at \$50 per month plus \$25 per hour of use. Would you save money by switching to this company?

2.20 Researchers studying acid rain measured the acidity of precipitation in an isolated wilderness area in Colorado for 150 consecutive weeks in the years 1975 to 1978. The acidity of a solution is measured by pH, with lower pH values indicating that the solution is more acid. The acid rain researchers observed a linear pattern over time. They reported that the least-squares line

$$\text{pH} = 5.43 - (.0053 \times \text{weeks})$$

fit the data well. (See William M. Lewis and Michael C. Grant, "Acid precipitation in the western United States," *Science*, 207 (1980), pp. 176–177.)

(a) Draw a graph of this line. Note that the linear change is decreasing rather than increasing.
(b) According to the fitted line, what was the pH at the beginning of the study (weeks = 1)? At the end (weeks = 150)?
(c) What is the slope of the fitted line? Explain clearly what this slope says about the change in the pH of the precipitation in this wilderness area.

2.21 Warren heats his home with natural gas. The amount of gas required to heat the home depends on the outdoor temperature—the colder the weather, the more gas will be consumed. Warren wants to predict gas consumption from the outdoor temperature. He measures his natural gas consumption each month during one heating season, from October to the following June. The explanatory variable x is the average number of heating degree days each day during the month, obtained from local weather records. (One heating degree day is accumulated for each degree a day's average temperature falls below 65° F. An average temperature of 20°F, for example, corresponds to 45 degree days.) The response variable y is the average gas consumption per day during the month, in hundreds of cubic feet. Here are Warren's data (provided by Robert Dale, Purdue University):

Month	Oct.	Nov.	Dec.	Jan.	Feb.	Mar.	Apr.	May	June
x	15.6	26.8	37.8	36.4	35.5	18.6	15.3	7.9	.0
y	5.2	6.1	8.7	8.5	8.8	4.9	4.5	2.5	1.1

(a) Make a scatterplot of these data. There is a strongly linear pattern with no outliers.
(b) Warren uses the Minitab statistical software to calculate the least-squares regression line of y on x. He sees this output on his screen:

```
MTB> REGRESS 'GAS' ON 1, 'DDAYS'

The regression equation is
GAS = 1.233 + .20221 DDAYS
```

Draw the regression line on your scatterplot.

(c) Warren adds insulation in his attic during the summer, hoping to reduce his gas consumption. The next February, there is an average of 40 degree days per day and his gas consumption is 870 cubic feet per day. Predict from the regression equation how much gas the house would have used at 40 degree days per day last winter before the extra insulation. Did the insulation reduce gas consumption?

2.22 Here are data on the the rate of water flow (liters per second) over an experimental soil bed and the amount of soil washed away (kilograms). These data first appeared in Exercise 2.5 (page 111), where a scatterplot showed a generally linear pattern.

Flow rate	.31	.85	1.26	2.47	3.75
Eroded soil	.82	1.95	2.18	3.01	6.07

(a) Find the equation of the least-squares regression line for predicting soil loss from water flow rate.
(b) For each flow rate given, compute the predicted soil loss using the least-squares line.
(c) Using the results of (b), compute the residuals. Verify that the residuals add to 0.
(d) Plot the residuals versus flow rate. Describe the pattern. What does the plot indicate about the adequacy of the linear fit?

2.23 Sarah's parents are concerned that she seems short for her age. Their pediatrician has the following record of Sarah's height:

Age (months)	36	48	51	54	57	60
Height (cm)	86	90	91	93	94	95

(a) Make a scatterplot of these data. Note the strong linear pattern.
(b) Use Equation 2.1 to compute the slope b and intercept a of the least-squares regression line of Sarah's height on age. What is the equation of the line?
(c) According to the regression line, how much does Sarah grow each month on the average? Normally growing girls gain about 6 cm in height between ages 4 (48 months) and 5 (60 months). What slope does this correspond to? Is Sarah's growth more rapid (larger slope) than normal or less rapid (smaller slope)? Sarah's pediatrician noticed the unusual slope of Sarah's growth curve and decided to do some additional tests. Sarah was found to have a growth hormone deficiency that can be treated with doses of a synthetic hormone.
(d) Use your equation from (b) to predict Sarah's height at 40 months and at 65 months. Then use this information to draw the least-squares line on the scatterplot.

2.24 A student who waits on tables at a Chinese restaurant in a college neighborhood records the cost of meals and the tip left by single diners. Here are some of the data:

Meal	\$4.50	\$5.79	\$6.24	\$4.62	\$6.35
Tip	\$0.50	\$0.75	\$0.85	\$0.60	\$1.00

(a) Compute the least-squares regression line for these data.
(b) Make a scatterplot of the data and draw the regression line on your plot.
(c) The next diner orders a meal costing \$4.89. Use your regression line to predict the tip.

2.25 Use your regression line from Exercise 2.23 to predict Sarah's height at each of 36, 48, 51, 54, 57, and 60 months. Then compute the residuals by subtracting these predicted heights from the actually observed heights at these ages. Verify that the residuals have sum 0 (up to roundoff error). Make a plot of the residuals against age and briefly describe the pattern of residuals.

2.26 We will now analyze the residuals for the Chinese restaurant data of Exercise 2.24. Compute the five residuals for these data, using the regression line that you have already computed. Verify that the residuals have sum 0 (up to roundoff error). Plot the residuals against the cost of the meals and comment on the pattern of the residuals.

2.27 Runners are concerned about their form when racing. One measure of form is the stride rate, defined as the number of steps taken per second. A runner is inefficient when the rate is either too high or too low. Of course, as the speed increases, the stride rate should also increase. In a study of 21 of the best American female runners, researchers measured the stride rate for different speeds. The following table gives the speeds (in feet per second) and the average stride rates for these women. (Data from R. C. Nelson, C. M. Brooks, and N. L. Pike, "Biomechanical comparison of male and female distance runners," in P. Milvy (ed.), *The Marathon: Physiological, Medical, Epidemiological, and Psychological Studies*, New York Academy of Sciences, 1977, pp. 793–807.)

Speed	15.86	16.88	17.50	18.62	19.97	21.06	22.11
Stride rate	3.05	3.12	3.17	3.25	3.36	3.46	3.55

(a) Plot the data with speed on the x axis and stride rate on the y axis. Does a straight line adequately describe these data?
(b) Compute the slope and intercept for the least-squares line, using Equation 2.1 or a statistical calculator or software. Graph the least-squares line on your plot from (a).

(c) For each of the speeds given, compute the predicted value using the least-squares line. Use your results to compute the residuals. Verify that the residuals add to 0.

(d) Plot the residuals versus speed. Describe the pattern. What does the plot indicate about the adequacy of the linear fit? Can you plot the residuals against the time at which the observations were made?

2.28 Research on digestion requires accurate measurements of blood flow through the lining of the stomach. A promising way to make such measurements easily is to inject mildly radioactive microscopic spheres into the blood stream. The spheres lodge in tiny blood vessels at a rate proportional to blood flow; their radioactivity allows blood flow to be measured from outside the body. Medical researchers compared blood flow in the stomachs of dogs, measured by use of microspheres, with simultaneous measurements taken using a catheter inserted into a vein. The data, in milliliters of blood per minute (ml/minute), appear below. (Based on L. H. Archibald, F. G. Moody, and M. Simons, "Measurement of gastric blood flow with radioactive microspheres," *Journal of Applied Physiology*, 38 (1975), pp. 1051–1056.)

Spheres	4.0	4.7	6.3	8.2	12.0	15.9	17.4	18.1	20.2	23.9
Vein	3.3	8.3	4.5	9.3	10.7	16.4	15.4	17.6	21.0	21.7

(a) Make a scatterplot of these data, with the microsphere measurement as the explanatory variable. There is a strongly linear pattern.

(b) Calculate the least-squares regression line of venous flow on microsphere flow. Draw your regression line on the scatterplot.

(c) Predict the venous measurement for microsphere measurements 6, 12, and 18 ml/minute. If the microsphere measurements are within about 10% to 15% of the predicted venous measurements, the researchers will simply use the microsphere measurements in future studies. Is this condition satisfied over this range of blood flow?

2.29 Suppose that the last customer in Exercise 2.24 had ordered a large meal and left no tip. The data are now as follows:

Meal	\$4.50	\$5.79	\$6.24	\$4.62	\$16.79
Tip	\$0.50	\$0.75	\$0.85	\$0.65	\$0.00

(a) Make a scatterplot of these data. Which observation will be most influential? Why?

(b) The least-squares regression line fitted to all five observations is $\hat{y} = 0.97 - 0.056x$. Draw this line on your graph. How well does the regression line describe the pattern of the first four points?

(c) Does it appear from the graph that the influential observation has a larger residual from the least-squares line given in (b) than the other observations? (You need not actually compute the residuals.)

2.30 Here are the golf scores of 11 members of a women's golf team in two rounds of tournament play:

Player	1	2	3	4	5	6	7	8	9	10	11
Round 1	89	90	87	95	86	81	105	83	88	91	79
Round 2	94	85	89	89	81	76	89	87	91	88	80

(a) Plot the data with the round 1 scores on the x axis and the round 2 scores on the y axis. There is a generally linear pattern except for one influential observation. Circle this observation on your graph.
(b) Computer software gives the following two least-squares lines, one calculated from all 11 data points and one omitting the influential observation:

$$\hat{y} = 20.49 + .754x$$

$$\hat{y} = 50.01 + .410x$$

Draw both lines on your scatterplot. Which line omits the influential observation? Explain how you identified this line.

2.31 A study of nutrition in developing countries collected data from the Egyptian village of Nahya. Here are the mean weights for 170 infants in Nahya who were weighed each month during their first year of life. (Data from Zeinab E. M. Afifi, "Principal components analysis of growth of Nahya infants: Size, velocity and two physique factors," *Human Biology*, 57 (1985), pp. 659–669.)

Age (months)	1	2	3	4	5	6	7	8	9	10	11	12
Weight (kg)	4.3	5.1	5.7	6.3	6.8	7.1	7.2	7.2	7.2	7.2	7.5	7.8

(a) Plot the mean weight against time.
(b) A hasty user of statistics enters the data into the Minitab computing system and computes the least-squares line without plotting the data. The result is

```
MTB> REGRESS 'WEIGHT' ON 1, 'AGE'

The regression equation is
WEIGHT = 4.88 + 0.267 AGE
```

Plot this line on your graph. Is it an acceptable summary of the overall pattern of growth? (Remember that you can calculate the least-squares line for *any* set of two-variable data. It's up to you to decide if it makes sense to fit a line.)

(c) Fortunately for the hasty user, the computing system prints out the residuals from the least-squares line. In order of age along the rows, they are

−.85	−.31	.02	.35	.58	.62
.45	.18	−.08	−.35	−.32	−.28

Plot these residuals against age. Describe carefully the pattern that you see.

2.32 One component of air pollution is airborne particulate matter such as dust and smoke. Particulate pollution is measured by using a vacuum motor to draw air through a filter for 24 hours. The filter is weighed at the beginning and end of the period and the weight gained is a measure of the concentration of particles in the air. In a study of pollution, measurements were made every 6 days with identical instruments in the center of a small city and at a rural location 10 miles southwest of the city. Because the prevailing winds blow from the west, it was suspected that the rural readings would be generally lower than the city readings, but that the city readings could be predicted from the rural readings. The table gives readings taken every 6 days over a 7-month period. The entry NA means that the reading for that date is not available, usually because of equipment failure. (Data provided by Matthew Moore.)

Rural	NA	67	42	33	46	NA	43	54	NA	NA	NA	NA
City	39	68	42	34	48	82	45	NA	NA	60	57	NA
Rural	38	88	108	57	70	42	43	39	NA	52	48	56
City	39	NA	123	59	71	41	42	38	NA	57	50	58
Rural	44	51	21	74	48	84	51	43	45	41	47	35
City	45	69	23	72	49	86	51	42	46	NA	44	42

To assess the success of predicting the city particulate reading from the rural reading, the 26 complete cases (both readings present) are analyzed. Computer work finds the least-squares regression line of the city reading y on the rural reading x to be

$$\hat{y} = -2.580 + 1.0935x$$

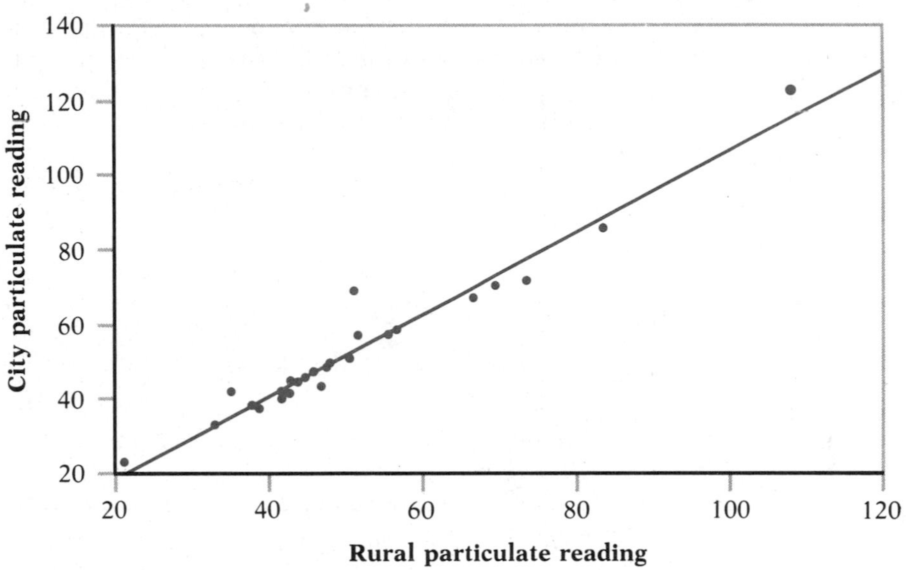

FIGURE 2.22 The regression of particulate concentration in a city center on rural particulate concentration for the same days (Exercise 2.32).

Figure 2.22 is a scatterplot for the 26 complete cases with the regression line.

(a) Which observation in Figure 2.22 appears to be the most influential? Is this the observation with the largest residual (vertical distance from the line)?

(b) Locate in the table the observation you chose from the graph in (a) and compute its residual.

(c) Do the data suggest that using the least-squares line for prediction will give approximately correct results over the range of values appearing in the data? (The incompleteness of the data does not seriously weaken this conclusion if equipment failures are independent of the variables being studied.)

(d) On the fourteenth date in the series, the rural reading was 88 and the city reading was not available. What do you estimate the city reading to be for that date?

2.33 Table 1.4 (page 36) gives the calories and sodium content for each of 17 brands of meat hot dogs. Figure 2.23 is a scatterplot of the amount of sodium in a hot dog of each brand versus the number of calories.

(a) Describe the main features of the relationship. (The discussion following Example 1.13 on page 44 may help you.)

(b) The plot shows two least-squares regression lines. One was calculated using all of the observations; the other omitted the brand of veal

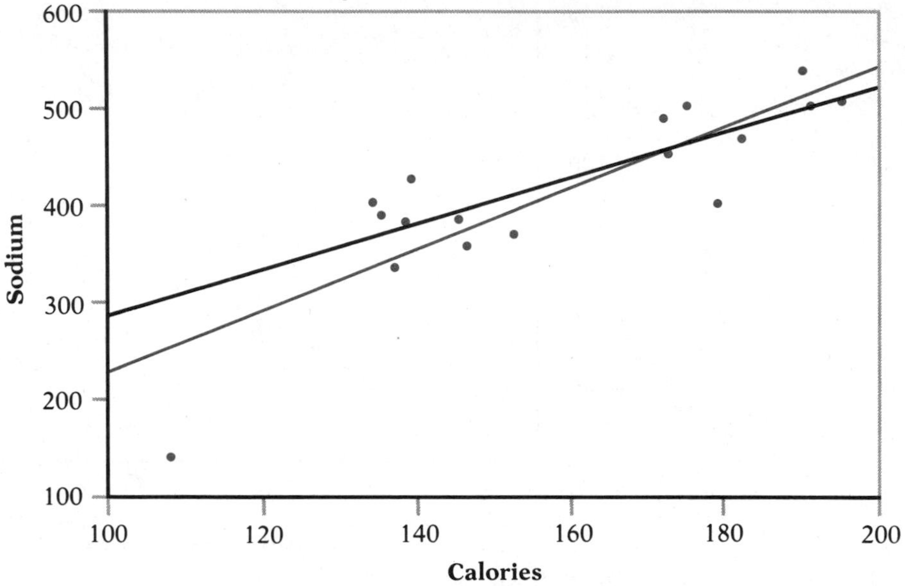

FIGURE 2.23 The regression of sodium content on calories for brands of hot dogs, calculated with and without an outlier (Exercise 2.33).

hot dogs that is an outlier in both variables measured. Which line (colored or black) was calculated from all of the data? Explain your answer.

(c) The regression line that ignores the outlier is

$$\hat{y} = 46.90 + 2.401x$$

A new brand of meat hot dog (not made with veal) has 150 calories per frank. How many milligrams of sodium do you estimate that one of these hot dogs contains?

2.34 Are baseball players paid according to their performance? To study this question, a statistician analyzed the salaries of over 260 major league hitters along with such explanatory variables as career batting average, career home runs per time at bat, and years in the major leagues. This is a *multiple regression* with several explanatory variables. More detail on multiple regression appears in Chapter 9, but the fit of the model is assessed just as we have done in this chapter, by calculating and plotting the residuals

$$\text{residual} = \text{observed } y - \text{predicted } y$$

(This analysis was done by Crystal Richard.)

(a) Figure 2.24(a) is a plot of the residuals versus the predicted salary.

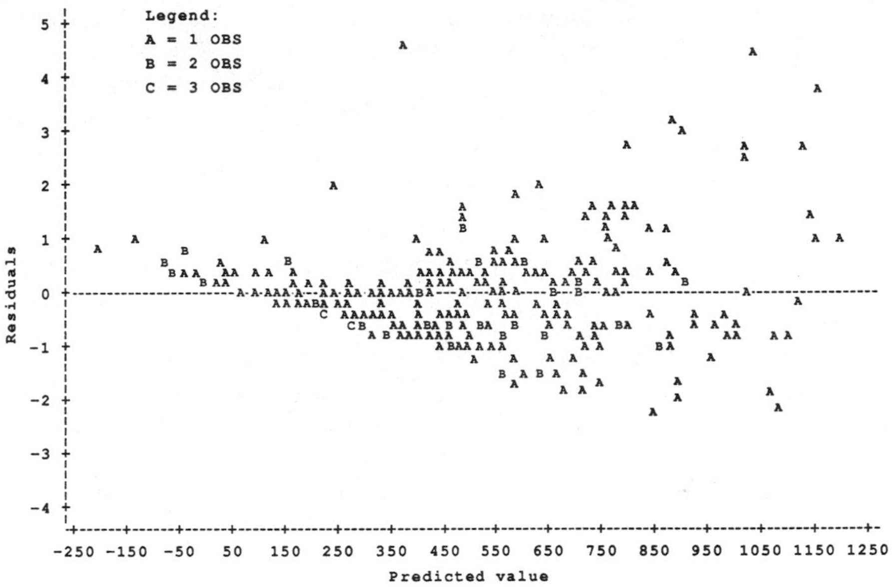

(a)

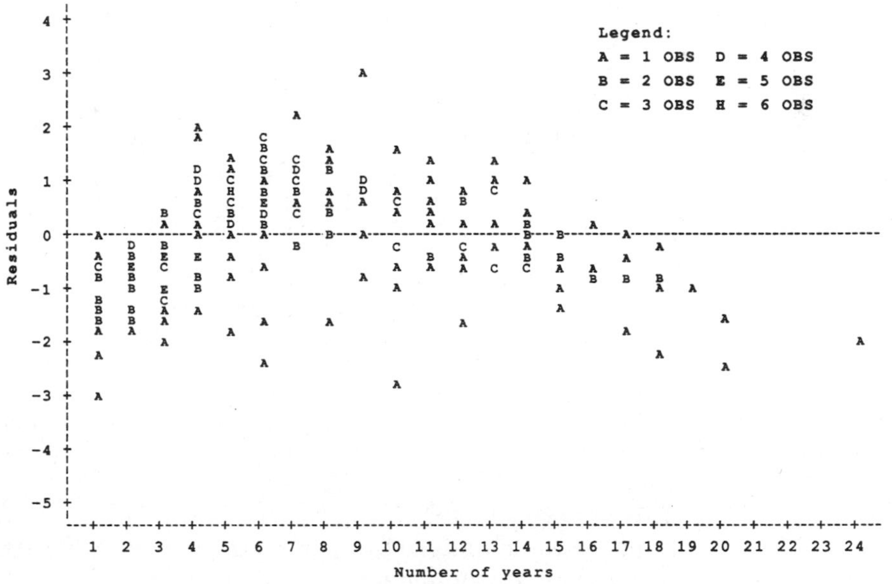

(b)

FIGURE 2.24 Two residual plots for the regression of baseball players' salaries on their performance (Exercise 2.34).

This plot was produced by the SAS statistical software system that was used to analyze the data. Notice that when points are too close together to plot separately, SAS uses letters of the alphabet to show how many points there are at each position. Describe the pattern that appears on this residual plot. Will the regression model predict high or low salaries more precisely?

(b) After studying the residuals in more detail, the statistician decided to predict the logarithm of salary rather than the salary itself. One reason was that although salaries are not normally distributed (the distribution is skewed to the right), their logarithms are nearly normal. When the response variable is the logarithm of salary, a plot of the residuals against the predicted value is satisfactory—it looks like Figure 2.15(a). Figure 2.24(b) is a plot of the residuals against the number of years the player has been in the major leagues. Describe the pattern that you see. Will the model overestimate or underestimate the salaries of players who are new to the major leagues? Of players who have been in the major leagues about 8 years? Of players with more than 15 years in the majors?

2.35 The following table gives the results of a study of a sensitive chemical technique called gas chromatography, which is used to detect very small amounts of a substance. Five measurements were taken for each of four amounts of the substance being investigated. The explanatory variable x is the amount of substance in the specimen, measured in nanograms (ng), or units of 10^{-9} g. The response variable y is the output reading from the gas chromatograph. The purpose of the study is to calibrate the apparatus by relating y to x. (Data from D. A. Kurtz (ed.), *Trace Residue Analysis*, American Chemical Society Symposium Series No. 284, 1985, appendix.)

Amount (ng)	Response				
.25	6.55	7.98	6.54	6.37	7.96
1.00	29.7	30.0	30.1	29.5	29.1
5.00	211	204	212	213	205
20.0	929	905	922	928	919

(a) Make a scatterplot of these data. The relationship appears to be approximately linear, but the wide variation in the response values makes it hard to see detail in this graph.
(b) Compute the least-squares regression line of y on x, and plot this line on your graph.
(c) Now compute the residuals and make a plot of the residuals against x. It is much easier to see deviations from linearity in the residual plot. Describe carefully the pattern displayed by the residuals.

2.3 AN APPLICATION: EXPONENTIAL GROWTH*

Petroleum has become the most important single source of energy for the developed nations in this century and recently the cause of economic dislocation and even war. Table 2.5 and Figure 2.25 show the growth of annual world crude oil production, measured in millions of barrels per year.[8] We have connected the points in the plot with lines to make the trend clearer. There is an increasing trend, but the overall pattern is not linear. Oil production has increased much faster than linear growth, but the pattern of growth follows a smooth curve until 1973, when a Mideast war touched off a vast price increase and a change in the previous pattern of production. Can we describe the growth of oil production from 1880 to 1973 by a simple mathematical model? We can indeed, and the resulting model for *exponential growth* is second only to straight lines as a useful simple pattern for data. A fitted line does not describe exponential growth, but we will see that a simple transformation of the data enables us to use a regression line and residuals to examine oil production over the past century.

The nature of exponential growth

exponential growth

A variable grows linearly over time if it *adds* a fixed increment in each equal period. *Exponential growth* occurs when a variable is *multiplied* by a fixed number in each time period. To grasp the effect of multiplicative growth, consider a population of bacteria in which each bacterium splits into two each hour. Beginning with a single bacterium, we have 2 after one

TABLE 2.5 Annual world crude oil production, 1880–1988 (millions of barrels)

Year	Mbbl	Year	Mbbl	Year	Mbbl
1880	30	1940	2150	1972	18,584
1890	77	1945	2595	1974	20,389
1900	149	1950	3803	1976	20,188
1905	215	1955	5626	1978	21,922
1910	328	1960	7674	1980	21,722
1915	432	1962	8882	1982	19,411
1920	689	1964	10,310	1984	19,837
1925	1069	1966	12,016	1986	20,246
1930	1412	1968	14,104	1988	21,338
1935	1655	1970	16,690		

*This material is important in biology, business, and some other areas of application, but can be omitted without loss of continuity.

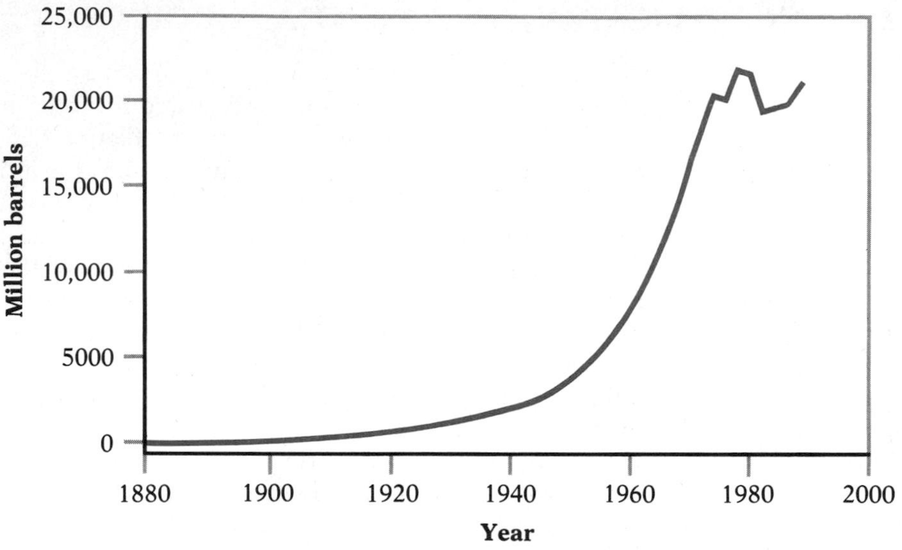

FIGURE 2.25 Annual world crude oil production, 1880 to 1988, from Table 2.5.

hour, 4 at the end of two hours, 8 after three hours, then 16, 32, 64, 128, and so on. These first few numbers are deceiving. After 1 day of doubling each hour, there are 2^{24}, or 16,777,216, bacteria in the population. That number then doubles the next hour! Try successive multiplications by 2 on your calculator to see for yourself the very rapid increase after a slow start. Figure 2.26 shows the growth of the bacteria population over 24 hours. For the first 15 hours, the population is too small to rise visibly above the zero level on the graph.

It is characteristic of exponential growth that the increase appears slow for a long period and then seems to explode. Both Figures 2.25 and 2.26 display explosive growth after a long period of gradual increase. Our first reaction to the apparently sudden surge is to search for some explanation, something new that distinguishes the high-growth present from the low-growth past. But the nature of the increase has not changed at all. It is our understanding of the long-term consequences of exponential growth that needs correcting. An old story tells how a king of Persia learned this lesson. There was once, the story goes, a courtier who asked the king for a simple reward: a grain of rice on the first square of a chess board, 2 grains on the second square, then 4, 8, 16, and so on through all 64 squares. The king foolishly granted the request. We now know better. Similarly, we should know better than to risk our prosperity on the assumption that oil production will grow exponentially forever.

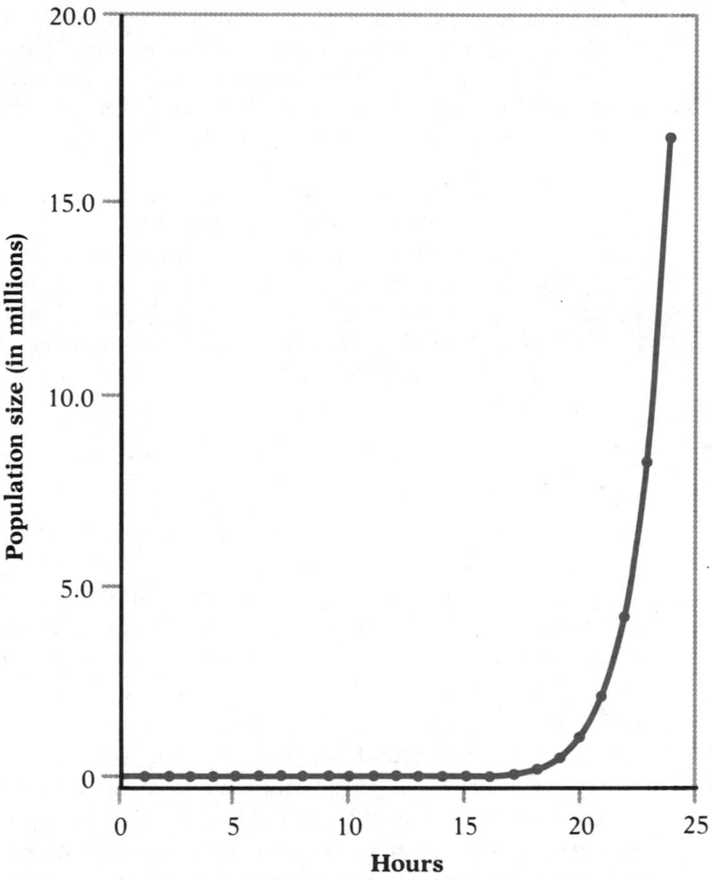

FIGURE 2.26 Growth of a bacteria population that doubles each hour.

> LINEAR VERSUS EXPONENTIAL GROWTH
>
> Linear growth increases by a fixed *amount* in each time period; exponential growth increases by a fixed *percentage* of the previous total.

Populations of living things—like bacteria—tend to grow exponentially if not restrained by outside limits such as lack of food or space. More pleasantly, money also displays exponential growth when returns to an investment are compounded. Compounding means that last period's income earns income this period.

EXAMPLE 2.13

A dollar invested at an annual rate of 6% turns into $1.06 in a year. The original dollar remains and has earned $0.06 in interest. That is, 6% annual interest means that any amount on deposit for the entire year is multiplied by 1.06. If the $1.06 remains invested for a second year, the new amount is therefore 1.06×1.06, or 1.06^2. That is only $1.12, but this in turn is multiplied by 1.06 during the third year, and so on.

If the Native Americans who sold Manhattan Island for $24 in 1626 had deposited the $24 in a savings account at 6% annual interest, they would now have over $40 billion. Our savings accounts don't make us billionaires because we don't stay around long enough. A century of growth at 6% per year turns $24 into $8143. That's 1.06^{100} times $24. By 1826, two centuries after the sale, the account would hold a bit over $2.7 million. Only after a patient 302 years do we finally reach $1 billion. That's real money, but 302 years is a long time. ●

The logarithm transformation

We can see multiplication at work in the growth of a biological population and of money deposited at compound interest. It is not surprising that the exponential model fits these cases. But there is no obvious multiplication going on in the case of world oil production. Is this really exponential growth? The shape of the growth curve for oil production in Figure 2.25 does indeed generally resemble the exponential curve in Figure 2.26. But our eyes are not very good at comparing curves of roughly similar shape. We need a better way to check whether growth is exponential. Our eyes are quite good at judging whether or not points lie along a straight line. So we will apply a mathematical transformation that changes exponential growth into linear growth—and patterns of growth that are not exponential into something other than linear.

logarithm

The necessary transformation is carried out by taking the *logarithms* of the data points. Use a calculator with a LOG button to compute logarithms. Better yet, most statistical software systems will calculate the logarithms of an entire data set in response to a single command. The essential property of the logarithm for our purposes is that it straightens an exponential growth curve. *If a variable grows exponentially, its logarithm grows linearly.*

EXAMPLE 2.14

Figure 2.26 shows the exponential growth of a population of bacteria that doubles each hour. The logarithms of the number of bacteria should show straight-line growth. Use your calculator to find some of the logarithms. The population counts for the first few hours are 2, 4, 8, ... The logarithms are approximately

$$\log 2 = .30$$

$$\log 4 = .60$$

$$\log 8 = .90$$

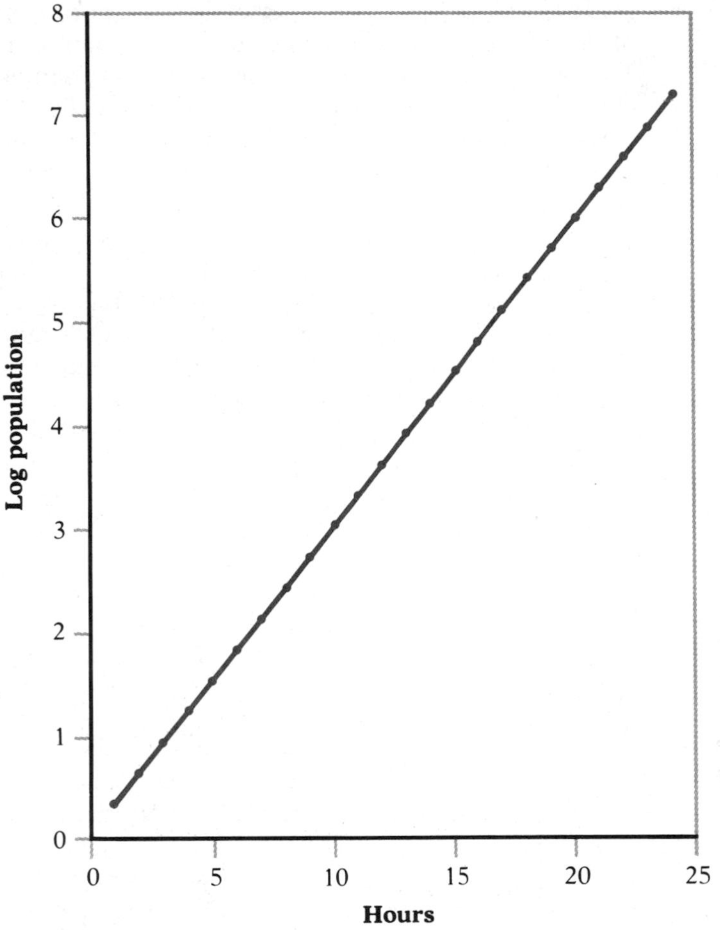

FIGURE 2.27 The logarithms of the bacteria population size, from Figure 2.26 (Example 2.14).

Figure 2.27 graphs the logarithms of the bacteria population counts from Figure 2.26 against time. The promised straight line has indeed appeared. After 15 hours, for example, the population contains 2^{15}, or 32,768, bacteria. The logarithm of 32,768 is 4.52, and this point appears above the 15-hour mark in Figure 2.27. (We have calculated and plotted common logarithms, or logarithms with base 10. There are other types of logarithms as well. Since all logarithms transform exponential growth into a straight line, you need not be concerned about which logarithm to use.) ●

Our artificial example of bacterial growth was exactly exponential. Real data will not fit the exponential model so perfectly, so applying logarithms will not produce a perfectly straight line. Figure 2.28 is a graph of the logarithm of the annual world crude oil production from Table 2.5. The individual points have been connected by lines to make the pattern clear. The graph is quite close to a straight line from 1880 to 1973. A clear deviation from the line begins in 1973. Closer examination shows a smaller deviation during the 1930s, when growth slowed during the Great Depression. The increase in oil production has indeed been close to exponential over most of the past century, but with important deviations.

The logarithmic transformation enables us to judge the appropriateness of a somewhat complicated mathematical model by reducing the question to judging whether points lie on a straight line. Comparison of Figures 2.25 and 2.28 reveals another advantage: Logarithms compress the vertical scale of the plot, making it easier to see deviations from fit in the earlier stages of exponential growth. The slowdown in the oil production during the Depression is visible in Figure 2.28, but not in Figure 2.25. On the other hand, the post-1973 disturbance, occurring at the top of the curve, is clearer in Figure 2.25.

Residuals again

Just as in the case of linear growth, calculating and plotting residuals allow a closer examination of the fit of the exponential growth model to data. Before we can find the residuals, however, we must fit to the data

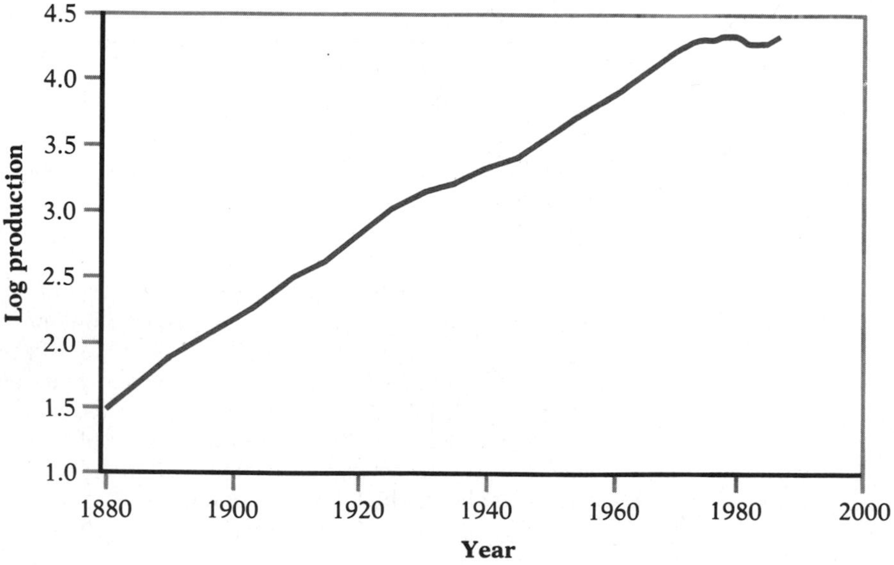

FIGURE 2.28 **The logarithms of annual world crude oil production, from Table 2.5.**

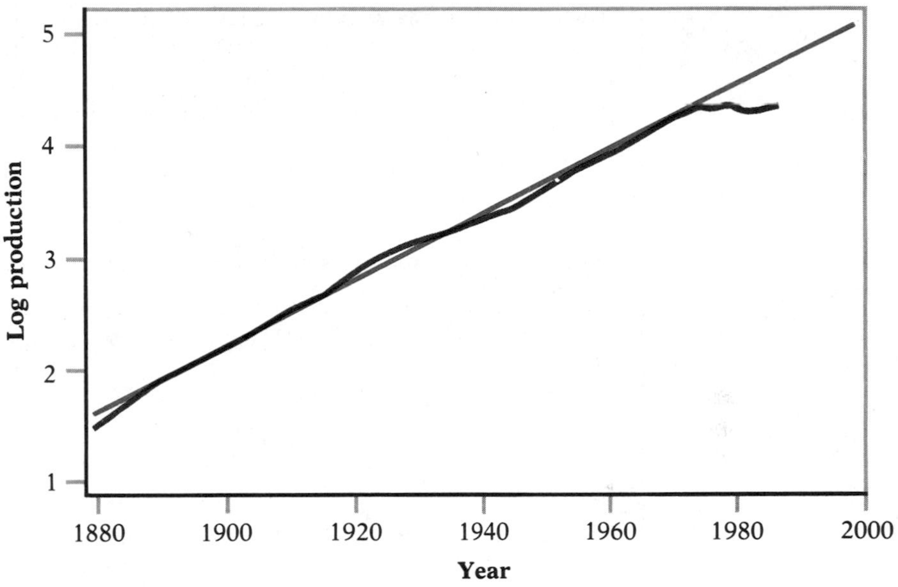

FIGURE 2.29 The least-squares line for world oil production based on data for 1880 to 1972.

an explicit model described by an equation. Since taking logarithms of data that display exponential growth produces a straight line, we can fit a straight line to the logarithms and obtain residuals as the deviations of the actual logarithms from this line. Figure 2.29 shows the result of fitting the least-squares regression line to the logarithms of annual oil production from 1880 to 1972. We did not allow data for years after 1972 to enter into the calculation, because we know that the pattern of exponential growth ended in 1973.

EXAMPLE 2.15

We enter the annual oil production for the years 1880 to 1972 into the Minitab computing system as values of the variable OIL, and the corresponding years as the variable YEAR. Minitab then computes the logarithms and finds the least-squares line for the logarithms as follows:

```
MTB> LET C3 = LOGT('OIL')
MTB> REGRESS C3 ON 1, 'YEAR'

The regression line is
C3 = -52.686 + 0.02888 YEAR
```

That is, the computer finds the regression line to be

$$\text{log production} = -52.686 + (.02888 \times \text{year})$$

The residuals of the data from the fitted line are computed in the usual way, as

$$\text{residual} = \text{observed value} - \text{predicted value}$$

For the year 1945, the value of the logarithm predicted by the regression line is

$$\begin{aligned}\log \text{production} &= -52.686 + (.02888 \times 1945) \\ &= 3.4856\end{aligned}$$

Since 2595 million barrels of oil were actually produced in 1945, the residual for that year is

$$\begin{aligned}\text{residual} &= \text{observed value} - \text{predicted value} \\ &= \log 2595 - 3.4856 \\ &= 3.4141 - 3.4856 \\ &= -.0715\end{aligned}$$

Most computer regression routines will compute and print a table of the residuals, and many will also produce residual plots. For example, a table of the residuals is part of the output from the Minitab command REGRESS. ●

The deviations from the overall pattern of exponential growth in oil production are easier to see in the residual plot (Figure 2.30) than in Figure 2.29. The pattern of residuals about the horizontal line at 0 shows

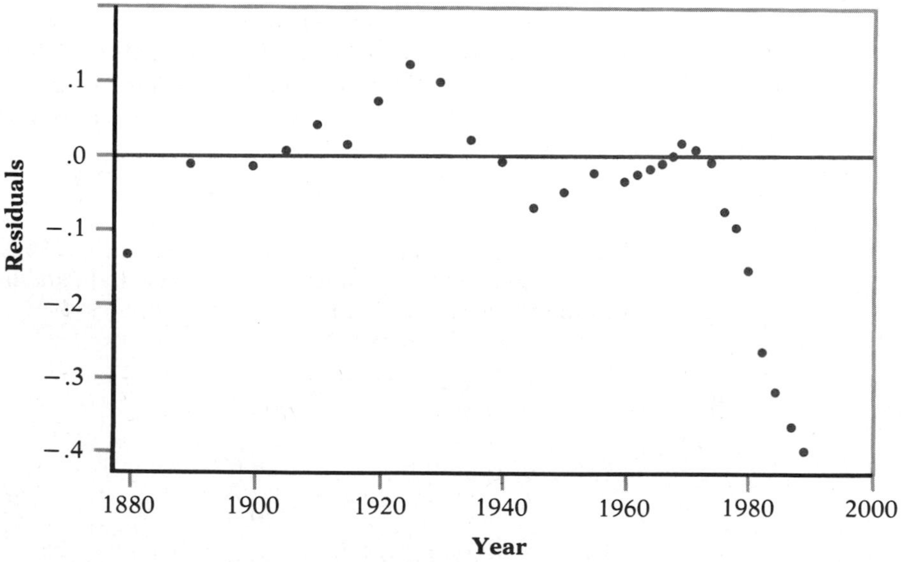

FIGURE 2.30 **Residual plot for the least-squares line for world oil production.**

systematic departures from the overall fit. Because we are plotting logarithms, straight-line patterns of residuals have a specific interpretation. The line of fit (zero residual) represents exponential growth at the average rate observed in the entire period 1880 to 1972. A straight-line rising pattern of residuals shows a period of growth at a faster rate, while a declining pattern shows a slower rate of growth.

The most dramatic departure of the residuals from 0 marks the end of exponential growth after 1972. But we can now see other systematic deviations. Oil production was increasing more rapidly than the long-term rate in the years between 1900 and the beginning of the Depression in 1929. Production increased more slowly not only during the Depression but also during World War II. That is, the turning points in the residual plot occur in 1925 (the last point before 1929) and 1945 (the end of World War II). Only after the war did oil production return to an above-average growth rate, which lasted until 1973. (The production of oil was of course increasing during the entire period up to 1973—the runs of declining residuals show periods of slower growth, not an actual drop in oil production.)

Our tools for data analysis have given us a quite detailed understanding of the history of world oil production. The long-term picture is one of exponential growth from the beginnings of commercial petroleum use to the Arab boycott of 1973. But the long-term rate of growth is an average over several periods of somewhat slower or somewhat faster growth, periods that coincide with major events in the economic history of the twentieth century.

The use of logarithms to obtain a linear pattern from exponential growth does complicate prediction a bit. Because the regression line is fitted to the logarithms of the observations, it also predicts the logarithm. To express the prediction in easily understood terms, we must restate it in terms of the original units of measurement. To do this, use the fact that any number y can be obtained from its common (base 10) logarithm $\log y$ by

$$y = 10^{\log y}$$

EXAMPLE 2.16

The regression line obtained in Example 2.15 predicts that in 1977

$$\begin{aligned} \log \text{production} &= -52.686 + (.02888 \times 1977) \\ &= 4.4098 \end{aligned}$$

The predicted production in the original units (millions of barrels of oil) is therefore

$$\text{production} = 10^{4.4098} = 25{,}692$$

The actual production in 1977 was 21,787 million barrels. The prediction is much too high because the exponential growth model fails to fit post-1973 oil production. ●

Could we have avoided logarithms by fitting an exponential growth curve to the actual production data displayed in Figure 2.25? Yes. But there are several reasons for transforming before fitting the model and calculating residuals. First, straight lines are simple, while the explicit mathematical form of the exponential growth model is more involved. A second reason for working with the logarithms is that it is easy to calculate the least-squares line. Fitting an exponential curve to a set of data is more difficult. Many statistical computing systems (not all) will carry out such nonlinear fits, but the calculations are not feasible without a computer.

SUMMARY

When a variable is multiplied by a fixed amount greater than 1 in each equal time period, **exponential growth** results.

A plot against time of exponential growth becomes linear (a straight line) if we take the **logarithm** before plotting. This fact makes it possible to detect exponential growth easily.

Deviations from the overall pattern of exponential growth are most easily examined by fitting a line to the logarithms of the data and plotting the **residuals** from this line.

SECTION 2.3 EXERCISES

2.36 (Exact exponential growth) The common intestinal bacterium *E. coli* is one of the fastest-growing bacteria. Under ideal conditions, the number of *E. coli* in a colony doubles about every 15 minutes until restrained by lack of resources. Starting from a single bacterium, how many *E. coli* will there be in 1 hour? In 5 hours?

2.37 (Exact exponential growth) A clever courtier, offered a reward by an ancient king of Persia, asked for a grain of rice on the first square of a chess board, 2 grains on the second square, then 4, 8, 16, and so on.

(a) Make a table of the number of grains on each of the first 10 squares of the board.

(b) Plot the number of grains on each square against the number of the square for squares 1 to 10, and connect the points with a smooth curve. This is an exponential curve.

(c) How many grains of rice should the king deliver for the 64th (and final) square?

(d) Take the logarithm of each of your numbers of grains from (a). Plot these logarithms against the number of squares from 1 to 10. You should get a straight line.

(e) From your graph in (d) find the approximate values of the slope b and the intercept a for the line. Use the equation $y = a + bx$ to predict the logarithm of the amount for the 64th square. Check your result by taking the logarithm of the amount you found in (c).

2.38 (Exact exponential growth) Alice is given a savings bond at birth. The bond is initially worth \$500 and earns interest at 7.5% each year. This means that the value is multiplied by 1.075 each year.

(a) Find the value of the bond at the end of 1 year, 2 years, and so on up to 10 years.

(b) Plot the value y against years x on graph paper. Connect them with a smooth curve. This is an exponential curve.

(c) Take the logarithm of each of the values y that you found in (a). Plot the logarithm $\log y$ against years x on graph paper. You should obtain a straight line.

2.39 Fred and Alice were born the same year, and each began life with \$500. Fred added \$100 each year, but earned no interest. Alice added nothing, but earned interest at 7.5% annually. After 25 years, Fred and Alice are getting married. Who has more money?

2.40 Biological populations can grow exponentially if not restrained by predators or lack of food. The gypsy moth outbreaks that occasionally devastate the forests of the Northeast illustrate approximate exponential growth. It is easier to count the number of acres defoliated by the moths than to count the moths themselves. Here are data on an outbreak in Massachusetts. (Data provided by Chuck Schwalbe, U. S. Department of Agriculture.)

Year	Acres
1978	63,042
1979	226,260
1980	907,075
1981	2,826,095

(a) Plot the number of acres defoliated y against the year x. The pattern of growth appears exponential.

(b) Verify that y is being multiplied by about 4 each year by calculating the ratio of acres defoliated each year to the previous year. (Start with 1979 to 1978, when the ratio is $226{,}260/63{,}042 = 3.6$.)

(c) Take the logarithm of each number y and plot the logarithms against the year x. The linear pattern confirms that the growth is exponential.

(d) The least-squares line fitted to the four points is

$$\log y = -1094.51 + (.55577 \times \text{year})$$

Use this line to predict the number of acres defoliated in 1982. (Predict $\log y$ by substituting $x = 1982$ in the equation. Then use the fact that $y = 10^{\log y}$ to predict y.) The actual number for 1982 was 1,383,265, far less than the prediction. The exponential growth of the gypsy moth population was cut off by a viral disease, and the population quickly collapsed back to a low level.

2.41 Federal expenditures on social insurance (chiefly social security and Medicare) increased rapidly after 1960. Here are the amounts spent, in millions of dollars:

Year	1960	1965	1970	1975	1980
Spending	14,307	21,807	45,246	99,715	191,162

(a) Plot social insurance expenditures against time. Does the pattern appear closer to linear growth or to exponential growth?

(b) Take the logarithm of the amounts spent. Plot these logarithms against time. Do you think that the exponential growth model fits well?

(c) After entering the data into the Minitab statistical system, with year as C1 and expenditures as C2, we obtain the least-squares line for the logarithms as follows:

```
MTB> LET C3 = LOGT(C2)
MTB> REGRESS C3 ON 1, C1

The regression equation is
C3 = -110.04 + 0.05824 C1
```

That is, the least-squares line is

$$\log y = -110.04 + (.05824 \times \text{year})$$

Draw this line on your graph from (b).

(d) Use this line to predict the logarithm of social insurance outlays for 1988. Then compute

$$y = 10^{\log y}$$

to predict the amount y spent in 1988.

(e) The actual amount (in millions) spent in 1988 was $358,412. Take the logarithm of this amount and add the 1988 point to your graph in (b). Does it fall close to the line? In 1981, President Reagan took office with a policy of slowing growth in spending on social programs. Did the trend of exponential growth in spending for social insurance change in a major way during the Reagan years 1981 to 1988?

2.42 The following table shows the growth of the population of Europe (millions of persons) between 400 BC and 1950:

Date	Population	Date	Population	Date	Population
400 B.C.	23	1200	61	1600	90
A.D.1	37	1250	69	1650	103
200	67	1300	73	1700	115
700	27	1350	51	1750	125
1000	42	1400	45	1800	187
1050	46	1450	60	1850	274
1100	48	1500	69	1900	423
1150	50	1550	78	1950	594

(a) Plot population against time.
(b) The graph shows that the population of Europe dropped at the collapse of the Roman Empire (A.D. 200–500) and at the time of the Black Death (A.D. 1348). Growth has been uninterrupted since 1400. Plot the logarithm of population against time, beginning in 1400.
(c) Was growth exponential between 1400 and 1950? If not, what overall pattern do you see?

2.43 The following table gives the resident population of the United States from 1790 to 1990, in millions of persons:

Date	Population	Date	Population	Date	Population
1790	3.9	1860	31.4	1930	122.8
1800	5.3	1870	39.8	1940	131.7
1810	7.2	1880	50.2	1950	151.3
1820	9.6	1890	62.9	1960	179.3
1830	12.9	1900	76.0	1970	203.3
1840	17.1	1910	92.0	1980	226.5
1850	23.2	1920	105.7	1990	248.7

(a) Plot population against time. The growth of the American population appears roughly exponential.
(b) Plot the logarithms of population against time. The pattern of growth is now clear. An expert says that "the population of the United States increased exponentially from 1790 to about 1880. After 1880 growth was still approximately exponential, but at a slower rate." Explain how this description is obtained from the graph.
(c) Lay a transparent straightedge over the graph of the logarithm of population from 1900 to 1990. Use this line to predict the population of the United States in the year 2000.

2.44 The number of motor vehicles (cars, trucks, and buses) registered in the United States grew as follows:

Year	Vehicles (millions)	Year	Vehicles (millions)
1940	32.4	1965	90.4
1945	31.0	1970	108.4
1950	49.2	1975	132.9
1955	62.7	1980	155.8
1960	73.9	1985	171.7

(a) Plot the number of vehicles against time.
(b) Compute the logarithm of the number of vehicles and plot the logarithms against time.
(c) Look at the years 1950 to 1980. Was the growth in motor vehicle registrations more nearly linear or more nearly exponential during this period? (Use a straightedge to assess the two graphs.)
(d) The year 1940 fits the pattern well, but 1945 and 1985 do not. The regular pattern of growth ended after 1980, probably due to oil price increases and other economic problems. All years in the 1980s would have negative residuals. But 1945 is a single unusual point. Can you suggest an explanation for the negative residual in 1945?
(e) Fit a line by eye to the graph you chose in (c) as more linear. Use this line to approximately predict the number of vehicles registered in 1989. The actual number for 1989 was 188.7 million. Did your extrapolation over- or under-predict the true value?

2.45 The least-squares line fitted to the 1950 to 1980 data from the previous exercise is approximately

$$\log y = -30.62 + (.01657 \times \text{year})$$

The computing system gives the residuals from the exact least-squares line for the years 1950 to 1980 as

1950	1955	1960	1965	1970	1975	1980
−.0116	.0109	−.0006	.0040	.0000	.0056	−.0082

(a) Use this line to predict motor vehicle registrations in 1945 and 1989.
(b) Make a residual plot for 1950 to 1980. Are there any signs of systematic deviation from exponential growth?
(c) Add the residuals for 1945 and 1989 to your plot. As usual, a plot of residuals magnifies deviations.

2.46 The productivity of American agriculture has grown rapidly due to improved technology (crop varieties, fertilizers, mechanization). Here are data on the output per hour of labor on American farms. The variable is an "index number" that gives productivity as a percent of the 1967 level.

Year	1940	1945	1950	1955	1960	1965	1970	1975	1980	1985
Productivity	21	27	35	47	67	91	113	137	166	217

Plot these data and also plot the logarithms of the productivity values against year. Then briefly describe the pattern of growth that your plots suggest.

2.4 CORRELATION

We have to this point concentrated on analyzing data having a clear explanatory-response structure. In order to fit a regression line, we must know which is the explanatory variable and which is the response. Figure 2.31 shows some familiar data, from the study in Example 2.12 (page 132), of the ages at which children spoke their first word and their later scores on a test of mental ability. The purpose of the study was to ascertain if age at first word x could predict the test score y. It is also plausible, however, to try to guess the age when a child first spoke from the test score. The figure displays *both* the regression line of y on x and the regression line of x on y. *These regression lines are very different:* One minimizes vertical deviations; the other minimizes horizontal deviations. Suppose, however, that we are interested in the relation or association between the two variables but do not wish to predict one from the other. We need numerical

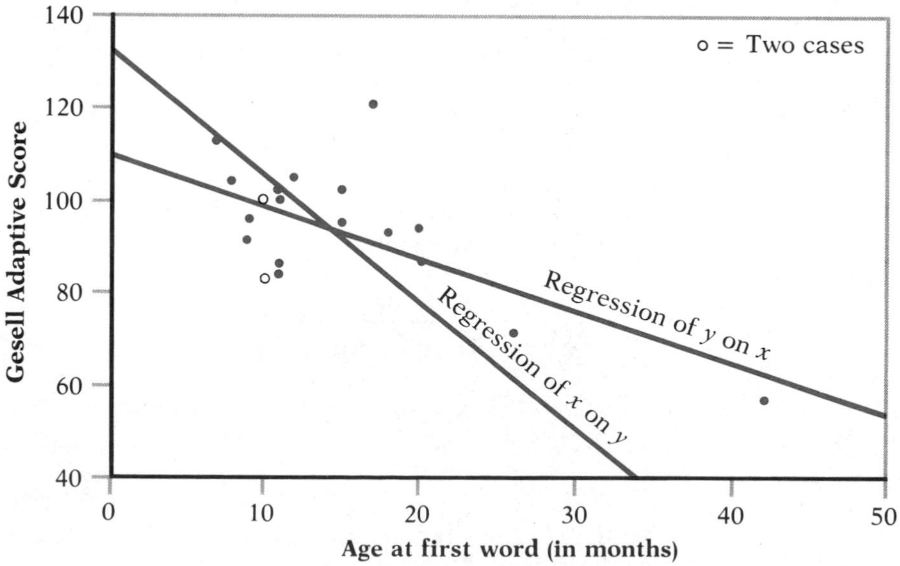

FIGURE 2.31 **The two least-squares regression lines of Gesell score on age at first word and of age at first word on Gesell score.**

measures that—unlike regression lines—do not depend on the distinction between explanatory and response variables.

A scatterplot portrays the direction, form, and strength of any relationship between two quantitative variables. But the interpretation of a scatterplot by eye is surprisingly subjective. Changing the horizontal or vertical scale, for example, greatly affects our perception of the strength of a linear or other pattern. Even the amount of white space around the point cloud in a scatterplot can fool us. As Figure 2.32 illustrates, a scatterplot appears to show a stronger relationship when the point cloud is surrounded by empty space. The two scatterplots in this figure are identical in every respect except that the lower plot is drawn smaller in a large field and therefore appears to show a stronger linear pattern.[9] We need a numerical measure to supplement the scatterplot.

The *correlation coefficient* measures the strength of the linear association between two quantitative variables. It does not distinguish explanatory from response variables and is not affected by changes in the unit of measurement of either or both variables.* The word *correlation* is often used as a vague synonym for "association." Because correlation is a specific numerical measure that applies only to linear association and only to quantitative variables, we will use the word only in this sense. There is a positive association between educational level and income in Example 2.5 (page 108), but correlation as we use the word is not meaningful because educational level in these data is a categorical variable.

correlation

Computing the correlation

We have n observations on two variables x and y, denoted by

$$(x_1, y_1), (x_2, y_2), \ldots, (x_n, y_n)$$

Unlike in the regression setting, x and y are not necessarily explanatory and response variables, though they may be. Here is the definition of the correlation coefficient.

> **CORRELATION COEFFICIENT**
>
> The correlation coefficient r for variables x and y computed from n cases is
>
> $$r = \frac{1}{n-1} \sum \left(\frac{x - \bar{x}}{s_x}\right)\left(\frac{y - \bar{y}}{s_y}\right) \qquad (2.2)$$

*Correlation and its relation to regression were also introduced by Sir Francis Galton, in 1888.

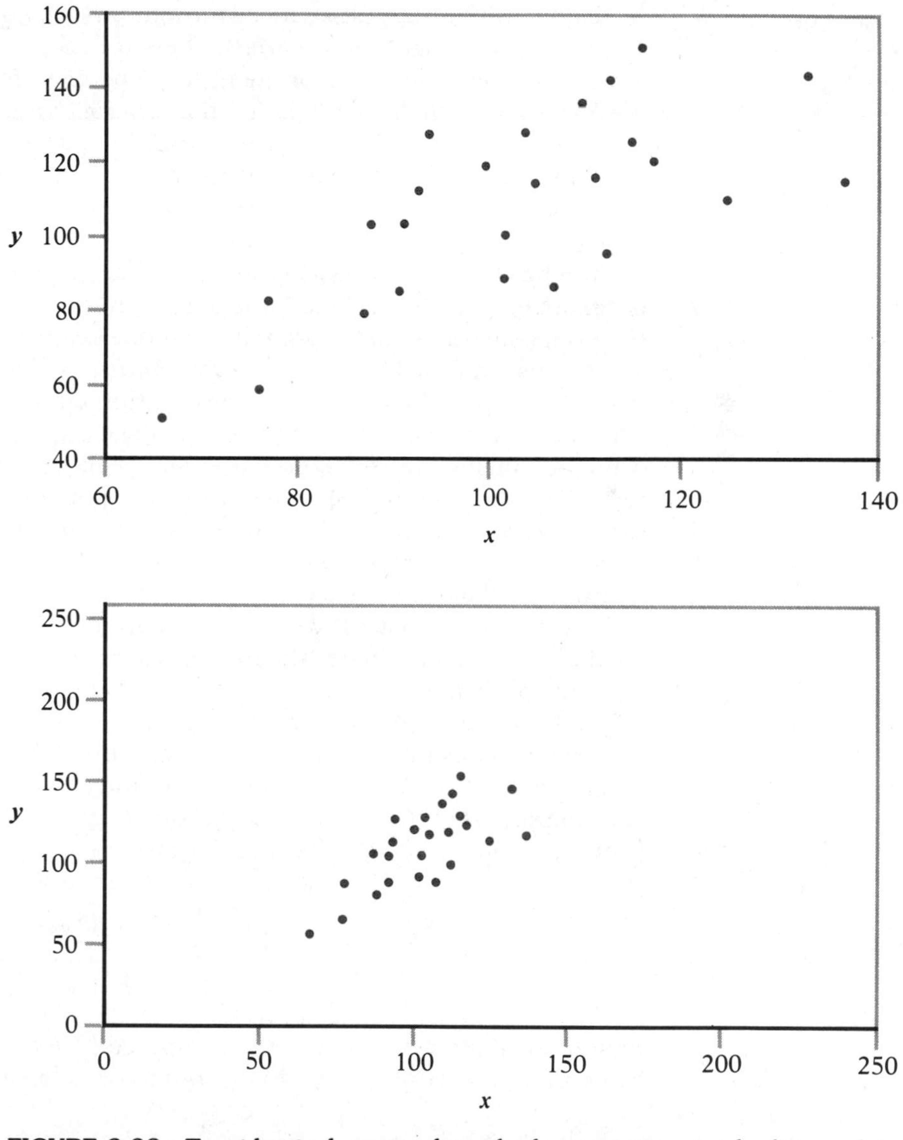

FIGURE 2.32 Two identical scatterplots; the linear pattern in the lower plot appears stronger because of the surrounding white space.

Here $\overline{x}$ and s_x are the mean and standard deviation of the x observations alone, and similarly $\overline{y}$ and s_y refer to the y observations. When we have several variables, we will regularly use subscripts on the standard deviation s to remind us which variable s refers to. As usual, the sum in Equation 2.2 runs over all of the cases for which the variables x and y have been measured.

Equation 2.2 suggests why r is a measure of association between x and y. Suppose that x and y are the height and the weight of a person. The n cases represent measurements on n people. Height and weight are positively associated, so that larger than average values of x tend to occur together with larger than average values of y. Therefore, the deviations $x - \bar{x}$ and $y - \bar{y}$ from the means will tend to either both be positive (for larger people) or both be negative (for smaller people). In both cases, the product $(x - \bar{x})(y - \bar{y})$ will be positive. Therefore r will be positive and will be larger as the positive association grows stronger. When the association is negative, on the other hand, the deviations $x - \bar{x}$ and $y - \bar{y}$ will tend to have opposite signs, so the sign of r will be negative.

The use in Equation 2.2 of the *standardized deviations* $(x - \bar{x})/s_x$ and $(y - \bar{y})/s_y$ implies that r measures association between x and y when both variables are measured in standard deviation units about the mean as origin. Changing the unit of measurement of either variable—for example, recording weight in kilograms rather than pounds—does not affect the value of r, because both variables are reduced to a standard scale before r is calculated. The properties of correlation will be explored in detail after an example that illustrates the calculation of r.

computing formula for r

Calculation is easier if we use a *computing formula* for the correlation coefficient r that eliminates the need to calculate the deviations $x - \bar{x}$ and $y - \bar{y}$ from the means. Like the earlier computing formulas Equation 1.3 (page 48) for the variance and Equation 2.1 (page 123) for the slope of the least-squares line, this formula is built up from basic sums. Because it is good practice to compute the means and standard deviations of each variable as part of the overall description, we give a form of the computing formula that uses the standard deviations:

$$r = \frac{\sum xy - \frac{1}{n}(\sum x)(\sum y)}{(n-1)s_x s_y} \tag{2.3}$$

In statistical practice, r is usually computed by a statistical calculator or by computer software rather than from a formula such as Equation 2.3.

EXAMPLE 2.17

Archaeopteryx is an extinct beast having flight feathers like a bird but teeth and a long bony tail like a reptile. Only six fossil specimens are known. Because these specimens differ greatly in size, they have sometimes been classified as different species rather than as individuals from the same species. Correlation can help decide the question. If the specimens belong to the same species and differ in size because they are at different stages of growth, there should be a strong straight-line relationship between the lengths of a pair of bones from all individuals. Outliers from this relationship would suggest a different species. Here are the lengths in millimeters of the femur (a leg bone) and the humerus (a bone in the upper arm) for the five specimens that preserve both bones.[10]

Femur	38	56	59	64	74
Humerus	41	63	70	72	84

There are $n = 5$ cases. Figure 2.33 shows strong positive linear association. This is evidence that all five specimens belong to the same species. The correlation coefficient r is a numerical measure of the strength of this linear association. To find r, first calculate the building block sums.

x	x^2	y	y^2	xy
38	1444	41	1681	1558
56	3136	63	3969	3528
59	3481	70	4900	4130
64	4096	72	5184	4608
74	5476	84	7056	6216
$\sum x = 291$	$\sum x^2 = 17,633$	$\sum y = 330$	$\sum y^2 = 22,790$	$\sum xy = 20,040$

Second, calculate the means and standard deviations of both variables. The means and standard deviations are useful in their own right as descriptions of the distributions of x and y. A calculator that gives you the mean and standard deviation from keyed-in data will speed this arithmetic. ●

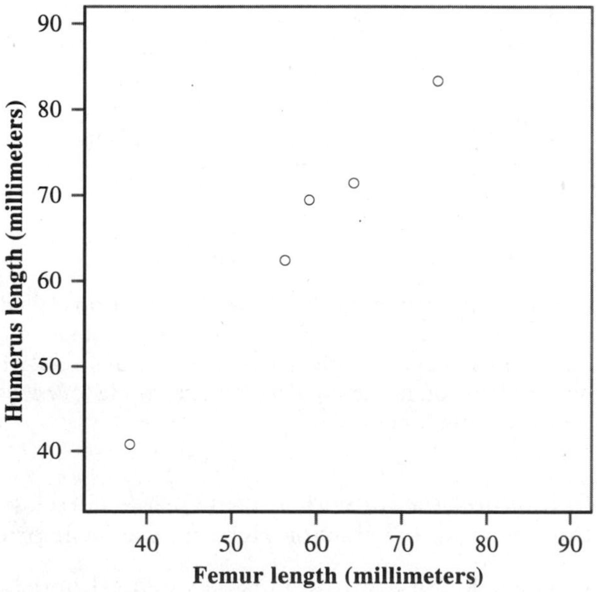

FIGURE 2.33 Humerus length versus femur length for five specimens of archaeopteryx.

$$\bar{x} = \frac{1}{n}\sum x = \frac{291}{5} = 58.2$$

$$\bar{y} = \frac{1}{n}\sum y = \frac{330}{5} = 66$$

$$s_x^2 = \frac{1}{n-1}\left[\sum x^2 - \frac{1}{n}(\sum x)^2\right]$$

$$= \frac{1}{4}\left(17,633 - \frac{291^2}{5}\right)$$

$$= \frac{1}{4}(17,633 - 16,936.2) = 174.2$$

$$s_x = \sqrt{174.2} = 13.198$$

$$s_y^2 = \frac{1}{n-1}\left[\sum y^2 - \frac{1}{n}(\sum y)^2\right]$$

$$= \frac{1}{4}\left(22,790 - \frac{330^2}{5}\right)$$

$$= \frac{1}{4}(22,790 - 21,780) = 252.5$$

$$s_y = \sqrt{252.5} = 15.890$$

Finally, substitute into the computing formula Equation 2.3 for the correlation coefficient.

$$r = \frac{\sum xy - \frac{1}{n}(\sum x)(\sum y)}{(n-1)s_x s_y}$$

$$= \frac{20,040 - \frac{1}{5}(291)(330)}{(4)(13.198)(15.890)}$$

$$= \frac{834}{838.865} = .994$$

As in other multistep calculations, the exact answer depends on how many decimal places are carried in the intervening steps. A computer or statistical calculator will generally give more accurate answers. After a few practice runs to ensure that you understand what the formula for r says, you should automate your arithmetic if possible. ●

To interpret the numerical value of the correlation coefficient r, you must understand its behavior. Here are the basic properties of r:

1. The value of r always falls between -1 and 1. Positive r indicates positive association between the variables, and negative r indicates negative association.

2. The extreme values $r = -1$ and $r = 1$ occur only in the case of perfect linear association, when the points in a scatterplot lie exactly along a straight line. Values of r close to 1 or -1 indicate that the points lie close to a straight line.
3. The value of r is not changed when the unit of measurement of x, y, or both changes. The correlation r has no unit of measurement; it is a dimensionless number between -1 and 1.
4. Correlation measures only the strength of *linear* association between two variables. Curved relationships between variables, no matter how strong, need not be reflected in the correlation.

The standardization of x and y in Equation 2.2 serves to constrain r to the range -1 to 1. The linear association increases in strength as r moves away from 0 toward either -1 or 1. The sign of r indicates only the direction of the association, so that $r = -0.7$ and $r = 0.7$ indicate linear association of the same strength but opposite directions. Understanding that r measures association in a standard scale helps avoid misinterpretation of scatterplots. Stretching or compressing the x or y scale can dramatically alter the appearance of a scatterplot, but it does not change the correlation. It is therefore not always easy to guess the value of r from visual inspection of a scatterplot. The scatterplots in Figure 2.34 illustrate how values of r closer to 1 or -1 correspond to stronger linear association.

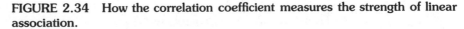

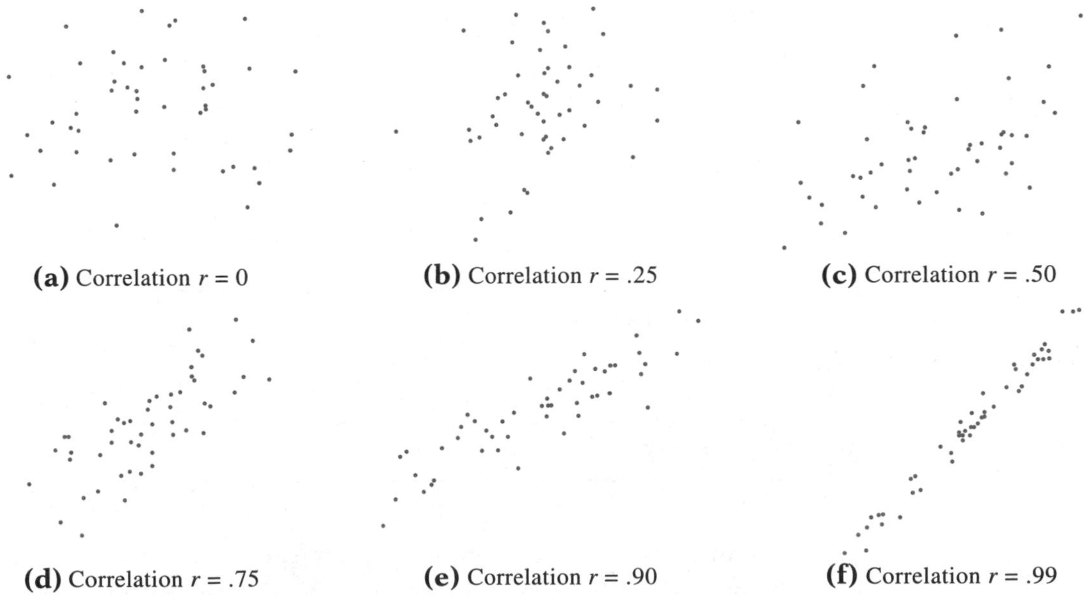

FIGURE 2.34 How the correlation coefficient measures the strength of linear association.

To make the essential meaning of r clear, the standard deviations of both variables in these plots are equal and the horizontal and vertical scales are the same. In general, it is not so easy to guess the value of r from the appearance of a scatterplot.

The nature of the correlation as a measure of linear association is also illustrated by the real data that we have examined. The very linear archaeopteryx data in Example 2.17 have correlation close to 1 ($r = 0.994$). The linear relationship between the percent of votes for Democrats in 1980 and 1984 (Figure 2.1 on page 100) is positive but less strong; the correlation is $r = 0.703$. Figure 2.3 (page 102) shows a negative association of similar strength ($r = -0.698$) between state mean SAT scores and the percent of each state's high school seniors who take the SAT. The association between birth date and draft lottery number in Figure 2.5 (page 106) is weak and slightly negative; the correlation is $r = -0.226$.

Remember that the correlation coefficient measures the strength of *linear* association only. It is possible to create examples of strong nonlinear association in which the correlation coefficient is small or even 0 (see Exercise 2.54). For example, we noted the strong nonlinear dependence of corn yield on planting rate in Example 2.3 (page 102). The correlation between these variables is $r = 0.135$, showing a very small linear association. Correlation is therefore not a general measure of all the types of association that may be visible in a scatterplot.

Correlation in the regression setting

Correlation does not require an explanatory-response relationship as regression does. The correlation coefficient r is nonetheless important in the regression setting, where y is the response to an explanatory variable x. In fact, the numerical value of r is most clearly interpreted from the following fact about regression.

> **r^2 IN REGRESSION**
>
> The square of the correlation coefficient, r^2, is the fraction of the variation in the values of y that is explained by the least-squares regression of y on x.

To understand this fact intuitively, consider again the scatterplot of the mean height y of toddlers versus their age x in months, which appears in Figure 2.35(a). The horizontal dashed line marks the mean height $\bar{y}$. Height shows considerable variation over time, as indicated by the vertical

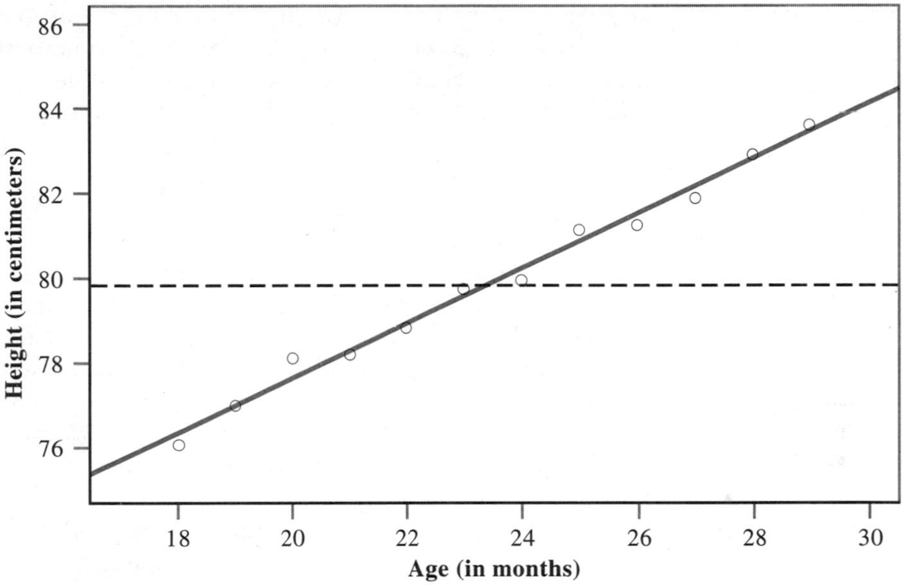

FIGURE 2.35 (a) High r^2: The variation in y about the regression line is much less than the variation in y about the mean $\bar{y}$; most of the variation in y is explained by the linear relationship of y and x.

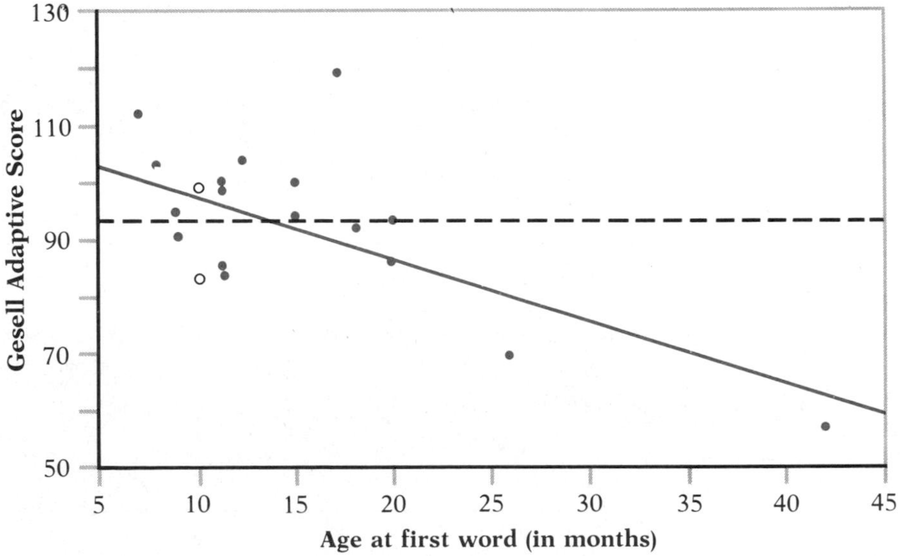

FIGURE 2.35 (b) Low r^2: The variation of y about the regression line remains large; less of the variation in y is explained by the linear relationship between y and x.

deviations of the points from the dashed line. The colored line is the least-squares regression line. The vertical deviations of the points from this line are small. That is, when x changes, y changes with it, and this linear relationship accounts for almost all of the variation in y. In this example $r = 0.994$, so that $r^2 = 0.988$, and we can say that the linear regression explains 98.8% of the observed variation in toddlers' mean heights over this age range.

On the other hand, the scatterplot of Gesell score y versus age at first word x in Figure 2.35(b) shows a weaker linear relationship. The vertical variation about the colored regression line, although smaller than the variation about the mean $\bar{y}$ (dashed line), is still large. The linear tie between x and y explains a smaller fraction of the observed variation in y. In fact, since $r = -0.640$ and $r^2 = 0.410$, we can say that 41% of the variation *in either variable* is explained by linear regression on the other variable. As Figure 2.31 illustrates, the regression lines of y on x and of x on y are quite different. But there is only a single correlation r between x and y (in either order), and r^2 helps interpret both regressions. The correlation between Gesell score and age at first word is negative, since late talkers tend to have lower aptitude scores. The interpretation of r through r^2 makes it clear that the magnitude of r, not its sign, measures the strength of a linear association. Because the goal of regression is to explain y by linear dependence on x, r^2 is a direct measure of the success of the regression and is almost always reported along with the regression results. This close connection with correlation is specifically a property of least-squares regression and is not shared by more resistant methods of fitting a line to a scatterplot.

Correlation and regression are also connected in a second way. The correlation coefficient r is closely related to the slope b of the least-squares regression line $\hat{y} = a + bx$. Some algebra based on Equations 2.1 and 2.3 establishes the following fact:

REGRESSION SLOPE

If s_x and s_y are the standard deviations of the observed x_i and y_i, and r is the correlation coefficient, then the slope of the least-squares regression line of y on x is

$$b = r\frac{s_y}{s_x} \tag{2.4}$$

Equation 2.4 is the formula for the slope b that is easiest to understand and remember. This equation says that along the regression line, *a change of one standard deviation in x corresponds to a change of r standard deviations in y*. When the variables are perfectly correlated ($r = 1$ or $r = -1$), the change in the response y is the same (in standard deviation units) as the change in x. Otherwise, since $-1 \leq r \leq 1$, the change in y is less than the change in x. As the correlation grows less strong, y moves less in response to changes in x.

It turns out that *the least-squares regression line always passes through the point* $(\bar{x}, \bar{y})$ on the graph of y versus x. So the least-squares regression line of y on x is the line with slope rs_y/s_x that passes through the point $(\bar{x}, \bar{y})$. Regression can therefore be described entirely in terms of the basic descriptive measures $\bar{x}$, s_x, $\bar{y}$, s_y and r. If both x and y are standardized variables, so that their means are 0 and their standard deviations are 1, then the regression line has slope r and passes through the origin.

EXAMPLE 2.18

In Example 2.17 we found that for the femur length x and humerus length y of archaeopteryx fossils,

$$\bar{x} = 58.2 \text{ and } s_x = 13.198$$

$$\bar{y} = 66 \text{ and } s_y = 15.890$$

$$r = .994$$

Because femur length is often used as an indicator of body size in birds, we can consider x an explanatory variable. The slope of the regression line of y on x is

$$b = .994 \frac{15.890}{13.198} = 1.197$$

The regression line passes through the point $(\bar{x}, \bar{y})$, which is (58.2, 66). Along the line, y increases by $b = 1.197$ millimeters when x increases by 1 millimeter. In terms of correlation, y increases by $r = 0.994$ standard deviation when x increases by one standard deviation. ●

Interpreting correlation and regression

Limitations of correlation and regression Correlation and regression are powerful tools for measuring the association between two variables and for expressing the dependence of one variable on the other. These tools must be used with an awareness of their limitations, beginning with the fact that they apply to *only linear* association or dependence. Also remember that *neither r nor the least-squares regression line is resistant*. One influential observation or incorrectly entered data point can greatly change these measures.

EXAMPLE 2.19

We saw in Example 2.12 (page 132) that case 18 is an influential observation in the regression of the Gesell score y on age at first word x for young children. The correlation based on all 21 children is $r = -0.640$. Because $(-0.64)^2 = 0.41$, age at first word appears to explain 41% of the variation in Gesell score among these children. But if case 18 is omitted, the correlation for the remaining 20 children falls to $r = -0.335$. Only 11% of the variation in aptitude score among these 20 children is explained by the age at which they first spoke. A decision to exclude case 18 as not belonging to the same population as the other children will weaken the study's conclusion that Gesell score can be partially predicted from age at first word. Just as calculation often adds to the information provided by a scatterplot, a plot is essential if calculation is not to be blind. Without a plot to help us spot the influential observation, numerical calculations for these data can be seriously misleading. ●

In the regression setting, points that lie apart from the other observations but (unlike case 18) close to the overall linear pattern of the other points are not called influential, because removing such a point does not greatly change the least-squares line. But removing such a point will often reduce the correlation substantially. Because regression and correlation are not resistant, give special attention to any unusual data point.

Lurking variables We have seen repeatedly that the effect of variables not included in a study can render a correlation or regression misleading. To give another example, there is a strong positive correlation over time between teachers' salaries and sales of liquor. Both increase with rising price levels and general prosperity, creating a strong association. Such correlations are sometimes called "nonsense correlations," but the corre-

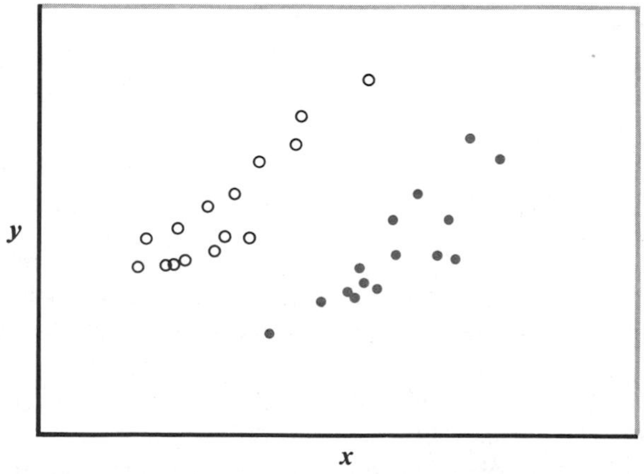

FIGURE 2.36 This scatterplot has a low r^2 even though there is a strong correlation within each of the two clusters.

lation is perfectly real. What *is* nonsense is the conclusion that because the correlation exists, teachers must be spending their salary increases on liquor. *Even a strong correlation does not imply a cause-and-effect relationship*. The question of causation is important enough to merit separate treatment in Section 2.6. For now, just remember that a correlation between two variables x and y can reflect many types of relationship between x, y, and other variables not explicitly recorded.

The effect of lurking variables can hide a true relationship between x and y as well as create an apparent relationship, as the following example illustrates.

EXAMPLE 2.20

A study of housing conditions and health in the city of Hull, England, measured a large number of variables for each of the wards in the city. (A ward is a small, relatively homogeneous geographic area.) Two of the variables were an index x of overcrowding and an index y of the lack of indoor toilets. Because x and y are both measures of inadequate housing, we expect a high correlation. In fact the correlation was only $r = 0.08$. How can this be? Investigation disclosed that some poor wards were dominated by public housing (called council housing in England). These wards had high values of x but low values of y because council housing always includes indoor toilets. Other deprived wards lacked council housing, and in these wards high values of x were accompanied by high values of y. Because the relationship between x and y differed in council and noncouncil wards, analyzing all wards together obscured the nature of the relationship.[11] ●

Figure 2.36 shows in simplified form how groups formed by a categorical lurking variable as in the housing example can make the correlation r misleading. The groups appear as clusters of points in the scatterplot. There is a strong relationship between x and y within each of the clusters; in fact, $r = 0.85$ and $r = 0.91$ in the two clusters. However, because similar values of x correspond to quite different values of y in the two clusters, x alone is of little value for predicting y. The correlation for all points displayed is therefore low; in fact, $r = 0.14$. This example is another reminder to plot the data rather than simply calculate numerical measures such as the correlation.

Prediction A regression line is often used to *predict* the response y to a given value of the explanatory variable x. This is clearly valid when the regression reflects a cause and effect relationship and r^2 is high enough to give us confidence that changes in x explain most of the variation in y. However, *successful prediction does not require a causal relationship*. If both x and y respond to the same underlying unmeasured variables, it may be possible to predict y from x even though x has no direct influence on y.

EXAMPLE 2.21

Decisions on which applicants to admit to graduate school are based on a number of criteria, such as the applicant's undergraduate grades. Another criterion is scores on the Graduate Record Examinations (GRE), national tests of both aptitude and knowledge administered by the Educational Testing Service. There is no causal relationship between the score x a college senior achieves on the GRE and the student's grade point average (GPA) y as a first-year graduate student the following year. But because both x and y respond to the student's level of ability and knowledge, it is plausible to predict y from x.

The success of this prediction depends on the strength of the association between GRE scores and later GPA in graduate school. A study[12] of a large number of first-year students at many graduate schools in many disciplines showed that for students of economics

- The correlation between the GRE verbal aptitude score and graduate GPA was $r = 0.09$.
- The correlation between the GRE quantitative aptitude score and graduate GPA was $r = 0.34$.
- The correlation between the GRE economics advanced test score and graduate GPA was $r = 0.45$.
- The correlation between undergraduate GPA and graduate GPA was $r = 0.27$.

These results show that the verbal aptitude test bears little relation to success as a graduate student of economics; but they also show that a student's score on the GRE economics advanced test is a better predictor of success than the undergraduate GPA. However, even the GRE economics test accounts for only 20% of the variation in first-year graduate GPA, because $r^2 = (0.45)^2 = 0.20$. Although prediction from a regression line makes sense in this setting, prediction based on the GRE advanced score alone will be quite unreliable. ●

For the purpose of assessing whether GRE scores are helpful in making admissions decisions, the data in Example 2.21 are incomplete in a systematic way. We have GPA information only for students who were admitted to graduate school. Because many students with low GRE scores were not admitted, the data refer primarily to students who did well on the GRE. The reported correlations describe the relation between GRE scores and graduate GPAs for students who make it to graduate school. They do not tell us whether students with GRE scores too low to allow them admission would in fact have earned low grades if they had entered graduate school.

extrapolation

Even when prediction is logically justified and r^2 is high, several additional cautions are in order. Remember the danger of *extrapolation*, the use of a regression line to predict y at values of x removed from the range of x-values used to fit the line. Most relationships remain linear only over a restricted range of x, so extrapolation can yield silly results.

Using averaged data Many regression or correlation studies work with averages or other measures that combine information from many individuals. You should note this carefully and resist the temptation to apply the results of such studies to individuals. There is, for example, a linear relationship between the natural gas consumption of a household that heats with gas and the outdoor temperature. If we regress the average gas consumption per day for a month on the average outdoor temperature for the month (measured in heating degree days), we find r^2 is over 95% for typical homes. (Exercise 2.21 on page 137 gives some actual data.) But the very high correlation observed does not apply to individual days. Prediction of gas usage from outside temperature on a single day would be much less reliable. The reason is that averaging over an entire month smooths out much of the day-to-day variation due to doors left open, house guests using more gas to heat water, and so on. Figure 2.37 shows the gas consumption and degree days for several individual days in April (A), May (M), and November (N). There is considerable variation within each month and the correlation for these individual observations would be moderate. But when we record only the three monthly averages (marked with •), the result is three of the data points from Exercise 2.21, with a correlation near 1. *Correlations based on averages are usually too high when applied to individuals*. This is another reminder that it is important to note exactly what variables were measured in a statistical study.

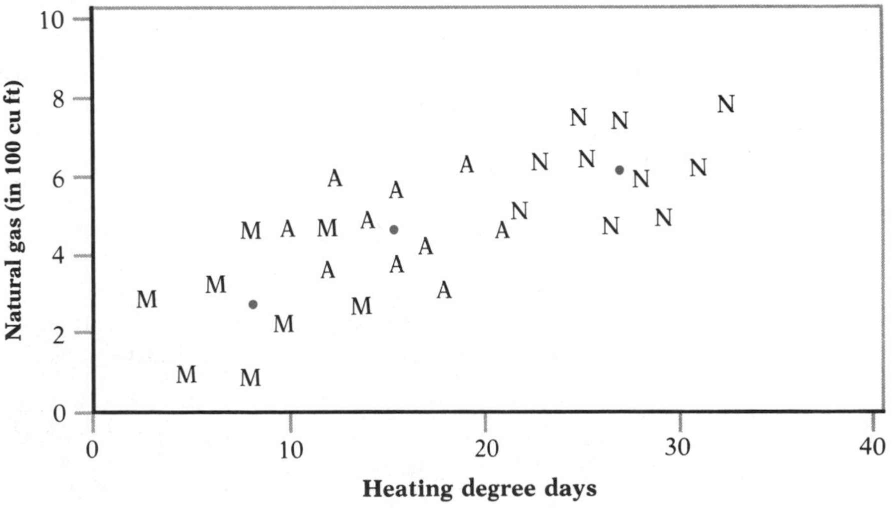

FIGURE 2.37 Natural gas consumption plotted against degree days for individual days in April (A), May (M), and November (N) and the averages of both variables in the 3 months (•). The correlation is much higher for the averaged data.

Correlation is not the whole story Finally, remember that *correlation is not a complete description of two-variable data*. The means and standard deviations of both x and y should be given along with the correlation. (Because correlation and regression make use of means and standard deviations, these measures are the proper choice to accompany a correlation.) Conclusions based on correlations alone may require rethinking in the light of a more complete description of the data.

EXAMPLE 2.22

Competitive divers are scored on their form by a panel of judges who use a scale from 1 to 10. The subjective nature of the scoring often results in controversy. We have the scores awarded by two judges, Ivan and George, on a large number of dives. How should we assess their agreement? Some computation shows that the correlation between their scores is $r = 0.9$. But the mean of Ivan's scores is 3 points lower than George's mean.

These facts do not contradict each other. They are simply different kinds of information. Ivan awards much lower scores than George, as the mean scores reveal. But because Ivan gives *every* dive a score about 3 points lower than George, the correlation remains high. Remember that adding or subtracting the same number to all values of either x or y does not change the correlation. If Ivan and George both rate several divers, the contest is consistently scored because, as the high correlation shows, Ivan and George agree on which dives are better than others. But if Ivan scores one diver and George another, we must add 3 points to Ivan's scores to arrive at a fair comparison. ●

Many statistical studies of important issues rely on correlation and regression to describe complex relationships among many variables. The examples in this section demonstrate that even when only two variables are involved, the numerical results of correlation or regression must be interpreted in the light of an understanding of both the behavior of these statistical procedures and the full background of the issue under study. The numerical work is easily automated, and you should let a computer do it. Interpretation requires an informed and skeptical human mind.

SUMMARY

The **correlation coefficient** r measures the strength and direction of the linear association between two quantitative variables x and y.

Correlation always satisfies $-1 \leq r \leq 1$, and $r = \pm 1$ only in the case of perfect linear association. The value of r is not affected by changes in the unit of measurement of either variable.

Correlation and regression are closely connected. The squared correlation coefficient r^2 is the fraction of the variance of one variable that is explained by least-squares regression on the other variable.

The least-squares regression line of y on x is the line with slope $b = rs_y/s_x$ that passes through the point $(\bar{x}, \bar{y})$.

A correlation or regression should be **interpreted with caution.** Be aware of the possible effects of lurking variables, the lack of resistance of these procedures, the danger of extrapolation, and the fact that correlations based on averages are usually too high for individuals. Remember that correlation and regression measure only linear relationships to the exclusion of other important aspects of the data.

SECTION 2.4 EXERCISES

2.47 Selective liberal arts colleges produce many graduates who go on to obtain doctorate degrees. Here are the percents of all doctorates earned between 1980 and 1989 by male and female graduates of six such colleges that were in the natural sciences. (Data compiled by Norean Radke-Sharpe, Department of Mathematics, Bowdoin College.)

College	Women	Men
Bowdoin	45	48
Carlton	38	61
Grinnell	35	47
Middlebury	36	46
Oberlin	20	34
Swarthmore	30	46

In each case, a higher percent of doctorates earned by men are in the natural sciences. We are interested in the relation between the percents of female and male doctorates that are in science.

(a) Make a scatterplot of these data. Take the female percent to be x.

(b) Describe the pattern of the data. Is there a generally linear relationship? Does any college produce an unusually high or low percent of female doctorates in science relative to the percent of male doctorates in science?

(c) Calculate the correlation coefficient r between the female and male percents. What percent of the variation among these colleges in the percent of female doctorates that are in science can be explained by straight-line dependence on the percent of male doctorates that are in science?

2.48 A student wonders if people of similar heights tend to date each other. She measures herself, her dormitory roommate, and the women in the adjoining rooms; then she measures the next man each woman dates. Here are the data (heights in inches):

Women	66	64	66	65	70	65
Men	72	68	70	68	71	65

(a) Make a scatterplot of these data. Based on the scatterplot, do you expect the correlation to be positive or negative? Near ± 1 or not?
(b) Use Equation 2.3 to compute the correlation r between the heights of the men and women.
(c) How would r change if all the men were 6 inches shorter than the heights given in the table? Does the correlation help answer the question of whether women tend to date men taller than themselves?
(d) If every woman dated a man exactly 3 inches taller than she, what would be the correlation between male and female heights?

2.49 The growth of young children is nearly linear. Here again are data on Sarah's height at several ages. (We found the regression line in Exercise 2.23 on page 138.)

Age (months)	36	48	51	54	57	60
Height (cm)	86	90	91	93	94	95

(a) From a scatterplot of height versus age, explain why you would expect the correlation to be close to 1.
(b) Compute the correlation coefficient r between height and age, using Equation 2.3.
(c) If Sarah were 4 cm taller at every age, how would the value of r change?

2.50 Compute the mean and standard deviation for the heights of the men and women in Exercise 2.48. Use your results and the correlation found in Exercise 2.48 to compute the slope of the regression line of male height on female height. What is the slope of the regression of female height on male height (when male height is on the x axis and female height on the y axis)? If both lines were drawn on the same graph, with female height on the x axis, at what point would they intersect?

2.51 Compute the mean and the standard deviation of both height and age in Exercise 2.49. Use these values and the correlation from Exercise 2.49 to find the slope of the regression line of height on age. (Compare your result with the slope you found in Exercise 2.23 on page 138.) What is the slope of the regression line of age on height?

2.52 Here are the golf scores of 11 members of a women's golf team in two rounds of college tournament play:

Player	1	2	3	4	5	6	7	8	9	10	11
Round 1	89	90	87	95	86	81	105	83	88	91	79
Round 2	94	85	89	89	81	76	89	87	91	88	80

The correlation between the round 1 and round 2 scores is $r = 0.55$. Remove player 7's scores and find the correlation for the remaining 10 players. Explain carefully why removing this single case substantially increases the correlation.

2.53 The British government conducts regular surveys of household spending. The following table shows the average weekly household spending on tobacco products and alcoholic beverages for each of the 11 regions of Britain. (Data from British official statistics, *Family Expenditure Survey*, Department of Employment, 1981.)

Region	Alcohol	Tobacco
North	£6.47	£4.03
Yorkshire	6.13	3.76
Northeast	6.19	3.77
East Midlands	4.89	3.34
West Midlands	5.63	3.47
East Anglia	4.52	2.92
Southeast	5.89	3.20
Southwest	4.79	2.71
Wales	5.27	3.53
Scotland	6.08	4.51
Northern Ireland	4.02	4.56

(a) Make a scatterplot of spending on tobacco y against spending on alcohol x.
(b) Describe the pattern of the plot. Circle the most influential observation.
(c) The correlation is only $r = 0.224$. Compute the correlation for the 10 regions omitting Northern Ireland. Explain why this r differs so greatly from the r for all 11 cases.

2.54 The gas mileage of an automobile first increases and then decreases as the speed increases. Suppose that this relationship is very regular, as shown by the following data on speed (miles per hour) and mileage (miles per gallon):

Speed	20	30	40	50	60
Mileage	24	28	30	28	24

Make a scatterplot of mileage versus speed. Show that the correlation between speed and mileage is $r = 0$. (Note that to show that $r = 0$, you need only compute the numerator in Equation 2.3.) Explain why the correlation is 0 even though there is a strong association between speed and mileage.

2.55 A college newspaper interviews a psychologist about a proposed system for rating the teaching ability of faculty members. The psychologist says, "The evidence indicates that the correlation between a faculty member's research productivity and teaching rating is close to zero." The paper reports this as, "Professor McDaniel said that good researchers tend to be poor teachers, and vice versa." Explain why the paper's report is wrong. Write a statement in plain language (don't use the word "correlation") explaining the psychologist's meaning.

2.56 Each of the following statements contains a blunder. Explain in each case what is wrong.

(a) "There is a high correlation between the sex of American workers and their income."

(b) "We found a high correlation ($r = 1.09$) between students' ratings of faculty teaching and ratings made by other faculty members."

(c) "The correlation between planting rate and yield of corn was found to be $r = 0.23$ bushel."

2.57 A study of class attendance and grades among first-year students at a state university showed that in general students who attended a higher percent of their classes earned higher grades. Class attendance explained 16% of the variation in grade index among the students studied. What is the numerical value of the correlation between percent of classes attended and grade index?

2.58 Suppose that the heights of the men in Exercise 2.48 were measured in centimeters rather than in inches, but that the heights of the women remained in inches. (There are 2.54 cm to an inch.)

(a) What would now be the correlation between male and female height? (Use information from Exercise 2.48—don't compute the new r directly.)

(b) What would be the slope of the regression line of male height on female height? (Use your calculations from Exercise 2.50—don't compute a new regression line. Hint: Use Equation 2.4 for the slope b.)

2.59 Changing the units of measurement can dramatically alter the appearance of a scatterplot. Consider the following data:

x	−4	−4	−3	3	4	4
y	.5	−.6	−.5	.5	.5	−.6

(a) Draw x and y axes, each extending from −6 to 6. Plot the data on these axes. Then plot $x^* = x/10$ against $y^* = 10y$ on the same axes using a different plotting symbol. The two plots are very different in appearance.

(b) The correlation between x and y is about $r = 0.25$. What must be the correlation between x^* and y^*?

(c) Will the regression line of y^* on x^* have the same slope as the regression line of y on x? Explain your answer. (Hint: Look at Equation 2.4 for the slope.)

2.60 Return to the scatterplot Figure 2.2 (page 100), which shows the percent of presidential votes cast for Democrats in 1980 and 1984, with the South emphasized. The correlation for all 50 states is $r = 0.703$. Would r be higher or lower if the 10 southern states were omitted? Why?

2.61 The full Minitab computer output for Example 2.11 (page 129) contains the entry `R-sq = 69.4%`. Explain what this means in this specific example, in language that can be understood by someone who knows no statistics.

2.62 A study of erosion produced the following data on the rate (in liters per second) at which water flows across a soil test bed and the weight (in kilograms) of soil washed away:

Flow rate	.31	.85	1.26	2.47	3.75
Eroded soil	.82	1.95	2.18	3.01	6.07

What percent of the variation in the amount of erosion can be explained by the fact that as the flow rate increases, erosion increases with it in a linear manner?

2.63 The mean height of American women in their early twenties is about 65.5 inches and the standard deviation is about 2.5 inches. The mean height of men the same age is about 68.5 inches, with standard deviation about 2.7 inches. If the correlation between the heights of husbands and wives is about $r = 0.5$, what is the slope of the regression line of the husband's height on the wife's height for young couples? Draw a graph of this regression line. Predict the height of the husband of a woman who is 67 inches tall.

2.64 In a large economics class, the correlation between a student's total score prior to the final examination and the final examination score is $r = 0.6$. The pre-exam totals for all students in the course have mean 280 and standard deviation 30. The final exam scores have mean 75 and standard deviation 8. The professor has lost Julie's final exam but knows that her total before the exam was 300. He decides to predict her final exam score from her pre-exam total.

(a) What is the slope of the regression line of final exam scores on pre-exam total scores in this course?

(b) Draw a graph of this regression line and use it to predict Julie's final examination score.

2.65 The price of seafood varies with species and time. The following table gives the prices in cents per pound received in 1970 and 1980 (PR70 and PR80) by fishers and vessel owners for several species:

Species	PR70	PR80
Cod	13.1	27.3
Flounder	15.3	42.4
Haddock	25.8	38.7
Menhaden	1.8	4.5
Ocean perch	4.9	23.0
Salmon, chinook	55.4	166.3
Salmon, coho	39.3	109.7
Tuna, albacore	26.7	80.1
Clams, soft	47.5	150.7
Clams, blue, hard	6.6	20.3
Lobsters, American	94.7	189.7
Oysters, eastern	61.1	131.3
Sea scallops	135.6	404.2
Shrimp	47.6	149.0

(a) Plot the data with PR70 on the x axis and PR80 on the y axis.
(b) Describe the overall pattern. Are there any points that lie away from the bulk of the data? If so, circle them on your graph. Are these unusual points outliers in the regression sense of having large residuals from a fitted line? Are they influential in the sense that removing them would change the fitted line? Or are they neither outliers nor influential?
(c) Compute the correlation for the entire set of data.
(d) What percent of the variation in 1980 prices is explained by the 1970 prices?
(e) Recompute the correlation discarding the cases that you circled in (b). Do these observations have a strong effect on the correlation? Explain why or why not.
(f) To what extent do you think the correlation provides a good measure of the relationship between the 1970 and 1980 prices for this set of data? Explain your answer.

2.66 Example 2.17 (page 164) illustrates the computation of the correlation coefficient r with a basic calculator. The result of that and similar calculations will vary as you carry more or fewer significant digits in the intermediate steps. Redo the calculation of Example 2.17 using the building block sums given there but rounding these sums and each later calculation to the nearest whole number. Is the resulting value of r reasonable?

2.67 There is a strong positive correlation between years of education and income for economists employed by business firms. (In particular, economists with doctorates earn more than economists with only a bachelor's degree.) There is also a strong positive correlation between years of education and income for economists employed by colleges and universities. But when all economists are considered, there is a *negative* correlation between education and income. The explanation for this is that business pays high salaries and employs mostly economists with bachelor's degrees, while colleges pay lower salaries and employ mostly economists with doctorates. Sketch a scatterplot with two groups of cases (business and academic) to illustrate how a strong positive correlation within each group and a negative overall correlation can occur together. (Hint: Begin by studying Figure 2.36.)

2.68 The data in Exercise 2.27 (page 139) give the average stride rate for a group of 21 elite female runners at each of several running speeds. There is a high positive correlation between average stride rate and speed. Suppose that you had complete data, recording stride rates for each runner separately at each speed. If you plotted each individual observation on stride rate and speed and computed the correlation, would you expect the correlation to be lower than, about the same as, or higher than the correlation for the published data? Sketch a scatterplot of stride rate versus speed for individual runners to illustrate your answer.

2.5 RELATIONS IN CATEGORICAL DATA

Up to this point we have focused on relationships between quantitative variables, although categorical variables played an important role in Section 2.1. Now we will shift to describing relationships between two or more categorical variables. Some variables—such as sex, race, and occupation—are inherently categorical. In other cases, categorical variables are created by grouping values of a quantitative variable into classes. Published data are often reported in this form to save space. Analysis of categorical data is based on the counts or percents of the cases that fall into various categories.

EXAMPLE 2.23

two-way table

Table 2.6 presents Census Bureau data on the educational attainment of Americans of different ages.[13] Because many persons under 25 years of age have not completed their education, they are not included in the table. Both variables, age and education, have been grouped into categories. The entries in this *two-way table* are the frequencies, or counts, of persons in each age by education class. Although both age and education as presented in this table are categorical variables, both have a natural order from least to most. The order of the rows and the columns in Table 2.6 reflects the order of the categories. ●

TABLE 2.6 Educational attainment by age, 1988 (thousands of persons)

	Age group					
Education	25–34	35–44	45–54	55–64	≥ 65	Total
Did not complete high school	5836	4841	5230	7024	13,183	36,114
Completed high school	17,889	13,200	9860	8580	9412	58,921
College, 1–3 years	9069	7309	3698	2793	2915	25,784
College, 4 or more years	10,174	9332	5008	3246	3018	30,781
Total	42,968	34,682	23,796	21,643	28,528	151,616

Analyzing two-way tables

How can we best grasp the information contained in Table 2.6 about the educational attainment of Americans? Notice first the abundance of information in a two-way table. The "Total" column at the right of the table gives the distribution of all Americans over 25 years of age by education. The "Total" row at the bottom gives their distribution by age. The bottom row contains the column totals, and the rightmost column contains the row totals. The distributions of both variables can therefore be calculated from the body of the table. These are often called the *marginal distributions* because they appear at the bottom and right margins of the table. If the row and column totals are missing, the first thing to do in studying a two-way table is to calculate the marginal distributions by summing each row and each column.

marginal distributions

The row and column sums in Table 2.6 show some discrepancies. The sum of the "College, 4 or more years" row, for example, is 30,778. But the total is given in the table as 30,781. The explanation is *roundoff error:* The table entries are in *thousands* of persons, and each entry is rounded to the nearest thousand. The Census Bureau obtained the row total of persons with at least 4 years of college by rounding the exact number of such persons to the nearest thousand: The result was 30,781,000. Adding the entries in that row, each of which is already rounded, gives a slightly different result.

roundoff error

Table 2.6 contains much more information than the two marginal distributions of age alone and education alone. The nature of the relationship between age and education cannot be deduced from the separate distributions, but requires the full table. *Relationships among categorical variables are described by calculating appropriate percentages from the counts given.*

EXAMPLE 2.24

We are interested in the association between age and college education among adult Americans. Table 2.6 shows that 10,174,000 persons aged 25 to 34 have completed college, but only 3,246,000 persons in the 55 to 64 age group have done so. These counts do not accurately describe the association, however, because there are nearly twice as many people in the younger age group. We must instead compare the *percent* in each age group who have finished college.

For persons aged 25 to 34, the percent who have completed college is the number of persons in this age group who completed college as a percent of the total number of 25- to 34-year-olds (the column total). This percent is

$$\frac{10,174}{42,968} = .237 = 23.7\%$$

Similar calculations for each age group give

Age group	25–34	35–44	45–54	55–64	≥ 65
Percent with 4 years of college	23.7	26.9	21.0	15.0	10.6

These percentages make it clear that a college education is much more common among younger Americans.

For the over-25 population as a whole, the percent who completed college is

$$\frac{30,781}{151,616} = .2030 = 20.3\%$$

This is the row total who completed college as a percent of the table total, which appears at the lower right corner of the table. ●

bar graph

If you are presenting the information in Example 2.24 to a general audience, a *bar graph* like Figure 2.38 may be helpful. Each bar represents one age group. The height of the bar is the percent of that age group with at least 4 years of college. Although bar graphs are somewhat similar to histograms, their details and uses are distinct. A histogram shows the distribution of frequencies or relative frequencies among the values of a single variable, while a bar graph compares the size of different items. The horizontal axis of a bar graph need not have any measurement scale but may simply identify the items being compared. The items compared in Figure 2.38 are the five age groups. Because each bar in a bar graph describes a different item, the bars are usually drawn with space between them.

If we wish to compare the complete distributions of education in the five age groups, we must go farther. Consider the 25 to 34 age group, represented by the first column in Table 2.6. The percent distribution of educational level within this age group is found by computing each count as a percent of the column total. The results are as follows:

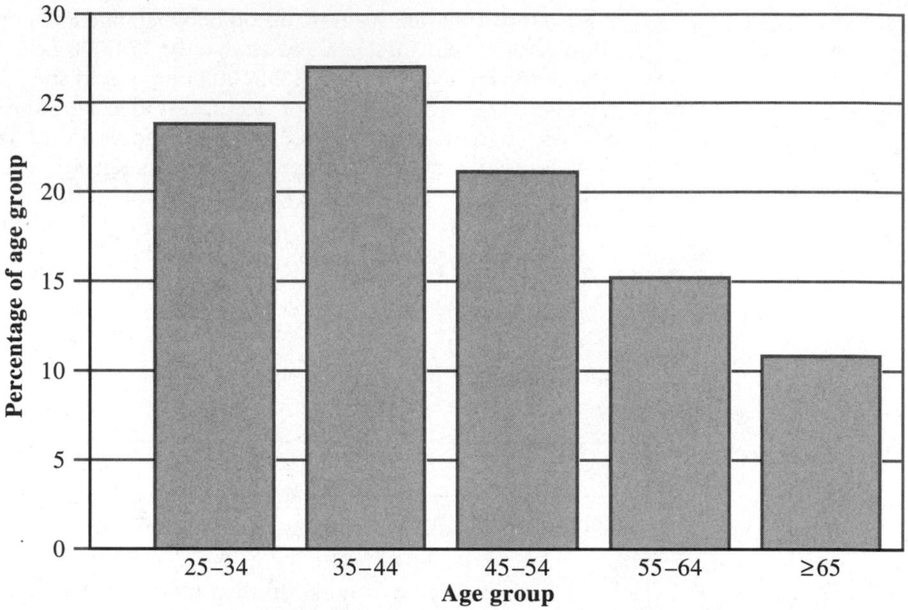

FIGURE 2.38 A bar graph comparing the percents in five age groups who have completed college (Example 2.24).

Education	1–3 years of high school	4 years of high school	1–3 years of college	≥4 years of college
Percent	13.6	41.6	21.1	23.7

conditional distribution

These percents add to 100%, since all 25- to 34-year-olds fall in one of the educational categories. The four percents together are the *conditional distribution* of education, given that a person is 25 to 34 years of age. The term "conditional" is used because the distribution refers only to people who satisfy the condition that they be 25 to 34 years old.

segmented bar graph

The conditional distributions of education in the different age groups differ from one another. They also differ from the marginal (overall) distribution of education computed from the rightmost column in Table 2.6. The influence of age on education is seen by comparing the five conditional distributions. Figure 2.39 does this graphically by means of a *segmented bar graph*, which compares all of the age groups in Table 2.6 as follows:

1. Each bar describes one age group, that is, one column in Table 2.6. The segments indicate the breakdown of that age group by educational attainment.

2. Each bar has height 100%. The division of the bar into segments shows what percent of that age group falls into each educational category.
3. Because both age and education have a clear order from least to most, the bars are arranged in order of increasing level of age from left to right; the segments in each bar are arranged in order of increasing level of education from bottom to top.

Figure 2.39 presents the relationship between age and education in visual form. We can see, for example, that the percent of Americans who did not complete high school (the bottom segment in each bar) rises rapidly with age. The percent who have completed college (the top segment) remains stable until over 44 years old, and then declines. These two comparisons are clear-cut because the segments compared all have a common starting point, either at the bottom or at the top of the bars. In order to compare the percent in each age group who completed high school, we must compare the lengths of segments that both start and end at different points, a task that our eyes do less well. Segmented bar graphs do not present information as vividly as most of the other graphs used in statistics, primarily because of the large amount of information given in a

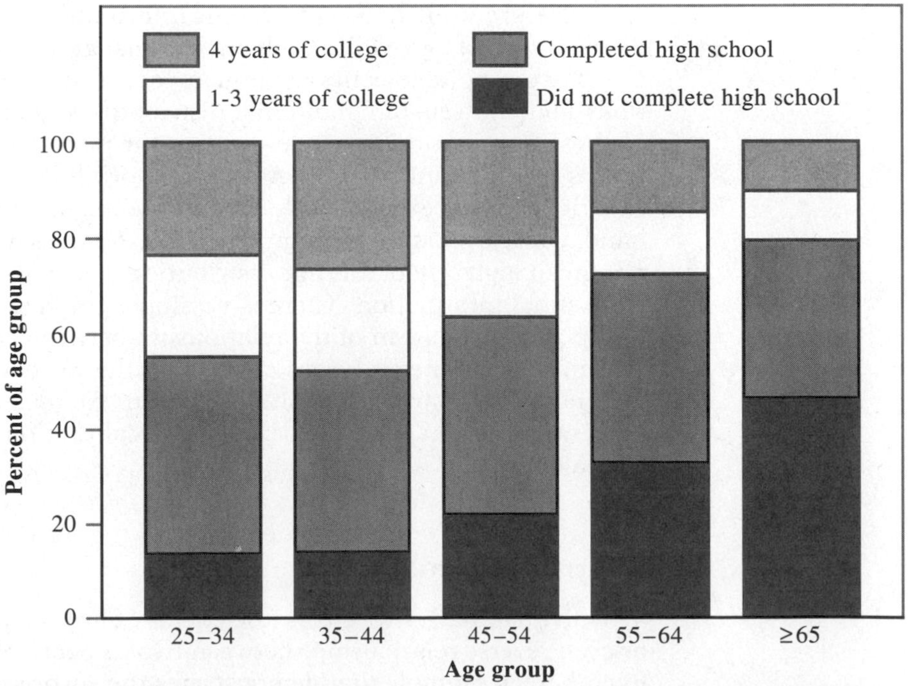

FIGURE 2.39 A segmented bar graph comparing the distributions of education level within five age groups.

two-way table.[14] The graph does not present all of the information in the table. For example, lengths in Figure 2.39 represent percents, not counts. This choice results in a clearer comparison of the five distributions of education within each age group. Table 2.6 shows the actual number of people in each age group by educational category, but Figure 2.39 does not present this information.

In calculating the percentages and constructing the segmented bar graph, we concentrated on the distribution of educational attainment in each age group and the way this distribution changes with age. That is, we regarded age as the explanatory variable and education as the response variable. We therefore considered each table entry as a part of its column total, and so obtained the conditional distribution of education for each age. We might also be interested in the distribution of age among persons having a certain level of education, that is, in the conditional distribution of age given education. Each table entry would then be considered as part of the corresponding *row total*. For example, the percentage of persons 65 years or older among those who did not complete high school is

$$\frac{13,183}{36,114} = .3650 = 36.5\%$$

Figure 2.39 summarizes the conditional distributions of education within each age group. It does not present information about the conditional distributions of age within each educational group.

A two-way table contains, in compact form, two marginal distributions and two sets of conditional distributions. You must decide which of these distributions are relevant to your purpose. If you are studying trends in the training of the American work force, conditioning on age reveals the more extensive education of younger people. If, on the other hand, you are planning a program to improve the skills of people who did not finish high school, the age distribution within this educational group is essential information. There is no single graph (such as a scatterplot) that portrays the form of the relationship between categorical variables, and no single numerical measure (such as the correlation coefficient) that summarizes the strength of an association. We must rely on well-chosen percentages, perhaps summarized graphically, or on more advanced statistical methods.[15]

Simpson's paradox

As with quantitative variables, the effects of lurking variables can change or even reverse relationships between two categorical variables. Here is a hypothetical example that demonstrates the surprises that can await the unsuspecting consumer of data.

EXAMPLE 2.25

In an attempt to help consumers make informed decisions about health care, the government releases data about patient outcomes in a large number of hospitals. You are interested in comparing Hospital A and Hospital B, which serve your community. Here are the data on the survival of patients after surgery in these two hospitals. All patients undergoing surgery in a recent time period are included; "survived" means that the patient lived at least 6 weeks following surgery.

	Hospital A	Hospital B
Died	63	16
Survived	2037	784
Total	2100	800

The evidence seems clear: Hospital A loses 3% (63/2100) of its surgery patients, whereas Hospital B loses only 2% (16/800). It seems that you should choose Hospital B when you next need surgery.

Not all surgery cases are equally serious, however. Later in the government report you find data on the outcome of surgery broken down by the condition of the patient before the operation. Patients are classified as being in either "poor" or "good" condition. Here are the more detailed data.

	Good condition			Poor condition	
	Hospital A	Hospital B		Hospital A	Hospital B
Died	6	8	Died	57	8
Survived	594	592	Survived	1443	192
Total	600	600	Total	1500	200

Aha! Hospital A beats Hospital B for patients in good condition: Only 1% (6/600) died in Hospital A, compared with 1.3% (8/600) in Hospital B. And Hospital A wins again for patients in poor condition, losing 3.8% (57/1500) to Hospital B's 4% (8/200). So Hospital A is safer both for patients in good condition and patients in poor condition. If you are facing surgery, you should choose Hospital A. ●

The result of Example 2.25 is surprising, and perhaps disturbing: Hospital A has a higher overall rate of patient deaths from surgery even though A has a lower death rate than B for each class of patients considered separately. This phenomenon is called Simpson's paradox.

SIMPSON'S PARADOX

Simpson's paradox refers to the reversal of the direction of a comparison or an association when data from several groups are combined to form a single group.

Although the apparent paradox is unexpected, a look at the data makes the explanation clear. Hospital A is a university medical center that attracts seriously ill patients from a wide region, and Hospital B is a local hospital that handles relatively few major surgery cases. Hospital A operated on 1500 patients in poor condition, but Hospital B saw only 200 such cases. Because patients in poor condition are more likely to die, Hospital A has a higher death rate despite its superior performance on each class of patients. The original two-way table, which did not take account of the condition of the patients, was misleading. The death rates compiled and released by the Federal Health Care Financing Agency do allow for the patient's condition and diagnosis. Hospitals continue to complain that there are differences among patients that can make the government's mortality data misleading.

three-way table

The table of outcome by hospital by patient condition in Example 2.25 is a *three-way table* that reports the frequencies of each combination of levels of three categorical variables. As the example illustrates, a three-way table can be presented as several two-way tables, one for each level of the third variable. In Example 2.25, there is a separate table of patient outcome by hospital for each patient condition. We can obtain the original two-way table by adding the corresponding entries in these tables, that is, by *aggregating* the data over patient conditions. Aggregating has the effect of ignoring patient condition, which then becomes a lurking variable. The peril of drawing incorrect conclusions from rates based on aggregated categorical data is so great—and so little understood—that special caution is required. Here are some real-life examples of this phenomenon.

- In 1979 the National Science Foundation conducted a study of persons who received a degree in science or engineering in 1977/78. The study found that at the bachelor's degree level the average salary of women with full-time jobs was only 77% of the average male salary. However, women's salaries in every individual field of science and engineering were at least 92% of male salaries. The aggregate data suggested a wide sex difference in salaries because women were concentrated in the life and social sciences, where salaries are lower than in the physical sciences and engineering.[16]
- Comparisons of the actual rate of federal income tax paid by individuals in different years often show that the average tax rate

has increased with time, even though the rate in every income category has decreased. Why? Because inflation has moved more taxpayers into higher income brackets, which are taxed at higher rates.[17]

- At one point during World War II, the percentage of women among U.S. industrial workers was rising in every individual industry but falling overall. Total employment was rising most rapidly in heavy industry, where few women were employed, so male employment rose faster even though women were gaining in every industry.[18]

The effects of lurking variables are often profound; sometimes they may even reverse the direction of an association based on aggregated data. How far to go in examining unreported variables is a matter of judgment, and even a matter for controversy. A charge of employment discrimination based on data showing adverse treatment of women or minorities, for example, will often be met by a defense claiming that other variables such as education and work experience explain the apparent discrimination. As Example 2.25 illustrates, the question is certainly not settled by first appearances. Careful examination of all relevant variables is needed to uncover the truth.

SUMMARY

A **two-way table** of counts or percents describes the relationship between two categorical variables. Two-way tables are often used to summarize large amounts of information by grouping outcomes into categories.

A two-way table allows us to calculate the **marginal distribution** of each variable alone from the row sums and column sums. It also allows us to obtain the **conditional distribution** of one variable given a specific level of the other by considering table entries as proportions of their row or column sums. A **segmented bar graph** provides a visual comparison for a set of conditional distributions.

Relationships among three categorical variables are described by a **three-way table,** which is printed as separate two-way tables for each level of the third variable. A comparison between two variables that holds for each level of a third variable can be changed or even reversed when the data are aggregated by summing over all levels of the third variable. **Simpson's paradox** refers to the reversal of a comparison by aggregation.

SECTION 2.5 EXERCISES

2.69 Business school researchers interested in the effect of company size on management practices conducted a survey of companies in their state. They classified companies as small, medium, or large and sent questionnaires to 200 randomly selected businesses of each size, for a total of 600 questionnaires. Not all the questionnaires were returned, so the survey team examined whether or not the response rate varied with the size of the business. The data are given in the following two-way table:

Size	Response	No response	Total
Small	125	75	200
Medium	81	119	200
Large	40	160	200

(a) Compute the response percentage for all 600 companies.
(b) Compute the response percentage for each of the three sizes of companies.
(c) Draw a bar graph that compares the response percentage for the three sizes of companies.
(d) Using the total number of responses as a base, compute the percent of responses that come from each of small, medium, and large businesses.
(e) The sampling plan was designed to obtain equal numbers of responses from small, medium, and large companies. In preparing an analysis of the survey results, do you think it would be reasonable to proceed as if the responses represented companies of each size equally?

2.70 A survey of commercial poultry-raising operations assessed the extent to which rodents are a problem in poultry houses. The operations surveyed were classified into two types (egg and turkey), and the extent of rodent infestation was classified as mild, moderate, or severe. The numbers of responses in each category are given in the following two-way table. (Data from R. M. Corrigan and R. E. Williams, "The house mouse in poultry operations: Pest significance and a novel baiting strategy for its control," *Proceedings of the Twelfth Vertebrate Pest Conference*, 1986, pp. 120–126.)

Type	Mild	Moderate	Severe
Egg	34	33	7
Turkey	22	22	4

(a) Give the overall (marginal) percentages of mild, moderate, and severe infestations.
(b) Give the percentages of each level of infestation for egg and for turkey operations separately. (These are the conditional distributions of the level of infestation, given the type of operation.)
(c) Construct a segmented bar graph showing how the extent of rodent infestation varies with the type of operation.

(d) Do you think it is reasonable to ignore the type of operation in discussing rodent problems in poultry houses? Explain your answer.

2.71 The following two-way table shows the relationship between the age of American civilians and their participation in the 1988 presidential election. Each column refers to an age group and contains the percent of that age group who voted in 1988, the percent who registered but did not vote, and the percent who did not register. That is, the percents in each column add to 100% of an age group.

	Age group				
Voting status	18–20	21–24	25–44	45–64	≥ 65
Voted	33.2	38.3	54.0	67.9	68.8
Registered only	11.7	12.3	9.0	7.6	9.6
Not registred	55.1	49.4	37.0	24.5	21.6

(a) Display these data in a segmented bar graph that compares the voting behavior of the five age groups.
(b) Based on your graph, write a brief description of the relationship between voting behavior and age.
(c) If you were planning a political campaign, you would be interested in the percent of each age group among registered voters. Can that percent be computed from the data in the table? If so, do it. If not, what information is lacking?

2.72 The following two-way table shows the relationship between the income of American households in 1987 and the educational level of the householder. The educational categories are the same as in Figure 2.39. Household income is recorded in thousands of dollars. The table entries are the counts of households (in thousands) in each education by income class.

	Income class				
Education	<15	15–34	35–49	≥ 50	Total
<4 years of high school	11,668	7217	1909	1180	21,976
4 years of high school	8088	12,417	5776	4279	30,561
1–3 years of college	2626	5263	3230	3173	14,294
≥ 4 years of college	1597	5189	4334	7888	19,007

(a) Find the sum of the counts in the first row of the table. Why do you think that this sum differs slightly from the total given for that row?
(b) Compute (in percents) the conditional distribution of income among persons who did not complete high school.

(c) Compute the conditional distribution of income in each of the remaining three educational classes.

(d) Make a segmented bar graph that shows how income varies with educational level.

(e) Write a brief description of the relationship between household income and education.

2.73 You are studying low-income households, defined as those with income less than $15,000. Use the data in the previous exercise to compute the distribution (in percents) of educational levels in low-income households. (This is the conditional distribution of education, given that the household has low income.) Describe your results in words.

2.74 Use the data in Exercise 2.72 to compute (in percents) the distribution of householder education among all households. Then compute the distribution of income among all households. (These are the two marginal distributions.) Which income class included the greatest number of American households in 1987?

2.75 In a study of the link between high blood pressure and cardiovascular disease, a group of white males aged 35 to 64 was followed for 5 years. At the beginning of the study, each man had either "low" systolic blood pressure (less than 140 mm Hg) or "high" blood pressure (140 mm Hg or higher). The following table gives the number of men in each blood pressure category and the number of deaths from cardiovascular disease during the 5-year period. (Data taken from J. Stamler, "The mass treatment of hypertensive disease: Defining the problem," *Mild Hypertension: To Treat or Not to Treat*, New York Academy of Sciences, 1978, pp. 333–358.)

Blood pressure	Deaths	Total
Low	21	2676
High	55	3338

(a) Calculate the mortality rate (deaths as a fraction of the total) for each group of men.

(b) Do these data support the idea that there is a link between high blood pressure and death from cardiovascular disease? Explain your answer.

2.76 The following two-way table categorizes suicides committed in 1988 by the sex of the victim and the method used ("hanging" also includes strangulation and suffocation). Based on these data, write a brief account of differences in suicide between males and females. Be sure to cite appropriate counts or percents to justify your statements.

Method	Male	Female
Firearms	15,656	2513
Poison	3403	2422
Hanging	3588	787
Other	1431	607
Total	24,078	6329

2.77 Upper Wabash Tech has two professional schools: business and law. Here is a three-way table of applicants to these professional schools, categorized by sex, school, and admission decision. (Although these data are fictitious, similar, though less simple, situations occur in reality. See P. J. Bickel and J. W. O'Connell, "Is there a sex bias in graduate admissions?" *Science*, 187 (1975), pp. 398–404.)

	Business			Law	
	Admit	Deny		Admit	Deny
Male	480	120	Male	10	90
Female	180	20	Female	100	200

(a) Make a two-way table of sex by admission decision for the combined professional schools by summing entries in the three-way table.
(b) Compute separately the percents of male and female applicants admitted from your two-way table. Upper Wabash Tech's professional schools admit a higher percent of male applicants than of female applicants.
(c) Now compute separately the percents of male and female applicants admitted by the business school and by the law school. Each school admits a higher percent of female applicants.
(d) Explain carefully, as if speaking to a skeptical reporter, how it can happen that Upper Wabash appears to favor males when each school individually favors females.

2.78 In 1987 Reggie Jackson retired from baseball after a career of 21 years in the major leagues. He had a reputation for playing particularly well in crucial situations such as World Series games. The following table gives the numbers of Reggie's times at bat and hits for regular-season play and World Series games.

	Hits	At bats
Regular season	2584	9864
World Series	35	98

(a) Compute Reggie's batting average for all games combined. (A player's batting average is the proportion of times that a player gets a hit, that is, "Hits" divided by "At bats.")
(b) Compute Reggie's batting average for regular-season and World Series games separately. Do the results support the idea that he played better than usual in World Series games?

2.79 A study of the salaries of full professors at Upper Wabash Tech shows that the median salary for female professors is considerably less than the median male salary. Further investigation shows that the median salaries for male and female full professors are about the same in every department (English, physics, etc.) of the university. Explain how equal salaries in every department can still result in a higher overall median salary for men.

2.80 Most baseball hitters perform differently against right-handed and left-handed pitching. Consider two players, Joe and Moe, both of whom bat right-handed. The table below records their performance against right-handed and left-handed pitchers:

Player	Pitcher	Hits	At bats
Joe	Right	40	100
	Left	80	400
Moe	Right	120	400
	Left	10	100

(a) Construct a two-way table of player (Joe or Moe) versus outcome (hit or no hit) summed over the different types of pitching.
(b) Compute the overall batting average (hits divided by total times at bat) for each player. On the basis of these calculations, who is the better hitter?
(c) For each of the two types of pitching, construct a two-way table of player by outcome (hit or no hit). When arranged side by side, these two two-way tables form a three-way table of player by pitcher type by outcome.
(d) Using the table in (c) compute batting averages to compare the two players for each type of pitching. Who is the better hitter against right-handed pitching? Who is the better hitter against left-handed pitching?
(e) Summarize the results of your calculations and explain them as if you were speaking to a skeptical manager. Which hitter would you prefer to have on your team?

2.81 The influence of race on imposition of the death penalty for murder has been much studied and contested in the courts. The following three-way table classifies 326 cases in which the defendant was convicted of murder. The three variables are the defendant's race, the victim's race, and whether the defendant was sentenced to death. (Data from M. Radelet, "Racial characteristics and imposition of the death penalty," *American Sociological Review*, 46 (1981), pp. 918–927.)

	White defendant: Death penalty			Black defendant: Death penalty	
	Yes	No		Yes	No
White victim	19	132	White victim	11	52
Black victim	0	9	Black victim	6	97

(a) From these data construct a two-way table of defendant's race by death penalty.

(b) Show that Simpson's paradox holds: A higher percent of white defendants are sentenced to death overall, but for both black and white victims a higher percent of black defendants are sentenced to death.

(c) Basing your reasoning on the data, explain why the paradox holds in language that a judge could understand.

2.6 THE QUESTION OF CAUSATION

In many studies of the relationship between two variables, the goal is to establish that changes in the explanatory variable *cause* changes in the response variable. Even when a strong association is present, the conclusion that this association is due to a causal link between the variables is often elusive. Traffic fatalities dropped after the United States adopted a 55-mph speed limit in 1974—but did the reduced speed limit *cause* the decline in road deaths? Nations with strict gun controls generally have much lower murder rates than does the United States—but do gun controls *cause* a drop in murders? Such questions are difficult to answer because many variables other than the proposed explanatory variable influence the response. The reduced highway speed limit coincided with a manyfold increase in gasoline prices which discouraged driving and so lessened our exposure to road accidents. Nations differ profoundly in so many ways that isolating the effect of one factor such as gun control is probably impossible. In this section we will explore the issues involved in establishing causation in the context of an important example, the connection between cigarette smoking and death from lung cancer.

Smoking and lung cancer

The relationship between smoking and lung cancer has been observed and contested for over 50 years. Here is an example of the connection that has been observed.

EXAMPLE 2.26

Table 2.7 summarizes a study carried out by government statisticians in England.[19] The data concern 25 occupational groups and are condensed from data on thousands of individual men. The explanatory variable x is a smoking ratio—that is, a measure of the number of cigarettes smoked per day by men in each occupation relative to the number smoked by all men of the same age. This smoking ratio is 100 if men in an occupation are exactly average in their smoking, it is below 100 if they smoke less than average, and above 100 if they smoke more than average. The response variable y is the standardized mortality ratio for deaths from lung cancer. It, too, is measured relative to the

TABLE 2.7 Indexes of cigarette smoking and lung cancer mortality for English males in several occupational groups

Occupational group	Smoking	Mortality
Farmers, foresters, fishermen	77	84
Miners and quarrymen	137	116
Gas, coke, and chemical makers	117	123
Glass and ceramics makers	94	128
Furnace, forge, foundry, rolling mill workers	116	155
Electrical and electronic workers	102	101
Engineering and allied trades	111	118
Woodworkers	93	113
Leather workers	88	104
Textile workers	102	88
Clothing workers	91	104
Food, drink, and tobacco workers	104	129
Paper and printing workers	107	86
Makers of other products	112	96
Construction workers	113	144
Painters and decorators	110	139
Drivers of stationary engines, cranes, etc.	125	113
Laborers not included elsewhere	133	146
Transport and communications workers	115	128
Warehousemen, storekeepers, packers, bottlers	105	115
Clerical workers	87	79
Sales workers	91	85
Service, sport, and recreation workers	100	120
Administrators and managers	76	60
Professional, technical workers, artists	66	51

entire population of men of the same ages as those studied and is greater or less than 100 when there are more or fewer deaths from lung cancer than would be expected based on the experience of all English men.

A scatterplot (Figure 2.40) of the data in Table 2.7 shows a moderately strong linear association. The correlation is $r = 0.716$. Because $r^2 = 0.51$, differences in smoking behavior among occupations explain 51% of the variation in deaths from lung cancer. The least-squares regression line also appears in Figure 2.40. The residual plot in Figure 2.41 may show a slight curved pattern, but this is by no means strong enough to lead us to abandon the linear relationship. There are no obvious outliers or influential observations. The data demonstrate a clear connection between smoking and death from lung cancer. How should this connection be interpreted? ●

Because the data of Example 2.26 do not refer to individuals but are averaged over occupational groups, the correlation may be higher than would be the case for individuals (recall Figure 2.37 on page 175). Many other studies, however, have shown a strong tie between cigarette smoking and the risk of death from lung cancer. The most convincing are *prospective studies* in which smokers and nonsmokers are identified and then followed for many years to observe their medical history and eventual cause of death. Prospective studies have consistently found that heavy smokers are 20 or more times as likely to contract lung cancer as nonsmokers.

prospective study

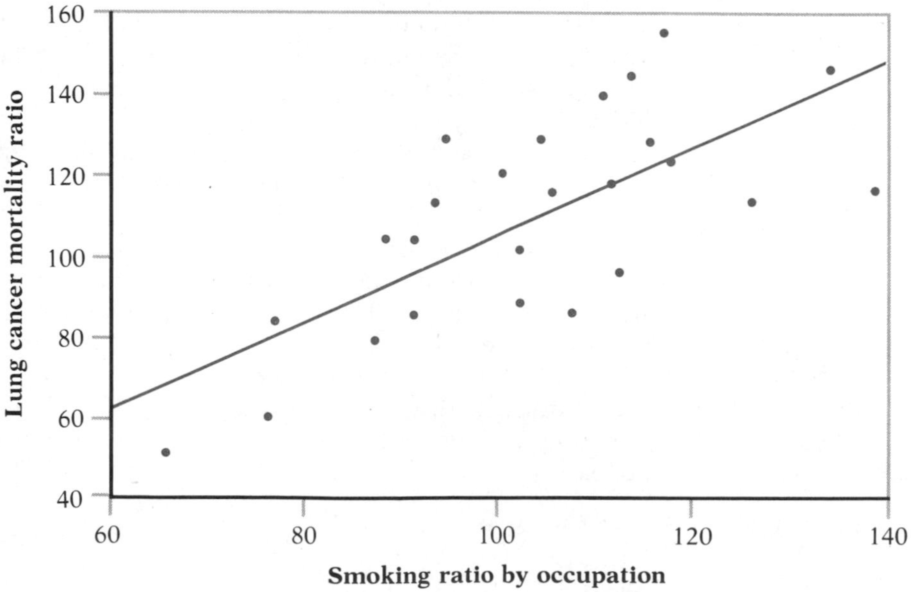

FIGURE 2.40 **The regression of a measure of deaths from lung cancer on a measure of smoking for British occupational groups, from Table 2.7.**

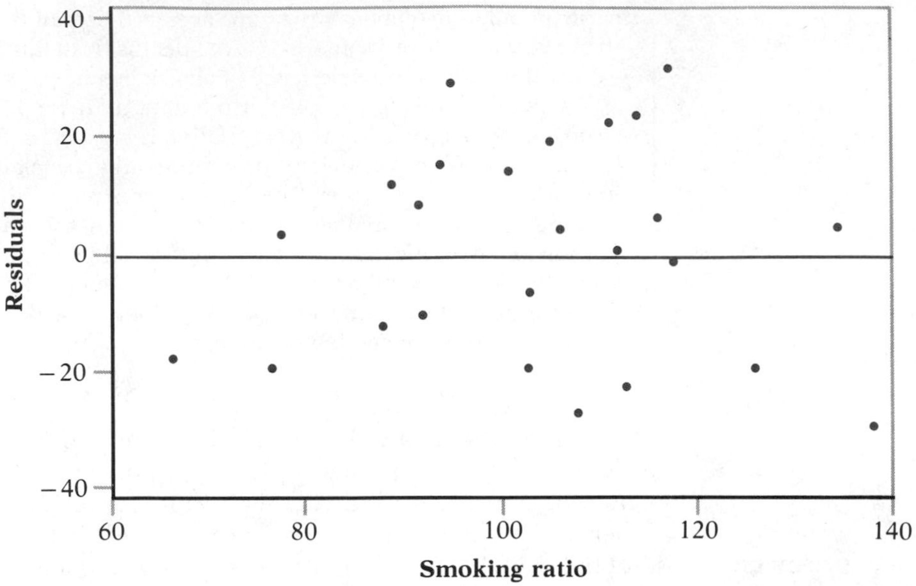

FIGURE 2.41 Residual plot for the regression of lung cancer deaths on smoking.

Because the evidence linking smoking and lung cancer is so strong, we will not dwell on the peculiarities of Example 2.26 but instead seek explanations for a connection that has been clearly demonstrated to exist. The obvious explanation is that smoking causes lung cancer. However, we have seen enough of the effects of lurking variables to know that we must look further.

What unrecorded variables might explain the strong association between smoking and lung cancer? An old favorite is the "genetic hypothesis." According to this hypothesis, there may be a hereditary trait that predisposes some people both to nicotine addiction and to lung cancer. These people would have high rates of lung cancer even if they never smoked. Another possibility, less favored by the tobacco industry because of its unflattering picture of smokers, is the "sloppy lifestyle" hypothesis. Perhaps smoking is most prevalent among people who also drink too much, don't exercise, eat unhealthy food, and so on. If some other component of a sloppy lifestyle caused lung cancer, smoking would wrongly appear to be the culprit.

These alternative explanations remind us that a strong association between two variables x and y can reflect any of several underlying relationships.

- *Causation:* Changes in x cause changes in y—for example, a drop in outdoor temperature causes an increase in natural gas consumption for heating. If we can change x, we can bring about a change in y. Quitting smoking will reduce a person's chance of getting lung cancer if causation holds.

- *Common response:* Both x and y respond to changes in some unobserved variable or variables—for example, both GRE scores and graduate GPA respond to a student's ability and level of knowledge. In this case y can sometimes be predicted from x, but intervening to change x would not bring about a change in y. The genetic hypothesis claims that smoking behavior and lung cancer are both responses to a genetic predisposition; quitting smoking does not change heredity and will have no effect on future lung cancer.
- *Confounding:* The effect on y of the explanatory variable x is hopelessly mixed up with the effects on y of other variables. Minority students, for example, have lower average scores on college entrance exams such as the SAT than do whites; but minorities (again on the average) grew up in poorer households and attended poorer schools than did whites. The effects of social and economic conditions on test scores are mixed together in a way that makes any cause-and-effect conclusion suspect. The "sloppy lifestyle" hypothesis claims that smoking is confounded with other types of behavior, so that we have no information about the effect of smoking alone on health.

Figure 2.42 illustrates these three kinds of relationship in schematic form. Both common response and confounding involve the influence of a lurking variable (or variables) z on the response variable y. We will not

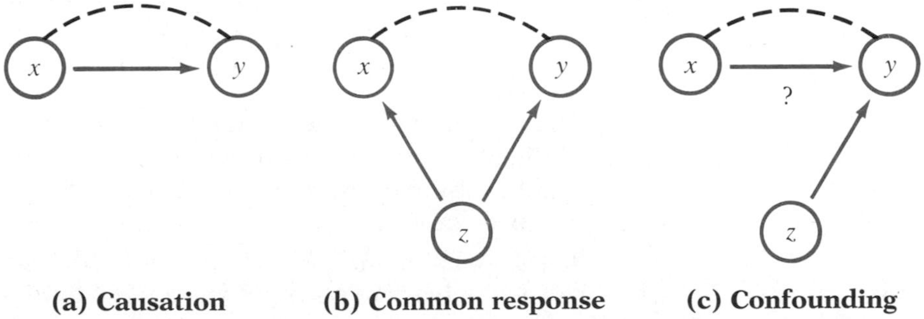

FIGURE 2.42 Variables x and y show a strong association (*dashed line*). This association may be the result of any of several causal relationships (*solid arrows*). (a) Causation: Changes in x cause changes in y. (b) Common response: Changes in both x and y are caused by changes in a lurking variable z. (c) Confounding: The effect (if any) of x on y is confounded with the effect of a lurking variable z.

usually attempt to distinguish common response from confounding. Both point to a lurking variable that explains the observed association at least in part.

EXAMPLE 2.27

An article in a women's magazine reported that women who nurse their babies feel more receptive toward their infants than mothers who bottle-feed. The author concluded that breast-feeding (x) leads to a more positive attitude (y) toward the child. But women choose whether to nurse or bottle-feed, and this choice may reflect already-existing attitudes (z) toward their infants. Mothers who already feel more positive about the child may choose to nurse, for example, while those to whom the baby is a nuisance more often choose the bottle. The mothers' already-established attitude is a lurking variable that prevents conclusions about whether breast-feeding itself changes mothers' attitudes. ●

Establishing causation

How can a direct causal link between x and y be established? The best method—indeed, the only fully compelling method—of establishing causation is to conduct a carefully designed *experiment* in which the effects of possible lurking variables are controlled. To experiment means to actively change x and observe the response y. Much of Chapter 3 is devoted to the art of designing convincing experiments. We can imagine selecting people at birth, forcing some to smoke and others to abstain and observing the subjects until death. Such an experiment would settle the issue, since smoking or not smoking would be imposed on the subjects independent of their heredity or lifestyle. It is because experiments such as this cannot be done—because people *choose* whether or not to smoke—that the question of whether cigarette smoking causes lung cancer remains arguable.

experiment

When experiments are not possible, good evidence for causation is less direct and requires a combination of several factors.

- The association between x and y must be observed in many studies of different types among different groups. This reduces the chance that the observed association is due to a defect in one type of study or a peculiarity in one group of subjects.
- The association must continue to hold when the effects of plausible third variables are taken into account—for example, by looking at three-way tables.
- There must be a plausible explanation for direct influence of x on y, so that the causal link does not depend on the observed association alone. Equally plausible explanations linking y to other variables that might be involved in confounding or common response should be absent.

The claim that cigarette smoking causes lung cancer meets these criteria. A strong association has been observed in numerous studies of many types in many countries. Consistent results are given by prospective studies that follow individuals over time and cross-sectional studies (such as Example 2.26) of diverse groups at a fixed point in time. Many possible sources of common response or confounding have been examined in these studies and have not been found to explain the association between smoking and lung cancer. In particular, the "sloppy lifestyle" hypothesis is not supported by the evidence. Data about nonsmokers who are exposed to tobacco smoke from nearby smokers contradict the genetic hypothesis. Lifelong nonsmokers who live with smokers get lung cancer 35% more often than those who do not live with smokers.[20] The genetic hypothesis cannot explain this effect. Nor can it account for the fact that lung cancer among women, once rare, has increased in step with smoking among women. There is a lag of about 30 years between increased smoking and increased lung cancer. Long the leading cause of cancer deaths among men, lung cancer is about to surpass breast cancer as the cancer that kills the most women (see Exercise 1.106 on page 86 for data).

Moreover, animal experiments have demonstrated conclusively that tobacco smoke contains substances that cause tumors. Some of these experiments have even attached animals to "smoking machines" so that they breathe cigarette smoke. Because the smoke contains substances that in high concentrations cause tumors in animals exposed to it for a short time, it is plausible that lower concentrations over many years cause tumors in human lungs. There is therefore a known pathway by which smoking could cause lung cancer. No similar pathway has been discovered for alternative explanations.

This evidence is quite conclusive. Most physicians in particular are now convinced that smoking causes lung cancer and has other serious consequences as well. The evidence for causation here is about as strong as nonexperimental evidence can be. Though association does not always indicate causation, sometimes it does. Nonetheless, the evidence linking cigarette smoking to lung cancer is not as strong as the evidence for causation provided by well-designed experiments.

Many important issues in which statistical evidence plays a central role concern relationships, like that between smoking and health, which cannot be clarified by experiments. It is easy to claim either too little or too much for the role of statistics in understanding these issues. This chapter has begun to equip you both with effective tools for understanding relationships and with cautions about the use of these tools.

SUMMARY

An observed association between two variables can be due to a **cause-and-effect** relationship between these variables, but it can also be due to the effect of **lurking variables** on the response.

That an association is due to causation is best established by an **experiment** in which the explanatory variable is directly changed and other influences on the response are controlled.

In the absence of experimental evidence, causation should be only cautiously accepted. Good evidence of causation requires that the association be observed in many varied studies, that examination of the effects of other variables not remove the association, and that a clear explanation for the alleged causation exist.

SECTION 2.6 EXERCISES

Each of Exercises 2.82 through 2.86 reports an observed association. In each case, suggest possible explanations for the association other than a causal relationship between x and y. (Note that more than one explanation can contribute to a single association. Even if x does cause changes in y, a second explanation can also be present.)

2.82 A study of elementary school children, ages 6 to 11, finds a high positive correlation between shoe size x and score y on a test of reading comprehension. What explains this correlation?

2.83 There is a strong positive association between the education x and the income y of American adults. A 1987 Census Bureau study found that the median monthly income of high school dropouts was $693 per month, high school graduates averaged $1045 a month, and holders of bachelor's degrees earned $1841. Is this entirely because more education opens the door to better jobs? (From an article in the *New York Times*, October 3, 1987.)

2.84 The National Halothane Study was a major investigation of the safety of the anesthetics used in surgery. Records of over 850,000 operations performed in 34 major hospitals showed the following death rates y for four common anesthetics x. (See L. E. Moses and F. Mosteller, "Safety of anesthetics," in J. M. Tanur et al. (eds.), *Statistics: A Guide to the Unknown*, 3rd ed., Wadsworth, Pacific Grove, Calif., 1989, pp. 15–24.)

Anesthetic	A	B	C	D
Death rate	1.7%	1.7%	3.4%	1.9%

Do these data prove that anesthetic C is causing more deaths than the others?

2.85 The Bureau of Labor Statistics conducts extensive surveys of the earnings of workers in various job categories. One such survey found that in the

South Bend, Indiana metropolitan area the median earnings of Class D secretaries (low-level clerical workers) were higher than the median for Class B secretaries, who have more skill and more responsible jobs. This negative association between job level x and pay y seems puzzling, since each individual employer pays Class B secretaries more. How can we explain it? (From V. L. Ward, "Measuring wage relationships among selected occupations," *Monthly Labor Review*, May 1980.)

2.86 A study shows that there is a clear positive correlation between the size of a hospital (measured by number of beds x) and the median number of days y that patients remain in the hospital. Are the large hospitals padding their bills by keeping patients longer?

2.87 A group of college students believes that herb tea has remarkable restorative powers. To test this belief, they make weekly visits to a local nursing home, visiting with the residents and serving them herb tea. The nursing home staff reports that after several months many of the residents are more cheerful and healthy. A skeptical sociologist commends the students for their good deeds but scoffs at the idea that herb tea helped the residents. It's all confounding, says the sociologist. Identify the explanatory and response variables in this informal study. Then explain what other variables are confounded with the explanatory variable.

2.88 Members of a high school language club believe that study of a foreign language improves a student's command of English. From school records, they obtain the scores on an English achievement test given to all seniors. The average score of seniors who had studied a foreign language for at least two years is much higher than the average score of seniors who studied no foreign language. The club's advisor says that these data are not good evidence that language study strengthens English skills. Identify the explanatory and response variables in this study. Then explain what lurking variable prevents the conclusion that language study improves students' English scores.

2.89 There is an observed association between the level of cholesterol in the blood and the formation of deposits in the arteries, which in turn increases the risk of a heart attack. Blood cholesterol levels may be influenced by many factors, including heredity and diet. It is suspected that diets high in red meat, eggs, and dairy products may cause high cholesterol levels. Outline the kinds of information you would want in order to produce evidence for or against the claim that this kind of diet raises blood cholesterol.

2.90 Return to the study of the safety of anesthetics from Exercise 2.84. Suppose that, for ethical reasons, surgeons are unwilling to do an experiment in which similar patients are given different anesthetics.

However, you have very detailed records for the 850,000 operations that were investigated in the study cited in Exercise 2.84. What kinds of information would you seek in these records in order to establish that anesthetic C does or does not cause extra deaths among patients?

CHAPTER 2 EXERCISES

2.91 Studies of disease often ask people about their diet in years past in order to discover links between diet and disease. But how well do people remember their past diet? Can we predict actual past diet as well or better from what subjects eat now as from their memory of past habits? Data on actual past diet are available for 91 people who were part of a study that started in the 1930s and were asked about their diet when they were 18 years old and again when they were 30. Researchers asked them recently (at age about 55) about their eating habits at ages 18 and 30 and about their current diet. The study results appear in J. T. Dwyer et al., "Memory of food intake in the distant past," *American Journal of Epidemiology*, 130 (1989), pp. 1033–1046. The authors say (page 1038), "The first study aim, to determine how accurately this group of participants remembered past consumption, was addressed by correlations between recalled and historical consumption in each time period. To evaluate the second study aim, that is, whether recalled intake or current intake more accurately predicts historical intake of food groups at age 30 years, we performed regression analysis."

(a) You must present the results of this paper to a group of people who know no statistics. Tell them in nontechnical language what "correlation" means, why correlation suits the first aim of the study, what "regression" means, and why regression fits the second study aim. Be sure to point out the distinction between correlation and regression.

(b) The study looked at the correlations between actual intake of many foods at age 18 and the intake the subjects now remember for age 18. The median correlation was $r = 0.217$. The authors "conclude that memory of food intake in the distant past is fair to poor." Explain to your audience why $r = 0.217$ points to this conclusion.

(c) The authors used regression to predict the intake of a number of foods at age 30 from current intake of those foods and from what the subjects now remember about their intake at age 30. They conclude that "recalled intake more accurately predicted historical intake at age 30 years than did current diet." As evidence, they present r^2-values for the regressions. Explain to your audience why comparing r^2 is one way to compare how well different explanatory variables predict a response.

2.92 Table 1.4 (page 36) gives the sodium content (in milligrams) and number of calories in each of 20 brands of beef hot dogs. Are sodium and calories related? In particular, we suspect that hot dogs that are high in calories are also high in sodium. The data are entered into the Minitab statistical system, calories as the variable CAL and sodium as SOD. We then compute descriptive measures as follows:

```
MTB> MEAN 'CAL'
   MEAN = 156.85

MTB> STD 'CAL'
   ST. DEV. = 22.642

MTB> MEAN 'SOD'
   MEAN = 401.15

MTB> STD 'SOD'
   ST. DEV. = 102.435

MTB> COR 'CAL' 'SOD'
   CORRELATION OF CAL AND SOD = 0.8871
```

(a) Plot sodium y against calories x for the 20 brands of beef hot dogs in Table 1.4. Describe the overall pattern of the relation and any major deviations from the pattern. In particular, do brands that are high in calories tend also to be high in sodium?
(b) Use the information in the Minitab output to find the equation of the least-squares regression line of sodium on calories. Add this line to your graph in (a). Are there any conspicuous outliers or influential observations? What percent of the variation in sodium level among brands can be explained by the linear relation between sodium and calories?
(c) A new brand of beef hot dogs has 180 calories per frank. Predict its sodium content.

2.93 What is the relationship between the number of home runs that a major league baseball team hits and its team batting average? On the one hand, we might expect better-hitting teams to have both more home runs and higher batting averages. For individual players this is true: The correlation is about $r = 0.2$. On the other hand, teams have different styles of play and play their home games in different stadiums. Some teams favor the home run and other teams rely on singles and speed. The table below gives the data for American League teams in the 1991 season. Analyze the data and report your conclusions.

Team	Home runs	Batting average
Baltimore	170	.253
Boston	126	.268
California	115	.255
Chicago	139	.261
Cleveland	79	.254
Detroit	209	.247
Kansas City	117	.264
Milwaukee	116	.271
Minnesota	140	.280
New York	147	.255
Oakland	159	.248
Seattle	126	.254
Texas	177	.269
Toronto	133	.257

2.94 The table below gives data on the relationship between running speed (feet per second) and stride rate (steps taken per second) for elite female runners. (See Exercise 2.27 on page 139 for the source of the data.)

Speed	15.86	16.88	17.50	18.62	19.97	21.06	22.11
Stride rate	3.05	3.12	3.17	3.25	3.36	3.46	3.55

Here are the corresponding data from the same source for male runners:

Speed	15.86	16.88	17.50	18.62	19.97	21.06	22.11
Stride rate	2.92	2.98	3.03	3.11	3.22	3.31	3.41

(a) Plot the data for both groups on one graph using different symbols to distinguish between the points for females and those for males.
(b) Suppose now that the data came to you without identification as to sex. Compute the least-squares line from all of the data and plot it on your graph.
(c) Compute the residuals from this line for each observation. Make a plot of the residuals against speed. How does the fact that the data come from two distinct groups show up in the residual plot?

2.95 To study the energy savings due to adding solar heating panels to a house, researchers measured the natural gas consumption of the house for more than a year and then installed solar panels and observed the natural gas consumption for almost 2 years. The variables are as in Exercise 2.21 (page 137): The explanatory variable x is degree days per day during the several weeks covered by each observation, and the response variable

y is gas consumption (in hundreds of cubic feet) per day during the same period. Figure 2.43 plots y against x, with separate symbols for observations taken before and after the installation of the solar panels. The least-squares regression lines were computed separately for the before and after data, and are drawn on the plot. The regression lines are

$$\text{Before: } \hat{y} = 1.089 + .189x$$

$$\text{After: } \hat{y} = 0.853 + .157x$$

(Data provided by Robert Dale, Purdue University.)

(a) Does the scatterplot suggest that a linear model is appropriate for the relationship between degree days and natural gas consumption? Do any individual observations appear to be outliers (large residuals) or highly influential?

(b) About how much additional natural gas was consumed per day for each additional degree day before the panels were added? After the panels were added?

(c) The daily average temperature during January in this location is about 30° F, which corresponds to 35 degree days per day. Use the regression lines to predict daily gas usage for a day with 35 degree days before and after installation of the panels. If the price of natural gas is $0.75 per hundred cubic feet, how much money do the solar panels save in the 31 days of a typical January?

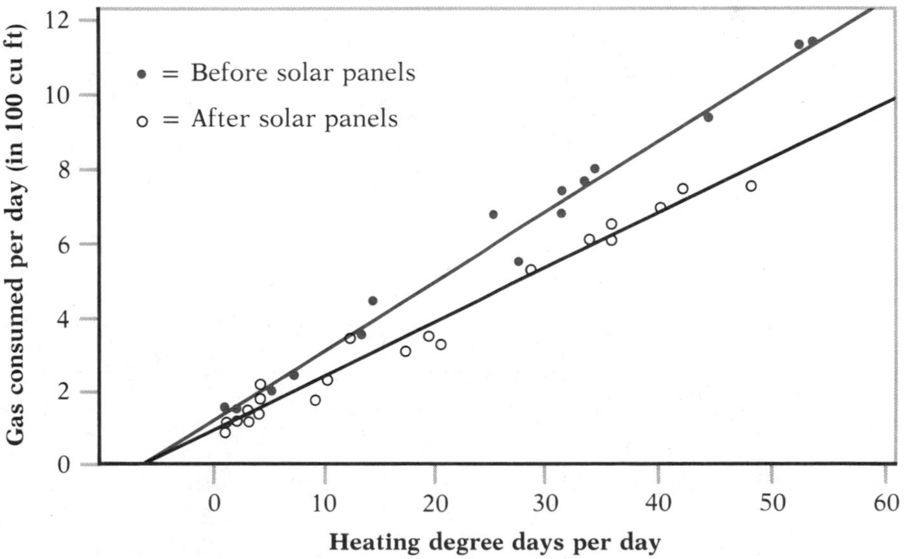

FIGURE 2.43 The regression of residential natural gas consumption on heating degree days before and after installation of solar heating panels (Exercise 2.95).

2.96 Wood scientists are conducting a research project to investigate the possibility of replacing solid wood construction material by products made from flakes of aspen and pine. They collected the following data to examine the relationship between the length (in inches) and the strength (in pounds per square inch) of beams made from wood flakes. (Data provided by Jim Bateman and Michael Hunt, Department of Forestry, Purdue University.)

Length	5	6	7	8	9	10	11	12	13	14
Strength	446	371	334	296	249	254	244	246	239	234

(a) Make a scatterplot that shows how the length of a beam affects its strength.
(b) Describe the pattern that you see in the plot, as well as any outliers that may be present.
(c) Fit a least-squares line to the entire set of data. Graph the line on your scatterplot. Does a straight line describe these data adequately?
(d) The scatterplot suggests that the relation between length and strength might be described by *two* straight lines, one for lengths of 5 to 9 inches and another for lengths of 9 to 14 inches. Fit least-squares lines to these two subsets of the data, and draw the lines on your plot. Do they describe the data adequately? What question would you now ask the wood experts?

2.97 Do mold colonies grow exponentially? In an investigation of the growth of molds, flasks containing a growth medium are inoculated with equal amounts of spores of the mold *Aspergillus nidulans*. The size of a colony is measured by analyzing how much remains of a radioactive tracer substance that is consumed by the mold as it grows. Each size measurement requires destroying that colony, so that the data below refer to 30 separate colonies. To smooth the pattern, we take the mean size of the three colonies measured at each time.

Hours	Colony size			Mean
0	1.25	1.60	.85	1.23
3	1.18	1.05	1.32	1.18
6	.80	1.01	1.02	.94
9	1.28	1.46	2.37	1.70
12	2.12	2.09	2.17	2.13
15	4.18	3.94	3.85	3.99
18	9.95	7.42	9.68	9.02
21	16.36	13.66	12.78	14.27
24	25.01	36.82	39.83	33.89
36	138.34	116.84	111.60	122.26

(Similar experiments are described by A. P. J. Trinci, "A kinetic study of the growth of *Aspergillus nidulans* and other fungi," *Journal of General Microbiology*, 57 (1969), pp. 11–24. These data were provided by Thomas Kuczek, Department of Statistics, Purdue University.)

(a) Graph the mean colony size against time. Then graph the logarithm of the mean colony size against time.

(b) On the basis of data like these, microbiologists divide the growth of mold colonies into three phases that follow each other in time. Exponential growth occurs during only one of these phases. Briefly describe the three phases, making specific reference to the graphs to support your description.

(c) The exponential growth phase for these data lasts from about 6 hours to about 24 hours. The least-squares regression line fitted to only the data between 6 and 24 hours is

$$\log \text{ size } = -.213 + (.0636 \times \text{ hours})$$

Use this line to predict the size of a colony 10 hours after inoculation. (Note that the line predicts the logarithm; you must then obtain the size from its logarithm.)

2.98 Recent studies have shown that earlier reports seriously underestimated the health risks associated with being overweight. The error was caused by overlooking important lurking variables. In particular, smoking tends both to reduce weight and to lead to earlier death. Illustrate Simpson's paradox by stating a simplified version of this situation. That is, make up a three-way table of overweight (yes or no) by early death (yes or no) by smoker (yes or no) such that

- Overweight smokers and overweight nonsmokers both tend to die earlier than those who are not overweight.
- But when smokers and nonsmokers are combined into a two-way table of overweight by early death, persons who are not overweight tend to die earlier.

2.99 The following table gives the U.S. resident population of voting age and the votes cast for president, both in thousands, for presidential elections between 1960 and 1988:

Year	Population	Votes
1960	109,672	68,838
1964	114,090	70,645
1968	120,285	73,212
1972	140,777	77,719
1976	152,308	81,556
1980	164,595	86,515
1984	174,467	92,653
1988	182,628	91,595

(a) For each year compute the percent of people who voted. Make a graph of the percent who voted against year and describe the change over time of participation in presidential elections.
(b) Before proposing political explanations for this change, we should examine possible lurking variables. The minimum voting age in presidential elections dropped from 21 to 18 years in 1970. Use this fact to propose a partial explanation for the trend you saw in (a).

CHAPTER 2 COMPUTER EXERCISES

2.100 Table 2.8 presents three sets of two-variable data prepared by the statistician Frank Anscombe to illustrate the dangers of calculating without first plotting the data. *All three sets have the same correlation and the same least squares regression line* to several decimal places.
(a) Calculate the correlation and the least-squares regression line for all three data sets and verify that they agree.
(b) Make a scatterplot for each of the three data sets and draw the regression line on each of the plots.
(c) In which of the three cases would you be willing to use the fitted regression line to predict y given that $x = 14$? Explain your answer in each case.

2.101 Return to the data on Gesell score and age at first word given in Table 2.4 (page 132). These data are plotted in Figure 2.19. Compute the correlation and the least-squares regression line for the data with (a) case 18 omitted, (b) case 19 omitted, (c) both cases 18 and 19 omitted. Write a brief discussion of the influence of cases 18 and 19. Which is more influential?

TABLE 2.8 Three data sets with the same correlation and least-squares regression line

	Data Set A										
x	10	8	13	9	11	14	6	4	12	7	5
y	8.04	6.95	7.58	8.81	8.33	9.96	7.24	4.26	10.84	4.82	5.68
	Data Set B										
x	10	8	13	9	11	14	6	4	12	7	5
y	9.14	8.14	8.74	8.77	9.26	8.10	6.13	3.10	9.13	7.26	4.74
	Data Set C										
x	8	8	8	8	8	8	8	8	8	8	19
y	6.58	5.76	7.71	8.84	8.47	7.04	5.25	5.56	7.91	6.89	12.50

2.102 With the aid of statistical software we can explore in more detail the relation between calories and sodium in hot dogs. Data for three types of hot dogs appear in Table 1.4 (page 36).

(a) Meat and beef hot dogs are quite similar. Combine the data for these types of hot dogs and make a plot of sodium y against calories x. In the discussion of Example 1.13 (page 44) we noted that the meat hot dog data contain an outlier in x, a veal hot dog that has unusually few calories. Mark the point for this brand on your scatterplot. Does it appear that it will be influential when the regression line is fitted?

(b) Compute the least-squares regression line of sodium on calories for all cases and then for all cases except the veal hot dog marked in (a). Draw both lines on your scatterplot. Was the case noted highly influential?

(c) Figure 1.11 (page 44) suggests that poultry hot dogs differ from the other types, at least in the distribution of calories. Make a new plot of sodium against calories, using one symbol for meat and beef hot dogs and a different symbol for poultry hot dogs. Describe the major differences (if any) in the calorie-sodium relationship for the two groups plotted.

(d) Compute the least-squares regression line of sodium on calories for poultry hot dogs. Draw this line and the regression line for meat and beef hot dogs (with the outlier omitted) on your graph from (c). Give the r^2-values for both regressions. If the regression lines are used to predict sodium from calories, for which group of hot dogs will the prediction be more reliable? Would you be willing to use a single regression line to predict sodium from calories for all types of hot dogs?

2.103 With a statistical computing system we can make a more thorough analysis of the particulate pollution data given in Exercise 2.32 (page 142).

(a) Do a regression of the city readings y on the rural readings x; verify the regression equation given in Exercise 2.32 and obtain the residuals.

(b) Plot the residuals both against x and against the time order of the observations, and comment on the results.

(c) Make a stemplot (or, if your system allows, a normal quantile plot) of the residuals. Is the distribution of the residuals nearly symmetric? Does it appear to be approximately normal?

2.104 Environmentalists, government officials, and vehicle manufacturers are all interested in studying the exhaust emissions produced by motor vehicles. The major pollutants in vehicle exhaust are hydrocarbons (HC), carbon monoxide (CO), and nitrogen oxides (NOX). Table 2.9 gives data for these three pollutants for a sample of 46 light-duty engines of the same type. In the table EN is an engine identifier. Figure 2.8 (page 110) is a plot of NOX against CO for these engines. With the aid of statistical

TABLE 2.9 Amounts of three pollutants (grams per mile) emitted by light-duty engines

EN	HC	CO	NOX	EN	HC	CO	NOX
1	.50	5.01	1.28	2	.65	14.67	.72
3	.46	8.60	1.17	4	.41	4.42	1.31
5	.41	4.95	1.16	6	.39	7.24	1.45
7	.44	7.51	1.08	8	.55	12.30	1.22
9	.72	14.59	.60	10	.64	7.98	1.32
11	.83	11.53	1.32	12	.38	4.10	1.47
13	.38	5.21	1.24	14	.50	12.10	1.44
15	.60	9.62	.71	16	.73	14.97	.51
17	.83	15.13	.49	18	.57	5.04	1.49
19	.34	3.95	1.38	20	.41	3.38	1.33
21	.37	4.12	1.20	22	1.02	23.53	.86
23	.87	19.00	.78	24	1.10	22.92	.57
25	.65	11.20	.95	26	.43	3.81	1.79
27	.48	3.45	2.20	28	.41	1.85	2.27
29	.51	4.10	1.78	30	.41	2.26	1.87
31	.47	4.74	1.83	32	.52	4.29	2.94
33	.56	5.36	1.26	34	.70	14.83	1.16
35	.51	5.69	1.73	36	.52	6.35	1.45
37	.57	6.02	1.31	38	.51	5.79	1.51
39	.36	2.03	1.80	40	.48	4.62	1.47
41	.52	6.78	1.15	42	.61	8.43	1.06
43	.58	6.02	.97	44	.46	3.99	2.01
45	.47	5.22	1.12	46	.55	7.47	1.39

software, a detailed study is feasible. (Data taken from T. J. Lorenzen, "Determining statistical characteristics of a vehicle emissions audit procedure," *Technometrics*, 22 (1980), pp. 483–493.)

(a) Plot HC against CO, HC against NOX, and NOX against CO.

(b) For each plot, describe the nature of the association. In particular, which pollutants are positively associated and which are negatively associated?

(c) Inspect your plots carefully and circle any points that appear to be outliers or influential observations. Next to the circles, indicate the EN identifier. Does any consistent pattern of unusual points appear in all three plots?

(d) Compute correlations for each pair of pollutants. If you found unusual points, compute the correlations both with and without these points.

(e) Compute the least-squares lines corresponding to the three plots; use CO to predict HC and NOX, and NOX to predict HC.

(f) Compute the least-squares lines omitting the unusual points as a check on the influence of these points.

(g) Summarize the results of your analysis. In particular, is any of these pollutants a good predictor of another pollutant?

2.105 There are different ways to measure the amount of money spent on education. Average salary paid to teachers and expenditures per pupil are two commonly used measures. Table 2.10, based on information from the 1991 *Statistical Abstract of the United States*, gives the 1990 values for these variables by state. The states are classified according to region: NE (New England), MA (Middle Atlantic), ENC (East North Central), WNC (West North Central), SA (South Atlantic), ESC (East South Central), WSC (West South Central), MN (Mountain), and PA (Pacific).

(a) Make a stemplot or histogram for teachers' pay. Label any outlying cases with the state identifier. Do the same for spending per pupil. Do the distributions appear symmetric or skewed? Are the same states outliers in both distributions?

TABLE 2.10 Average teachers' pay and per-pupil educational spending by state (thousands of dollars)

State	Region	Pay	Spending	State	Region	Pay	Spending
Maine	NE	26.9	5.58	N.H.	NE	29.0	5.15
Vt.	NE	28.8	5.42	Mass.	NE	34.2	6.17
R.I.	NE	36.1	6.52	Conn.	NE	40.5	7.93
N.Y.	MA	38.9	8.09	N.J.	MA	35.7	8.44
Pa.	MA	33.3	5.67	Ohio	ENC	31.2	4.39
Ind.	ENC	30.5	4.13	Ill.	ENC	32.8	4.85
Mich.	ENC	36.0	5.07	Wis.	ENC	31.9	5.70
Minn.	WNC	32.2	5.11	Iowa	WNC	26.7	4.59
Mo.	WNC	27.2	4.23	N. Dak.	WNC	23.0	3.58
S. Dak.	WNC	21.3	3.31	Nebr.	WNC	25.5	3.87
Kans.	WNC	28.7	4.71	Del.	SA	33.4	5.85
Md.	SA	36.6	5.89	D.C.	SA	38.0	7.41
Va.	SA	30.9	5.00	W.Va.	SA	22.8	4.15
N.C.	SA	27.8	4.39	S.C.	SA	27.2	3.73
Ga.	SA	27.9	5.05	Fla.	SA	28.8	5.05
Ky.	ESC	26.3	3.82	Tenn.	ESC	27.1	3.31
Ala.	ESC	25.5	3.31	Miss.	ESC	24.4	3.15
Ark.	WSC	22.0	3.85	La.	WSC	24.3	3.46
Okla.	WSC	23.1	3.48	Tex.	WSC	27.5	4.06
Mont.	MN	25.1	4.15	Idaho	MN	23.9	3.04
Wyo.	MN	28.2	5.28	Colo.	MN	30.8	4.58
N. Mex.	MN	25.1	4.18	Ariz.	MN	29.4	3.85
Utah	MN	23.7	2.73	Nev.	MN	30.6	4.39
Wash.	PA	30.5	4.64	Oreg.	PA	30.8	5.09
Calif.	PA	36.4	4.60	Alaska	PA	43.2	7.25
Hawaii	PA	32.0	4.50				

(b) Make a scatterplot of pay y versus spending x. Describe the pattern of the relationship between pay and spending. Is there a strong association? If so, is it positive or negative? Explain why you might expect to see an association of this kind.

(c) On your plot, circle any outlying points found in (a) and any additional points that appear to be outliers or influential in the scatterplot. Label the circled points with the state identifier. Are the points found in (a) outliers from the overall relationship?

(d) Compute the least-squares regression line for predicting pay from spending. Graph the line on your plot. Give a numerical measure of the success of overall spending on education in explaining variations in teachers' pay among states.

(e) Repeat (d) excluding Alaska. Do any of your conclusions change substantially?

2.106 Continue the analysis of teachers' pay and education spending begun in the previous exercise by looking for regional effects. Group the states from Table 2.10 into four regions as follows:

- Northeast: NE and MA
- Heartland: ENC and WNC
- South: ESC, SA, and WSC
- Mountain/Pacific: MN and PA

(a) Construct side-by-side modified boxplots for education spending in the four regions. For each region, label any outliers (points plotted individually in the modified boxplot) with the state identifier.

(b) Repeat part (a) for teachers' pay.

(c) Do you see important differences in spending and pay by region? Are the differences consistent for the two variables, that is, are regions that are high in spending also high in pay and vice versa?

(d) Using the least-squares line from the previous exercise, compute the residuals. Display the residuals by region using side-by-side boxplots. Draw a line across your plot at 0. Do the residuals appear to differ by region? Describe any patterns that are noteworthy.

2.107 Can college grade point average be predicted from SAT scores and high school grades? The CSDATA data set described in the data appendix contains information on this issue for a large group of computer science students. We will look only at SAT mathematics scores as a predictor of later GPA, using the variables SATM and GPA from CSDATA. Make a scatterplot, obtain r and r^2, and draw on your plot the least-squares regression line for predicting GPA from SATM. Then write a brief discussion of the ability of SATM alone to predict GPA. (In Chapter 9 we will see how combining several explanatory variables improves our ability to predict.)

2.108 The WOOD data set given in the data appendix originates in a study of the strength of wood products. The researchers measured the modulus of elasticity for each of 50 strips of wood two times. If no strong relationship exists between the two measurements, the measurement process is not repeatable. In many fields the correlation is used to assess the strength of the relationship between measurements and is reported as the *reliability* of the measurement process.

(a) Plot the data with T1 (first measurement) on the x axis and T2 (second measurement) on the y axis. Describe the relationship. Why would you expect it to be linear and positive?

(b) Calculate the correlation. Is this measurement process highly reliable?

(c) We have two ways to calculate the least-squares regression line for predicting T2 from T1. First, find the line using your system's regression command. Then find the means and standard deviations for T1 and T2, and use these numbers along with the correlation to find the slope of the least-squares line. Verify that both methods give the same slope.

2.109 We will investigate the effect of adding an unusual observation to the WOOD data set. Add the observation T1 = 1.6, T2 = 1.9 to the data set and make a scatterplot of T2 versus T1.

(a) The new point is an outlier in regression. To confirm this, regress T2 on T1 and plot the residuals versus T1.

(b) The new point is not an outlier in T1 alone and it is also not an outlier in T2 alone. Verify these statements by constructing stemplots, boxplots, or normal quantile plots for both variables.

(c) How does the new point change the regression equation? Find the equation with and without the new point. Plot both regression lines on your scatterplot. Is this point influential?

(d) How does the new point change the correlation? Report the correlation with and without the new point.

2.110 Once again we will add an unusual observation to the WOOD data set. Add the observation T1 = 2.2, T2 = 1.2 to the original data set (do not keep the extra observation from the previous exercise).

(a) The new point is somewhat extreme in T1 and T2, but it is not far away from other values of these variables. Verify these statements by constructing stemplots, boxplots, and/or normal quantile plots for both T1 and T2.

(b) Calculate the means and standard deviations for T1 and T2 with and without the new point. Does the new point affect these summary measures substantially for either variable?

(c) Plot T2 versus T1 with the new data point included. Does the new point appear to be influential? Find the equation for the regression line of T2 on T1 both with and without the new point and draw both lines on your plot. Is the new point in fact influential?

(d) Calculate the correlation with and without the new point. Does the new point change the correlation substantially?

NOTES

1. The data are from the *Statistical Abstract of the United States*, 1987.

2. A sophisticated treatment of improvements and additions to scatterplots is W. S. Cleveland and R. McGill, "The many faces of a scatterplot," *Journal of the American Statistical Association*, 79 (1984), pp. 807–822.

3. The data are from W. L. Colville and D. P. McGill, "Effect of rate and method of planting on several plant characters and yield of irrigated corn," *Agronomy Journal*, 54 (1962), pp. 235–238.

4. The draft lottery is analyzed in detail by S. E. Fienberg, "Randomization and social affairs: The 1970 draft lottery," *Science*, 171 (1971), pp. 255–261. The design of the 1971 lottery is described in another article in that same issue.

5. Data provided by Linda McCabe and the Agency for International Development.

6. Data provided by Peter Cook, Department of Mathematics, Purdue University.

7. These data were originally collected by L. M. Linde of the University of California—Los Angeles but were first published by M. R. Mickey, O. J. Dunn, and V. Clark, "Note on the use of stepwise regression in detecting outliers," *Computers and Biomedical Research*, 1 (1967), pp. 105–111. The data have been used by several authors. We found them in N. R. Draper and J. A. John, "Influential observations and outliers in regression," *Technometrics*, 23 (1981), pp. 21–26.

8. Data from the Energy Information Administration, recorded in Robert H. Romer, *Energy: An Introduction to Physics*, W. H. Freeman, San Francisco, 1976, for 1880 to 1972, and in *The World Almanac and Book of Facts 1991*, Newspaper Enterprise Association, New York, 1990, for more recent years.

9. A careful study of this phenomenon is W. S. Cleveland, P. Diaconis, and R. McGill, "Variables on scatterplots look more highly correlated when the scales are increased," *Science*, 216 (1982), pp. 1138–1141.

10. Data from Marilyn A. Houck et al., "Allometric scaling in the earliest fossil bird, *Archaeopteryx lithographica*," *Science*, 247 (1990), pp. 195–198. The authors conclude from a variety of evidence that all specimens represent the same species.

11. This example is drawn from M. Goldstein, "Preliminary inspection of multivariate data," *The American Statistician*, 36 (1982), pp. 358–362.

12. Kenneth M. Wilson, *The Validation of GRE Scores as Predictors of First-Year Performance in Graduate Study*, Educational Testing Service, Princeton, N.J., 1979, p. 21.

13. The data are from the U.S. Bureau of the Census as reported in *The World Almanac and Book of Facts*, 1990.

14. An alternative to segmented bar graphs is presented in W. S. Cleveland, *The Elements of Graphing Data*, Wadsworth, Monterey, Calif., 1985, pp. 259–262.

15. A clear and comprehensive discussion of numerical measures of association for categorical data appears in Chapter 3 of A. M. Liebetrau, *Measures of Association*, Sage Publications, Beverly Hills, Calif., 1983.

16. National Science Foundation, *Science Indicators 1980*, U.S. Government Printing Office, 1981, pp. 152–153.

17. A specific example comparing the years 1974 and 1978 appears in C. H. Wagner, "Simpson's paradox in real life," *The American Statistician*, 36 (1982), pp. 46–48.

18. This example is reported by Nathan Mantel in a letter to *The American Statistician*, 36 (1982), p. 395.

19. The original source for these data is *Occupational Mortality: The Registrar General's Decennial Supplement for England and Wales, 1970–1972*, Her Majesty's Stationery Office, London, 1978. We found the data in a course of The Open University, *Statistics in Society*, Unit C4, The Open University Press, Milton Keynes, England, 1983.

20. See "Involuntary smokers face health risks," *Science*, 234 (1986), p. 1066, for this and other evidence on the risks of passive smoking drawn from a report of the National Research Council.

CHAPTER

3

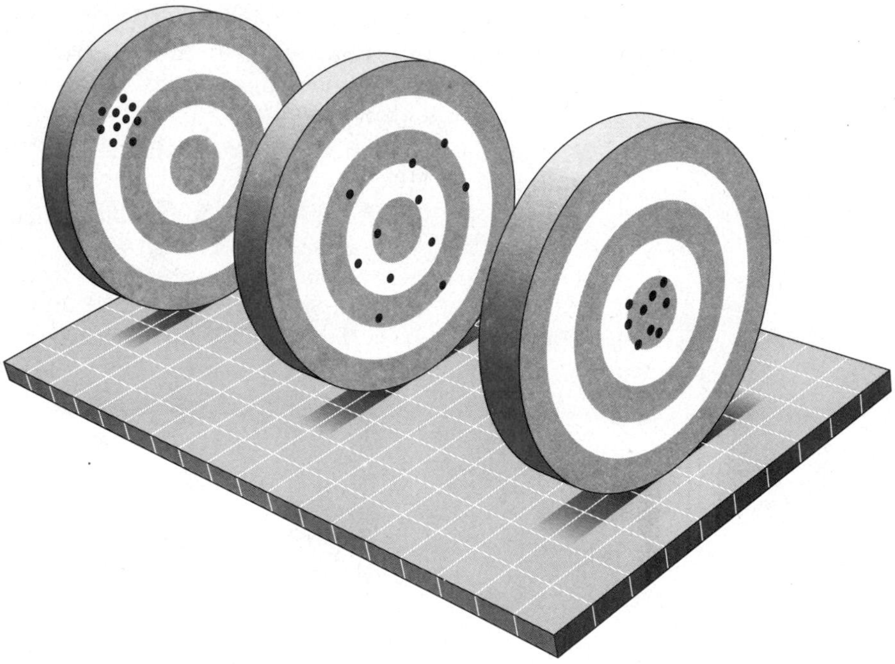

Producing Data

PRELUDE

Statistical designs for producing trustworthy data are perhaps the single most influential contribution of statistics to the advance of knowledge. In this century, random sampling and randomized comparative experiments have revolutionized the practice of many fields of applied science. This chapter presents the basic ideas of statistical designs for choosing a sample that represents some wider population and for laying out an experiment to study the effects of explanatory variables on a response. The deliberate use of chance is central to these designs, so we conclude the chapter with a preliminary look at the consequences of using chance in producing data. Here are some of the examples we will encounter:

- How many homeless people are there in Chicago?
- Does taking aspirin regularly reduce the risk of a heart attack?
- Ann Landers says that 70% of parents would not have children if they could make the decision again. Is this sense or nonsense?
- The Gallup Poll finds that 58% of its sample of adults did not attend church or synagogue last week. If Gallup took a second poll on the same question at the same time, how different would the outcome be?

• • •

Numerical data are raw material for the growth of knowledge. Human efforts to understand the world have been most successful in areas of knowledge where we have learned what to measure and how to measure it. Statistics is a tool that helps measurement produce knowledge rather than confusion. As such, it must be concerned with producing data as well as with interpreting already available data.

Chapters 1 and 2 explored the art of data analysis. They showed how to uncover the characteristics of a set of data by applying numerical and graphical techniques. We guide the application of these techniques by strategies such as seeking an overall pattern and deviations from it, moving from graphs to numerical descriptions to mathematical models, and advancing from the distributions of individual variables to relationships among several variables.

It is helpful to distinguish between two purposes in analyzing data. Sometimes we are presented with data that seem interesting and deserve careful exploration. We do not know what the data may reveal, and we may have no specific questions in mind before we begin our examination. Our inquiry into the behavior of world oil production over time (Section 2.3) was of this kind. Such work is often called *exploratory data analysis*, on the analogy of an explorer reporting the nature of unknown lands. Exploratory analysis relies heavily on graphical methods and seeks patterns in the data that suggest novel conclusions or questions for further study. However, exploratory analysis alone can rarely provide convincing evidence for its conclusions because striking patterns in data can arise from many sources.

exploratory data analysis

A second purpose in analyzing data is to provide clear answers to specific questions. "What is the speed of light?" (Section 1.1); "How many plants per acre give the highest yield of corn?" (Section 2.1); "Does smoking cigarettes cause lung cancer?" (Section 2.6). Data that answer such questions do not arise on their own. They are as much the product of intelligent effort as hybrid tomatoes and compact disc players. This chapter is devoted to developing the skills needed for producing trustworthy data and for judging the quality of data produced by others. The techniques for producing data discussed in this chapter are simple and involve no formulas or graphs, but they are among the most important ideas in statistics.

Statistical techniques for producing data open the door to a further advance in data analysis, *formal statistical inference*, which not only answers specific questions but also provides a measure of the reliability of the conclusions. Although statistical inference uses the descriptive tools we have already presented, it involves new kinds of reasoning. It is primarily numerical rather than graphical. The later chapters of this book are devoted to inference. Even in the setting of inference, however, exploratory data analysis remains important. Large data sets will contain errors that exploration can uncover, and surprising results can lie hidden in the most routine study. The wise statistician always explores the data before proceeding to a formal analysis.

statistical inference

3.1 FIRST STEPS

The first questions you must answer when you plan to produce data are "What shall I measure?" and "How shall I measure it?" That is, what variables (weight, number of tumors, aggressiveness) will you record, and what instruments or methods (balance scale, pathologist's examination, written personality questionnaire) will you use to give numerical values to these variables? These are central questions in the development of any science but they are not questions to which statistics gives answers. Statistics is not completely a bystander in the discussion of measurement—it can help to describe the accuracy of measurements and often helps to clarify the meaning of a variable by showing its relation to other variables. But by and large, questions of measurement belong to the substantive fields of science, not to the methodological field of statistics. We will therefore take for granted that all the variables we work with have specific definitions and are satisfactorily measured.

design for producing data

Statistics begins to play an essential role when the focus turns from measurements on individual cases (whether ball bearings, rats, or people) to arrangements for collecting data from many individuals. These arrangements or patterns for producing data are called *designs*. A design addresses issues such as: How many individuals shall we collect data from? How shall we select the individuals to be studied? If, as in many experiments, several groups of individuals are to receive different treatments, how shall we form the groups? If there is no systematic design for producing data, we may be misled by haphazard or incomplete data or by confounding.

The need for design

How many homeless people are there in the United States? The homeless are visible in any large city, sleeping on heat vents and clutching the shopping bags that hold their few belongings. But is homelessness a massive problem or a matter affecting relatively few people? We need data. In the absence of data, it is tempting to draw conclusions based on observation of a few individual cases.

> **ANECDOTAL EVIDENCE**
>
> Anecdotal evidence is based on haphazardly selected individual cases, which often come to our attention because they are striking in some way. These cases need not be representative of any larger group of cases.

We can see and talk to individual street people. Recognizing the suffering of these individuals, advocates for the homeless urge large programs of subsidized housing to make homes available to the very poor. But quite different conclusions are also possible. Many individual street people appear to be mentally disturbed. Some experts therefore trace the problem of homelessness to the program of "deinstitutionalizing" the mentally ill, releasing them from large institutions into communities that were not prepared to offer care. What is needed, they say, is not a massive housing program but better mental health facilities. As these conflicting recommendations illustrate, anecdotal evidence is a poor guide to policy.

EXAMPLE 3.1

An effective response to the problem of homelessness requires data on the number of the homeless and their condition. Such data are not available in the usual sources. The U.S. Census, for example, relies primarily on questionnaires sent to dwelling units and has had only limited success in its attempts to count the homeless. In the absence of carefully designed data collection, estimates of the number of homeless people in the United States have ranged from about 250,000 to over 3 million. Guesses about the proportion of mentally disturbed people among the homeless have varied from 20% to 90%.

The first attempt to collect reliable data was made in Chicago. The Chicago Homeless Study searched a sample of several hundred city blocks during the night, counting homeless people and interviewing them. The researchers also counted and interviewed people spending the night in shelters. Though the data apply only to Chicago, they support the view that the homeless are relatively few in number and often unable to function in society. The number of homeless in Chicago on any one night was estimated at 2000 to 3000 persons. About 80% of those interviewed had been in a mental institution, jail, or drug treatment center. The data reveal a small population with many severe problems that simple provision of housing will not address.[1] ●

Example 3.1 illustrates the weakness of anecdotal evidence and also the fact that available data may be inadequate.

AVAILABLE DATA

Available data are data that were produced in the past for some other purpose but that may help answer a present question.

Available data from the library are the only data used in most student reports. Because producing new data is expensive, we all use available data whenever possible. Census data, though unable to reveal the size of the homeless population, are a staple source of information for social science, economics, and market research. However, the clearest answers to current questions usually require the production of new data to answer those specific questions. Statistical designs for producing data rely on either *sampling* or *experiments*.

Sampling

The Chicago Homeless Study set out to collect data to answer specific questions about the number and condition of homeless people. It was not possible to search the entire city of Chicago in a single night to count all homeless people. Instead, the researchers produced their data by using a statistical design to choose a part of the city to be searched. That is, they selected a *sample* of city blocks to represent the larger *population* of all blocks in the city. The idea of *sampling* is to study a part in order to gain information about the whole. Data are often produced by sampling a population of people or things. Opinion polls, for example, report the views of the entire country based on interviews with a sample of about 1500 people. Government reports on employment and unemployment are produced from a monthly sample of about 60,000 households. The quality of manufactured items is monitored by inspecting small samples on a regular basis.

sampling

In all of these cases, as in the Chicago Homeless Study, the expense of examining every item or person in the population makes sampling a practical necessity. Timeliness is another reason for preferring a sample to a *census*, which is an attempt to contact every individual in the entire population. We want information on current unemployment and public opinion next week, not next year. Finally, a carefully conducted sample is often more accurate than a census. Accountants, for example, sample a firm's inventory to verify the accuracy of the records. Attempting to count every last part in the warehouse would be not only expensive but inaccurate. Bored people do not count carefully.

census

If conclusions based on a sample are to be valid for the entire population, a sound design for selecting the sample is required. Sampling designs are discussed in Section 3.3.

Experiments

The purpose of sampling is to collect information about some population by selecting and measuring a sample from the population. The goal is a picture of the population disturbed as little as possible by the act of gathering information. An *experiment*, in contrast, deliberately imposes some treatment on the experimental units or subjects in order to observe the response. A merely observational study, even one based on a sound statistical sample, is a poor way to measure the effect of an intervention. To see how nature responds to a change, we must actually impose the change.

experiment

EXAMPLE 3.2

A majority of adult recipients of welfare are mothers of young children. Observational studies of welfare mothers show that many individuals are able to increase their earnings and leave the welfare system even though they have preschool children. Some of them take advantage of voluntary job training programs to improve their skills. Should participation in job-training and job-search programs be required of all able-bodied welfare mothers? Observational studies cannot tell us what the effects of such a policy would be. Even if the mothers studied are a properly chosen sample of all welfare recipients, those who seek out training and find jobs may differ in many ways from those who do not. They are observed to have more education, for example, but they may also differ in values and motivation, things that cannot be observed. To see if a required jobs program will help mothers escape welfare, such a program must actually be tried. Choose two similar groups of mothers when they apply for welfare. Require one group to participate in a job-training program, but do not offer the program to the other group. Comparing the income and work record of the two groups after several years will show whether requiring job programs has the desired effect.[2] ●

confounding

In Example 3.2, the effect of voluntary training programs on success in finding work is *confounded* with the special character of mothers who seek out training. We emphasized in Section 2.6 that observational studies of the effect of one variable on another often fail because the explanatory variable is confounded with lurking variables. When our goal is to understand cause and effect, experiments are the only source of fully convincing data. Because experiments allow us to pin down the effects of specific variables of interest to us, they are the preferred method for gaining knowledge in science, medicine, and industry.

Experiments, like samples, require careful design. Because of the unique importance of experiments as a method for producing data, we begin the discussion of statistical designs for data collection in Section 3.2 with the principles underlying the design of experiments.

SUMMARY

Data analysis is sometimes **exploratory** in nature. Exploratory analysis asks what the data tell us about the variables and their relations to each other. The conclusions of an exploratory analysis may not generalize beyond the specific data studied.

Statistical inference produces answers to specific questions, along with a statement of how confident we can be that the answer is correct. The conclusions of statistical inference are usually intended to apply beyond the specific cases studied. Successful statistical inference usually requires **production of data** intended to answer the specific questions posed.

Anecdotal evidence based on a few individual cases is rarely trustworthy. **Available data** collected for other purposes, such as census data, are sometimes helpful. Data intended to answer specific questions are usually produced by sampling or experimentation.

Sampling selects a part of a population of interest to represent the whole. **Experiments** are distinguished from observational studies by the active imposition of some treatment on the subjects of the experiment.

SECTION 3.1 EXERCISES

3.1 A letter to the editor of *Organic Gardening* magazine (August 1980) said, "Today I noticed about eight stinkbugs on the sunflower stalks. Immediately I checked my okra, for I was sure that they'd be under attack. There wasn't one stinkbug on them. I'd never read that stinkbugs are attracted to sunflowers, but I'll surely interplant them with my okra from now on."

Explain briefly why this anecdote does not provide good evidence that sunflowers attract stinkbugs away from okra. In your explanation, suggest some factors that might account for all of the bugs being on the sunflowers.

3.2 When the discussion turns to the pros and cons of wearing automobile seat belts, Herman always brings up the case of a friend who survived an accident because he was not wearing seat belts. The friend was thrown out of the car and landed on a grassy bank, suffering only minor injuries, while the car burst into flame and was destroyed. Explain briefly why this anecdote does not provide good evidence that it is safer not to wear seat belts.

3.3 Explain carefully why each of the following studies is *not* an experiment.

(a) The question of whether a radical mastectomy (removal of breast, chest muscles, and lymph nodes) is more effective than simple mastectomy (removal of the breast only) in prolonging the life of women with breast cancer has been debated intensely. To study this question, a medical team examines the records of five large hospitals and compares the survival times after surgery of all women who have had either operation.

(b) It has been suggested that there is a "gender gap" in political party preference in the United States, with women more likely than men to prefer Democratic candidates. A political scientist selects a large sample of registered voters, both men and women, and asks them whether they voted for the Democratic or Republican candidate in the last congressional election.

3.4 Even though the studies in Exercise 3.3 are not experiments, they have explanatory and response variables. What are these variables in each case?

3.5 A study of the effect of living in public housing on family stability and other variables in poverty-level households was carried out as follows. The researchers obtained a list of all applicants for public housing during the previous year. Some applicants had been accepted, and others had been turned down by the housing authority. Both groups were interviewed and compared. Was this study an experiment? Why or why not? What are the explanatory and response variables in the study?

3.6 A study of the effect of abortions on the health of subsequent children was conducted as follows. The names of women who had had abortions were obtained from medical records in New York City hospitals. Birth records were then searched to locate all women in this group who bore a child within 5 years of the abortion. Then hospital records were examined again for information about the health of the newborn child. Was this study an experiment? Why or why not?

3.7 Some people believe that exercise raises the body's metabolic rate for as long as 12 to 24 hours, enabling us to continue to burn off fat after the workout has ended. In a study of this effect, subjects were asked to walk briskly on a treadmill for several hours. Their metabolic rates were measured before, immediately after, and 12 hours after the exercise. The study was criticized because eating raises the metabolic rate, and no record was kept of what the subjects ate after exercising. Was this study an experiment? Why or why not? What are the explanatory and response variables?

3.8 Before a new variety of frozen muffins is put on the market, it is subjected to extensive taste testing. People are asked to taste the new muffin and a competing brand and to say which they prefer. (Both muffins are unidentified in the test.) Is this an experiment? Why or why not?

3.2 DESIGN OF EXPERIMENTS

A study is an experiment when we actually do something to people, animals, or objects in order to observe the response. Here is the basic vocabulary of experiments.

> **EXPERIMENTAL UNITS, SUBJECTS, TREATMENT**
>
> The objects on which the experiment is performed are the experimental units. When the units are human beings, they are called subjects. A specific experimental condition applied to the units is called a treatment.

Because the purpose of an experiment is to reveal the response of one variable to changes in other variables, the distinction between explanatory and response variables is important. The explanatory variables in an experiment are often called *factors*. Many experiments study the joint effects of several factors; in such an experiment, each treatment is formed by combining a specific value (often called a *level*) of each of the factors.

factor

level

EXAMPLE 3.3

In a study of the absorption of a drug into the bloodstream, an injection of the drug (the treatment) is given to 25 persons (the experimental subjects). The response variable is the concentration of the drug in the subjects' blood, measured 30 minutes after the injection. This experiment has a single factor with only one level. If three different doses of the drug are injected, there is still a single factor (the dosage of the drug), now with three levels. ●

EXAMPLE 3.4

A mathematics education researcher is studying where in high school mathematics texts it is most effective to insert questions. She wants to know whether it is better to present questions as motivation before the text passage or as review after the passage. The result may depend on the type of question asked: simple fact, computation, or word problem. The researcher therefore prepares six versions of an instructional unit in elementary algebra. All versions have the same text except that questions are added to each version in a different fashion, according to the scheme of Figure 3.1.

Six groups of students are selected and each group studies a different version of the unit. The students are the subjects in the experiment. All of them are given the same homework assignment for the material and then all take the same test. The response variable is the test score. There are two factors, the position and the type of questions added to the reading material. Position has two levels (before, after) and type has three levels (simple fact, computation, word problem). Each of the six combinations of one level of each factor is a treatment. ●

Both of these examples illustrate our basic vocabulary for talking about experiments. They also suggest the advantages of experiments over

		Factor B—Question type		
		Simple fact	Computation	Word problem
Factor A—Question position	Before	1	2	3
	After	4	5	6

FIGURE 3.1 The treatments in a two-factor experimental design (see Example 3.4).

observational studies, in which the researcher observes subjects and measures variables of interest but does not impose any treatment. Experimentation allows us to study the effects of the specific treatments that are of interest. Moreover, we can control the environment of the experimental units to hold constant factors that are of no interest to us, such as the text content and homework assignment in Example 3.4. In the ideal laboratory case, we control all outside factors. Like most ideals, such control is realized only rarely in practice and certainly not in the schoolroom. Nonetheless, a well-designed experiment makes it possible to draw conclusions about the effect of one variable on another.

A less obvious advantage of experiments is that we can study the combined effects of several factors simultaneously. The interaction of several factors can produce effects that could not be predicted from looking at the effects of each factor alone. In the absence of careful planning, the researcher of Example 3.4 might well perform a single-factor experiment, simply comparing the effectiveness of asking factual questions before versus after a reading assignment. Later, when she realized that the type of question might be important, she might repeat the experiment with word problems. These repeated single-factor experiments, done at different times and perhaps in different schools, are less reliable and require more effort than a single two-factor experiment.

Comparative experiments

The design of an experiment begins with a description of the response variable or variables, the factors (explanatory variables), and what specific treatments will be administered. Figure 3.1 illustrates the factors and treatments in Example 3.4. Laboratory experiments in the sciences and engineering often have a simple design with only a single treatment, which is applied to all of the experimental units. The design of such an experiment can be displayed as

$$\text{Treatment} \longrightarrow \text{Observation} \tag{3.1}$$

or, if before-and-after measurements are made,

$$\text{Observation 1} \longrightarrow \text{Treatment} \longrightarrow \text{Observation 2} \tag{3.2}$$

For example, we may subject a steel beam to a load (treatment) and measure its deflection (observation). Alas, when experiments are conducted in the field or with living subjects, simple designs such as (3.1) and (3.2) often yield invalid data. That is, we cannot tell whether or not the treatment had an effect on the units. An example will show what can go wrong.

EXAMPLE 3.5

Ulcers in the upper intestine are unfortunately common in modern society; they are the product of stress and of acids secreted in the stomach. Here is a clever treatment for ulcer patients: The patient swallows a deflated balloon with tubes attached and then a refrigerated solution is pumped through the balloon for an hour. This "gastric freezing" therapy promised to reduce acid secretion by cooling the stomach and so relieve ulcers. An experiment reported in the *Journal of the American Medical Association* showed that gastric freezing did reduce acid secretion and relieve ulcer pain. The treatment was safe and easy and was widely used for several years.

placebo effect

Unfortunately, the design of the experiment was defective. The response of the patients may have been due to the *placebo effect*. A placebo is a dummy treatment, such as a sugar pill. Many patients respond favorably to *any* treatment, even a placebo, presumably because of trust in the physician and expectations of a cure. This response to a dummy treatment is the placebo effect. The placebo effect is quite strong. It extends to measurable physical responses, such as acid secretion in the stomach, as well as to subjective responses, such as lessening of pain. A better-designed experiment, done several years later, divided ulcer patients into two groups. One group was treated by gastric freezing as before, while the other group received a placebo treatment in which the solution in the balloon was at body temperature rather than freezing. The results: 34% of the 82 patients in the treatment group improved versus 38% of the 78 patients in the control (placebo) group. This and other properly designed experiments showed that gastric freezing had no effect, and its use was abandoned.[3] ●

The original gastric freezing experiment had a design of the form (3.2):

$$\text{Observe pain} \longrightarrow \text{Gastric freezing} \longrightarrow \text{Observe pain}$$

The resulting data were misleading because of confounding with the placebo effect. The data also reflect any special features of that particular study, such as a physician with a soothing manner. Confounding of the treatment with these outside variables occurs because the environment cannot be completely controlled, as it might in a chemistry laboratory experiment. Fortunately, the remedy is simple: Experiments should *compare* treatments rather than attempt to assess a single treatment in isolation. When the two groups of patients in the second experiment are compared, the placebo effect and environmental variables operate on both groups. The only difference between the groups is the actual effect of gastric freezing. If the freezing group had experienced more relief from ulcer pain, the explanation must be that gastric freezing is effective. The group of patients who were given a sham treatment is called a *control group*, because it enables us to control the effects of environmental variables on the outcome. *Control of the effects of outside variables is the first principle of statistical design of experiments*. Comparison of several treatments is the simplest form of control.

control group

Without comparison of treatments, experimental results in medicine and the behavioral sciences can be dominated by such influences as the details of the experimental arrangement, the selection of subjects, and the placebo effect. The result is often bias.

> **BIAS**
>
> The design of a study is biased if it systematically favors certain outcomes.

An uncontrolled study of a new medical therapy, for example, is biased in favor of finding the treatment effective because of the placebo effect. It should not surprise you to learn that uncontrolled studies in medicine give new therapies a markedly higher success rate than proper comparative experiments. Medical researchers who trust their professional judgment to assess the value of a treatment from a series of patients without control are overconfident.

EXAMPLE 3.6

Lack of adequate controls can also destroy the credibility of nonexperimental studies. The Love Canal neighborhood near Niagara Falls was the site of the first major controversy over the health effects of toxic chemicals dumped decades earlier. The emotional climax of the Love Canal affair came in May 1980 when the U.S. Environmental Protection Agency reported evidence of chromosome damage among Love Canal residents. Already worried, the residents demanded that they be relocated, and the federal government soon did so.

But when the chromosome study was reviewed by an expert panel, it appeared that the study's author had compared the chromosomes of Love Canal residents with those of a population he had studied earlier. This, the panel knew, is no more valid than comparing patients treated by gastric freezing with past ulcer patients who received no treatment. They pointed out that everyone has some chromosome damage, which varies with age, sex, geographic location, and other variables. Moreover, the way in which the cells are processed in the laboratory affects the count of damaged chromosomes. Finally, whether a chromosome is damaged or not is sometimes hard to judge. Adequate control of these factors requires that Love Canal chromosomes be compared with those of a control population similar in age, sex, and location; that cells from both populations be processed in the same way and at the same time; and that the person counting the damaged chromosomes do all the counts without knowing which cells are from the "suspect" population. Lack of such control made the Love Canal chromosome study meaningless.[4] ●

Randomization

The design of an experiment first identifies the variables and describes the layout of treatments, with comparison as the leading principle. The second essential aspect of the design is the rule by which the experimental units are assigned to the treatments. Comparison of the effects of several treatments is valid only when all treatments are applied to similar groups of experimental units. If one corn variety is planted on more fertile ground, or if one cancer drug is given to less seriously ill patients, comparisons among treatments are meaningless. Systematic differences among the groups of experimental units in a comparative experiment are another source of bias. Allocating the available units or subjects among the treatments must therefore be done with care.

matching

Experimenters often attempt to *match* groups by elaborate balancing acts. Medical researchers, for example, try to match the patients in a "new drug" experimental group and a "standard drug" control group by age, sex, physical condition, smoker or not, and so on. Such attempts are helpful but not adequate—there are too many lurking variables that might affect the outcome. The experimenter is unable to measure some of these variables and will not think of others until after the experiment. Some important variables, such as how advanced a cancer patient's disease is, are so subjective that an experimenter might bias the study by, for example, assigning less advanced cancer cases to a promising new treatment in the subconscious hope that it will help them.

The remedy is to rely on chance to make an assignment that does not depend on any characteristic of the experimental unit and does not rely on the judgment of the experimenter in any way. The use of chance can be combined with matching, but the simplest design creates groups by chance alone. Here is an example.

EXAMPLE 3.7

A food company assesses the nutritional quality of a new "instant breakfast" product by feeding it to newly weaned male white rats and measuring their weight gain over a 28-day period. A control group of rats is fed a standard diet for comparison. This nutrition experiment has a single factor (the diet) with two levels. The researchers use 30 rats for the experiment and so must divide them into two groups of 15. To do this in a completely unbiased fashion, they put the cage numbers of the 30 rats in a hat, mix them up, and draw 15. These rats form the experimental group and the remaining 15 make up the control group. ●

randomization

The use of chance to allocate experimental units into groups is called randomization. *Randomization is the second major principle of statistical* *design of experiments*.* Combining comparison and randomization, we arrive at the simplest randomized comparative design:

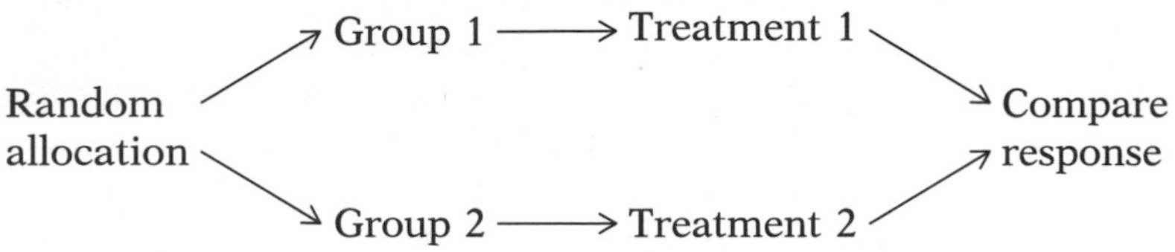

(3.3)

The logic behind a randomized comparative design is straightforward. Randomization produces groups of experimental units that should be similar in all respects before the treatments are applied. Comparative design ensures that influences other than the experimental treatments operate equally on all groups. Therefore, differences in the response variable must be due to the effects of the treatments—that is, the treatments are not only associated with the observed differences in the response, but must actually cause them.

Experimenters were slow to accept randomized comparative designs when they were first put forward in the 1920s. Many researchers are overly confident of their ability to assign units to groups without bias and even to judge the effectiveness of a treatment without controls. They dislike randomization because it makes the outcome of the study depend on chance. If the researchers in Example 3.7 drew names out of the hat a second time, they would get a different 15 rats in the experimental group and no doubt a somewhat different result when they compare the average weight gains of the two groups.

These objections have little justification. Many studies have demonstrated that researchers cannot in fact avoid bias in using their expert judgment. As for randomization, it does indeed render the outcomes dependent on chance, but in a regular manner that can be described by the laws of probability. "Random" in statistics does not mean "haphazard." Feeding the new breakfast food to the first 15 rats that the technician could catch would be haphazard, and perhaps biased if these rats were smaller, slower, or friendlier than the others. Randomization, on the other hand, gives each rat an equal chance to be chosen.

The presence of chance variation does require us to look more closely at the logic of randomized comparative experiments. We cannot say that *any* difference in average weight gain between the two groups of rats, however small, must be due to the diets. Some differences will appear even if

*The use of randomization in experimental design was pioneered by Sir R. A. Fisher (1890–1962). Fisher worked with agricultural field experiments and genetics at the Rothamsted Experiment Station in England. He invented statistical design of experiments and formal statistical methods for analyzing experimental data, and developed them to an advanced state.

the diets are identical, because the rats are not exactly alike. Some rats will grow faster than others, and by chance more of the faster-growing rats may end up in one group than in the other. Even though there are no systematic differences between the groups, there will still be chance differences. It is the business of statistical inference, using the laws of probability, to say whether the observed difference is too large to occur plausibly as a result of chance alone. A difference too large to attribute plausibly to chance is called *statistically significant*. If we observe significant differences among the groups after a comparative randomized experiment, we can conclude that they are caused by the treatments.

statistical significance

We will explore statistical significance in detail in Chapter 6. One important point should be made immediately, however: Experiments with many subjects are better able to detect differences among the effects of the treatments than similar experiments with fewer subjects. You would not trust the results of an experiment that fed each diet to only one rat—the role of chance is too large if we use two rats and toss a coin to decide which is fed the new diet. The more rats used, the more likely it is that randomization will create groups that are alike on the average. When differences among the rats are averaged out, only the effects of the different treatments will remain. Here is a third principle of statistical design of experiments, called *replication:* Repeat each treatment on a large enough number of experimental units or subjects to allow the systematic effects of the treatments to be seen.

replication

PRINCIPLES OF EXPERIMENTAL DESIGN

The basic principles of statistical design of experiments are:

1. **Control** of the effects of lurking variables on the response, most simply by comparing several treatments.
2. **Randomization,** the use of chance to assign subjects to treatments to eliminate bias due to systematic differences among the groups.
3. **Replication** of the experiment on many subjects to reduce chance variation in the results.

How to randomize

The idea of randomization is to assign subjects to treatments by drawing names from a hat. In practice we randomize more quickly by using a *table of random digits*. Table B at the back of the book and on the back endpaper is such a table. A table of random digits is a long string of the digits 0, 1, 2, 3, 4, 5, 6, 7, 8, 9 with these two properties:

table of random digits

1. Each entry in the table is equally likely to be any of the 10 digits 0, 1, 2, 3, 4, 5, 6, 7, 8, 9.
2. The entries are independent of each other. That is, knowledge of one part of the table gives no information about any other part.

You can think of Table B as the result of asking an assistant (or a computer) to mix the digits 0 to 9 in a hat, draw one, then replace the digit drawn, mix again, draw a second digit, and so on. The assistant's mixing and drawing saves us the work of mixing and drawing when we need to randomize.

In Table B, each entry is equally likely to be any of the 10 possibilities. Each pair of entries is equally likely to be any of the 100 possible pairs 00, 01, . . . , 99. Each triple of entries is equally likely to be any of the 1000 possibilities 000, 001, . . . , 999, and so on. The division of the table entries into groups of five is only a device to make the table easier to read, and the numbered rows merely make it easier to say where you started in the table. These conveniences do not affect the nature of the table as a long string of randomly chosen digits. How to use Table B for experimental randomization is best learned from an example.

EXAMPLE 3.8

In the nutrition experiment of Example 3.7, we must divide 30 rats at random into two groups of 15 rats each. First, give each rat a numerical label, using as few digits as possible. Two digits are needed to label each of 30 rats, so we use labels

01, 02, 03, . . ., 29, 30

Next, enter Table B anywhere and read two-digit groups. Suppose we enter at line 130, which is

69051 64817 87174 09517 84534 06489 87201 97245

The first 10 two-digit groups in this line are

69 05 16 48 17 87 17 40 95 17

Each successive two-digit group is a label. The labels 00 and 31 to 99 are not used in this example and we ignore them. The first 15 labels between 01 and 30 that we encounter in the table choose rats for the experimental group. Of the first 10 labels in line 130, five are ignored because they are too high (over 30). The others are 05, 16, 17, 17, and 17. The rats labeled 05, 16, and 17 go into the experimental group. The second and third 17s are ignored because that rat is already in the group. Now run your finger across line 130 (and continue to lines 131, 132, and so on) until 15 rats are chosen. They are the rats labeled

05, 16, 17, 20, 19, 04, 25, 29, 18, 07, 13, 02, 23, 27, 21

These rats form the experimental group; the remaining 15 are the control group. ●

In practice, randomization consists of two steps: Assign labels to the experimental units and then use Table B to select labels at random. Don't try to scramble the labels as you assign them. Table B will do the required randomizing, so assign labels in any convenient manner, such as in alphabetical order for human subjects. You can assign each subject several labels to speed use of the table. Just be certain that all subjects have the same number of labels and that all labels have the same number of digits. Only then will all subjects have the same chance to be chosen. You can read digits from Table B in any order—along a row, down a column, and so on—because the table has no order. As an easy standard practice, we recommend reading along rows.

EXAMPLE 3.9

Many electronic devices contain circuit packs, which are small boards on which many hundreds of components are mounted. During production of a circuit pack, about 2000 connections must be soldered to electrically connect the components. These connections are soldered simultaneously by passing the packs through a standing wave of molten solder. If the conveyor carrying the circuit packs moves either too quickly or too slowly, the quality of the soldered connections will deteriorate. The engineer charged with setting up the production process for a new circuit pack conducts an experiment to determine the best speed for the conveyor in the wave soldering operation. The candidate speeds are 5, 6, and 7 feet per minute. The engineer chooses to use 15 circuit packs for the experiment.

In this example, the experimental unit is a circuit pack, the explanatory variable is the conveyor speed, and the response variable is the count of imperfectly soldered connections in the pack. There are three treatments to be compared. The outline of the design is

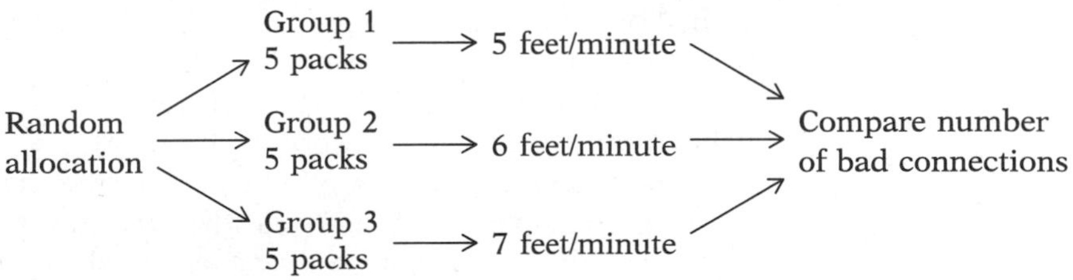

To carry out the random assignment of circuit packs to treatments, label the 15 packs 01 to 15. Enter Table B at any row, say line 110. The first 5 labels encountered select Group 1. This group consists of the packs labeled 06, 08, 12, 04, and 11. Continue through the table until 5 more packs are selected to make up Group 2. They are 02, 14, 03, 09, and 15. The remaining 5 packs form Group 3. Because there are only 15 packs, most labels are unused and the randomization requires searching through several lines in Table B. Fortunately, this takes little time. ●

The diagram of the experiment in Example 3.9 is more specific than the generic diagram 3.3 (page 234). The response variable (count of bad connections) is indicated in the diagram, and the number of units in each group (5 packs) is also shown. We will see later that for statistical inference there are some advantages to having equal numbers of units in all groups. Groups of equal size should usually be chosen unless there is some compelling reason not to do so.

completely randomized design

When all experimental units are allocated at random among all treatments, the experimental design is *completely randomized*. Design 3.3 is the simplest such design, comparing only two treatments. Completely randomized designs can compare more than two treatments; this is done by allocating the experimental units at random to as many groups as there are treatments, as in Example 3.9. The process is tedious but straightforward.

Cautions about experimentation

Although experiments are the preferred means of gaining knowledge about the effects of explanatory variables on a response, you should examine even experimental evidence with a critical eye.

hidden bias

First, the way the experiment is conducted may produce *hidden bias* despite the use of comparison and randomization. Great care must be taken to deal with all experimental units or subjects in the exact same way, so that the treatments are the only systematic differences present. Sometimes violations of this principle are obvious. A study of the roasting of meat in large commercial ovens once found a mysteriously large weight loss in the roasts cooked at the right front corner of the oven. The reason turned out to be that in every run of the experiment, a meat thermometer was thrust into the roast at that location, allowing juice to escape.

Less obvious violations of equal treatment can occur in medical and behavioral experiments. Suppose that in the comparative gastric freezing experiment of Example 3.5 the person interviewing the subjects knew which patients received the "real" therapy and which received "only" the placebo. This knowledge may subconsciously affect the interviewer's attitude toward the patient and recording of the patient's report of the degree of pain relief experienced. The experiment should therefore be *double-blind*—that is, neither the subjects nor the evaluators should know which treatment a subject received. The gastric freezing experiment was double-blind, with attention to such details as ensuring that the tube in the mouth of each subject was cold, whether or not the fluid in the balloon was refrigerated. Careful planning and attention to detail are crucial to avoiding hidden bias.

double-blind experiment

lack of realism in experiments

The most serious potential weakness of experiments is *lack of realism*. Here are some examples. A behavioral experiment on the effects of stress exposes student subjects to an artificial situation for a few hours—do the results apply to prolonged stress on the job? An industrial experiment uses a small-scale pilot production process to find the choices of catalyst concentration and temperature that maximize yield—will these be the best choices for the operation of a large plant? A new computer-based teaching method works well in a few classes taught by the designers of the computer program—will the method succeed in typical schools staffed by overworked teachers? Lack of realism can limit our ability to apply the conclusions of an experiment to the settings of greatest interest.

Most experimenters want to generalize their conclusions to some setting wider than that of the actual experiment. Statistical analysis of the original experiment cannot tell us how far the results will generalize. Rather, the experimenter must argue based on an understanding of psychology or chemical engineering or education that the experimental results do describe the wider world. Other psychologists or engineers or educators may disagree. This is one reason why a single experiment is rarely completely convincing, despite the compelling logic of experimental design. The true scope of a new finding must usually be explored by a number of experiments in various settings.

A convincing case that an experiment is sufficiently realistic to produce useful information is based not on statistics but on the experimenter's knowledge of the subject matter of the experiment. The attention to detail required to avoid hidden bias also rests on knowledge of the subject matter. Good experiments combine statistical principles with understanding of a specific field of study.

Other experimental designs

To this point, we have discussed completely randomized experimental designs comparing levels of a single factor. Completely randomized designs can have more than one factor. The education experiment of Example 3.4 has two factors—type and placement of questions in mathematics textbooks—and their combinations form six treatments. A completely randomized design assigns students at random to these six treatments. Once we determine the treatments to be studied, the randomization needed for a completely randomized design remains straightforward.

Completely randomized designs are the simplest statistical designs for experiments; they illustrate clearly the principles of control, randomization, and replication. However, completely randomized designs are often inferior to more elaborate statistical designs. In particular, matching the subjects in various ways can produce more precise results than simple randomization.

EXAMPLE 3.10

Are cereal leaf beetles more strongly attracted by the color yellow or by the color green? Agriculture researchers want to know, because they detect the presence of the pests in farm fields using sticky boards to trap insects that land on them. The board color should attract beetles as strongly as possible. We must design an experiment to compare yellow and green by mounting sticky boards on poles in a large field of oats.

The experimental units are locations within the field far enough apart to represent independent observations. We erect a pole at each location to hold the boards. We might employ a completely randomized design in which we randomly select half the poles to receive a yellow board while the remaining poles receive green. However, there is wide variation among the locations in the number of beetles present. For example, the alfalfa that borders the oats on one side is a natural host of the beetles, so locations near the alfalfa will have more beetles. This variation among experimental units can hide the systematic effect of the board color. It is more efficient to use a *matched pairs* design that mounts boards of both colors on each pole. The observations (numbers of beetles trapped) are matched in pairs from the same poles. We compare the number of trapped beetles on a yellow board with the number trapped by the green board on the same pole. Because the boards are mounted one above the other, we select the color of the top board at random. Just toss a coin for each board—if the coin falls heads, the yellow board is mounted above the green board. ●

matched pairs design

The matched pairs design of Example 3.10 uses the principles of comparison of treatments, randomization, and replication on several experimental units. However, the randomization is not complete (all locations randomly assigned to treatment groups) but restricted to assigning the order of the boards at each location. Each location serves as its own control. The matched pairs design reduces the effect of variation among locations in the field and is able to detect the effect of board color with fewer locations. Matched pairs are an example of block designs.

BLOCK DESIGN

A block is a group of experimental units or subjects that are known before the experiment to be similar in some way that is expected to affect the response to the treatments. In a block design, the random assignment of units to treatments is carried out separately within each block.

In Example 3.10, each location forms a separate block. In general, matched pairs designs compare two treatments using blocks that contain either a single experimental unit or a matched pair of units. For example, a block in a matched pairs experiment to compare two methods of teaching reading might contain two students having the same age, race, sex, socioeconomic status, and achievement test scores. We choose at random one student from each matched pair to receive the new teaching method; the other student in each pair receives the standard method.

Block designs can have blocks of any size. A block design combines the old idea of creating equivalent treatment groups by matching with the newer idea of forming treatment groups at random. Blocks are another form of *control*. They control the effects of some outside variables by bringing those variables into the experiment to form the blocks. Here are some typical examples of block designs.

- The progress of a type of cancer differs in women and men. A clinical experiment to compare three therapies for this cancer therefore treats sex as a blocking variable. Two separate randomizations are done, the first assigning the male subjects to the treatments and the second assigning the female subjects. Figure 3.2 outlines the design of this experiment. Note that there is no randomization involved in making up the blocks. They are groups of subjects that differ in some way (sex in this case) that is apparent before the experiment begins.
- The soil type and fertility of farmland differs by location. Because of this, a test of the effect of tillage type (two types) and pesticide application (three application schedules) on soybean yields uses small fields as blocks. Each block is divided into six plots, and the

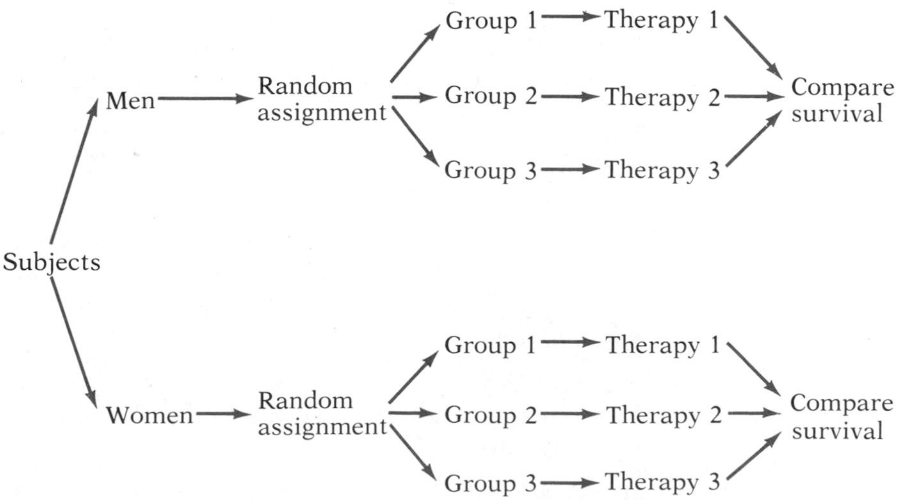

FIGURE 3.2 The outline of a randomized block design. The blocks consist of male and female subjects; the treatments are three therapies for cancer.

six treatments are randomly assigned to plots separately within each block.

- A social policy experiment will assess the effect on family income of several proposed new welfare systems and compare them with the present welfare system. Because the income of a family under any welfare system is strongly related to its present income, the families who agree to participate are divided into blocks of similar income levels. The families in each block are then allocated at random among the welfare systems.

Experiments with blocks and several factors can be quite complicated to design and analyze. A detailed discussion would quickly take us beyond the basic practice of statistics. The design of complex experiments is work for a professional statistician. However, the idea of blocking is an important additional principle of statistical design of experiments. A wise experimenter will form blocks based on the most important unavoidable sources of variability among the experimental units. Randomization will then average out the effects of the remaining variation and allow an unbiased comparison of the treatments.

SUMMARY

In an experiment, one or more **treatments** are imposed on the experimental **units** or **subjects.** Each treatment is a combination of levels of the explanatory variables, or **factors.**

The **design** of an experiment refers to the choice of treatments and the manner in which the experimental units or subjects are assigned to the treatments.

The basic principles of statistical design of experiments are **control, randomization,** and **replication.**

Experiments should **compare** several treatments in order to avoid **confounding** of the effect of a treatment with other influences, such as outside variables.

Randomization creates treatment groups that are similar (except for chance variation) before the treatments are applied. Randomization and comparison together prevent **bias,** or systematic favoritism, in experiments.

Randomization can be carried out by giving numerical labels to the experimental units and using a **table of random digits** to choose treatment groups.

Replication of the treatments on many units reduces the role of chance variation and makes the experiment more sensitive to differences among the treatments.

The validity of experimental results may be threatened by **hidden bias** or **lack of realism.** Statistics alone does not protect against these threats.

In addition to comparison, a second form of control is to restrict randomization by forming **blocks** of experimental units that are similar in some way that is important to the response. Randomization is then carried out separately within each block.

SECTION 3.2 EXERCISES

For each of the experimental situations described in Exercises 3.9 to 3.11, identify the experimental units, the explanatory variable(s) (factors), and the response variable.

3.9 A manufacturer of food products uses package liners that are sealed at the top by applying heated jaws after the package is filled. The customer peels the sealed pieces apart to open the package. What effect does the temperature of the jaws have on the force required to peel the liner? To answer this question, the engineers obtain 20 pairs of pieces of package liner. Five pairs are sealed at each of 250° F, 275° F, 300° F, and 325° F. Then the peel strength of each seal is measured.

3.10 Can aspirin help prevent heart attacks? The Physicians' Health Study, a large medical experiment involving 22,000 male physicians, attempted to answer this question. One group of about 11,000 physicians took an aspirin every second day, while the rest took a placebo. After several years it was found that subjects in the aspirin group had significantly fewer heart attacks than subjects in the placebo group.

3.11 New varieties of corn with altered amino acid patterns may have higher nutritive value than standard corn, which is low in the amino acid lysine. An experiment compares two new varieties, called opaque-2 and floury-2, with normal corn. Corn-soybean meal diets using each type of corn are prepared at three different protein levels: 12%, 16%, and 20%. There are thus nine diets in all. Researchers assign 10 one-day-old male chicks to each diet and record their weight gains after 21 days. The weight gain of the chicks is a measure of the nutritive value of their diet.

3.12 List the specific treatments studied in the Physicians' Health Study (Exercise 3.10). Then draw a diagram to outline an appropriate completely randomized design for this study. (When using a diagram to outline the design of an experiment, be sure to indicate the size of the treatment groups and the response variable. The diagram in Example 3.9 is a model.)

3.13 List the specific treatments studied in the package liner experiment of Exercise 3.9. Then draw a diagram to describe a completely randomized experimental design that is appropriate for this study.

3.14 List the specific treatments compared in the high-lysine corn experiment of Exercise 3.11. Draw a diagram to describe a completely randomized design for this experiment.

3.15 The following situations were not experiments. Can experiments be performed to answer the questions raised? If so, briefly outline the design for each.

(a) The comparison of two surgical procedures for breast cancer in Exercise 3.3(a) (page 227).

(b) The "gender gap" issue of Exercise 3.3(b).

3.16 You are testing a new medication for relief of migraine headache pain. You intend to give the drug to migraine sufferers and ask them 1 hour later to estimate what percent of their pain has been relieved. You have 40 patients available to serve as subjects.

(a) Outline an appropriate design for the experiment, taking the placebo effect into account.

(b) The names of the subjects are given below. Use Table B beginning at line 131 to do the randomization required by your design. List the subjects to whom you will give the drug.

Abrams	Danielson	Gutierrez	Lippman	Rosen
Adamson	Durr	Hale	Martinez	Solomon
Afifi	Edwards	Hwang	McNeill	Thompson
Brown	Fluharty	Iselin	Morse	Travers
Cansico	Garcia	Janle	Ng	Turing
Chen	Gerson	Kaplan	Obramowitz	Ullmann
Cranston	Green	Kim	Rivera	Williams
Curzakis	Gupta	Lattimore	Roberts	Wong

3.17 A horticulturist is comparing two methods (call them A and B) of growing potatoes. Standard potato cuttings will be planted in small plots of ground; the response variables are number of tubers per plant and fresh weight (weight when just harvested) of vegetable growth per plant. There are 20 plots available for the experiment. Sketch the outline of a field divided into 20 small plots. Then diagram the experimental design and use line 145 of Table B to do the required randomization. Mark on your sketch which growing method you will use in each plot.

3.18 We want to determine the best color for attracting cereal leaf beetles to boards on which they will be trapped. We will compare four colors (treatments): green, orange, yellow, and red. The response variable is the count of beetles trapped. One board will be mounted on each of 16 poles evenly spaced in a square field, with four poles in each of four rows. Sketch the field with the locations of the 16 poles. Outline the design of a completely randomized experiment to compare the colors. Then use Table B, starting at line 115, to randomly assign colors to the poles. Mark on your sketch the color assigned to each pole.

3.19 Use Table B, starting at line 120, to do the randomization required by your design for the package liner experiment in Exercise 3.13.

3.20 You decide to use a completely randomized design in the two-factor mathematics education experiment described in Example 3.4 (page 229). Because it is disruptive to assign schoolchildren at random to the treatment groups, you will use 12 classes of the same grade level instead. Diagram the design of an experiment in which you assign these classes at random to the six treatments. Then use Table B, beginning at line 101, to do the randomization required.

3.21 The horticulturist of Exercise 3.17 realizes that different genetic lines of potatoes may respond differently to the two growing methods. He therefore enlarges the experiment to include five genetic lines as a second factor. Use a diagram like Figure 3.1 to describe the treatments in this two-factor design. Outline a completely randomized design based on 20 plots of ground. Return to Table B at line 145 to do the randomization required to allocate the 20 plots among treatments. Make a sketch of the field divided into 20 plots and mark the treatment assigned to each plot.

3.22 Briefly discuss the scientific advantages of a well-designed experiment comparing breast cancer treatments over the observational study of Exercise 3.3(a) (page 227).

3.23 An experiment that claimed to show that meditation lowers anxiety proceeded as follows. The experimenter interviewed the subjects and rated their level of anxiety. Then the subjects were randomly assigned to two groups. The experimenter taught one group how to meditate and they meditated daily for a month. The other group was simply told to relax more. At the end of the month, the experimenter interviewed all the subjects again and rated their anxiety level. The meditation group now had less anxiety. Psychologists said that the results were suspect because the ratings were not blind. Explain what this means and how lack of blindness could bias the reported results.

3.24 A chemist wants to compare the variability of a new and simpler assay method with the standard method. She prepares a batch of solution, divides it into 40 specimens, and then selects 20 at random. She asks her technician to analyze these 20 specimens using the new method and the remaining 20 using the standard method. Because all of the specimens come from the same solution, all assays should give the same answer except for variation caused by differences in carrying out the chemical analysis. After performing the analyses with the new method, the technician falls ill, so the chemist herself performs the 20 assays using the old method. The results using the new method are much more variable than those using the old, and the chemist therefore rejects the new assay technique. This experiment is fatally flawed despite a randomized comparative design. Explain why the results cannot be trusted.

3.25 Is the right hand generally stronger than the left in right-handed people? You can crudely measure hand strength by placing a bathroom scale on a shelf with the end protruding and then squeezing the scale between the thumb below and the four fingers above. The reading of the scale shows the force exerted. Describe the design of a matched pairs experiment to compare the strength of the right and left hands, using 10 right-handed people as subjects. Use a coin to do the required randomization.

3.26 Twenty overweight females have agreed to participate in a study of the effectiveness of four reducing regimens: A, B, C, and D. The researcher first calculates how overweight each subject is by comparing the subject's actual weight with her "ideal" weight. The subjects and their excess weights in pounds are

Birnbaum	35	Hernandez	25	Moses	25	Smith	29
Brown	34	Jackson	33	Nevesky	39	Stall	33
Brunk	30	Kendall	28	Obrach	30	Tran	35
Dixon	34	Loren	32	Rodriguez	30	Wilansky	42
Festinger	24	Mann	28	Santiago	27	Williams	22

The response variable is the weight lost after 8 weeks of treatment. Because the initial amount overweight will influence the response variable, a block design is appropriate.

(a) Arrange the subjects in order of increasing excess weight. Form five blocks by grouping the four least overweight, then the next four, and so on.

(b) Use Table B to do the required random assignment of subjects to the four reducing regimens separately within each block. Be sure to explain exactly how you used the table.

3.27 Return to the mathematics education experiment of Exercise 3.20. Six of the 12 available classes are in one school district and the other six are in another district. Differences between the districts, in curriculum and otherwise, may have a strong effect on the response. You therefore decide to use a block design with the two districts as blocks. The six treatments will be assigned at random to the six classes within each block separately. Outline the design with a diagram. Then use Table B, beginning at line 125, to do the randomization. Report your result in a table that lists the six classes in each district and the treatment you assigned to each class.

3.28 Return to the experiment of Exercise 3.18 on the attractiveness of several colors to cereal leaf beetles. The researchers decide to use two oat fields in different locations and to space eight poles equally within each field. Outline a randomized block design using the fields as blocks. Then use Table B, beginning at line 105, to carry out the random assignment of colors to poles. Report your results by means of a sketch of the two fields with the color at each pole noted.

3.3 SAMPLING DESIGN

A political scientist wants to know what fraction of the public consider themselves Democrats. A quality engineer must determine what fraction of the bearings rolling off an assembly line are defective. Government economists inquire about average household income. In all these cases, we want to gather information about a large group of people or things. We will not, as in an experiment, impose a treatment in order to observe the response. Time, cost, and inconvenience usually forbid inspecting every bearing or contacting every household. In such cases, we gather information about only part of the group in order to draw conclusions about the whole.

POPULATION, UNITS, SAMPLE

The entire group of objects or people about which information is wanted is called the population. Individual members of the population are called units.

A sample is a part of the population that is actually examined in order to gather information.

Notice that population is defined in terms of our desire for knowledge. If we wish to draw conclusions about all U.S. college students, that group is our population even if only local students are available for questioning. The sample is the part from which we draw conclusions about the whole. A poorly designed sampling procedure can produce misleading conclusions, as the following examples illustrate.

EXAMPLE 3.11

A coil mill produces large coils of thin steel for use in manufacturing home appliances. The quality engineer wants to submit a sample of 5-centimeter squares to detailed laboratory examination. She asks a technician to cut a sample of 10 such squares. Wanting to provide "good" pieces of steel, the technician carefully avoids the visible defects in the coil material when cutting the sample. The laboratory results are wonderful but the customers complain about the material they are receiving. ●

EXAMPLE 3.12

The advice columnist Ann Landers once asked her readers, "If you had it to do over again, would you have children?" A few weeks later, her column was headlined "70% OF PARENTS SAY KIDS NOT WORTH IT." Indeed, 70% of the nearly 10,000 parents who wrote in said they would not have children if they could make the choice again. The results of this sample are worthless as indicators of opinion in the population of all American parents. Because the sample was self-selected, it consisted of people who felt strongly enough to take the trouble to write Ann Landers. Many of them, as their letters showed, were angry at their children and alienated from them. *Voluntary response samples* of this kind overrepresent people with strong opinions, most often negative opinions. It is not surprising that a statistically designed opinion poll on the same issue a few months later found that 91% of parents *would* have children again. Ann Landers's poorly designed sampling method produced a 70% "No" result when the truth about the population was close to 90% "Yes." ●

voluntary response sample

In both Examples 3.11 and 3.12, the sample was selected in a manner that guaranteed that it would not be representative of the entire population. These sampling schemes display *bias*, or systematic error, in favoring some parts of the population over others. The remedy is to allow impersonal chance to choose the sample, so that there is neither favoritism by the sampler (as in Example 3.11) nor self-selection by respondents (as in Example 3.12). Random selection of a sample eliminates bias by giving all units an equal chance to be chosen, just as randomization eliminates bias in assigning experimental subjects.

bias in sampling

Simple random samples

A sampling design is the pattern of randomization used to select the sample. The simplest design amounts to placing names in a hat (the population) and drawing out a handful (the sample). This is simple random sampling.

> **SIMPLE RANDOM SAMPLE**
>
> A simple random sample (SRS) of size n consists of n units from the population chosen in such a way that every set of n units has an equal chance to be the sample actually selected.

Each treatment group in a completely randomized experimental design is just an SRS drawn from the available experimental units. An SRS is selected by labeling all units and then using a table of random digits to select a sample of the desired size, just as in experimental randomization. Notice that an SRS not only gives each unit an equal chance to be chosen (thus avoiding bias in the choice), but gives every possible sample an equal chance to be chosen. There are other random sampling designs that give each unit, but not each sample, an equal chance. One such design, systematic random sampling, is described in Exercise 3.37 (page 256).

EXAMPLE 3.13

An academic department wishes to choose a three-member advisory committee at random from the members of the department. To choose an SRS of size 3 from the 28 faculty listed below, first label the members of the population as shown.

00 Abbott	07 Goodwin	14 Pillotte	21 Theobald
01 Cicirelli	08 Haglund	15 Raman	22 Vader
02 Crane	09 Johnson	16 Riemann	23 Wang
03 Dunsmore	10 Keegan	17 Rodriguez	24 Wieczoreck
04 Engle	11 Lechtenberg	18 Rowe	25 Williams
05 Fitzpatrick	12 Martinez	19 Sommers	26 Wilson
06 Garcia	13 Nguyen	20 Stone	27 Zink

Now enter Table B, and read two-digit groups until you have chosen three committee members. If you enter at line 140, the committee consists of Martinez (12), Nguyen (13), and Engle (04). ●

Most statistical software systems will select an SRS for you, eliminating the need for Table B. In the Minitab system, for example, suppose that we have stored the names of 28 subjects in column C1. The commands

```
MTB> SAMPLE 3 C1 C2
MTB> PRINT C2
```

choose an SRS of 3 entries from column C1, place them in column C2, and print the names selected.

Other sampling designs

probability sampling

The general framework for sampling is a *probability sampling design*. Any such design must give each member of the population a known nonzero chance to be selected. Some probability sampling designs (such as an SRS) give each member of the population an equal chance to be selected. This may not be true in more elaborate sampling designs. In every case, however, the use of chance to select the sample is the essential principle of statistical sampling.

Designs for sampling from large geographically dispersed populations are usually much more complex than an SRS. For example, it is common to restrict the random selection by dividing the population into groups of similar units, called *strata,* and then selecting a separate SRS in each stratum. This is a *stratified random sample*. The strata, like the blocks in a block design for an experiment, are based on facts known before the sample is taken. A stratified design can produce more exact information than an SRS of the same size by taking advantage of the fact that units in the same stratum are similar to one another. If all units in each stratum are identical, for example, just one unit from each stratum is enough to completely describe the population. Once the strata are specified, you can obtain a stratified random sample by taking several SRSs, one from each stratum.

stratified random sample

EXAMPLE 3.14

A radio or television station that broadcasts a piece of music owes a royalty to the composer. ASCAP (the American Society of Composers, Authors and Publishers) sells licenses permitting broadcast of the works of any of its members. ASCAP must then pay the proper royalties to the composers whose music was actually played. The major television networks keep program logs of all music played, but local radio and television stations do not. Because there are over a billion ASCAP-licensed performances each year, a detailed accounting is too expensive and cumbersome. Here is a case for sampling.

ASCAP allocates royalties among its members by taping a random sample of broadcasts. The sample of local commercial radio stations, for example, consists of 60,000 hours of broadcast time each year. Radio stations are stratified by type of community (metropolitan, rural), geographic location (New England, Pacific, etc.), and the size of the license fee paid to ASCAP (which reflects the size of the audience). In all, there are 432 strata. Tapes are made at random hours for randomly selected members of each stratum. The tapes are reviewed by experts who can recognize almost every piece of music ever written, and the composers are then paid according to their popularity.[5] ●

Another common means of restricting random selection is to perform the selection in stages. This is often done for national samples of families, households, or individuals. For example, government data on employment and unemployment are gathered by the Current Population Survey, which conducts interviews in about 60,000 households each month. It is not practical to maintain a list of all U.S. households from which to select an SRS. Moreover, the cost of sending interviewers to the widely scattered households in an SRS would be excessive. The Current Population Survey therefore uses a *multistage sample design*. The final sample consists of clusters of nearby households. Most opinion polls and other national samples are also multistage, though interviewing in most national samples today is done by telephone rather than in person, eliminating the economic need for clustering.

multistage sample

A national multistage sample proceeds somewhat as follows:

Stage 1. Take a sample from the 3000 counties in the United States

Stage 2. Select a sample of townships within each of the counties chosen

Stage 3. Select a sample of blocks within each chosen township

Stage 4. Take a sample of households within each block

Lists of counties and townships are easily available. From that point, the sampling can be based on a local map if necessary. No national list of households is required. Moreover, the sample households occur in clusters in the same block and so can be contacted by a single interviewer. The sample at each stage of a multistage design may be an SRS. Stratified samples are also common—for example, counties might be grouped into rural and metropolitan strata before sampling.

Analysis of data from sampling designs more complex than an SRS takes us beyond basic statistics. But the SRS is the building block of more complex designs, and analysis of more complex data differs more in complexity of detail than in fundamental concepts.

Cautions about sample surveys

Random selection eliminates bias in the choice of a sample from a list of the population. When the population consists of human beings, however, accurate information from a sample requires much more than a good sampling design.[6] To begin, we need an accurate and complete list of the population. Because such a list is rarely available, most samples suffer from some degree of *undercoverage*. A sample survey of households, for example, will miss not only the homeless but prison inmates and students in dormitories. An opinion poll conducted by telephone will miss the 7% to 8% of the American population without residential phones. The results of national sample surveys therefore have some bias if the people not covered—who most often are poor people—differ from the rest of the population. The Census Bureau estimates that even the 1990 census, backed by the authority and resources of the federal government, missed about 2.1% of the total population. Because the undercount was greater in the poorer sections of large cities, the Census Bureau estimates that it failed to count 4.8% of blacks and 5.2% of Hispanics.

undercoverage

nonresponse

A more serious source of bias in most sample surveys is *nonresponse*, which occurs when a selected individual either cannot be contacted or refuses to cooperate. Nonresponse to nongovernmental surveys often reaches 30% or more, even with careful planning and several callbacks. Because nonresponse is much higher in urban areas, most sample surveys substitute other accessible people in the same area to avoid favoring rural areas in the final sample. If accessible people differ from those who are rarely at home or who refuse to answer questions, some bias remains.

response bias

In addition, the behavior of the respondent or of the interviewer can cause *response bias* in sample results. Respondents may lie, especially if asked about illegal or antisocial behavior. The sample then underestimates the presence of such behavior in the population. An interviewer whose attitude suggests that some answers are more desirable than others will get these answers more often. The race or sex of the interviewer may influence responses to questions about race relations or attitudes toward feminism. Answers to questions that ask respondents to recall past events are often inaccurate because of faulty memory. In particular, many people will "telescope" events in the past, bringing them forward in memory to more recent time periods. "Have you visited a dentist in the last 6 months?" will often draw a "Yes" from someone who last visited a dentist 8 months ago.[7] Because of these effects, careful training of interviewers and careful supervision to avoid variation among the interviewers are important to good sample survey practice. Because it is so difficult to train large numbers of interviewers, a sample survey is often at least as accurate as a census.

wording questions

The *wording of questions* is the most important influence on the answers given to a sample survey. Opinion polls on nuclear arms reduction, for example, have produced strongly differing conclusions depending on the questions asked. Suppose that you want to show by a statistically sound opinion poll that most people oppose nuclear disarmament. First, because more people favor arms reduction in general than agree with any one proposal, make your question as specific as possible. Second, word the question to suggest some unpleasant effect of arms reduction. In 1982, this question drew a 58% response opposing a nuclear freeze at a time when other questions showed much more favorable attitudes.[8]

> Do you agree or disagree with the following statement: A freeze in nuclear weapons should be opposed because it would do nothing to reduce the danger of thousands of nuclear weapons already in place and would leave the Soviet Union in a position of nuclear superiority.

Never trust the results of a sample survey until you have read the exact questions posed. The sampling design, the amount of nonresponse, and the date of the survey are also important. Good statistical design is a part, but only a part, of a trustworthy survey.

SUMMARY

In a sample survey, a **sample** is selected from the **population** of all people or things about which we desire information. Conclusions about the population are based on examination of the sample.

The **design** of a sample refers to the method used to select the sample from the population. **Probability sampling designs** use random selection to give each member of the population a known nonzero chance to be selected for the sample.

The basic probability sample is a **simple random sample (SRS),** which gives every possible sample of a given size the same chance of being selected. Simple random samples are chosen by labeling the members of the population and using random digits to select the sample.

In **stratified random sampling** the population is divided into **strata,** groups of units that are similar in some way that is important to the response. A separate SRS is then selected from each stratum. **Multistage samples** select successively smaller regions within the population in stages. Each stage may employ an SRS, a stratified sample, or another type of sample.

Failure to use probability sampling often results in **bias,** or systematic errors in the way the sample represents the population. **Voluntary response** samples, in which the respondents choose themselves, are particularly prone to large bias.

In human populations, even probability samples can suffer from bias due to **undercoverage** or **nonresponse,** from **response bias** due to the behavior of the interviewer or the respondent, or from misleading results due to poorly worded questions.

SECTION 3.3 EXERCISES

3.29 For each of the following situations, identify the population as exactly as possible—that is, identify the basic unit and specify which units fall in the population. If the information given is not complete, complete the description of the population in a reasonable way.

(a) Each week, the Gallup Poll questions a sample of about 1500 adult U.S. residents in order to discover the opinions of Americans on a wide variety of issues.

(b) The 1990 census tried to gather basic information from every household in the United States. But a "long form" requesting much additional information was sent to a sample of about 17% of U.S. households.

(c) A machinery manufacturer purchases voltage regulators from a supplier. There are reports that variation in the output voltage of the regulators is affecting the performance of the finished products. To assess the quality of the supplier's production, the manufacturer subjects a sample of 5 regulators from the last shipment to careful laboratory analysis.

3.30 Follow the instructions given in Exercise 3.29 for each of the following situations.

(a) A sociologist wants to study the attitudes of American male college students toward marriage and husband-wife relations. She gives a questionnaire to an SRS of the men enrolled in Sociology 101 at her college.

(b) A member of Congress wants to know what his constituents think of proposed legislation on health insurance. His staff reports that 228 letters have been received on the subject, of which 193 oppose the legislation.

(c) An insurance company wants to monitor the quality of its procedures for handling loss claims from its auto insurance policyholders. Each month the company selects an SRS of all auto insurance claims filed that month to examine them for accuracy and promptness.

3.31 The author Shere Hite undertook a study of women's attitudes toward sex and love by distributing 100,000 questionnaires through women's groups. Only 4.5% of the questionnaires were returned. Based on this sample of women, Hite wrote *Women and Love*, a best-selling book claiming that women are fed up with men. For example, 91% of the divorced women in the sample said that they had initiated the divorce, and 70% of the married women said that they had committed adultery.

Explain briefly why Hite's sampling method is nearly certain to produce a strong bias. Are the sample results cited (91% and 70%) much higher or much lower than the truth about the population of all adult American women?

3.32 Some television stations take quick polls of public opinion by announcing a question on the air and asking viewers to call one of two telephone numbers to register their opinion as "Yes" or "No." Telephone companies make available "900" numbers for this purpose; dialing such a number results in a small charge to your telephone bill. One such call-in poll finds that 73% of those who called are opposed to a proposed local gun control ordinance. Explain why this sampling method is biased. Is the percent of the population who oppose gun control probably higher or lower than the 73% of the sample who are opposed?

3.33 A manufacturer of specialty chemicals chooses 3 from each lot of 25 containers of a reagent to be tested for purity and potency. Below are the control numbers stamped on the bottles in the current lot. Use Table B at line 111 to choose an SRS of 3 of these bottles.

A1096	A1097	A1098	A1101	A1108
A1112	A1113	A1117	A2109	A2211
A2220	B0986	B1011	B1096	B1101
B1102	B1103	B1110	B1119	B1137
B1189	B1223	B1277	B1286	B1299

3.34 Figure 3.3 is a map of a census tract in Cedar Rapids, Iowa. Census tracts are small, homogeneous areas with an average population of about 4000. Each block in the tract is marked with a Census Bureau identifying

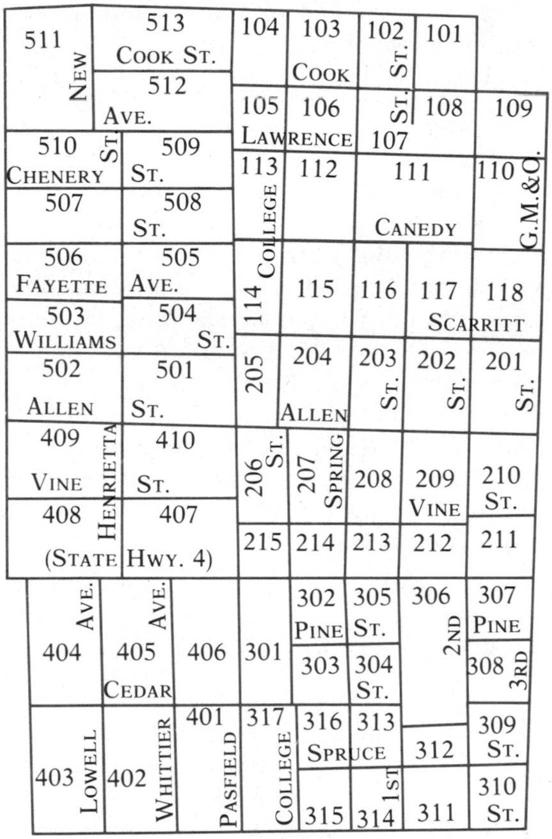

FIGURE 3.3 **Map of a census tract in Cedar Rapids, Iowa (Exercise 3.34).**

number. A sample of blocks from a census tract is often the next-to-last stage in a multistage sample. Use Table B at line 125 to choose an SRS of 5 blocks from the tract illustrated.

3.35 The people listed below are enrolled in a statistics course taught by means of television. Use Table B at line 139 to choose 6 to be interviewed in detail about the quality of the course.

Agarwal	Dewald	Hixson	Puri
Anderson	Fernandez	Klassen	Rodriguez
Baxter	Frank	Mihalko	Rubin
Bowman	Fuhrmann	Moser	Santiago
Bruvold	Goel	Naber	Shen
Casella	Gupta	Petrucelli	Shyr
Cote	Hicks	Pliego	Sundheim

3.36 In using Table B repeatedly to choose samples or do experimental randomization, you should not always begin at the same place, such as line 101. Why not?

3.37 *Systematic random samples* are often used to choose a sample of apartments in a large building or dwelling units in a block at the last stage of a multistage sample. An example will illustrate the idea of a systematic sample. Suppose that we must choose 4 addresses out of 100. Because $100/4 = 25$, we can think of the list as four lists of 25 addresses. Choose 1 of the first 25 at random, using Table B. The sample contains this address and the addresses 25, 50, and 75 places down the list from it. If 13 is chosen, for example, then the systematic random sample consists of the addresses numbered 13, 38, 63, and 88.

(a) Use Table B to choose a systematic random sample of 5 addresses from a list of 200. Enter the table at line 120.

(b) Like an SRS, a systematic sample gives all units the same chance to be chosen. Explain why this is true, and then explain carefully why a systematic sample is nonetheless *not* an SRS.

3.38 A club contains 30 student members and 10 faculty members. The students are

Abel	Fisher	Huber	Moran	Reinmann
Carson	Golomb	Jimenez	Moskowitz	Santos
Chen	Griswold	Jones	Neyman	Shaw
David	Hein	Kiefer	O'Brien	Thompson
Deming	Hernandez	Klotz	Pearl	Utts
Elashoff	Holland	Liu	Potter	Vlasic

and the faculty members are

Andrews	Fernandez	Kim	Moore	Rabinowitz
Besicovitch	Gupta	Lightman	Phillips	Yang

The club can send four students and two faculty members to a convention and decides to choose those who will go by random selection. Use Table B to choose a stratified random sample of 4 students and 2 faculty members.

3.39 A university has 2000 male and 500 female faculty members. The equal opportunity employment officer wants to poll the opinions of a random sample of faculty members. In order to give adequate attention to female faculty opinion, he decides to choose a stratified random sample of 200 males and 200 females. He has alphabetized lists of female and male faculty members. Explain how you would assign labels and how you would use Table B to choose the desired sample. Enter Table B at line 122 and give the labels of the first 5 females and the first 5 males in the sample.

3.40 A university has 2000 male and 500 female faculty members. A stratified random sample of 50 female and 200 male faculty members gives each member of the faculty one chance in 10 to be chosen. This sample design gives every unit in the population the same chance to be chosen for the sample. Is it an SRS? Explain your answer.

3.41 The list of units from which a sample is actually selected is called the *sampling frame*. Ideally, the frame should list every unit in the population, but in practice this is often difficult. A frame that is not representative of the population is a common source of bias.

(a) Suppose that a sample of households in a community is selected at random from the telephone directory. What households are omitted from this frame? What types of people do you think are likely to live in them? These people will probably be underrepresented in the sample.

(b) It is more common in telephone surveys to use random digit dialing equipment that selects the last four digits of a telephone number at random after being given the exchange (the first three digits). Which of the households mentioned in your answer to (a) will be included in the sampling frame by random digit dialing?

3.42 Bias is present in each of the following cases. Identify the source of the bias and specify the direction of the bias (that is, whether the sample result will be systematically above or below the true population result).

(a) A flour company wants to know what fraction of Minneapolis households bake their own bread. An SRS of 500 residential addresses is drawn and interviewers are sent to these addresses. The interviewers are employed during regular working hours on weekdays and they interview only during those hours.

(b) The Miami Police Department wants to know if black residents of Miami are satisfied with police service in their neighborhood. A questionnaire is prepared. An SRS of 300 mailing addresses in predominantly black neighborhoods is chosen, and a uniformed police officer is sent to each address to interview an adult resident.

3.43 Comment on each of the following as a potential sample survey question. Is the question clear? Is it slanted toward a desired response?

(a) "Does your family use food stamps?"

(b) "Which of the following best represents your opinion on gun control?

1. The government should confiscate our guns.
2. We have the right to keep and bear arms."

(c) "A freeze in nuclear weapons should be favored because it would begin a much-needed process to stop everyone in the world from building nuclear weapons now and reduce the possibility of nuclear war in the future. Do you agree or disagree?"

(d) "In view of escalating environmental degradation and incipient resource depletion, would you favor economic incentives for recycling of resource-intensive consumer goods?"

3.4 TOWARD STATISTICAL INFERENCE

The deliberate introduction of chance into the process of producing data is a central idea of statistical designs for both sampling and experimentation. In both cases, randomization eliminates favoritism, or bias, in allocating units to treatments or choosing units to be in the sample. Reducing bias is not the only justification for randomization, however. We usually produce data in order to draw conclusions about some wider population, a process called *statistical inference* because we infer conclusions about the wider population from data on selected units. The reasoning of inference relies on the laws of probability, which describe random behavior. Any use of statistical inference therefore requires that the data arise from some process governed by probability. The most direct way to meet this requirement is to actually use randomization in producing our data.

statistical inference

Inference will occupy the rest of this book, beginning with the necessary probability background in the next chapter. We can see more clearly how statistical inference works and why probability is important by a preliminary study of the most straightforward setting: inference about a population on the basis of a simple random sample drawn from that population. Because statistics deals with numerical variables, we commonly use a number computed from the data to make inferences about an unknown number that describes the population.

PARAMETER, STATISTIC

A parameter is a number describing the population.
A statistic is a number that can be computed from the data without making use of any unknown parameters.

EXAMPLE 3.15

The Gallup Poll has collected data about churchgoing in the United States over several decades. The population is all U.S. residents age 18 and over. A recent poll interviewed a sample containing 1785 randomly selected adults and asked them, "Did you, yourself, happen to attend church or synagogue in the last seven days?" Of the respondents, 1035, or 58%, said "No." The number 58% is a *statistic*. The corresponding *parameter* is the percent of all adult U.S. residents who would have said "No" if asked the same question. ●

Sampling distributions

The opinion poll of Example 3.15 offers information on the churchgoing habits of Americans. Let us examine it in more detail. We want to estimate the proportion p of the population that did not attend church or synagogue the week of the poll. The poll reports that of 1785 randomly selected adults, 1035 had not attended church or synagogue the previous week. The sample proportion

$$\hat{p} = \frac{1035}{1785} = 0.58$$

is a statistic that we can use to estimate the unknown parameter p.

How can $\hat{p}$, based on only 1785 of the more than 180 million American adults, be an accurate estimate of p? After all, a second random sample taken at the same time would select different respondents and no doubt produce a different value of $\hat{p}$. This basic fact is called *sampling variability:* The value of a statistic varies in repeated random sampling.

sampling variability

To understand why sampling variability need not discredit the results of sampling, we will produce some data ourselves. Suppose that an opinion poll takes an SRS of size 100 from the population of all adult U.S. residents. Suppose also that in fact 60% of the population did not attend church or synagogue last week. (So $p = 0.6$.) We can imitate the population by a huge table of random digits, with each digit standing for a person. Six of the digits (say 0 to 5) stand for people who did not attend church. The remaining four digits 6 to 9 stand for those who attended. Because all digits in a random number table are equally likely, this assignment produces a population proportion of nonchurchgoers equal to $p = 0.6$. We then imitate taking an SRS of 100 people from the population by taking 100 consecutive digits from Table B. The statistic $\hat{p}$ is the proportion of 0s to 5s in the sample. For example, the first 100 entries in Table B contain 63 digits between 0 and 5, so $\hat{p} = 63/100 = 0.63$. A second SRS based on the second 100 entries in Table B gives a different result, $\hat{p} = 0.56$. That's sampling variability.

simulation

We used the table of random digits to *simulate* drawing an SRS from a large population. Simulation is a powerful tool for studying random phenomena. It is much faster to use Table B than to actually draw repeated SRSs, and much faster yet to use a computer programmed to produce random digits. Figure 3.4 displays the results of simulating 1000 separate SRSs of size 100 from a population with $p = 0.6$. This frequency histogram shows the sampling distribution of the sample proportion $\hat{p}$. Strictly speaking, the sampling distribution is the ideal pattern that would emerge if we considered all possible samples of size 100 from our population. A distribution obtained from a fixed number of trials, like that in Figure 3.4, is only an approximation to the sampling distribution. One of

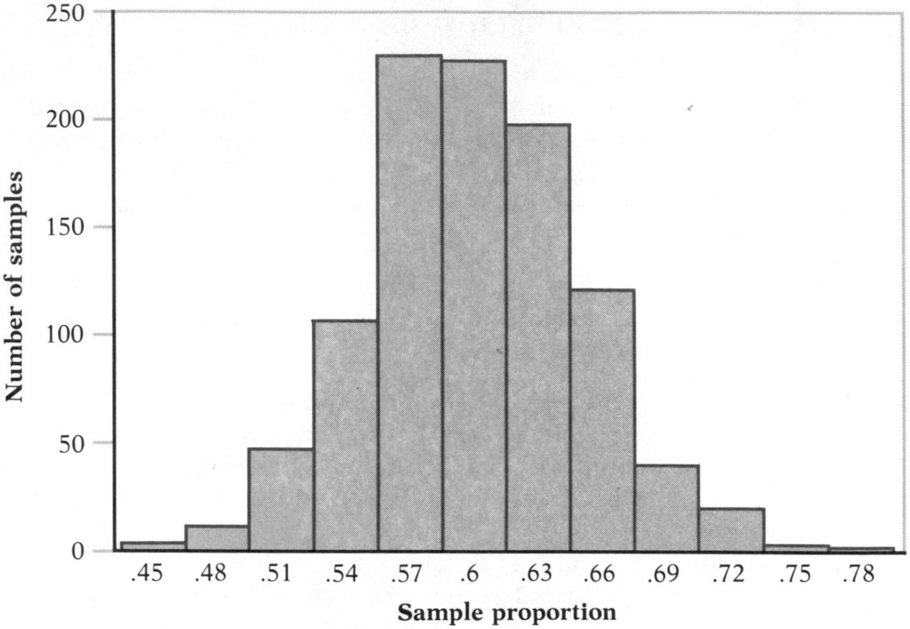

FIGURE 3.4 The sampling distribution of the sample $\hat{p}$ in 1000 repetitions of an SRS of size 100 from a population with population proportion $p = 0.6$.

the uses of probability theory in statistics is to obtain exact sampling distributions without simulation. Simulation has the advantage that it shows clearly how a sampling distribution arises from the results of many samples.

> SAMPLING DISTRIBUTION
>
> The sampling distribution of a statistic is the distribution of values taken by the statistic in all possible samples of the same size from the same population.

Although the values of $\hat{p}$ vary from sample to sample, the behavior of the statistic in repeated sampling is regular and predictable. The sampling distribution describes this regular pattern of behavior. Statistical inference is based on the fact that statistics follow a regular and predictable pattern if we repeatedly sample from the same population. Sampling distributions are therefore fundamental to inference.

We can apply to Figure 3.4 our familiar tools for describing distributions. There are no gaps or outliers. The distribution is symmetric and

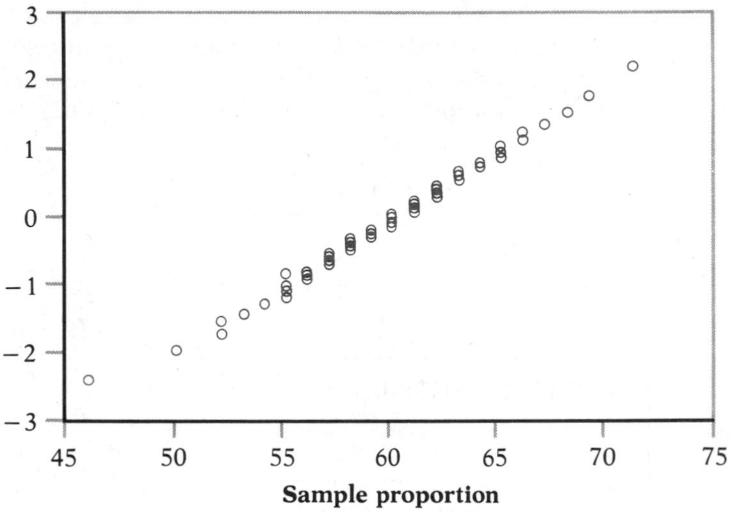

FIGURE 3.5 **Normal quantile plot of the sample proportions in Figure 3.4.**

appears to be approximately normal. Figure 3.5 is a normal quantile plot of the 1000 observed $\hat{p}$'s, with only every twentieth point plotted because of the large number of observations. The plot confirms that the sampling distribution is very close to normal. The only important deviation is slight granularity caused by the fact that $\hat{p}$ from a sample of size 100 can take only values that are multiples of 0.01. It is a remarkable fact that the sampling distributions of several common statistics are approximately normal. This is another reason why the normal distributions are so important in statistics.

The appearance of the sampling distribution in Figures 3.4 and 3.5 is a consequence of the random selection of samples. Haphazard sampling does not give such regular and predictable results. When randomization is used in a design for producing data, statistics computed from the data have a definite pattern of behavior over many repetitions, even though the result of a single repetition is uncertain. The regular pattern of outcomes described by the sampling distribution is the basis for understanding how trustworthy a statistic is as an estimator of a parameter.

Bias

The fraction of all adults who did not attend church or synagogue last week is a parameter p, which in practice is unknown. The sample proportion $\hat{p}$ is a statistic used to estimate the parameter p. Bias refers to systematic error in the estimation process that causes $\hat{p}$ to regularly miss the true p in the same direction. We describe bias more precisely in terms of the center of the sampling distribution of $\hat{p}$. Suppose that in fact $p = 0.6$. Then the sampling distribution of $\hat{p}$ appears in Figure 3.4. The center of this distribution is very close to 0.6. In fact, the exact sampling distribution has mean exactly equal to 0.6. In repeated sampling, some samples produce a $\hat{p}$ greater than 0.6, and some produce a $\hat{p}$ less than 0.6. But the center of the distribution of values produced by repeated samples is 0.6, the true value of the parameter. We will see in Chapter 5 that this is not an accident: The mean of the sampling distribution of the sample proportion $\hat{p}$ from a simple random sample is always equal to the population proportion p. We measure bias numerically as the difference between the mean of the sampling distribution and the true value of the parameter.

UNBIASED ESTIMATOR

A statistic used to estimate a parameter is an unbiased estimator of the parameter if the mean of its sampling distribution is equal to the true value of the parameter.

Unbiasedness says that the mean of the distribution reflects the truth about the population. If we draw an SRS from a population in which 50% did not attend church or synagogue, the center of the sampling distribution of $\hat{p}$ would be 0.5 rather than 0.6. An unbiased statistic computed from an individual sample will sometimes fall above the true value of the parameter and sometimes below. Because its sampling distribution is centered at the true value, however, there is no systematic tendency to overestimate or underestimate the parameter.

Figure 3.6 illustrates the idea of bias as systematic deviation from the truth. Think of the parameter as the bull's-eye on a target and the sample estimate as an arrow shot at the target. Unbiasedness means that our aim is properly aligned. Even though the arrows do not all hit the target at the same point they are centered around the bull's-eye. Bias occurs when the arrows fall regularly away from the bull's-eye in a single direction. Avoiding bias requires both a proper sampling design and a proper choice of the statistic used to estimate the unknown parameter. Common statistics such as the sample proportion $\hat{p}$ will be biased if the sampling procedure is biased in the sense of favoring some part of the population. The proportion $\hat{p}$ of parents who would not have children again in the Ann Landers voluntary response sample of Example 3.12, for example, is strongly biased as an estimate of the population proportion p.

Variability

The goal of archery is to shoot arrows as close as possible to the bull's-eye. To achieve this the archer needs small bias, but also small variability so

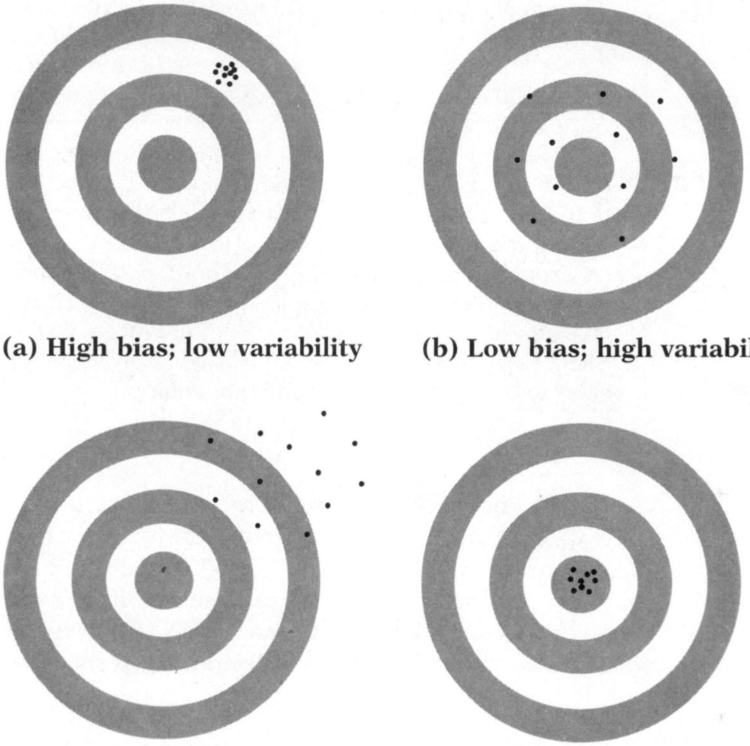

(a) High bias; low variability **(b) Low bias; high variability**

(c) High bias; high variability **(d) Low bias; low variability**

FIGURE 3.6 Bias and variability.

that the arrows fall close together. As Figure 3.6 reminds us, bias and variability are separate properties. Both large and small bias can appear in combination with either large or small variability. Properly chosen statistics computed from random samples have little or no bias. But Figure 3.4 shows that estimates from an SRS can have quite a bit of variability. The variability of a statistic in repeated samples is measured by the spread of its sampling distribution. Our 1000 $\hat{p}$'s range from 0.46 to 0.77. Because the distribution is nearly normal, the variability is better described by the standard deviation, which is 0.049. Remember the 68–95–99.7 rule for normal distributions? In this case, it says that in the long run, 95% of all samples will give a $\hat{p}$ within 0.098 (two standard deviations) of the mean. Lack of bias puts the mean at the true $p = 0.6$, so 95% of all samples give an estimate $\hat{p}$ between 0.502 and 0.698. That is, if in fact 60% of the population does not attend church or synagogue, the estimates from repeated SRSs of size 100 will usually fall between 50.2% and 69.8%. That's not very satisfactory.

Randomization produces a sampling distribution and eliminates bias. *Variability is controlled by the size of the sample*. Larger samples produce statistics with less variability.

EXAMPLE 3.16

The actual opinion poll in Example 3.15 interviewed not 100 people, but 1785. We can simulate repeated SRSs of size 1785 from the same population ($p = 0.6$). Figure 3.7 shows the sampling distribution of the sample proportions $\hat{p}$ from 1000 such samples. The scale is the same as that of Figure 3.4 to make comparison easy.

The center remains at 0.6 (no bias). The sampling variability is much reduced. In fact, the smallest and largest among the $\hat{p}$'s are 0.569 and 0.623, so that all of the 1000 samples gave a result close to the truth. The sampling distribution is again very close to normal. Figure 3.8 displays the distribution on an expanded scale that makes the normal shape more visible. The standard deviation of the distribution is about 0.01. Therefore 95% of all samples will give an estimate within about 0.02 of the mean, or between 0.58 and 0.62. An SRS of size 1785 can in fact be trusted to give sample estimates that are very close to the truth about the entire population. ●

More knowledge of probability will soon enable us to give the standard deviation of the sampling distribution for any size sample. We will then see Example 3.16 as part of a general rule that shows exactly how the variability of sample results decreases for larger samples. One important and surprising aspect of this conclusion is that the spread of the sampling distribution does *not* depend very much on the size of the *population*.

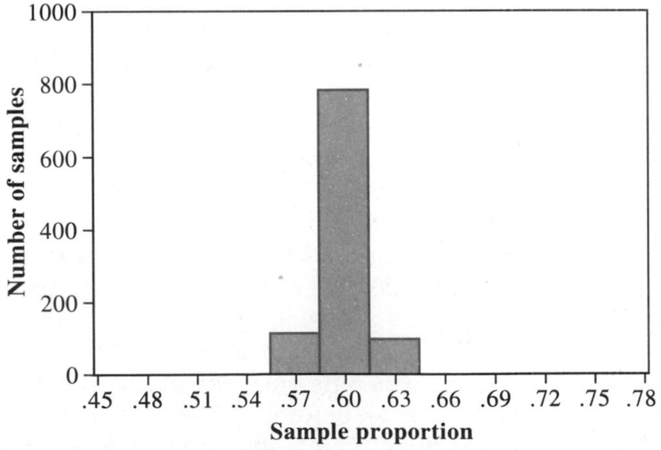

FIGURE 3.7 The sampling distribution of the sample proportion $\hat{p}$ in 1000 repetitions of a SRS of size 1785 from the same population as in Figure 3.4. The graph has the same scale as Figure 3.4, so that the smaller variability in the larger sample is clearly visible (see Example 3.16).

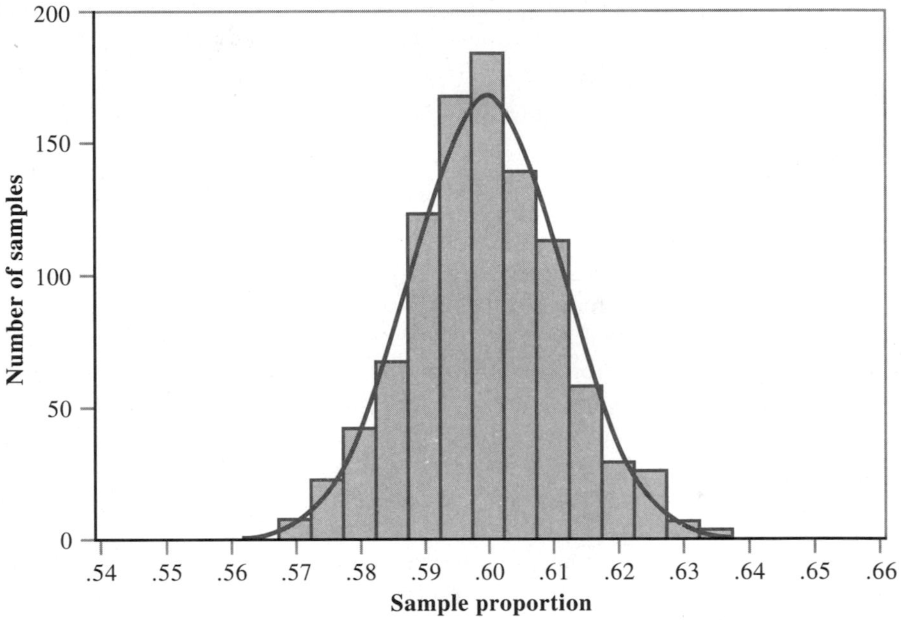

FIGURE 3.8 The same sampling distribution as in Figure 3.7, on a scale that better displays the normal shape of the distribution.

> **VARIABILITY OF A STATISTIC**
>
> The variability of a statistic from a probability sampling design is described by the spread of its sampling distribution. This spread is determined by the sampling design and the size of the sample. As long as the population is much larger than the sample (say, 100 times as large), the spread of the sampling distribution is the same for any population size

Why does the size of the population have little influence on the behavior of statistics from random samples? Imagine sampling harvested corn by thrusting a scoop into a lot of corn kernels. The scoop doesn't know whether it is surrounded by a bag of corn or by an entire truckload. As long as the corn is well mixed (so that the scoop selects a random sample), the variability of the result depends only on the size of the scoop.

The fact that the variability of sample results is controlled by the size of the sample has important consequences for sampling design. A statistic from an SRS of size 1785 from the more than 250,000,000 residents of the United States is just as precise as an SRS of size 1785 from the 740,000 inhabitants of San Francisco. This is good news for designers of national samples but bad news for those who want accurate information about the citizens of San Francisco. If both use an SRS, both must use the same size sample to obtain equally trustworthy results.

What about experiments?

The idea of a sampling distribution is clearest when we are drawing a sample from a population that actually exists, such as all adult residents of the United States or all unsold items in a firm's inventory. Statisticians use the language of sample and population, statistic and parameter, and sampling distribution much more widely in discussing inference. Experimenters, for example, usually do not actually select experimental units from the entire population that they want information about. Yet most experiments do aim to make inferences about a population beyond the units actually involved in the experiment. We tacitly assume that the experimental units are representative of this wider population. An experimenter who compares the performance on a dexterity test of college students before and after they take a stimulant, for example, will argue that results for college students extend to a wider population of adults. Even though a population did not enter into the mechanics of the experimental design, the experiment aims to make inferences about a population just as sampling does. In some experiments, the population exists only in our minds, as the following example illustrates.

EXAMPLE 3.17

A semiconductor logic chip is manufactured in circular wafers containing 120 chips. One stage in the process is to bake a chemical coating on the wafer; the next is to etch in the circuits. Will baking at 90° F or 105° F lead to fewer defective chips per wafer? The experimenters divide a batch of wafers at random into one group to be baked at 90° F and another group to be baked at 105° F. The wafers are baked individually in a random order. The first group averages 57.3 defective chips per wafer, the second, 63.8 defective chips.

The *statistic* of interest is the difference $63.8 - 57.3 = 6.5$. That is, baking at 90° F produced 6.5 fewer defective chips per wafer on the average. The *parameter* is the average difference in the counts of defective chips between the population of all wafers baked at 90° F and the population of all wafers baked at 105° F. These populations are hypothetical, because in production the wafers will be baked at one temperature only. It is common in experiments to seek information about a "What if?" population that cannot be randomly sampled because its units do not all exist in actual fact. ●

Conclusion

Why randomize? Because the act of randomizing guarantees that the results of analyzing our data are subject to the laws of probability. The behavior of a statistic is then described by its sampling distribution. The form of the distribution is known, and in many cases is approximately normal. Often the center of the distribution lies at the true parameter value, so that the notion that randomization eliminates bias is made more precise. The spread of the distribution describes the variability of the statistic and can be made as small as we wish at the expense of a larger sample. In a randomized experiment, variability can be similarly reduced at the expense of larger groups of subjects for each treatment.

These facts are at the heart of formal statistical inference. Later chapters will have much to say in more technical language about sampling distributions and the way statistical conclusions are based on them. What any user of statistics must understand is that all the technical talk has its basis in a simple question: What would happen if the sample or the experiment were repeated many times? The reasoning applies not only to an SRS but also to the complex sampling designs actually used by opinion polls and other national sample surveys. The same conclusions hold as well for randomized experimental designs. The details vary with the design, but the basic facts are true whenever randomization is used to produce data.

Good statistical design, such as use of simple random sampling, is an essential aspect of a good sample. But it is not the whole story. We must avoid badly worded questions and take steps to reduce nonresponse, for example. The sampling distribution, although it is the basis for statistical inference, is similarly not the whole story. It displays the variation in the statistic due to choosing the subjects and shows that an SRS, unlike a voluntary response sample, has no bias due to the choice of subjects. But the sampling distribution reveals nothing about possible bias due to high nonresponse and other practical difficulties. The true distance of the statistic from the parameter can be much larger than the sampling distribution suggests. What is worse, there is no way to say how large the added error is. The inference procedures that we will learn assume that practical difficulties like nonresponse in sample surveys and lack of realism in experiments have been overcome. This is not always easy in practice.

SUMMARY

A number that describes a population is called a **parameter.** A number that can be computed from the data is called a **statistic.** The purpose of sampling or experimentation is usually to use statistics to make statements about unknown parameters.

A statistic from a probability sample or randomized experiment has a **sampling distribution** that describes how the statistic varies in repeated data collection. Formal statistical inference is based on the sampling distributions of statistics.

A statistic as an estimator of a parameter may suffer from **bias** or from high **variability.** Bias means that the center of the sampling distribution is not equal to the true value of the parameter. The variability of the statistic is described by the spread of its sampling distribution.

Properly chosen statistics from randomized data production designs have no bias resulting from the way the sample is selected or the way the experimental units are assigned to treatments. The variability of the statistic is determined by the size of the sample or by the size of the experimental groups.

SECTION 3.4 EXERCISES

State whether each boldface number in Exercises 3.44 to 3.47 is a parameter or a statistic.

3.44 The Bureau of Labor Statistics last month interviewed 60,000 members of the U.S. labor force, of whom **7.2%** were unemployed.

3.45 A carload lot of ball bearings has a mean diameter of **2.5003** cm. This is within the specifications for acceptance of the lot by the purchaser. By chance, however, the acceptance sampling procedure inspects 100 bearings from the lot with a mean diameter of **2.5009** cm. Because this is outside the specified limits, the lot is mistakenly rejected.

3.46 A telemarketing firm in Los Angeles uses a device that dials residential telephone numbers in that city at random. Of the first 100 numbers dialed, **23** are unlisted. This is not surprising because **38%** of all Los Angeles residential phones are unlisted.

3.47 A researcher investigating the effects of a toxic compound in food conducts a randomized comparative experiment with young male white rats. A control group is fed a normal diet, and the experimental group receives a diet with 2500 parts per million of the toxic material. After 8 weeks, the mean weight gain is **335** grams for the control group and **289** grams for the experimental group.

3.48 Coin tossing can illustrate the concept of a sampling distribution. The population is all outcomes (heads or tails) we would get if we tossed a coin forever. The parameter p is the proportion of heads in this population. We suspect that p is close to 0.5; that is, we think the coin will show about one-half heads in the long run. The sample is the outcomes of 20 tosses, and the statistic $\hat{p}$ is the proportion of heads in these 20 tosses.

(a) Toss a coin 20 times and record the value of $\hat{p}$.

(b) Repeat this sampling process 10 times. Make a stemplot of the 10 values of $\hat{p}$. Is the center of this distribution close to 0.5? (Ten repetitions give only a crude approximation to the sampling distribution. If possible, pool your work with that of other students to obtain several hundred repetitions and make a histogram of the values of $\hat{p}$.)

3.49 Let us illustrate the idea of a sampling distribution in the case of a very small sample from a very small population. The population is the scores of 10 students on an exam:

Student	0	1	2	3	4	5	6	7	8	9
Score	82	62	80	58	72	73	65	66	74	62

The parameter of interest is the mean score, which is 69.4. The sample is an SRS of size $n = 4$ drawn from this population. The students are labeled 0 to 9 so that a single random digit from Table B chooses one student for the sample.

(a) Use Table B to draw an SRS of size 4 from this population. Write the four scores in your sample and calculate the mean $\bar{x}$ of the sample scores. This statistic is an estimate of the population parameter.

(b) Repeat this process 10 times. Make a stemplot of the 10 values of $\bar{x}$. You are constructing the sampling distribution of $\bar{x}$. Is the center of your histogram close to 69.4? (Ten repetitions give only a crude approximation to the sampling distribution. If possible, pool your work with that of other students—using different parts of Table B—to obtain several hundred repetitions and make a histogram of the values of $\bar{x}$. This histogram is a better approximation to the sampling distribution.)

3.50 An entomologist samples a field for egg masses of a harmful insect by placing a yard-square frame at random locations and examining the ground within the frame carefully. He wishes to estimate the proportion of square yards in which egg masses are present. Suppose that in a large field egg masses are present in 20% of all possible yard-square areas—that is, $p = 0.2$ in this population.

(a) Use Table B to simulate the presence or absence of egg masses in each square yard of an SRS of 10 square yards from the field. Be sure to explain clearly which digits you used to represent the presence and the absence of egg masses. What proportion of your 10 sample areas had egg masses?

(b) Repeat (a) with different lines from Table B, until you have simulated the results of 20 SRSs of size 10. What proportion of the square yards in each of your 20 samples had egg masses? Make a stemplot from these 20 values to display the sampling distribution of $\hat{p}$ in this case. What is the mean of this distribution? What is its shape?

3.51 An opinion poll asks, "Are you afraid to go outside at night within a mile of your home because of crime?" Suppose that the proportion of all adult

U.S. residents who would say "Yes" to this question is $p = 0.4$.

(a) Use Table B to simulate the result of an SRS of 20 adults. Be sure to explain clearly which digits you used to represent each of "Yes" and "No." What proportion of your 20 responses were "Yes"?

(b) Repeat (a) using different lines in Table B until you have simulated the results of 10 SRSs of size 20 from the same population. Compute the proportion of "Yes" responses in each sample. Find the mean of these 10 proportions. Is it close to p?

3.52 The table below contains the results of simulating on a computer 100 repetitions of the drawing of an SRS of size 200 from a large lot of bearings, 10% of which do not conform to the specifications. The numbers in the table are the counts of nonconforming bearings in each sample of 200.

17	23	18	27	15	17	18	13	16	18	20	15	18	16	21
17	18	19	16	23	20	18	18	17	19	13	27	22	23	26
17	13	16	14	24	22	16	21	24	21	30	24	17	14	16
16	17	24	21	16	17	23	18	23	22	24	23	23	20	19
20	18	20	25	16	24	24	24	15	22	22	16	28	15	22
9	19	16	19	19	25	24	20	15	21	25	24	19	19	20
28	18	17	17	25	17	17	18	19	18					

(a) Make a frequency table for the counts, and list for each count of defectives the corresponding value of the sample proportion of defectives

$$\hat{p} = \frac{\text{count of defectives}}{200}$$

Then draw a frequency histogram for the values of the statistic $\hat{p}$.

(b) Where is the center of this sampling distribution? Does the shape appear to be approximately normal? Does the statistic $\hat{p}$ appear to have large or small bias as an estimate of the population proportion, which is $p = 0.10$ in this particular population?

(c) If we had simulated the repeated selection of SRSs of size 1000 instead of 200 from this same population, where would be the center of the sampling distribution of the sample proportion $\hat{p}$? Would the spread be larger, smaller, or about the same when compared with the spread of your histogram in (b)?

3.53 Figure 3.9 shows histograms of four sampling distributions of statistics intended to estimate the same parameter. Label each distribution relative to the others as large or small bias and as large or small variability.

3.54 A management student is planning to take a survey of student attitudes toward part-time work while attending college. He develops a question-

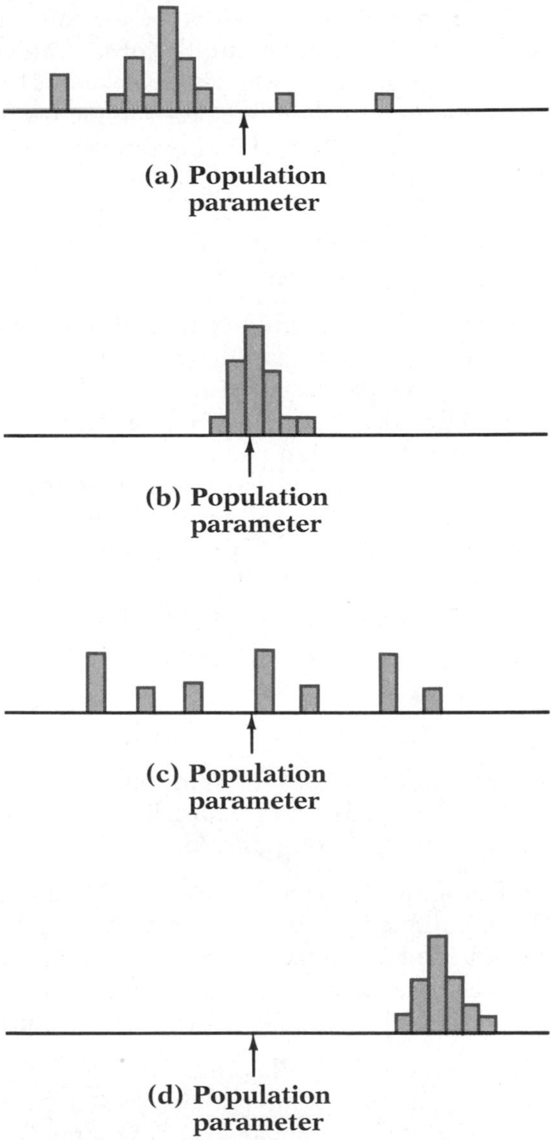

FIGURE 3.9 Which of these sampling distributions displays high or low bias and high or low variability (Exercise 3.53)?

naire and plans to ask 25 randomly selected students to fill it out. His faculty advisor approves the questionnaire but urges that the sample size be increased to at least 100 students. Why is the larger sample helpful?

3.55 For a study of state and federal taxation, the Internal Revenue Service is planning to select an SRS of individual federal income tax returns from each state. One variable of interest is the proportion of returns claiming itemized deductions. The total number of tax returns in a state varies from over 12 million in California to fewer than 210,000 in Wyoming.

(a) Will the sampling variability of the sample proportion change from state to state if an SRS of 2000 tax returns is selected in each state? Explain your answer.

(b) Will the sampling variability of the sample proportion change from state to state if an SRS of 1% of all tax returns is selected in each state? Explain your answer.

3.56 A national opinion poll recently estimated that 44% ($\hat{p} = 0.44$) of all American adults agree that parents of school-age children should be given vouchers good for education at any public or private school of their choice. The polling organization used a probability sampling method for which the sample proportion has a normal distribution with standard deviation about 0.015. The poll therefore announced a "margin of error" of 0.03 (two standard deviations) for its result. If a sample were drawn by the same method from the state of New Jersey (population 7.8 million) instead of from the entire United States (population 250 million), would this margin of error be larger or smaller? Explain your answer.

CHAPTER 3 EXERCISES

3.57 Do consumers prefer the taste of Pepsi or Coke in a blind test in which neither cola is identified? Describe briefly the design of a matched pairs experiment to investigate this question.

3.58 Is the number of days a letter takes to reach another city affected by the time of day it is mailed and whether or not the zip code is used? Describe briefly the design of a two-factor experiment to investigate this question. Be sure to specify the treatments exactly and to tell how you will handle lurking variables such as the day of the week on which the letter is mailed.

3.59 The previous two exercises illustrate the use of statistically designed experiments to answer questions that arise in everyday life. Select a question of interest to you that an experiment might answer and carefully discuss the design of an appropriate experiment.

3.60 There are several psychological tests available to measure the extent to which Mexican Americans are oriented toward Mexican/Spanish or Anglo/English culture. Two such tests are the Bicultural Inventory (BI) and the Acculturation Rating Scale for Mexican Americans (ARSMA). To study the correlation between the scores on these two tests, researchers will give both tests to a group of 22 Mexican Americans.

(a) Briefly describe a matched pairs design for this study. In particular, how will you use randomization in your design?

(b) You have an alphabetized list of the subjects (numbered 1 to 22). Carry out the randomization required by your design and report the result.

3.61 A medical study of heart surgery investigates the effect of drugs called beta-blockers on the pulse rate of the patient during surgery. The pulse rate will be measured at a specific point during the operation. The investigators decide to use as subjects 30 patients facing heart surgery who have consented to take part in the study. You have a list of these patients, numbered 1 to 30 in alphabetical order.

(a) Diagram a completely randomized experimental design for this study.

(b) Enter Table B at line 125 to carry out the randomization required by your design and report the result.

3.62 In a study of the relationship between physical fitness and personality, middle-aged college faculty who have volunteered for an exercise program are divided into low-fitness and high-fitness groups on the basis of a physical examination. All subjects then take the Cattell Sixteen Personality Factor Questionnaire (a 187-item multiple-choice test often used by psychologists), and the results for the two groups are compared. Is this study an experiment? Explain your answer.

3.63 A university's financial aid office wants to know how much it can expect students to earn from summer employment. This information will be used to set the level of financial aid. The population contains 3,478 students who have completed at least 1 year of study but have not yet graduated. A questionnaire will be sent to an SRS of 100 of these students, drawn from an alphabetized list.

(a) Describe how you will label the students in order to select the sample.

(b) Use Table B, beginning at line 105, to select the *first five* students in the sample.

3.64 A labor organization wants to study the attitudes of college faculty members toward collective bargaining. These attitudes appear to be different depending on the type of college. The American Association of University Professors classifies colleges as follows:

Class I. Offer doctorate degrees and award at least 15 per year

Class IIA. Award degrees above the bachelor's but are not in Class I

Class IIB. Award no degrees beyond the bachelor's

Class III. Two-year colleges

Discuss the design of a sample of faculty from colleges in your state, with total sample size about 200.

3.65 You want to investigate the attitudes of students at your school toward the faculty's commitment to teaching. The student government will pay the costs of contacting about 500 students.

(a) Specify the exact population for your study; for example, will you include part-time students?
(b) Describe your sample design. Will you use a stratified sample?
(c) Briefly discuss the practical difficulties that you anticipate; for example, how will you contact the students in your sample?

3.66 You are participating in the design of a medical experiment to investigate whether a calcium supplement in the diet will reduce the blood pressure of middle-aged men. Preliminary work suggests that calcium may be effective and that the effect may be greater for black men than for white men.
(a) Diagram the design of an appropriate experiment.
(b) Choosing the sizes of the treatment groups requires more statistical expertise. We will learn more about this aspect of design in later chapters. Explain in plain language the advantage of using larger groups of subjects.

3.67 A chemical engineer is designing the production process for a new product. The chemical reaction that produces the product may have higher or lower yield, depending on the temperature and the stirring rate in the vessel in which the reaction takes place. The engineer decides to investigate the effects of combinations of two temperatures (50° C and 60° C) and three stirring rates (60 revolutions per minute (rpm), 90 rpm, and 120 rpm) on the yield of the process. Two batches of the feedstock will be processed at each combination of temperature and stirring rate.
(a) How many factors are there in this experiment? How many treatments? Identify each of the treatments. How many experimental units (batches of feedstock) does the experiment require?
(b) Diagram the design of an appropriate experiment.
(c) The randomization in this experiment determines the order in which batches of the feedstock will be processed according to each treatment. Use Table B starting at line 128 to carry out the randomization, and state the result.

3.68 A psychologist is interested in the effect of room temperature on the performance of tasks requiring manual dexterity. She chooses temperatures of 70° F and 90° F as treatments. The response variable is the number of correct insertions, during a 30-minute period, in an elaborate peg-and-hole apparatus that requires the use of both hands simultaneously. Each subject is trained on the apparatus and then asked to make as many insertions as possible in 30 minutes of continuous effort.
(a) Outline a completely randomized design to compare dexterity at 70° and 90°. Twenty subjects are available.
(b) Because individuals differ greatly in dexterity, the wide variation in individual scores may hide the systematic effect of temperature unless there are many subjects in each group. Describe in detail the

design of a matched pairs experiment in which each subject serves as his or her own control.

3.69 The requirement that human subjects give their informed consent to participate in an experiment can greatly reduce the number of available subjects. In trials of treatments for cancer, for example, patients must agree to be randomly assigned to the standard therapy or to an experimental therapy. The patients who do not wish their treatment to be decided by a coin toss are dropped from the experiment and given the standard therapy. Why is it not correct to keep these patients in the experiment as part of the control group, since the control group receives the standard therapy?

3.70 You are on the staff of a member of Congress who is considering a controversial bill that would provide for government-sponsored insurance to cover care in nursing homes. You report that 1128 letters dealing with the issue have been received, of which 871 oppose the legislation. "I'm surprised that most of my constituents oppose the bill. I thought it would be quite popular," says the congresswoman. Are you convinced that a majority of the voters oppose the bill? State briefly how you would explain the statistical issue to the congresswoman.

3.71 A study of the effects of running on personality involved 231 male runners who each ran about 20 miles a week. The runners were given the Cattell Sixteen Personality Factor Questionnaire. A news report (*New York Times*, Feb. 15, 1988) stated, "The researchers found statistically significant personality differences between the runners and the 30-year-old male population as a whole." A headline on the article said, "Research has shown that running can alter one's moods." Explain carefully, to someone who knows no statistics, why the headline is misleading.

3.72 To demonstrate how randomization reduces confounding, return to the nutrition experiment described in Example 3.7 (page 233). Suppose that the 30 rats are labeled 01 to 30. Suppose also that, unknown to the experimenter, the 10 rats labeled 01 to 10 have a genetic defect that will cause them to grow more slowly than normal rats. If the experimenter simply put rats 01 to 15 in the experimental group and rats 16 to 30 in the control group, this lurking variable will bias the experiment against the new food product.

Use Table B to assign 15 rats at random to the experimental group as in Example 3.8 (page 236). Record how many of the 10 rats with genetic defects are placed in the experimental group and how many are in the control group. Repeat the randomization using different lines in Table B until you have done five random assignments. What is the mean number of genetically defective rats in experimental and control groups in your five repetitions?

CHAPTER 3 COMPUTER EXERCISES

3.73 In addition to selecting simple random samples, statistical software can randomly assign experimental subjects as follows. If there are n subjects, label them 1 to n, and then arrange the whole numbers from 1 to n in a random order; a treatment group of (say) 10 subjects contains the first 10 in this random order, a second group of 10 contains the second 10, and so on. For example, Minitab randomly orders the numbers 1 to 30 in response to these commands:

```
MTB> SET C1
 DATA> 1:30
 DATA> END
MTB> SAMPLE 30 C1 C2
```

Column C1 has the numbers 1 to 30, and column C2 has these same numbers in random order.

Use your software to randomly assign 150 subjects among 6 treatments.

Most statistical software packages allow you to do simulations. This makes it practical to explore sampling variability more thoroughly than if you use Table B for simulation. Suppose, for example, that you wish to simulate 100 repetitions of an SRS of size 50 from a large population of which the proportion $p = 0.25$ would say "Yes" to a certain question. The following Minitab commands carry out the simulation, store the 100 counts of "Yes" results as column C1, calculate the 100 values of $\hat{p}$ (count/50), and store these sample proportions in column C2:

```
MTB> RANDOM 100 C1;
 SUBC> BINOMIAL N = 50, P = .25.
MTB> LET C2=C1/50
```

Chapter 5 will explain why "binomial" is the magic word here. The next two exercises involve simulations to be carried out using a computer.

3.74 Draw 100 samples of size $n = 50$ from populations with $p = 0.1$, $p = 0.3$, and $p = 0.5$. Make a stemplot of the 100 values of $\hat{p}$ obtained in each simulation. Compare your three stemplots. Do they show about the same variability? How does changing the parameter p affect the sampling distribution? If your software permits, make a normal quantile plot of the values of $\hat{p}$ from the population with $p = 0.5$. Is the sampling distribution approximately normal?

3.75 Draw 100 samples of each of the sizes $n = 50$, $n = 200$ and $n = 800$ from a population with $p = 0.6$. Prepare a frequency histogram of the values of $\hat{p}$ for each simulation, using the same horizontal and vertical scales so that the three graphs can be compared easily. How does increasing the size of an SRS affect the sampling distribution of $\hat{p}$?

NOTES

1. P. H. Rossi, J. D. Wright, G. A. Fisher, and G. Wills, "The urban homeless: Estimating composition and size," *Science*, 235 (1987), pp. 1336–1339.
2. Both observational studies and experiments on the issue of requiring job programs for welfare recipients are discussed in Constance Holden, "Is the time ripe for welfare reform?" *Science*, 238 (1987), pp. 607–609.
3. L. L. Miao, "Gastric freezing: An example of the evaluation of medical therapy by randomized clinical trials," in J. P. Bunker, B. A. Barnes, and F. Mosteller (eds.), *Costs, Risks and Benefits of Surgery*, Oxford University Press, New York, 1977, pp. 198–211.
4. G. B. Kolata, "Love Canal: False alarm caused by botched study," *Science*, 208 (1980), pp. 1239–1242.
5. The information in this example is taken from *The ASCAP Survey and Your Royalties*, ASCAP, New York, undated.
6. For more detail on the material of this section and complete references, see P. E. Converse and M. W. Traugott, "Assessing the accuracy of polls and surveys," *Science*, 234 (1986), pp. 1094–1098.
7. For more detail on the limits of memory in surveys, see N. M. Bradburn, L. J. Rips and S. K. Shevell, "Answering autobiographical questions: The impact of memory and inference on surveys," *Science*, 236 (1987), pp. 157–161.
8. This question and the response are reported in the *New York Times*, May 6, 1982. A different question, given in Exercise 3.43(c) (page 257), asked at the same time drew a 56% response in favor of a nuclear freeze.

CHAPTER

4

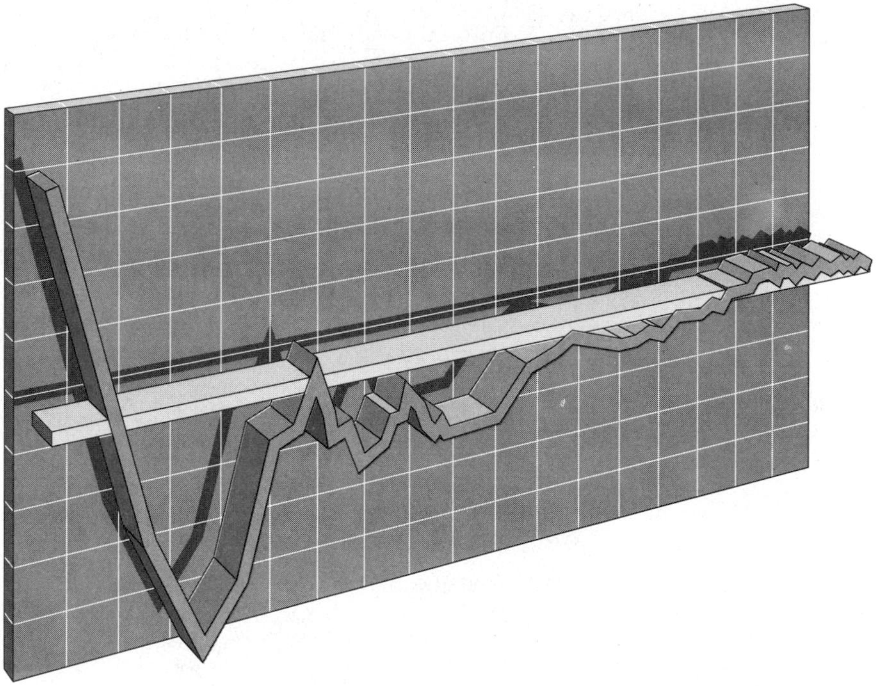

Probability: The Study of Randomness

PRELUDE

Chance is all around us. Sometimes chance results from human design, as in the casino's games of chance and the statistician's random samples. Sometimes nature uses chance, as in choosing the sex of a child. Sometimes the reasons for chance behavior are mysterious, as when the number of deaths each year in a large population is as regular as the number of heads in many tosses of a coin. Probability is the branch of mathematics that describes the pattern of chance outcomes. Probability is the topic of this chapter, but not for its own sake. We will look only at the ideas from probability that we need to go more deeply into statistics. We move quickly from general facts about probability in Section 4.1 to tools for describing the pattern of numerical chance outcomes in Sections 4.2 and 4.3. Later chapters apply these tools to describe the behavior of statistics from random samples and randomized comparative experiments. Even our brief acquaintance with probability will enable us to answer questions like these:

- If we know the blood types of a man and a woman, what can we say about the blood types of their future children?
- Give a test for the AIDS virus to the employees of a small company. What is the chance of at least one positive test if all the people tested are free of the virus?
- An opinion poll asks a sample of 1500 adults what they consider the most serious problem facing our schools. How often will the poll percent who answer "drugs" come within two percentage points of the truth about the entire population?
- If you buy a ticket to Vermont's Green Mountain Numbers state lottery game every day for many years, how much will each ticket win on the average?

• • •

Many things in the world can be predicted, given a sufficiently advanced science and a sufficiently large computer. Drop an apple from a known distance above the ground and elementary physics can tell you how long it will take to fall. Observe the position and motion of an asteroid and rather advanced physics can tell you where it will be a year from now. Other things in the world are unpredictable—the toss of a coin, the time between emissions of alpha particles by a radioactive source, the sexes of the next litter of lab rats. But there is a regularity to these particular phenomena, and to many others, that places them between strict predictability and unanalyzable haphazardness. Although an individual outcome is indeed uncertain, there is a regular *pattern* of outcomes that emerges in many repetitions. The next toss of the coin cannot be foretold, but in the long run very close to half the tosses will be heads and half will be tails.

RANDOM PHENOMENON

We call a phenomenon random if individual outcomes are uncertain but there is nonetheless a regular distribution of relative frequencies in a large number of repetitions.

probability theory

"Random" in the vocabulary of statistics does not mean haphazard but rather refers to a kind of order that emerges only in the long run. The long-run regularity of random phenomena can be described mathematically, just as the fall of an apple or the motion of an asteroid can be predicted. The mathematical study of randomness is called *probability theory*.

We are most interested in the orderly behavior of sample statistics used to gain information about unknown population parameters. If our data come from a probability sample or from a randomized experiment, the values of a statistic are random. Probability theory describes how a statistic will vary in repeated samples or repeated experiments when the state of the underlying population remains unchanged. Suppose, for example, that in fact a proportion $p = 0.6$ of all adults did not attend church or synagogue last week. We poll an SRS of 1785 persons from this population and find that a proportion $\hat{p}$ of the sample were not in church last week. As the histogram in Figure 3.8 (page 265) reminds us, the statistic $\hat{p}$ would vary if we selected repeated samples. We obtained the histogram by simulating 1000 samples. Probability theory allows us to replace simulation by calculation and to obtain the exact distribution, not just the results of a particular 1000 samples.

If the parameter p has a value different from 0.6, the statistic $\hat{p}$ has a different distribution. In an actual statistical application, we would not know that the value of the parameter p is 0.6. That is an essential distinction between probability and statistical inference. Probability describes how the statistic $\hat{p}$ would behave in repeated samples if p were known. Inference starts from one sample that produced one value of $\hat{p}$ and draws conclusions about the value of the unknown parameter p. Probability calculations are nonetheless the basis for inference. This chapter presents the fundamentals of probability theory. With probability in hand we will turn to inference in Chapters 5 and 6.

The idea of probability

The mathematics of probability begins with the observed fact that some phenomena are random—that is, the relative frequencies of their outcomes seem to settle down to fixed values in the long run. Consider tossing a single coin. The relative frequency of heads is quite erratic in 2 or 5 or 10 tosses. But after several thousand tosses it remains stable, changing very little over further thousands of tosses. Some diligent people have in fact made thousands of tosses:

- The French naturalist Buffon (1707–1788) tossed a coin 4040 times. Result: 2048 heads, or relative frequency $2048/4040 = 0.5069$ for heads.
- Around 1900, the English statistician Karl Pearson heroically tossed a coin 24,000 times. Result: 12,012 heads, a relative frequency of 0.5005.
- While imprisoned by the Germans during World War II, the English mathematician John Kerrich tossed a coin 10,000 times. Result: 5067 heads, a relative frequency of 0.5067.

long-term relative frequency

If in many tosses of a coin the proportion of heads observed becomes close to 1/2, we say that the probability of a head on any single toss is 1/2. In intuitive terms, *probability is long-term relative frequency*. From this point of view, correct probabilities can only be determined empirically, by actually watching many tosses of the coin to see if about 1/2 the outcomes are heads. Of course, we can never observe a probability exactly, because we could always continue tossing the coin. Mathematical probability is an idealization based on imagining what would happen to the relative frequencies in an indefinitely long series of trials.

EXAMPLE 4.1

To illustrate the fact that relative frequencies stabilize in the long run, here is an example slightly less obvious than tossing a single coin repeatedly. We used a computer to simulate tossing *four* coins 1000 times and to record the number of heads among the four coins on each trial. We then found the relative frequency of the outcome "exactly 2 heads" after each trial. The results of the first few trials were

Trial	Outcome	Relative frequency of exactly 2 heads
1	3 heads	$\frac{0}{1} = 0$
2	0 heads	$\frac{0}{2} = 0$
3	0 heads	$\frac{0}{3} = 0$
4	2 heads	$\frac{1}{4} = .25$
5	4 heads	$\frac{1}{5} = .20$
6	2 heads	$\frac{2}{6} = .33$

Figure 4.1 plots the relative frequency of the outcome "2 heads" against the number of trials. This relative frequency is 0 for the first three trials, but then jumps up to $1/4 = 0.25$ when the fourth toss yields 2 heads. The relative frequency is quite variable during the first 100 repetitions but stabilizes as we make more trials. In the next section, we will find that the mathematical probability of 2 heads in four tosses of a balanced coin is 0.375. This probability is marked on the graph as a black horizontal line. The gradual approach of the relative frequency to the probability is clearly visible. (The horizontal scale in Figure 4.1 is not linear but logarithmic. This choice makes it easier to see the erratic early fluctuations in the relative frequency while still allowing 1000 tosses to fit on the graph.) ●

personal probability

Long-term relative frequency is not the only intuitive interpretation of probability. From another point of view, probability describes a personal opinion about the chance that the next toss will produce a head. My personal opinion is reflected, for example, in the odds I am willing to accept in betting on the outcome. If I will accept even odds on the toss of a coin, my personal probability of a head is $1/2$. Your opinion, and therefore your personal probability of a head, may differ from mine. *Probability as personal opinion* frees us from having to imagine many repetitions. This freedom has the great advantage that we can assign a probability even to one-time events (for example, "I think the probability that the Redskins will win the Superbowl this year is $1/4$"). On the other hand, the broad scope of personal probability tends to lead us away from long-term regularity of relative frequencies as an observed fact that we want to describe mathematically.

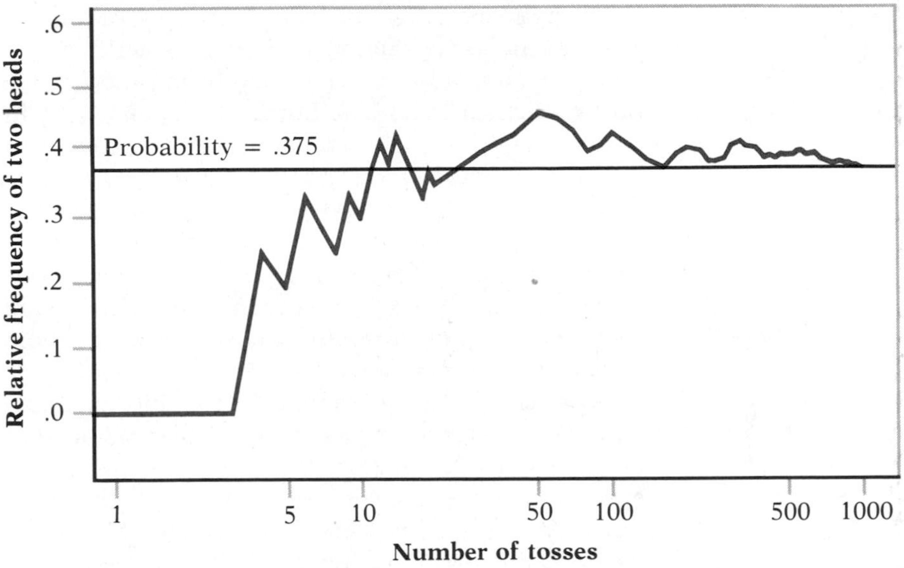

FIGURE 4.1 The long-term behavior of the relative frequency of two heads in many tosses of four coins. The relative frequency approaches the probability 0.375.

The mathematics of probability is motivated by these commonsense notions of what probability means. The mathematics makes more sense if you keep in mind long-term relative frequency or personal assessment of chance. Statistical designs for sampling or experimentation deliberately introduce randomness so that there will be a regular long-term pattern. We will therefore concentrate on the relative frequency interpretation of probability.

The uses of probability

Probability theory originated in the study of games of chance. Tossing dice, dealing shuffled cards, and spinning a roulette wheel are examples of deliberate randomization that are similar to random sampling. Although games of chance are ancient, they were not studied by mathematicians until the sixteenth and seventeenth centuries. It is only a mild simplification to say that probability theory as a branch of mathematics arose when seventeenth-century French gamblers asked the mathematicians Blaise Pascal and Pierre de Fermat for help. Gambling is still with us, in casinos and state lotteries. We will make use of games of chance as simple examples that illustrate the principles of probability. Careful measurements in astronomy and surveying led to further advances in probability in the eighteenth and nineteenth centuries because the results of repeated measurements are random and can be described by relative frequency distributions much like those arising from random sampling. Similar distributions appear in data on human life span (mortality tables) and in data on lengths or weights in a population of skulls, leaves, or cockroaches.[1] In the twentieth century, probability is used to describe the flow of traffic through a highway system, a telephone interchange, or a computer processor; the genetic makeup of individuals or populations; the energy states of subatomic particles; the spread of epidemics or rumors; and the rate of return on risky investments. Although we are interested in probability because of its usefulness in statistics, the mathematics of chance is important in many fields of study.

Statistical inference applies probability calculations to the analysis of data. Our goal is to replace the uncertainties of "eyeball appraisal" by the exactness and power of mathematics. You or I might guess that the beam above our heads is strong enough to keep the building from collapsing, but we feel safer knowing that an engineer has calculated it. Similarly, we might guess that the observed mean difference between the responses of the treatment and control groups indicates a real difference between treatments, but we will feel more confident if we can back our eyeball appraisal with calculations.

4.1 PROBABILITY MODELS

Earlier chapters gave mathematical models for linear relationships (in the form of equations) and for some distributions of data (in the form of normal density curves). Now we must give a mathematical description or model for randomness. To see how to proceed, think first about a very simple random phenomenon, tossing a coin once. When we toss a coin, we cannot know the outcome in advance. What do we know? We are willing to say that the outcome will be either heads or tails. Because the coin appears to be balanced, we believe that each of these outcomes has probability $1/2$. This description of coin tossing has two parts:

- A list of possible outcomes
- A probability for each outcome

Such a description is the basis for all probability models. We will begin by describing the outcomes of a random phenomenon and then learn how to assign probabilities to the outcomes.

Sample spaces

The first element of a probability model is a statement of the outcomes that are possible.

SAMPLE SPACE

The sample space S of a random phenomenon is the set of all possible outcomes.

The name "sample space" is natural in random sampling, where each possible outcome is a sample and the sample space contains all possible samples.

To specify S, we must state what constitutes an individual outcome and then state which outcomes can occur. We often have some freedom in defining the sample space, so the choice of S is a matter of convenience as well as correctness. The idea of a sample space, and the freedom we may have in specifying it, are best illustrated by examples.

EXAMPLE 4.2

Toss a coin. There are only two possible outcomes, and the sample space is

$$S = \{\text{Heads, Tails}\}$$

or, abbreviated, $S = \{\text{H, T}\}$. ●

EXAMPLE 4.3

Let your pencil point fall blindly into Table B of random digits and record the value of the digit you land on. Plainly,

$$S = \{0, 1, 2, 3, 4, 5, 6, 7, 8, 9\}$$ ●

EXAMPLE 4.4

Toss a coin four times and record the results. That's a bit vague. To be exact, record the results of each of the four tosses in order. A typical outcome is then HTTH. Counting shows that there are 16 possible outcomes. The sample space S is the set of all 16 strings of four H's and T's.

Suppose that our only interest is the number of heads in four tosses. Now we can be exact in a simpler fashion: The random phenomenon is to toss a coin four times and count the number of heads. The sample space contains only five outcomes:

$$S = \{0, 1, 2, 3, 4\}$$

This example illustrates the importance of carefully specifying what constitutes an individual outcome. ●

Although these examples seem remote from the practice of statistics, the connection is surprisingly close. Suppose that in the course of conducting an opinion poll you select four people at random from a large population and ask each if he or she favors reducing federal spending on low-interest student loans. The answers are "Yes" or "No." The possible outcomes—the sample space—are exactly as in Example 4.4 if we replace heads by "Yes" and tails by "No." Similarly, the possible outcomes of an SRS of 1500 people are the same in principle as the possible outcomes of tossing a coin 1500 times. One of the great advantages of mathematics is that the essential features of quite different phenomena can be described by the same mathematical model.

EXAMPLE 4.5

Many computing systems have a function that will generate a random number between 0 and 1. For example, the Minitab command

```
MTB> RANDOM 1 C1;
 SUBC> UNIFORM 0 1.
```

puts such a number into C1. Because the computer routine generates a random number between 0 and 1,

$$S = \{\text{all numbers between 0 and 1}\}$$

This S is a mathematical idealization. Any specific random number generator produces numbers with some limited number of decimal places so that, strictly speaking, not all numbers between 0 and 1 are possible outcomes. The entire interval from 0 to 1 is easier to think about. It also has the advantage of being a suitable sample space for different computers that produce random numbers with different numbers of significant digits. ●

Assigning probabilities

A sample space S lists the possible outcomes of a random phenomenon. To complete a mathematical model for the random phenomenon, we must also give the probability with which these outcomes occur.

The true long-term relative frequency of any outcome—say, "exactly 2 heads in four tosses of a coin"—can be found only by experiment, and then only approximately. How then can we describe probability mathematically? Rather than immediately attempt to give "correct" probabilities, let's confront the easier task of laying down rules that any assignment of probabilities must satisfy. We need to assign probabilities not only to single outcomes but also to sets of outcomes.

> **EVENT**
>
> An event is a set of outcomes of a random phenomenon, that is, a subset of the sample space.

EXAMPLE 4.6

Take the sample space S for four tosses of a coin to be the list of all possible outcomes in the form HTHH. Then "exactly 2 heads" is an event. Call this event A. The event A expressed as a set of outcomes is

$$A = \{\text{HHTT, HTHT, HTTH, THHT, THTH, TTHH}\}$$

Listing the outcomes in an event such as A is tedious; the verbal description "exactly 2 heads" is shorter. If we let X be the number of heads, we have an even shorter description: A is the event that X=2, which we write as $\{X = 2\}$. The event "2 or more heads" can similarly be described as $\{X \geq 2\}$. We will follow up this natural way of describing numerical outcomes in the next section. ●

In a probability model, events have probabilities. We will write the probability of the event A as $P(A)$. What properties must any assignment of probabilities to events have? A relative frequency is always a number between 0 and 1. The sample space S contains all possible outcomes and therefore always has relative frequency 1. If we think of probability as long-run relative frequency, we are led to two rules that any assignment of probabilities must obey.

BASIC PROBABILITY RULES

Rule 1. Any probability $P(A)$ is a number between 0 and 1. That is, $0 \leq P(A) \leq 1$.

Rule 2. The collection S of all possible outcomes has probability 1. That is, $P(S) = 1$.

There are other rules that any legitimate assignment of probabilities to events must obey. Before we meet these rules, let us look at actual assignments of probability in Examples 4.2 and 4.3.

EXAMPLE 4.7

In tossing a coin (Example 4.2) there are only two outcomes in S. We simply assign probabilities to each of them. If we believe that the coin is balanced, then we assign

$$P(H) = .5$$
$$P(T) = .5$$

If observation convinces us that the coin is unbalanced so as to land tails more often, we might assign probabilities

$$P(H) = .4$$
$$P(T) = .6$$

It is *not* true that the probability of each outcome must be 1/2 just because there are two outcomes. Some coins are unbalanced. ●

In Example 4.7, any assignment of two probabilities (numbers between 0 and 1) with sum 1 satisfies Rules 1 and 2. We can determine which assignment is correct for a particular coin only by observing many tosses. This reinforces the lesson that our mathematics only says which assignments of probability make sense, not which is actually correct.

EXAMPLE 4.8

The successive digits in Table B (Example 4.3) were produced by a careful randomization that makes each entry equally likely to be any of the 10 candidates. Because the total probability must be 1, the probability of each of the 10 outcomes must be 1/10. This assignment of probabilities to individual outcomes can be summarized in a table as follows:

Outcome	0	1	2	3	4	5	6	7	8	9
Probability	.1	.1	.1	.1	.1	.1	.1	.1	.1	.1

We must assign probability to all events, not just to individual outcomes. The probability of an event in this example is simply the sum of the probabilities of the outcomes making up the event. For example, the probability that an odd digit is chosen is

$$P(\text{odd outcome}) = P(1) + P(3) + P(5) + P(7) + P(9) = .5$$

This assignment of probability satisfies Rules 1 and 2. ●

Examples 4.7 and 4.8 illustrate one way of assigning probabilities to events when a random phenomenon has only a finite number of possible outcomes.

PROBABILITIES IN A FINITE SAMPLE SPACE

Assign a probability to each individual outcome. These probabilities must be numbers between 0 and 1, and must have sum 1. The probability of any event is the sum of the probabilities of the outcomes making up the event.

Assigning correct probabilities to individual outcomes often requires long observation of the random phenomenon. In some special circumstances, however, we are willing to assume that individual outcomes are equally likely because of some balance in the phenomenon we are observing. Ordinary coins have a physical balance that should make heads and tails equally likely, and the table of random digits comes from a deliberate randomization. In Example 4.8 all outcomes have the same probability. Because there are 10 equally likely outcomes, each must have probability 1/10. Because exactly 5 of the 10 equally likely outcomes are odd, the probability of an odd outcome is 5/10, or 0.5. In the special situation where all outcomes are equally likely, we have a simpler rule for assigning probabilities to events.

EQUALLY LIKELY OUTCOMES

If a random phenomenon has k possible outcomes, all equally likely, then each individual outcome has probability $1/k$. The probability of any event A is

$$P(A) = \frac{\text{count of outcomes in } A}{\text{count of outcomes in } S} = \frac{\text{count of outcomes in } A}{k}$$

Most random phenomena do not have equally likely outcomes, so the general rule for finite sample spaces is more important than the special rule for equally likely outcomes. Here is an example in which the outcomes are not equally likely.

EXAMPLE 4.9

Select one M&M candy at random from a large bag and record its color. The sample space is

$$S = \{\text{brown, green, orange, red, tan, yellow}\}$$

The probability of each outcome will be the proportion of all M&M's made that have that color. (As usual this is an idealization that imagines the result of looking at one candy after another forever.) From the manufacturer we can learn that the correct assignment of probabilities to the outcomes is

Color	Brown	Red	Yellow	Green	Orange	Tan
Probability	.3	.2	.2	.1	.1	.1

Check that these probabilities have sum 1. We assign probabilities to events by adding the outcome probabilities. For example,

$$P(\text{red or orange}) = P(\text{red}) + P(\text{orange}) = .2 + .1 = .3$$

●

Addition and multiplication rules

Rules 1 and 2 concern the assignment of probabilities to single events. To complete a basic description of probability, we must learn a few additional rules that govern how probabilities of events combine. Three such rules are important for statistics. In each case, we first describe a particular relation between two events and then the probability rule that applies to this relation.

Disjoint events Suppose that a large dormitory contains 40% freshmen, 20% seniors, and 40% belonging to other classes. The relative frequency of students who are *either* freshmen *or* seniors is 60%. We can add the relative frequencies of freshmen and seniors because no student can be both a freshman and a senior. This "addition rule" is one of the most important facts about probability.

> **DISJOINT EVENTS**
>
> Events that have no outcomes in common are called disjoint events.

Venn diagram

Select a student at random from the inhabitants of the dormitory and record the student's class. "Freshman" and "senior" are disjoint events. Figure 4.2 illustrates disjoint events A and B as nonoverlapping areas. Diagrams like Figure 4.2 that picture the sample space S as an area in the plane and events as areas within S are called *Venn diagrams*. Venn diagrams are helpful in showing relations such as disjointness.

Two disjoint events cannot both occur on the same trial of a random phenomenon. The long-term relative frequency that "one or the other" occurs is therefore the sum of their individual relative frequencies. Any

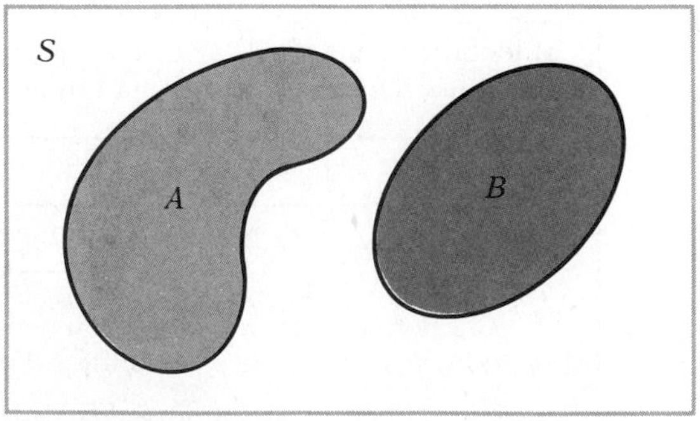

FIGURE 4.2 Venn diagram showing disjoint events A and B.

legitimate assignment of probability must therefore satisfy the addition rule for disjoint events.

> **ADDITION RULE FOR DISJOINT EVENTS**
>
> **Rule 3.** If events A and B are disjoint, then
>
> $$P(A \text{ or } B) = P(A) + P(B)$$

EXAMPLE 4.10

The addition rule is satisfied when we assign probability by summing the probabilities of individual outcomes. Example 4.8 gives an assignment of probabilities for choosing a random digit between 0 and 9. Consider the events

$$A = \text{the digit chosen is odd}$$
$$B = \text{the digit chosen is a multiple of 4}$$

The outcomes in A are 1, 3, 5, 7, and 9, so $P(A) = 0.5$. The event B contains outcomes 0, 4, and 8, with $P(B) = 0.3$. What is the probability of the event $\{A \text{ or } B\}$ that the random digit is *either* odd *or* a multiple of 4? This event contains all of the outcomes in either A or B, namely, 0, 1, 3, 4, 5, 7, 8, and 9. By the rule for equally likely outcomes, these 8 outcomes have probability 8/10. Because A and B have no outcomes in common, the same result is obtained by adding $P(A)$ and $P(B)$:

$$\begin{aligned} P(A \text{ or } B) &= P(A) + P(B) \\ &= .5 + .3 = .8 \end{aligned}$$

●

Complements The addition rule leads to another important law of probability, the complement rule. If 20% of the students in a dormitory are seniors, then 80% are not seniors. That is, the relative frequency that a student is a senior and the relative frequency that a student is *not* a senior must add to 100%. This is true because every student either is or is not a senior and no student is both.

> **COMPLEMENT OF AN EVENT**
>
> If A is any event, the event that A does *not* occur is called the complement of A, written A^c.

Figure 4.3 is a Venn diagram that illustrates how an event and its complement partition the sample space between them. Notice that A and

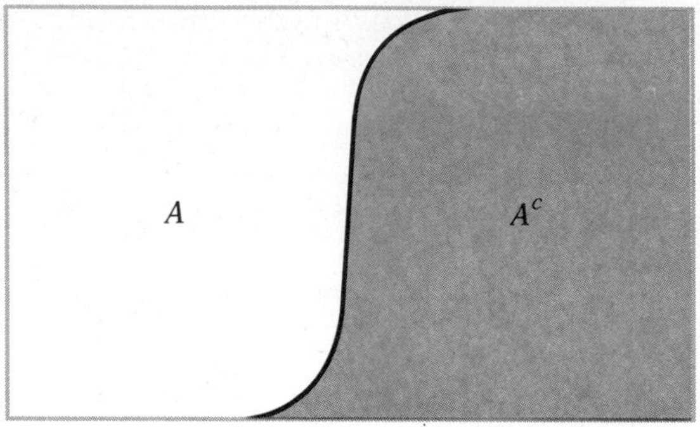

FIGURE 4.3 Venn diagram showing the complement A^c of an event A.

A^c are always disjoint events. The complement A^c contains exactly those outcomes that are not in A. The long-term relative frequency of A^c is therefore 1 minus the relative frequency of A because A^c occurs exactly when A does not. This fact is another of our basic rules for probability.

COMPLEMENT RULE

Rule 4. For any event A, the probability that A does not occur is

$$P(A^c) = 1 - P(A)$$

EXAMPLE 4.11

A sociologist studies social mobility in England by recording the social class of a large sample of fathers and their sons. Social class is determined by such factors as education and occupation. The social classes are ordered from Class 1 (lowest) to Class 5 (highest). Here are the probabilities that the son of a lower-class (Class 1) father ends up in each social class:

Son's class	1	2	3	4	5
Probability	.48	.38	.08	.05	.01

Consider the events

A = son remains in Class 1

B = son reaches one of the two highest classes

From the table of probabilities,

$$P(A) = .48$$
$$P(B) = .05 + .01 = .06$$

What is the probability that the son of a Class 1 father does *not* remain in Class 1? This is

$$P(A^c) = 1 - P(A)$$
$$= 1 - .48 = .52$$

The events A and B are disjoint, so the probability that the son of a lower-class father either remains in the lower class or reaches one of the two top classes is

$$P(A \text{ or } B) = P(A) + P(B)$$
$$= .48 + .06 = .54$$

●

Independence Rule 3, the addition rule for disjoint events, describes the probability that *one or the other* of two events A and B will occur in the special situation when A and B cannot occur together because they are disjoint. Our final rule describes the probability that *both* events A and B occur, again only in a special situation. More general rules appear in Section 4.4, but in our study of statistics we will need only the rules that apply to special situations.

Suppose that you toss a balanced coin twice. You are counting heads, so two events of interest are

$$A = \text{first toss is a head}$$
$$B = \text{second toss is a head}$$

The events A and B are not disjoint. They occur together whenever both tosses give heads. We want to compute the probability of the event $\{A \text{ and } B\}$ that *both* tosses are heads. Figure 4.4 is a Venn diagram that illustrates the event $\{A \text{ and } B\}$ as the overlapping area that is common to both A and B.

The coin-tossing experiments of Buffon, Pearson, and Kerrich described at the beginning of this chapter make us willing to assign

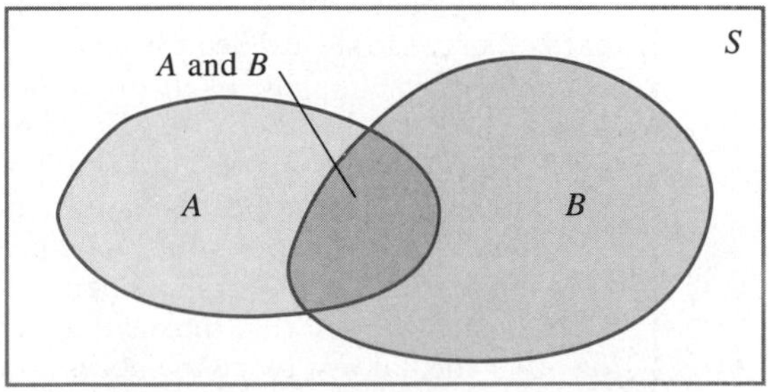

FIGURE 4.4 **Venn diagram showing the event $\{A$ and $B\}$.**

probability 1/2 to a head when we toss a coin. So

$$P(A) = .5$$
$$P(B) = .5$$

What is $P(A \text{ and } B)$? Our common sense says that it is 1/4. The first coin will give a head half the time and then the second will give a head on half of those trials, so both coins will give heads on $1/2 \times 1/2 = 1/4$ of all trials in the long run. This reasoning assumes that the second coin still has probability 1/2 of a head after the first has given a head. This is true—we can verify it by tossing two coins many times and observing the relative frequency of heads on the second toss after the first toss has produced a head. We say that the events "head on the first toss" and "head on the second toss" are independent.

INDEPENDENCE

Events A and B are independent if knowing whether A occurs does not change the probability that B occurs.

This definition is rather informal. A more precise definition appears in Section 4.4. In practice, though, we rarely need a precise definition of independence, because independence is usually *assumed* as part of a probability model when we want to describe random phenomena that seem to be physically unrelated to each other.

EXAMPLE 4.12

Because a coin has no memory and most coin-tossers cannot influence the fall of the coin, it is safe to assume that successive coin tosses are independent. For a balanced coin this means that after we see the outcome of the first toss, we still assign probability 1/2 to heads on the second toss.

On the other hand, the colors of successive cards dealt from the same deck are not independent. A standard 52-card deck contains 26 red and 26 black cards. For the first card dealt from a shuffled deck, the probability of a red card is $26/52 = 0.50$ (equally likely outcomes again). Once we see that the first card is red, we know that there are only 25 reds among the remaining 51 cards. The probability that the second card is red is therefore only $25/51 = 0.49$. Knowing the outcome of the first deal changes the probabilities for the second.

If a doctor measures your blood pressure twice, it is reasonable to assume that the results X and Y are independent because the first result does not influence the instrument that makes the second reading. But if you take an IQ test or other mental test twice in succession, the test scores X and Y are not independent. The learning that occurs on the first attempt influences your second attempt. ●

We concluded in the case of tossing two coins that the probability that *both* of two independent events occur is the product of their individual probabilities. This fact is always true and is our final basic probability rule.

MULTIPLICATION RULE FOR INDEPENDENT EVENTS

Rule 5. If events A and B are independent, then

$$P(A \text{ and } B) = P(A)P(B)$$

EXAMPLE 4.13

A person's blood type is inherited from the parents. Inheritance operates randomly. Each parent carries two genes for blood type, and each of these genes has probability 1/2 of being passed to the child. The two genes that the child receives, one from each parent, determine the child's blood type. The parents contribute their genes independently of each other.

Suppose that both parents carry the O and the A genes. The child will have blood type O if both parents contribute an O gene; otherwise the child's blood type will be A. If M is the event that the mother contributes her O gene and F is the event that the father contributes his O gene, then the probability that the child has O-type blood is

$$\begin{aligned} P(M \text{ and } F) &= P(M)P(F) \\ &= (.5)(.5) = .25 \end{aligned}$$

In the long run, one-fourth of all children born to such parents will have type O blood. ●

EXAMPLE 4.14

Choose at random a U.S. resident at least 25 years of age. We are interested in the events

A = the person chosen completed 4 years of college

B = the person chosen is between 25 and 34 years old

Government data recorded in Table 2.6 (page 184) allow us to assign probabilities $P(A) = 0.203$ and $P(B) = 0.283$ to these events. These are the relative frequencies of events A and B in the entire population, so they are also the probabilities of observing the events when we choose one person at random. The probability that the person chosen is 25 to 34 years old *and* has 4 years of college education is $P(A \text{ and } B) = 0.067$. This is not equal to the product $P(A)P(B) = 0.057$ because A and B are not independent events. Younger adults tend to have more education than the overall adult population, so the probability of choosing a young college graduate is higher than the multiplication rule suggests. ●

The multiplication rule $P(A \text{ and } B) = P(A)P(B)$ is valid if A and B are independent but not otherwise. The addition rule $P(A \text{ or } B) = P(A)+P(B)$ is valid if A and B are disjoint but not otherwise. Resist the temptation to use these simple formulas when the circumstances that justify them are not present. You must also be certain not to confuse disjointness and independence. If A and B are disjoint, then the fact that A occurs tells us that B cannot occur. So disjoint events are not independent. Unlike disjointness or complements, independence cannot be pictured by a Venn diagram, because it involves the probabilities of the events rather than just the outcomes that make up the events.

Like the rule for complements, the multiplication rule for independence can be derived from the first three rules of probability. We are not concerned with such derivations, but it is useful to note that we need check only the first three rules to be sure that an assignment of probabilities to events is legitimate. All other probability laws follow from these three.*

If two events A and B are independent, then their complements A^c and B^c are also independent, and A^c is independent of B. Suppose for example that 75% of all registered voters in a rural district are Republicans. If an opinion poll interviews two voters chosen independently, the probability that the first is a Republican and the second is not a Republican is $(0.75)(0.25) = 0.1875$. The multiplication rule also extends to collections of more than two events, provided that all are independent. Independence of events A, B, and C means that no information about any one or any two can change the probability of the remaining events. The formal definition is a bit messy. Fortunately, independence is usually assumed in setting up a probability model. We can then use the multiplication rule freely, as in this example.

EXAMPLE 4.15

A transatlantic telephone cable contains repeaters at regular intervals to amplify the signal. If a repeater fails, it must be replaced by fishing the cable to the surface at great expense. Each repeater has probability 0.999 of functioning without failure for 10 years. Repeaters fail independently of each other. (This assumes that there are no "common causes" such as earthquakes that would affect several repeaters at once.) Denote by A_i the event that the ith repeater operates successfully for 10 years.

The probability that two repeaters both last 10 years is

$$\begin{aligned} P(A_1 \text{ and } A_2) &= P(A_1)P(A_2) \\ &= .999 \times .999 = .998 \end{aligned}$$

*Although probability is an old branch of mathematics, the idea of deriving all probability laws from a few basic rules is rather recent. The Russian mathematician A. N. Kolmogorov did this in 1933. (Actually, a more elaborate version of Rule 3 is needed to obtain all of the usual theory of probability.)

For a cable with 10 repeaters the probability of no failures in 10 years is

$$\begin{aligned} P(A_1 \text{ and } A_2 \text{ and} \ldots \text{and } A_{10}) &= P(A_1)P(A_2)\cdots P(A_{10}) \\ &= .999 \times .999 \times \cdots \times .999 \\ &= .999^{10} = .990 \end{aligned}$$

Cables with 2 or 10 repeaters are quite reliable. Unfortunately, a transatlantic cable has 300 repeaters. The probability that all 300 work for 10 years is

$$P(A_1 \text{ and } A_2 \text{ and} \ldots \text{and } A_{300}) = .999^{300} = .741$$

There is therefore about one chance in four that the cable will have to be fished up for replacement of a repeater sometime during the next 10 years. Repeaters are in fact designed to be much more reliable than 0.999 in 10 years. Some transatlantic cables have served for more than 20 years with no failures. ●

By combining the rules we have learned, we can compute probabilities for rather complex events. Here is an example.

EXAMPLE 4.16

A diagnostic test for the presence of the AIDS virus has probability 0.005 of producing a false positive. That is, when a person free of the AIDS virus is tested, the test has probability 0.005 of falsely indicating that the virus is present. If the 140 employees of a medical clinic are tested and all 140 are free of AIDS, what is the probability that at least one false positive will occur?

It is reasonable to assume as part of the probability model that the test results for different individuals are independent. The probability that the test is positive for a single person is 0.005, so the probability of a negative result is $1 - 0.005 = 0.995$ by the complement rule. The probability of at least one false positive among the 140 people tested is therefore

$$\begin{aligned} P(\text{at least one positive}) &= 1 - P(\text{no positives}) \\ &= 1 - P(140 \text{ negatives}) \\ &= 1 - .995^{140} \\ &= 1 - .496 = .504 \end{aligned}$$

The probability is greater than 1/2 that at least one of the 140 people will test positive for AIDS, even though no one has the virus. ●

SUMMARY

Probability can be viewed intuitively as relative frequency in many repeated trials of a random phenomenon. Alternatively, a probability can express a personal assessment of chance. Because we are familiar with sampling distributions, we emphasize the interpretation of probability as idealized long-term relative frequency.

A **probability model** for a random phenomenon consists of a **sample space** S and an assignment of probabilities P. S is the set of all possible outcomes of the random phenomenon. Sets of outcomes are called **events.** P assigns a number $P(A)$ to an event A as its probability.

Events A and B are called **disjoint** if they have no outcomes in common. The **complement** A^c of an event A consists of exactly the outcomes that are not in A. Events A and B are called **independent** if knowing whether one event occurs does not change the probability we would assign to the other event.

Any assignment of probability must obey the rules that state the basic properties of probability:

1. $0 \leq P(A) \leq 1$ for any event A.
2. $P(S) = 1$.
3. **Addition rule:** If A and B are disjoint events, then $P(A \text{ or } B) = P(A) + P(B)$.
4. **Complement rule:** For any event A, $P(A^c) = 1 - P(A)$.
5. **Multiplication rule:** If events A and B are independent, then $P(A \text{ and } B) = P(A)P(B)$

SECTION 4.1 EXERCISES

4.1 Suppose that instead of tossing a penny, you spin it on a smooth surface and observe which face is up when the coin finally falls. Assign approximate probabilities to the outcomes {heads, tails} by making at least 20 trials and using the relative frequencies to estimate the probabilities. (Hold the penny upright with a finger and snap it with a finger of the other hand to start it spinning. Do not count trials in which you fail to spin the coin well or in which the coin hits an object before falling.) If possible, combine your results with those of other students to obtain long-term relative frequencies that are closer to the probabilities.

4.2 In the game of *heads or tails*, Betty and Bob toss a coin four times. Betty wins a dollar from Bob for each head and pays Bob a dollar for each tail—that is, she wins or loses the difference between the number of heads and the number of tails. For example, if there are one head and three tails, Betty loses \$2. You can check that Betty's possible outcomes are

$$\{-4, -2, 0, 2, 4\}$$

Assign probabilities to these outcomes by playing the game 20 times and using the relative frequencies of the outcomes as estimates of the probabilities. If possible, combine your trials with those of other students to obtain long-run relative frequencies that are closer to the probabilities.

4.3 In each of the following situations, describe a sample space S for the indicated random phenomenon. In some cases, you have some freedom in your choice of S.

(a) A seed is planted in the ground. It either germinates or fails to grow.
(b) A patient with a usually fatal form of cancer is given a new treatment. The response variable is the length of time that the patient lives after treatment.
(c) A student enrolls in a statistics course and at the end of the semester receives a letter grade.
(d) A basketball player shoots two free throws.
(e) A year after knee surgery, a patient is asked to rate the amount of pain in the knee. A 7-point scale is used, with 1 corresponding to no pain and 7 corresponding to extreme discomfort.

4.4 In each of the following situations, describe a sample space S for the indicated random phenomenon. In some cases you have some freedom in specifying S, especially in setting the largest and smallest value in S.

(a) Choose a student in your class at random. Ask how much time that student spent studying during the past 24 hours.
(b) The Physicians' Health Study asked 11,000 physicians to take an aspirin every other day and observed how many of them had a heart attack in a 5-year period.
(c) In a test of a new package design, you drop a carton of a dozen eggs from a height of 1 foot and count the number of broken eggs.
(d) Choose a student in your class at random. Ask how much cash that student is carrying.
(e) A nutrition researcher feeds a new diet to a young male white rat. The response variable is the weight (in grams) that the rat gains in 8 weeks.

4.5 Let X be the maximum daily rainfall (in inches) recorded next year at South Bend, Indiana. Give a reasonable sample space S for the possible values of X. (Exercise 1.20 on page 27 has past values of X to guide you.)

4.6 Let X be the number of calories in a hot dog. Give a reasonable sample space S for the possible values of X. (Table 1.4 on page 36 contains some typical values to guide you.)

4.7 All human blood can be typed as either O, A, B, or AB, but the distribution of the types varies a bit with race. Here is the distribution of the blood type of a randomly chosen black American. If this is to be a legitimate assignment of probabilities, what must be the probability of type AB blood?

Blood type	O	A	B	AB
Probability	.49	.27	.20	

4.8 The distribution of colors in M&M peanut candies differs from the distribution given in Example 4.9 (page 289) for plain M&M's. The table below gives the probability that a randomly chosen peanut M&M has each color. What must the probability for tan candies be to make this a legitimate assignment of probabilities?

Color	Brown	Red	Yellow	Green	Orange	Tan
Probability	.3	.2	.2	.2	.1	

4.9 Here are several assignments of probabilities to the six faces of a die. We can learn which assignment is actually *accurate* for a particular die only by rolling the die many times. However, some of the assignments violate Rules 1 or 2, and so are not *legitimate* assignments of probability. Which are legitimate and which are not? In the case of the illegitimate models, explain what is wrong.

Outcome	Model 1	Model 2	Model 3	Model 4
⚀	$\frac{1}{3}$	$\frac{1}{6}$	$\frac{1}{7}$	$\frac{1}{3}$
⚁	0	$\frac{1}{6}$	$\frac{1}{7}$	$\frac{1}{3}$
⚂	$\frac{1}{6}$	$\frac{1}{6}$	$\frac{1}{7}$	$-\frac{1}{6}$
⚃	0	$\frac{1}{6}$	$\frac{1}{7}$	$-\frac{1}{6}$
⚄	$\frac{1}{6}$	$\frac{1}{6}$	$\frac{1}{7}$	$\frac{1}{3}$
⚅	$\frac{1}{3}$	$\frac{1}{6}$	$\frac{1}{7}$	$\frac{1}{3}$

4.10 In each of the following situations, state whether or not the given assignment of probabilities to individual outcomes is legitimate, that is, satisfies Rules 1 and 2. If not, give specific reasons for your answer.

(a) When a coin is spun, $P(\text{H}) = 0.55$ and $P(\text{T}) = 0.45$.
(b) When two coins are tossed, $P(\text{HH}) = 0.4, P(\text{HT}) = 0.4, P(\text{TH}) = 0.4$, and $P(\text{TT}) = 0.4$.
(c) The mixture of colors for M&M's given in Example 4.9 (page 289) is quite new. Previously, there were no red candies, and the other five colors had the same probabilities that are given in Example 4.9.

4.11 When asked to predict the Atlantic Coast Conference basketball champion, Las Vegas Zeke follows the modern practice of giving probabilistic predictions. He says, "North Carolina's probability of winning is twice Duke's. North Carolina State and Virginia each have probability 0.1 of winning, but Duke's probability is three times that. Nobody else has a chance." Has Zeke given a legitimate assignment of probabilities to the eight teams in the conference? Explain your answer.

4.12 You draw an M&M from a bag, with an outcome governed by the probabilities given in Example 4.9 (page 289). Find the probability of each of the following events.
(a) You select brown or red.
(b) You select green, red, or tan.
(c) The M&M you draw is not yellow.
(d) The M&M you select is neither orange nor tan.
(e) You select brown, red, yellow, green, orange, or tan.

4.13 A sociologist studying social mobility in Denmark finds that the probability that the son of a lower-class father remains in the lower class is 0.46. What is the probability that the son moves to one of the higher classes?

4.14 The probability that a randomly chosen American woman aged 20 to 24 is married is 0.36. What is the probability that she is not married?

4.15 Government data assign a single cause for each death that occurs in the United States. The data show that the probability is 0.45 that a randomly chosen death was due to cardiovascular (mainly heart) diseases, and 0.22 that it was due to cancer. What is the probability that a death was due either to cardiovascular disease or to cancer? What is the probability that the death was due to some other cause?

4.16 Choose an acre of land in the United States at random. The probability is 0.20 that it is forested and 0.28 that it is pasture (including rangeland). What is the probability that the land chosen is either forest or pasture? What is the probability that the land is anything other than forest or pasture?

4.17 Select a first-year college student at random and ask what his or her academic rank was in high school. Here are the probabilities, based on relative frequencies from a large sample survey:

Outcome	Top 20%	Second 20%	Third 20%	Fourth 20%	Lowest 20%
Probability	.41	.23	.29	.06	.01

Let A be the event that the student selected ranked in the top 40% in high school, and let B be the event that the student fell in the lower 40%.

(a) Find $P(A)$ and $P(B)$.

(b) Describe the event A^c in words. Find $P(A^c)$ in two ways, first by adding the probabilities of outcomes and then by the complement rule.

(c) The events A and B are disjoint. Find $P(A \text{ or } B)$ in two ways, first by adding the probabilities of outcomes and then by the addition rule.

4.18 Choose an American farm at random and measure its size in acres. Here are the probabilities that the farm chosen falls in several acreage categories:

Acres	< 10	10–49	50–99	100–179	180–499	500–999	1000–1999	≥ 2000
Probability	.08	.20	.15	.16	.24	.09	.05	.03

Let A be the event that the farm is less than 50 acres in size, and let B be the event that it is 500 acres or more.

(a) Find $P(A)$ and $P(B)$.

(b) Describe A^c in words and find $P(A^c)$ by the complement rule.

(c) Describe $\{A \text{ or } B\}$ in words and find its probability by the addition rule.

4.19 Choose an American worker at random and classify his or her occupation into one of the following classes. These classes are used in government employment data.

A	Managerial and professional
B	Technical, sales, administrative support
C	Service occupations
D	Precision production, craft, and repair
E	Operators, fabricators, and laborers
F	Farming, forestry, and fishing

The table below gives the probabilities that a randomly chosen worker falls into each of 12 sex-by-occupation classes.

Class	A	B	C	D	E	F
Male	.14	.11	.05	.11	.12	.025
Female	.11	.20	.08	.01	.04	.005

(a) Verify that this is a legitimate assignment of probabilities to these outcomes.
(b) What is the probability that the worker is female?
(c) What is the probability that the worker is not engaged in farming, forestry, or fishing?
(d) Classes D and E include most mechanical and factory jobs. What is the probability that the worker holds a job in one of these classes?
(e) What is the probability that the worker does not hold a job in classes D or E?

4.20 A roulette wheel has 38 slots, numbered 0, 00, and 1 to 36. The slots 0 and 00 are colored green, 18 of the others are red, and 18 are black. The dealer spins the wheel and at the same time rolls a small ball along the wheel in the opposite direction. The wheel is carefully balanced so that the ball is equally likely to land in any slot when the wheel slows. Gamblers can bet on various combinations of numbers and colors.
(a) What is the probability that the ball will land in any one slot?
(b) If you bet on "red," you win if the ball lands in a red slot. What is the probability of winning?
(c) The slot numbers are laid out on a board on which gamblers place their bets. One column of numbers on the board contains all multiples of 3, that is, 3, 6, 9, . . . , 36. You place a "column bet" that wins if any of these numbers comes up. What is your probability of winning?

4.21 Abby, Deborah, Jim, Julie, and Sam work in a firm's public relations office. Their employer must choose two of them to attend a conference in Paris. To avoid unfairness, the choice will be made by drawing two names from a hat. (This is an SRS of size 2.)
(a) Write down all possible choices of two of the five names. This is the sample space.
(b) The random drawing makes all choices equally likely. What is the probability of each choice?
(c) What is the probability that Julie is chosen?
(d) What is the probability that neither of the two men (Jim and Sam) is chosen?

4.22 A general can plan a campaign to fight one major battle or three small battles. He believes that he has probability 0.6 of winning the large battle and probability 0.8 of winning each of the small battles. Victories or defeats in the small battles are independent. The general must win either the large battle or all three small battles to win the campaign. Which strategy should he choose?

4.23 An automobile manufacturer buys computer chips from a supplier. The supplier sends a shipment containing 5% defective chips. Each chip chosen from this shipment has probability 0.05 of being defective, and each automobile uses 12 chips selected independently. What is the probability that all 12 chips in a car will work properly?

4.24 Government data show that 26% of the civilian labor force have at least 4 years of college and that 15% of the labor force work as laborers or operators of machines or vehicles. Can you conclude that because $(0.26)(0.15) = 0.039$ about 4% of the labor force are college-educated laborers or operators? Explain your answer.

4.25 A string of Christmas lights contains 20 lights. The lights are wired in series, so that if any light fails, the whole string will go dark. Each light has probability 0.02 of failing during a 3-year period. The lights fail independently of each other. What is the probability that the string of lights will remain bright for 3 years?

4.26 A six-sided die has four green and two red faces and is balanced so that each face is equally likely to come up. The die will be rolled several times. You must choose one of the following three sequences of colors; you will win $25 if the first rolls of the die give the sequence that you have chosen.

RGRRR
RGRRRG
GRRRRR

Which sequence do you choose? Explain your choice. (In a psychological experiment, 63% of 260 students who had not studied probability chose the second sequence. This is evidence that our intuitive understanding of probability is not very accurate. This and similar experiments are reported by A. Tversky and D. Kahneman, "Extensional versus intuitive reasoning: The conjunction fallacy in probability judgment," *Psychological Review*, 90 (1983), pp. 293–315.)

4.27 Suppose that both parents carry genes for blood types A and B. Each parent passes one of these genes to a child and is equally likely to pass either gene. The two parents pass genes independently. The child will have blood type A if both parents pass their A genes, type B if both pass their B genes, and type AB if one A and one B gene are passed. What are the probabilities that a child of these parents has type A blood? Type B? Type AB?

4.28 The "random walk" theory of securities prices holds that price movements in disjoint time periods are independent of each other. Suppose that we record only whether the price is up or down each year, and that the probability that our portfolio rises in price in any one year is 0.65. (This probability is approximately correct for a portfolio containing equal dollar amounts of all common stocks listed on the New York Stock Exchange.)

(a) What is the probability that our portfolio goes up for 3 consecutive years?
(b) If you know that the portfolio has risen in price 2 years in a row, what probability do you assign to the event that it will go down the next year?
(c) What is the probability that the portfolio's value moves in the same direction in both of the next 2 years?

4.29 Here is a two-way table of the composition of the 101st Congress (elected in 1988) by party and seniority. The entries in the body of the table should be the probabilities that a randomly chosen member of Congress has both the stated seniority and party affiliation. Only the two sets of probabilities for party alone and seniority alone are given. If party and seniority were independent, what would be the probabilities in the body of the table?

Seniority	Democratic	Republican	Total
< 2 years			.090
2–9 years			.478
≥ 10 years			.432
Total	.614	.386	1

4.30 The type of medical care a patient receives may vary with the age of the patient. A large study of women who had lumps on their breasts investigated whether or not each woman received a mammogram and a biopsy when the lump was discovered. Here are some probabilities estimated by the study. The entries in the table are the probabilities that *both* of two events occur; for example, 0.321 is the probability that a patient is under 65 years of age *and* the tests were done.

Age	Tests done	Tests not done
Under 65	.321	.124
65 or over	.365	.190

(a) What is the probability that a patient in this study is under 65? That a patient is 65 or over?
(b) What is the probability that the tests were done for a patient? That they were not done?
(c) Are the events $A = \{$the patient was 65 or older$\}$ and $B = \{$the tests were done$\}$ independent? Were the tests omitted on older patients more or less frequently than would be the case if testing were independent of age?

4.2 RANDOM VARIABLES

Sample spaces need not consist of numbers. When we toss four coins, we can record the outcome as a string of heads and tails, such as HTTH. In statistics, however, we are most often interested in numerical outcomes such as the count of heads in the four tosses. It is convenient to use a shorthand notation: Let X be the number of heads. If our outcome is HTTH, then $X = 2$. If the next outcome is TTTH, the value of X changes to $X = 1$. The possible values of X are 0, 1, 2, 3, and 4. Tossing a coin four times will give X one of these possible values. Tossing four more times will give X another and probably different value. We call X a random variable because its values vary when the coin tossing is repeated.

> **RANDOM VARIABLE**
>
> A random variable is a variable whose value is a numerical outcome of a random phenomenon.

We usually denote random variables by capital letters near the end of the alphabet, such as X or Y. A statistic calculated from a random sample or a randomized comparative experiment is a random variable. As we progress from general rules of probability toward statistical inference, we will concentrate on random variables. When a random phenomenon is described by a random variable X, the sample space S just lists the possible values of the random variable. We rarely mention S separately. There remains the second part of any probability model, the assignment of probabilities to events. In this section, we will learn two ways of assigning probabilities to the values of a random variable. The two types of probability models that result will dominate our application of probability to statistical inference.

Discrete random variables

We have learned several rules of probability, but only one method of assigning probabilities: State the probabilities of the individual outcomes and assign probabilities to events by summing over the outcomes. The outcome probabilities must be between 0 and 1 and have sum 1. When the outcomes are numerical, they are values of a random variable. We will now attach a name to random variables having probability assigned in this way.[2]

DISCRETE RANDOM VARIABLE

A discrete random variable X takes a finite number of values, call them $x_1, x_2, \ldots, x_k$. A probability model for X is given by assigning probabilities p_i to these outcomes,

$$P(X = x_i) = p_i$$

The probabilities p_i must satisfy

$$\begin{aligned} &1.\ 0 \leq p_i \leq 1 \text{ for each } i \\ &2.\ p_1 + p_2 + \cdots + p_k = 1 \end{aligned} \tag{4.1}$$

The probability $P(X \text{ in } A)$ of any event is found by summing the p_i for the outcomes x_i making up A.

probability distribution

The assignment of probabilities to the values of a random variable X is called the *probability distribution* of X. When X is discrete, its probability distribution must satisfy the requirements 1 and 2 of definition 4.1. If a random variable takes only a few values, its probability distribution can be given as a table. Here is an example.

EXAMPLE 4.17

The instructor of a large class gives 15% each of A's and D's, 30% each of B's and C's, and 10% F's. If a student is selected at random from this course, his grade on a 4-point scale (A = 4.0) is a discrete random variable X having the distribution

Outcome x_i	0	1	2	3	4
Probability p_i	.10	.15	.30	.30	.15

The probability of the event "the student got a B or better" is found from the table of probabilities as follows:

$$\begin{aligned} P(X \geq 3) &= P(X = 3) + P(X = 4) \\ &= .30 + .15 = .45 \end{aligned}$$

●

probability histogram

The probability distribution of a discrete random variable can be presented graphically in a *probability histogram*. Figure 4.5 displays two probability histograms. Part (a) is the distribution of an entry in the table of random digits, with probability distributed uniformly over all 10 digits. Part (b) is the distribution of grades in Example 4.17. The horizontal scale shows the possible values of X, and the height of each bar is the probability for the value at its base. A probability histogram is in effect a relative frequency histogram for a very large number of trials.

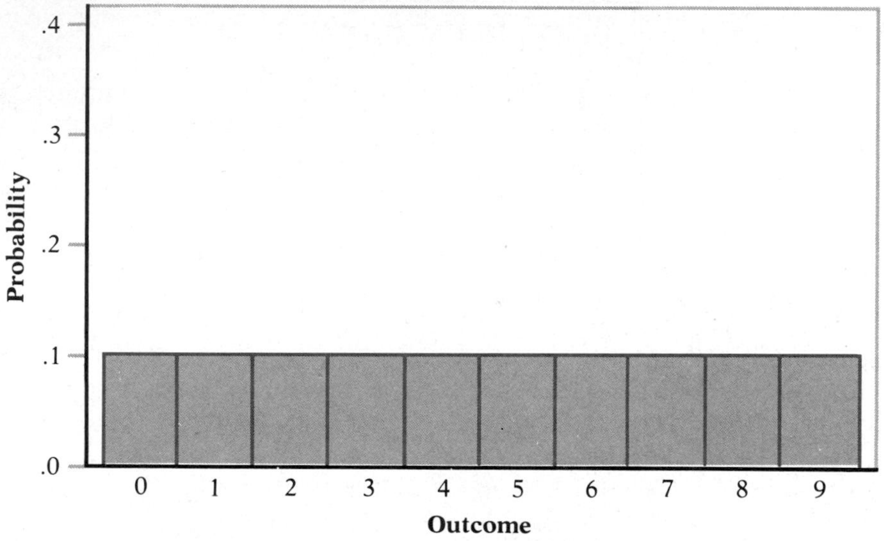

(a)

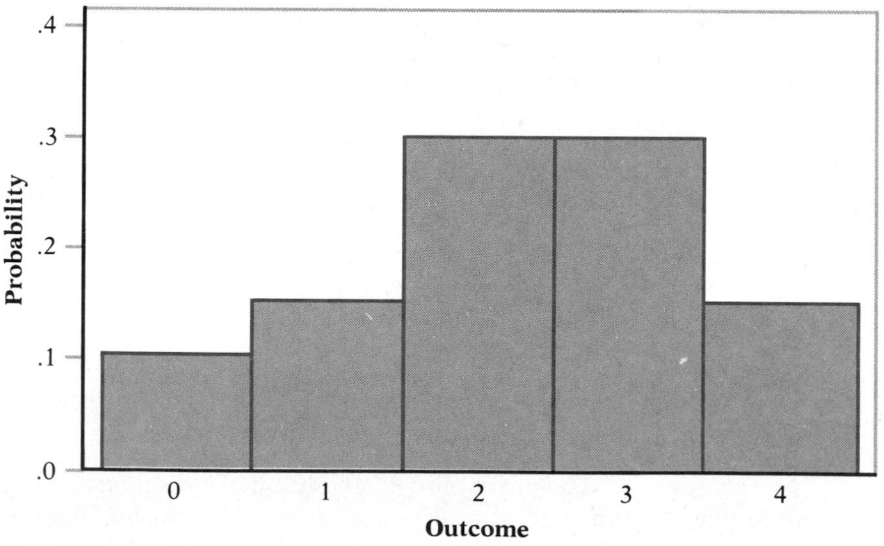

(b)

FIGURE 4.5 Probability histograms for the distributions of discrete random variables: (a) generating a random digit, and (b) grades in a large course (Example 4.17).

EXAMPLE 4.18

What is the probability distribution of the discrete random variable X that counts the number of heads in four tosses of a coin? We can derive this distribution if we make two reasonable assumptions: The coin is balanced, so each toss is equally likely to give H or T; the coin has no memory, so tosses are independent.

The outcome of four tosses is a sequence of heads and tails such as HTTH. There are 16 possible outcomes in all. They are listed in Figure 4.6, along with the value of X for each outcome. The multiplication rule for independent events tells us, for example, that

$$P(\text{HTTH}) = \frac{1}{2} \times \frac{1}{2} \times \frac{1}{2} \times \frac{1}{2} = \frac{1}{16}$$

Each of the 16 possible outcomes similarly has probability 1/16. That is, these outcomes are equally likely.

The number of heads X has possible values 0, 1, 2, 3, and 4. These values are *not* equally likely. As Figure 4.6 shows, there is only one way that $X = 0$ can occur, namely, when the outcome is TTTT. So $P(X = 0) = 1/16$. But the event $\{X = 2\}$ can occur in six different ways. So

$$\begin{aligned} P(X = 2) &= \frac{\text{count of ways } X = 2 \text{ can occur}}{16} \\ &= \frac{6}{16} \end{aligned}$$

We can find the probability of each value of X from Figure 4.6 in the same way. Here is the result:

$$\begin{aligned} P(X = 0) &= \frac{1}{16} = .0625 \\ P(X = 1) &= \frac{4}{16} = .25 \\ P(X = 2) &= \frac{6}{16} = .375 \\ P(X = 3) &= \frac{4}{16} = .25 \\ P(X = 4) &= \frac{1}{16} = .0625 \end{aligned}$$

These probabilities have sum 1, so this is a legitimate probability distribution. In table form the distribution is

Outcome X_i	0	1	2	3	4
Probability P_i	.0625	.25	.375	.25	.0625

Figure 4.7 is a probability histogram for this distribution. The probability distribution is exactly symmetric. It is an idealization of the relative frequency distribution of the number of heads after many tosses of four coins, which would be nearly symmetric but is unlikely to be exactly symmetric.

		HTTH		
		HTHT		
	HTTT	THTH	HHHT	
	THTT	HHTT	HHTH	
	TTHT	THHT	HTHH	
TTTT	TTTH	TTHH	THHH	HHHH
$X = 0$	$X = 1$	$X = 2$	$X = 3$	$X = 4$

FIGURE 4.6 **Possible outcomes in four tosses of a coin. The random variable *X* is the number of heads.**

Any event involving the number of heads observed can be expressed in terms of X, and its probability can be found from the distribution of X. For example, the probability of tossing at least two heads is

$$P(X \geq 2) = .375 + .25 + .0625 = .6875$$

The probability of at least one head is most simply found by use of the complement rule:

$$P(X \geq 1) = 1 - P(X = 0)$$
$$= 1 - .0625 = .9375$$

The assignment of probabilities to four tosses of a balanced coin gives the same probability to all 16 possible outcomes. Rules 1 and 2 then demand that each outcome have probability 1/16. If the coin is not balanced, say $P(\text{H}) = 0.4$, the outcomes are not equally likely and our work is a bit

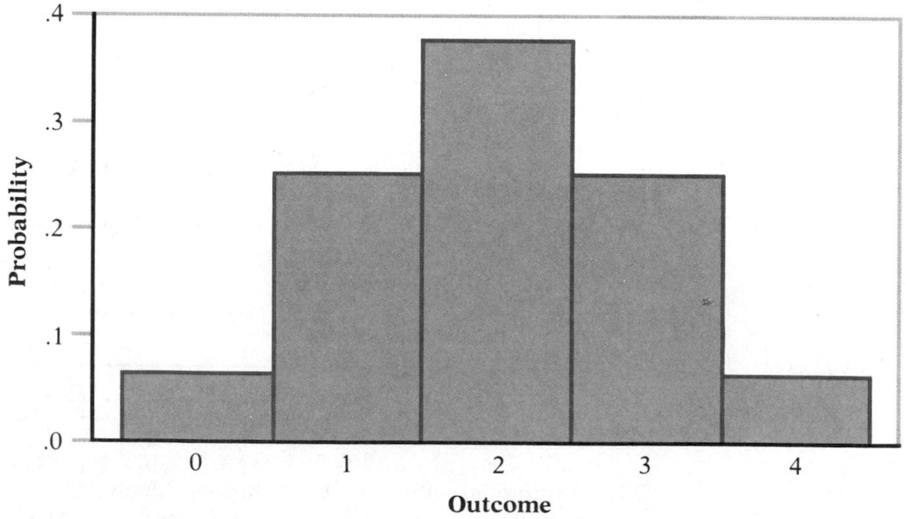

FIGURE 4.7 **Probability histogram for the number of heads in four tosses of a coin.**

harder. Recall that tossing a coin n times is similar to choosing an SRS of size n from a large population and asking a yes-or-no question. If in fact 40% of the population favor reduced spending on student loans, the responses to a question about student loans can be likened to repeated tosses of a coin with probability 0.4 of a head on each toss. A model with probability 0.5 of a head may be reasonable for many coins, but in the sampling situation it represents the unusual case in which exactly half of the population would respond "Yes." The next chapter will explain how to produce a more general discrete distribution for a count X of "Yes" responses in a yes-or-no situation.

Continuous random variables

When we use the table of random digits to select a digit between 0 and 9, the result is a discrete random variable. The probability model assigns probability 1/10 to each of the 10 possible outcomes, as Figure 4.5(a) shows. Suppose that we want to choose a number at random between 0 and 1, allowing *any* number between 0 and 1 as the outcome. You can visualize such a random number by thinking of a spinner (Figure 4.8) that turns freely on its axis and slowly comes to a stop. The pointer can come to rest anywhere on a circle that is marked from 0 to 1. The sample space allows the random number X to take any value between 0 and 1—that is, the possible values of X occupy an entire interval of numbers. We call X a *continuous* random variable because its values are not isolated numbers but an entire interval of numbers. How can we assign probabilities to such events as $\{0.3 \leq X \leq 0.7\}$? As in the case of selecting a random digit, we would like all possible outcomes to be equally likely. But we

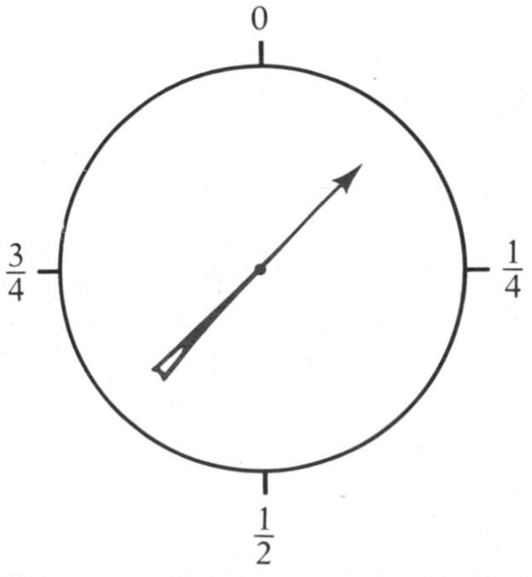

FIGURE 4.8 **A spinner that generates a random number between 0 and 1.**

cannot assign probabilities to each individual value of X and then sum, because the spinner can stop at any of an infinite number of points around the circle.

Instead, we use a new way of assigning probabilities directly to events—as *areas under a curve*. Probability histograms picture probabilities by the areas of their bars; we now use area to assign probability. Because the entire sample space must have probability 1, a curve with area 1 beneath it describes a probability distribution.

EXAMPLE 4.19

The random number generator will spread its output uniformly across the entire interval from 0 to 1 as we allow it to generate a long sequence of numbers. We therefore draw a line at height 1 above this interval (Figure 4.9). The area under this line is 1, and the probability of any event is the area under the line and above the event in question. Compare Figure 4.5(a), where 10 bars of equal height show the equal probabilities of the 10 digits, with Figure 4.9, where a line of fixed height shows probability spread evenly across an interval.

Because the height of the line is 1 and the area of a rectangle is the product of height and length, the probability of any interval of outcomes is just the length of the interval. As Figure 4.9(a) illustrates, the probability that the idealized random number generator produces a number between 0.3 and 0.7 is

$$P(.3 \leq X \leq .7) = .4$$

because the area under the line and above the interval from 0.3 to 0.7 is 0.4.

Similarly,

$$P(X \leq .5) = .5$$
$$P(X > .8) = .2$$
$$P(X \leq .5 \text{ or } X > .8) = .7$$

Notice that the last event consists of two nonoverlapping intervals, so the total area above the event is found by adding two areas, as illustrated by Figure 4.9(b). ●

The assignment of probabilities in Example 4.19 satisfies Rules 1 through 3, and therefore the other rules of probability. The entire sample space S has probability $P(0 \leq X \leq 1) = 1$ because the total area under the line is 1. For this same reason, the probability of any event is always a number between 0 and 1. Finally, the total area of two disjoint areas under the line is the sum of their individual areas. Probability as area under a curve is a second important way of assigning probabilities to events. Here is a general statement of the method.

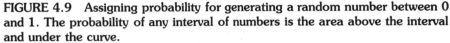

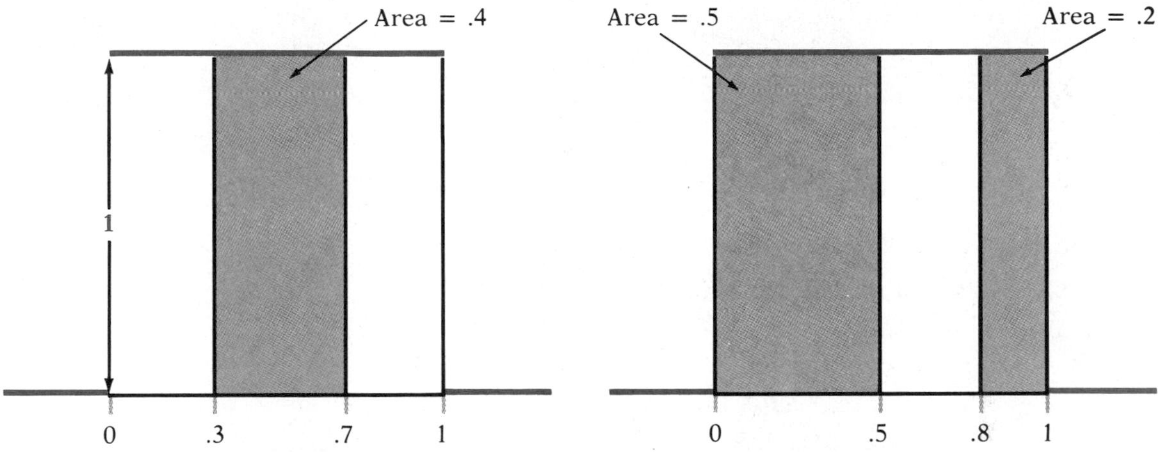

(a) $P(.3 \leq X \leq .7)$ (b) $P(X \leq .5 \text{ or } X > .8)$

FIGURE 4.9 Assigning probability for generating a random number between 0 and 1. The probability of any interval of numbers is the area above the interval and under the curve.

> **CONTINUOUS RANDOM VARIABLE**
>
> A continuous random variable X takes all values in an interval of real numbers. A probability model for X is given by assigning to a set of outcomes A the probability $P(A)$ equal to the area above A and under a curve. The curve is the graph of a function $p(x)$ that satisfies
>
> 1. $p(x) \geq 0$ for all x.
> 2. The total area under the graph of $p(x)$ is 1. (4.2)

Figure 4.10 illustrates this definition. In Example 4.19, the function $p(x)$ takes the value 1 for $0 \leq x \leq 1$ and the value 0 everywhere else. In general, $p(x)$ is 0 in regions where the random variable X cannot take values. When the function $p(x)$ is at all complicated, we cannot find areas under it by simple geometry. In such cases, we find probabilities with the aid of tables or computer software. Note carefully that the values of $p(x)$ are *not* probabilities—it is areas under the curve that are probabilities.

Continuous probability distributions have the same advantage over discrete distributions that density curves have over histograms—a single smooth curve is easier to work with than a large number of individual probabilities. In fact, the connection between continuous distributions and density curves is very close. Any density curve must lie on or above the horizontal axis and must have total area 1 underneath it. These are

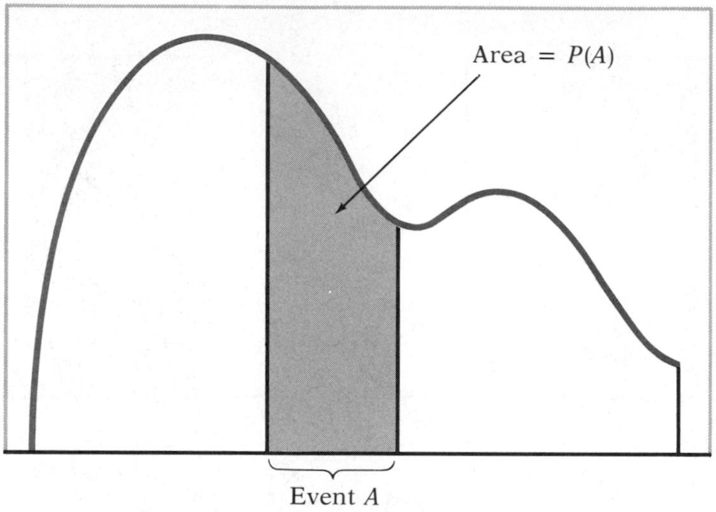

FIGURE 4.10 The probability distribution of a continuous random variable assigns probabilities as areas under a density curve.

exactly the requirements on $p(x)$ in definition 4.2 of a continuous random variable. So $p(x)$ is the equation of a density curve. *Any density curve describes the probability distribution of some continuous random variable.*

The probability model for a continuous random variable assigns probabilities to intervals of outcomes rather than to individual outcomes. In fact, *all continuous probability distributions assign probability 0 to every individual outcome.* Only intervals of values have positive probability. To see that this is true, consider a specific outcome such as $P(X = 0.8)$ in Example 4.19. The probability of any interval is the same as its length. The point 0.8 has no length, so its probability is 0. Although this fact may seem odd at first glance, it does make intuitive as well as mathematical sense. The random number generator produces a number between 0.79 and 0.81 with probability 0.02. An outcome between 0.799 and 0.801 has probability 0.002, and a result between 0.7999 and 0.8001 has probability 0.0002. Continuing to home in on 0.8, we can see why an outcome *exactly* equal to 0.8 should have probability 0. Because there is no probability exactly at $X = 0.8$, the two events $\{X > 0.8\}$ and $\{X \geq 0.8\}$ have the same probability. We can ignore the distinction between $>$ and $\geq$ when finding probabilities for continuous (but not discrete) random variables.

Normal distributions The density curves that are most familiar to us are the normal curves. So *normal distributions are probability distributions*. That is, a normal curve assigns probabilities to intervals of outcomes, and we find these probabilities by using Table A of areas under the standard normal curve. To employ Table A we must standardize the original scale of measurement. Recall that $N(\mu, \sigma)$ is our shorthand notation for the normal distribution having mean μ and standard deviation σ. In the language of random variables, if X has the $N(\mu, \sigma)$ distribution, then the standardized variable

$$Z = \frac{X - \mu}{\sigma}$$

has the standard normal distribution $N(0, 1)$. Here is another example of finding probabilities as areas under a density curve, this time a normal density curve.

EXAMPLE 4.20

An opinion poll asks an SRS of 1500 American adults what they consider to be the most serious problem facing our schools. Suppose that if we could ask all adults this question, 30% would say "drugs." The proportion $p = 0.3$ is a population parameter. The proportion $\hat{p}$ of the sample who answer "drugs" is a statistic used to estimate p. We will see in the next chapter that $\hat{p}$ is a random variable that has approximately the $N(0.3, 0.0118)$ distribution. The mean 0.3 of this distribution is the same as the population parameter because $\hat{p}$ is an unbiased estimate of p. The standard deviation is controlled mainly by the sample size, which is 1500 in this case.

What is the probability that the poll result differs from the truth about the population by more than two percentage points? Figure 4.11 shows this probability as an area under a normal density curve. By the addition rule for disjoint events, the desired probability is

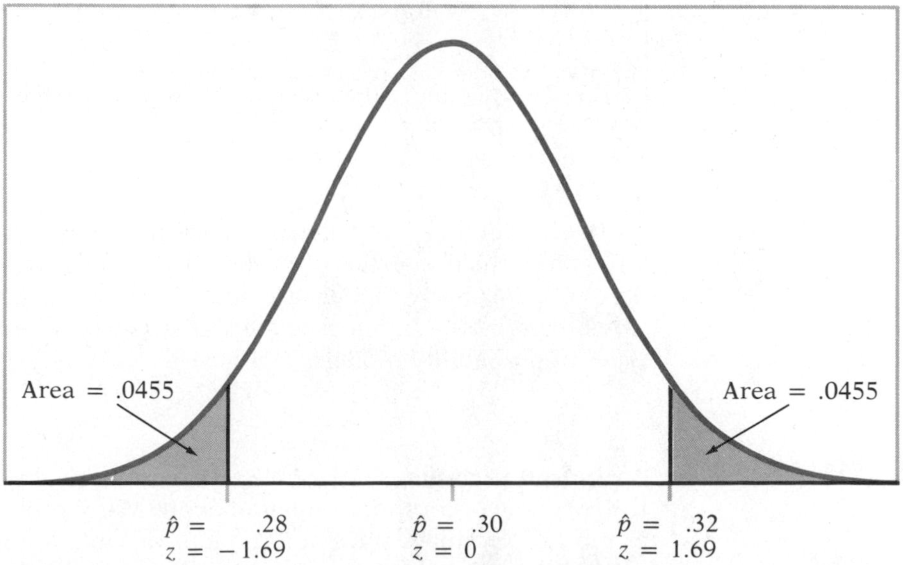

FIGURE 4.11 Probability in Example 4.20 as area under a normal density curve.

$$P(\hat{p} < .28 \text{ or } \hat{p} > .32) = P(\hat{p} < .28) + P(\hat{p} > .32)$$

The individual probabilities are found by standardizing and using Table A.

$$\begin{aligned} P(\hat{p} < .28) &= P\left(Z < \frac{.28 - .3}{.0118}\right) \\ &= P(Z < -1.69) = .0455 \\ P(\hat{p} > .32) &= P\left(Z > \frac{.32 - .3}{.0118}\right) \\ &= P(Z > 1.69) = .0455 \end{aligned}$$

Therefore,

$$P(\hat{p} < .28 \text{ or } \hat{p} > .32) = .0455 + .0455 = .0910$$

The probability that the sample result will miss the truth by more than two percentage points is .091. The arrangement of this calculation is familiar from our earlier work with normal distributions. Only the language of probability is new.

We could also do the calculation by first finding the probability of the complement:

$$\begin{aligned} P(.28 \le \hat{p} \le .32) &= P\left(\frac{.28 - .3}{.0118} \le Z \le \frac{.32 - .3}{.0118}\right) \\ &= P(-1.69 \le Z \le 1.69) \\ &= .9545 - .0455 = .9090 \end{aligned}$$

Then by the complement rule,

$$\begin{aligned} P(\hat{p} < .28 \text{ or } \hat{p} > .32) &= 1 - P(.28 \le \hat{p} \le .32) \\ &= 1 - .9090 = .0910 \end{aligned}$$

There is often more than one correct way to use the rules of probability to answer a question. ●

We began this chapter with a general discussion of the idea of probability and the properties of probability models. Two very useful specific types of probability models are distributions of discrete and continuous random variables. In our study of statistics we will employ only these two types of probability models.

SUMMARY

A **random variable** is a variable taking numerical values determined by the outcome of a random phenomenon. Many probability models take S to be all possible values of a random variable X and define P from the **probability distribution** of X. A random variable X and its distribution can be discrete or continuous.

A **discrete random variable** takes a finite collection of values. The probability distribution assigns each of these values a probability between 0 and 1 such that the sum of all the probabilities assigned is 1. The probability of an event A is then the sum of the probabilities of all the values that make up A.

A **continuous random variable** takes all values in some interval of real numbers. The probability distribution assigns to an event A the area above A and under a **density curve.** The height of a density curve must be 0 or positive everywhere, and the curve must have total area 1 beneath it.

Normal distributions are one type of continuous probability distribution.

SECTION 4.2 EXERCISES

4.31 Studies of social mobility in England can be summarized in the language of random variables. Social classes are numbered from 1 (low) to 5 (high). Take the random variable X to be the class of a randomly chosen son of a lower-class father. The distribution of X is as follows:

Son's class x_i	1	2	3	4	5
Probability p_i	.48	.38	.08	.05	.01

(a) Verify that the table gives a legitimate probability distribution for X. Draw a probability histogram to display the distribution.
(b) What is $P(X \leq 3)$?
(c) What is $P(X < 3)$?
(d) Write the event "a son of a lower-class father reaches one of the two highest classes" in terms of values of X. What is the probability of this event?
(e) The father's social class is Class 1. Write the event "the son of a lower-class father reaches a class higher than his father's" in terms of values of X. What is the probability of this event?

4.32 Choose an American household at random and let the random variable X be the number of persons living in the household. If we ignore the few households with more than seven inhabitants, the probability distribution of X is as follows:

Outcome x_i	1	2	3	4	5	6	7
Probability p_i	.240	.322	.177	.155	.067	.024	.015

(a) Verify that this is a legitimate discrete probability distribution and draw a probability histogram to display it.
(b) What is $P(X \geq 5)$?
(c) What is $P(X > 5)$?
(d) What is $P(2 < X \leq 4)$?
(e) What is $P(X \neq 1)$?
(f) Write the event that a randomly chosen household contains more than two persons in terms of the random variable X. What is the probability of this event?

4.33 A study selected a sample of children who were then in fifth grade and recorded how many years of school they eventually completed. Let X be the highest year of school that a randomly chosen fifth grader completes. (Students who go on to college are included in the outcome $X = 12$.) The probability distribution of X is as follows:

x_i	4	5	6	7	8	9	10	11	12
p_i	.010	.007	.007	.013	.032	.068	.070	.041	.752

(a) Verify that this is a legitimate discrete probability distribution and draw a probability histogram to display it.
(b) Find $P(X \geq 6)$.
(c) Find $P(X > 6)$.
(d) What values of X make up the event "the student completed at least 1 year of high school?" (High school begins with the ninth grade.) What is the probability of this event?
(e) Express the event that a fifth-grade student will not complete twelfth grade in terms of X and find its probability.

4.34 A couple plans to have four children. If we consider only the sex of the children, having four children is much like tossing a coin four times. There are 16 possible arrangements of girls and boys, and all 16 are (approximately) equally likely.
(a) Let the discrete random variable X be the number of girls among the four children. Following the pattern of Figure 4.6, list all 16 outcomes and give the value of X for each. The distribution of X is the same as that given in Example 4.18 (page 309) for the number of heads in four tosses of a coin.
(b) What is the probability that the couple will have at least one girl?
(c) What is the probability that the couple will have two girls and two boys?
(d) What is the probability that the first child is a girl? (Notice that this event cannot be described in terms of X. You must return to the original outcomes.)

4.35 Some games of chance rely on tossing two dice. Each die has six faces, marked with 1, 2, . . . , 6 spots called pips. The dice used in casinos are carefully balanced so that each face is equally likely to come up. When two dice are tossed, each of the 36 possible pairs of faces is equally likely to come up. The outcome of interest to a gambler is the sum of the pips on the two up faces. Call this random variable X.

(a) Write down all 36 possible pairs of faces.

(b) If all pairs have the same probability, what must be the probability of each pair?

(c) Write the value of X next to each pair of faces and use this information with the result of (b) to give the probability distribution of X. Draw a probability histogram to display the distribution.

(d) One bet available in craps wins if a 7 or an 11 comes up on the next roll of two dice. What is the probability of rolling a 7 or an 11 on the next roll?

(e) Several bets in craps lose if a 7 is rolled; if any outcome other than 7 occurs, these bets either win or continue to the next roll. What is the probability that anything other than a 7 is rolled?

4.36 Weary of the low turnout in student elections, a college administration decides to choose an SRS of three students to form an advisory board that represents student opinion. Suppose that 40% of all students oppose the use of student fees to fund student interest groups. Then the opinions of the three students on the board are independent, and the probability is 0.4 that each opposes the funding of interest groups.

(a) Call the three students A, B, and C. What is the probability that A and B support funding and C opposes it?

(b) List all possible combinations of opinions that can be held by students A, B, and C. (Hint: There are eight possibilities.) Then give the probability of each of these outcomes. Note that they are not equally likely.

(c) Let the random variable X be the number of student representatives who oppose the funding of interest groups. Give the probability distribution of X.

(d) Express the event "a majority of the advisory board opposes funding" in terms of X and find its probability.

4.37 Let X be a random number between 0 and 1 produced by the idealized uniform random number generator described in Example 4.19 and Figure 4.9 (page 313). Find the following probabilities:

(a) $P(0 \leq X \leq 0.4)$

(b) $P(0.4 \leq X \leq 1)$

(c) $P(0.3 \leq X \leq 0.5)$

(d) $P(0.3 < X < 0.5)$

(e) $P(0.226 \leq X \leq 0.713)$

4.38 Let the random variable X be a random number with the density curve in Figure 4.9, as in the previous exercise. Find the following probabilities:
(a) $P(X \leq 0.49)$
(b) $P(X \geq 0.27)$
(c) $P(0.27 < X < 1.27)$
(d) $P(0.1 \leq X \leq 0.2 \text{ or } 0.8 \leq X \leq 0.9)$
(e) The probability that X is not in the interval 0.3 to 0.8.
(f) $P(X = 0.5)$

4.39 Most random number generators allow users to specify the range of the random numbers to be produced. In Minitab, for example, a random number X between 0 and 2 is placed in C1 by the command

```
MTB> RANDOM 1 C1;
 SUBC> UNIFORM 0 2.
```

The distribution of X spreads its probability uniformly between 0 and 2. The density curve of X should have constant height between 0 and 2 and height 0 elsewhere.
(a) What is the height of the density curve between 0 and 2? Draw a graph of the density curve of X.
(b) Use your graph from (a) and the fact that probability is area under the curve to find $P(X \leq 1)$.
(c) Find $P(0.5 < X < 1.3)$.
(d) Find $P(X \geq 0.8)$.
(e) Find $P(1.4 \leq X \leq 2.4)$.
(f) Find $P(X = 1)$.

4.40 Suppose that the random variable X is normally distributed with mean 0 and standard deviation 1. Use Table A to find the following probabilities:
(a) $P(X \leq 1)$
(b) $P(X \geq -1)$
(c) $P(X = 1)$

4.41 The random variable X has the standard normal $N(0, 1)$ distribution. Find each of the following probabilities:
(a) $P(-1 \leq X \leq 1)$
(b) $P(1 \leq X \leq 2)$
(c) $P(1 < X < 2)$

4.42 An SRS of 400 American adults is asked, "What do you think is the most serious problem facing our schools?" Suppose that in fact 30% of all adults would answer "drugs" if asked this question. That is, the population proportion is $p = 0.3$. The sample proportion $\hat{p}$ who answer "drugs" will vary in repeated sampling. The sampling distribution will be approximately normal with mean 0.3 and standard deviation 0.023. Using this approximation, find the probabilities of the following events:
(a) At least half of the sample believes that drugs are the schools' most serious problem.

(b) Less than 25% of the sample believes that drugs are the most serious problem.
(c) The sample proportion is between 0.25 and 0.35.
(d) $\{\hat{p} \leq 0.4 \text{ or } \hat{p} \geq 0.6\}$.

4.43 An opinion poll asks an SRS of 1500 adults, "Do you happen to jog?" Suppose (as is approximately correct) that the population proportion who jog is $p = 0.15$. Then the proportion $\hat{p}$ in the sample who answer "Yes" will be approximately normally distributed with mean $\mu = 0.15$ and standard deviation $\sigma = 0.0092$. Find the following probabilities:
(a) $P(\hat{p} = 0.16)$
(b) $P(\hat{p} \geq 0.16)$
(c) $P(0.14 \leq \hat{p} \leq 0.16)$

4.3 MEANS AND VARIANCES OF RANDOM VARIABLES

Probability is the mathematical language that describes the long-run regular behavior of random phenomena. The probability distribution of a random variable is an idealized relative frequency distribution. The probability histograms and density curves that picture probability distributions resemble our earlier pictures of distributions of data. In describing data, we moved from graphs to numerical measures such as means and standard deviations. Now we will make the same move to expand our descriptions of the distributions of random variables. We can speak of the mean winnings in a game of chance or the standard deviation of the randomly varying number of calls a travel agency receives in an hour. In this section we will learn more about how to compute these descriptive measures and about the laws they obey.

The mean of a random variable

The mean $\bar{x}$ of a set of observations is their ordinary average. The mean of a random variable X is also an average of the possible values of X, but with an essential change to take into account the fact that not all outcomes need be equally likely. An example will show what we must do.

EXAMPLE 4.21 State lotteries sell over \$15 billion worth of tickets each year and continue to expand rapidly. Here is one of the simplest available wagers: In Vermont's Green Mountain Numbers game, you choose a three-digit number; the state chooses a three-digit winning number at random and pays you \$500 if your number is chosen. Because there are 1000 three-digit numbers, you have probability 1/1000 of winning. Taking X to be the amount you win, the probability distribution of X is

Outcome	0	500
Probability	.999	.001

What are your average winnings? The ordinary average of the two possible outcomes \$0 and \$500 is \$250, but \$250 is certainly not the average winnings, because \$500 is much less likely than \$0. In the long run you win \$500 once in every 1000 tickets and \$0 on the remaining 999 of 1000 tickets. Your long-run average winnings from a ticket is

$$\$500\frac{1}{1000} + \$0\frac{999}{1000} = \$.50$$

or 50 cents. That number is the mean of the random variable X. (Tickets cost \$1, so in the long run Vermont keeps half the money you spend on Green Mountain Numbers.) ●

If you play Green Mountain Numbers several times, we would as usual call the mean of the actual amounts you win $\bar{x}$. The mean in Example 4.21 is a different quantity—it is the long-run average winnings you expect if you play a very large number of times. Just as probabilities are an idealized description of long-run relative frequencies, so the mean of a probability distribution describes the long-run average outcome. We can't call this mean $\bar{x}$, so we need a different symbol. The common symbol for the mean of a probability distribution is μ, the Greek letter mu. We used μ in Chapter 1 for the mean of a normal distribution, so this is not a new notation. We will often be interested in several random variables, each having a different probability distribution with a different mean. To remind ourselves that we are talking about the mean of X we usually write μ_X rather than plain μ. In the example, $\mu_X = \$0.50$. You will often find the mean of a random variable X called the *expected value* of X. This term can be misleading, for we don't necessarily expect one observation on X to be close to its expected value.

expected value

The mean of any discrete random variable is found just as in Example 4.21. It is an average of the possible outcomes, but it is a *weighted* average, in which each outcome is weighted by its probability. Because the probabilities add to 1, we have total weight 1 to distribute among the outcomes. An outcome that occurs half the time has probability 1/2 and so gets one-half the weight in calculating the mean. Here is the general definition.

MEAN OF A DISCRETE RANDOM VARIABLE

If X is a discrete random variable taking the values $x_1, x_2, \ldots, x_k$ with probabilities $p_1, p_2, \ldots, p_k$, then the mean of X is found by multiplying each outcome by its probability and adding over all the outcomes,

$$\begin{aligned}\mu_X &= x_1p_1 + x_2p_2 + \cdots + x_kp_k \\ &= \sum x_ip_i\end{aligned} \qquad (4.3)$$

EXAMPLE 4.22

The distribution of the count X of heads in four tosses of a balanced coin was found in Example 4.18 to be

Outcome x_i	0	1	2	3	4
Probability p_i	.0625	.25	.375	.25	.0625

The mean of X is therefore

$$\mu_X = (0)(.0625) + (1)(.25) + (2)(.375) + (3)(.25) + (4)(.0625)$$
$$= 2$$

This discrete distribution is symmetric, as the probability histogram Figure 4.12(a) reminds us. The mean therefore falls at the center of symmetry. ●

EXAMPLE 4.23

What is the mean size of an American family? Here is the distribution of the size of American families according to Census Bureau studies:

Persons in family	2	3	4	5	6	7
Fraction of families	.413	.236	.211	.090	.032	.018

If we imagine selecting a single family at random, the size of the family drawn is a random variable X with probability distribution given by the table. The mean μ_X is the mean family size in the population. This mean is

$$\mu_X = (2)(.413) + (3)(.236) + (4)(.211) + (5)(.090) + (6)(.032) + (7)(.018)$$
$$= 3.146$$

Figure 4.12(b) locates the mean on a probability histogram. (In this example, we have ignored the few families with 8 or more members. The Census Bureau reports that the actual mean size of American families is 3.17 people when these few large families are included.) ●

What about continuous random variables? The probability distribution of a continuous random variable X is described by a density curve. Chapter 1 showed how to find the mean of the distribution: It is the point at which the area under the density curve would balance if it were made out of solid material. The mean lies at the center of symmetric density curves such as the normal curves. Exact calculation of the mean of a distribution with a skewed density curve requires advanced mathematics. The idea that the mean is the balance point of the distribution applies to discrete random variables as well, but in the discrete case we have the recipe 4.3 that calculates this point.

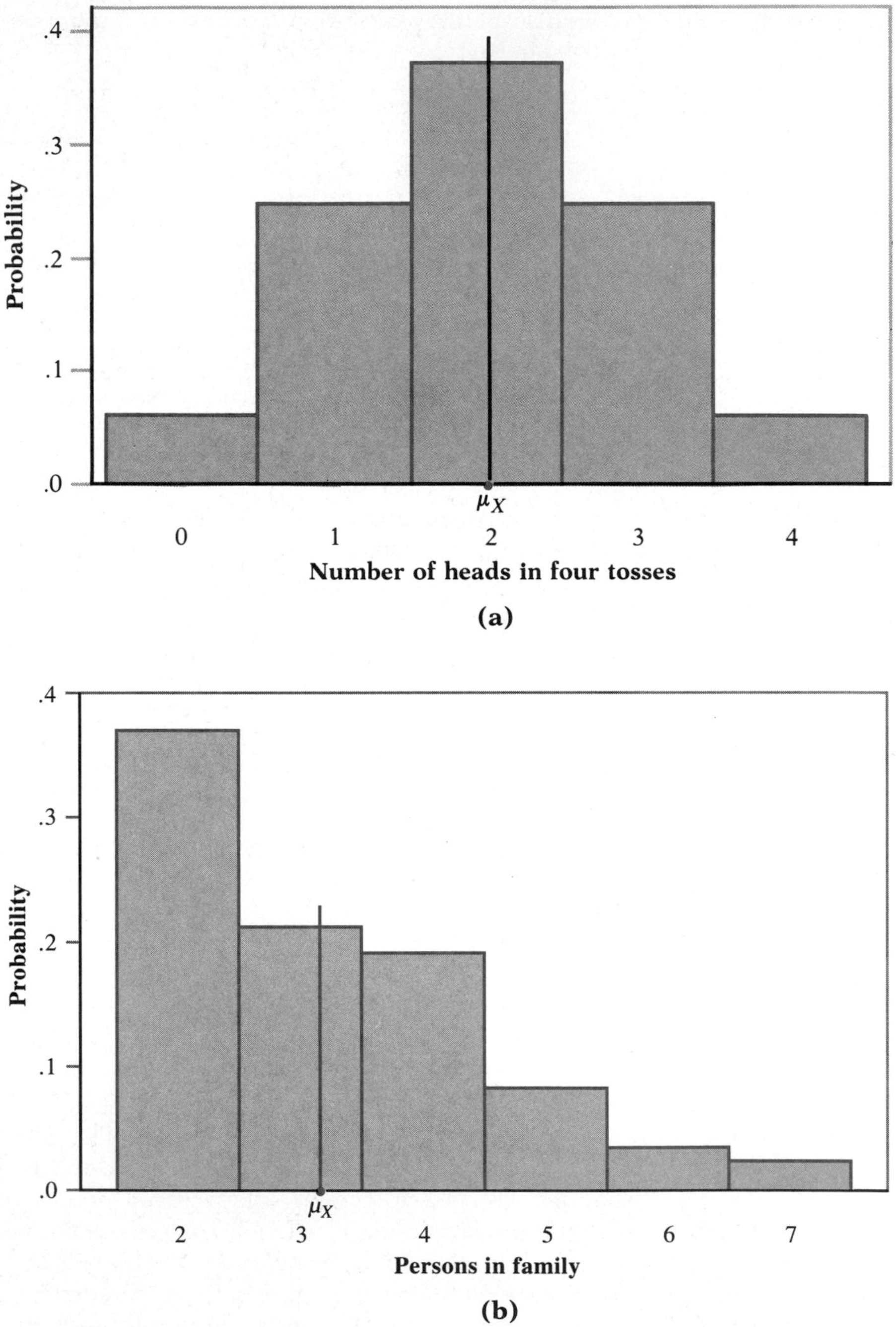

FIGURE 4.12 Locating the mean of a discrete random variable on the probability histogram for (a) the number of heads in four tosses of a coin (Example 4.22) and (b) the number of persons in a family (Example 4.23).

The law of large numbers

The mean of a random variable is by its definition the average outcome of the random variable as computed from its distribution. The mean of a random variable is also the average outcome in a second sense. If we actually observe a large number of outcomes of a random variable and calculate their mean (arithmetic average), this random mean will be close to the fixed mean of the distribution. This fact deserves closer study.

Suppose that we repeatedly toss four coins and record the number X of heads obtained on each trial. We did this in Example 4.1. The first toss gave $X = 3$, the second toss $X = 0$, then another 0, then 2, 4, 2, and so on. The relative frequency of each outcome in many repetitions will be close to its probability. The probabilities of all possible outcomes make up the probability distribution of the discrete random variable X. In Example 4.22 we found that the mean of this distribution is $\mu_X = 2$. What happens to the average number of heads observed? The numbers of heads obtained on each of the first n trials form n observations. The average number of heads in the first n trials is the familiar mean $\bar{x}$ of these observations. The mean $\bar{x}$ is a random variable that takes different values if we repeat the tosses.[3] Here are the values of the observations and the mean $\bar{x}$ of all observations to date after each of the first few trials:

Trial	Outcome	Average number of heads per trial
1	3	$\bar{x} = \frac{3}{1} = 3.00$
2	0	$\bar{x} = \frac{3}{2} = 1.50$
3	0	$\bar{x} = \frac{3}{3} = 1.00$
4	2	$\bar{x} = \frac{5}{4} = 1.25$
5	4	$\bar{x} = \frac{9}{5} = 1.80$
6	2	$\bar{x} = \frac{11}{6} = 1.83$

Figure 4.13 shows the behavior of $\bar{x}$ as the number of trials n increases. At first the value of $\bar{x}$ is unstable, but *in the long run the observed mean $\bar{x}$ approaches and remains close to the distribution mean $\mu_X = 2$.* After 100 trials $\bar{x} = 1.86$; at the end of our series of 1000 tosses of four coins $\bar{x} = 2.012$. If we repeated our work, we would obtain a different sequence of outcomes, and the graph of the progress of $\bar{x}$ would differ from Figure

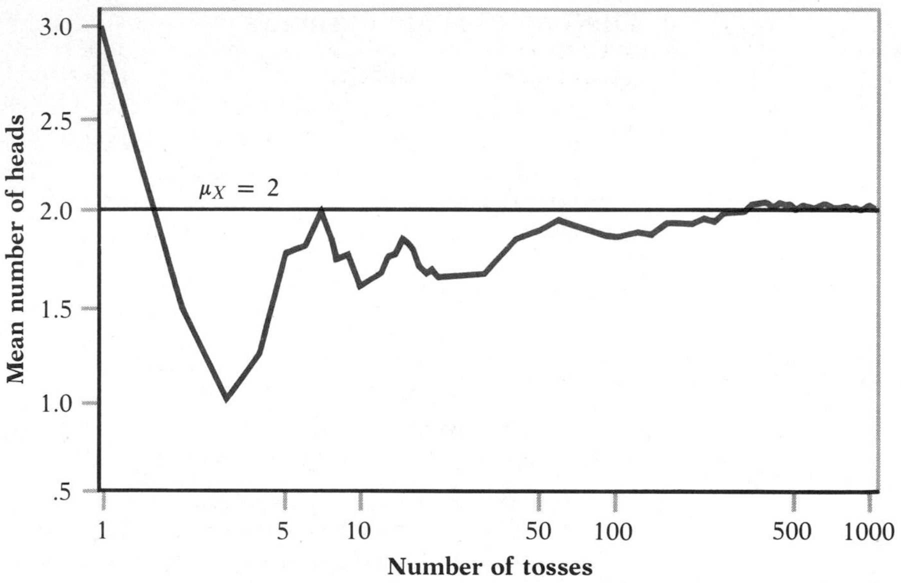

FIGURE 4.13 The long-term behavior of the mean outcome of many trials. The mean number of heads $\bar{x}$ observed when four coins are tossed many times approaches the mean $\mu_X = 2$ of the probability distribution.

4.13. But $\bar{x}$ will always approach $\mu_X = 2$ ever more closely as the number of trials grows.

Starting from the basic laws of probability, it can be *proved* that the sample mean outcome calculated from repeated independent observations of a random variable must approach the mean of the distribution. The behavior of $\bar{x}$ is similar to the idea of probability—in the long run, relative frequencies of outcomes get close to the probability distribution, and the average outcome gets close to the distribution mean. These facts, especially when they are considered as mathematical results that can be derived from basic laws of probability, are called the *law of large numbers*.*

law of large numbers

The law of large numbers is the foundation of such business enterprises as gambling casinos and insurance companies. The winnings (or losses) of a gambler on a single play are uncertain—that's why gambling is exciting. The average winnings of the house on tens of thousands of plays will be very close to the mean of the distribution of winnings—that's why gambling can be a business, for this distribution guarantees the house a profit. In Green Mountain Numbers, for example, we saw that the mean payout for a $1 ticket is 50 cents. The law of large numbers guarantees

*The earliest version of the law of large numbers was proved by the Swiss mathematician Jacob Bernoulli (1654–1705). The Bernoullis were a remarkably mathematical family; five of them contributed to the early study of probability.

that in the long run Vermont will pay out an average of very close to 50 cents per ticket sold and keep the other 50 cents.

The law of averages The law of large numbers is the precise version of the popular "law of averages." Both say that random fluctuations eventually average out. But psychologists have discovered that the popular understanding of randomness is quite different from the true laws of chance.[4] Most people believe in an incorrect "law of small numbers." That is, we expect even short sequences of random events to show the kind of average behavior that in fact appears only in the long run.

Try this experiment: Write down a sequence of heads and tails that you think imitates 10 tosses of a balanced coin. How long was the longest string (called a *run*) of consecutive heads or consecutive tails in your tosses? Most people will write a sequence with no runs of more than two consecutive heads or tails. Longer runs don't seem "random" to us. But in fact the probability of a run of three or more consecutive heads or tails in 10 tosses is greater than 0.8, and the probability of *both* a run of three or more heads and a run of three or more tails is almost 0.2.[5] This and other probability calculations suggest that a short sequence of coin tosses will often not appear random to us. The runs of consecutive heads or consecutive tails that appear in real coin tossing (and that are predicted by probability theory) seem surprising to us. Because we don't expect to see long runs, we may conclude that the coin tosses are not independent or that some influence is disturbing the random behavior of the coin.

run

Belief in the law of small numbers influences behavior. If a basketball player makes several consecutive shots, both the fans and his teammates believe that he has a "hot hand" and is more likely to make the next shot. This is doubtful. Careful study suggests that runs of baskets made or missed are no more frequent in basketball than would be expected if each shot were independent of the player's previous shots. Players perform consistently, not in streaks. (Of course, some players make a higher percent of their shots in the long run than others.) Our perception of hot or cold streaks simply shows that we don't perceive random behavior very well.[6]

Gamblers often follow the hot hand theory, betting that a run will continue. At other times, however, they draw the opposite conclusion when confronted with a run of outcomes. If a coin gives 10 straight heads, some gamblers feel that it must now produce some extra tails to get back to the average of half heads and half tails. Not so. If the next 10,000 tosses give about 50% tails, those 10 straight heads will be swamped by the later thousands of heads and tails. No compensation is needed to get back to the average in the long run. Remember that it is *only* in the long run that the regularity described by probabilities and means takes over.

Our inability to accurately distinguish random behavior from systematic influences points out once more the need for statistical inference to supplement exploratory analysis of data. Probability calculations can help verify that what we see in the data is more than a random pattern.

Rules for means

Means of random variables are average outcomes (in two senses). It is not surprising that means obey rules that reflect the behavior of averages. Suppose, for example, that you are discussing flaws in the finish of refrigerators made by your firm. Dimples and paint sags are two kinds of surface flaw. Not all refrigerators have the same number of dimples: many have none, some have one, some two, and so on. You ask for the average number of imperfections on a refrigerator. The inspectors report finding an average of 0.7 dimples and 1.4 sags per refrigerator. How many total imperfections of both kinds (on the average) are there on a refrigerator? Easy: If the average number of dimples is 0.7 and the average number of sags is 1.4, then counting both gives an average $0.7 + 1.4 = 2.1$ flaws.

Putting this example into more formal language, the number of dimples on a refrigerator is a random variable X that takes values 0, 1, 2, and so on. X varies as we inspect one refrigerator after another. Only the mean number of dimples $\mu_X = 0.7$ was reported to you. The number of paint sags is a second random variable Y having mean $\mu_Y = 1.4$. (You see how the subscripts keep straight which variable we are talking about.) The total number of both dimples and sags is the sum $X + Y$. That sum is another random variable that varies from refrigerator to refrigerator. Its mean μ_{X+Y} is the average number of dimples and sags together and is just the sum of the individual means μ_X and μ_Y. That is an important rule for how means of random variables behave.

Another important rule says how the mean of a random variable changes when we multiply every outcome by the same fixed number or add the same fixed number to every outcome. Suppose X is the length in centimeters of a cockroach randomly chosen from those living in a dormitory and that the mean length is $\mu_X = 2.2$ centimeters. If we decide to measure in millimeters, we multiply every value of X by 10 because there are 10 millimeters in a centimeter. The length in millimeters is the new random variable $10X$ formed by multiplying X by 10 to convert centimeters to millimeters. What is the mean μ_{10X} of this new variable? If we multiply every value of X by 10, surely we also multiply the mean (the average value) by 10. That's right: The mean length is $2.2 \times 10 = 22$ millimeters. In more formal language, the mean μ_{10X} of $10X$ is $10\mu_X$. Similarly, if we add the same fixed number to every value of a random variable X, we add that same number to the mean. The first rule in the box below combines the results of multiplying by a fixed number and adding a fixed number.

RULES FOR MEANS

If X is a random variable and a and b are fixed numbers, then

$$\mu_{a+bX} = a + b\mu_X$$

If X and Y are random variables, then (4.4)

$$\mu_{X+Y} = \mu_X + \mu_Y$$

EXAMPLE 4.24

Gain Communications sells aircraft communications units to both the military and the civilian markets. Next year's sales depend on market conditions that cannot be predicted exactly. Gain follows the modern practice of using probability estimates of sales. The military division estimates its sales as follows:

Units sold	1000	3000	5000	10,000
Probability	.1	.3	.4	.2

These are personal probabilities that express the informed opinion of Gain's executives. The corresponding sales estimates for the civilian division are

Units sold	300	500	750
Probability	.4	.5	.1

Take X to be the number of military units sold and Y the number of civilian units. From the probability distributions we compute that

$$\begin{aligned}\mu_X &= (1000)(.1) + (3000)(.3) + (5000)(.4) + (10,000)(.2)\\ &= 5000 \text{ units}\\ \mu_Y &= (300)(.4) + (500)(.5) + (750)(.1)\\ &= 445 \text{ units}\end{aligned}$$

Gain makes a profit of \$2000 on each military unit sold and \$3500 on each civilian unit. Next year's profit from military sales will be $2000X$, \$2000 times the number X of units sold. By the first rule in 4.4, the mean military profit is

$$\mu_{2000X} = 2000\mu_X = (2000)(5000) = \$10,000,000$$

Similarly, the civilian profit is 3500Y, and the mean profit from civilian sales is

$$\mu_{3500Y} = 3500\mu_Y = (3500)(445) = \$1,557,500$$

The total profit Z is the sum of the military and civilian profits,

$$Z = 2000X + 3500Y$$

The second rule in 4.4 says that the mean of this sum of two variables is the sum of the two individual means:

$$\begin{aligned}\mu_Z &= \mu_{2000X} + \mu_{3500Y} \\ &= 10,000,000 + 1,557,500 \\ &= \$11,557,500\end{aligned}$$

This mean is the company's best estimate of next year's profit, combining the probability estimates of the two divisions. ●

Once you have gained some experience in applying the rules for means of random variables, the calculation of mean total profit in Example 4.24 can be done more quickly by applying both rules at once as follows:

$$\begin{aligned}\mu_Z &= \mu_{2000X+3500Y} \\ &= 2000\mu_X + 3500\mu_Y \\ &= (2000)(5000) + (3500)(445) = \$11,557,500\end{aligned}$$

The variance of a random variable

The mean is a measure of the center of a distribution. Even the most basic numerical description requires in addition a measure of the spread or variability of the distribution. The variance and the standard deviation are the measures of spread that accompany the choice of the mean to measure center. As for the mean, we need a distinct symbol to distinguish the variance of a random variable from the variance s^2 of a data set. We denote the variance of a random variable X by σ_X^2. Once again the subscript just reminds us which variable we have in mind. The definition of the variance σ_X^2 of a random variable is similar to the definition of the sample variance s^2 given in Chapter 1. That is, the variance is an average of the squared deviation $(X - \mu_X)^2$ of the variable X from its mean μ_X. As for the mean, the average we use is a weighted average in which each outcome is weighted by its probability in order to take account of outcomes that are not equally likely. Calculating this weighted average is straightforward for discrete random variables but requires advanced mathematics in the continuous case. Here is the resulting definition.

VARIANCE OF A DISCRETE RANDOM VARIABLE

If X is a discrete random variable taking values $x_1, x_2, \ldots, x_k$ with probabilities $p_1, p_2, \ldots, p_k$ the variance of X is given by

$$\begin{aligned}\sigma_X^2 &= (x_1 - \mu_X)^2 p_1 + (x_2 - \mu_X)^2 p_2 + \cdots + (x_k - \mu_X)^2 p_k \qquad (4.5) \\ &= \sum (x_i - \mu_X)^2 p_i\end{aligned}$$

The standard deviation σ_X of X is the square root of the variance.

EXAMPLE 4.25

In Example 4.24 we saw that the number X of communications units sold by the Gain Communications military division has distribution

Units sold	1000	3000	5000	10,000
Probability	.1	.3	.4	.2

We can find the mean and variance of X by arranging the calculation in the form of a table. Both μ_X and σ_X^2 are sums of columns in this table.

x_i	p_i	$x_i p_i$	$(x_i - \mu_Y)^2 p_i$
1000	.1	100	$(1000 - 5000)^2(.1) = 1{,}600{,}000$
3000	.3	900	$(3000 - 5000)^2(.3) = 1{,}200{,}000$
5000	.4	2000	$(5000 - 5000)^2(.4) = 0$
10,000	.2	2000	$(10,000 - 5000)^2(.2) = 5{,}000{,}000$
		$\mu_X = 5000$	$\sigma_X^2 = 7{,}800{,}000$

We see that $\sigma_X^2 = 7,800,000$. The standard deviation of X is $\sigma_X = \sqrt{7,800,000} = 2792.8$. The standard deviation is a measure of how variable the number of units sold is. As in the case of distributions for data, the standard deviation of a probability distribution is most meaningful when the distribution has a specific form—particularly when the distribution is normal. ●

Rules for variances

Rules for variances similar to the rules 4.4 for means are often needed to understand statistical questions. Although the mean of a sum of random variables is always the sum of their means, this addition rule is not always true for variances. To understand why, take X to be the percent of a family's after-tax income that is spent and Y the percent that is saved. When X increases, Y decreases by the same amount. Though X and Y may vary widely from year to year, their sum $X + Y$ is always 100% and does not vary at all. It is the association between the variables X and Y that prevents their variances from adding. If random variables are *independent*, this kind of association between their values is ruled out, and their variances do add. Two random variables X and Y are independent if any event involving X alone is independent of any event involving Y alone. Probability models often assume independence when the random variables describe outcomes that appear unrelated to each other. You should ask in each instance whether the assumption of independence seems reasonable.

RULES FOR VARIANCES

If X is a random variable and a and b are fixed numbers, then

$$\sigma^2_{a+bX} = b^2\sigma^2_X$$

If X and Y are *independent* random variables, then (4.6)

$$\sigma^2_{X+Y} = \sigma^2_X + \sigma^2_Y$$
$$\sigma^2_{X-Y} = \sigma^2_X + \sigma^2_Y$$

Notice that because a variance is the average of *squared* deviations from the mean, multiplying X by a constant b multiplies σ^2_X by the *square* of the constant. Adding a constant a to a random variable changes its mean but does not change its variability. The variance of $X + a$ is therefore the same as the variance of X. Because the square of -1 is 1, the addition rule says that the variance of a difference $X - Y$ is the *sum* of the variances. The difference $X - Y$ is more variable than either X or Y alone, because variations in both X and Y contribute to variation in their difference.

As with data, we often prefer the standard deviation to the variance as a measure of variability. The addition rule for variances implies that standard deviations do *not* generally add. Standard deviations are most easily combined by using the rules for variances rather than by giving separate rules for standard deviations. For example, the standard deviations of $2X$ and $-2X$ are both equal to $2\sigma_X$ because this is the square root of the variance $4\sigma^2_X$.

EXAMPLE 4.26

In Example 4.21 we found that the payoff X of a \$1 ticket in Vermont's Green Mountain Numbers game has mean $\mu_X = \$0.50$. Here is the combined calculation of mean and variance:

x_i	p_i	$x_i p_i$	$(x_i - \mu_Y)^2 p_i$
0	.999	0	$(0 - .5)^2(.999) = .24975$
500	.001	.5	$(500 - .5)^2(.001) = 249.50025$
		$\mu_X = .5$	$\sigma^2_X = 249.75000$

The standard deviation is $\sigma_X = \sqrt{249.75} = \$15.80$. It is usual for games of chance to have large standard deviations, because large variability makes gambling exciting.

If you buy a Green Mountain Numbers ticket, your winnings are $W = X - 1$ because the dollar you paid for the ticket must be subtracted from the payoff. By the rules 4.4 for means, the mean amount you win is

$$\mu_W = \mu_X - 1 = -\$.50$$

That is, you lose an average of 50 cents per ticket. The rules 4.6 for variances remind us that the variance and standard deviation of the winnings $W = X - 1$ are the same as those of X. Subtracting a fixed number changes the mean but not the variance.

Suppose now that you buy a \$1 ticket on each of two different days. The payoffs X and Y on the two tickets are independent because separate drawings are held each day. Your total payoff $X + Y$ has mean

$$\mu_{X+Y} = \mu_X + \mu_Y = \$.50 + \$.50 = \$1.00$$

Because X and Y are independent, the variance of $X + Y$ is

$$\sigma^2_{X+Y} = \sigma^2_X + \sigma^2_Y = 249.75 + 249.75 = 499.5$$

The standard deviation of the total payoff is

$$\sigma_{X+Y} = \sqrt{499.5} = \$22.35$$

This is not the same as the sum of the individual standard deviations, which is \$15.80 + \$15.80 = \$31.60. Variances of independent random variables add; their standard deviations do not. ●

EXAMPLE 4.27

A college uses Scholastic Aptitude Test (SAT) scores as one criterion for admission. Experience has shown that the distribution of SAT scores among the entire population of applicants is such that

SAT math score X	$\mu_X = 625$	$\sigma_X = 90$
SAT verbal score Y	$\mu_Y = 590$	$\sigma_Y = 100$

What are the mean and standard deviation of the total score $X + Y$ for students applying to this college?

The mean overall SAT score is

$$\mu_{X+Y} = \mu_X + \mu_Y = 625 + 590 = 1215$$

The variance and standard deviation of the total *cannot be computed* from the information given. The verbal and math scores are clearly not independent, because students who score high on one exam will tend to score high on the other also. Therefore, the addition rule for variances does not apply. ●

EXAMPLE 4.28

Tom and George play golf at the same club. Tom's score X varies from round to round but has

$$\mu_X = 110 \text{ and } \sigma_X = 10$$

George's score Y also varies, with

$$\mu_Y = 100 \text{ and } \sigma_Y = 8$$

Tom and George are playing the first round of the club tournament. Because they are not playing together, we will assume that their scores vary independently of each other. The difference between their scores on this first round has mean

$$\mu_{X-Y} = \mu_X - \mu_Y = 110 - 100 = 10$$

The variance of the difference between the scores is

$$\sigma^2_{X-Y} = \sigma^2_X + \sigma^2_Y = 10^2 + 8^2 = 164$$

The standard deviation is found from the variance:

$$\sigma_{X-Y} = \sqrt{164} = 12.8$$

The variation in the difference between the scores is greater than the variation in the score of either player, because the difference contains two independent sources of variation. It also makes the tournament interesting, because George will not always finish ahead of Tom even though he has the lower mean score. Computations of the variance of sums and differences of statistics play an important role in statistical inference. ●

SUMMARY

The probability distribution of a random variable X, like a distribution of observations, has a **mean** μ_X and a **standard deviation** σ_X.

If X is a discrete random variable taking values $x_1, x_2, \ldots, x_k$ with probabilities $p_1, p_2, \ldots, p_k$ the mean and variance can be computed from its distribution as follows:

$$\mu_X = x_1p_1 + x_2p_2 + \cdots + x_kp_k$$
$$\sigma^2_X = (x_1 - \mu_X)^2 p_1 + (x_2 - \mu_X)^2 p_2 + \cdots + (x_k - \mu_X)^2 p_k$$

The mean and variance of a continuous random variable can be computed from the density curve, but to do so requires more advanced mathematics.

The **law of large numbers** states that the actually observed mean outcome in many independent trials must approach the mean of the distribution of outcomes. This statement includes the fact that observed relative frequencies of events must approach the probabilities of the events. Short sequences of trials will often not display the regularity that appears in the long run.

The means and variances of random variables obey the following rules. If a and b are fixed numbers, then

$$\mu_{a+bX} = a + b\mu_X$$
$$\sigma^2_{a+bX} = b^2\sigma^2_X$$

If X and Y are any two random variables, then

$$\mu_{X+Y} = \mu_X + \mu_Y$$

and if X and Y are independent, then

$$\sigma^2_{X+Y} = \sigma^2_X + \sigma^2_Y$$
$$\sigma^2_{X-Y} = \sigma^2_X + \sigma^2_Y$$

SECTION 4.3 EXERCISES

4.44 Keno is a favorite game in the casinos of Nevada and Atlantic City. Balls numbered 1 to 80 are tumbled in a machine as the bets are placed, then 20 of the balls are chosen at random. Players select numbers by marking a card. The simplest of the many wagers available is "Mark 1 Number." The payoff is \$3 on a \$1 bet if the number you select is one of those chosen. Because 20 of 80 numbers are chosen, your probability of winning is 20/80, or 0.25.

(a) What is the probability distribution (the outcomes and their probabilities) of the payoff X on a single play?

(b) What is the mean payoff μ_X?

(c) In the long run, how much does the casino keep from each dollar bet?

4.45 In government data a household consists of all people who live together in a dwelling unit, whether or not they are related. The distribution of the size of American households is

Outcome	1	2	3	4	5	6	7
Probability	.240	.322	.177	.155	.067	.024	.015

Find the mean number of people living in a household.

4.46 The distribution of the highest grade X completed by schoolchildren who reach the fifth grade is

x_i	4	5	6	7	8	9	10	11	12
p_i	.010	.007	.007	.013	.032	.068	.070	.041	.752

Find the mean μ_X.

4.47 Example 4.17 (page 307) gives the distribution of grades (A = 4, B = 3, and so on) in a large course as

Outcome	0	1	2	3	4
Probability	.10	.15	.30	.30	.15

Find the average (that is, the mean) grade in this course.

4.48 A life insurance company sells a term insurance policy to a 21-year-old male that pays \$100,000 if the insured dies within the next 5 years. The probability that a randomly chosen male will die each year can be found in mortality tables. The company collects a premium of \$250 each year as payment for the insurance. The amount X that the company earns on this policy is \$250 per year, less the \$100,000 that it must pay if the insured dies. Here is the distribution of X. Fill in the missing probability in the table and calculate the mean earnings μ_X.

Age at death	21	22	23	24	25	≥ 26
Payout	−\$99,750	−\$99,500	−\$99,250	−\$99,000	−\$98,750	\$1250
Probability	.00183	.00186	.00189	.00191	.00193	

4.49 It would be quite risky for you to insure the life of a 21-year-old friend under the terms of the previous exercise. There is a high probability that your friend would live and you would gain \$1250 in premiums. But if he were to die, you would lose almost \$100,000. Explain carefully why selling insurance is not risky for an insurance company that insures many thousands of 21-year-old men.

4.50 Green Mountain Numbers, a part of the Vermont state lottery, offers a choice of several bets. You choose a three-digit number. The lottery commission announces the winning three-digit number, chosen at random, at the end of each day. The "box" pays \$83.33 if the number you choose has the same digits as the winning number, in any order. Find the expected payoff for a \$1 bet on the box. (Assume that you chose a number having three different digits.)

4.51 For each of the following situations, would you expect the random variables X and Y to be independent? Explain your answers.

(a) X is the rainfall (in inches) on November 6 of this year and Y is the rainfall at the same location on November 6 of next year.

(b) X is the amount of rainfall today and Y is the rainfall at the same location tomorrow.

(c) X is today's rainfall at the Orlando, Florida, airport, and Y is today's rainfall at Disney World just outside Orlando.

4.52 In which of the following games of chance would you be willing to assume independence of X and Y in making a probability model? Explain your answer in each case.

(a) In blackjack, you are dealt two cards and examine the total points X on the cards (face cards count 10 points). You can choose to be dealt another card and compete based on the total points Y on all three cards.

(b) In craps, the betting is based on successive rolls of two dice. X is the sum of the faces on the first roll, and Y the sum of the faces on the next roll.

4.53 (a) A gambler knows that red and black are equally likely to occur on each spin of a roulette wheel. He observes five consecutive reds and bets heavily on red at the next spin. Asked why, he says that "red is hot" and that the run of reds is likely to continue. Explain to the gambler what is wrong with this reasoning.

(b) After hearing you explain why red and black remain equally probable after five reds on the roulette wheel, the gambler moves to a poker game. He is dealt five straight red cards. He remembers what you said and assumes that the next card dealt in the same hand is equally likely to be red or black. Is the gambler right or wrong? Why?

4.54 A time and motion study measures the time required for an assembly line worker to perform a repetitive task. The data show that the time required to bring a part from a bin to its position on an automobile chassis varies from car to car with mean 11 seconds and standard deviation 2 seconds. The time required to attach the part to the chassis varies with mean 20 seconds and standard deviation 4 seconds.

(a) What is the mean time required for the entire operation of positioning and attaching the part?

(b) If the variation in the worker's performance is reduced by better training, the standard deviations will decrease. Will this decrease change the mean you found in (a) if the mean times for the two steps remain as before?

(c) The study finds that the times required for the two steps are independent. A part that takes a long time to position, for example, does not take more or less time to attach than other parts. Would your answer in (a) change if the two variables were dependent?

4.55 Laboratory data show that the time required to complete two chemical reactions in a production process varies. The first reaction has a mean time of 40 minutes and a standard deviation of 2 minutes; the second has a mean time of 25 minutes and a standard deviation of 1 minute. The two reactions are run in sequence during production. There is a fixed period of 5 minutes between them as the product of the first reaction is pumped into the vessel where the second reaction will take place. What is the mean time required for the entire process?

4.56 Find the standard deviation σ_X of the distribution of grades in Exercise 4.47.

4.57 In an experiment on the behavior of young children, each subject is placed in an area with five toys. The response of interest is the number of toys that the child plays with. Past experiments with many subjects have shown that the probability distribution of the number X of toys played with is as follows:

Number of toys	0	1	2	3	4	5
Probability	.03	.16	.30	.23	.17	.11

Calculate the mean μ_X and the standard deviation σ_X.

4.58 In government data, a household consists of all occupants of a dwelling unit, but a family consists of two or more persons who live together and are related by blood, marriage, or adoption. Here are the distributions of household size and of family size in the United States:

Number of persons	1	2	3	4	5	6	7
Household probability	.240	.322	.177	.155	.067	.024	.015
Family probability	0	.413	.236	.211	.090	.032	.018

Compare the two distributions using probability histograms, means, and standard deviations. Then write a brief comparison, using your calculations to back up your statements.

4.59 Find the standard deviation of the time required to complete the chemical production process described in Exercise 4.55. The times for the two reactions are independent.

4.60 Find the standard deviation of the time required for the two-step assembly operation studied in Exercise 4.54.

4.61 Examples 4.24 and 4.25 (page 329) concern a probabilistic projection of sales and profits by an electronics firm, Gain Communications.

(a) Find the variance and standard deviation of the estimated sales Y of Gain's civilian division, using the distribution and mean from Example 4.24.

(b) Because the military budget and the civilian economy are not closely linked, Gain is willing to assume that its military and civilian sales vary independently. Combine your result from (a) with the results for the military division from Example 4.25 to obtain the standard deviation of the total sales $X + Y$.

(c) Find the standard deviation of the estimated profit, $Z = 2000X + 3500Y$.

4.62 The academic motivation and study habits of female students as a group are better than those of males. The Survey of Study Habits and Attitudes (SSHA) is a psychological test that measures these factors. The distribution of SSHA scores among the women at a college has mean 120 and standard deviation 28, and the distribution of scores among men students has mean 105 and standard deviation 35. You select a single male student and a single female student at random and give them the SSHA test.

(a) Explain why it is reasonable to assume that the scores of the two students are independent.
(b) What are the mean and standard deviation of the difference (female minus male) between their scores?
(c) From the information given, can you find the probability that the woman chosen scores higher than the man? If so, find this probability. If not, explain why you cannot.

4.63 In a process for manufacturing glassware, glass stems are sealed by heating them in a flame. The temperature of the flame varies a bit. Here is the distribution of the temperature X measured in degrees Celsius:

Temperature	540°	545°	550°	555°	560°
Probability	.1	.25	.3	.25	.1

(a) Find the mean temperature μ_X and the standard deviation σ_X.
(b) The target temperature is 550° C. What are the mean and standard deviation of the number of degrees off target $X - 550$?
(c) A manager asks for results in degrees Fahrenheit. The conversion of X into degrees Fahrenheit is given by

$$Y = \frac{9}{5}X + 32$$

What are the mean μ_Y and standard deviation σ_Y of the temperature of the flame in the Fahrenheit scale?

4.64 You have two scales for measuring weights in a chemistry lab. Both scales give answers that vary a bit in repeated weighings of the same item. If the true weight of a compound is 2 grams (g), the first scale produces readings X that have mean 2.000 g and standard deviation 0.002 g. The second scale's readings Y have mean 2.001 g and standard deviation 0.001 g.

(a) What are the mean and standard deviation of the difference $Y - X$ between the readings? (The readings X and Y are independent.)
(b) You measure once with each scale and average the readings. Your result is $Z = (X + Y)/2$. What are μ_Z and σ_Z? Is the average Z more or less variable than the reading Y of the less variable scale?

4.65 The risk of an investment is often measured by the standard deviation of the return on the investment. The more variable the return is (the larger σ is), the riskier the investment. We can measure the great risk of insuring a single person's life in Exercise 4.48 by computing the standard deviation of the income X that the insurer will receive. Find σ_X, using the distribution and mean found in Exercise 4.48.

4.66 The risk of insuring one person's life is reduced if we insure many people. Use the result of the previous exercise and rules 4.4 and 4.6 for means and variances to answer the following questions.

(a) Suppose that we insure two 21-year-old males and that their ages at death are independent. If X and Y are the insurer's income from the two insurance policies, the total income is $T = X + Y$. The mean and variance of each of X and Y are as you computed them in the previous exercise. What are the mean μ_T and the standard deviation σ_T of the insurer's total income T?

(b) The insurer's *average income* on the two policies is

$$Z = \frac{X+Y}{2} = \frac{1}{2}X + \frac{1}{2}Y$$

Find the mean and standard deviation of Z. You see that the mean income is the same as for a single policy, but the standard deviation is less.

(c) If four 21-year-old men are insured, the insurer's average income is

$$Z = \frac{1}{4}(X_1 + X_2 + X_3 + X_4)$$

where X_i is the income from insuring one man. The X_i are independent and each has the same distribution as before. Find the mean and standard deviation of Z. Compare your results with the results of (b).

4.4 PROBABILITY LAWS*

Our study of probability has concentrated on random variables and their distributions. Now we return to the laws that govern any assignment of probabilities. The purpose of learning more laws of probability is to be able to give probability models for more complex random phenomena. We have already met and used five basic rules.

*The content of this section is important for an understanding of probability. However, it is not needed to understand the statistical methods in later chapters and can therefore be omitted without loss of continuity.

RULES OF PROBABILITY

Rule 1. $0 \leq P(A) \leq 1$ for any event A
Rule 2. $P(S) = 1$
Rule 3. Addition rule: If A and B are disjoint events, then

$$P(A \text{ or } B) = P(A) + P(B)$$

Rule 4. Complement rule: For any event A,

$$P(A^c) = 1 - P(A)$$

Rule 5. Multiplication rule: If A and B are independent events, then

$$P(A \text{ and } B) = P(A)P(B)$$

The first three rules state the fundamental properties of probability. The next two rules can be derived from these three, and so can the other laws of probability that we will study. We will not emphasize the derivations, but it is important to note that the three commonsense facts in Rules 1 to 3 carry all of the other laws with them. Any assignment of probabilities to events that satisfies Rules 1 to 3 automatically satisfies the other laws. This is true in particular of the distributions of discrete and continuous random variables.

General addition rules

Probability has the property that if A and B are disjoint events, then $P(A \text{ or } B) = P(A) + P(B)$. But what if there are more than two events, or if the events are not disjoint? These circumstances are covered by more general addition rules for probability.

UNION

The union of any collection of events is the event that at least one of the collection occurs.

For two events A and B, the union is the event $\{A \text{ or } B\}$ that contains all outcomes in A, in B, or in both A and B. From the addition rule for disjoint events we can obtain rules for more general unions. Suppose first that we have several events—say, A, B, and C—that are disjoint in pairs. That is,

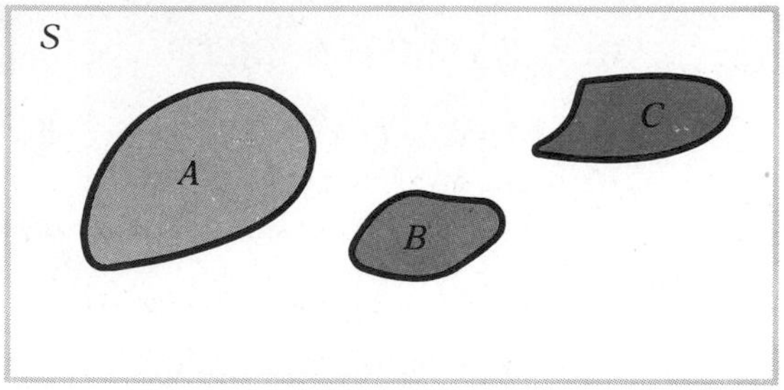

FIGURE 4.14 **The addition rule for disjoint events: $P(A \text{ or } B \text{ or } C) = P(A) + P(B) + P(C)$ when events A, B, and C are disjoint.**

no two can occur simultaneously. Figure 4.14 illustrates this case with a Venn diagram. The basic addition rule for two disjoint events extends to the following rule:

ADDITION RULE FOR DISJOINT EVENTS

If events A, B, and C are disjoint in the sense that no two have any outcomes in common, then

$$P(\text{one or more of } A, B, C) = P(A) + P(B) + P(C)$$

This rule extends to any number of disjoint events.

EXAMPLE 4.29

Generate a random number X between 0 and 1. What is the probability that the first digit will be odd? The random number X is a continuous random variable whose density curve has constant height 1 between 0 and 1 and is 0 elsewhere. The event that the first digit of X is odd is the union of five disjoint events. These events are

$$.10 \le X < .20$$
$$.30 \le X < .40$$
$$.50 \le X < .60$$
$$.70 \le X < .80$$
$$.90 \le X < 1.00$$

Figure 4.15 illustrates the probabilities of these events as areas under the density curve. Each has probability 0.1 equal to its length. The union of the five therefore has probability equal to the sum, or 0.5. As we should expect, a random number is equally likely to begin with an odd or an even digit. ●

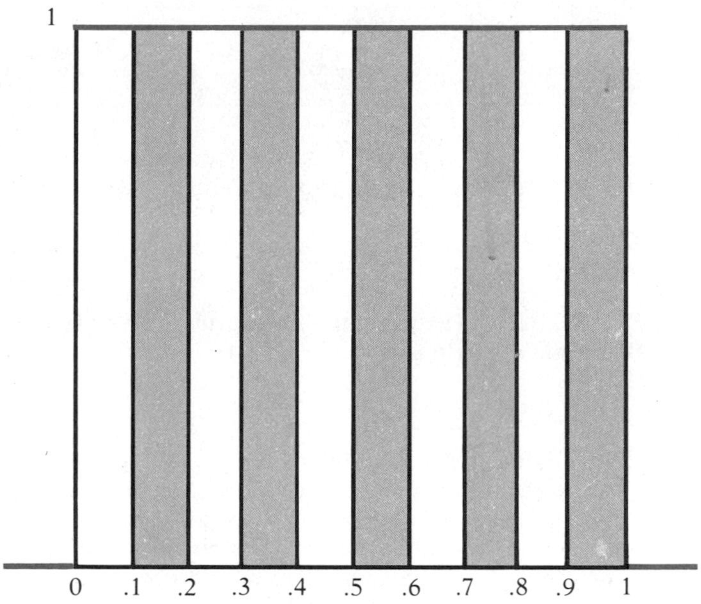

FIGURE 4.15 The probability that the first digit of a random number is odd is the sum of the probabilities of the 5 disjoint events shown.

If events A and B are not disjoint, they can occur simultaneously and the probability of their union is less than the sum of their probabilities. As Figure 4.16 suggests, the outcomes common to both are counted twice when we add probabilities, so we must subtract this probability once. Here is the addition rule for the union of any two events, disjoint or not.

> **GENERAL ADDITION RULE FOR UNIONS OF TWO EVENTS**
>
> For any two events A and B
>
> $$P(A \text{ or } B) = P(A) + P(B) - P(A \text{ and } B)$$

empty event

If A and B are disjoint, the event $\{A \text{ and } B\}$ that both occur has no outcomes in it. As you might guess, this *empty event* must have probability 0. So the general addition rule includes the earlier addition rule for disjoint events.

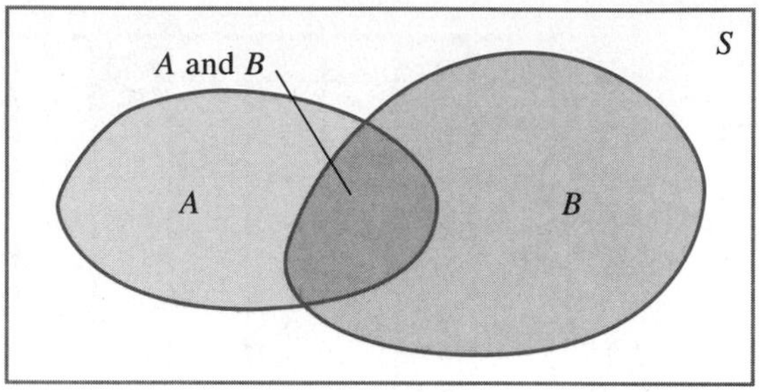

FIGURE 4.16 The general addition rule: $P(A \text{ or } B) = P(A) + P(B) - P(A \text{ and } B)$ for any events A and B.

EXAMPLE 4.30

Deborah and Matthew are anxiously awaiting word on whether they have been made partners of their law firm. Deborah guesses that her probability of making partner is 0.7 and that Matthew's is 0.5. (These are personal probabilities reflecting Deborah's assessment of chance.) This assignment of probabilities does not give us enough information to compute the probability that at least one of the two is promoted. In particular, adding the individual probabilities of promotion gives the impossible result 1.2. If Deborah also guesses that the probability that *both* she and Matthew are made partners is 0.3, then by the addition rule for unions

$$P(\text{at least one is promoted}) = .7 + .5 - .3 = .9$$

The probability that *neither* is promoted is then 0.1 by the complement rule.

Venn diagrams are a great help in finding probabilities for unions, because you can just think of adding and subtracting areas. Figure 4.17 shows some events and their probabilities for this example. Suppose we want the probability that Deborah is promoted and Matthew is not. The Venn diagram shows that this is the probability that Deborah is promoted minus the probability that both are promoted, $0.7 - 0.3 = 0.4$. Similarly, the probability that Matthew is promoted and Deborah is not is $0.5 - 0.3 = 0.2$. The four probabilities that appear in the figure add to 1 because they refer to four disjoint events whose union is the entire sample space. ●

Conditional probabilities and general multiplication rules

Harry Edwards is an authority on the sociology of sports. He has found that only 5% of male high school basketball, baseball, and football players go on to play at the college level. Of these, only 1.7% enter major league professional sports. The proportion of high school athletes who will have both a college and a professional career is therefore 1.7% of 5%. That's $0.017 \times 0.05 = 0.00085$, or about eight out of 10,000.[7]

We will express this commonsense relative frequency calculation in terms of a new rule of probability theory. Our sample space consists of all

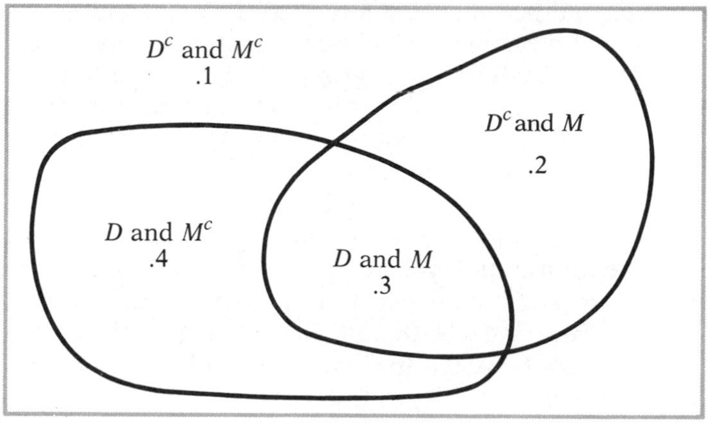

D = Deborah is made partner
M = Matthew is made partner

FIGURE 4.17 Venn diagram and probabilities for Example 4.30.

male athletes in the three major high school sports. The events of interest are

$$A = \text{competes in college}$$
$$B = \text{competes professionally}$$

We wish to compute the probability of the event $\{A \text{ and } B\}$ that the athlete competes both in college and professionally. This is called the intersection of the events A and B.

> **INTERSECTION**
>
> The intersection of any collection of events is the event that *all* of the events occur.

We know that $P(A) = 0.05$. We are also told that the probability that an athlete *who has already advanced to college sports* goes on to the pros is 0.017. This is not $P(B)$ but rather the probability that event B occurs given that the event A is known to occur. We call this the *conditional probability* of event B given the event A; we write it as $P(B|A) = 0.017$.

conditional probability

Whether one event has or has not occurred often influences the probability of another event—that is, the events are not independent. The conditional probability that a high school athlete plays professionally without having played in college, $P(B|A^c)$, is different from both $P(B|A)$ and the unconditional probability $P(B)$. Like unconditional probabilities, conditional probabilities can be thought of as relative frequencies. $P(B)$ is the proportion of a large group of high school athletes who go on to play professionally. $P(B|A)$ is found by looking only at the high school athletes who competed in college (event A) and finding the proportion of these who enter professional sports.

Be sure to keep in mind the distinct roles in $P(B|A)$ of the event B, whose probability we are computing, and the event A, which represents the information we are given. The conditional probability $P(B|A)$ makes no sense if the event A can never occur, so we require that $P(A) > 0$ whenever we talk about $P(B|A)$. A rule for the probability of intersections can now be stated in terms of conditional probability.

GENERAL MULTIPLICATION RULE FOR INTERSECTIONS

The probability of the intersection of two events is found by

$$P(A \text{ and } B) = P(A)P(B|A)$$

In words, this rule says that for both of two events to occur, first one must occur $[P(A)]$ and then, given that the first event has occurred, the second must occur $[P(B|A)]$. In the case of Edwards's study, the probability that an athlete competes both in college (event A) and professionally (event B) is

$$\begin{aligned} P(A \text{ and } B) &= P(A)P(B|A) \\ &= (.05)(.017) = .00085 \end{aligned}$$

as we computed earlier.

If $P(A)$ and $P(A \text{ and } B)$ are given, we can rearrange the multiplication rule to produce a definition of the conditional probability $P(B|A)$ in terms of unconditional probabilities.

CONDITIONAL PROBABILITY

When $P(A) > 0$, the conditional probability of B given A is

$$P(B|A) = \frac{P(A \text{ and } B)}{P(A)}$$

EXAMPLE 4.31 Select at random a U.S. resident age 25 or older. The census data in Table 2.6 gives us the following information about the age and education of this population. (Counts in the table are in thousands of persons.)

Age	<35	≥35	Total
Completed college	10,174	20,607	30,781
Did not complete college	32,794	88,041	120,835
Total	42,968	108,648	151,616

Let

$$A = \text{the person selected is under 35}$$
$$B = \text{the person selected has completed college}$$

Then because all 151,616,000 people are equally likely to be selected,

$$P(A) = \frac{42,968}{151,616} = .2834$$
$$P(B) = \frac{30,781}{151,616} = .2030$$
$$P(A \text{ and } B) = \frac{10,174}{151,616} = .0671$$

The conditional probability $P(B|A)$ is the proportion of people who have completed college (event B) among those who are under 35 (event A). We can find this probability directly from the counts as

$$\frac{10,174}{42,968} = .237$$

Now we can also use the formal definition of conditional probability:

$$P(B|A) = \frac{P(A \text{ and } B)}{P(A)} = \frac{.0671}{.2834} = .237$$

●

The definition of conditional probability reminds us that in principle all probabilities, including conditional probabilities, can be found from the assignment of probabilities to events that describes a random phenomenon. More often, however, conditional probabilities are part of the information given to us in a probability model, and the multiplication rule is used to compute $P(A \text{ and } B)$.

EXAMPLE 4.32

Slim is a professional poker player. At the moment, he wants very much to draw two diamonds in a row. As he sits at the table looking at his hand and at the upturned cards on the table, Slim sees 11 cards. Of these, 4 are diamonds. The full deck contains 13 diamonds among its 52 cards so 9 of the 41 unseen cards are diamonds. Because the deck was carefully shuffled, each card that Slim draws is equally likely to be any of the cards that he has not seen.

To find Slim's probability of drawing two diamonds, first calculate

$$P(\text{first card diamond}) = \frac{9}{41}$$
$$P(\text{second card diamond} \mid \text{first card diamond}) = \frac{8}{40}$$

Both probabilities were found by counting cards. The probability that the first card drawn is a diamond is 9/41 because 9 of the 41 unseen cards are diamonds. If the first card is a diamond, that leaves 8 diamonds among the 40 remaining cards. So the *conditional* probability of another diamond is 8/40. The multiplication rule now says that

$$P(\text{both cards diamonds}) = \left(\frac{9}{41}\right)\left(\frac{8}{40}\right) = .044$$

Slim will need luck to draw his diamonds. ●

The multiplication rule extends to the probability that all of several events occur. The key is to condition each event on the occurrence of *all* of the preceding events. For example, the intersection of three events A, B, and C has probability

$$P(A \text{ and } B \text{ and } C) = P(A)P(B \mid A)P(C \mid A \text{ and } B)$$

EXAMPLE 4.33

Take C to be the event that a major league professional athlete has a career that lasts longer than 3 years. Harry Edwards also finds that about 40% of the athletes who compete in college and then reach the pros have a career of more than 3 years. Taking A as the event that a high school athlete competes in college and B as the event that the athlete competes professionally, we know from Edwards's studies that

$$P(A) = .05$$
$$P(B \mid A) = .017$$
$$P(C \mid A \text{ and } B) = .4$$

The probability that a high school athlete competes in college and then goes on to have a professional career of more than 3 years is therefore

$$\begin{aligned} P(A \text{ and } B \text{ and } C) &= P(A)P(B \mid A)P(C \mid A \text{ and } B) \\ &= (.05)(.017)(.40) = .00034 \end{aligned}$$

Only about 3 of every 10,000 high school athletes can expect to compete in college and have a professional career of more than 3 years. Edwards concludes that high school students should concentrate on studies rather than on unrealistic hopes of fortune from professional sports. ●

Probability problems often require us to combine several of the basic rules into a more elaborate calculation. Here is an example that combines the addition and multiplication rules.

EXAMPLE 4.34

What is the probability that a high school athlete will go on to professional sports? In the notation of Example 4.33, this is $P(B)$. Since every athlete either does or does not compete in college, B is the union of $\{B \text{ and } A\}$ (the athlete competes both in college and professionally) and $\{B \text{ and } A^c\}$ (the athlete competes professionally but not in college). These two events are disjoint because no one can both compete and not compete in college. Figure 4.18 shows the relations among events. By the addition rule for disjoint events

$$P(B) = P(B \text{ and } A) + P(B \text{ and } A^c)$$

Now apply the multiplication rule to obtain each term in the sum separately. We have already seen that

$$\begin{aligned} P(B \text{ and } A) &= P(A)P(B \mid A) \\ &= (.05)(.017) = .00085 \end{aligned}$$

To apply the multiplication rule to the second term in the sum, we must know $P(B \mid A^c)$, the conditional probability that a high school athlete reaches a major professional league given that he does not compete in college. (Baseball players, for example, can reach the major leagues through the minor leagues without attending college.) If $P(B \mid A^c) = 0.0001$, then

$$\begin{aligned} P(B \text{ and } A^c) &= P(A^c)P(B \mid A^c) \\ &= (.95)(.0001) = .000095 \end{aligned}$$

The final result is

$$P(B) = .00085 + .000095 = .000945$$

About 9 high school athletes out of 10,000 will play professional sports. ●

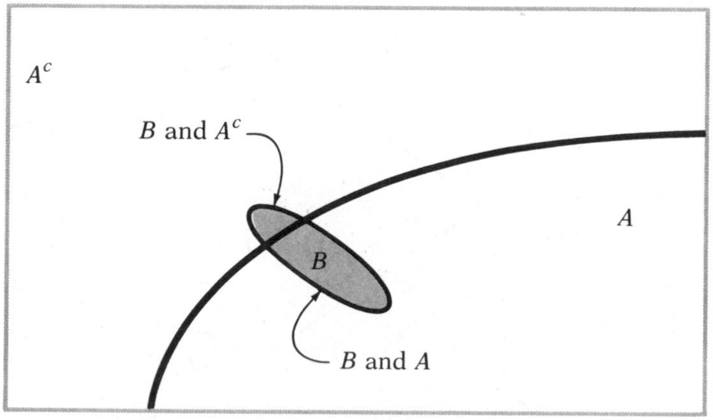

S = Male high school athletes in major sports
A = Competes in college
B = Competes professionally (colored)

FIGURE 4.18 Venn diagram for Example 4.34.

tree diagram

Tree diagrams are helpful in organizing calculations that involve several stages. Figure 4.19 is a tree diagram for Example 4.34. Each segment in the tree is one stage of the problem. Each complete branch shows a path that an athlete can take. The probabilities written by the segments are conditional probabilities that an athlete follows that segment given that he has reached the point from which it branches. The multiplication rule says that the probability of reaching the end of any complete branch is the product of the probabilities written by its segments. The probability of any outcome, such as the event B that an athlete reaches professional sports, is then found by adding the probabilities of all branches that are part of that event.

Bayes's formula There is another kind of probability question that we might ask in the context of Edwards's studies of athletes. Our earlier calculations look forward toward professional sports as the final stage of an athlete's career. Now let's concentrate on professional athletes and look back at their earlier careers.

EXAMPLE 4.35

What proportion of professional athletes competed in college? In the notation of Examples 4.33 and 4.34 this is the conditional probability $P(A \mid B)$. We start from the definition of conditional probability and then apply the results of

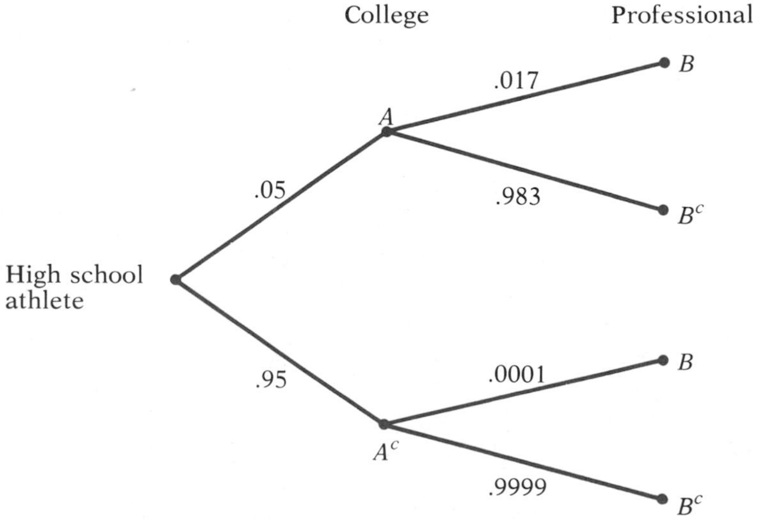

FIGURE 4.19 Tree diagram for Example 4.34. The probability $P(B)$ is the sum of the probabilities of the two branches ending at B.

Example 4.34:

$$P(A \mid B) = \frac{P(A \text{ and } B)}{P(B)} = \frac{.00085}{.000945} = .8995$$

Almost 90% of professional athletes competed in college. ●

We know from Edwards's studies the probabilities $P(A)$ and $P(A^c)$ that a high school athlete does and does not compete in college. We also know the conditional probabilities $P(B \mid A)$ and $P(B \mid A^c)$ that an athlete from each group reaches professional sports. Example 4.34 shows how to use this information to calculate $P(B)$. The method can be summarized in a single expression:

$$P(B) = P(B \mid A)P(A) + P(B \mid A^c)P(A^c)$$

In Example 4.35, we calculated the "reverse" conditional probability $P(A \mid B)$. The method can again be summarized in a single expression that builds on the formula just given for $P(B)$:

$$P(A \mid B) = \frac{P(B \mid A)P(A)}{P(B \mid A)P(A) + P(B \mid A^c)P(A^c)}$$

Bayes's formula

This expression is called *Bayes's formula*, after Thomas Bayes, who wrestled with arguing from outcomes like B back to antecedents like A in a book published in 1763. It is better to think your way through problems like Examples 4.34 and 4.35 rather than memorize these expressions.

Independence The conditional probability $P(B \mid A)$ is generally not equal to the unconditional probability $P(B)$. That is because the occurrence of event A generally gives us some additional information about whether or not event B occurs. If knowing that A occurs gives no additional information about B, then A and B are independent events. The precise definition of independence is expressed in terms of conditional probability.

INDEPENDENT EVENTS

Two events A and B that both have positive probability are independent if

$$P(B \mid A) = P(B)$$

This is a more exact description of independence than that given in Section 4.1. We now see that the multiplication rule for independent events, $P(A \text{ and } B) = P(A)P(B)$, is a special case of the general multiplication rule, $P(A \text{ and } B) = P(A)P(B|A)$, just as the addition rule for disjoint events is a special case of the general addition rule.

Decision analysis Perhaps the most straightforward view of statistical inference regards the goal of inference as *making decisions in the presence of uncertainty*. One kind of decision making in the presence of uncertainty seeks to ensure that the decision made will be the one with the highest probability of a favorable outcome. Here is an example that illustrates how the multiplication and addition rules, organized with the help of a tree diagram, apply to a decision problem.

EXAMPLE 4.36

The kidneys of a patient with end-stage kidney disease will not support life. If the patient is to survive, the available choices are a kidney transplant or regular hemodialysis (use of a kidney machine several times a week). Lynn is faced with this choice. Both treatments are risky. Her doctor gives her the following information for patients in her condition: About 68% of dialysis patients survive for 5 years. Of transplant patients, about 48% survive with the transplanted kidney for 5 years, 43% must undergo regular dialysis because the transplanted kidney fails, and the remaining 9% do not survive the transplant. Of those transplant patients who return to dialysis, about 42% survive 5 years. Which treatment should Lynn choose to give herself the best chance of living for 5 years?[8] ●

We can organize the information provided by the doctor in a tree diagram (Figure 4.20). Each path through the tree represents a possible outcome of Lynn's case. The probability written beside each branch after the first stage is the conditional probability of the next step given that Lynn has reached this point. For example, 0.68 is the probability that a dialysis patient survives 5 years. The conditional probability that a patient survives 5 years on dialysis given that the patient first had a transplant and then returned to dialysis because the transplant failed is different. It is 0.42 and appears on a different branch of the tree. Study the tree to convince yourself that it organizes all the information available.

The multiplication rule for probabilities says that the probability of reaching the end of a branch in the tree is the product of the probabilities of all the steps along that branch. These probabilities are written at the end of each branch in Figure 4.20. For example, the event that a transplant patient returns to dialysis and then survives 5 years is

$$A = \{D \text{ and } S\}$$

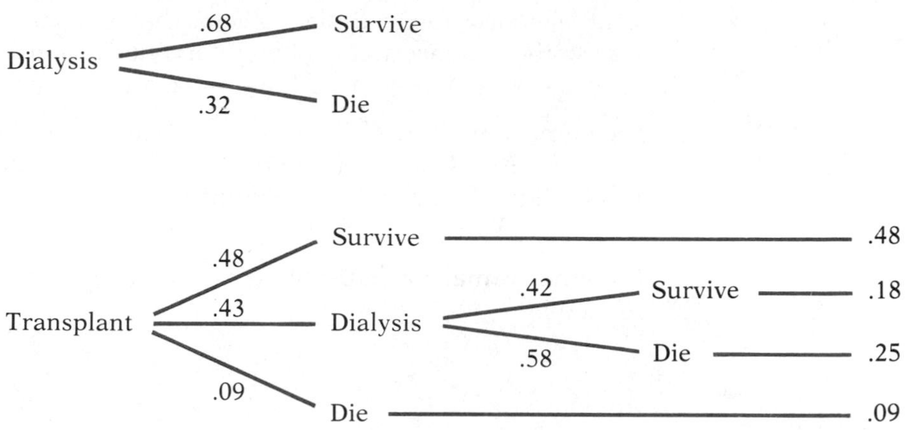

FIGURE 4.20 Tree diagram for the kidney failure decision problem in Example 4.36.

where

$$D = \text{transplant patient returns to dialysis}$$
$$S = \text{patient survives 5 years}$$

Therefore,

$$\begin{aligned} P(A) &= P(D \text{ and } S) \\ &= P(D)P(S|D) \\ &= (.43)(.42) = .18 \end{aligned}$$

Lynn can now compare her probability of surviving 5 years with regular dialysis with the probability of surviving 5 years with a transplant. For patients who choose to undergo dialysis, the survival probability is 0.68. A transplant patient can survive in either of two ways: with the transplant functioning or, if the transplant has failed, with a return to dialysis. These are *disjoint* events, represented by separate branches of the tree; they have probabilities 0.48 and 0.18, respectively. The addition rule for probability says that the overall probability of surviving 5 years is the sum, $0.48 + 0.18 = 0.66$. This is slightly less than the survival probability under dialysis, so dialysis is the better choice. In this example, there are only two decisions to be compared, dialysis or transplant. Other decision problems may have more alternatives.

Where do the conditional probabilities in Example 4.36 come from? They are based in part on data—that is, on studies of many patients with kidney disease. But an individual's chances of survival depend on her age, general health, and other factors. Lynn's doctor considered her individual situation before giving her these particular probabilities. It is characteristic of most decision analysis problems that *personal probabilities* are used to describe the uncertainty of an informed decision maker.

SUMMARY

The **complement** A^c of an event A contains all outcomes that are not in A. The **union** $\{A \text{ or } B\}$ of events A and B contains all outcomes in A, in B, or in both A and B. The **intersection** $\{A \text{ and } B\}$ contains all outcomes that are in both A and B, but not outcomes in A alone or B alone.

The **conditional probability** $P(B|A)$ of an event B given an event A is defined by

$$P(B|A) = \frac{P(A \text{ and } B)}{P(A)}$$

when $P(A) > 0$, but in practice is most often found from directly available information.

The essential general laws of elementary probability are

Legitimate values: $0 \leq P(A) \leq 1$ for any event A, and $P(S) = 1$.
Complement rule: $P(A^c) = 1 - P(A)$.
Addition rule: $P(A \text{ or } B) = P(A) + P(B) - P(A \text{ and } B)$.
Multiplication rule: $P(A \text{ and } B) = P(A)P(B|A)$.

If A and B are **disjoint,** then $P(A \text{ and } B) = 0$. The general addition rule for unions then becomes the special addition rule, $P(A \text{ or } B) = P(A) + P(B)$.

A and B are **independent** when $P(B|A) = P(B)$. The multiplication rule for intersections then becomes $P(A \text{ and } B) = P(A)P(B)$.

In problems with several stages, use of the multiplication and addition rules is aided by drawing a **tree diagram.**

SECTION 4.4 EXERCISES

4.67 Call a household prosperous if its 1987 income exceeded \$50,000. Call the household educated if the householder completed college. Select an American household at random, and let A be the event that the selected household is prosperous and B the event that it is educated. According to the Census Bureau, $P(A) = 0.192$, $P(B) = 0.221$, and the probability that a household is both prosperous and educated is $P(A \text{ and } B) = 0.092$. What is the probability $P(A \text{ or } B)$ that the household selected is either prosperous or educated?

4.68 Consolidated Builders has bid on two large construction projects. The company president believes that the probability of winning the first contract (event A) is 0.6, that the probability of winning the second (event

B) is 0.4, and that the probability of winning both jobs (event $\{A \text{ and } B\}$) is 0.2. What is the probability of the event $\{A \text{ or } B\}$ that Consolidated will win at least one of the jobs?

4.69 Draw a Venn diagram that shows the relation between the events A and B in Exercise 4.67. Indicate each of the following events on your diagram and use the information in Exercise 4.67 to calculate the probability of each event. Finally, describe in words what each event is.
(a) $\{A \text{ and } B\}$
(b) $\{A \text{ and } B^c\}$
(c) $\{A^c \text{ and } B\}$
(d) $\{A^c \text{ and } B^c\}$

4.70 Draw a Venn diagram that illustrates the relation between events A and B in Exercise 4.68. Write each of the following events in terms of A, B, A^c, and B^c. Indicate the events on your diagram and use the information in Exercise 4.68 to calculate the probability of each.
(a) Consolidated wins both jobs.
(b) Consolidated wins the first job but not the second.
(c) Consolidated does not win the first job but does win the second.
(d) Consolidated does not win either job.

4.71 Here is the assignment of probabilities that describes the age (in years) and the sex of a randomly selected American college student:

Age	14–17	18–24	25–34	≥35
Male	.01	.30	.12	.04
Female	.01	.30	.13	.09

(a) What is the probability that the student is a female?
(b) What is the conditional probability that the student is a female given that the student is at least 35 years old?
(c) What is the probability that the student is either a female or at least 35 years old?

4.72 Here is a two-way table of all suicides committed in a recent year by sex of the victim and method used:

Method	Male	Female
Firearms	15,518	2635
Poison	3516	2520
Hanging	3761	845
Other	1431	678
Total	24,226	6678

Suppose that a suicide victim is selected at random. Answer the following questions from the table.

(a) What is the probability that the suicide victim is male?

(b) What is the probability that the suicide victim used a firearm?

(c) What is the conditional probability that the suicide used a firearm given that the victim is male?

(d) Are the events "victim is male" and "a firearm was used in the suicide" independent? Why or why not?

4.73 Here is a two-way table of American households classified by education and income (the table entries are in thousands of households):

	Income class (thousands of dollars)				
Education	<15	15–34	35–49	≥ 50	Total
<4 years of high school	11,668	7217	1909	1180	21,976
4 years of high school	8088	12,417	5776	4279	30,561
1–3 years of college	2626	5263	3230	3173	14,294
$\geq$4 years of college	1597	5189	4334	7888	19,007

Answer the following questions from this table.

(a) What is the probability that a randomly selected household has an income of at least \$50,000?

(b) What is the conditional probability that a household earns over \$50,000 given that the householder completed at least 4 years of college?

(c) What is the conditional probability that the householder completed at least 4 years of college given that the household income is at least \$50,000?

(d) Are the random variables X (household income in dollars) and Y (years of education for the householder) independent? Why or why not?

4.74 Common sources of caffeine in the diet are coffee, tea, and cola drinks. Suppose that

55% of adults drink coffee
25% of adults drink tea
45% of adults drink cola

and also that

15 % drink both coffee and tea
5 % drink all three beverages
25 % drink both coffee and cola
5 % drink only tea

Draw a Venn diagram marked with this information. Use it along with the addition rules to answer the following questions.

(a) What percent of adults drink only cola?

(b) What percent drink none of these beverages?

4.75 Choose an employed person at random. Let A be the event that the person chosen is a woman and B the event that the person holds a managerial or professional job. As of 1990, $P(A) = 0.45$, and the probability of managerial and professional jobs among women is $P(B|A) = 0.26$. Find the probability that a randomly chosen employed person is a woman holding a managerial or professional position.

4.76 Functional Robotics Corporation buys electrical controllers from a Japanese supplier. The company's treasurer feels that there is probability 0.4 that the dollar will fall in value against the Japanese yen in the next month. The treasurer also believes that *if* the dollar falls, there is probability 0.8 that the supplier will demand renegotiation of the contract. What probability has the treasurer assigned to the event that the dollar falls and the supplier demands renegotiation?

4.77 In the language of government statistics, the "labor force" includes all civilians over 16 years of age who are working or looking for work. Select a member of the U.S. labor force at random. Let A be the event that the person selected is white and B the event that he or she is employed. In 1990, 85.9% of the labor force was white. Of the whites in the labor force, 95.3% were employed. Among nonwhite members of the labor force, 89.9% were employed.

(a) Express each of the percents given as a probability involving the events A and B—for example, $P(A) = 0.859$.

(b) Draw a tree diagram for the outcomes of recording first the race (white or nonwhite) of a randomly chosen member of the labor force and then whether or not the person is employed.

(c) Find the probability that the person chosen is an employed white. Also find the probability that an employed nonwhite is chosen. What is the probability $P(B)$ that the person chosen is employed?

4.78 Suppose that in Exercise 4.76 the treasurer also feels that if the dollar does not fall, there is probability 0.2 that the Japanese supplier will demand that the contract be renegotiated. What is the probability that the supplier will demand renegotiation?

4.79 Use your results from Exercise 4.77 and the definition of conditional probability to find the probability $P(A|B)$ that a randomly selected member of the labor force is white, given that he or she is employed. (Alternatively, you can use Bayes's formula.)

4.80 An examination consists of multiple-choice questions, each having five possible answers. Linda estimates that she has probability 0.75 of knowing the answer to any question that may be asked. If she does not know the answer, she will guess, with conditional probability 1/5 of being correct. What is the probability that Linda gives the correct answer to a question? (Draw a tree diagram to guide the calculation.)

4.81 The voters in a large city are 40% white, 40% black, and 20% Hispanic. (Hispanics may be of any race in official statistics, but in this case we are speaking of political blocks.) A black mayoral candidate anticipates attracting 30% of the white vote, 90% of the black vote, and 50% of the Hispanic vote. Draw a tree diagram with probabilities for the race (white, black, or Hispanic) and vote (for or against the candidate) of a randomly chosen voter. What percent of the overall vote does the candidate expect to get?

4.82 In the setting of Exercise 4.80, find the conditional probability that Linda knows the answer, given that she supplies the correct answer. (You can use the result of Exercise 4.80 and the definition of conditional probability, or you can use Bayes's formula.)

4.83 Choose a point at random in the square with sides $0 \le x \le 1$ and $0 \le y \le 1$. This means that the probability that the point falls in any region within the square is the area of that region. Let X be the x coordinate and Y the y coordinate of the point chosen. Find the conditional probability $P(Y < 1/2 \mid Y > X)$. (Hint: Draw a diagram of the square and the events $Y < 1/2$ and $Y > X$.)

4.84 Psychologists use probability models to describe learning in animals. In one experiment, a rat is placed in a dark compartment and a door leading to a light compartment is opened. The rat will not move to the light compartment without reason. A bell is rung and if after 5 seconds the rat has not moved, it gets a shock through the floor of the dark compartment. To avoid the shock, rats soon learn to move when the bell is rung. A simple model for learning says that the rat can only be in one of two states:

- In *state A* the rat will not move from the dark compartment until it receives a shock.
- In *state B* the rat has learned to respond to the bell and moves immediately when the bell rings.

A rat starts in state A and eventually changes to state B. The change from state A to state B is the result of learning. A rat in state A gets a shock every time the bell rings; after it changes to state B, it never gets a shock. Suppose that a rat has probability 0.2 of learning each time it is shocked. Let the random variable X be the number of shocks that this rat receives.[9]

(a) If a rat receives exactly 4 shocks, which state was the rat in at the end of each of trials 1, 2, 3, and 4?

(b) Use the result of (a) and the multiplication rule to find $P(X = 4)$, the probability that the rat receives exactly 4 shocks.

(c) Based on your work in (a) and (b), give the probability distribution of X. That is, for any positive whole number x, what is $P(X = x)$?

4.85 John has coronary artery disease. He and his doctor must decide between medical management of the disease and coronary bypass surgery. Because John has been quite active, he is concerned about his quality of life as well as the length of life. He wants to make the decision that will maximize the probability of the event A that he survives for 5 years and is able to carry on moderate activity during that time. The doctor makes the following probability estimates for patients of John's age and condition:

- Under medical management, $P(A) = 0.7$.
- There is probability 0.05 that John will not survive bypass surgery, probability 0.10 that he will survive with serious complications, and probability 0.85 that he will survive the surgery without complications.
- If he survives with complications, the conditional probability of the desired outcome A is 0.73. If there are no serious complications, the conditional probability of A is 0.76.

Draw a tree diagram that summarizes this information. Then calculate $P(A)$ assuming that John chooses the surgery. Does surgery or medical management offer him the better chance of achieving his goal? (Based loosely on M. C. Weinstein, J. S. Pliskin, and W. B. Stason, "Coronary artery bypass surgery: Decision and policy analysis," in J. P. Bunker, B. A. Barnes, and F. W. Mosteller (eds.), *Costs, Risks and Benefits of Surgery*, Oxford University Press, New York, 1977, pp. 342–371.)

4.86 Zipdrive, Inc., has developed a new disk drive for small computers. The demand for the new product is uncertain but can be described as "high" or "low" in any one year. After 4 years, the product is expected to be obsolete. Management must decide whether to build a plant or to contract with a factory in Hong Kong to manufacture the new drive. Building a plant will be profitable if demand remains high but could lead to a loss if demand drops in future years.

After careful study of the market and of all relevant costs, Zipdrive's planning office provides the following information. Let A be the event that the first year's demand is high and B the event that the following 3 years' demand is high. The marketing division's best estimate of the probabilities is

$$P(A) = .9$$
$$P(B \mid A) = .36$$
$$P(B \mid A^c) = 0$$

The probability that building a plant is more profitable than contracting the production to Hong Kong is 0.95 if demand is high all 4 years, 0.3 if demand is high only in the first year, and 0.1 if demand is low all 4 years.

Draw a tree diagram that organizes this information. Each branch will have three segments: first year demand, next 3 years' demand, and whether building or contracting is more profitable. Which decision has the higher probability of being more profitable? (When decision analysis is used for investment decisions like this, firms in fact compare the mean profits rather than the probability of a profit. We ignore this complication.)

CHAPTER 4 EXERCISES

4.87 The original simple form of the Connecticut state lottery (ignoring a few gimmicks) awarded the following prizes for each 100,000 tickets sold. The winners were chosen by drawing tickets at random.

1	\$5000 prize
18	\$200 prizes
120	\$25 prizes
270	\$20 prizes

If you hold one ticket in this lottery, what is your probability of winning anything? What is the mean amount of your winnings?

4.88 Rotter Partners is planning a major investment. The amount of profit X is uncertain but a probabilistic estimate gives the following distribution (in millions of dollars):

Profit	1	1.5	2	4	10
Probability	.1	.2	.4	.2	.1

(a) Find the mean profit μ_X and the standard deviation of the profit.

(b) Rotter Partners owes its source of capital a fee of \$200,000 plus 10% of the profits X. So the firm actually retains

$$Y = .9X - .2$$

from the investment. Find the mean and standard deviation of Y.

4.89 A grocery store gives its customers cards that may win them a prize when matched with other cards. The back of the card announces the following probabilities of winning various amounts if a customer visits the store 10 times:

Amount	\$1000	\$200	\$50	\$10
Probability	1/10,000	1/1000	1/100	1/20

(a) What is the probability of winning nothing?
(b) What is the mean amount won?
(c) What is the standard deviation of the amount won?

4.90 The time required for a telephone operator to handle a directory assistance call is normally distributed with mean 25 seconds and standard deviation 5 seconds.
(a) What is the probability that it takes more than 30 seconds to handle such a call?
(b) The telephone company plans to monitor the longest 1% of directory assistance calls to study the reasons for the delay. You must instruct the computer how long to wait before beginning to monitor the call. What is the length of time that only 1% of directory assistance calls exceed?

4.91 A study of the weights of the brains of Swedish men found that the weight X was a random variable with mean 1400 grams and standard deviation 20 grams. Find positive numbers a and b such that $Y = a + bX$ has mean 0 and standard deviation 1.

4.92 In assigning a character's intelligence in the game "Dungeons and Dragons," the player rolls three dice and adds the spots on the up faces. Assume that the dice are all balanced so that each face is equally likely and that the three dice fall independently.
(a) Give a sample space for the sum X of the spots.
(b) Find $P(X = 5)$.
(c) If X_1, X_2, and X_3 are the number of spots on the up faces of the three dice, then $X = X_1 + X_2 + X_3$. Use this fact to find the mean μ_X and the standard deviation σ_X without finding the distribution of X. (Start with the distribution of each of the X_i.)

4.93 The real return on an investment is its rate of increase corrected for the effects of inflation. You believe that the annual real return X on a portfolio of stocks will vary in the future with mean $\mu_X = 0.11$ and standard deviation $\sigma_X = 0.28$. (That is, you expect stocks to give an average return of 11% in the future, but with large variation from year to year.) You further think that the annual real return Y on Treasury bills will vary with mean $\mu_Y = 0.02$ and standard deviation $\sigma_Y = 0.05$. Even though this is *not* realistic, assume that returns on stocks and Treasury bills vary independently.
(a) If you put half of your assets into stocks and half into Treasury bills, your overall return will be $Z = 0.5X + 0.5Y$. Calculate μ_Z and σ_Z.

(b) You decide that you are willing to take more risk (greater variation in the return) in exchange for a higher mean return. Choose an allocation of your assets between stocks and Treasury bills that will accomplish this and calculate the mean and standard deviation of the overall return. (If you put a proportion α of your assets into stocks, the total return is $Z = \alpha X + (1 - \alpha)Y$.)

4.94 The most popular game of chance in Roman times was tossing four *astragali*. An astragalus is a small bone from the heel of an animal that comes to rest on one of four sides when tossed. (The other two sides are rounded.) The table gives the probabilities of the outcomes for a single astragalus based on modern experiments. The names "Broad convex," etc., describes the four sides of the heel bone. The best throw was the "Venus," with all four uppermost sides different. What is the probability of rolling a Venus? (From Florence N. David, *Games, Gods and Gambling*, Charles Griffin, London, 1962, p. 7.)

Side	Broad convex	Broad concave	Narrow flat	Narrow hollow
Probability	.4	.4	.1	.1

4.95 Ann and Bob are playing the game "two-finger Morra." Each player shows either one or two fingers and at the same time calls out a guess for the number of fingers the other player will show. If a player guesses correctly and the other player does not, the player wins a number of dollars equal to the total number of fingers shown by both players. If both or neither guesses correctly, no money changes hands. On each play both Ann and Bob choose one of the following options:

Choice	Show	Guess
A	1	1
B	1	2
C	2	1
D	2	2

(a) Give the sample space S by writing all possible choices for both players on a single play of this game.

(b) Let X be Ann's winnings on a play. (If Ann loses \$2, then $X = -2$; when no money changes hands, $X = 0$.) Write the value of the random variable X next to each of the outcomes you listed in (a). This is another choice of sample space.

(c) Now assume that Ann and Bob choose independently of each other. Moreover, they both play so that all four choices listed above are equally likely. Find the probability distribution of X.

(d) If the game is fair, X should have mean 0. Does it? What is the standard deviation of X?

The following exercises require familiarity with the material presented in the optional Section 4.4.

4.96 Here is the distribution of the adjusted gross income X (in thousands of dollars) reported on individual federal income tax returns in 1984.

Income	< 11	11–24	25–49	50–99	≥ 100
Probability	.369	.316	.248	.057	.010

(a) What is the probability that a randomly chosen return shows an adjusted gross income of $50,000 or more?
(b) Given that a return shows an income of at least $50,000, what is the conditional probability that the income is at least $100,000?

4.97 You have torn a tendon and are facing surgery to repair it. The orthopedic surgeon explains the risks to you: Infection occurs in 3% of such operations, the repair fails in 14%, and both infection and failure occur together in 1%. What percent of these operations succeed and are free from infection?

4.98 The distribution of blood types among white Americans is approximately as follows: 37% type A, 13% type B, 44% type O, and 6% type AB. Suppose that the blood types of married couples are independent and that both the husband and wife follow this distribution.
(a) An individual with type B blood can safely receive transfusions only from persons with type B or type O blood. What is the probability that the husband of a woman with type B blood is an acceptable blood donor for her?
(b) What is the probability that in a randomly chosen couple the wife has type *B* blood and the husband has type A?
(c) What is the probability that one of a randomly chosen couple has type A blood and the other has type B?
(d) What is the probability that at least one of a randomly chosen couple has type O blood?

4.99 Exercise 4.19 (page 302) gives the probability distribution of the sex and occupation of a randomly chosen American worker. Use this distribution to answer the following questions:
(a) Given that the worker chosen holds a managerial (class A) job, what is the conditional probability that the worker is female?
(b) Classes D and E include most mechanical and factory jobs. What is the conditional probability that a worker is female given that he or she holds a job in one of these classes?

4.100 It is difficult to conduct sample surveys on sensitive issues because many people will not answer questions if the answers might embarrass them. "Randomized response" is an effective way to guarantee anonymity while collecting information on topics such as student cheating or sexual behavior. Here is the idea: To ask a sample of students whether they have plagiarized a term paper while in college, have each student toss a coin in private. If the coin lands heads *and* they have not plagiarized, they are to answer "No." Otherwise, they are to give "Yes" as their answer. Only the student knows whether the answer reflects the truth or just the coin toss, but the researchers can use a proper random sample with follow-up for nonresponse and other good sampling practices.

Suppose that in fact the probability is 0.3 that a randomly chosen student has plagiarized a paper. Draw a tree diagram in which the first segment of each branch is tossing the coin and the second segment is the truth about plagiarism. The outcome at the end of each branch is the answer given to the randomized response question. What is the probability of a "No" answer in the randomized response poll? If the probability of plagiarism were 0.2, what would be the probability of a "No" response on the poll? Now suppose that you get 39% "No" answers in a randomized response poll of a large sample of students at your college. What do you estimate to be the percent of the population who have plagiarized a paper?

4.101 ELISA tests are used to screen donated blood for the presence of the AIDS virus. The test actually detects antibodies, substances that the body produces when the virus is present. When antibodies are present, ELISA is positive with probability about 0.997 and negative with probability 0.003. When the blood tested is not contaminated with AIDS antibodies, ELISA gives a positive result with probability about 0.015 and a negative result with probability 0.985.[10] (Because ELISA is designed to keep the AIDS virus out of blood supplies, the higher probability 0.015 of a false positive is acceptable in exchange for the low probability 0.003 of failing to detect contaminated blood. These probabilities depend on the expertise of the particular laboratory doing the test.) Suppose that 1% of a large population carries the AIDS antibody in their blood.

(a) Draw a tree diagram for selecting a person from this population (outcomes: the person does or does not carry the AIDS antibody) and for testing his or her blood (outcomes: positive or negative).

(b) What is the probability that the ELISA test for AIDS is positive for a randomly chosen person from this population?

(c) What is the probability that a person has the antibody given that the ELISA test is positive?

(This exercise illustrates a fact that is important when considering proposals for widespread testing for AIDS or illegal drugs: If the condition being tested is uncommon in the population, most positives will be false positives.)

CHAPTER 4 COMPUTER EXERCISES

4.102 Example 4.9 (page 289) gives the distribution of colors of plain M&M candies as

Color	Brown	Red	Yellow	Green	Orange	Tan
Probability	.3	.2	.2	.1	.1	.1

Many statistical software systems allow you to simulate observations from any discrete distribution. The DISCRETE subcommand of the RANDOM command in Minitab does this, for example. A small bag contains 25 M&M's. Simulate the contents of 10 bags and record the number of candies of each color in each of the bags. Combine your data with those of other students so that you have information on at least 100 bags.

(a) What was the overall proportion of orange candies in your data? Was the observed proportion close to the probability, which is .1?

(b) What was the mean number $\bar{x}$ of orange candies in the bags? The theoretical mean is $\mu = 2.5$. Was the observed mean close to the theoretical value?

(c) How many bags contained no orange candies? Do you expect bags of 25 M&Ms to often, sometimes, or almost never contain no orange candies?

4.103 Toss a balanced coin 10 times. What is the probability of a run of 3 or more consecutive heads? What is the distribution of the length of the longest run of heads? What is the mean length of the longest run of heads? These are quite difficult questions if we must rely on mathematical calculations of probability. Computer simulation of the probability model can provide approximate answers.

First, simulate 50 repetitions of tossing a balanced coin 10 times. Here is the Minitab command that places the results of 10 tosses in each of columns 1 to 50; heads are represented by 1s and tails by 0s:

```
MTB > RANDOM 10 C1-C50;
 SUBC> BERNOULLI .5.
```

Now examine your 50 repetitions (using the PRINT command). Record the length of the longest run of heads (1s) in each trial. Combine your results with those of other students so that you have the results of several hundred repetitions.

(a) Make a table of the (approximate) probability distribution of the length X of the longest run of heads in 10 coin tosses. (The relative frequency of each outcome in many repetitions is approximately equal to its probability.) Draw a probability histogram of this distribution.
(b) What is your estimate of the probability of a run of 3 or more heads?
(c) Find the mean μ_X from your table of probabilities in (a). Then find the average of the 50 values of X in your 50 repetitions. How close to the mean μ_X was the average obtained in 50 repetitions?

NOTES

1. An informative and entertaining account of the origins of probability theory is Florence N. David, *Games, Gods and Gambling*, Charles Griffin, London, 1962.
2. We will consider only the case in which X takes a finite number of possible values. The same ideas, implemented with more advanced mathematics, apply to random variables with an infinite but still countable collection of values.
3. The notation $\bar{x}$ is used both for the random variable, which takes different values in repeated sampling, and for the numerical value of the random variable in a particular sample. Similarly, s and $\hat{p}$ stand both for random variables and for specific values. This notation is mathematically imprecise but statistically convenient.
4. See A. Tversky and D. Kahneman, "Belief in the law of small numbers," *Psychological Bulletin*, 76 (1971), pp. 105–110, and other writings of these authors for a full account of our misperception of randomness.
5. Probabilities involving runs can be quite difficult to compute. That the probability of a run of three or more heads in 10 independent tosses of a fair coin is $(1/2) + (1/128) = 0.508$ can be found by clever counting, as can the other results given in the text. A general treatment using advanced methods appears in Section XIII.7 of William Feller, *An Introduction to Probability Theory and Its Applications*, vol. 1, 3d ed., Wiley, New York, 1968.
6. R. Vallone and A. Tversky, "The hot hand in basketball: On the misperception of random sequences," *Cognitive Psychology*, 17 (1985), pp. 295–314. A later series of articles that debate the independence question is A. Tversky and T. Gilovich, "The cold facts about the 'hot hand' in basketball," *Chance*, 2 (1989), no. 1, pp. 16–21; P. D. Larkey, R. A. Smith, and J. B. Kadane, "It's OK to believe in the 'hot hand,'" *Chance*, 2 (1989), no. 4, pp. 22–30; and A. Tversky and T. Gilovich, "The 'hot hand': Statistical reality or cognitive illusion?" *Chance*, 2 (1989), no. 4, pp. 31–34.
7. Professor Edwards's findings are reported in the *New York Times*, February 25, 1986.
8. The probabilities in this example are based on the article by Benjamin A. Barnes, "An overview of the treatment of end-stage renal disease and a consideration of some of the consequences," in J. P. Bunker, B. A. Barnes, and F. W. Mosteller (eds.), *Costs, Risks and Benefits of Surgery*, Oxford University

Press, New York, 1977, pp. 325–341. See this article for a more realistic analysis.

9. This model is too simple to fit experimental data well. Probability models for learning that give more realistic descriptions of experiments with rats are discussed in R. C. Atkinson, G. H. Bower, and E. J. Crothers, *An Introduction to Mathematical Learning Theory*, Wiley, New York, 1965.

10. These probabilities are estimated from a large national study reported in E. M. Sloand et al., "HIV testing: State of the Art," *Journal of the American Medical Association*, 266 (1991), pp. 2861–2866.

CHAPTER

5

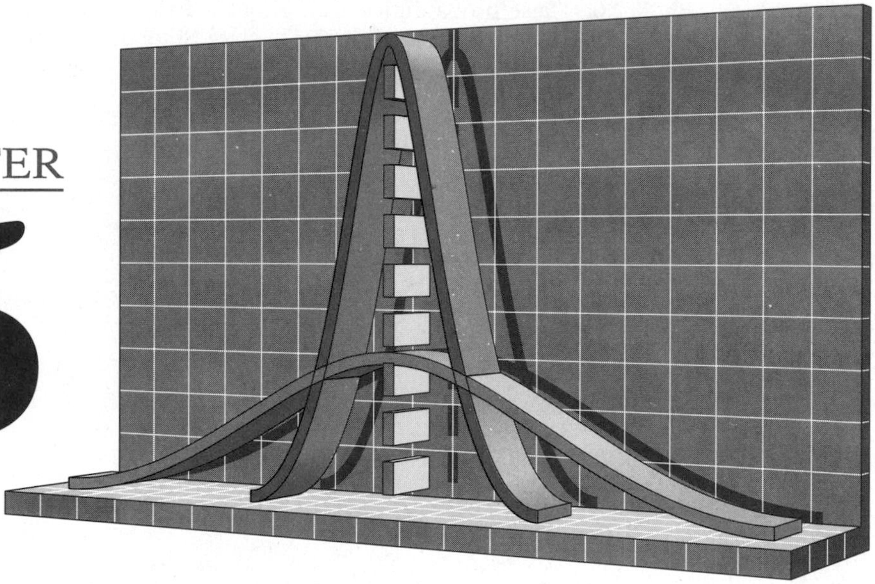

From Probability to Inference

PRELUDE

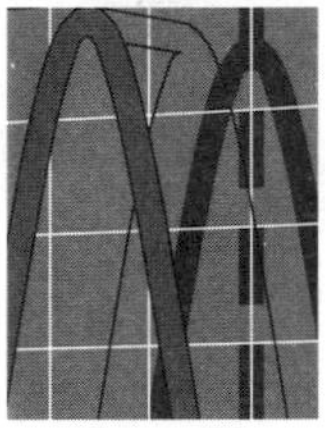

When we examine data, we often come to some conclusion about the individuals represented by the data. Can we draw conclusions not only about these individuals but about a wider population or process? Our ability to generalize beyond the individuals we actually measure depends on designs for data production and on the behavior of numerical summaries—statistics—from the data produced. In this chapter we enlarge our knowledge of probability and prepare for the study of statistical inference by looking at the probability distributions of some very common statistics: sample counts, sample proportions, and sample means. Section 5.3 presents a simple but important application of probability distributions to quality control. Here are some questions we will soon answer:

- A random drawing to assign the top line on a county ballot to either the Democratic or the Republican candidate gave the line to the Democrat 40 times in 41 tries. Is this just chance, or is the drawing rigged?
- A study of the effect of a cholesterol-reducing drug on heart attacks plans to give the drug to 2000 men and a placebo to another 2000. Do we expect to see enough heart attacks among these men to draw conclusions about the drug, or are larger samples needed?
- It is good practice to repeat laboratory measurements several times and report the mean result. What is the advantage of the mean over a single measurement?

• • •

Statistical inference draws conclusions about a population or process on the basis of data. The data are summarized by *statistics* such as means, proportions, and the slopes of least-squares regression lines. When the data are produced by random sampling or randomized experimentation, a statistic is a random variable that obeys the laws of probability theory. The link between probability and data is formed by the *sampling distributions* of statistics. A sampling distribution shows how a statistic would vary in repeated data production. That is, a sampling distribution is a probability distribution. In Section 3.4 we looked at several sampling distributions empirically by actually simulating 1000 simple random samples (SRSs). Considered as a probability distribution, a sampling distribution is an idealized mathematical description of the results of an indefinitely large number of samples rather than the results of a particular 1000 samples.

STATISTIC

A statistic from a probability sample or randomized experiment is a random variable. The probability distribution of the statistic is its sampling distribution.

Distributions also play a second role in statistical inference. Any quantity that can be measured for each member of a population is described by its distribution of values, which we take to be a probability distribution.

EXAMPLE 5.1

The distribution of heights of women between the ages of 18 and 24 is approximately normal with mean 65.5 inches and standard deviation 2.5 inches. Select women at random and measure their heights. The height X of the next woman to be chosen is a random variable. We don't know the height of the next woman, but we do know that in repeated sampling X will have the same $N(65.5, 2.5)$ distribution that describes the pattern of heights in the entire population. We call $N(65.5, 2.5)$ the population distribution. ●

The population of all women between the ages of 18 and 24 actually exists, so that we can in principle draw an SRS from it. In another situation, we might have an industrial process producing rods with lengths that vary according to a normal distribution. The population contains all rods that would be produced if the process continued forever in its present state. That population does not actually exist, and we cannot draw an SRS from it. Yet we are often justified in treating the rods as if they were independent observations randomly chosen from this hypothetical population.

Once again we describe the population (or the process, if you prefer) by the normal distribution that governs individual observations. In statistical inference, probability distributions therefore play two important roles:

population distribution

- The distribution of a variable in the population is described by a probability distribution. This is the context in which we first met distributions, as models for data. The language of probability is used for convenience and is introduced by thinking of selecting a unit at random from the population. We will simply speak of the *population distribution* without emphasizing that it is a probability distribution.
- The sampling distribution of a statistic is a probability distribution. This distribution describes the behavior of the statistic in repeated sampling or repeated experimentation.

To progress from discussing probability as a topic in itself to probability as a foundation for inference, we must study the sampling distributions of some common statistics. That is the topic of Sections 5.1 and 5.2. We will meet other sampling distributions in the study of inference. In each case, the nature of the sampling distribution depends on both the nature of the population distribution and the way in which the data are collected from the population.

Section 5.3 presents a simple and direct use of sampling distributions for inference that is also of considerable practical importance: *control charts* for assessing whether a process remains stable over time.

5.1 COUNTS AND PROPORTIONS

count

sample proportion

An opinion poll selects a sample of 1785 persons and asks each if they attended church or synagogue in the past week; the number who say "No" is a random variable X. A quality control engineer selects a sample of 10 switches from a supplier's shipment for detailed inspection; of these, a number X fail to meet the specifications. In both of these examples, the random variable X is a *count* of the occurrences of some outcome in a fixed number of observations. There are 1785 observations in the first example and 10 observations in the second. If the number of observations is n, then the *sample proportion* is $\hat{p} = X/n$. For example, if 1000 of the 1785 people interviewed by the opinion poll say "No," the sample proportion is

$$\hat{p} = \frac{1000}{1785} = .56$$

Sample counts and sample proportions are common statistics. Finding their sampling distributions is the goal of this section.

The binomial distributions

The distribution of a count X depends on how the data are produced. Here is a simple but common situation.

> **THE BINOMIAL SETTING**
>
> **1.** There are a fixed number n of observations.
>
> **2.** The n observations are all independent.
>
> **3.** Each observation falls into one of just two categories, which for convenience we call "success" and "failure."
>
> **4.** The probability of a success, call it p, is the same for each observation.

Think of tossing a coin n times as an example of the binomial setting. Each toss gives either heads or tails; the outcomes of successive tosses are independent; if we call heads a success, then p is the probability of a head and remains the same as long as we toss the same coin. The number of heads we count is a random variable X. The distribution of X, and more generally of the count of successes in any binomial setting, is completely determined by the number of observations n and the success probability p.

> **BINOMIAL DISTRIBUTION**
>
> The distribution of the count X of successes in the binomial setting is called the binomial distribution with parameters n and p. As an abbreviation, we say that X is $B(n, p)$.

The binomial distributions are an important class of discrete probability distributions. Later in this section we will learn how to assign probabilities to outcomes and how to find the mean and standard deviation of binomial distributions. But the first, and most important, aspect of using binomial distributions is the ability to recognize situations to which they apply.

EXAMPLE 5.2

(a) Toss a balanced coin 10 times and count the number X of heads. There are $n = 10$ observations. Successive tosses are independent. If the coin is truly balanced, the probability of a head is $p = 0.5$ on each toss. So the number of heads we observe has the binomial distribution $B(10, 0.5)$.

(b) Deal 10 cards from a shuffled deck and count the number X of red cards. There are 10 observations, and each gives either a red or a black card. But the observations are *not* independent. If the first card is black, the second is more likely to be red because there are now more red cards than black cards remaining in the deck. The count X does *not* have a binomial distribution.

(c) Genetics says that offspring independently receive genes from their parents. If both parents carry genes for the O and A blood types, each offspring has probability 0.25 of getting two O genes and so of having blood type O. The number of O blood types among 5 children of these parents is the count X of successes in 5 independent trials with probability 0.25 of a success on each trial. So X has the $B(5, 0.25)$ distribution.

(d) Engineers define reliability as the probability that an item will perform its function under specific conditions for a specific period of time. If an aircraft engine turbine has probability 0.999 of performing properly for an hour of flight, the number of turbines in a fleet of 350 engines that fly for an hour without failure has the $B(350, 0.999)$ distribution. This binomial distribution is obtained by assuming, as seems reasonable, that the turbines fail independently of each other. A common cause of failure, such as sabotage, would destroy the independence and make the binomial model inappropriate. ●

The binomial distributions are important in statistics when we wish to make inferences about the proportion p of "successes" in the population. Here is a typical example.

EXAMPLE 5.3

A quality engineer chooses an SRS of 10 switches from a large shipment. Unknown to the engineer, 10% of the switches in the shipment do not conform to the specifications. The engineer counts the number X of nonconforming switches in the sample.

Is this a binomial situation? Not quite. Removing one switch changes the proportion of bad switches in the remaining population, so the state of the second switch chosen is not independent of the first. If the shipment is large, however, removing a few items has a very small effect on the composition of the remaining population. Successive inspection results are very nearly independent. Suppose, for example, that the shipment contains 10,000 switches, of which 1000 are nonconforming (1000/10,000 is 0.1, or 10%). If the first switch chosen is bad, the proportion of bad switches remaining is $999/9999 = 0.0999$; if the first switch is good, the proportion of bad switches left in the shipment is $1000/9999 = 0.10001$. These proportions are so close to 0.1 that for practical purposes we act as if removing one switch has no effect on the proportion of bad switches remaining. We act as if the count X of switches in the sample that fail inspection has the binomial distribution $B(10, 0.1)$. ●

In Example 5.3 and similar settings, the population distribution is just the distribution of successes and failures in the population, described by the single parameter p. The sampling distribution of the count of successes in an SRS from the population is nearly binomial if the population is large enough to allow us to ignore the small dependence between observations.

> **SAMPLING DISTRIBUTION OF A COUNT**
>
> When the population is much larger than the sample, a count of successes in an SRS of size n has approximately the $B(n, p)$ distribution if the population proportion of successes is p.

The accuracy of this approximation improves as the size of the population increases relative to the size of the sample. As a rule of thumb, we will use the binomial sampling distribution for counts when the population is at least 10 times as large as the sample. This was the case in Example 5.3.

Finding binomial probabilities We will give a formula for the probability that a binomial random variable takes any of its values. In practice, you should avoid using this formula whenever possible. Some calculators and most statistical software packages calculate binomial probabilities. If you do not have suitable computing facilities, you can still shorten the work of calculating binomial probabilities for some values of n and p by looking up probabilities in Table C in the back of this book. The entries in the table are the probabilities $P(X = k)$ of individual outcomes for a binomial random variable X.

EXAMPLE 5.4

A quality engineer selects an SRS of 10 switches from a large shipment for detailed inspection. Unknown to the engineer, 10% of the switches in the shipment fail to meet the specifications. What is the probability that no more than 1 of the 10 switches in the sample fails inspection?

Example 5.3 reminds us that the count X of bad switches in the sample has (approximately) the $B(10, 0.1)$ distribution. Figure 5.1 is a probability histogram for this distribution. The distribution is strongly skewed. Although X can take any whole number value from 0 to 10, the probabilities of values larger than 5 are so small that they do not appear in the histogram.

We want to calculate

$$P(X \leq 1) = P(X = 1) + P(X = 0)$$

when X is $B(10, 0.1)$. The probability $P(X = 1)$ can be obtained in Minitab as follows:

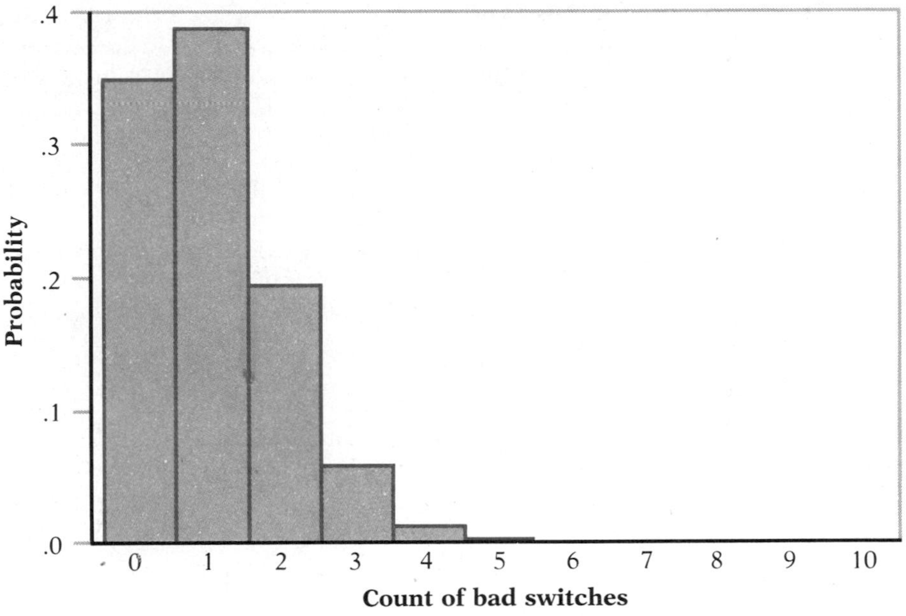

FIGURE 5.1 **Probability histogram for the binomial distribution with $n = 10$ and $p = 0.1$.**

```
MTB> PDF 1;
 SUBC> BINOMIAL n = 10, p = .1.
```

Other computing systems have similar commands.

To use Table C for this calculation, look opposite $n = 10$ and under $p = .10$. This part of the table appears at the left. The entry opposite each k is $P(X = k)$. We find

		p
n	k	.10
10	0	.3487
	1	.3874
	2	.1937
	3	.0574
	4	.0112
	5	.0015
	6	.0001
	7	.0000
	8	.0000
	9	.0000
	10	.0000

$$P(X \le 1) = P(X = 1) + P(X = 0)$$
$$= .3874 + .3487 = .7361$$

About 74% of all samples will contain no more than 1 bad switch. In fact, 35% of the samples will contain no bad switches. A sample of size 10 cannot be trusted to alert the engineer to the presence of unacceptable items in the shipment. ●

The excerpt from Table C contains the full $B(10,0.1)$ distribution. The probabilities are rounded to four decimal places; values above 6 do not have probability 0, but their probabilities are so small that the rounded values are .0000. Check that the sum of the probabilities given is 1, as it should be. The values of p that are present in Table C are all 0.5 or smaller. When the probability of a success is greater than 0.5, restate the problem in terms of the number of failures. The probability of a failure is less than 0.5 when the probability of a success exceeds 0.5. When using the table, always stop to ask which outcomes we must count.

EXAMPLE 5.5

Rick is a basketball player who makes 75% of his free throws over the course of a season. In a key game, Rick shoots 12 free throws and misses 5 of them. The fans think that he failed because he was nervous. Is it unusual for Rick to perform this poorly?

To answer this question, assume that free throws are independent with probability 0.75 of a success on each shot. (Studies of long sequences of free throws have found no evidence that they are dependent, so this is a reasonable assumption.) Because the probability of making a free throw is greater than 0.5, we count misses in order to use Table C. The probability of a miss is $1 - 0.75$, or 0.25. The number X of misses in 12 attempts has the $B(12, 0.25)$ distribution.

We want the probability of missing 5 or more. This is

$$\begin{aligned} P(X \geq 5) &= P(X = 5) + P(X = 6) + \cdots + P(X = 12) \\ &= .1032 + .0401 + \cdots + .0000 = .1576 \end{aligned}$$

Rick will miss 5 or more out of 12 free throws about 16% of the time, or roughly one of every six games. Although below his average level, this performance is well within the range of the usual chance variation in his shooting. ●

Binomial probabilities*

We will illustrate by an example how to obtain a general formula for binomial probabilities. The argument uses the multiplication rule for the probability that all of a number of independent events occur.

EXAMPLE 5.6

Nicholas Caputo is the clerk of Essex County, New Jersey. The top line on county ballots is supposed to be assigned by a random drawing to either the Republican or the Democratic candidate. Mr. Caputo, who is a Democrat, conducts the drawings. He has assigned the top line to the Democrats 40 out of 41 times. What is the probability that the Democrats would do so well in a truly random lottery?

In a properly conducted lottery, the successive drawings are independent and each has probability $1/2$ of giving the top line to the Democrats. The count X of Democrats on the top line in 41 drawings is therefore a binomial random variable with $n = 41$ and $p = 1/2$. The observed count of successes for Democrats was $X = 40$. We will calculate

$$P(X \geq 40) = P(X = 40) + P(X = 41)$$

which is the probability that the Democrats win at least 40 of the 41 drawings. ●

By the multiplication rule, the probability that the Democrats are awarded the top spot in all 41 drawings is

*The formula for binomial probabilities is useful in many settings, but we will not use it in our study of statistical inference. This section can therefore be omitted if desired.

$$P(X = 41) = \left(\frac{1}{2}\right)^{41}$$

Similarly, the probability that the Democrats win the first 40 drawings is $(1/2)^{40}$. So 40 straight Democratic wins followed by a Republican win has probability

$$\left(\frac{1}{2}\right)^{40}\left(\frac{1}{2}\right) = \left(\frac{1}{2}\right)^{41} \tag{5.1}$$

This is *not* the probability $P(X = 40)$, because the event $\{X = 40\}$ allows the single Republican win to fall anywhere in the sequence of 41 drawings. We argue as follows: Any specific sequence of 40 Democratic and 1 Republican wins has the same probability given by Equation 5.1, by a similar use of the multiplication rule. There are 41 such outcomes because the single Republican success can occur on any one of the 41 drawings. These 41 outcomes are disjoint. The event $\{X = 40\}$ therefore has probability given by the sum of 41 probabilities, each equal to the expression in Equation 5.1. We see at last that

$$P(X = 40) = (41)\left(\frac{1}{2}\right)^{40}\left(\frac{1}{2}\right)$$

These are very small probabilities. Their sum $P(X \geq 40)$ is about 0.00000000002, or 1 in 50 billion. There is almost no chance that the Democrats would have won a truly random drawing so consistently. The New Jersey Supreme Court made calculations like this one and told Mr. Caputo to change his ways.[1]

The idea of Example 5.6 extends to binomial situations in general. Suppose that we have n independent observations, each a "success" or a "failure." Suppose also that a success has the same probability p on each observation. Then any arrangement of k successes and $n - k$ failures in a specific order has probability $p^k(1-p)^{n-k}$ by the multiplication rule. The probability $P(X = k)$ of exactly k successes is therefore $p^k(1-p)^{n-k}$ times the number of different ways we can distribute k successes among n observations. We use the following fact to find this number.

BINOMIAL COEFFICIENT

The number of ways in which k successes can be chosen from among n observations is given by the binomial coefficient

$$\binom{n}{k} = \frac{n!}{k!\,(n-k)!} \tag{5.2}$$

for $k = 0, 1, 2, \ldots, n$.

factorial The binomial coefficients are expressed using the *factorial* notation: For any positive whole number n,

$$n! = n \times (n-1) \times (n-2) \times \cdots \times 3 \times 2 \times 1$$

and $0! = 1$. Notice that the larger of the two factorials in the denominator of a binomial coefficient will cancel much of the $n!$ in the numerator. For example,

$$\begin{aligned}\binom{8}{3} &= \frac{8!}{3!\,5!} \\ &= \frac{(8)(7)(6)}{(3)(2)(1)} = 56\end{aligned}$$

The notation $\binom{n}{k}$ is *not* related to the fraction $\frac{n}{k}$. A helpful way to remember its meaning is to read it as "binomial coefficient n choose k." Binomial coefficients have many uses in mathematics, but we are interested in them only as an aid to finding binomial probabilities. The binomial coefficient $\binom{n}{k}$ counts the number of ways in which k successes can be distributed among n observations. The binomial probability $P(X = k)$ is this count multiplied by the probability of any specific arrangement of the k successes. Here is the formula we seek.

BINOMIAL PROBABILITY

If X is $B(n, p)$, then

$$P(X = k) = \binom{n}{k} p^k (1-p)^{n-k} \qquad (5.3)$$

for $k = 0, 1, \ldots, n$.

In Example 5.6, X is $B(41, 1/2)$ and $k = 40$. The probability

$$P(X = 40) = \binom{41}{40} \left(\frac{1}{2}\right)^{40} \left(\frac{1}{2}\right)^{1}$$

obtained from the binomial formula 5.3 agrees with the result obtained in Example 5.6 because

$$\binom{41}{40} = \frac{41!}{40!\,1!} = 41$$

Here is another example of the use of the binomial probability formula.

EXAMPLE 5.7

The number X of switches that fail inspection in Example 5.4 has the $B(10, 0.1)$ distribution. The probability that no more than 1 switch fails is

$$\begin{aligned} P(X \le 1) &= P(X = 1) + P(X = 0) \\ &= \binom{10}{1}(.1)^1(.9)^9 + \binom{10}{0}(.1)^0(.9)^{10} \\ &= \frac{10!}{1!\,9!}(.1)(.3874) + \frac{10!}{0!\,10!}(1)(.3487) \\ &= (10)(.1)(.3874) + (1)(1)(.3487) \\ &= .3874 + .3487 = .7361 \end{aligned}$$

The calculation used the facts that $0! = 1$ and that $a^0 = 1$ for any number $a \ne 0$. The result agrees with that obtained from Table C in Example 5.4. ●

Binomial mean and variance

If a count X is $B(n, p)$, what are the mean μ_X and the standard deviation σ_X? We can guess the mean. If a basketball player makes 80% of her free throws, the mean number made in 10 tries should be 80% of 10, or 8. That's μ_X when X is $B(10, 0.8)$. Intuition suggests more generally that the mean of the $B(n, p)$ distribution should be np. Can we show that this is correct and also obtain a short formula for the standard deviation? Because binomial distributions are discrete probability distributions, we could find the mean and variance by using definitions 4.3 and 4.5 in Section 4.3. For example, we could obtain the mean by multiplying each outcome by its probability and adding over all outcomes. Here is an easier way.

A binomial random variable X is the count of successes in n independent observations that each have the same probability p of success. Let the random variable Z_i indicate whether the ith observation is a success or failure by taking the values $Z_i = 1$ if a success occurs and $Z_i = 0$ if the outcome is a failure. The Z_i are independent because the observations are, and each Z has the same simple distribution:

Outcome	1	0
Probability	p	$1 - p$

Definition 4.3 (page 322) of the mean of a discrete random variable says that

$$\mu_Z = (1)(p) + (0)(1 - p) = p$$

Similarly, definition 4.5 (page 330) for the variance shows that $\sigma_Z^2 = p(1 - p)$. The count X is just the number of 1s among the Z_i. That is,

$$X = Z_1 + Z_2 + \cdots + Z_n$$

Now apply the addition rules for means and variances to this sum. The mean of X is the sum of the means of the variables Z_i:

$$\begin{aligned}\mu_X &= \mu_{Z_1} + \mu_{Z_2} + \cdots + \mu_{Z_n} \\ &= n\mu_Z = np\end{aligned}$$

Similarly, the variance is n times the variance of a single Z, so that $\sigma_X^2 = np(1-p)$. The standard deviation σ_X is the square root of the variance. Here is the result.

BINOMIAL MEAN AND STANDARD DEVIATION

If X has the $B(n, p)$ distribution, then

$$\begin{aligned}\mu_X &= np \\ \sigma_X &= \sqrt{np(1-p)}\end{aligned} \tag{5.4}$$

EXAMPLE 5.8

The Helsinki Heart Study asks whether the anticholesterol drug gemfibrozil will reduce heart attacks. In planning such an experiment, the researchers must be confident that the sample sizes are large enough to enable them to observe enough heart attacks. The Helsinki study plans to give gemfibrozil to 2000 men and a placebo to another 2000. The probability of a heart attack during the 5-year period of the study for men this age is about 0.04. What are the mean and standard deviation of the number of heart attacks that will be observed in one group if the treatment does not change this probability?

There are 2000 independent observations, each having probability $p = 0.04$ of a heart attack. The count X of heart attacks is $B(2000, 0.04)$ and

$$\begin{aligned}\mu_X &= np = (2000)(.04) = 80 \\ \sigma_X &= \sqrt{np(1-p)} = \sqrt{(2000)(.04)(.96)} = 8.76\end{aligned}$$

The expected number of heart attacks is large enough to permit conclusions about the effectiveness of the drug. ●

Sample proportions

In the binomial setting we often wish to estimate the proportion p of successes in the population. Our estimator is the sample proportion of successes,

$$\begin{aligned}\hat{p} &= \frac{\text{count of successes in sample}}{\text{size of sample}} \\ &= \frac{X}{n}\end{aligned}$$

The count X has a binomial distribution. We do probability calculations about the sample proportion $\hat{p}$ by restating them in terms of the count X and using binomial methods. As an example, we return to the opinion poll that we studied by simulation in Example 3.16 (page 264). We must develop probability methods to support statistical inference in settings like this.

EXAMPLE 5.9

An opinion poll asks an SRS of 1785 adults whether they attended church or synagogue during the past week. Suppose that 60% of the adult population did not attend. What is the probability that the sample proportion who did not attend is at least 58%?

The sample proportion does *not* have a binomial distribution because it is not a count. We therefore translate any question about the sample proportion $\hat{p}$ into a question about the sample count X, which *does* have the binomial distribution $B(1785, 0.6)$. Because 58% of 1785 is 1035.3,

$$\begin{aligned} P(\hat{p} \geq .58) &= P(X \geq 1035.3) \\ &= P(X = 1036) + P(X = 1037) + \cdots + P(X = 1785) \end{aligned}$$

●

To find the probability in Example 5.9 we must add more than 700 binomial probabilities, each with $n = 1785$. Tables are not available for such large values of n, and most computer packages are unable to calculate binomial probabilities when n is this large. Even the optional formula 5.3 for binomial probabilities is not practical in this case. We can state a probability question about the opinion poll in terms of the binomial distribution, but we cannot easily find the numerical value of the probability using this distribution. We shall soon see that a normal distribution can replace the binomial distribution in this example.

The mean and standard deviation of the sample proportion can be obtained from the mean and standard deviation of a sample count using the rules from Section 4.3 for the mean and variance of a constant times a random variable. Here is the result.

MEAN AND STANDARD DEVIATION OF A SAMPLE PROPORTION

Let $\hat{p}$ be the sample proportion of successes in an SRS of size n drawn from a large population having population proportion p of successes. The mean and standard deviation of $\hat{p}$ are

$$\mu_{\hat{p}} = p \tag{5.5}$$

$$\sigma_{\hat{p}} = \sqrt{\frac{p(1-p)}{n}}$$

EXAMPLE 5.10 The mean and standard deviation of the proportion of the opinion poll sample in Example 5.9 who did not attend church the week of the poll are

$$\mu_{\hat{p}} = p = .6$$

$$\sigma_{\hat{p}} = \sqrt{\frac{p(1-p)}{n}} = \sqrt{\frac{(.6)(.4)}{1785}} = .0116$$

●

unbiased estimator

The fact that the mean of $\hat{p}$ is p states in statistical language that the sample proportion $\hat{p}$ in an SRS is an *unbiased estimator* of the population proportion p. When a sample is drawn from a new population having a different value of the population proportion p, the sampling distribution of the unbiased estimator $\hat{p}$ changes so that its mean moves to the new value of p. We observed this fact empirically in Section 3.4 and have now verified it from the laws of probability.

The variability of $\hat{p}$ about its mean, as described by the variance or standard deviation, decreases as the sample size increases. So a sample proportion from a large sample will usually lie quite close to the population proportion p. We observed this in the simulation study of Example 3.16. Now we have discovered exactly how the variability decreases: The standard deviation is $\sqrt{p(1-p)/n}$, or $\sqrt{p(1-p)}/\sqrt{n}$. The $\sqrt{n}$ in the denominator means that the sample size must be multiplied by 4 if we wish to divide the standard deviation in half.

Normal approximation for proportions and counts

We discovered empirically in Section 3.4 that the sampling distribution of a sample proportion $\hat{p}$ is close to normal. Now we have deduced from basic probability that when the population is much larger than the sample, the distribution of $\hat{p}$ is that of a binomial count divided by the sample size n. This seems to be a contradiction but it is not. The exact distribution is indeed based on the binomial, but it is approximately normal when n is large. Here are the facts.

> **NORMAL APPROXIMATION FOR SAMPLE PROPORTIONS**
>
> Let $\hat{p}$ be the sample proportion of successes in an SRS of size n drawn from a large population having population proportion p of successes. When n is large, the sampling distribution of $\hat{p}$ is approximately normal with mean p and standard deviation $\sqrt{p(1-p)/n}$.

Because the binomial count X is just a fixed number n times the proportion, X is also approximately normal, with mean np and standard deviation $\sqrt{np(1-p)}$.

Both of these normal approximations are easy to remember because they simply say that $\hat{p}$ and X are normal with their usual means and standard deviations given by Equations 5.4 and 5.5. The accuracy of the normal approximations improves as the sample size n increases. They are most accurate for any fixed n when p is close to $1/2$, and least accurate when p is near 0 or 1. Whether or not you use the normal approximations should depend on how accurate your calculations need to be. For most statistical purposes great accuracy is not required. We will therefore use the approximations for values of n and p that satisfy $np \geq 10$ and $n(1-p) \geq 10$.

EXAMPLE 5.11

The normal approximation makes the difficult calculation of Example 5.9 easy. We wished to calculate $P(\hat{p} \geq 0.58)$ when the sample size was $n = 1785$ and the population proportion was $p = 0.6$. Example 5.10 shows that

$$\mu_{\hat{p}} = .6$$
$$\sigma_{\hat{p}} = .0116$$

Act as if $\hat{p}$ were normal with mean 0.6 and standard deviation 0.0116. The approximate probability, as illustrated in Figure 5.2, is

$$\begin{aligned} P(\hat{p} \geq .58) &= P\left(\frac{\hat{p} - .6}{.0116} \geq \frac{.58 - .6}{.0116}\right) \\ &\doteq P(Z \geq -1.72) = .9573 \end{aligned}$$

That is, 96% of all samples have a sample proportion that is at least 0.58. Because the sample was large, this normal approximation is quite accurate. ●

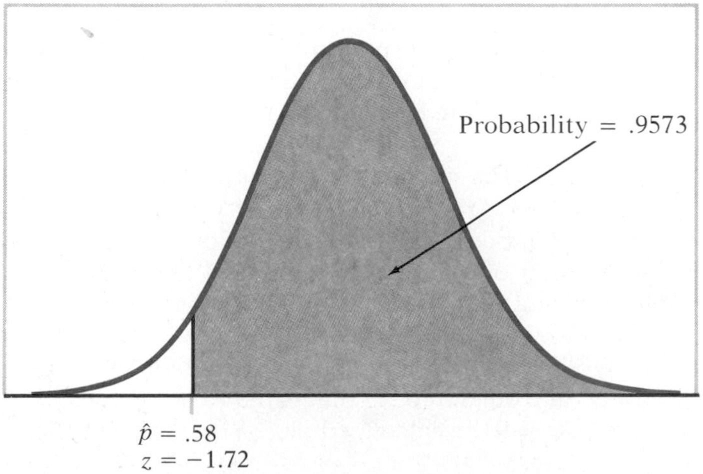

FIGURE 5.2 The normal probability calculation for Example 5.11.

EXAMPLE 5.12

The quality engineer of Example 5.4 realizes that a sample of 10 switches is inadequate and decides to inspect 100 switches from a large shipment. If in fact 10% of the switches in the shipment fail to meet the specifications, then the count X of bad switches in the sample has approximately the $B(100, 0.1)$ distribution. The distribution of the count of bad switches in a sample of 10 is distinctly nonnormal, as Figure 5.1 showed. When the sample size is increased to 100, the shape of the binomial distribution becomes approximately normal.

The probability $P(X \leq 9)$ that no more than 9 of the switches in the sample fail inspection can be found from the binomial probability formula (Equation 5.3) or from a computer program. The exact probability is $P(X \leq 9) = 0.4513$.

According to the normal approximation to the binomial distributions, the count X is approximately normal with mean and standard deviation

$$\mu = np = (100)(.1) = 10$$
$$\sigma = \sqrt{np(1-p)} = \sqrt{(100)(.1)(.9)} = 3$$

Figure 5.3 displays the probability histogram of the binomial distribution with the density curve of the approximating normal distribution superimposed. Both distributions have the same mean and standard deviation, and both the area under the histogram and the area under the curve are 1. (Although the binomial random variable X can take values as large as 100, the probabilities of outcomes larger than 20 are so small that the bars have no visible height in the figure.)

The normal approximation to the probability of no more than 9 bad switches is found from Table A, as illustrated by Figure 5.4:

$$P(X \leq 9) = P\left(Z \leq \frac{9-10}{3}\right)$$
$$= P(Z \leq -.33) = .3707$$

The approximation 0.37 to the binomial probability 0.45 is not very accurate. Notice that $np = (100)(0.1) = 10$, so that this combination of n and p is on the border of the values for which we are willing to use the approximation. ●

The continuity correction* Figure 5.3 suggests an idea that greatly improves the accuracy of the normal approximation to binomial probabilities. The discrete binomial distribution in Example 5.12 puts probability exactly on $X = 9$ and $X = 10$ and no probability between these whole numbers. The normal distribution spreads its probability continuously. The bar for $X = 9$ in the probability histogram of Figure 5.3 extends from 8.5 to 9.5, but the normal calculation of $P(X \leq 9)$ includes only the area to the left of the center of this bar. The normal approximation is more accurate if we consider $X = 9$ to extend from 8.5 to 9.5, $X = 10$ to extend from 9.5 to 10.5, and so on.

When we want to include the outcome $X = 9$, we include the entire interval from 8.5 to 9.5 that is the base of the $X = 9$ bar in the histogram. So $P(X \leq 9)$ is calculated as $P(X \leq 9.5)$. On the other hand, $P(X < 9)$

*This material can be omitted if desired.

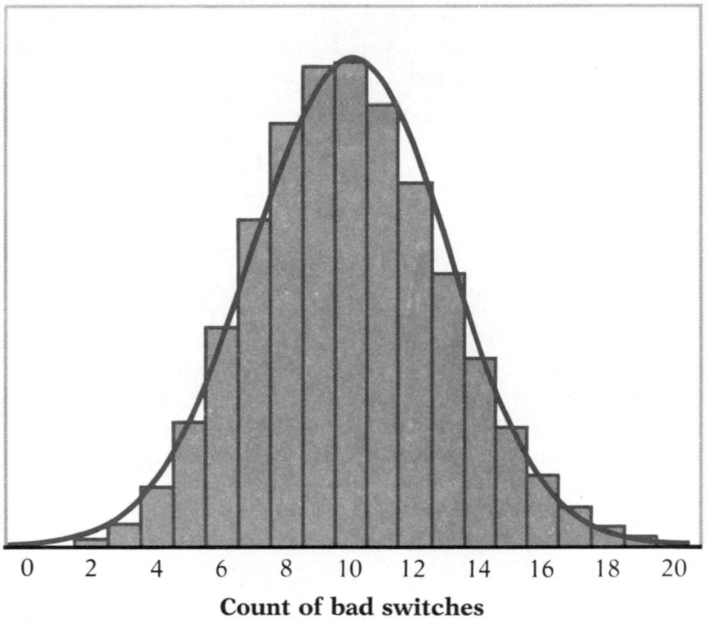

FIGURE 5.3 Probability histogram and normal approximation for the binomial distribution with $n = 100$ and $p = 0.1$.

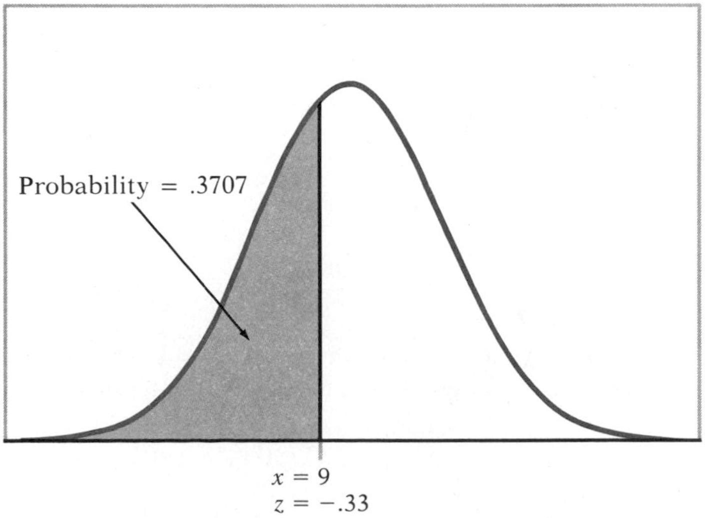

FIGURE 5.4 Area under the normal approximation curve for the probability in Example 5.12.

excludes the outcome $X = 9$, so we exclude the entire interval from 8.5 to 9.5 and calculate $P(X \leq 8.5)$ from the normal table. Here is the result of the normal calculation in Example 5.12 improved in this way:

$$\begin{aligned} P(X \leq 9) &= P(X \leq 9.5) \\ &= P\left(Z \leq \frac{9.5 - 10}{3}\right) \\ &= P(Z \leq -.17) = .4325 \end{aligned}$$

The improved approximation 0.43 is much closer to the binomial probability 0.45. Acting as though a whole number occupies the interval from 0.5 below to 0.5 above the number is called the *continuity correction* to the normal approximation. If you need accurate values for binomial probabilities, try to use computer software to do exact calculations. If no software is available, use the continuity correction unless n is very large. Because most statistical purposes do not require extremely accurate probability calculations, we do not emphasize use of the continuity correction.

continuity correction

SUMMARY

A count X of successes has the **binomial distribution** $B(n, p)$ in the **binomial setting:** There are n trials, all independent, each resulting in a success or a failure, and each having the same probability p of a success.

Binomial probabilities can be found in Table C or calculated from the formula

$$P(X = k) = \binom{n}{k} p^k (1-p)^{n-k}$$

where the possible values of X are $k = 0, 1, \ldots, n$.

The binomial probability formula uses the **binomial coefficient**

$$\binom{n}{k} = \frac{n!}{k!\,(n-k)!}$$

Here the **factorial** $n!$ is

$$n! = n \times (n-1) \times (n-2) \times \cdots \times 3 \times 2 \times 1$$

for positive whole numbers n and $0! = 1$. The binomial coefficient counts the number of ways of distributing k successes among n trials.

The binomial distribution gives a good approximation to the sampling distribution of a count from an SRS of size n when the population is at least 10 times larger than the sample.

The means and standard deviations of a binomial count X and a sample proportion of successes $\hat{p} = X/n$ are

$$\mu_X = np \qquad \mu_{\hat{p}} = p$$

$$\sigma_X = \sqrt{np(1-p)} \qquad \sigma_{\hat{p}} = \sqrt{\frac{p(1-p)}{n}}$$

The sample proportion $\hat{p}$ is therefore an unbiased estimator of the population proportion p.

If X has the binomial distribution $B(n, p)$ and n is large, then X has approximately the normal distribution $N(np, \sqrt{np(1-p)})$ and the sample proportion $\hat{p} = X/n$ has approximately the normal distribution $N(p, \sqrt{p(1-p)/n})$. We will use these approximations when $np \geq 10$ and $n(1-p) \geq 10$. If the **continuity correction** is employed to improve the accuracy, the normal approximations can be used more freely.

SECTION 5.1 EXERCISES

All of the binomial probability calculations required in these exercises can be done by using Table C. Your instructor may request that you use the binomial probability formula Equation 5.3 or computer software. In exercises requiring the normal approximation, you should use the continuity correction only if you studied that topic.

5.1 For each of the following situations, indicate whether a binomial distribution is a reasonable probability model for the random variable X. Give your reasons in each case.

(a) You observe the sex of the next 50 children born at a local hospital; X is the number of girls among them.

(b) A couple decides to continue to have children until their first girl is born; X is the total number of children the couple has.

(c) You want to know what percent of married people believe that mothers of young children should not be employed outside the home. You plan to interview 50 people, and for the sake of convenience you decide to interview both the husband and the wife in 25 married couples. The random variable X is the number among the 50 persons interviewed who think mothers should not be employed.

5.2 In each situation below, is it reasonable to use a binomial distribution for the random variable X? Give reasons for your answer in each case.

(a) An auto manufacturer chooses one car from each hour's production for a detailed quality inspection. One variable recorded is the count X of finish defects (dimples, ripples, etc.) in the car's paint.

(b) The pool of potential jurors for a murder case contains 100 persons chosen at random from the adult residents of a large city. Each person in the pool is asked whether he or she opposes the death penalty; X is the number who say "Yes."

(c) Joe buys a state lottery ticket every week; X is the number of times in a year that he wins a prize.

5.3 In each of the following cases, decide whether or not a binomial distribution is an appropriate model, and give your reasons.
(a) Fifty students are taught about binomial distributions by a television program. After completing their study, all students take the same examination. The number of students who pass is counted.
(b) A student studies binomial distributions using computer-assisted instruction. After the initial instruction is completed, the computer presents 10 problems. The student solves each problem and enters the answer; the computer gives additional instruction between problems if the student's answer is wrong. The number of problems that the student solves correctly is counted.
(c) A chemist repeats a solubility test 10 times on the same substance. Each test is conducted at a temperature 10° higher than the previous test.

5.4 A factory employs over 3000 workers, of whom 30% are black. If the 15 members of the union executive committee were chosen from the workers without regard to race, the number of blacks on the committee would have the $B(15, 0.3)$ distribution.
(a) What is the probability that 3 or fewer members of the committee are black?
(b) What is the probability that 10 or more members of the committee are white?

5.5 A university that is better known for its basketball program than for its academic strength claims that 80% of its basketball players get degrees. An investigation examines the fate of all 20 players who entered the program over a period of several years that ended 5 years ago. Of these players, 10 graduated and the remaining 10 are no longer in school. If the university's claim is true, the number of players who graduate among the 20 studied should have the $B(20, 0.8)$ distribution.
(a) Find the probability that exactly 10 players graduate under these assumptions.
(b) Find the probability that 10 or fewer players graduate. This probability is so small that it casts doubt on the university's claim.

5.6 Suppose that both parents in a family carry genes for blood types A and B. The blood types of their children are independent and each child has probability 1/4 of having blood type A. There are 4 children in the family. Let X be the number of children who have blood type A. What are n and p in the binomial distribution of X? Find the probability of each possible value of X and draw a probability histogram for this distribution.

5.7 A believer in the "random walk" theory of the behavior of stock prices thinks that an index of stock prices has probability 0.65 of increasing in any year. Moreover, the change in the index in any given year is not influenced by whether it rose or fell in earlier years. Let X be the number of years among the next 6 years in which the index rises. What are n and p in the binomial distribution of X? Give the possible values that X can take and the probability of each value. Draw a probability histogram for the distribution of X.

5.8 According to government data, 25% of employed women have never been married.

(a) If 10 employed women are selected at random, what is the probability that exactly 2 have never been married?
(b) What is the probability that 2 or fewer have never been married?
(c) What is the probability that at least 8 have been married?

5.9 Use the definition of binomial coefficients to show that each of the following facts is true. Then restate each fact in words in terms of the number of ways that k successes can be distributed among n observations.

(a) $\binom{n}{n} = 1$ for any whole number $n \geq 1$.
(b) $\binom{n}{n-1} = n$ for any whole number $n \geq 1$.
(c) $\binom{n}{k} = \binom{n}{n-k}$ for any n and k with $k \leq n$.

5.10 Find the mean number of children with type A blood in the family of Exercise 5.6, and mark the location of the mean on your probability histogram for that problem. Then find the standard deviation of X.

5.11 Find the mean of the number X of years in which the stock price index rises according to the random walk stock price model of Exercise 5.7, and mark the mean on your probability histogram for this distribution. Then compute the standard deviation of X. What is the probability that X takes a value within one standard deviation of its mean?

5.12 In a test for ESP (extrasensory perception), a subject is told that cards the experimenter can see but he cannot contain either a star, a circle, a wave, or a square. As the experimenter looks at each of 20 cards in turn, the subject names the shape on the card.

(a) If a subject simply guesses the shape on each card, what is the probability of a successful guess on a single card? Because the cards are independent, the count of successes in 20 cards has a binomial distribution.
(b) What is the probability that a subject correctly guesses at least 10 of the 20 shapes?
(c) In many repetitions of this experiment with a subject who is guessing, how many cards will the subject guess correctly on the average? What is the standard deviation of the number of correct guesses?

(d) A standard ESP deck actually contains 25 cards, with five different shapes each appearing on five cards. The subject knows that the deck has this makeup. Is a binomial model still appropriate for the count of correct guesses in one pass through this deck? If so, what are n and p? If not, why not?

5.13 A college is considering a new core curriculum that would require all undergraduates to study a foreign language. The student newspaper plans to interview several faculty members and report their opinions. Suppose that in fact 60% of the faculty support the language requirement.

(a) If the newspaper interviews 5 faculty members, what is the probability that a majority (that is, 3 or more) oppose the language requirement? (Assume that the faculty members are chosen at random.)

(b) If 15 faculty are interviewed instead, what is the probability that a majority (8 or more) oppose the requirement? Larger samples make it less likely that majority opinion in the population is incorrectly reported.

5.14 A study by a federal agency concludes that polygraph (lie detector) tests given to people telling the truth have probability about 0.2 of suggesting that the person is deceptive. (Office of Technology Assessment, *Scientific Validity of Polygraph Testing: A Research Review and Evaluation*, Government Printing Office, Washington, D.C., 1983.)

(a) A firm asks 12 job applicants about thefts from previous employers, using a polygraph to assess their truthfulness. Suppose that all 12 answer truthfully. What is the probability that the polygraph says at least 1 is deceptive?

(b) Among 12 truthful persons, what is the mean number who will be classified as deceptive? What is the standard deviation of this number?

(c) What is the probability that the number classified as deceptive is less than the mean?

5.15 According to government data, 22% of American children under the age of 6 live in households with incomes less than the official poverty level. A random sample of 300 children is selected for a study of learning in early childhood.

(a) What is the mean number of children in the sample who come from poverty-level households? What is the standard deviation of this number?

(b) Use the normal approximation to calculate the probability that at least 80 of the children in the sample live in poverty.

5.16 One way of checking the effect of undercoverage, nonresponse, and other sources of error in a sample survey is to compare the sample with known demographic facts about the population. About 11% of American adults are black. The number X of blacks in a random sample of 1500 adults should therefore vary with the $B(1500, 0.11)$ distribution.

(a) What are the mean and standard deviation of X?
(b) Use the normal approximation to find the probability that the sample will contain 100 or fewer blacks.

5.17 A selective college would like to have an entering class of 1200 students. Because not all students who are offered admission accept, the college admits more than 1200 students. Past experience shows that about 70% of the students admitted will accept. The college decides to admit 1500 students. Assuming that students make their decisions independently, the number who accept has the $B(1500, 0.7)$ distribution. If this number is less than 1200, the college will admit students from its waiting list.
(a) What are the mean and the standard deviation of the number X of students who accept?
(b) Use the normal approximation to find the probability that at least 1000 students accept.
(c) The college does not want more than 1200 students. What is the probability that more than 1200 will accept?
(d) If the college decides to increase the number of admission offers to 1700, what is the probability that more than 1200 will accept?

5.18 The Gallup Poll once asked a random sample of 1540 adults, "Do you happen to jog?" Suppose that in fact 15% of all American adults jog.
(a) Find the mean and standard deviation of the proportion of the sample who jog. (Assume that an SRS was used.)
(b) Use the normal approximation to find the probability that between 13% and 17% of the sample jog.
(c) What sample size would be required to reduce the standard deviation of the sample proportion to one-half the value you found in (a)?

5.19 Here is a simple probability model for multiple-choice tests. Suppose that each student has probability p of correctly answering a question chosen at random from a universe of possible questions. (A strong student has a higher p than a weak student.) The correctness of answers to different questions are independent. Julie is a good student for whom $p = 0.75$.
(a) Use the normal approximation to find the probability that Julie scores 70% or lower on a 100-question test.
(b) If the test contains 250 questions, what is the probability that Julie will score 70% or lower?
(c) How many questions must the test contain in order to reduce the standard deviation of Julie's proportion of correct answers on a 100-item test to half its value?
(d) Laura is a weaker student for whom $p = 0.6$. Does the answer you gave in (c) for the standard deviation of Julie's score apply to Laura's standard deviation also?

5.20 When the ESP study of Exercise 5.12 discovers a subject whose performance appears to be better than guessing, the study continues at greater length. The experimenter looks at many cards bearing one of 5 shapes (star, square, circle, wave, and cross) in an order determined by random numbers. The subject cannot see the experimenter as he looks at each card in turn, in order to avoid any possible nonverbal clues. The answers of a subject who does not have ESP should be independent observations, each with probability 1/5 of success. We record 1000 attempts.

(a) What are the mean and the standard deviation of the count of successes?
(b) What are the mean and standard deviation of the proportion of successes among the 1000 attempts?
(c) What is the probability that a subject without ESP will be successful in at least 24% of 1000 attempts?
(d) The researcher considers evidence of ESP to be a proportion of successes so large that there is only probability 0.01 that a subject could do this well or better by guessing. What proportion of successes must a subject have to meet this standard? (Hint: Example 1.24 on page 73 shows how to do a normal calculation of the type required here.)

5.21 ELISA tests are used to screen donated blood for the presence of antibodies to the AIDS virus. When presented with AIDS-contaminated blood, ELISA gives a positive result (that is, signals that an antibody is present) in about 99% of all cases. Suppose that among the many units of blood that pass through a blood bank in a year there are 20 units containing AIDS antibodies.

(a) What is the probability that ELISA will detect all of these cases?
(b) What is the probability that more than 1 of the 20 contaminated units will escape detection?
(c) What is the mean number of units among the 20 that will be detected by ELISA? What is the standard deviation of the number detected?

5.22 Because ELISA (see the previous exercise) is a screening test designed to keep the AIDS virus out of blood supplies, it is conservative in the sense that it signals the presence of AIDS antibodies too often. In fact, when presented with uncontaminated blood, ELISA is positive (claims that antibodies are present) about 2% of the time. Positive test results for uncontaminated blood are called *false positives*.

(a) A blood bank tests 12,000 units of uncontaminated blood. What is the mean number of false positives among these units? What is the standard deviation of the number of false positives?
(b) Suppose that the blood bank screens 12,000 units of uncontaminated blood and 20 units of blood that carry AIDS antibodies. What is the mean number of positive ELISA results among the 12,020 units? What is the standard deviation of the number of positive results? (Use the information in both this and the previous exercise along with the rules for the mean and variance of a sum of random variables.)

(c) What is the mean number of false positives as a percent of the overall mean number of positive results? This gives a rough answer to the following question: When ELISA is applied to 12,020 units of blood of which 20 carry AIDS antibodies, about what percent of positive test results are false positives? (A more exact way of answering such a question appears in Exercise 4.101 on page 364. The conclusion is important: If a condition is uncommon, most positive test results will be false positives even if the test has a high probability of giving the correct result on any one trial. A detailed statistical analysis appears in J. L. Gastwirth, "The statistical precision of medical screening procedures: Application to polygraph and AIDS antibodies test data," *Statistical Science* 2 (1987), pp. 213–222.)

5.2 SAMPLE MEANS

Counts and proportions are discrete random variables that describe categorical data. The statistics most often used to describe measured data, on the other hand, are continuous random variables. The sample mean, percentiles, and standard deviation are examples of statistics based on measured data. Statistical theory describes the sampling distributions of these statistics. In this section we will concentrate on the sample mean. Because sample means are just averages of observations, they are among the most common statistics.

EXAMPLE 5.13

A basic principle of investment is that diversification reduces risk. That is, buying several securities rather than just one reduces the variability of the return on an investment. Figure 5.5 illustrates this principle in the case of common stocks listed on the New York Stock Exchange. Figure 5.5(a) shows the distribution of returns for all 1815 stocks listed on the Exchange for the entire year 1987. This was a year of extreme swings in stock prices, including a record loss of over 20% in a single day. The mean return for all 1815 stocks was −3.5%, and the distribution shows a very wide spread.

Figure 5.5(b) shows the distribution of returns for all possible portfolios that invested equal amounts in each of 5 stocks. A portfolio is just a sample of 5 stocks, and its return is the average return for the 5 stocks chosen. The mean return for all portfolios is still −3.5%, but the variation among portfolios is much less than the variation among individual stocks. For example, 11% of all individual stocks had a loss of more than 40%, but only 1% of the portfolios had a loss that large. ●

The histograms in Figure 5.5 illustrate a fact that we will make precise in this section: Averages are less variable than individual observations. More detailed examination of the distributions, by normal quantile plots, would point to a second fact: Averages are more normal than individual observations. These two facts contribute to the popularity of sample means in statistical inference.

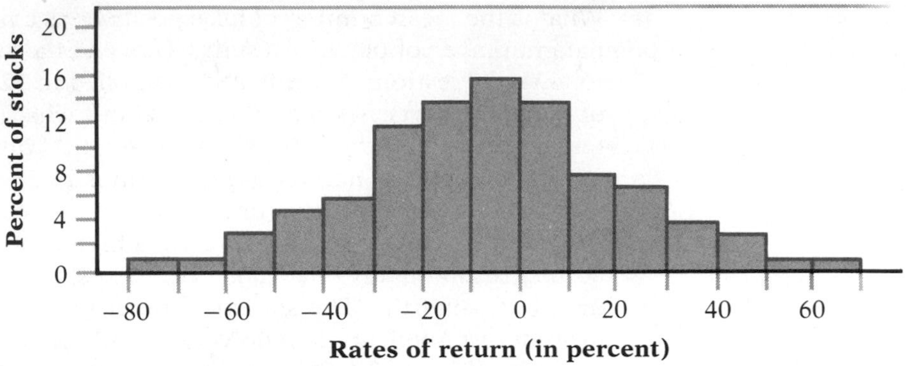

FIGURE 5.5(a) The distribution of returns for New York Stock Exchange common stocks in 1987.

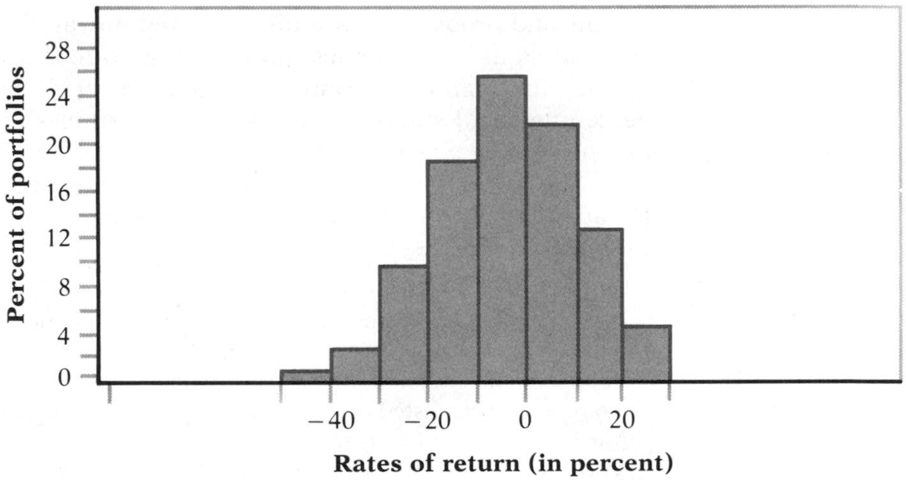

FIGURE 5.5(b) The distribution of returns for all possible portfolios of five stocks in 1987. Figure 5.5 is reprinted with permission from John K. Ford, "A method for grading 1987 stock recommendations," *American Association of Individual Investors Journal,* March 1988, pp. 16–17.

The distribution of a sample mean

The sample mean $\bar{x}$ from a sample or an experiment is an estimate of the mean μ of the underlying population, just as a sample proportion $\hat{p}$ is an estimate of a population proportion p. The sampling distribution of $\bar{x}$ is determined by the design used to produce the data, the sample size n, and the population distribution.

Select an SRS of size n from a population, and measure a variable X on each unit in the sample. The data consist of observations on n random variables $X_1, X_2, \ldots, X_n$. A single X_i is a measurement on one unit selected at random from the population, and therefore has the distribution of the population. If the population is large relative to the sample, we can consider $X_1, X_2, \ldots, X_n$ to be independent random variables each having the same distribution. This is our probability model for measurements on each unit in an SRS.

The sample mean of an SRS of size n is

$$\bar{x} = \frac{1}{n}(X_1 + X_2 + \cdots + X_n)$$

If the population has mean μ, then μ is the mean of each observation X_i. Therefore, by the addition rule for means of random variables,

$$\begin{aligned}\mu_{\bar{x}} &= \frac{1}{n}\left(\mu_{X_1} + \mu_{X_2} + \cdots + \mu_{X_n}\right)\\ &= \frac{1}{n} n\mu = \mu\end{aligned}$$

That is, *the mean of $\bar{x}$ is the same as the mean of the population*. The sample mean $\bar{x}$ is therefore an unbiased estimator of the unknown population mean μ.

The observations are independent, so the addition rule for variances also applies:

$$\begin{aligned}\sigma^2_{\bar{x}} &= \left(\frac{1}{n}\right)^2 \left(\sigma^2_{X_1} + \sigma^2_{X_2} + \cdots + \sigma^2_{X_n}\right)\\ &= \left(\frac{1}{n}\right)^2 n\sigma^2\\ &= \frac{\sigma^2}{n}\end{aligned}$$

Just as in the case of a sample proportion $\hat{p}$, the variability of the sampling distribution of a sample mean decreases as the sample size grows. *The mean of several observations is less variable than a single observation.* Because the standard deviation of $\bar{x}$ is $\sigma/\sqrt{n}$, it is again true that the standard deviation of the statistic decreases in proportion to the square root of the sample size. Here is a summary of these facts.

MEAN AND STANDARD DEVIATION OF A SAMPLE MEAN

If $\bar{x}$ is the mean of an SRS of size n from a population having mean μ and standard deviation σ, then

$$\begin{aligned}\mu_{\bar{x}} &= \mu\\ \sigma_{\bar{x}} &= \frac{\sigma}{\sqrt{n}}\end{aligned}$$

EXAMPLE 5.14

The height X of a single randomly chosen young woman varies according to the $N(65.5, 2.5)$ distribution. If a medical study asked the height of an SRS of 100 young women, the sampling distribution of the sample mean height $\bar{x}$ would have mean and standard deviation

$$\mu_{\bar{x}} = \mu = 65.5 \text{ inches}$$
$$\sigma_{\bar{x}} = \frac{\sigma}{\sqrt{n}} = \frac{2.5}{\sqrt{100}} = 0.25 \text{ inch}$$

The heights of individual women vary widely about the population mean, but the average height of a sample of 100 women has a standard deviation only one-tenth as large. ●

EXAMPLE 5.15

The fact that a mean of several measurements is less variable than a single measurement is also important in science. When Simon Newcomb set out to measure the speed of light, he took repeated measurements of the time required for a beam of light to travel a known distance. Newcomb's 66 measurements appear in Table 1.1 (page 3). We saw in Chapter 1 that the data contain two outliers that may not be observations from the same population. Disregarding these leaves 64 observations. These are the values of 64 independent random variables, each with a probability distribution that describes the population of all measurements made with Newcomb's apparatus. If Newcomb's procedures were correct, the population mean μ is the true passage time of light. The population variability reflects the random variation in the measurements due to small changes in the environment, the equipment, and the procedure. Suppose that the standard deviation for this population is $\sigma = 5$. (The units of the coded data in Table 1.1 are billionths of a second, that is, seconds $\times 10^{-9}$.)

If Newcomb had taken a single measurement, the standard deviation of the result would be 5, so that a second observation might have a quite different value. Taking 10 measurements and reporting their mean $\bar{x}$ reduces the standard deviation to

$$\sigma_{\bar{x}} = \frac{5}{\sqrt{10}} = 1.58$$

The sample mean of any number of measurements is an unbiased estimator of the population mean. Averaging over more measurements reduces the variability and makes it more likely that our result is close to the truth. The mean $\bar{x}$ of all 64 measurements has standard deviation

$$\sigma_{\bar{x}} = \frac{5}{\sqrt{64}} = .625$$

Newcomb's observed value of this random variable, $\bar{x} = 27.75$, is a much more reliable estimate of the true passage time than is a single measurement. Increasing the number of measurements from 1 to 64 divides the standard deviation by 8, the square root of 64. ●

We have described the mean and standard deviation of the probability distribution of a sample mean $\bar{x}$, but not the distribution itself. The shape of the distribution of $\bar{x}$ depends on the shape of the population distribution. In particular, if the population distribution is normal, then so is the distribution of the sample mean.

> **SAMPLING DISTRIBUTION OF A SAMPLE MEAN**
>
> If a population has the $N(\mu, \sigma)$ distribution, then the sample mean of n independent observations has the $N(\mu, \sigma/\sqrt{n})$ distribution.

EXAMPLE 5.16

A normal quantile plot of Newcomb's 64 observations (Figure 1.21(d) on page 76) shows that their distribution is close to normal. The relative frequency distribution of repeated observations reflects the population distribution, so we can be confident that the population of all measurements made with Newcomb's apparatus is close to normal. A single measurement has the $N(\mu, 5)$ distribution, where μ is the unknown population mean. By the 95 part of the 68–95–99.7 rule, 95% of these measurements will lie within 2×5, or 10, of the unknown μ.

Now we combine the calculations of $\sigma_{\bar{x}}$ from Example 5.15 with the fact that $\bar{x}$ follows a normal distribution. The mean $\bar{x}$ of 10 measurements follows the $N(\mu, 1.58)$ distribution, so 95% of the time the observed mean falls within $2 \times 1.58 = 3.16$ of μ. Figure 5.6 compares the distributions of a single measurement and the sample mean of 10 measurements. The picture makes dramatically clear the advantages of averaging over many observations. Newcomb actually made 64 measurements. Their sample mean $\bar{x}$ has the $N(\mu, 0.625)$ distribution. In repetitions of the experiment, 95% of the values of $\bar{x}$ would lie within $2 \times 0.625 = 1.25$ of the population mean μ. Newcomb could be confident that his announced result was close to μ. (But if there was bias in his experimental method, the mean μ of the population of measurements may not be equal to the true passage time of light. This is a question of physics, not of statistics.) ●

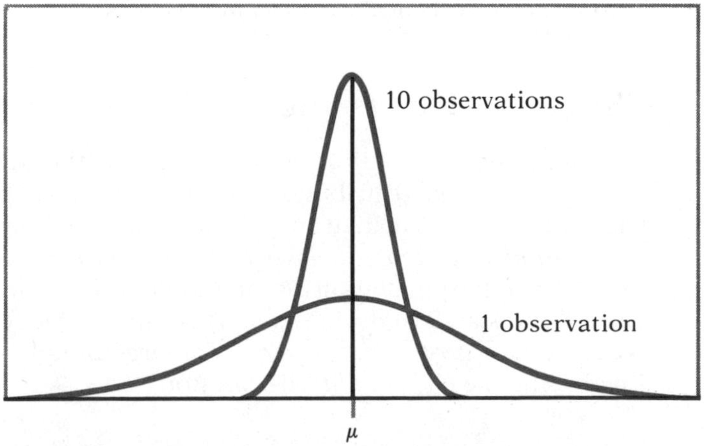

FIGURE 5.6 **The sampling distribution of $\bar{x}$ for samples of size 10 compared with the distribution of a single observation.**

The fact that the sample mean of an SRS from a normal population has a normal distribution is a special case of a more general fact: *Any linear combination of independent normal random variables is also normally distributed*. That is, if X and Y are independent normal random variables and a and b are constants, $aX + bY$ is also normally distributed, and so it is for any number of normal variables. In particular, the sum or difference of independent normal random variables has a normal distribution. The mean and standard deviation of $aX + bY$ are found as usual from the addition rules for means and variances. These facts are often used in statistical calculations.

EXAMPLE 5.17

Tom and George are playing in the club golf tournament. Their scores vary as they play the course repeatedly. Tom's score X has the $N(110, 10)$ distribution, and George's score Y varies from round to round according to the $N(100, 8)$ distribution. If they play independently, what is the probability that Tom will score lower than George and thus do better in the tournament? The difference $X - Y$ between their scores is normally distributed, with mean and variance

$$\mu_{X-Y} = \mu_X - \mu_Y = 110 - 100 = 10$$
$$\sigma^2_{X-Y} = \sigma^2_X + \sigma^2_Y = 10^2 + 8^2 = 164$$

Because $\sqrt{164} = 12.8$, $X - Y$ has the $N(10, 12.8)$ distribution. Figure 5.7 illustrates the probability computation:

$$\begin{aligned} P(X < Y) &= P(X - Y < 0) \\ &= P\left(\frac{(X - Y) - 10}{12.8} < \frac{0 - 10}{12.8}\right) \\ &= P(Z < -.78) = .2177 \end{aligned}$$

Although George's score is 10 strokes lower on the average, Tom will have the lower score in about one of every five matches. ●

The central limit theorem

The sampling distribution of $\bar{x}$ is normal if the underlying population itself has a normal distribution. What happens when the population distribution is not normal? It turns out that *as the sample size increases, the distribution of $\bar{x}$ becomes closer to a normal distribution*. This is true no matter what the population distribution may be, as long as the population has a finite standard deviation σ. This famous fact of probability theory is called the *central limit theorem*.* For large sample size n, we can regard $\bar{x}$ as having the $N(\mu, \sigma/\sqrt{n})$ distribution.

central limit theorem

*The first general version of the central limit theorem was established in 1810 by the French mathematician Pierre Simon Laplace (1749–1827).

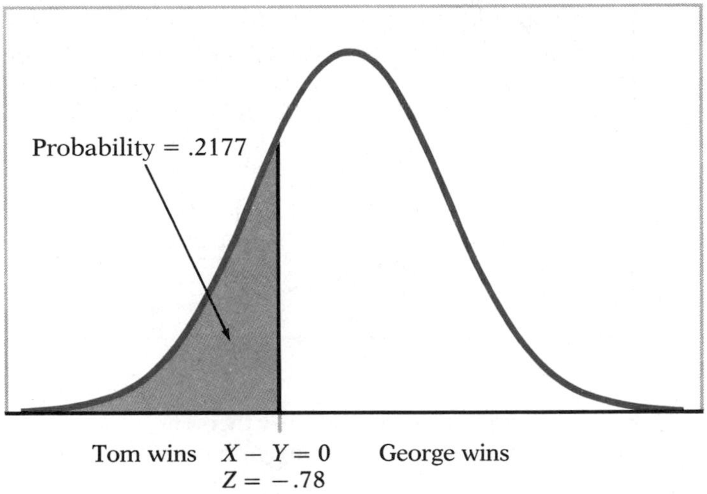

FIGURE 5.7 The normal probability calculation for Example 5.17.

More generally, the central limit theorem says that the distribution of a sum or average of many small random quantities is close to normal. This is true even if the quantities are not independent (as long as they are not too strongly associated) and even if they have different distributions (as long as no one random quantity is so large that it dominates the others). The central limit theorem suggests why the normal distributions are common models for observed data. Any variable that is a sum of many small influences will have approximately a normal distribution.

How large a sample size n is needed for $\bar{x}$ to be close to normal depends on the population distribution. More observations are required if the shape of the population distribution is far from normal.

EXAMPLE 5.18

Figure 5.8 shows the central limit theorem in action in the case of a strongly nonnormal population. Figure 5.8(a) displays the density curve of a single observation, that is, of the population. The distribution is strongly right skewed, and the most probable outcomes are near 0 at one end of the range of possible values. The mean μ of this distribution is 1 and its standard deviation σ is also 1. This particular continuous distribution is called an exponential distribution from the shape of its density curve. Exponential distributions are used as models for the lifetime in service of electronic components and for the time required to serve a customer or repair a machine.

Figure 5.8(b) and Figure 5.8(c) are the density curves of the sample mean of 2 and 10 observations from this population. As n increases, the shape becomes more normal. The mean remains at $\mu = 1$ and the standard deviation decreases, taking the value $1/\sqrt{n}$. The density curve for 10 observations is still somewhat skewed to the right but already resembles a normal curve with $\mu = 1$ and $\sigma = 1/\sqrt{10} = 0.32$. The contrast between the shape of the population distribution and the distribution of the mean of 10 observations is striking. ●

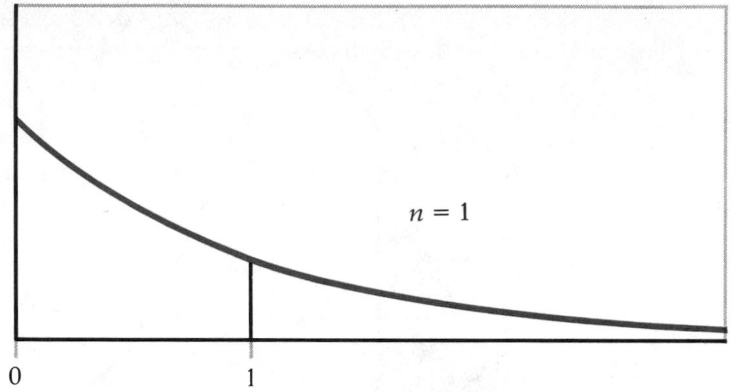

FIGURE 5.8(a) The central limit theorem in action: The density curve for a single observation from a strongly nonnormal population.

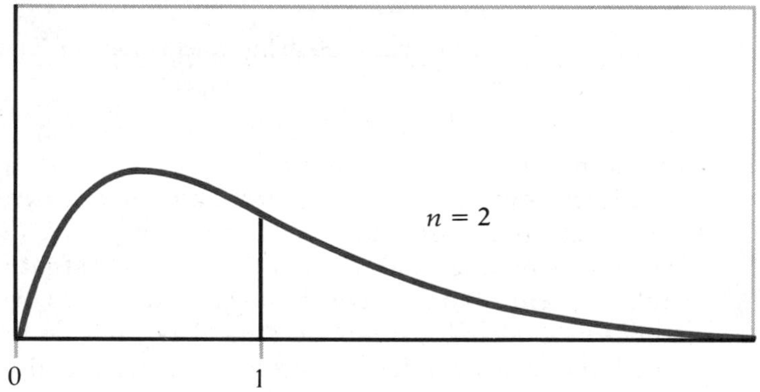

FIGURE 5.8(b) The central limit theorem in action: The density curve for the sample mean of 2 observations.

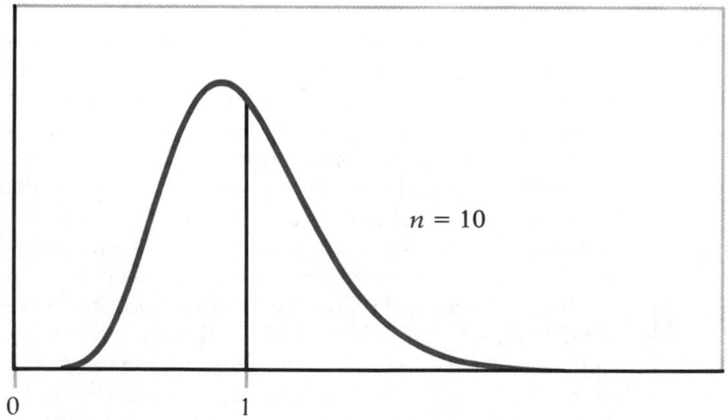

FIGURE 5.8(c) The central limit theorem in action: The density curve for the sample mean of 10 observations.

The central limit theorem allows us to use normal probability calculations to answer questions about sample means from many observations even when the population distribution is not normal.

EXAMPLE 5.19

The time X that a technician requires to perform preventive maintenance on an air conditioning unit is governed by the exponential distribution whose density curve appears in Figure 5.8(a). The mean time is $\mu = 1$ hour and the standard deviation is $\sigma = 1$ hour. Your company operates 70 of these units. What is the probability that their average maintenance time exceeds 50 minutes?

The central limit theorem says that the sample mean time $\bar{x}$ spent working on 70 units has approximately the $N(\mu, \sigma/\sqrt{70})$ distribution. In this case,

$$\frac{\sigma}{\sqrt{70}} = \frac{1}{\sqrt{70}} = .12$$

The distribution of $\bar{x}$ is therefore $N(1, 0.12)$. Because 50 minutes is 50/60 of an hour, or 0.83 hour, the probability we want is

$$\begin{aligned} P(\bar{x} > .83) &= P\left(\frac{\bar{x} - 1}{.12} > \frac{.83 - 1}{.12}\right) \\ &= P(Z > -1.42) = .9222 \end{aligned}$$

Figure 5.9 illustrates this calculation. ●

The normal approximation for sample proportions and counts is an important example of the central limit theorem. This is true because a sample proportion can be thought of as a sample mean. Recall the idea that we used to find the mean and variance of a $B(n, p)$ random variable

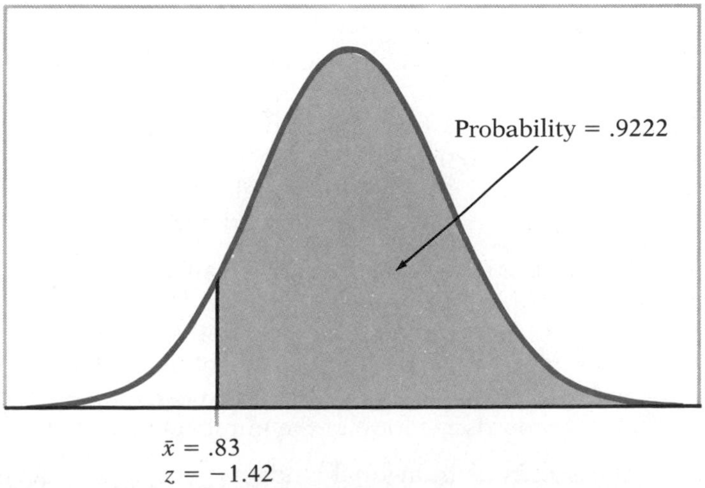

FIGURE 5.9 **The normal probability calculation for Example 5.19.**

X. We wrote the count X as a sum

$$X = Z_1 + Z_2 + \cdots + Z_n$$

of random variables Z_i that take the value 1 if a success occurs on the ith trial and the value 0 otherwise. The variables Z_i take only the values 0 and 1 and are far from normal. The proportion $\hat{p} = X/n$ is the sample mean of the Z_i and, like all sample means, is approximately normal when n is large.

SUMMARY

The sample mean $\bar{x}$ of an SRS of size n drawn from a large population with mean μ and standard deviation σ has a sampling distribution with mean and standard deviation

$$\mu_{\bar{x}} = \mu$$
$$\sigma_{\bar{x}} = \frac{\sigma}{\sqrt{n}}$$

The sample mean $\bar{x}$ is therefore an unbiased estimator of the population mean μ and is less variable than a single observation.

Linear combinations of independent normal random variables have normal distributions. In particular, if the population has a normal distribution, so does $\bar{x}$.

The **central limit theorem** states that for large n the sampling distribution of $\bar{x}$ is approximately $N(\mu, \sigma/\sqrt{n})$ for any population with finite standard deviation σ.

SECTION 5.2 EXERCISES

5.23 The law requires coal mine operators to test the amount of dust in the atmosphere of the mine. A laboratory carries out the test by weighing filters that have been exposed to the air in the mine. The test has a standard deviation of $\sigma = 0.08$ milligram in repeated weighings of the same filter. The laboratory weighs each filter 3 times and reports the mean result. What is the standard deviation of the reported result?

5.24 An automatic grinding machine in an auto parts plant prepares axles with a target diameter $\mu = 40.125$ millimeters (mm). The machine has some variability, so the standard deviation of the diameters is $\sigma = 0.002$ mm. A sample of 4 axles is inspected each hour for process control purposes, and records are kept of the sample mean diameter. What will be the mean and standard deviation of the numbers recorded?

5.25 The scores of individual students on the American College Testing (ACT) Program composite college entrance examination have a normal distribution with mean 18.6 and standard deviation 5.9. At Northside High, 76 seniors take the test. If the scores at this school have the same distribution as national scores, what are the mean and standard deviation of the average (sample mean) score for the 76 students?

5.26 Investors remember 1987 as the year stocks lost 20% of their value in a single day. For 1987 as a whole, the mean return of all common stocks on the New York Stock Exchange was $\mu = -3.5\%$ (that is, these stocks lost 3.5% of their value in 1987). The standard deviation of the returns was about $\sigma = 26\%$. A student of finance forms all possible portfolios that invested equal amounts in 5 of these stocks and records the return for each portfolio. This return is the average of the returns of the 5 stocks chosen. What are the mean and the standard deviation of the portfolio returns?

5.27 The scores of students on the ACT college entrance examination in a recent year had the normal distribution with mean $\mu = 18.6$ and standard deviation $\sigma = 5.9$.

(a) What is the probability that a single student randomly chosen from all those taking the test scores 21 or higher?

(b) The average score of the 76 students at Northside High who took the test was $\bar{x} = 20.4$. What is the probability that the mean score for 76 students randomly selected from all who took the test nationally is 20.4 or higher?

5.28 A bottling company uses a filling machine to fill plastic bottles with a popular cola. The bottles are supposed to contain 300 milliliters (ml). In fact, the contents vary according to a normal distribution with mean $\mu = 298$ ml and standard deviation $\sigma = 3$ ml.

(a) What is the probability that an individual bottle contains less than 295 ml?

(b) What is the probability that the mean contents of the bottles in a six-pack is less than 295 ml?

5.29 Judy's doctor is concerned that she may suffer from hypokalemia (low potassium in the blood). There is variation both in the actual potassium level and in the blood test that measures the level. Judy's measured potassium level varies according to the normal distribution with $\mu = 3.8$ and $\sigma = 0.2$. A patient is classified as hypokalemic if the potassium level is below 3.5.

(a) If a single potassium measurement is made, what is the probability that Judy is diagnosed as hypokalemic?

(b) If measurements are made instead on 4 separate days and the mean result is compared with the criterion 3.5, what is the probability that Judy is diagnosed as hypokalemic?

5.30 A laboratory weighs filters from a coal mine to measure the amount of dust in the mine atmosphere. Repeated measurements of the weight of dust on the same filter vary normally with standard deviation $\sigma = 0.08$ milligram (mg) because the weighing is not perfectly precise. The dust on a particular filter actually weighs 123 mg. Repeated weighings will then have the $N(123, 0.08)$ distribution.

(a) The laboratory reports the mean of 3 weighings. What is the distribution of this mean?

(b) What is the probability that the laboratory reports a result of 124 mg or higher for this filter?

5.31 A company that owns and services a fleet of cars for its sales force has found that the service lifetime of disc brake pads varies from car to car according to a normal distribution with mean $\mu = 55,000$ miles and standard deviation $\sigma = 4500$ miles. The company installs a new brand of brake pads on 8 cars.

(a) If the new brand has the same lifetime distribution as the previous brand, what is the distribution of the sample mean lifetime for the 8 cars?

(b) The average life of the pads on these 8 cars turns out to be $\bar{x} = 51,800$ miles. What is the probability that the sample mean lifetime is 51,800 miles or less if the lifetime distribution is unchanged? The company takes this probability as evidence that the average lifetime of the new brand of pads is less than 55,000 miles.

5.32 The design of an electronic circuit calls for a 100-ohm resistor and a 250-ohm resistor connected in series so that their resistances add. The components used are not perfectly uniform, so that the actual resistances vary independently according to normal distributions. The resistance of 100-ohm resistors has mean 100 ohms and standard deviation 2.5 ohms, while that of 250-ohm resistors has mean 250 ohms and standard deviation 2.8 ohms.

(a) What is the distribution of the total resistance of the two components in series?

(b) What is the probability that the total resistance lies between 345 and 355 ohms?

5.33 The clearance between a pin and the collar around it is important for the proper performance of a disc drive for small computers. The specifications call for the pin to have diameter 0.525 centimeter (cm) and for the collar to have diameter 0.526 cm. The clearance will then be 0.001 cm. In practice, both diameters vary from part to part independently of each other. The diameter X of the pin has the $N(0.525, 0.0003)$ distribution, and the distribution of the diameter Y of the collar is $N(0.526, 0.0004)$.

(a) What is the distribution of the clearance $Y - X$?

(b) What is the probability $P(Y - X \leq 0)$ that the pin will not fit inside the collar?

5.34 An experiment to compare the nutritive value of normal corn and high-lysine corn divides 40 chicks at random into two groups of 20. One group is fed a diet based on normal corn while the other receives high-lysine corn. At the end of the experiment, inference about which diet is superior is based on the difference $\bar{y} - \bar{x}$ between the mean weight gain $\bar{y}$ of the 20 chicks in the high-lysine group and the mean weight gain $\bar{x}$ of the 20 in the normal-corn group. Because of the randomization, the two sample means are independent.

(a) Suppose that $\mu_X = 360$ grams (g) and $\sigma_X = 55$ g in the population of all chicks fed normal corn and that $\mu_Y = 385$ g and $\sigma_Y = 50$ g in the high-lysine population. What are the mean and standard deviation of $\bar{y} - \bar{x}$?

(b) The weight gains are normally distributed in both populations. What is the distribution of $\bar{x}$? Of $\bar{y}$? What is the distribution of $\bar{y} - \bar{x}$?

(c) What is the probability that the mean weight gain in the high-lysine group exceeds the mean weight gain in the normal-corn group by 25 g or more?

5.35 An experiment on the teaching of reading compares two methods, A and B. The response variable is the Degree of Reading Power (DRP) score. The experimenter uses method A in a class of 26 students and method B in a comparable class of 24 students. The classes are assigned to the teaching methods at random. Suppose that in the population of all children of this age the DRP score has the $N(34,12)$ distribution if method A is used and the $N(37,11)$ distribution if method B is used.

(a) What is the distribution of the mean DRP score $\bar{x}$ for the 26 students in the A group? (Assume that this group can be regarded as an SRS from the population of all children of this age.)

(b) What is the distribution of the mean score $\bar{y}$ for the 24 students in the B group?

(c) Use the results of (a) and (b), keeping in mind that $\bar{x}$ and $\bar{y}$ are independent, to find the distribution of the difference $\bar{y} - \bar{x}$ between the mean scores in the two groups.

(d) What is the probability that the mean score for the B group will be at least 4 points higher than the mean score for the A group?

5.36 The two previous exercises illustrate a common setting for statistical inference. This exercise gives the general form of the sampling distribution needed in this setting. We have a sample of n observations from a treatment group and an independent sample of m observations from a control group. Suppose that the response to the treatment has the $N(\mu_X, \sigma_X)$ distribution and that the response of control subjects has the $N(\mu_Y, \sigma_Y)$ distribution. Inference about the difference $\mu_Y - \mu_X$ between the population means is based on the difference $\bar{y} - \bar{x}$ between the sample means in the two groups.

(a) Under the assumptions given, what is the distribution of $\overline{y}$? Of $\overline{x}$?
(b) What is the distribution of $\overline{y} - \overline{x}$?

5.37 A mechanical assembly consists of a shaft with a bearing at each end. The total length of the assembly is the sum $X + Y + Z$ of the shaft length X and the lengths Y and Z of the bearings. These lengths vary from part to part in production, independently of each other and with normal distributions. The shaft length X has mean 11.2 inches and standard deviation 0.002 inch. Each bearing length Y and Z has mean 0.4 inch and standard deviation 0.001 inch.

(a) According to the 68–95–99.7 rule, about 95% of all shafts have lengths in the range $11.2 \pm d_1$ inches. What is the value of d_1? Similarly, about 95% of the bearing lengths fall in the range $0.4 \pm d_2$. What is the value of d_2?
(b) It is common practice in industry to state the "natural tolerance" of parts in the form used in (a). An engineer who knows no statistics thinks that tolerances add, so that the natural tolerance for the total length of the assembly (shaft and two bearings) is $12 \pm d$ inches, where $d = d_1 + 2d_2$. Find the standard deviation of the total length $X+Y+Z$. Then find the value d such that about 95% of all assemblies have lengths in the range $12 \pm d$. Was the engineer correct?

5.38 The amount of nitrogen oxides (NOX) present in the exhaust of a particular type of car varies from car to car according to the normal distribution with mean 1.4 grams per mile (g/mile) and standard deviation 0.3 g/mile. Two cars of this type are tested. One has 1.1 g/mile of NOX; the other, 1.9. The test station attendant finds this much variation between two similar cars surprising. If X and Y are independent NOX levels for cars of this type, find the probability

$$P(X - Y \geq .8 \text{ or } X - Y \leq -.8)$$

that the difference is at least as large as the value the attendant observed.

5.39 Leona and Fred are friendly competitors in high school. Both are about to take the ACT college entrance examination. They agree that if one of them scores 5 or more points better than the other, the loser will buy the winner a meal. Suppose that in fact Fred and Leona have equal ability, so that each score varies normally with mean 24 and standard deviation 2. (The variation is due to luck in guessing and the accident of the specific questions being familiar to the student.) The two scores are independent. What is the probability that the scores differ by 5 or more points in either direction?

5.40 The study habits portion of the Survey of Study Habits and Attitudes (SSHA) psychological test consists of two sets of questions. One set measures "delay avoidance" and the other measures "work methods."

A subject's study habits score is the sum $X + Y$ of the delay avoidance score X and the work methods score Y. The distribution of X in a large population of first-year college students is close to $N(25, 10)$, and the distribution of Y in the same population is close to $N(25, 9)$.

(a) If a subject's X and Y scores were independent, what would be the distribution of the study habits score $X + Y$?

(b) Using the distribution you found in (a), what percent of the population have a study habits score of 60 or higher?

(c) In fact, the X and Y scores are strongly correlated. In this case, does the mean of $X + Y$ still have the value you found in (a)? Does the standard deviation still have the value you found in (a)?

5.41 A study of working couples measures the income X of the husband and the income Y of the wife in a large number of couples in which both partners are employed. Suppose that you knew the means μ_X and μ_Y and the variances σ_X^2 and σ_Y^2 of both variables in the population.

(a) Is it reasonable to take the mean of the total income $X + Y$ to be $\mu_X + \mu_Y$? Explain your answer.

(b) Is it reasonable to take the variance of the total income to be $\sigma_X^2 + \sigma_Y^2$? Explain your answer.

5.42 The number of flaws per square yard in a type of carpet material varies, with mean 1.6 flaws per square yard and standard deviation 1.2 flaws per square yard. The distribution is not normal—in fact, it is discrete. An inspector studies 200 square yards of the material, records the number of flaws found in each square yard, and calculates $\bar{x}$, the mean number of flaws per square yard inspected. Use the central limit theorem to find the approximate probability that the mean number of flaws exceeds 2 per square yard.

5.43 The number of accidents per week at a hazardous intersection varies with mean 2.2 and standard deviation 1.4. This distribution is discrete and so is certainly not normal.

(a) Let $\bar{x}$ be the mean number of accidents per week at the intersection during a year (52 weeks). What is the approximate distribution of $\bar{x}$ according to the central limit theorem?

(b) What is the approximate probability that $\bar{x}$ is less than 2?

(c) What is the approximate probability that there are fewer than 100 accidents at the intersection in a year? (Hint: Restate this event in terms of $\bar{x}$.)

5.44 The distribution of annual returns on common stocks is roughly symmetric, but extreme observations are more frequent than in a normal distribution. Because the distribution is not strongly nonnormal, the mean return over even a moderate number of years is close to normal. In the long run, annual real returns on common stocks have varied with mean about 9% and standard deviation about 28%. Andrew plans to retire in 45 years and is considering investing in stocks. What is the probability (assuming that the past pattern of variation continues) that the mean annual return on common stocks over the next 45 years will exceed 15%? What is the probability that the mean return will be less than 5%?

5.45 The level of nitrogen oxide (NOX) in the exhaust of a particular car model varies with mean 1.4 g/mile and standard deviation 0.3 g/mile. A company has 125 cars of this model in its fleet. If $\bar{x}$ is the mean NOX emission level for these cars, what is the level L such that the probability that $\bar{x}$ is greater than L is only 0.01?

5.46 Children in kindergarten are sometimes given the Ravin Progressive Matrices Test (RPMT) to assess their readiness for learning. Experience at Southwark Elementary School suggests that the RPMT scores for its kindergarten pupils have mean 13.6 and standard deviation 3.1. The distribution is close to normal. Mr. Lavin has 22 children in his kindergarten class this year. He suspects that their RPMT scores will be unusually low because the test was interrupted by a fire drill. To check this suspicion, he wants to find the level L such that there is probability only 0.05 that the mean score of 22 children falls below L when the usual Southwark distribution remains true. What is the value of L?

5.3 CONTROL CHARTS*

There are many situations in which our goal is to hold a variable constant over time. You may monitor your weight or blood pressure, and plan to modify your behavior if either changes. Manufacturers watch the results of regular measurements made during production and plan to take action if quality deteriorates. Statistics plays a central role in these situations because of the presence of variation. All processes have variation. Your weight fluctuates from day to day; the critical dimension of a machined part varies a bit from item to item. An engineer may hope to maintain the dimension of a part exactly at the design specification, but in practice we expect variation due to small variations in the raw material, the adjustment of the machine, the behavior of the operator, and even the temperature in the plant. The statistical description of stability over time requires that the pattern of variation remain stable, not that there be no variation in the variable measured.

*Control charts are important in industry and also illustrate the use of sampling distributions in inference. Nonetheless, this section is not required for an understanding of later material.

> **STATISTICAL CONTROL**
>
> A variable that continues to be described by the same distribution when observed over time is said to be in statistical control or, simply, in control.

Control charts are statistical tools that monitor the control of a process and alert us when the process has been disturbed. Control charts record the natural variation that is inherent in the process; they sound an alarm when variation beyond the usual range suggests that the process has changed. The most common application of control charts is to monitor the performance of an industrial process. The same methods, however, can be used to check the stability of quantities as varied as the ratings of a television show and the level of ozone in the atmosphere. Control charts combine graphical and numerical descriptions of data with probability computations. They therefore provide a natural bridge between exploratory analysis and the use of probability in formal statistical inference.

$\bar{x}$ control charts

Control charts were invented in the 1920s by Walter Shewhart at the Bell Telephone Laboratories.[2] The most common control chart plots the means $\bar{x}$ of small samples taken from the process at regular intervals over time. The process mean μ is the mean of the distribution of individual measurements on items produced by the process in its present state; μ describes the center or aim of the process. Because each sample mean $\bar{x}$ is an unbiased estimator of μ, an $\bar{x}$ control chart can detect changes in the aim of the process.

EXAMPLE 5.20

A manufacturer of high-resolution video terminals must control the tension on the mesh of fine wires that lies behind the surface of the viewing screen. Too much tension will tear the mesh and too little will allow wrinkles. The tension is measured by an electrical device with output readings in millivolts (mV). A careful study shows that the proper tension is 275 mv. Some variation is inherent in the production process, and the study shows that when the production process is operating properly, the standard deviation of the tension readings is $\sigma = 43$ mV. The quality engineer measures the tension on a sample of 4 terminals each hour. The mean $\bar{x}$ of each sample estimates the mean tension μ for the process at the time of the sample.

Table 5.1 shows the observed $\bar{x}$'s for 20 consecutive hours of production. How can we use these data to monitor the process? ●

TABLE 5.1 $\bar{x}$ from 20 samples of size 4

Sample	$\bar{x}$	Sample	$\bar{x}$
1	269.5	11	264.7
2	297.0	12	307.7
3	269.6	13	310.0
4	283.3	14	343.3
5	304.8	15	328.1
6	280.4	16	342.6
7	233.5	17	338.8
8	257.4	18	340.1
9	317.5	19	374.6
10	327.4	20	336.1

center line

A plot against time helps us see whether or not the process is stable. Figure 5.10 presents a plot of the successive sample means against the order in which the samples were taken. Because the desired value for the process mean is $\mu = 275$ mV, we draw a *center line* at that level across the plot. The means from the later samples fall above this line and are consistently higher than those from earlier samples. This suggests that the process mean μ may have shifted upward, away from its target value of 275 mV. We are aware, however, how easily our intuitive judgment can

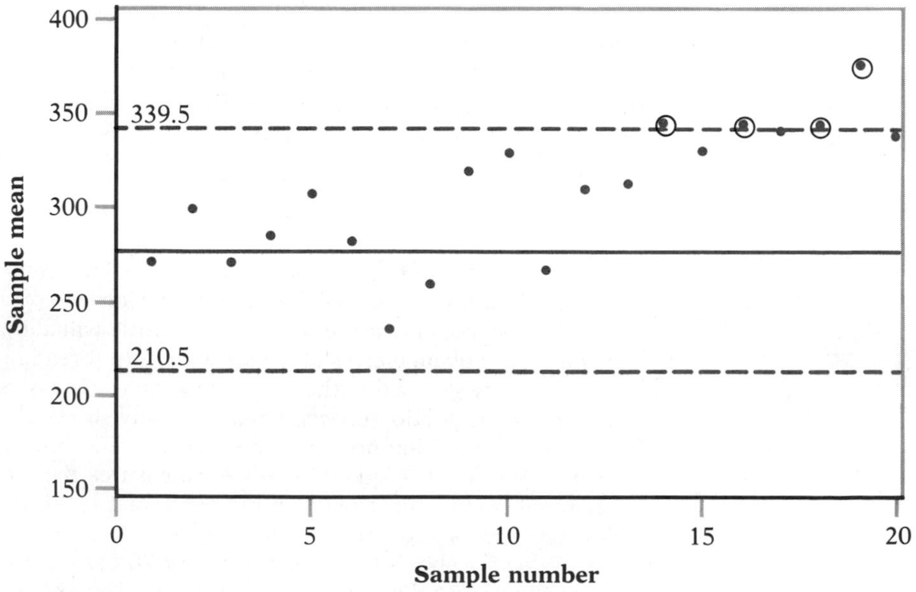

FIGURE 5.10 The $\bar{x}$ control chart for the data of Table 5.1.

be misled by short random sequences. Can we back up our perception by calculation?

We expect $\bar{x}$ to have a distribution that is nearly normal. Not only are the tension measurements roughly normal, but also the central limit theorem effect suggests that the sample mean will be closer to normal than the individual measurements. Because a control chart is a warning device, it is not necessary that our probability calculations be exactly correct. Approximate normality is good enough. Control charts therefore use the approximate normal probabilities given by the 68–95–99.7 rule rather than more exact calculations from Table A.

If the standard deviation of the individual screens remains at $\sigma = 43$ mV, the standard deviation of $\bar{x}$ from 4 screens is

$$\sigma_{\bar{x}} = \frac{\sigma}{\sqrt{n}} = \frac{43}{\sqrt{4}} = 21.5$$

As long as the mean remains at its target value $\mu = 275$ mV, the 99.7 part of the 68–95–99.7 rule says that almost all values of $\bar{x}$ will lie between

$$\mu - 3\sigma_{\bar{x}} = 275 - (3)(21.5) = 210.5$$
$$\mu + 3\sigma_{\bar{x}} = 275 + (3)(21.5) = 339.5$$

control limits

We therefore draw dashed *control limits* at these two levels on the plot.

> **$\bar{x}$ CONTROL CHART**
>
> To evaluate the control of a process with given standards μ and σ, plot the means of regular samples of size n against time, with a center line at μ and control limits at $\mu \pm 3\sigma/\sqrt{n}$.

Four points, which are circled in Figure 5.10, lie above the upper control limit of the control chart. It is unlikely (probability less than 0.003) that a particular point would fall outside the control limits if μ and σ remain at their target values. These points are therefore good evidence that the distribution of the production process has changed. It appears that the process mean moved up at about sample number 12. In practice, the operators search for a disturbance in the process as soon as the first out-of-control point is noticed, that is, after sample number 14. Lack of control might be caused by a new operator, a new batch of mesh, or a breakdown in the tensioning apparatus. The out-of-control signal alerts us to the change before a large number of defective screens are produced.

$\bar{x}$ chart

An $\bar{x}$ control chart is often called simply an $\bar{x}$ *chart*. Points $\bar{x}$ that vary between the control limits of an $\bar{x}$ chart represent the chance variation that is present in a normally operating process. Points that are out of control suggest that some source of additional variability has disturbed the stable operation of the process. Such a disturbance makes out-of-control points probable rather than unlikely. For example, if the process mean μ in Example 5.20 shifts from 275 mV to 339.5 mV, the probability that the next point falls above the upper control limit increases from about 0.0015 to 0.5.

Statistical process control The purpose of a control chart is not to ensure good quality by inspecting most of the items produced. Control charts focus on the manufacturing process itself rather than on the products. By examining a few items at regular intervals, we can detect disturbances and correct them before they become serious. This is called *statistical process control*. Process control achieves high quality at a lower cost than inspecting all of the products. Small samples of 4 or 5 items are usually adequate for process control.

statistical process control

A process that is in control is stable over time, but stability alone does not guarantee good quality. The natural variation in the process may be so large that many of the products are unsatisfactory. Nonetheless, establishing control has a number of advantages. First, in order to assess whether the process is adequate, we must observe the process operating in control, that is, free of breakdowns and other disturbances. A process in control is doing as well as it can. If the process is not capable of producing adequate quality even when undisturbed, some major change in the process (new machines, retraining the operators, and so on) is needed. Second, a process in control is predictable. We can predict both the quantity and the quality of items produced. Third, when a process is in control, we can easily see the effects of attempts to improve the process because they will not be hidden by the unpredictable variation that characterizes lack of statistical control.

Comments on control charts A control chart gets it name from the statistic that is plotted against time—an $\bar{x}$ chart plots $\bar{x}$, and a $\hat{p}$ chart plots the sample proportion $\hat{p}$. (By an accident of history that cannot now be undone, a $\hat{p}$ chart is called a p chart in most writing on quality control.) We find the center line and control limits from the sampling distribution of the statistic. Because you know that $\hat{p}$ from a sample of size n has approximately the $N(p, \sqrt{p(1-p)/n})$ distribution, you should be able to set up a $\hat{p}$ chart to control a population proportion at a target value p. Such $\hat{p}$ charts are most often used to monitor the proportion p of items produced that fail to conform to a set of specifications.

There are many variations on even the $\bar{x}$ chart. In practice we rarely know the process mean μ and the process standard deviation σ. We must then base control limits on estimates of μ and σ from past samples. There are some fine points in using control limits based on past data—for example, we must first check that the process was already in control when the data were gathered. We will content ourselves with the case in which values μ and σ are given.

The probability calculations in a control chart assume that the individual observations are a random sample from the population of interest. If the usual 4 or 5 items in a sample are an SRS from an hour's production, the population is all items produced that hour. It is more common, however, to sample 4 or 5 consecutively produced items once each hour. In that case the population exists only in our minds. It contains all items that would be produced by the process as it was operating at the time of sampling. The control chart monitors the state of the process once each hour to see if a change has taken place.

Out-of-control signals

The basic signal for lack of control in an $\bar{x}$ chart is a single point beyond the control limits. In practice, however, other signals are used as well. In each case we can compute the probability that a process in control (that is, with μ and σ at their target values) gives a false signal. The probability of a false signal is about 0.003 for the "one point beyond the limits" signal, by the 99.7 part of the 68–95–99.7 rule. Figure 5.11 illustrates several popular signals.

EXAMPLE 5.21

One common out-of-control signal is a *run* of 9 consecutive points above or below the center line (Figure 5.11b). What is the probability of a false signal of this type when the process is in fact in control?

When the process is in control, the center line is the true process mean μ. Because of the symmetry of the normal distributions, a sample mean $\bar{x}$ is equally likely to fall above or below the line. The means of successive samples are independent because the samples are separated in time. So by the multiplication rule, the probability of a run of 9 points above μ is

$$\left(\frac{1}{2}\right)^9 \doteq .002$$

The probability of a run of 9 points below the center line is the same. The probability that the next 9 points show a run of either kind is therefore about 0.004. The probability that such a run occurs not in the next 9 points but somewhere in a longer sequence of observations would also be useful to know. Unfortunately, the calculation of such probabilities is quite difficult. ●

In the $\bar{x}$ chart of Figure 5.10, the run criterion does not give an out-of-control signal until sample number 20. The "one point out" criterion

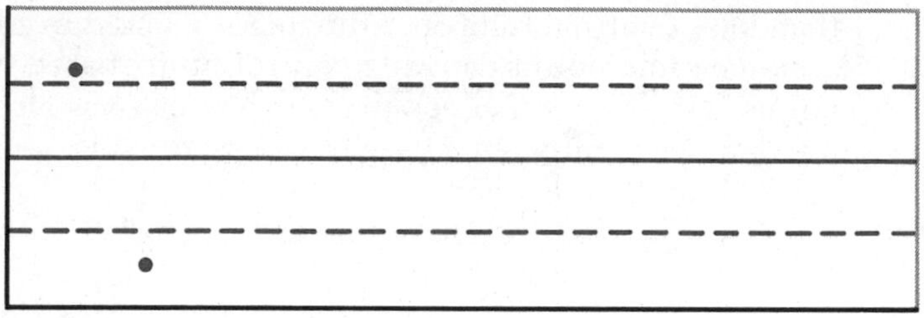

FIGURE 5.11(a) Out-of-control signals for $\bar{x}$ charts: One point beyond the 3σ level.

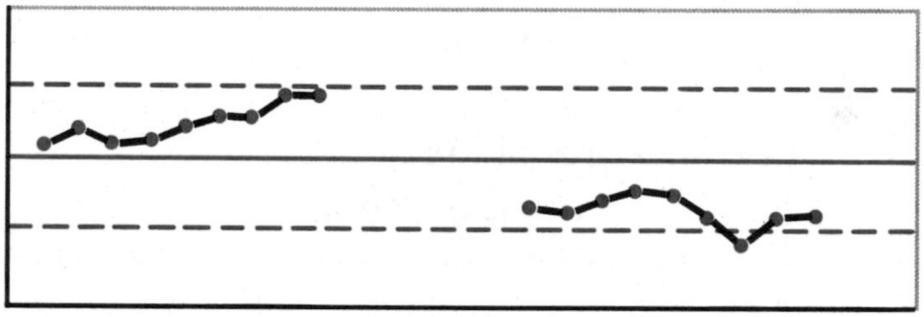

FIGURE 5.11(b) Out-of-control signals for $\bar{x}$ charts: Run of 9 points on one side of the center line.

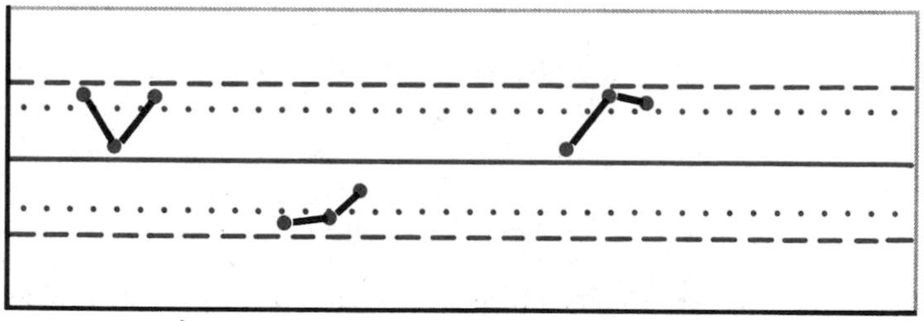

FIGURE 5.11(c) Out-of-control signals for $\bar{x}$ charts: 2 out of 3 points beyond the 2σ level on the same side of the center line.

alerts us at sample 14. In this case the run criterion was slow to detect lack of control. In cases where the process mean slowly drifts away from its target value, however, the run criterion will often give an out-of-control signal before any individual point falls outside the control limits. That is why it is common practice to use the "one point out" criterion and the run criterion simultaneously.

EXAMPLE 5.22

Another common out-of-control signal is "2 out of 3 points beyond the 2σ level on the same side of the center line." That is, at least 2 of the next 3 $\bar{x}$'s fall either below $\mu - 2\sigma/\sqrt{n}$ or above $\mu + 2\sigma/\sqrt{n}$. Figure 5.11(c) illustrates several configurations that give this signal.

The probability that a process in control produces an $\bar{x}$ larger than $\mu + 2\sigma/\sqrt{n}$ is about 0.025 (that's half of the 0.05 left over from the 95 part of the 68–95–99.7 rule). The count of $\bar{x}$'s that fall above this level in the next 3 samples is a count of successes in 3 independent trials. This count X therefore has the $B(3, 0.025)$ distribution. The binomial formula gives the probability of a false signal as

$$\begin{aligned} P(X \geq 2) &= P(X = 2) + P(X = 3) \\ &= \binom{3}{2}(.025)^2(.975) + \binom{3}{3}(.025)^3(.975)^0 \\ &= .00183 + .00002 \doteq .002 \end{aligned}$$

The probability that the next 3 samples give a false signal on the low side is the same. So the overall probability of a false signal by this criterion is about 0.004. ●

These examples illustrate the application of probability calculations in inference. They confirm that each of these out-of-control signals is unlikely to occur as long as the process remains in control. So when a signal *does* occur, we suspect that the process has been disturbed. Our probability calculations assumed that μ and σ were given. The results would be a bit different in the more common case in which σ is unknown and must be estimated from past observations. The same out-of-control signals are nonetheless used in both cases. Control charts are an informal tool designed for easy use. We do probability calculations in the simplest case to see roughly how likely false signals are and do not attempt to refine these calculations for different conditions. We will see in Chapter 7 that more formal inference does adapt its probability calculations to the specific situation.

SUMMARY

A process that can be measured over time is **in control** if the variable measured has the same probability distribution at all times. A process that is in control is operating under stable conditions.

An $\bar{x}$ **control chart** is a graph of sample means plotted against the time order of the samples, with a solid center line at the target value μ of the process mean and dashed **control limits** at $\mu \pm 3\sigma/\sqrt{n}$. An $\bar{x}$ chart helps us decide if a process is in control with mean μ and standard deviation σ.

The probability that the next point lies outside the control limits on an $\bar{x}$ chart is about 0.003 if the process is in control. Such a point is evidence that the process is **out of control**—that is, that the distribution of the process has changed due to some disturbance. A cause for the change in the process should be sought.

There are other common signals for lack of control, such as a run of 9 consecutive points on the same side of the center line. In each case, we can compute the probability that a process in control will give the signal. If this probability is small, we can regard the signal as good evidence of lack of control.

SECTION 5.3 EXERCISES

5.47 A maker of auto air conditioners checks a sample of 4 thermostatic controls from each hour's production. The thermostats are set at 75° and then placed in a chamber where the temperature is raised gradually. The temperature at which the thermostat turns on the air conditioner is recorded. The standard for the process mean is $\mu = 75°$. Past experience indicates that the response temperature of properly adjusted thermostats varies with $\sigma = 0.5°$. The mean response temperature $\bar{x}$ for each hour's sample is plotted on an $\bar{x}$ control chart. Calculate the center line and control limits for this chart.

5.48 The width of a slot cut by a milling machine is important to the proper functioning of a hydraulic system for large tractors. The manufacturer checks the control of the milling process by measuring a sample of 5 consecutive items during each hour's production. The mean slot width for each sample is plotted on an $\bar{x}$ control chart. The target width for the slot is $\mu = 0.8750$ inch. When properly adjusted, the milling machine should produce slots with mean width equal to the target value and standard deviation $\sigma = 0.0012$ inch. What center line and control limits should be drawn on the $\bar{x}$ chart?

5.49 A pharmaceutical manufacturer forms tablets by compressing a granular material that contains the active ingredient and various fillers. The hardness of a sample from each lot of tablets is measured in order to control the compression process. The target values for the hardness are $\mu = 11.5$ and $\sigma = 0.2$. Table 5.2 gives three sets of data, each representing $\bar{x}$ for 20 successive samples of $n = 4$ tablets. One set remains in control at the target value. In a second set, the process mean μ shifts suddenly to a new value. In a third, the process mean drifts gradually.

(a) What are the center line and control limits for an $\bar{x}$ chart for this process?

(b) Draw a separate $\bar{x}$ chart for each of the three data sets. Circle any points that are beyond the control limits. Also, check for runs of 9

TABLE 5.2 Three sets of $\bar{x}$ from 20 samples of size 4

Sample	Data set A	Data set B	Data set C
1	11.602	11.627	11.495
2	11.547	11.613	11.475
3	11.312	11.493	11.465
4	11.449	11.602	11.497
5	11.401	11.360	11.573
6	11.608	11.374	11.563
7	11.471	11.592	11.321
8	11.453	11.458	11.533
9	11.446	11.552	11.486
10	11.522	11.463	11.502
11	11.664	11.383	11.534
12	11.823	11.715	11.624
13	11.629	11.485	11.629
14	11.602	11.509	11.575
15	11.756	11.429	11.730
16	11.707	11.477	11.680
17	11.612	11.570	11.729
18	11.628	11.623	11.704
19	11.603	11.472	12.052
20	11.816	11.531	11.905

points above or below the center line and mark the ninth point of any run as being out of control.

(c) Based on your work in (b) and the appearance of the control charts, which set of data comes from a process that is in control? In which case does the process mean shift suddenly, and at about which sample do you think that the mean changed? Finally, in which case does the mean drift gradually?

5.50 The diameter of a bearing deflector in an electric motor is supposed to be 2.205 cm. Experience shows that when the manufacturing process is properly adjusted, it produces items with mean 2.2050 cm and standard deviation 0.0010 cm. A sample of 5 consecutive items is measured once each hour. The sample means $\bar{x}$ for the past 12 hours are given below:

Hour	1	2	3	4	5	6
$\bar{x}$	2.2047	2.2047	2.2050	2.2049	2.2053	2.2043

Hour	7	8	9	10	11	12
$\bar{x}$	2.2036	2.2042	2.2038	2.2045	2.2026	2.2040

Make an $\bar{x}$ control chart for the deflector diameter. Use both the "one point out" and "run of nine" signals to assess the control of the process. At what point should action have been taken to correct the process as the hourly point was added to the chart?

5.51 Ceramic insulators are baked in lots in a large oven. After the baking, 3 insulators are selected at random from each lot and tested for breaking strength. The mean breaking strength for these samples is plotted on a control chart. The specifications call for a mean breaking strength of at least 10 pounds per square inch (psi). Past experience suggests that if the ceramic is properly formed and baked, the standard deviation in the breaking strength is about 1.2 psi. Here are the sample means from the last 15 lots.

Lot	1	2	3	4	5	6	7	8
$\bar{x}$	12.94	11.45	11.78	13.11	12.69	11.77	11.66	12.60

Lot	9	10	11	12	13	14	15
$\bar{x}$	11.23	12.02	10.93	12.38	7.59	13.17	12.14

(a) Find the center line and control limits for an $\bar{x}$ chart with standards $\mu = 10$ and $\sigma = 1.2$. Plot the $\bar{x}$ points on a control chart with these lines drawn on it.

(b) In this case, a process mean breaking strength greater than 10 psi is acceptable, so points out of control in the high direction do not call for remedial action. With this in mind, use both the "one point out" and "run of nine" signals to assess the control of the process and recommend action.

5.52 A manager who knows no statistics asks you, "What does it mean to say that a process is in control? Is being in control a guarantee that the quality of the product is good?" Answer these questions in plain language that the manager can understand.

5.53 There are other out-of-control signals that are sometimes used with $\bar{x}$ charts. One is "4 out of 5 points beyond the 1σ level" on the same side of the center line. That is, at least 4 of 5 consecutive points lie above $\mu + \sigma/\sqrt{n}$, or at least 4 of 5 lie below $\mu - \sigma/\sqrt{n}$. Find the probability that the next 5 points give this signal when the process remains in control at the given μ and σ. Use the binomial distribution and approximate probabilities from the 68–95–99.7 rule.

5.54 Another out-of-control signal that is sometimes used is "15 points in a row within the 1σ level." That is, 15 consecutive points fall between $\mu - \sigma/\sqrt{n}$

and $\mu + \sigma/\sqrt{n}$. This signal suggests either that the value of σ used for the chart is too large or that careless measurement is producing results that are suspiciously close to the target. Find the probability that the next 15 points will give this signal when the process remains in control with the given μ and σ.

5.55 The usual American and Japanese practice in making $\bar{x}$ charts is to place the control limits at $\mu \pm 3\sigma/\sqrt{n}$. The probability that a particular $\bar{x}$ falls outside these limits when the process is in control is about 0.003, using the 99.7 part of the 68–95–99.7 rule. European practice, on the other hand, places the control limits at $\mu \pm c\sigma/\sqrt{n}$, where the constant c is chosen to give exactly probability 0.001 of a point $\bar{x}$ falling above $\mu + c\sigma/\sqrt{n}$ when the target μ and σ remain true. (The probability that $\bar{x}$ falls below $\mu - c\sigma/\sqrt{n}$ is also 0.001 because of the symmetry of the normal distributions.) Use Table A to find the value of c.

Sometimes we want to make a control chart for a single measurement x at each time period. Control charts for individual measurements are just $\bar{x}$ charts with the sample size $n = 1$. We cannot estimate the short-term process standard deviation σ from the individual samples because a sample of size 1 has no variation. Even with advanced methods[3] of combining the information in several samples, the estimated σ will include some long-term variation and so will be too large. To compensate for this, it is common to use 2σ rather than 3σ control limits, that is, control limits $\mu \pm 2\sigma$. The next two exercises deal with control charts for individual measurements.

5.56 Joe has recorded his weight, measured at the gym after a workout, for several years. The mean is 162 pounds and the standard deviation 1.5 pounds, with no signs of lack of control. An injury keeps Joe away from the gym for several months. The data below give his weight, measured once each week for the first 16 weeks after he returns from the injury.

Week	1	2	3	4	5	6	7	8
Weight	168.7	167.6	165.8	167.5	165.3	163.4	163.0	165.5

Week	9	10	11	12	13	14	15	16
Weight	162.6	160.8	162.3	162.7	160.9	161.3	162.1	161.0

The short-term variation in Joe's weight, estimated from these measurements by advanced methods, is about $\sigma = 1.3$ pounds. Joe has a target of $\mu = 162$ pounds for his weight. Make a control chart for his measurements, using control limits $\mu \pm 2\sigma$. Comment on individual points out of control and on runs. Is Joe's weight stable or does it change systematically over this period?

5.57 Professor Moore, who lives a few miles outside a college town, keeps a record of his commuting time. The data (in minutes) appear in Table 5.3. They cover most of the fall semester, although on some dates the professor was out of town or forgot to set his stopwatch. He also noted unusual occurrences on his record sheet: On October 27, a truck backing into a loading dock delayed him, and on December 5, ice on the windshield forced him to stop and clear the glass.

(a) Find $\bar{x}$ and s for the driving times in Table 5.3.

(b) Plot the driving times against the order in which the observations were made. Add a center line and the control limits $\bar{x} \pm 2s$ to your chart. (The standard deviation s of all observations includes any long-term variation, so these limits are a bit crude.)

(c) Comment on the control of the process. Can you suggest explanations for individual points that are out of control? Is there any indication of an upward or downward trend in driving time?

5.58 A manufacturer of compact disc players uses statistical process control to monitor the quality of the circuit board that contains most of the player's electronic components. Every circuit board is tested for proper function by a computer-directed test bed after assembly. The plant produces 400 circuit boards per shift, and the proportion $\hat{p}$ of the 400 boards that fail the test is recorded each day. Company standards call for a failure rate of no more than 10%, or $p = 0.1$.

(a) If the process is in control with proportion $p = 0.1$ of defective boards, what are the mean and standard deviation of the sample proportion $\hat{p}$?

TABLE 5.3 Commuting time (minutes), September 10 to December 19

Date	Time	Date	Time	Date	Time
Sept. 10	8.25	Oct. 10	7.75	Nov. 22	7.75
Sept. 11	7.83	Oct. 13	7.92	Nov. 24	7.42
Sept. 12	8.30	Oct. 16	8.00	Nov. 28	6.75
Sept. 15	8.42	Oct. 17	8.08	Nov. 29	7.42
Sept. 17	8.50	Oct. 22	8.42	Dec. 2	8.50
Sept. 22	8.67	Oct. 23	8.75	Dec. 4	8.67
Sept. 23	8.17	Oct. 24	8.08	Dec. 5	10.17
Sept. 24	9.00	Oct. 27	9.75	Dec. 8	8.75
Sept. 25	9.00	Oct. 28	8.33	Dec. 10	8.58
Sept. 29	8.17	Nov. 3	7.83	Dec. 11	8.67
Sept. 30	7.92	Nov. 5	7.92	Dec. 12	9.17
Oct. 3	9.00	Nov. 19	8.58	Dec. 15	9.08
Oct. 8	8.50	Nov. 20	7.83	Dec. 17	8.83
Oct. 9	9.00	Nov. 21	8.42	Dec. 19	8.67

(b) According to the normal approximation, what is the distribution of $\hat{p}$ when the process is in control?

(c) Give the center line (at the mean of $\hat{p}$) and control limits (at three standard deviations of $\hat{p}$ above and below the mean).

(d) The table below lists the proportions defective in 16 consecutive days' production. Make a $\hat{p}$ control chart with center line and control limits from (c). Plot these values of $\hat{p}$ on the chart. Are any points out of control?

Day	1	2	3	4	5	6	7	8
$\hat{p}$	.1150	.1600	.1300	.1225	.1000	.1225	.1900	.1150

Day	9	10	11	12	13	14	15	16
$\hat{p}$	.1000	.1600	.1675	.1225	.1375	.1975	.1525	.1675

5.59 In a process control setting, successive samples of n items are inspected. Each item is classified as conforming to specifications or as failing to conform. The proportion $\hat{p}$ of nonconforming items in each sample is recorded, and the values of $\hat{p}$ are plotted against the order of the samples. There is a target value p for the population proportion of nonconforming items. Generalize the results of the previous exercise to give formulas in terms of p for the center line and control limits of a $\hat{p}$ control chart. Use the normal approximation for $\hat{p}$ and follow the model of the $\bar{x}$ chart.

5.60 A producer of agricultural machinery purchases bolts from a supplier in lots of several thousand. The producer submits 80 bolts from each lot to a shear test and accepts the lot if no more than 3 of them fail the test. Past records show that an average of 1.8 bolts per lot have failed the test. This is an opportunity to keep a control chart to monitor the quality level of the supplier over time. Because 1.8/80 =0.0225, a target proportion of nonconforming bolts is set at $p = 0.0225$ based on past data. The proportion of nonconforming bolts among the 80 tested from each incoming lot will be plotted on a control chart. Give the center line and control limits to be used on this chart. (Because of the informal nature of process control procedures, the chart is based on the normal approximation to the distribution of the sample proportion $\hat{p}$ even though the approximation is not very accurate for this n and p.)

CHAPTER 5 EXERCISES

5.61 Ray is a basketball player who has made about 70% of his free throws over several years. In a tournament game he makes only 2 of 6 free throws. Ray's coach says this was just bad luck. Suppose that Ray's free throws are independent trials with probability 0.7 of a success on each trial. What is the probability that he makes 2 or fewer in 6 attempts? Do you think that his tournament performance is just chance variation?

5.62 The distribution of scores for persons over 16 years of age on the Wechsler Adult Intelligence Scale (WAIS) is approximately normal with mean 100 and standard deviation 15. The WAIS is one of the most common "IQ tests" for adults.

(a) What is the probability that a randomly chosen individual has a WAIS score of 105 or higher?
(b) What are the mean and standard deviation of the average WAIS score $\bar{x}$ for an SRS of 60 people?
(c) What is the probability that the average WAIS score of an SRS of 60 people is 105 or higher?
(d) Would your answers to any of (a), (b), or (c) be affected if the distribution of WAIS scores in the adult population were distinctly nonnormal?

5.63 High school dropouts make up 14.1% of all Americans aged 18 to 24. A vocational school that wants to attract dropouts mails an advertising flyer to 25,000 persons between the ages of 18 and 24.

(a) If the mailing list can be considered a random sample of the population, what is the mean number of high school dropouts who will receive the flyer? What is the standard deviation of this number?
(b) What is the probability that at least 3500 dropouts will receive the flyer?

5.64 A political activist is gathering signatures on a petition by going door to door asking citizens to sign. She wants 100 signatures. Suppose that the probability of getting a signature at each household is 1/10, and let the random variable X be the number of households visited to collect exactly 100 signatures. Does X have a binomial distribution? If so, give n and p. If not, explain why not.

5.65 According to genetic theory, the blossom color in the second generation of a certain cross of sweet peas should be red or white in a 3:1 ratio. That is, each plant has probability 3/4 of having red blossoms, and the blossom colors of separate plants are independent.

(a) What is the probability that exactly 6 out of 8 of these plants have red blossoms?
(b) What is the mean number of red-blossomed plants when 80 plants of this type are grown from seeds?
(c) What is the probability of obtaining at least 50 red-blossomed plants when 80 plants are grown from seeds?

5.66 The weight of the eggs produced by a certain breed of hen is normally distributed with mean 65 g and standard deviation 5 g. If cartons of such eggs can be considered to be SRSs of size 12 from the population of all eggs, what is the probability that the weight of a carton falls between 750 g and 825 g?

5.67 According to a market research firm, 52% of all residential telephone numbers in Los Angeles are unlisted. A telephone sales firm uses random digit dialing equipment that dials residential numbers at random, regardless of whether they are listed in the telephone directory. The firm calls 500 numbers in Los Angeles.

(a) What is the exact distribution of the number X of unlisted numbers that are called?

(b) Use a suitable approximation to calculate the probability that at least half of the numbers called are unlisted.

5.68 A study of rush hour traffic in San Francisco records the number of people in each car entering a freeway at a suburban interchange. Suppose that this number X has mean 1.5 and standard deviation 0.75 in the population of all cars that enter at this interchange during rush hours.

(a) Does the count X have a binomial distribution? Why or why not?

(b) Could the exact distribution of X be normal? Why or why not?

(c) Traffic engineers estimate that the capacity of the interchange is 700 cars per hour. According to the central limit theorem, what is the approximate distribution of the mean number of persons $\bar{x}$ in 700 randomly selected cars at this interchange?

(d) The count of people in 700 cars is $700\bar{x}$. Use your result from (c) to give an approximate distribution for the count. What is the probability that 700 cars will carry more than 1075 people?

5.69 An opinion poll asks a sample of 500 adults whether they favor giving parents of school-age children vouchers that can be exchanged for education at any public or private school of their choice. Each school would be paid by the government on the basis of how many vouchers it collected. Suppose that in fact 45% of the population favor this idea. What is the probability that at least half of the sample are in favor? (Assume that the sample is an SRS.)

5.70 A machine fastens plastic screw-on caps onto containers of motor oil. If the machine applies more torque than the cap can withstand, the cap will break. Both the torque applied and the strength of the caps vary. The capping machine torque has the normal distribution, with mean 7 inch-pounds and standard deviation 0.9 inch-pounds. The cap strength (the torque that would break the cap) has the normal distribution with mean 10 inch-pounds and standard deviation 1.2 inch-pounds.

(a) Explain why it is reasonable to assume that the cap strength and the torque applied by the machine are independent.

(b) What is the probability that a cap will break while being fastened by the capping machine?

5.71 The process that molds the plastic caps referred to in the previous exercise is monitored by testing a sample of 6 caps every 20 minutes. The breaking strength of the caps is measured and an $\bar{x}$ chart is kept. Use the information in the previous exercise to find the center line and control limits for this $\bar{x}$ chart.

5.72 An important step in the manufacture of integrated circuit chips is etching the lines that will conduct current between components on each chip. Each chip contains a line width test pattern that is used for process control measurements. The target width is 3.0 micrometers (μm). Past data show that the line width varies in production according to a normal distribution with mean 2.829 μm and standard deviation 0.1516 μm.

(a) What is the probability that the line width of a randomly chosen chip falls outside the acceptable range 3.0 ± 0.2 μm?

(b) What are the control limits for an $\bar{x}$ chart for line width if samples of size 5 are selected at regular intervals during production? (Use the target value 3.0 as your center line.)

5.73 In an experiment on learning foreign languages, researchers studied the effect of delaying oral practice when beginning language study. The researchers randomly assigned 23 beginning students of Russian to an experimental group and another 23 to a control group. The control group began speaking practice immediately, but the experimental group delayed speaking for 4 weeks. At the end of the semester both groups took a standard test of comprehension of spoken Russian. Suppose that in the population of all beginning students, the test scores under the control method vary according to the $N(32, 6)$ distribution. The population distribution when oral practice is delayed is $N(29, 5)$.

(a) What is the sampling distribution of the mean score $\bar{x}$ in the control group in many repetitions of the experiment? What is the sampling distribution of the mean score $\bar{y}$ in the experimental group?

(b) If the experiment were repeated many times, what would be the sampling distribution of the difference $\bar{y} - \bar{x}$ between the mean scores in the two groups?

(c) What is the probability that the experiment will find (misleadingly) that the experimental group has a mean at least as large as that of the control group?

CHAPTER 5 COMPUTER EXERCISES

5.74 We can observe the distribution of a statistic in repeated sampling by simulation. Consider a large population of young women whose heights vary according to the normal distribution with mean $\mu = 65.5$ inches and standard deviation $\sigma = 2.5$ inches. We will draw repeated SRSs of size $n = 9$ from this population and observe the behavior of the sample mean $\bar{x}$ in repeated sampling. The sampling distribution of $\bar{x}$ is the theoretical distribution that would result from an indefinitely large number of samples; the distribution we observe in this exercise is an approximation based on a few samples.

(a) You can simulate an SRS of size 9 by generating 9 observations from the $N(65.5, 2.5)$ distribution. Do this 100 times and save the 100 SRSs that result. (Most statistical software allows you to generate 100 samples with a single command.)

(b) Now calculate the sample mean $\bar{x}$ for each sample and save the 100 values of $\bar{x}$. (Again, most software can find the 100 means with a single command.)

(c) Investigate the distribution of the 100 values of $\bar{x}$. What are the mean and standard deviation of these data? Compare the results of your simulation with the theoretical results on page 395.

5.75 Some statistical software systems will make $\bar{x}$ charts. In the Minitab system, for example, the XBARCHART command does this. In addition to saving time, the software incorporates methods of estimating unknown process μ and σ from the data. Use software to produce a control chart for the driving times given in Table 5.3. Compare the control limits calculated by the software with those used in Exercise 5.57. The software uses a more sophisticated method to estimate σ.

NOTES

1. Joseph F. Sullivan, "Ruling in Jersey upholds idea of equal odds for all," *New York Times*, August 13, 1985.

2. See his classic book, W. A. Shewhart, *Economic Control of Quality of Manufactured Product*, Van Nostrand, New York, 1931.

3. These methods are discussed in all texts on statistical quality control. See, for example, Section 6-4 of Douglas C. Montgomery, *Introduction to Statistical Quality Control*, 2nd ed., Wiley, New York, 1991.

CHAPTER

6

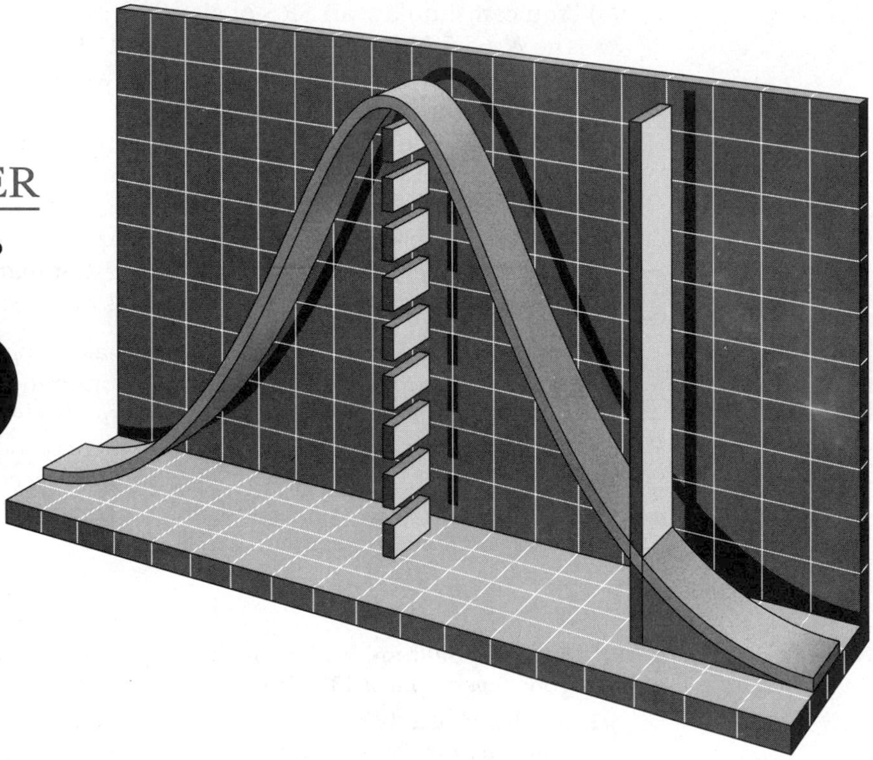

Introduction to Inference

PRELUDE

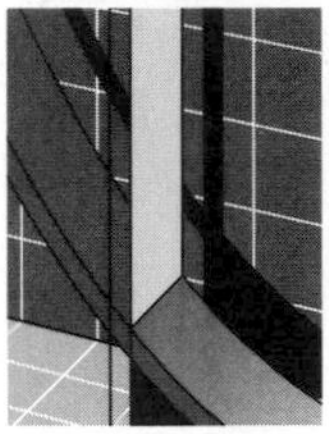

Statistical inference draws conclusions about a population or process based on sample data. Statistical inference also provides a statement, expressed in the language of probability, of how much confidence we can place in the conclusions. Although there are many specific recipes for inference in specific settings, there are only a few general types of statistical inference. This chapter introduces the two most common types: confidence intervals and tests of significance.

We concentrate in this chapter on the reasoning of inference. We will examine when confidence intervals and significance tests are appropriate, what kinds of conclusions they draw and how, and what their limitations are. Because of the emphasis on understanding the reasoning of inference, we look only at a single simple setting: inference about the mean of a normal population whose standard deviation we know. Later chapters will present recipes for inference in many other situations. The reasoning is more important, first, because it applies to all of the recipes and, second, because a computer can carry out the recipes but not the reasoning. We will deal with questions like these:

- If we have the Scholastic Aptitude Test scores of a random sample of 500 California high school seniors, what can we conclude about the average score of all seniors in the state?
- If you find that the mean contents of the cola bottles in a six-pack is 299 milliliters, is this good evidence that the mean for all bottles is less than the 300 milliliters claimed on the label?
- Psychologists compared a group of mentally ill subjects with a group of healthy subjects on 77 different variables describing their childhood and family background. There were "statistically significant" differences between the groups for two of these variables. Is this a sensible scientific procedure?

• • •

The purpose of statistical inference is to draw conclusions from data. We have examined data and arrived at conclusions many times previously. Formal inference adds an emphasis on substantiating our conclusions by probability calculations.

Probability theory allows us to take chance variation into account and so to correct our judgment by calculation. Was the first Vietnam-era draft lottery biased? The correlation between birth date and draft number was $r = -0.226$. Because any two variables will have some chance association in practice, this rather small correlation does not seem convincing. Calculation that a correlation this far from 0 has probability less than .001 in a truly random lottery gives strong evidence that the lottery was in fact unfair. Our unaided judgment can also err in the opposite direction, seeing a systematic effect when only chance variation is at work. Give a new drug and a placebo to 20 patients each; 12 of those taking the drug show improvement, but only 8 of the placebo patients improve. Is the drug more effective than the placebo? Perhaps, but a difference this large or larger between the results in the two groups would occur about one time in five simply because of chance variation. An effect that could so easily be just chance is not convincing.

In this chapter we introduce the two most prominent types of formal statistical inference. Section 6.1 concerns *confidence intervals* for estimating the value of a population parameter. Section 6.2 presents *tests of significance*, which assess the evidence for a claim. Both types of inference are based on the sampling distributions of statistics. That is, both report probabilities that state *what would happen if we used the inference method many times*. This kind of probability statement is characteristic of standard statistical inference. Users of statistics should understand from the first the nature of the reasoning employed and the meaning of the probability statements that appear, for example, on computer output for statistical procedures.

Because the methods of formal inference are based on sampling distributions, they require a probability model for the data. Trustworthy probability models can arise in many ways, but a model is most secure and inference is most reliable when the data are produced by a properly randomized design. When you use statistical inference you are acting as if the data are a random sample or result from a randomized experiment. If this is not true, your conclusions may be open to challenge. Do not be overly impressed by the complex details of formal inference. This elaborate machinery cannot remedy basic flaws in producing the data such as voluntary response samples and confounded experiments. Use the common sense developed in your study of the first three chapters of this book, and proceed to detailed formal inference only when you are satisfied that the data deserve such analysis.

The primary purpose of this chapter is to describe the reasoning used in statistical inference. We will discuss only a few specific inference techniques, and these require rather unrealistic assumptions. Later chapters will present inference methods for use in most of the settings we met in learning to explore data. There are libraries—both of books and of computer software—full of more elaborate statistical techniques. Informed use of any of these methods requires an understanding of the underlying reasoning. A computer will do the arithmetic, but you must still exercise judgment based on understanding.

6.1 ESTIMATING WITH CONFIDENCE

The Scholastic Aptitude Tests (SAT) are widely used measures of readiness for college study. There are two tests, one for verbal ability (SAT-V) and one for mathematical ability (SAT-M). The scores on each test are adjusted so that the mean is 500 and the standard deviation is 100 in a large "standardization group" on which the tests were developed. This scale is maintained from year to year so that scores have a constant interpretation. The scores of the students who actually take the SATs in a given year have a mean that is lower than 500. For example, the mean scores in 1991 were 422 on the verbal examination and 474 for mathematics.

EXAMPLE 6.1

You want to estimate the mean SAT-M score for the more than 250,000 high school seniors in California. You know better than to trust data only from the students who choose to take the SATs. About 45% of California students take the SATs. These self-selected students are all planning to attend college and are not representative of all California seniors. At considerable effort and expense, you give the test to a simple random sample (SRS) of 500 California high school seniors. The mean score for your sample is $\bar{x} = 461$. What can you say about the mean score μ in the population? ●

The sample mean $\bar{x}$ is the natural estimator of the unknown population mean μ. We know that $\bar{x}$ is an unbiased estimator of μ. More important, the law of large numbers says that the sample mean must approach the population mean as the size of the sample grows. The value $\bar{x} = 461$ therefore appears to be a reasonable estimate of the mean score μ that all 250,000 students would achieve if they took the test. But how reliable is this estimate? A second sample would surely not give 461 again. Unbiasedness says only that there is no systematic tendency to underestimate or overestimate the truth. Could we plausibly get a sample mean of 530 or 410 on repeated samples? An estimate without an indication of its variability is of little value.

Statistical confidence

Just as unbiasedness of an estimator concerns the center of its sampling distribution, questions about variation are answered by looking at the spread. We know that if the entire population of SAT scores has mean μ and standard deviation σ, then in repeated samples of size 500 the sample mean $\bar{x}$ follows the $N(\mu, \sigma/\sqrt{500})$ distribution. Let us suppose that we know that the standard deviation σ of SAT-M scores in our California population is the same $\sigma = 100$ that is true of the standardization group. (This is not realistic. We will see in the next chapter how to proceed when σ is not known. For now, we are more interested in statistical reasoning than in details of realistic methods.) In repeated sampling, the sample mean $\bar{x}$ follows the normal distribution centered at the unknown population mean μ and having standard deviation

$$\sigma_{\bar{x}} = \frac{100}{\sqrt{500}} \doteq 4.5$$

Now we are in business. Consider this line of thought, which is illustrated by Figure 6.1:

- The 68–95–99.7 rule says that the probability is about 0.95 that $\bar{x}$ will be within 9 points (two standard deviations of $\bar{x}$) of the population mean score μ.
- To say that $\bar{x}$ lies within 9 points of μ is the same as saying that μ is within 9 points of $\bar{x}$.

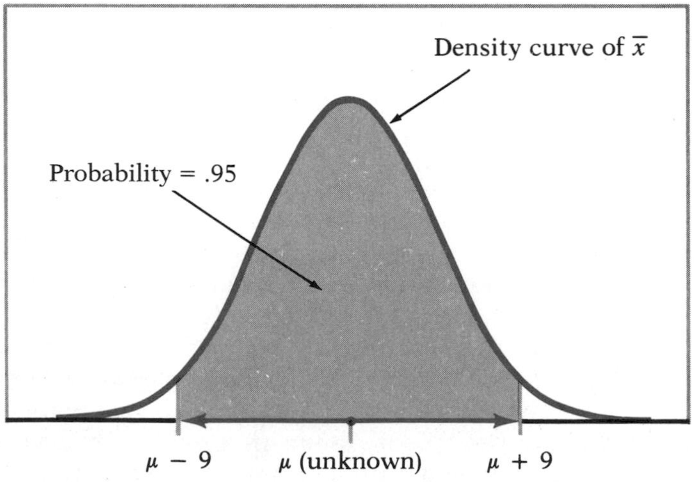

FIGURE 6.1 **$\bar{x}$ lies within $\pm$ 9 of μ in 95% of all samples, so μ also lies within $\pm$ 9 of $\bar{x}$ in those samples.**

- So 95% of all samples will capture the true μ in the interval from $\bar{x} - 9$ to $\bar{x} + 9$.

We have simply restated a fact about the sampling distribution of $\bar{x}$. *The language of statistical inference uses this fact about what would happen in the long run to express our confidence in the results of any one sample.* Our sample gave $\bar{x} = 461$. We say that we are *95% confident* that the unknown mean score for all California seniors lies between

$$\bar{x} - 9 = 461 - 9 = 452$$

and

$$\bar{x} + 9 = 461 + 9 = 470$$

Be sure you understand the grounds for our confidence. There are only two possibilities:

1. The interval between 452 and 470 contains the true μ.
2. Our SRS was one of the few samples for which $\bar{x}$ is not within 9 points of the true μ. Only 5% of all samples give such inaccurate results.

We cannot know whether our sample is one of the 95% for which the interval $\bar{x} \pm 9$ captures μ or one of the unlucky 5%. The statement that we are 95% confident that the unknown μ lies between 452 and 470 is shorthand for saying, "We arrived at these numbers by a method that gives correct results 95% of the time."

The interval of numbers between the values $\bar{x} \pm 9$ is called a *95% confidence interval* for μ. Like most confidence intervals we will meet, this one has the form

$$\text{estimate} \pm \text{margin of error}$$

margin of error

The estimate ($\bar{x}$ in this case) is our guess for the value of the unknown parameter. The *margin of error* ± 9 shows how accurate we believe our guess is, based on the variability of the estimate. The confidence level shows how confident we are that the procedure will catch the true population mean μ.

Figure 6.2 illustrates the behavior of 95% confidence intervals in repeated sampling. The center of each interval is at $\bar{x}$ and therefore varies from sample to sample. The sampling distribution of $\bar{x}$ appears at the top of the figure to show the long-term pattern of this variation. The 95% confidence intervals $\bar{x} \pm 9$ from 25 SRSs appear below. The center $\bar{x}$ of each interval is marked by a dot. The arrows on either side of the dot span the confidence interval. All except one of the 25 intervals cover the true value

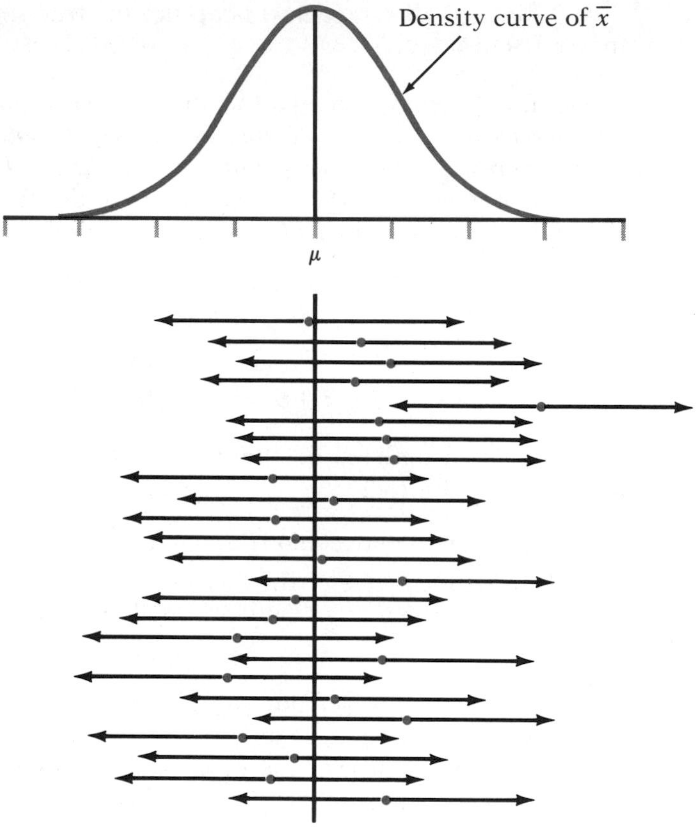

FIGURE 6.2 Twenty-five samples from the same population gave these 95% confidence intervals. In the long run, 95% of all samples give an interval that covers μ.

of μ. In a very large number of samples, 95% of the confidence intervals would contain μ.*

Confidence intervals

Statisticians have constructed confidence intervals for many different parameters based on a variety of designs for data collection. We will meet a number of these in later chapters. Any confidence interval has two aspects: an *interval* computed from the data and a *confidence level* giving

confidence level

*Confidence intervals as a systematic method were invented in 1937 by Jerzy Neyman (1894–1981). Neyman was a Pole who moved to England in 1934 and spent the last half of his long life at the University of California at Berkeley. He remained active until his death, almost doubling the list of his publications after his official retirement in 1961.

the probability that the method produces an interval that covers the parameter. Users can choose the confidence level, most often 90% or higher because we most often want to be quite sure of our conclusions. We will use C to stand for the confidence level in decimal form. For example, a 95% confidence level corresponds to $C = 0.95$. Here is the general definition of a confidence interval for an unknown parameter, which we call θ, the Greek letter theta. In our examples, θ is the mean μ of the population, but in other settings it might be the median, the standard deviation, or any other parameter.

CONFIDENCE INTERVAL

A level C confidence interval for a parameter θ is an interval computed from sample data by a method that has probability C of producing an interval containing the true value of θ.

We will now construct a level C confidence interval for the mean μ of a population when the data are an SRS of size n. The construction is based on the sampling distribution of the sample mean $\bar{x}$. This distribution is exactly $N(\mu, \sigma/\sqrt{n})$ when the population has the $N(\mu, \sigma)$ distribution. The central limit theorem says that this same sampling distribution is approximately correct for large samples whenever the population mean and standard deviation are μ and σ.

Our construction of a 95% confidence interval for the mean SAT score began by noting that any normal distribution has probability about 0.95 within ± 2 standard deviations of its mean. To construct a level C confidence interval we first catch the central area C under a normal curve. That is, we must find the number z^* such that any normal distribution has probability C within $\pm z^*$ standard deviations of its mean. We can find z^* from Table A of standard normal probabilities, using Figure 6.3 as a guide. The value z^* for confidence C catches the central area C between $-z^*$ and z^*, and so omits area $1 - C$. Half the omitted area lies in each tail, so z^* is the point with area $(1 - C)/2$ to its right under the standard normal curve. You can carry through the reasoning without the notation, as the following example illustrates.

EXAMPLE 6.2

A 90% confidence interval is based on catching the central 90% of the normal sampling distribution of $\bar{x}$. In catching the central 90% we leave out 10%, or 5% in each tail. So z^* is the point with area 5% above it under the standard normal curve. (In more formal language, $C = 0.9$, and $(1 - C)/2 = 0.05$.) Table A shows that this z^* lies between 1.64 and 1.65. A more exact calculation of areas under the standard normal curve gives $z^* = 1.645$. ●

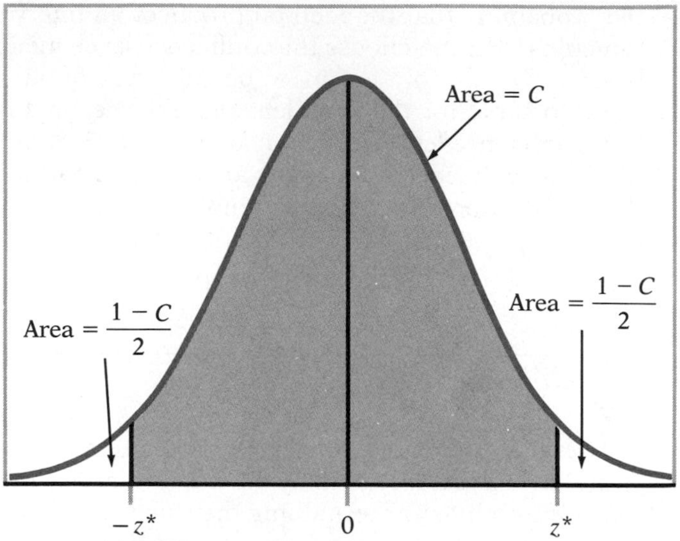

FIGURE 6.3 **The area between $-z^*$ and z^* under the standard normal curve is C. Because z^* has area $(1 - C)/2$ to its right under the curve, it is called the upper $(1 - C)/2$ critical value.**

> **CRITICAL VALUE**
>
> The number z^* with probability p lying to its right under the standard normal density curve is called the upper p critical value of the standard normal distribution.

Table D in the back of the book gives the upper p critical values of the standard normal distribution for several values of p. It is the usual practice to arrange such tables by the upper tail area p, and we will need this arrangement in the following section. But to find the appropriate critical value z^* for a confidence interval, you can just look opposite the confidence level you have chosen. For a 90% confidence interval, Table D shows that $z^* = 1.645$. Here is a brief tabulation taken from Table D of the normal critical values needed for the most common levels of confidence.

Confidence level	p	z^*
90%	.05	1.645
95%	.025	1.960
99%	.005	2.576

Any normal curve has probability C between the point z^* standard deviations below the mean and the point z^* standard deviations above the mean, as Figure 6.3 reminds us. The sample mean $\bar{x}$ has the normal distribution with mean μ and standard deviation $\sigma/\sqrt{n}$. So the probability is C that $\bar{x}$ lies between

$$\mu - z^* \frac{\sigma}{\sqrt{n}} \quad \text{and} \quad \mu + z^* \frac{\sigma}{\sqrt{n}}$$

This is exactly the same as saying that the unknown population mean μ lies between

$$\bar{x} - z^* \frac{\sigma}{\sqrt{n}} \quad \text{and} \quad \bar{x} + z^* \frac{\sigma}{\sqrt{n}}$$

That is, there is probability C that the interval $\bar{x} \pm z^*\sigma/\sqrt{n}$ contains μ. That is our confidence interval. The estimator of the unknown μ is $\bar{x}$ and the margin of error is $z^*\sigma/\sqrt{n}$.

CONFIDENCE INTERVAL FOR A POPULATION MEAN

Suppose that an SRS of size n is drawn from a population having unknown mean μ and known standard deviation σ. A level C confidence interval for μ is

$$\bar{x} \pm z^* \frac{\sigma}{\sqrt{n}}$$

where z^* is the upper $(1-C)/2$ critical value for the standard normal distribution. This interval is exact when the population distribution is normal and is approximately correct for large n in other cases.

EXAMPLE 6.3

A laboratory analyzes specimens of a pharmaceutical product to determine the concentration of the active ingredient. Such chemical analyses are not perfectly precise. Repeated measurements on the same specimen will give slightly different results. The results of repeated measurements follow a normal distribution quite closely. The analysis procedure has no bias, so that the mean μ of the population of all measurements is the true concentration in the specimen. The standard deviation of this distribution is a property of the analytical procedure and is known to be $\sigma = 0.0068$ grams per liter. The laboratory analyzes each specimen three times and reports the mean result.

Three analyses of one specimen give concentrations

.8403 .8363 .8447

Give a 99% confidence interval for the true concentration μ.

The sample mean of these readings is

$$\overline{x} = \frac{.8403 + .8363 + .8447}{3} = .8404$$

For 99% confidence, we see from Table D that $z^* = 2.576$. A 99% confidence interval for μ is therefore

$$\begin{aligned}\overline{x} \pm z^* \frac{\sigma}{\sqrt{n}} &= .8404 \pm 2.576 \frac{.0068}{\sqrt{3}} \\ &= .8404 \pm .0101 \\ &= (.8303, .8505)\end{aligned}$$

We are 99% confident that the true concentration lies between 0.8303 and 0.8505. ●

Suppose that a single measurement gave $x = 0.8404$, the same value that the sample mean took in Example 6.3. Repeating the calculation with $n = 1$ shows that the 99% confidence interval based on a single measurement is

$$\begin{aligned}\overline{x} \pm z^* \frac{\sigma}{\sqrt{1}} &= .8404 \pm (2.576)(.0068) \\ &= .8404 \pm .0175 \\ &= (.8229, .8579)\end{aligned}$$

The mean of three measurements gives a smaller margin of error and therefore a shorter interval than a single measurement. Figure 6.4 illustrates the gain from using three observations.

The argument leading to the form of confidence intervals for the population mean μ rests on the fact that the statistic $\overline{x}$ used to estimate μ has a normal distribution. Because many sample estimates have normal distributions (at least approximately), it is useful to notice that the confidence interval has the form

$$\text{estimate} \pm z^* \sigma_{\text{estimate}}$$

The estimate based on the sample is the center of the confidence interval. The margin of error is $z^* \sigma_{\text{estimate}}$. The desired confidence level determines z^* from Table D. The standard deviation of the estimate is found from

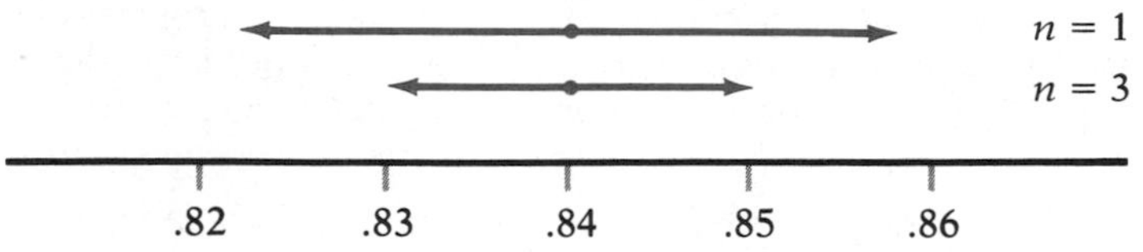

FIGURE 6.4 **Confidence intervals for $n = 3$ and $n = 1$ for Example 6.3.**

a knowledge of the sampling distribution in a particular case. When the estimate is $\overline{x}$ from an SRS, the standard deviation of the estimate is $\sigma/\sqrt{n}$.

How confidence intervals behave

The confidence interval $\overline{x} \pm z^*\sigma/\sqrt{n}$ for the mean of a normal population illustrates several important properties that are shared by all confidence intervals in common use. The user chooses the confidence level and the margin of error follows from this choice. High confidence is desirable and so is a small margin of error. High confidence says that our method almost always gives correct answers. A small margin of error says that we have pinned down the parameter quite precisely. There is a trade-off between the confidence level and the margin of error. To obtain higher confidence from the same data, you must be willing to accept a larger margin of error.

In the case of the confidence interval for a population mean, these facts reflect the behavior of the critical value z^*. A look at Figure 6.3 will convince you that z^* must be larger for higher confidence (larger C). Table D shows that this is indeed the case. If n and σ are unchanged, a larger z^* leads to a larger margin of error. On the other hand, increasing the sample size n reduces the margin of error for any fixed confidence level. The square root in the formula implies that we must multiply the number of observations by 4 in order to halve the margin of error. The standard deviation σ measures the variation in the population. You can think of the variation among individuals in the population as noise that obscures the average value μ. It is harder to pin down the mean μ of a highly variable population; that is why the margin of error of a confidence interval increases with σ.

EXAMPLE 6.4

Suppose that the laboratory in Example 6.3 is content with 90% confidence rather than 99%. Table D gives the critical value for 90% confidence as $z^* = 1.645$. The 90% confidence interval for μ based on three repeated measurements with mean $\overline{x} = .8404$ is

$$\begin{aligned}\overline{x} \pm z^* \frac{\sigma}{\sqrt{n}} &= .8404 \pm 1.645 \frac{.0068}{\sqrt{3}} \\ &= .8404 \pm .0065 \\ &= (.8339, .8469)\end{aligned}$$

Settling for 90% rather than 99% confidence has reduced the margin of error from ± 0.0101 to ± 0.0065. Figure 6.5 compares the two intervals.

Increasing the number of measurements from 3 to 12 will also reduce the margin of error of the 99% confidence interval in Example 6.3. Check that replacing $\sqrt{3}$ by $\sqrt{12}$ cuts the ± 0.0101 margin of error in half, because we now have four times as many observations. ●

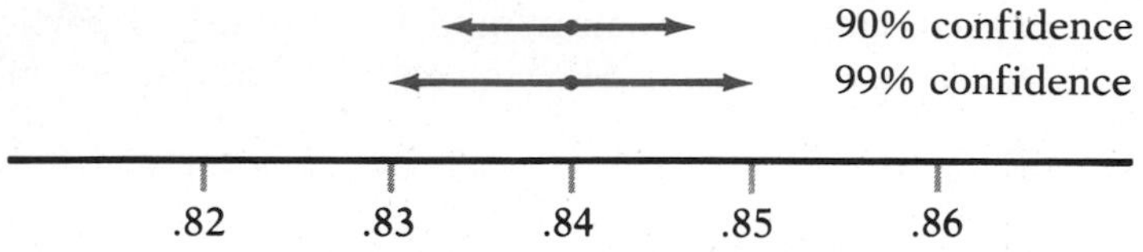

FIGURE 6.5 **90% and 99% confidence intervals for Example 6.4.**

Choosing the sample size A wise user of statistics never plans data collection without at the same time planning the inference. You can arrange to have both high confidence and a small margin of error. The margin of error of the confidence interval $\bar{x} \pm z^*\sigma/\sqrt{n}$ for the mean of a normal distribution is

$$z^* \frac{\sigma}{\sqrt{n}}$$

To obtain a desired margin of error m, just set this expression equal to m, substitute the value of z^* for your desired confidence level, and solve for the sample size n. The result is as follows:

SAMPLE SIZE FOR DESIRED MARGIN OF ERROR

The confidence interval for a population mean will have a specified margin of error m when the sample size is

$$n = \left(\frac{z^*\sigma}{m}\right)^2$$

This formula is not the proverbial free lunch. In practice, taking observations costs time and money. The required sample size may be impossibly expensive. Do notice once again that it is the size of the *sample* that determines the margin of error. The size of the *population* (as long as the population is much larger than the sample) does not influence the sample size we need.

EXAMPLE 6.5

A new customer of the laboratory of Example 6.3 wants results accurate to within ± 0.005 with 95% confidence. How many measurements must be averaged to comply with this request?

The desired margin of error is $m = 0.005$. For 95% confidence, Table D gives $z^* = 1.960$. (This is more exact than the $z^* = 2$ from the 68–95–99.7 rule.) Therefore

$$n = \left(\frac{z^*\sigma}{m}\right)^2 = \left[\frac{(1.96)(.0068)}{.005}\right]^2 = 7.1$$

Because 7 measurements will give a slightly wider interval than desired and 8 measurements a slightly narrower interval, the lab must take 8 measurements on each specimen to meet the customer's demand. (Always round *up* to the next higher whole number when finding n.) If this is too expensive, the lab can refuse the contract or raise the price. Signing a contract before doing the calculation of n would be foolish. ●

Some cautions We have already seen that small margins of error and high confidence can require large numbers of observations. You should also be keenly aware that *any formula for inference is correct only in specific circumstances*. If the government required statistical procedures to carry warning labels like those on drugs, most inference methods would have long labels indeed. Our handy formula $\bar{x} \pm z^*\sigma/\sqrt{n}$ for estimating a normal mean comes with the following list of warnings for the user:

- The data must be an SRS from the population. We are completely safe if we actually did a randomization and drew an SRS. We are not in great danger if the data can plausibly be thought of as independent observations from a population. That is the case in Examples 6.3 to 6.5, where we have in mind the population resulting from a very large number of repeated analyses of the same specimen.
- The formula is not correct for probability sampling designs more complex than an SRS. Correct methods for other designs are available. We will not discuss confidence intervals based on multistage or stratified samples. If you plan such samples, be sure that you (or your statistical consultant) know how to carry out the inference you desire.
- There is no correct method for inference from data haphazardly collected with bias of unknown size. Fancy formulas cannot rescue badly produced data.
- Because $\bar{x}$ is not resistant, outliers can have a large effect on the confidence interval. You should search for outliers and try to correct them or justify their removal before computing the interval. If the outliers cannot be removed, ask your statistical consultant about procedures that are not sensitive to outliers.
- If the sample size is small and the population is not normal, the true confidence level will be different from the value C used in computing the interval. Examine your data carefully for skewness and other signs of nonnormality. The interval relies only on the distribution of $\bar{x}$, which even for quite small sample sizes is much

closer to normal than that of the individual observations. When $n \geq 15$, the confidence level is not greatly disturbed by nonnormal populations unless extreme outliers or quite strong skewness are present. We will discuss this issue in more detail in the next chapter.

- You must know the standard deviation σ of the population. This unrealistic requirement renders the interval $\bar{x} \pm z^*\sigma/\sqrt{n}$ of little use in statistical practice. We will learn in the next chapter what to do when σ is unknown. If, however, the sample is large, the sample standard deviation s will be close to the unknown σ. The recipe $\bar{x} \pm z^*s/\sqrt{n}$ is then an approximate confidence interval for μ.

The most important caution concerning confidence intervals is a consequence of the first of these warnings. *The margin of error in a confidence interval covers only random sampling errors.* The margin of error is obtained from the sampling distribution and indicates how much error can be expected because of chance variation in randomized data production. Practical difficulties such as undercoverage and nonresponse in a sample survey can cause additional errors that may be larger than the random sampling error. Remember this unpleasant fact when reading the results of an opinion poll or other sample survey. The practical conduct of the survey influences the trustworthiness of its results in ways that are not included in the announced margin of error.

Every inference procedure that we will meet has its own list of warnings. Because many of the warnings are similar to those above, we will not print the full warning label each time. It is easy to state (from the mathematics of probability) conditions under which a method of inference is exactly correct. These conditions are *never* fully met in practice. For example, no population is exactly normal. Deciding when a statistical procedure should be used in practice often requires judgment assisted by exploratory analysis of the data. Mathematical facts are therefore only a part of statistics. The difference between statistics and mathematics can be stated thus: Mathematical theorems are true; statistical methods are often effective when used with skill.

Finally, you should understand what statistical confidence does not say. We are 95% confident that the mean SAT-M score for the California students in Example 6.1 lies between 452 and 470. This says that these numbers were calculated by a method that gives correct results in 95% of all possible samples. It does *not* say that the probability is 95% that the true mean falls between 452 and 470. No randomness remains after we draw a particular sample and get from it a particular interval. The true mean either is or is not between 452 and 470. Probability in its interpretation as long-term relative frequency makes no sense in this situation. The probability calculations of standard statistical inference describe how often the *method* gives correct answers.

SUMMARY

The purpose of a **confidence interval** is to estimate an unknown parameter with an indication of how accurate the estimate is and of how confident we are that the result is correct.

Any confidence interval has two parts: an interval computed from the data and a confidence level. The interval often has the form

$$\text{estimate} \pm \text{margin of error}$$

The **confidence level** states the probability that the method will give a correct answer. That is, if you use 95% confidence intervals often, in the long run 95% of your intervals will contain the true parameter value. You cannot know whether the result of applying a confidence interval to a particular set of data is correct.

A level C confidence interval for the mean μ of a normal population with known standard deviation σ, based on an SRS of size n, is given by

$$\bar{x} \pm z^* \frac{\sigma}{\sqrt{n}}$$

Here z^* is the **upper critical value** of the standard normal distribution for $p = (1 - C)/2$, given in Table D.

Other things being equal, the margin of error of a confidence interval decreases as

- the confidence level C decreases,
- the sample size n increases, and
- the population standard deviation σ decreases.

The sample size required to obtain a confidence interval of specified margin of error m for a normal mean is

$$n = \left(\frac{z^* \sigma}{m}\right)^2$$

where z^* is the critical value for the desired level of confidence.

A specific confidence interval recipe is correct only under specific conditions. The most important conditions concern the method used to produce the data. Other factors, such as the form of the population distribution, may also be important.

SECTION 6.1 EXERCISES

6.1 A closely contested presidential election in 1976 pitted Jimmy Carter against Gerald Ford. A poll taken immediately before the 1976 election showed that 51% of the sample intended to vote for Carter. The polling organization announced that they were 95% confident that the sample result was within ±2 points of the true percent of all voters who favored Carter.

(a) Explain in plain language to someone who knows no statistics what "95% confident" means in this announcement.
(b) The poll showed Carter leading. Yet the polling organization said the election was too close to call. Explain why.
(c) On hearing of the poll, a nervous politician asked, "What is the probability that over half the voters prefer Carter?" A statistician said in reply that this question not only can't be answered from the poll results, it doesn't even make sense to talk about such a probability. Explain why.

6.2 A *New York Times* poll on women's issues interviewed 1025 women and 472 men randomly selected from the United States excluding Alaska and Hawaii. The poll found that 47% of the women said they do not get enough time for themselves.

(a) The poll announced a margin of error of ±3 percentage points for 95% confidence in conclusions about women. Explain to someone who knows no statistics why we can't just say that 47% of all adult women do not get enough time for themselves.
(b) Then explain clearly what "95% confidence" means.
(c) The margin of error for results concerning men was ±4 percentage points. Why is this larger than the margin of error for women?

6.3 A student reads that a 95% confidence interval for the mean SAT math score of California high school seniors is 452 to 470. Asked to explain the meaning of this interval, the student says "95% of California high school seniors have SAT math scores between 452 and 470." Is the student right? Justify your answer.

6.4 A radio talk show invites listeners to enter a dispute about a proposed pay increase for city council members. "What yearly pay do you think council members should get? Call us with your number." In all, 958 people call. The mean pay they suggest is $\bar{x} = \$8740$ per year, and the standard deviation of the responses is $s = \$1125$. For a large sample such as this, s is very close to the unknown population σ. The station calculates that the 95% confidence interval for the mean pay μ that all citizens would propose for council members is \$8669 to \$8811. Is this result trustworthy? Explain your answer.

6.5 You measure the weights of a random sample of 24 male runners. The sample mean is $\bar{x} = 60$ kilograms (kg). Suppose that the standard deviation of the population is known to be $\sigma = 5$ kg.

(a) What is $\sigma_{\bar{x}}$, the standard deviation of $\bar{x}$?
(b) Give a 95% confidence interval for μ, the mean of the population

from which the sample is drawn. Are you quite sure that the average weight of the population of runners is less than 65 kg?

6.6 Crop researchers plant 50 plots with a new variety of corn. The average yield for these plots is $\bar{x}$=130 bushels per acre. Assume that $\sigma = 10$ bushels per acre.

(a) Find the 90% confidence interval for the mean yield μ for this variety of corn.

(b) Find the 95% confidence interval.

(c) Find the 99% confidence interval.

(d) How do the margins of error in (a), (b), and (c) change as the confidence level increases?

6.7 Find a 99% confidence interval for the mean weight μ of the population of male runners in Exercise 6.5. Is the 99% confidence interval wider or narrower than the 95% interval found in Exercise 6.5? Explain in plain language why this is true.

6.8 Suppose that the crop researchers in Exercise 6.6 obtained the same value of $\bar{x}$ from a sample of 100 plots rather than 50.

(a) Compute the 95% confidence interval for the mean yield μ.

(b) Is the margin of error larger or smaller than the margin of error found for the sample of 50 plots in Exercise 6.6? Explain in plain language why the change occurs.

(c) Will the 90% and 99% intervals for a sample of size 100 be wider or narrower than those for $n = 50$? (You need not actually calculate these intervals.)

6.9 A test for the level of potassium in the blood is not perfectly precise. Moreover, the actual level of potassium in a person's blood varies slightly from day to day. Suppose that repeated measurements for the same person on different days vary normally with $\sigma = 0.2$.

(a) Julie's potassium level is measured once. The result is $x = 3.2$. Give a 90% confidence interval for her mean potassium level.

(b) If three measurements were taken on different days and the mean result is $\bar{x} = 3.2$, what is a 90% confidence interval for Julie's mean blood potassium level?

6.10 The Acculturation Rating Scale for Mexican Americans (ARSMA) is a psychological test developed to measure the degree of Mexican/Spanish versus Anglo/English acculturation of Mexican Americans. The distribution of ARSMA scores in a population used to develop the test was approximately normal, with mean 3.0 and standard deviation 0.8. A further study gave ARSMA to 42 first-generation Mexican Americans. The mean of their scores is $\bar{x} = 2.13$. If the standard deviation for the first-generation population is also $\sigma = 0.8$, give a 95% confidence interval for the mean ARSMA score for first-generation Mexican Americans.

6.11 Here are measurements (in millimeters) of a critical dimension on a sample of auto engine crankshafts:

224.120	224.001	224.017	223.982	223.989	223.961
223.960	224.089	223.987	223.976	223.902	223.980
224.098	224.057	223.913	223.999		

The data come from a production process that is known to have standard deviation $\sigma = 0.060$ mm. A normal quantile plot shows that the distribution is very close to normal. The process mean is supposed to be $\mu = 224$ mm but can drift away from this target during production. Give a 95% confidence interval for the process mean at the time these crankshafts were produced.

6.12 Here are the Degree of Reading Power (DRP) scores for a sample of 44 third-grade students. We first examined these data in Chapter 1, where a normal quantile plot (Figure 1.25 on page 84) showed that the distribution is close to normal except for slightly short tails.

40	26	39	14	42	18	25	43	46	27	19
47	19	26	35	34	15	44	40	38	31	46
52	25	35	35	33	29	34	41	49	28	52
47	35	48	22	33	41	51	27	14	54	45

Suppose that the standard deviation of the population of DRP scores is known to be $\sigma = 11$. Give a 99% confidence interval for the population mean score.

6.13 Researchers planning a study of the reading ability of third-grade children want to obtain a 95% confidence interval for the population mean score on a reading test, with margin of error no greater than 5 points. They carry out a small pilot study to estimate the variability of test scores. The sample standard deviation is $s = 12$ points in the pilot study, so in preliminary calculations the researchers take the population standard deviation to be $\sigma = 12$.

(a) The study budget will allow as many as 100 students. Calculate the margin of error of the 95% confidence interval for the population mean based on $n = 100$.

(b) There are many other demands on the research budget. If all of these demands were met, there would be funds to measure only 10 children. What is the margin of error of the confidence interval based on $n = 10$ measurements?

(c) Find the smallest value of n that would satisfy the goal of a 95% confidence interval with margin of error 5 or less. Is this sample size within the limits of the budget?

6.14 How large a sample of the crankshafts in Exercise 6.11 would be needed to estimate the mean μ within ± 0.020 mm with 95% confidence?

6.15 In Exercises 6.6 and 6.8, we compared confidence intervals based on corn yields from 50 and 100 small plots of ground. These are large sample sizes for an agricultural field experiment. How large a sample is required to estimate the mean yield within ± 5 bushels per acre with 90% confidence?

6.16 To assess the accuracy of a laboratory scale, a standard weight known to weigh 10 grams (g) is weighed repeatedly. The scale readings are normally distributed with unknown mean (this mean is 10 grams if the scale has no bias). The standard deviation of the scale readings is known to be 0.0002 g.

(a) The weight is weighed five times. The mean result is 10.0023 g. Give a 98% confidence interval for the mean of repeated measurements of the weight.

(b) How many measurements must be averaged to get a margin of error of ± 0.0001 with 98% confidence?

6.17 The 1990 census "long form" asked the total 1989 income of the householder, the person in whose name the dwelling unit was owned or rented. This census form was sent to a random sample of 17% of the nation's households. Suppose (alas, it is too simple to be true) that the households that returned the long form are an SRS of the population of all households in each district. In Middletown, a city of 40,000 persons, 2621 householders reported their income. The mean of the responses was $\bar{x} = \$23,453$, and the standard deviation was $s = \$8721$. The sample standard deviation for so large a sample will be very close to the population standard deviation σ. Use these facts to give an approximate 99% confidence interval for the 1989 mean income of Middletown householders.

6.18 The Gallup Poll asked 1571 adults what they considered to be the most serious problem facing the nation's public schools; 30% said drugs. This sample percent is an estimate of the percent of all adults who think that drugs are the schools' most serious problem. The news article reporting the poll result adds, "The poll has a margin of error—the measure of its statistical accuracy—of three percentage points in either direction; aside from this imprecision inherent in using a sample to represent the whole, such practical factors as the wording of questions can affect how closely a poll reflects the opinion of the public in general" (*The New York Times*, August 31, 1987).

The Gallup Poll uses a complex multistage sample design, but the sample percent has approximately a normal distribution. Moreover, it is standard practice to announce the margin of error for a 95% confidence interval unless a different confidence level is stated.

(a) The announced poll result was $30\% \pm 3\%$. Can we be certain that the true population percent falls in this interval?

(b) Explain to someone who knows no statistics what the announced result $30\% \pm 3\%$ means.

(c) This confidence interval has the same form we have met earlier:

$$\text{estimate} \pm z^* \sigma_{\text{estimate}}$$

(Actually σ is estimated from the data, but we ignore this for now.) What is the standard deviation σ_{estimate} of the estimated percent?

(d) Does the announced margin of error include errors due to practical problems such as undercoverage and nonresponse?

6.19 When the statistic that estimates an unknown parameter has a normal distribution, a confidence interval for the parameter has the form

$$\text{estimate} \pm z^* \sigma_{\text{estimate}}$$

In a complex sample survey design, the appropriate unbiased estimate of the population mean and the standard deviation of this estimate may require elaborate computations. But when the estimate is known to have a normal distribution and its standard deviation is given, we can calculate a confidence interval for μ from complex sample designs without knowing the formulas that led to the numbers given.

A report based on the Current Population Survey estimates the 1991 median weekly earnings of families of wage and salary workers as $664 and also estimates that the standard deviation of this estimate is $3.50. The Current Population Survey uses an elaborate multistage sampling design to select a sample of about 60,000 households. The sampling distribution of the estimated median income is approximately normal. Give a 95% confidence interval for the 1991 median weekly earnings of all families of wage and salary workers.

6.20 As we prepare to take a sample and compute a 95% confidence interval, we know that the probability that the interval we compute will cover the parameter is 0.95. That's the meaning of 95% confidence. If we use several such intervals, however, our confidence that *all* give correct results is less than 95%.

In an agricultural field trial a corn variety is planted in seven separate locations, which may have different mean yields due to differences in soil and climate. At the end of the experiment, seven independent 95% confidence intervals will be calculated, one for the mean yield at each location.

(a) What is the probability that every one of the seven intervals covers the true mean yield at its location? This probability (expressed as a percent) is our overall confidence level for the seven simultaneous statements.

(b) What is the probability that at least six of the seven intervals cover the true mean yields?

6.2 TESTS OF SIGNIFICANCE

Confidence intervals are one of the two most common types of formal statistical inference. They are appropriate when our goal is to estimate a population parameter. The second common type of inference is directed at a quite different goal: to assess the evidence provided by the data in favor of some claim about the population. An example will illustrate the reasoning we use.

EXAMPLE 6.6

Does the result of the 1970 draft lottery provide strong evidence that the lottery was not truly random? We examined this question in Chapter 2. A scatterplot of draft numbers against birth dates (Figure 6.6) does not show a convincing nonrandom pattern. A statistic that measures the strength of the association between draft number (1 to 366) and birth date (1 to 366) is the correlation coefficient. The computer tells us that $r = -0.226$ for the 1970 lottery. Is this the evidence we seek?

Formal question: Suppose for the sake of argument that the lottery were truly random. What is the probability that a random lottery would produce an r at least as far from 0 as the observed $r = -0.226$?

Answer: The probability that a random lottery will produce an r this far from 0 is less than 0.001.

Conclusion: Because an r as far from 0 as that observed in 1970 would almost never occur in a random lottery, we have strong evidence that the 1970 lottery was not random. ●

In a random assignment of draft numbers to birth dates, we expect the correlation to be close to 0. The observed correlation for the 1970 lottery was $r = -0.226$, showing that men born later in the year tended to get lower draft numbers. Common sense cannot decide if $r = -0.226$ means that the lottery was not random. After all, the correlation in a random lottery will almost never be exactly 0. Perhaps $r = -0.226$ is within the range of values that could plausibly occur. As an aid to answering the informal question "Is this good evidence of a nonrandom lottery?" we state a formal question about probability. We ask just how often a random lottery will produce an r as far from 0 as the r observed in 1970. The answer: less than once in a thousand tries.[1] This convinces us that the 1970 lottery was biased.

Be sure you understand why this probability calculation is convincing. There are two possible explanations for that notorious $r = -0.226$.

1. The lottery was random, and by bad luck a very unlikely outcome occurred.
2. The lottery was biased, so the outcome is about what would be expected from such a lottery.

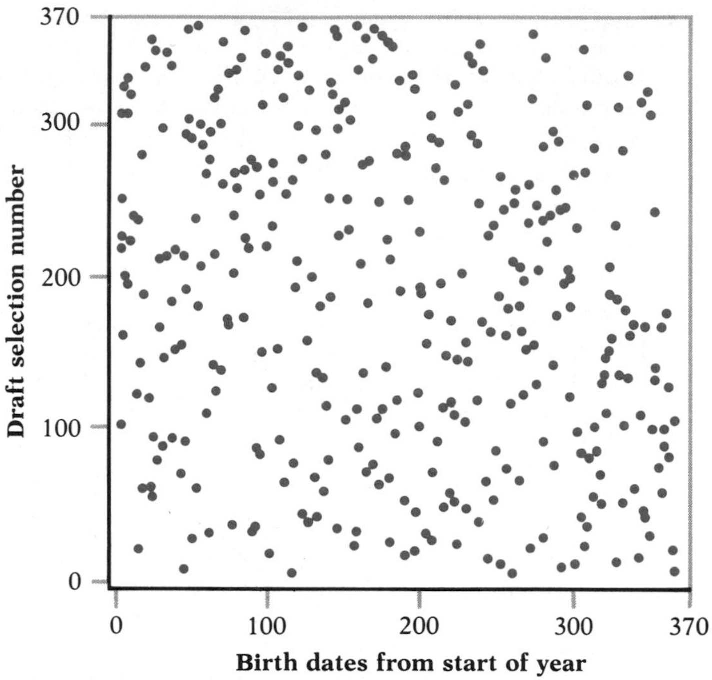

FIGURE 6.6 Scatterplot of birth date (1 to 366) and draft number (1 to 366) for the 1970 draft lottery.

We cannot be certain that the first explanation is untrue. The 1970 results *could* be due to chance alone. But the probability that such results will occur by chance in a random lottery is so small (0.001) that we are quite confident that the second explanation is correct. Here is a second example of this reasoning.

EXAMPLE 6.7

Cobra Cheese Company buys milk from several suppliers as the essential raw material for its cheese. Cobra suspects that some producers are adding water to their milk to increase their profits. Excess water can be detected by determining the freezing point of the milk. The freezing temperature of natural milk varies normally, with a mean of $\mu = -0.545°$ Celsius (C) and a standard deviation of $\sigma = 0.008°$ C. Added water raises the freezing temperature toward 0°, the freezing point of water. Cobra's laboratory manager measures the freezing temperature of five consecutive lots of milk from one producer. The mean measurement is $\bar{x} = -0.538°$ C. Is this good evidence that the producer is adding water to the milk?

Formal question: Suppose for the sake of argument that no water has been added, so that the mean freezing point of the population of all milk from this producer is $\mu = -0.545°$ C. What is the probability that five measurements would give a sample mean as high as $-0.538°$ C or higher?

Answer: An outcome this high or higher has probability 0.025 if natural milk is measured.

Conclusion: Because a mean freezing temperature as high as that observed would occur only 2.5 times per 100 samples of natural milk, there is evidence that the producer is watering the milk. ●

The evidence in Example 6.7 is less strong than the evidence in Example 6.6, because the observed effect is more likely to occur simply by chance.

The nature of significance testing

The reasoning used in Examples 6.6 and 6.7 is codified in *tests of significance*.* In both cases, we ask whether some effect is present—whether the draft lottery is biased in Example 6.6 and whether the freezing point of the milk is elevated in Example 6.7. To do this, we begin by supposing for the sake of argument that the effect is *not* present. In Example 6.6 we suppose that the lottery is random, not biased. In Example 6.7 we suppose that the freezing point of the suspect milk is the same as that of natural milk. We then ask whether the data provide evidence against the supposition we made. If so, we have evidence in favor of the effect we are seeking. The first step in a test of significance is to state a claim that we will try to find evidence *against*.

NULL HYPOTHESIS

The statement being tested in a test of significance is called the null hypothesis. The test of significance is designed to assess the strength of the evidence against the null hypothesis. Usually the null hypothesis is a statement of "no effect" or "no difference."

Stating hypotheses The term "null hypothesis" is abbreviated H_0. A null hypothesis is a statement about a population, expressed in terms of some parameter or parameters. For example, if μ is the mean freezing point of all milk shipped to the cheesemaker by the milk producer in Example 6.7,

*The reasoning of significance tests has been used sporadically at least since Laplace in the 1820s. A clear exposition by the English statistician Francis Y. Edgeworth (1845–1926) appeared in 1885 in the context of a study of social statistics. Edgeworth introduced the term "significant" as meaning "corresponds to a real difference in fact."

our null hypothesis is

$$H_0: \mu = -.545$$

This says that the mean freezing point of the milk shipped is the same as for unwatered milk.

alternative hypothesis

It is convenient to also give a name to the statement we hope or suspect is true instead of H_0. This is called the *alternative hypothesis* and is abbreviated by H_a. In Example 6.7, the alternative hypothesis states that the mean freezing point of the suspect milk is higher than that of unwatered milk. We write this as

$$H_a: \mu > -.545$$

The hypotheses in Example 6.6 refer to the randomization procedure used in the 1970 draft lottery, or if you prefer, to the hypothetical population of all lottery outcomes that would arise from repeated use of this procedure. Let ρ (the Greek letter rho) stand for the correlation between birth date and draft number in this population, averaged over all of the outcomes. The null hypothesis states that the lottery is truly random, so that on the average there is no correlation. That is,

$$H_0: \rho = 0$$

The alternative is a biased lottery with

$$H_a: \rho \neq 0$$

Hypotheses always refer to some population, not to a particular outcome. For this reason, we must state H_0 and H_a in terms of population parameters.

one-sided and two-sided alternatives

Because H_a expresses the effect that we hope to find evidence *for*, we often begin with H_a and then set up H_0 as the statement that the hoped-for effect is not present. Stating H_a is often the more difficult task. It is not always clear, in particular, whether H_a should be *one-sided* or *two-sided*. In the draft lottery example, the alternative H_a: $\rho \neq 0$ is two-sided. That is, it allows the lottery to give men born later in the year either higher ($\rho > 0$) or lower ($\rho < 0$) draft numbers than men with earlier birth dates. This H_a simply says that the lottery procedure is biased without specifying the direction of the bias. The alternative H_a: $\mu > -0.545$ in the cheesemaking example is one-sided. Because watering milk always increases the freezing point, we are only interested in detecting an upward shift in μ. The alternative hypothesis should express the hopes or suspicions we bring to the data. It is cheating to first look at the data and then frame H_a to fit what the data show. Thus the fact that the 1970 draft lottery produced a negative correlation between birth date and draft number should not influence our choice of H_a. If you do not have a specific direction firmly in mind in advance, use a two-sided alternative.

The choice of the hypotheses in Example 6.7 as

$$H_0: \mu = -.545$$
$$H_a: \mu > -.545$$

deserves a final comment. The cheesemaker is not concerned with the possibility that the milk may have a *lower*-than-normal freezing point, perhaps due to an unusually high cream content. However, we can allow for the possibility that μ is less than $-0.545°$ C by including this case in the null hypothesis. Then we would write

$$H_0: \mu \leq -.545$$
$$H_a: \mu > -.545$$

This statement is logically satisfying because the hypotheses account for all possible values of μ. However, only the parameter value in H_0 that is closest to H_a influences the form of the test in all common significance testing situations. We will therefore take H_0 to be the simpler statement that the parameter has a specific value, in this case H_0: $\mu = -0.545$.

The form of the test We will learn specific recipes for the form of significance tests in a number of common situations. Here are some principles that apply to most tests and that help in understanding the form of tests:

- The test is based on a statistic that estimates the parameter that appears in the hypotheses. Usually this is the same estimate we would use in a confidence interval for the parameter. When H_0 is true, we expect the estimate to take a value near the parameter value specified by H_0.
- Values of the estimate far from the parameter value specified by H_0 give evidence against H_0. The alternative hypothesis determines which directions count against H_0.

EXAMPLE 6.8

In the draft lottery, Example 6.6, the hypotheses concern ρ, the population correlation,

$$H_0: \rho = 0$$
$$H_a: \rho \neq 0$$

The test is based on the sample correlation r, which estimates the population correlation. When H_0 is true, we expect r to be near 0. Because H_a is two-sided, values of r away from 0 in either direction give evidence against the null hypothesis.

In the Cobra Cheese case, Example 6.7, the hypotheses are stated in terms of the population mean freezing point:

$$H_0: \mu = -.545$$
$$H_a: \mu > -.545$$

The estimate of μ is the sample mean $\bar{x}$. Because H_a is one-sided on the high side, only large values of $\bar{x}$ count as evidence against the null hypothesis. ●

***P*-values** A test of significance assesses the evidence against the null hypothesis in terms of probability. If the observed outcome is unlikely under the supposition that the null hypothesis is true, but is more probable if the alternative hypothesis is true, that outcome is evidence against H_0 in favor of H_a. The less probable the outcome is, the stronger the evidence that H_0 is false. Usually, any *specific* outcome has low probability. A random draft lottery is unlikely to give exactly $r = 0$, but if we observe $r = 0$, we certainly do not have evidence against the null hypothesis that the lottery is random. The alternative hypothesis determines what kinds of outcomes count as evidence against H_0 and in favor of H_a. In the draft lottery case, observed correlations r away from 0 in either direction count against the hypothesis of a random lottery. The farther from 0 the observed r is, the stronger the evidence. The probability that measures the strength of the evidence that the 1970 lottery was biased is therefore the probability that a random lottery would produce an *r at least as far from 0* as the 1970 lottery did. This is

$$P(r \leq -.226 \text{ or } r \geq .226)$$

We calculate this probability assuming that H_0 is true.

In general, a test of significance finds the probability of getting an outcome *as extreme or more extreme than the actually observed outcome.* "Extreme" means "far from what we would expect if H_0 were true." The direction or directions that count as "far from what we would expect" are determined by H_a as well as H_0. In the draft lottery example, an observed r away from 0 in either direction is evidence of a nonrandom lottery, because H_a is two-sided. In the cheesemaking example, we want to know if the mean freezing point has been raised, a one-sided H_a. So the evidence against H_0 is measured by the probability

$$P(\bar{x} \geq -.538)$$

that the sample mean freezing point would be *as high or higher* than the observed value. This probability is calculated under the assumption that the population mean remains at $\mu = -0.545$.

> ***P*-VALUE**
>
> The probability, computed assuming that H_0 is true, that the test statistic would take a value as extreme or more extreme than that actually observed is called the *P*-value of the test. The smaller the *P*-value, the stronger the evidence against H_0 provided by the data.

Computer software that carries out tests of significance usually calculates the *P*-value for us. In some cases we can find *P*-values from our knowledge of sampling distributions.

EXAMPLE 6.9

In Example 6.7 the observations are an SRS of size $n = 5$ from a normal population with $\sigma = 0.008$. The observed average freezing point was $\bar{x} = -0.538$. The P-value for testing

$$H_0: \mu = -.545$$
$$H_a: \mu > -.545$$

is therefore

$$P(\bar{x} \geq -.538)$$

calculated assuming that H_0 is true. When H_0 is true, $\bar{x}$ has the normal distribution with

$$\mu_{\bar{x}} = \mu = -.545$$
$$\sigma_{\bar{x}} = \frac{\sigma}{\sqrt{n}} = \frac{.008}{\sqrt{5}}$$

The P-value is found by a normal probability calculation as follows:

$$\begin{aligned} P(\bar{x} \geq -.538) &= P\left(\frac{\bar{x} - (-.545)}{.008/\sqrt{5}} \geq \frac{-.538 - (-.545)}{.008/\sqrt{5}}\right) \\ &= P(Z \geq 1.96) \\ &= 1 - .9750 = .025 \end{aligned}$$

This is the value that was reported in Example 6.7. Figure 6.7 illustrates the P-value in terms of the sampling distribution of $\bar{x}$ when H_0 is true. ●

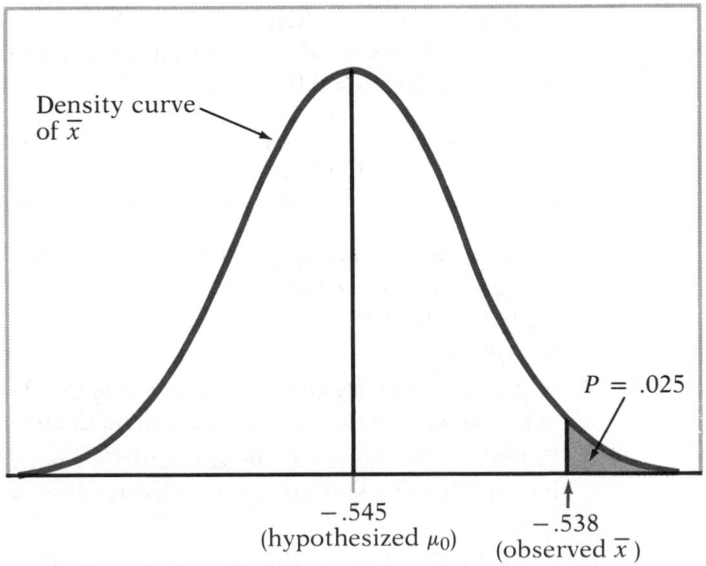

FIGURE 6.7 The P-value for Example 6.9. The P-value here is the probability (when H_0 is true) that $\bar{x}$ takes a value as large or larger than the actually observed value.

Statistical significance One final step is sometimes taken to assess the evidence against H_0. We can compare the P-value we obtained to a fixed value that we regard as decisive. This amounts to announcing in advance how much evidence against H_0 we will insist on. The decisive value of P is called the *significance level*. It is denoted by α, the Greek letter alpha. If we choose $\alpha = 0.05$, we are requiring that the data give evidence against H_0 so strong that it would happen no more than 5% of the time (1 time in 20) when H_0 is true. If we choose $\alpha = 0.01$, we are insisting on stronger evidence against H_0, evidence so strong that it would appear only 1% of the time (1 time in 100) if H_0 is in fact true.

significance level

> STATISTICAL SIGNIFICANCE
>
> If the P-value is as small or smaller than α, we say that the data are *statistically significant at level* α.

"Significant" in the statistical sense does not mean "important." The original meaning of the word is "signifying something." In statistics the term is used to indicate only that the evidence against the null hypothesis reached the standard set by α. Significance at level 0.01 is often expressed by the statement, "The results were significant ($P < 0.01$)." Here P stands for the P-value. The P-value is more informative than a statement of significance, because we can then assess significance at any level we choose. For example, a result with $P = 0.03$ is significant at the $\alpha = 0.05$ level but not significant at the $\alpha = 0.01$ level.

Tests of significance A test of significance is a recipe for assessing the significance of the evidence provided by data against a null hypothesis. The steps common to all tests of significance are as follows:

1. State the *null hypothesis* H_0 and the *alternative hypothesis* H_a. The test is designed to assess the strength of the evidence against H_0; H_a is the statement that we will accept if the evidence enables us to reject H_0.
2. (Optional) Specify the *significance level* α. This states how much evidence against H_0 we will regard as decisive.
3. Calculate the value of the *test statistic* on which the test will be based. This is a statistic that measures how well the data conform to H_0.
4. Find the *P-value* for the observed data. This is the probability, calculated assuming that H_0 is true, that the test statistic will weigh against H_0 at least as strongly as it does for the observed

data. If the P-value is less than or equal to α, the test result is *statistically significant at level* α.

We will learn the details of many tests of significance in the following chapters. In most cases, Steps 3 and 4 of our outline are almost automatic once you have practiced a bit. The proper test statistic is determined by the hypotheses and the data collection design, and its numerical value is found by computer software or with a calculator. The computation of the P-value is done by computer software or from tables. The computer will not formulate your hypotheses for you, however. Nor will it decide if significance testing is appropriate or help you to interpret the P-value that it presents to you. The reasoning of significance tests is a bit subtle. We have looked at this reasoning with little attention to the details of carrying out a test. Next we will examine the details of a simple significance test, the one that is appropriate in the cheesemaking example.

Tests for a population mean

We have an SRS of size n drawn from a normal population with unknown mean μ. We want to test the hypothesis that μ has a specified value. Call the specified value μ_0. The null hypothesis is

$$H_0: \mu = \mu_0$$

The test is based on the sample mean $\bar{x}$. Because normal calculations require standardized variables, we will use as our test statistic the *standardized* sample mean z,

$$z = \frac{\bar{x} - \mu_0}{\sigma/\sqrt{n}}$$

The statistic z has the standard normal distribution when H_0 is true. If the alternative hypothesis is one-sided on the high side

$$H_a: \mu > \mu_0$$

then the P-value is the probability that a standard normal random variable Z takes a value at least as large as the observed z. That is,

$$P = P(Z \geq z)$$

Example 6.9 applies this test. There, $\mu_0 = -0.545$, the standardized sample mean was $z = 1.96$, and the P-value was $P(Z \geq 1.96)$.

Similar reasoning applies when the alternative hypothesis states that the true μ lies below the hypothesized μ_0 (one-sided). When H_a states that μ is simply unequal to μ_0 (two-sided), values of z away from 0 in either direction count against the null hypothesis. The P-value is the probability that a standard normal Z is at least as far from 0 as the observed z. For example, if $z = 1.7$ is observed, the two-sided P-value is the probability that $Z \leq -1.7$ or $Z \geq 1.7$. Because the standard normal distribution is symmetric, we calculate this probability by finding $P(Z \geq 1.7)$ and *doubling* it:

$$\begin{aligned} P(Z \leq -1.7 \text{ or } Z \geq 1.7) &= 2P(Z \geq 1.7) \\ &= 2(1 - .9554) = .0892 \end{aligned}$$

We would make exactly the same calculation if we observed $z = -1.7$. It is the absolute value $|z|$ that matters, not whether z is positive or negative. Here is a statement of the test in general terms.

z TEST FOR A POPULATION MEAN

To test the hypothesis H_0: $\mu = \mu_0$ based on an SRS of size n from a population with unknown mean μ and known standard deviation σ, compute the test statistic

$$z = \frac{\bar{x} - \mu_0}{\sigma/\sqrt{n}}$$

In terms of a standard normal random variable Z, the P-value for a test of H_0 against

$$\begin{aligned} H_a&: \mu > \mu_0 \quad \text{is} \quad P(Z \geq z) \\ H_a&: \mu < \mu_0 \quad \text{is} \quad P(Z \leq z) \\ H_a&: \mu \neq \mu_0 \quad \text{is} \quad 2P(Z \geq |z|) \end{aligned}$$

These P-values are exact if the population distribution is normal and approximately correct for large n in other cases.

EXAMPLE 6.10

Do middle-aged male executives have different average blood pressure than the general population? The National Center for Health Statistics reports that the mean systolic blood pressure for males 35 to 44 years of age is 128 and the standard deviation in this population is 15. The medical director of a company looks at the medical records of 72 company executives in this age group and finds that the mean systolic blood pressure in this sample is $\bar{x} = 126.07$. Is this evidence that executive blood pressures differ from the national average?

The null hypothesis is "no difference" from the national mean $\mu_0 = 128$. The alternative is two-sided, because the medical director did not have a particular direction in mind before examining the data. So the hypotheses about the unknown mean μ of the executive population are

$$\begin{aligned} H_0&: \mu = 128 \\ H_a&: \mu \neq 128 \end{aligned}$$

As usual in this chapter, we make the unrealistic assumption that the population standard deviation is known, in this case that executives have the same $\sigma = 15$ as the general population of middle-aged males. The z test requires that the 72 executives in the sample are an SRS from the population of all middle-aged male executives in the company. We check this assumption by asking how the data were produced. If medical records are available only for executives with recent medical problems, for example, the data are of little value for our purpose. It turns out that all executives are given a free annual medical exam and that the medical director selected 72 exam results at random.

We can now proceed to compute the test statistic

$$z = \frac{\bar{x} - \mu_0}{\sigma/\sqrt{n}} = \frac{126.07 - 128}{15/\sqrt{72}}$$
$$= -1.09$$

Figure 6.8 illustrates the P-value, which is the probability that a standard normal variable Z takes a value at least 1.09 away from 0. From Table A we find that this probability is

$$2P(Z \geq |-1.09|) = 2P(Z \geq 1.09)$$
$$= 2(1 - 0.8621) = .2758$$

That is, more than 27% of the time an SRS of size 72 from the general male population would have a mean blood pressure at least as far from 128 as that of the executive sample. The observed $\bar{x} = 126.07$ is therefore not good evidence that executives differ from other men. ●

The data in Example 6.10 do *not* establish that the mean blood pressure μ for this company's young male executives is 128. We sought ev-

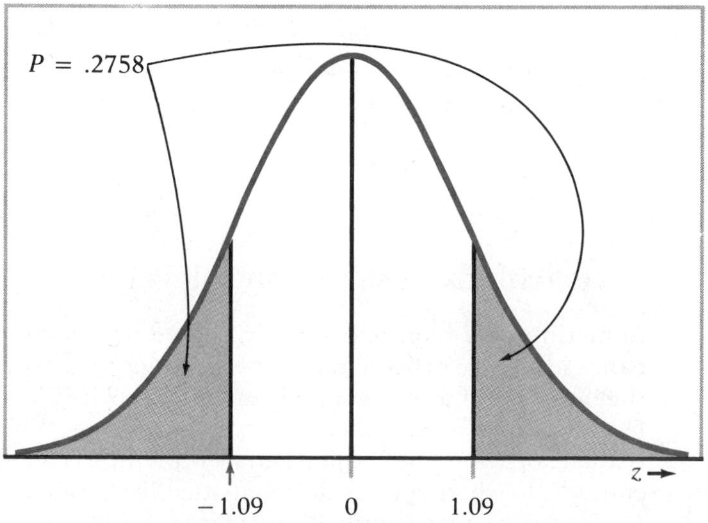

FIGURE 6.8 The *P*-value for the two-sided test in Example 6.10.

idence that μ differed from 128 and failed to find convincing evidence. That is all we can say. No doubt the mean blood pressure of the entire executive population is not exactly equal to 128. A large enough sample would give evidence of the difference, even if it is very small. Tests of significance assess the evidence *against* H_0. If the evidence is strong, we can confidently reject H_0 in favor of the alternative. Failing to find evidence against H_0 means only that the data are consistent with H_0, not that we have clear evidence that H_0 is true.

EXAMPLE 6.11

In a discussion of SAT scores, someone comments: "Because only a minority of high school students take the test, the scores overestimate the ability of typical high school seniors. The mean SAT mathematics score is about 475, but I think that if all seniors took the test, the mean score would be no more than 450." You gave the test to an SRS of 500 seniors from California (Example 6.1). These students had a mean score of $\bar{x} = 461$. Is this good evidence against the claim that the mean for all California seniors is no more than 450? The hypotheses are

$$H_0: \mu = 450$$
$$H_a: \mu > 450$$

As in Example 6.1, we assume that $\sigma = 100$. The z statistic is

$$z = \frac{\bar{x} - \mu_0}{\sigma/\sqrt{n}} = \frac{461 - 450}{100/\sqrt{500}} = 2.46$$

Because H_a is one-sided on the high side, large values of z count against H_0. From Table A, we find that the P-value is

$$P(Z \geq 2.46) = 1 - .9931 = .0069$$

Figure 6.9 illustrates this P-value. A mean score as large as that observed would occur fewer that seven times in 1000 samples if the population mean were 450. This is convincing evidence that the mean SAT-M score for all California high school seniors is higher than 450. ●

Tests with fixed significance level

Sometimes we require a specific degree of evidence, stated as a significance level α, in order to reject the null hypothesis. In terms of the P-value, the outcome of a test is significant at level α if $P \leq \alpha$. Significance at any level is easy to assess once we have the P-value. When we do not use statistical software, the P-value can be difficult to calculate. Fortunately, we can decide whether a result is statistically significant without calculating P. The following example illustrates how to assess significance at a fixed

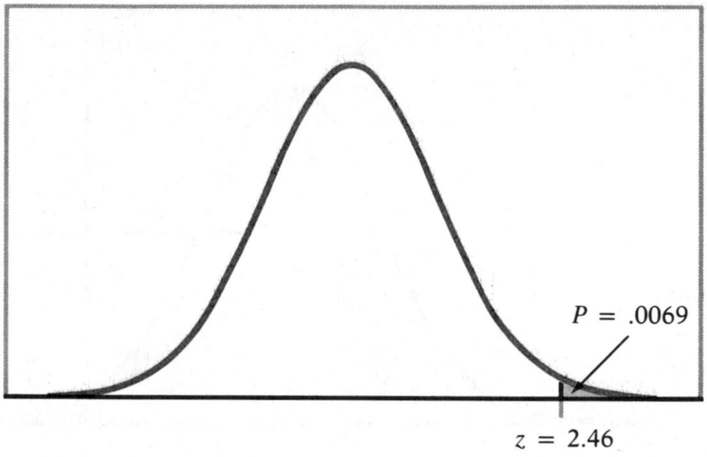

FIGURE 6.9 The *P*-value for the one-sided test in Example 6.11.

level α by using a table of critical values, the same table we used to obtain confidence intervals.

EXAMPLE 6.12

In Example 6.11, we examined whether the mean SAT-M score of California high school seniors is higher than 450. The hypotheses are

$$H_0: \mu = 450$$
$$H_a: \mu > 450$$

Is the evidence against H_0 statistically significant at the 1% level?

Compute the standardized value $z = 2.46$ of $\bar{x}$ as before. The *P*-value is the area to the right of 2.46 under the standard normal curve, shown in Figure 6.9. To be significant at the 1% level, this area must be no more than 1%. That is, z must lie in the upper 1% of the standard normal distribution. So to determine significance, we need only compare the observed $z = 2.46$ with the upper 0.01 critical value $z^* = 2.326$ from Table D that marks off the upper 1%. Because z is larger than z^*, the observed $\bar{x}$ is significant at level $\alpha = 0.01$. Figure 6.10 illustrates the procedure. ●

upper p critical value

The number z^* with probability p falling to the right of it under the standard normal density curve is the *upper p critical value* for the standard normal distribution. Table D gives these critical values for several choices of p. We made use of upper critical values for confidence intervals. Example 6.12 shows how they are used in tests with a fixed level α with a one-sided alternative hypothesis. Similar reasoning applies to other choices of the alternative hypothesis. Here is a summary of the procedures.

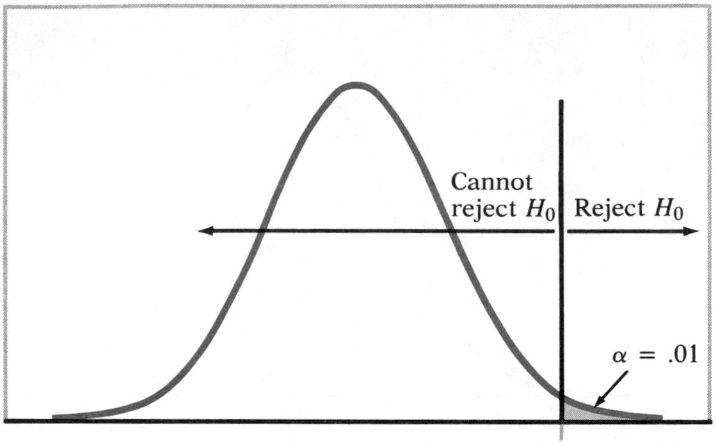

FIGURE 6.10 The level $\alpha = 0.01$ one-sided test in Example 6.12.

> **FIXED SIGNIFICANCE LEVEL z TESTS FOR A POPULATION MEAN**
>
> To test the hypothesis H_0: $\mu = \mu_0$ based on an SRS of size n from a population with unknown mean μ and known standard deviation σ, compute the z test statistic
>
> $$z = \frac{\bar{x} - \mu_0}{\sigma/\sqrt{n}}$$
>
> Reject H_0 at significance level α against a one-sided alternative
>
> $$H_a\colon \mu > \mu_0 \quad \text{if} \quad z \geq z^*$$
> $$H_a\colon \mu < \mu_0 \quad \text{if} \quad z \leq -z^*$$
>
> where z^* is the upper α critical value from Table D. Reject H_0 at significance level α against a two-sided alternative
>
> $$H_a\colon \mu \neq \mu_0 \quad \text{if} \quad |z| \geq z^*$$
>
> where z^* is the upper $\alpha/2$ critical value from Table D.

EXAMPLE 6.13

The analytical laboratory of Example 6.3 (page 435) is asked to evaluate the claim that the concentration of the active ingredient in a specimen is 0.86%. As in Example 6.3, the mean of three repeated analyses of the specimen is $\bar{x} = 0.8404$, and the standard deviation of the analysis process is known to be $\sigma = 0.0068$. The true concentration is the mean μ of the population of repeated analyses. The hypotheses are

$$H_0: \mu = 0.86$$
$$H_a: \mu \neq 0.86$$

The lab chooses the 1% level of significance, $\alpha = 0.01$. The z statistic is

$$z = \frac{.8404 - .86}{.0068/\sqrt{3}} = -4.99$$

Because the alternative is two-sided, we compare $|z| = 4.99$ with the $\alpha/2 = 0.005$ critical value from Table D. This critical value is $z^* = 2.576$. The values of z that lead to rejection are illustrated in Figure 6.11. Because $|z| > z^*$, we reject H_0. ●

The calculation in Example 6.13 for a 1% significance test is very similar to that in Example 6.3 for a 99% confidence interval. In fact, a two-sided test at significance level α can be carried out directly from a confidence interval with confidence level $C = 1 - \alpha$.

CONFIDENCE INTERVALS AND TWO-SIDED TESTS

A level α two-sided significance test rejects a hypothesis H_0: $\mu = \mu_0$ exactly when the value μ_0 falls outside a level $1-\alpha$ confidence interval for μ.

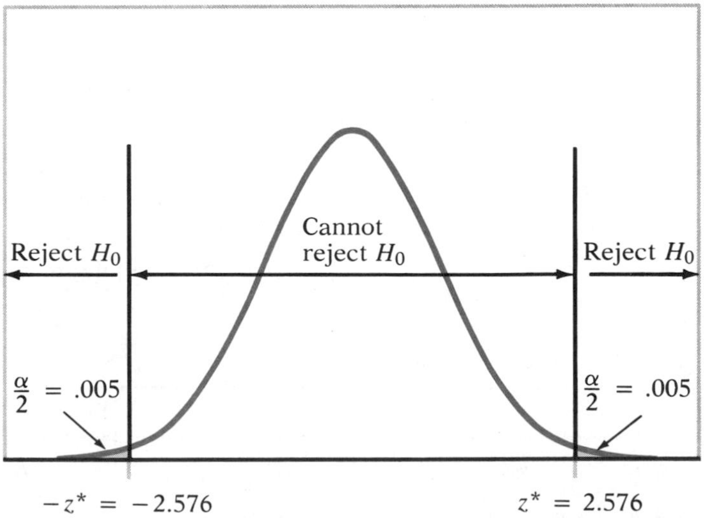

FIGURE 6.11 The level $\alpha = 0.01$ two-sided test in Example 6.13.

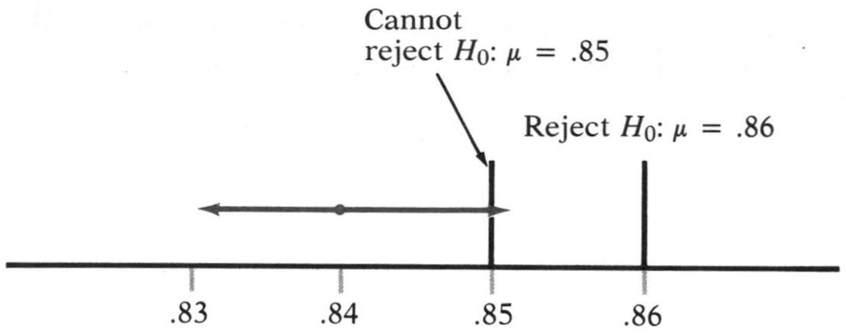

FIGURE 6.12 **Values of μ falling outside a 99% confidence interval can be rejected at the 1% significance level; values falling inside the interval cannot be rejected.**

EXAMPLE 6.14

The 99% confidence interval for μ in Example 6.3 is

$$\bar{x} \pm z^* \frac{\sigma}{\sqrt{n}} = .8404 \pm .0101$$
$$= (.8303, .8505)$$

The hypothesized value $\mu_0 = 0.86$ in Example 6.13 falls outside this confidence interval, so we reject

$$H_0\colon \mu = .86$$

at the 1% significance level. On the other hand, we cannot reject

$$H_0\colon \mu = .85$$

at the 1% level in favor of the two-sided alternative $H_a\colon \mu \neq 0.85$, because 0.85 lies inside the 99% confidence interval for μ. Figure 6.12 illustrates both cases. ●

***P*-values versus fixed level α** The observed result in Example 6.13 was $z = -4.99$. The conclusion that this result is significant at the 1% level does not tell the whole story. The observed z is far beyond the 1% critical value, and the evidence against H_0 is far stronger than 1% significance suggests. The *P*-value

$$2P(Z \geq 4.99) = .0000006$$

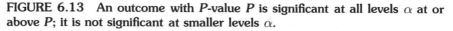

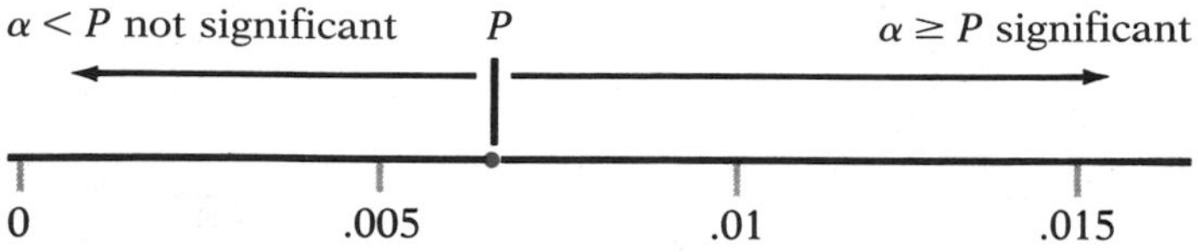

FIGURE 6.13 **An outcome with *P*-value *P* is significant at all levels α at or above *P*; it is not significant at smaller levels α.**

gives a better sense of how strong the evidence is. *The P-value is the smallest level α at which the data are significant.* Knowing the P-value allows us to assess significance at any level.

EXAMPLE 6.15

In Example 6.11, we tested the hypotheses

$$H_0: \mu = 450$$
$$H_a: \mu > 450$$

concerning the mean SAT mathematics score μ of California high school seniors. The test had the P-value $P = 0.0069$. This result is significant at the $\alpha = 0.01$ level because $0.0069 \leq 0.01$. It is not significant at the $\alpha = 0.005$ level, because the P-value is larger than 0.005. See Figure 6.13. ●

A P-value is more informative than a reject-or-not finding at a fixed significance level. But assessing significance at a fixed level α is easier, because no probability calculation is required. You need only look up a critical value in a table. Because the practice of statistics almost always employs computer software that calculates P-values automatically, the use of tables of critical values is becoming outdated. We include the usual tables of critical values (such as Table D) at the end of the book for learning purposes and to rescue students without good computing facilities. The tables can be used directly to carry out fixed-α tests. They also allow us to approximate P-values quickly without a probability calculation. The following example illustrates the use of Table D to find an approximate P-value.

EXAMPLE 6.16

Bottles of a popular cola drink are supposed to contain 300 milliliters (ml) of cola. There is some variation from bottle to bottle because the filling machinery is not perfectly precise. The distribution of the contents is normal with standard deviation $\sigma = 3$ ml. A student who suspects that the bottler is underfilling measures the contents of six bottles. The results are

299.4   297.7   301.0   298.9   300.2   297.0

Is this convincing evidence that the mean contents of cola bottles is less than the advertised 300 ml? The hypotheses are

$$H_0: \mu = 300$$
$$H_a: \mu < 300$$

The sample mean contents of the six bottles measured is $\bar{x} = 299.03$ ml. The z test statistic is therefore

$$z = \frac{\bar{x} - \mu_0}{\sigma/\sqrt{n}} = \frac{299.03 - 300}{3/\sqrt{6}}$$
$$= -.792$$

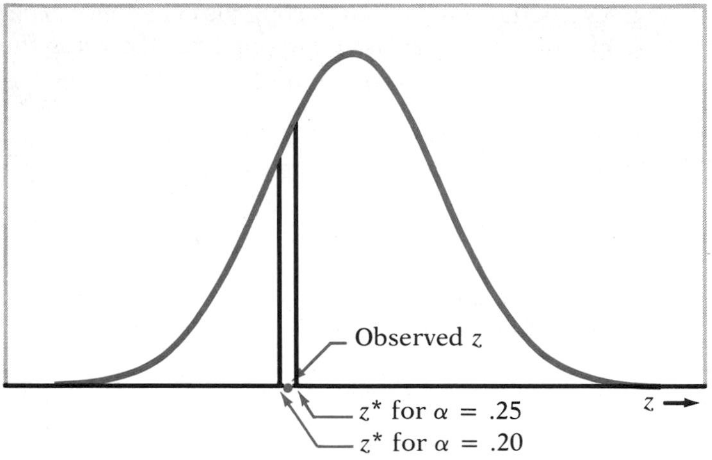

FIGURE 6.14 The *P*-value of a *z* statistic can be approximated by noting which levels from Table D it falls between. Here, *P* lies between 0.20 and 0.25.

Small values of z count against H_0, because H_a is one-sided on the low side. The *P*-value is

$$P(Z \leq -.792)$$

Rather than compute this probability, compare $z = -0.792$ with the critical values in Table D. According to this table, we would

- reject H_0 at level $\alpha = .25$ if $z \leq -.674$,
- reject H_0 at level $\alpha = .20$ if $z \leq -.841$

As Figure 6.14 illustrates, the observed $z = -0.792$ lies between these two critical points. We can reject at $\alpha = 0.25$ but not at $\alpha = 0.20$. That is the same as saying that the *P*-value lies between 0.25 and 0.20. A sample mean at least as small as that observed will occur in between 20% and 25% of all samples if the population mean is $\mu = 300$ ml. There is no convincing evidence that the mean is below 300, and there is no need to calculate the *P*-value more exactly. ●

SUMMARY

A **test of significance** assesses the evidence provided by data against a **null hypothesis** H_0 in favor of an **alternative hypothesis** H_a.

The hypotheses are stated in terms of population parameters. Usually H_0 is a statement that no effect is present and H_a says that a parameter differs from its null value, in a specific direction **(one-sided alternative)** or in either direction **(two-sided alternative).**

The test is based on a **test statistic.** The ***P*-value** is the probability, computed assuming that H_0 is true, that the test statistic will take a value at least as extreme as that actually observed. Small *P*-values indicate strong evidence against H_0. Calculating *P*-values requires knowledge of the sampling distribution of the test statistic when H_0 is true.

If the *P*-value is as small or smaller than a specified value α, the data are **statistically significant** at significance level α.

Significance tests for the hypothesis H_0: $\mu = \mu_0$ concerning the unknown mean μ of a population are based on the z **statistic**

$$z = \frac{\overline{x} - \mu_0}{\sigma/\sqrt{n}}$$

The z test assumes an SRS of size n, known population standard deviation σ, and either a normal population or a large sample. *P*-values are computed from the normal distribution (Table A). Fixed-α tests use the table of **standard normal critical values** (Table D).

SECTION 6.2 EXERCISES

6.21 Each of the following situations requires a significance test about a population mean μ. State the appropriate null hypothesis H_0 and alternative hypothesis H_a in each case.

(a) The mean area of the several thousand apartments in a new development is advertised to be 1250 square feet. A tenant group thinks that the apartments are smaller than advertised. They hire an engineer to measure a sample of apartments to test their suspicion.

(b) Larry's car averages 32 miles per gallon on the highway. He now switches to a new motor oil that is advertised as increasing gas mileage. After driving 3000 highway miles with the new oil, he wants to determine if his gas mileage actually has increased.

(c) The diameter of a spindle in a small motor is supposed to be 5 mm. If the spindle is either too small or too large, the motor will not perform properly. The manufacturer measures the diameter in a sample of motors to determine whether the mean diameter has moved away from the target.

6.22 In each of the following situations, a significance test for a population mean μ is called for. State the null hypothesis H_0 and the alternative hypothesis H_a in each case.

(a) Experiments on learning in animals sometimes measure how long it takes a mouse to find its way through a maze. The mean time is 18 seconds for one particular maze. A researcher thinks that a loud noise will cause the mice to complete the maze faster. She measures how long each of 10 mice takes with a noise as stimulus.

(b) The examinations in a large accounting class are scaled after grading so that the mean score is 50. A self-confident teaching assistant thinks that his students have a higher mean score than the class as a whole. His students this semester can be considered a sample from the population of all students he might teach, so he compares their mean score with 50.

(c) A university gives credit in French language courses to students who pass a placement test. The language department wants to know if students who get credit in this way differ in their understanding of spoken French from students who actually take the French courses. Some faculty think the students who test out of the courses are better, but others argue that they are weaker in oral comprehension. Experience has shown that the mean score of students in the courses on a standard listening test is 24. The language department gives the same listening test to a sample of 40 students who passed the credit examination to see if their performance is different.

6.23 In each of the following situations, state an appropriate null hypothesis H_0 and alternative hypothesis H_a. Be sure to identify the parameters that you use to state the hypotheses. (We have not yet learned how to test these hypotheses.)

(a) A sociologist asks a large sample of high school students which academic subject they like best. She suspects that a higher percent of males than of females will name mathematics as their favorite subject.

(b) An education researcher randomly divides sixth-grade students into two groups for physical education class. He teaches both groups basketball skills, using the same methods of instruction in both classes. He encourages Group A with compliments and other positive behavior but acts cool and neutral toward Group B. He hopes to show that positive teacher attitudes result in a higher mean score on a test of basketball skills than do neutral attitudes.

(c) An economist believes that among employed young adults there is a positive correlation between income and the percent of disposable income that is saved. To test this, she gathers income and savings data from a sample of employed persons in her city aged 25 to 34.

6.24 A randomized comparative experiment examined whether a calcium supplement in the diet reduces the blood pressure of healthy men. The subjects received either a calcium supplement or a placebo for 12 weeks. The statistical analysis was quite complex, but one conclusion was that "the calcium group had lower seated systolic blood pressure ($P = 0.008$) compared with the placebo group." Explain this conclusion, especially the P-value, as if you were speaking to a doctor who knows no statistics. (From R. M. Lyle et al.,"Blood pressure and metabolic effects of calcium supplementation in normotensive white and black men," *Journal of the American Medical Association*, 257 (1987), pp. 1772–1776.)

6.25 A social psychologist reports that "in our sample, ethnocentrism was significantly higher ($P < 0.05$) among church attenders than among nonattenders." Explain what this means in language understandable to someone who knows no statistics. Do not use the word "significance" in your answer.

6.26 The financial aid office of a university asks a sample of students about their employment and earnings. The report says that "for academic year earnings, a significant difference ($P = 0.038$) was found between the sexes, with men earning more on the average. No difference ($P = 0.476$) was found between the earnings of black and white students." Explain both of these conclusions, for the effects of sex and of race on mean earnings, in language understandable to someone who knows no statistics. (From a study by M. R. Schlatter et al., Division of Financial Aid, Purdue University.)

6.27 Statistics can help decide the authorship of literary works. Sonnets by an Elizabethan poet are known to contain an average of $\mu = 6.9$ new words (words not used in the poet's other works). The standard deviation of the number of new words is $\sigma = 2.7$. Now a manuscript with 5 new sonnets has come to light, and scholars are debating whether it is the poet's work. The new sonnets contain an average of $\bar{x} = 8.2$ words not used in the poet's known works. We expect poems by another author to contain more new words, so to see if we have evidence that the new sonnets are not by our poet we test

$$H_0: \mu = 6.9$$
$$H_a: \mu > 6.9$$

Give the z test statistic and its P-value. What do you conclude about the authorship of the new poems?

6.28 The Survey of Study Habits and Attitudes (SSHA) is a psychological test that measures the motivation, attitude toward school, and study habits of students. Scores range from 0 to 200. The mean score for U.S. college students is about 115, and the standard deviation is about 30. A teacher who suspects that older students have better attitudes toward school gives the SSHA to 25 students who are at least 30 years of age. Their mean score is $\bar{x} = 125.2$.

(a) Assuming that $\sigma = 30$ for the population of older students, carry out a test of

$$H_0: \mu = 115$$
$$H_a: \mu > 115$$

Report the P-value of your test, and state your conclusion clearly.

(b) Your test in (a) required two important assumptions in addition to the assumption that the value of σ is known. What are they? Which of these assumptions is most important to the validity of your conclusion in (a)?

6.29 The mean yield of corn in the United States is about 120 bushels per acre. A survey of 50 farmers this year gives a sample mean yield of $\bar{x} = 123.6$ bushels per acre. We want to know whether this is good evidence that the national mean this year is not 120 bushels per acre. Assume that the farmers surveyed are an SRS from the population of all commercial corn growers and that the standard deviation of the yield in this population is $\sigma = 10$ bushels per acre. Give the *P*-value for the test of

$$H_0\colon \mu = 120$$
$$H_a\colon \mu \neq 120$$

Are you convinced that the population mean is not 120 bushels per acre? Is your conclusion correct if the distribution of corn yields is somewhat nonnormal? Why?

6.30 In the past, the mean score of the seniors at South High on the American College Testing (ACT) college entrance examination has been 20. This year a special preparation course is offered, and all 43 seniors planning to take the ACT test enroll in the course. The mean of their 43 ACT scores is 21.1. The principal believes that the new course has improved the students' ACT scores.

(a) Assume that ACT scores vary normally with standard deviation 6. Is the outcome $\bar{x} = 21.1$ good evidence that the population mean score is greater than 20? State H_0 and H_a, compute the test statistic and the *P*-value, and answer the question by interpreting your result.

(b) The results are in any case inconclusive because of the design of the study. The effects of the new course are confounded with any change from past years, such as other new courses or higher standards. Briefly outline the design of a better study of the effect of the new course on ACT scores.

6.31 Here are measurements (in millimeters) of a critical dimension on a sample of automobile engine crankshafts:

224.120	224.001	224.017	223.982	223.989	223.961
223.960	224.089	223.987	223.976	223.902	223.980
224.098	224.057	223.913	223.999		

The manufacturing process is known to vary normally with standard deviation $\sigma = 0.060$ mm. The process mean is supposed to be 224 mm. Do these data give evidence that the process mean is not equal to the target value 224 mm?

(a) State the H_0 and H_a that you will test.
(b) Give the P-value of the test. Are you convinced that the process mean is not 224 mm?

6.32 The level of calcium in the blood in healthy young adults varies with mean about 9.5 milligrams per deciliter and standard deviation about $\sigma = 0.4$. A clinic in rural Guatemala measures the blood calcium level of 180 healthy pregnant women at their first visit for prenatal care. The mean is $\bar{x} = 9.57$. Is this an indication that the mean calcium level in the population from which these women come differs from 9.5?
(a) State H_0 and H_a.
(b) Carry out the test and give the P-value, assuming that $\sigma = 0.4$ in this population. Report your conclusion.
(c) Give a 95% confidence interval for the mean calcium level μ in this population. We are confident that μ lies quite close to 9.5. This illustrates the fact that a test based on a large sample ($n = 180$ here) will often declare even a small deviation from H_0 to be statistically significant.

6.33 Here are the Degree of Reading Power (DRP) scores for a sample of 44 third grade students:

40	26	39	14	42	18	25	43	46	27	19
47	19	26	35	34	15	44	40	38	31	46
52	25	35	35	33	29	34	41	49	28	52
47	35	48	22	33	41	51	27	14	54	45

These students can be considered to be an SRS of the third graders in a suburban school district. DRP scores are approximately normal. Suppose that the standard deviation of scores in this school district is known to be $\sigma = 11$. The researcher believes that the mean score μ of all third graders in this district is higher than the national mean, which is 32.
(a) State the appropriate H_0 and H_a to test this suspicion.
(b) Carry out the test. Give the P-value, and then interpret the result in plain language.

6.34 A computer has a random number generator designed to produce random numbers that are uniformly distributed in the interval from 0 to 1. If this is true, the numbers generated come from a population with $\mu = 0.5$ and $\sigma = 0.2887$. A command to generate 100 random numbers gives outcomes with mean $\bar{x} = 0.4365$. Assume that the population σ remains fixed. We want to test

$$H_0\colon \mu = .5$$
$$H_a\colon \mu \neq .5$$

(a) Calculate the value of the z test statistic.
(b) Is the result significant at the 5% level ($\alpha = 0.05$)?

(c) Is the result significant at the 1% level ($\alpha = 0.01$)?

6.35 To determine whether the mean nicotine content of a brand of cigarettes is greater than the advertised value of 1.4 milligrams, a health advocacy group tests

$$H_0: \mu = 1.4$$
$$H_a: \mu > 1.4$$

The calculated value of the test statistic is $z = 2.42$.
(a) Is the result significant at the 5% level?
(b) Is the result significant at the 1% level?

6.36 There are other z statistics that we have not yet studied. The significance of any z statistic is assessed from Table D. A study compares the habits of students who are on academic probation with students whose grades are satisfactory. One variable measured is the hours spent watching television last week. The null hypothesis is "no difference" between the means for the two populations. The alternative hypothesis is two-sided. The value of the test statistic is $z = -1.37$.
(a) Is this result significant at the 5% level?
(b) Is the result significant at the 1% level?

6.37 Explain in plain language why a significance test that is significant at the 1% level must always be significant at the 5% level.

6.38 Use Table D to find the approximate P-value for the test in Exercise 6.34 without doing a probability calculation. That is, find from the table two numbers that contain the P-value between them.

6.39 Use Table D to find the approximate P-value for the test in Exercise 6.35. That is, between what two numbers obtained from the table does the P-value lie?

6.40 Between what values from Table D does the P-value for the outcome $z = -1.37$ in Exercise 6.36 lie? (Remember that H_a is two-sided.) Calculate the P-value using Table A, and verify that it lies between the values you found from Table D.

6.41 Radon is a colorless, odorless gas that is naturally released by rocks and soils and may concentrate in tightly closed houses. Because radon is slightly radioactive, there is some concern that it may be a health hazard. Radon detectors are sold to homeowners worried about this risk, but the detectors may be inaccurate. University researchers placed 12 detectors in a chamber where they were exposed to 105 picocuries per liter (pCi/l) of radon over 3 days. Here are the readings given by the detectors. (Data provided by Diana Schellenberg, Purdue University School of Health Sciences.)

91.9	97.8	111.4	122.3	105.4	95.0
103.8	99.6	96.6	119.3	104.8	101.7

Assume (unrealistically) that you know that the standard deviation of readings for all detectors of this type is $\sigma = 9$.

(a) Give a 90% confidence interval for the mean reading μ for this type of detector.

(b) Is there significant evidence at the 10% level that the mean reading differs from the true value 105? State hypotheses, and base a test on your confidence interval from (a).

6.42 A sample of 24 male runners has average weight $\bar{x} = 60$ kilograms. Exercise 6.5 (page 442) asks you to find a 95% confidence interval for the mean weight of the population of all such runners, assuming that the population standard deviation is $\sigma = 5$ kg.

(a) Give the confidence interval from that exercise, or calculate the interval if you did not do the exercise.

(b) Based on this confidence interval, does a test of

$$H_0\colon \mu = 61.5$$
$$H_a\colon \mu \neq 61.5$$

reject H_0 at the 5% significance level?

(c) Would $H_0\colon \mu = 63$ be rejected at the 5% level if tested against a two-sided alternative?

6.43 Researchers studying the absorption of sugar by insects feed cockroaches a diet containing measured amounts of a particular sugar. After 10 hours, the cockroaches are killed and the concentration of the sugar in various body parts is determined by a chemical analysis. The paper that reports the research states that a 95% confidence interval for the mean amount (in milligrams) of the sugar in the hindguts of the cockroaches is 4.2 ± 2.3. (From D. L. Shankland et al., "The effect of 5-thio-D-glucose on insect development and its absorption by insects," *Journal of Insect Physiology*, 14 (1968), pp. 63–72.)

(a) Does this paper give evidence that the mean amount of sugar in the hindgut under these conditions is not equal to 7 mg? State H_0 and H_a and base a test on the confidence interval.

(b) Would the hypothesis that $\mu = 5$ mg be rejected at the 5% level in favor of a two-sided alternative?

6.44 An old farmer claims to be able to detect the presence of water with a forked stick. In a test of this claim, he is presented with five identical barrels, some containing water and some not. He is right in four of the five cases.

(a) Suppose the farmer has probability p of being correct. If he is just guessing, $p = 0.5$. State an appropriate H_0 and H_a in terms of p for a test of whether he does better than guessing.

(b) If the farmer is simply guessing, what is the distribution of the number X of correct answers in five tries?

(c) The observed outcome is $X = 4$. What is the P-value of the test that takes large values of X to be evidence against H_0?

6.3 USE AND ABUSE OF TESTS

Carrying out a test of significance is often quite simple, especially if a fixed significance level α is used or if the P-value is given effortlessly by a computer. Using tests wisely is not so simple. Each test is valid only in certain circumstances, with properly produced data being particularly important. The z test, for example, should bear the same warning label that was attached in Section 6.1 to the corresponding confidence interval. Similar warnings accompany the other tests that we will learn. There are additional caveats that concern tests more than confidence intervals, enough to warrant this separate section. Some hesitation about the unthinking use of significance tests is a sign of statistical maturity.

Using significance tests

The reasoning of significance tests has appealed to researchers in many fields, so that tests are widely used to report research results. In this setting H_a is a "research hypothesis" asserting that some effect or difference is present. The null hypothesis H_0 says that there is no effect or no difference. A low P-value represents good evidence that the research hypothesis is true. Here are some comments on the use of significance tests, with emphasis on their use in reporting scientific research.

Choosing a level of significance The goal of a test of significance is to give a clear statement of the degree of evidence provided by the sample against the null hypothesis. The P-value does this. But sometimes you will make some decision or take some action if your evidence reaches a certain standard. You can set such a standard by giving a level of significance α. Perhaps you will announce a new scientific finding if your data are significant at the $\alpha = 0.05$ level. Or perhaps you will recommend using a new method of teaching reading if the evidence of its superiority is significant at the $\alpha = 0.01$ level.

Making a decision is different in spirit from testing significance, though the two are often mixed in practice. Choosing a level α in advance makes sense if you must make a decision, but not if you wish only to describe the strength of your evidence. Using tests with fixed α for decision making is discussed at greater length at the end of this chapter.

If you do use a fixed-α significance test to make a decision, choose α by asking how much evidence is required to reject H_0. This depends first on how plausible H_0 is. If H_0 represents an assumption that everyone in your field has believed for years, strong evidence (small α) will be needed to reject it. Second, the level of evidence required to reject H_0 depends on the consequences of such a decision. If rejecting H_0 in favor of H_a means making an expensive changeover from one medical therapy or instructional method to another, strong evidence is needed. Both the plausibility of H_0 and H_a and the consequences of any action that rejection may lead to are somewhat subjective. Different persons may feel that different levels of significance are appropriate. It is better to report the *P*-value, which allows each of us to decide individually if the evidence is sufficiently strong.

Users of statistics have often emphasized certain standard levels of significance, such as 10%, 5%, and 1%. This emphasis reflects the time when tables of critical points rather than computer programs dominated statistical practice. The 5% level ($\alpha = 0.05$) is particularly common. Significance at that level is still a widely accepted criterion for meaningful evidence in research work. *There is no sharp border between "significant" and "insignificant," only increasingly strong evidence as the P-value decreases.* There is no practical distinction between the *P*-values 0.049 and 0.051. It makes no sense to treat $\alpha = 0.05$ as a universal rule for what is significant.

There is a reason for the common use of $\alpha = 0.05$—the great influence of Sir R. A. Fisher. Fisher did not originate tests of significance. But because his writings organized statistics, especially as a tool of scientific research, his views on tests were enormously influential. Here is his opinion on choosing a level of significance:

> ... it is convenient to draw the line at about the level at which we can say: "Either there is something in the treatment, or a coincidence has occurred such as does not occur more than once in twenty trials...."
>
> If one in twenty does not seem high enough odds, we may, if we prefer it, draw the line at one in fifty (the 2 percent point), or one in a hundred (the 1 percent point). Personally, the writer prefers to set a low standard of significance at the 5 percent point, and ignore entirely all results which fail to reach that level. A scientific fact should be regarded as experimentally established only if a properly designed experiment *rarely fails* to give this level of significance.[2]

There you have it. Fisher thought 5% was about right, and who was to disagree with the master? Fisher was of course not an advocate of blind use of significance at the 5% level as a yes-or-no criterion. The last sentence quoted above shows an experienced scientist's feeling for the repeated studies and variable results that mark the advance of knowledge.[3]

What statistical significance doesn't mean When a null hypothesis ("no effect" or "no difference") can be rejected at the usual levels, $\alpha = 0.05$ or $\alpha = 0.01$, there is good evidence that an effect is present. But that effect may be extremely small. When large samples are available, even tiny deviations from the null hypothesis will be significant. For example, suppose that we are testing the hypothesis of no correlation between two variables. With 1000 observations, an observed correlation of only $r =$

*We have met Fisher as the inventor of randomized experimental designs. He also originated many other statistical techniques, derived the distributions of many common statistics, and introduced such basic terms as "parameter" and "statistic" into statistical writing.

0.08 is significant evidence at the $\alpha = 0.01$ level that the correlation in the population is not zero but positive. The low significance level does not mean there is a strong association, only that there is strong evidence of some association. The true population correlation is probably quite close to the observed sample value, $r = 0.08$. We might well conclude that for practical purposes we can ignore the association between these variables, even though we are confident (at the 1% level) that the correlation is positive. Remember the wise saying: *Statistical significance is not the same thing as practical significance.* Exercise 6.47 demonstrates in detail the effect on P of increasing the sample size.

The remedy for attaching too much importance to statistical significance is to pay attention to the actual experimental results as well as to the P-value. Plot your data and examine them carefully. Are there outliers or other deviations from a consistent pattern? A few outlying observations can produce highly significant results if you blindly apply common tests of significance. Outliers can also destroy the significance of otherwise-convincing data. The foolish user of statistics who feeds the data to a computer without exploratory analysis will often be embarrassed. Is the effect you are seeking visible in your plots? If not, ask yourself if the effect is large enough to be practically important. It is usually wise to give a confidence interval for the parameter in which you are interested. A confidence interval actually estimates the size of an effect, rather than simply asking if it is too large to reasonably occur by chance alone. Confidence intervals are not used as often as they should be, while tests of significance are perhaps overused.

Don't ignore lack of significance Researchers typically have in mind the research hypothesis that some effect exists. Following the peculiar logic of tests of significance, they set up as H_0 the null hypothesis that no such effect exists and try their best to get evidence against H_0. A perverse legacy of Fisher's opinion on $\alpha = 0.05$ is that research in some fields has rarely been published unless significance at that level is attained. For example, a survey of four journals of the American Psychological Association showed that of 294 articles using statistical tests, only 8 reported results that did not attain the 5% significance level.[4]

Such a publication policy impedes the spread of knowledge, and not only by declaring that a P-value of 0.051 is "not significant." If a researcher has good reason to suspect that an effect is present and then fails to find significant evidence of it, that may be interesting news—perhaps more interesting than if evidence in favor of the effect at the 5% level had been found. If you follow the history of science, you will recall examples such as the Michelson-Morley experiment, which changed the course of physics by *not* detecting an expected change in the speed of light. Keeping silent about negative results may condemn other researchers to repeat the attempt to find an effect that isn't there.

Of course, an experiment that fails only causes a stir if it is clear that the experiment would have detected the effect if it were really there. An important aspect of planning a study is to verify that the test you plan to use does have high probability of detecting an effect of the size you hope to find. This probability is the *power* of the test. Power calculations are discussed later in this section.

Abuse of significance tests

Tests of statistical significance are routinely used to assess the results of research in agriculture, education, engineering, medicine, psychology, and sociology and increasingly in other fields as well. Any tool used routinely is often used unthinkingly. We therefore offer some comments for the thinking researcher on possible abuses of this tool. Thinking consumers of research findings (such as students) should also ponder these comments.

Statistical inference is not valid for all sets of data We learned long ago that badly designed surveys or experiments often produce invalid results. Formal statistical inference cannot correct basic flaws in the design. There is no doubt a significant difference in English vocabulary scores between high school seniors who have studied a foreign language and those who have not. But because the effect of actually studying a language is confounded with the differences between students who choose language study and those who do not, this statistical significance is hard to interpret. It does indicate that the difference in English scores is greater than would often arise by chance alone. That leaves unsettled the issue of *what* other than chance caused the difference. The most plausible explanation is that students who were already good at English chose to study another language. A randomized comparative experiment would isolate the actual effect of language study and so make significance meaningful.

Tests of significance and confidence intervals are based on the laws of probability. Randomization in sampling or experimentation ensures that these laws apply. But we must often analyze data that do not arise from randomized samples or experiments. To apply statistical inference to such data, we must have confidence in a probability model for the data. The diameters of successive holes bored in auto engine blocks during production, for example, may behave like independent observations on a normal distribution. We can check this probability model by examining the data. If the model appears correct, we can apply the recipes of this chapter to do inference about the process mean diameter μ. Do ask how the data were produced, and don't be too impressed by P-values on a printout until you are confident that the data deserve a formal analysis.

Beware of searching for significance Statistical significance is a commodity much sought after by researchers. It means (or ought to mean) that you have found an effect that you were looking for. *The reasoning behind statistical significance works well if you decide what effect you are seeking, design an experiment or sample to search for it, and use a test of significance to weigh the evidence you get*. But because a successful search for a new scientific phenomenon often ends with statistical significance, it is all too tempting to make significance itself the object of the search. There are several ways to do this, none of them acceptable in polite scientific society.

One tactic is to make many tests on the same data. Once upon a time three psychiatrists studied a sample of schizophrenic persons and a sample of nonschizophrenic persons. They measured 77 variables for each subject—religion, family background, childhood experiences, and so on. Their goal was to discover what distinguishes persons who later become schizophrenic. Having measured 77 variables, they performed 77 separate tests of the significance of the differences between the two groups of subjects. Pause for a moment of reflection. If you made 77 tests at the 5% level, you would expect a few of them to be significant by chance alone. After all, results significant at the 5% level do occur 5 times in 100 in the long run even when H_0 is true. The psychiatrists found 2 of their 77 tests significant at the 5% level and immediately published this exciting news.[5] Running one test and reaching the $\alpha = 0.05$ level is reasonably good evidence that you have found something; running 77 tests and reaching that level only twice is not.

The case of the 77 tests happened long ago. Such crimes are rarer now—or at least better concealed. The computer has freed us from the labor of doing arithmetic. This is surely a blessing in statistics, where the arithmetic can be long and complicated indeed. Comprehensive statistical software systems are everywhere available, so a few simple commands will set the machine to work performing all manner of complicated tests and operations on your data. The result can be much like the 77 tests of old. We will state it as a law that any large set of data—even several pages of a table of random digits—contains some unusual pattern. Sufficient computer time will discover that pattern, and when you test specifically for the pattern that turned up, the result will be significant. It also will mean exactly nothing.

One lesson here is not to be overawed by the computer. The computer has greatly extended the range of statistical inference, allowing us to handle larger sets of data and to carry out more complex analyses. But it has changed the logic of inference not one bit. Doing 77 tests and finding 2 significant at the $\alpha = 0.05$ level is not evidence of a real discovery. Neither is doing factor analysis followed by discriminant analysis followed by multiple regression and at last discovering a significant pattern in the data. Fancy computer programs are no remedy for bad scientific logic. It is convincing to hypothesize that an effect or pattern will be present, design a study to look for it, and find it at a low significance level. It is not convincing to search for any effect or pattern whatever and find one.

We do not mean that searching data for suggestive patterns is not proper scientific work. It certainly is. Many important discoveries have been made by accident rather than by design. Exploratory analysis of data is an essential part of statistics. We do mean that the usual reasoning of statistical inference does not apply when the search for a pattern is successful. You cannot legitimately test a hypothesis on the same data that first suggested that hypothesis. The remedy is clear. Once you have a hypothesis, design a study to search specifically for the effect you now think is there. If the result of this study is statistically significant, you have real evidence at last.

Power*

In examining the usefulness of a confidence interval, we are concerned with both the level of confidence and the margin of error. The confidence level tells us how reliable the method is in repeated use. The margin of error tells us how sensitive the method is, that is, how closely the interval pins down the parameter being estimated. Fixed level α significance tests are closely related to confidence intervals—in fact, we saw that a two-sided test can be carried out directly from a confidence interval. The significance level, like the confidence level, says how reliable the method is in repeated use. If we use 5% significance tests repeatedly when H_0 is in fact true, we will be wrong (the test will reject H_0) 5% of the time and right (the test will fail to reject H_0) 95% of the time.

High confidence is of little value if the interval is so wide that few values of the parameter are excluded. Similarly, it can happen that a test with small level α almost never rejects H_0 even when the true parameter value is far from the hypothesized value. We must be concerned with the ability of a test to detect that H_0 is false, just as we are concerned with the margin of error of a confidence interval. This ability is measured by the probability that the test will reject H_0 when an alternative is true. The higher this probability is, the more sensitive the test is. Because the probability that the test will reject H_0 is different for different values of the parameter, we must have in mind one of the parameter values covered by H_a when we compute power.

POWER

The probability that a fixed level α significance test will reject H_0 when a particular alternative value of the parameter is true is called the power of the test against that alternative.

*Although power is important in planning and interpreting significance tests, this section and all later comments on power can be omitted without loss of continuity.

EXAMPLE 6.17

The cheesemaker of Example 6.7 (page 448) determines that milk so heavily watered that the freezing point is raised to $-0.53°$ C will damage the quality of the cheese. Will a 5% significance test of the hypotheses

$$H_0: \mu = -.545$$
$$H_a: \mu > -.545$$

based on a sample of five lots usually detect a mean freezing point this high? To answer this question, we find the power of the test against the specific alternative $\mu = -0.53$.

The test measures the freezing point of five lots of milk from a producer and rejects H_0 when

$$z = \frac{\bar{x} - (-.545)}{.008/\sqrt{5}} \geq 1.645$$

This is the same as

$$\bar{x} \geq -.545 + 1.645\frac{.008}{\sqrt{5}}$$

or

$$\bar{x} \geq -.539$$

That the significance level is $\alpha = 0.05$ means that this event has probability 0.05 of occurring *when in fact the population mean μ is* -0.545. To help us keep in mind that the probability calculation assumes that $\mu = -0.545$, we will write

$$P(\bar{x} \geq -.539 \mid \mu = -.545) = .05$$

(This is not a conditional probability, because μ is not a random variable. It is just a notation to remind us what value of μ was assumed in calculating the probability.)

The power against the alternative -0.53 is the probability that H_0 will be rejected *when in fact* $\mu = -0.53$. This probability is

$$P(\bar{x} \geq -.539 \mid \mu = -.53)$$

We can calculate this probability by standardizing $\bar{x}$, but we must use the value $\mu = -0.53$ of the population mean in doing so. The population standard deviation $\sigma = 0.008$ does not change.

$$\begin{aligned} &P(\bar{x} \geq -.539 \mid \mu = -.53) \\ &= P\left(\frac{\bar{x} - (-.53)}{.008/\sqrt{5}} \geq \frac{-.539 - (-.53)}{.008/\sqrt{5}}\right) \\ &= P(Z \geq -2.52) = .9941 \end{aligned}$$

Figure 6.15 illustrates the power in terms of the sampling distribution of $\bar{x}$ when $\mu = -0.53$ is true. This significance test will almost always (probability more than 99%) reject H_0 when in fact $\mu = -0.53$. It is sensitive enough for the cheesemaker's purpose. ●

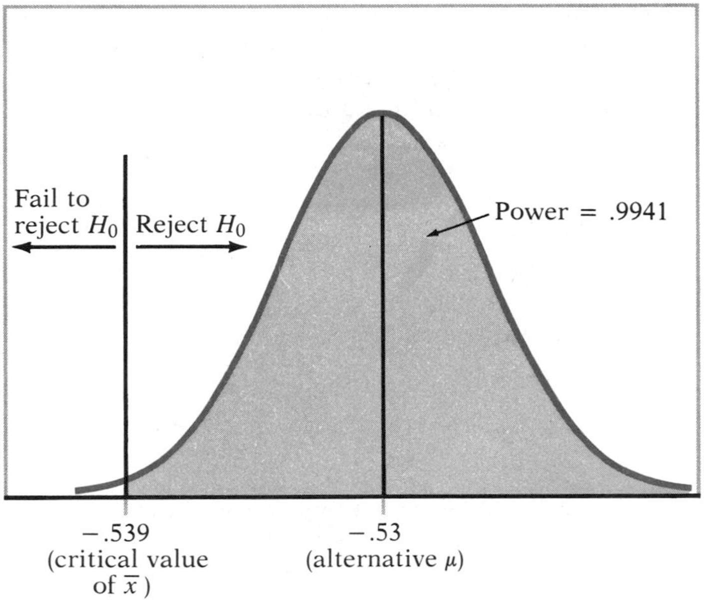

FIGURE 6.15 The power for Example 6.17. Power is the probability that the test rejects H_0 when the alternative is true.

High power is desirable. The numerical value of the power depends on which particular parameter value in H_a we are interested in. Values of the mean μ that are in H_a but lie close to the hypothesized value μ_0 are harder to detect (lower power) than values of μ that are far from μ_0.

In planning an investigation that will include a test of significance, a careful user of statistics should decide what alternatives the test should detect and check that the power is adequate. If the power is too low, a larger sample size will increase the power for the same significance level α. In order to calculate power, we must fix an α so that there is a fixed criterion for rejecting H_0. We prefer to report P-values rather than to use a fixed significance level. The usual practice is to calculate the power at common significance levels such as $\alpha = 0.05$ even though you intend to report a P-value.

Power calculations are important in planning studies. The use of a significance test with low power makes it unlikely that you will find a significant effect even if the truth is far from the null hypothesis. A null hypothesis that is in fact false can become widely believed if repeated attempts to find evidence against it fail because of low power. Consider for example the "efficient market hypothesis" for the time series of stock prices. This hypothesis says that future stock prices (when adjusted for inflation) show only random variation. No information available now will help us predict stock prices in the future, because the efficient working of the market has already incorporated all available information in the present price. Many studies have tested the claim that one or another kind of information is helpful. In these studies, the efficient market hypothesis is H_0 and the claim that prediction is possible is H_a. Almost all the studies have failed to find good evidence against H_0. As a result, the efficient market theory is quite popular. But an examination of the significance tests employed finds that the power is generally low. Failure to reject H_0 when using tests of low power is not evidence that H_0 is true. As one expert says, "The widespread impression that there is strong evidence for market efficiency may be due just to a lack of appreciation of the low power of many statistical tests."[6]

The outline of a power calculation is as follows:

- Write down the event that the test rejects H_0.
- Find the probability of this event under an alternative value of the parameter. This probability is the power against that alternative.

In the case of tests on a population mean, we use the test statistic z, the sample mean standardized assuming that H_0 is true. To calculate power, it is easiest to first restate the test in terms of $\bar{x}$, as we did in Example 6.17. Here is another example, this time for a two-sided z test.

EXAMPLE 6.18

Example 6.13 (page 460) presented a test of

$$H_0: \mu = .86$$
$$H_a: \mu \neq .86$$

at the 1% level of significance. What is the power of this test against the specific alternative $\mu = 0.845$?

The test in Example 6.13 rejects H_0 when $|z| \geq 2.576$. Because

$$z = \frac{\bar{x} - .86}{.0068/\sqrt{3}}$$

some arithmetic shows that the test rejects when either of the following is true

$$\begin{aligned} z &\geq 2.576 \quad \text{that is, } \bar{x} \geq .870 \\ z &\leq -2.576 \quad \text{that is, } \bar{x} \leq .850 \end{aligned}$$

These are disjoint events, so the power is the sum of their probabilities, *computed assuming that the alternative* $\mu = 0.845$ *is true*. We find that

$$\begin{aligned} P(\bar{x} \geq .87 \mid \mu = .845) &= P\left(\frac{\bar{x} - .845}{.0068/\sqrt{3}} \geq \frac{.87 - .845}{.0068/\sqrt{3}}\right) \\ &= P(Z \geq 6.37) \doteq 0 \\ P(\bar{x} \leq .85 \mid \mu = .845) &= P\left(\frac{\bar{x} - .845}{.0068/\sqrt{3}} \leq \frac{.85 - .845}{.0068/\sqrt{3}}\right) \\ &= P(Z \leq 1.27) = .8980 \end{aligned}$$

Figure 6.16 illustrates this calculation. Because the power is about 0.9, we are quite confident that the test will reject H_0 when this alternative is true. ●

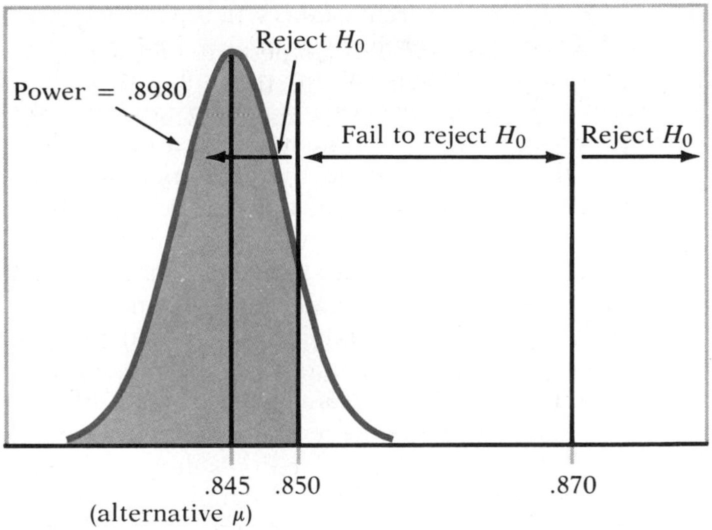

FIGURE 6.16 **The power for Example 6.18.**

Inference as decision*

We have presented tests of significance as methods for assessing the strength of evidence against the null hypothesis. This assessment is made by the *P*-value, which is a probability computed under the assumption that H_0 is true. The alternative hypothesis (the statement we seek evidence for) enters the test only to help us see what outcomes count against the null hypothesis. Such is the reasoning of tests of significance as advocated by Fisher and as practiced by many users of statistics.

But signs of another way of thinking were present in the discussion of significance tests with fixed level α. A level of significance α chosen in advance points to the outcome of the test as a *decision*. If the *P*-value is less than α, we reject H_0 in favor of H_a. Otherwise we do not reject H_0. The transition from measuring the strength of evidence to making a decision is not a small step. Many statisticians agree with Fisher that making decisions is too grand a goal, especially in scientific inference. A decision is reached only after the evidence of many studies is weighed. Indeed, the goal of research is not "decision" but a gradually evolving understanding. Statistical inference should content itself with confidence intervals and tests of significance. Many users of statistics are content with such methods. It is rare to set up a level α in advance as a rule for making a decision in a scientific problem. More commonly, users think of significance at level 0.05 as a description of good evidence. This is made clearer by giving the *P*-value.

*The purpose of this section is to clarify the reasoning of significance tests by contrast with a related type of reasoning. It can be omitted without loss of continuity.

acceptance sampling

Yet there are circumstances that call for a decision or action as the end result of inference. *Acceptance sampling* is one such circumstance. A producer of bearings and the consumer of the bearings agree that each carload lot shall meet certain quality standards. When a carload arrives, the consumer chooses a sample of bearings to be inspected. On the basis of the sample outcome, the consumer will either accept or reject the carload. Fisher agreed that this is a genuine decision problem. But he insisted that acceptance sampling is completely different from scientific inference. Other eminent statisticians have argued that if "decision" is given a broad meaning, almost all problems of statistical inference can be posed as problems of making decisions in the presence of uncertainty. We will not venture further into the arguments over how we ought to think about inference. We do want to show how a different concept—inference as decision—changes the reasoning used in tests of significance.

Two types of error Tests of significance concentrate on H_0, the null hypothesis. If a decision is called for, however, there is no reason to single out H_0. There are simply two hypotheses, and we must accept one and reject the other. It is convenient to call the two hypotheses H_0 and H_a, but H_0 no longer has the special status (the statement we try to find evidence against) that it had in tests of significance. In the acceptance sampling problem, we must decide between

- H_0: the lot of bearings meets standards,
- H_a: the lot does not meet standards

on the basis of a sample of bearings.

We hope that our decision will be correct, but sometimes it will be wrong. There are two types of incorrect decisions. We can accept a bad lot of bearings, or we can reject a good lot. Accepting a bad lot injures the consumer, while rejecting a good lot hurts the producer. To help distinguish these two types or error, we give them specific names.

TYPE I AND TYPE II ERRORS

If we reject H_0 (accept H_a) when in fact H_0 is true, this is a Type I error. If we accept H_0 (reject H_a) when in fact H_a is true, this is a Type II error.

The possibilities are summed up in Figure 6.17. If H_0 is true, our decision is either correct (if we accept H_0) or is a Type I error. If H_a is true, our decision is either correct or is a Type II error. Only one error

Decision based on sample		Truth about the population: H_0 true	H_a true
	Reject H_0	Type I error	Correct decision
	Accept H_0	Correct decision	Type II error

FIGURE 6.17 **The two types of error in testing hypotheses.**

is possible at one time. Figure 6.18 applies these ideas to the acceptance sampling example.

Error probabilities Any rule for making decisions is assessed in terms of the probabilities of the two types of error. This is in keeping with the idea that statistical inference is based on probability. We cannot (short of inspecting the whole lot) guarantee that good lots of bearings will never be rejected and bad lots never be accepted. But by random sampling and the laws of probability, we can say what the probabilities of both kinds of error are.

Significance tests with fixed level α give a rule for making decisions, because the test either rejects H_0 or fails to reject it. If we adopt the decision-making way of thought, failing to reject H_0 means deciding that H_0 is true. We can then describe the performance of a test by the probabilities of Type I and Type II errors.

Decision based on sample		Truth about the lot: Does meet standards	Does not meet standards
	Reject the lot	Type I error	Correct decision
	Accept the lot	Correct decision	Type II error

FIGURE 6.18 **The two types of error in the acceptance sampling setting.**

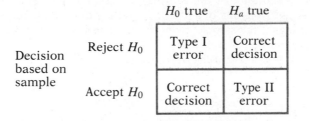

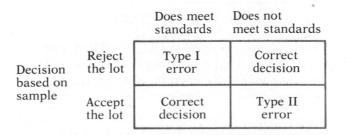

EXAMPLE 6.19

The mean diameter of a type of bearing is supposed to be 2.000 centimeters (cm). The bearing diameters vary normally with standard deviation $\sigma = 0.010$ cm. When a lot of the bearings arrives, the consumer takes an SRS of five bearings from the lot and measures their diameters. The consumer rejects the bearings if the sample mean diameter is significantly different from 2 at the 5% significance level.

This is a test of the hypotheses

$$H_0: \mu = 2$$
$$H_a: \mu \neq 2$$

To carry out the test, the consumer computes the z statistic

$$z = \frac{\bar{x} - 2}{.01/\sqrt{5}}$$

and rejects H_0 if

$$z < -1.96 \quad \text{or} \quad z > 1.96$$

A Type I error is to reject H_0 when in fact $\mu = 2$.

What about Type II errors? Because there are many values of μ in H_a, we will concentrate on one value. The producer and the consumer agree that a lot of bearings with mean 0.015 cm away from the desired mean 2.000 should be rejected. So a particular Type II error is to accept H_0 when in fact $\mu = 2.015$.

Figure 6.19 shows how the two probabilities of error are obtained from the two sampling distributions of $\bar{x}$, for $\mu = 2$ and for $\mu = 2.015$. When $\mu = 2$, H_0 is true and to reject H_0 is a Type I error. When $\mu = 2.015$, accepting H_0 is a Type II error. We will now calculate these error probabilities. ●

The probability of a Type I error is the probability of rejecting H_0 when it is really true. In Example 6.19, this is the probability that $|z| \geq 1.96$ when $\mu = 2$. But this is exactly the significance level of the test. The critical value 1.96 was chosen to make this probability 0.05, so we do not have to compute it again. The definition of "significant at level 0.05" is that sample outcomes this extreme will occur with probability 0.05 when H_0 is true.

SIGNIFICANCE AND TYPE I ERROR

The significance level α of any fixed level test is the probability of a Type I error. That is, α is the probability that the test will reject the null hypothesis H_0 when H_0 is in fact true.

The probability of a Type II error for the particular alternative $\mu = 2.015$ in Example 6.19 is the probability that the test will fail to reject

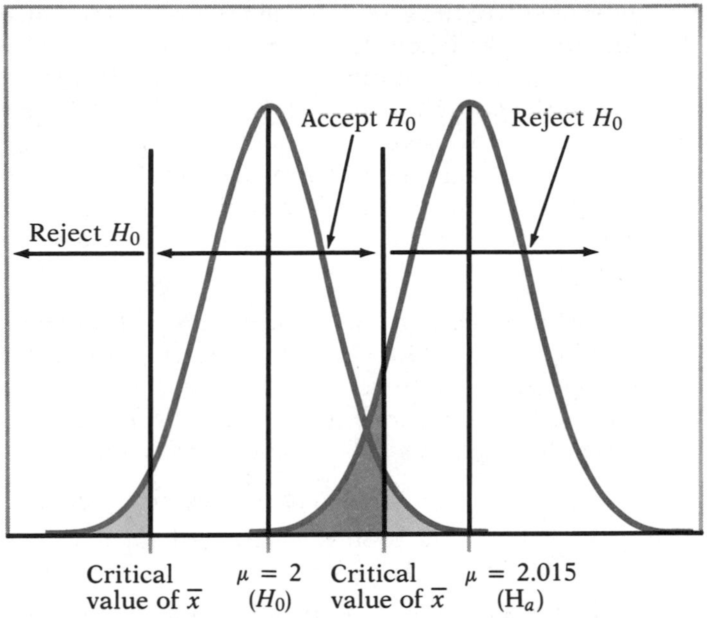

FIGURE 6.19 The two error probabilities for Example 6.19. The probability of a Type I error *(light shaded area)* is the probability of rejecting H_0: $\mu = 2$ when in fact $\mu = 2$. The probability of a Type II error *(dark shaded area)* is the probability of accepting H_0 when in fact $\mu = 2.015$.

H_0 when μ has this alternative value. The *power* of the test against the alternative $\mu = 2.015$ is just the probability that the test *does* reject H_0. By following the method of Example 6.18, we can calculate that the power is about 0.92. The probability of a Type II error is therefore $1 - 0.92$, or 0.08.

> POWER AND TYPE II ERROR
>
> The power of a fixed level test against a particular alternative is 1 minus the probability of a Type II error for that alternative.

The two types of error and their probabilities provide another interpretation of the significance level and power of a test. The distinction between tests of significance and tests as rules for deciding between two hypotheses does not lie in the calculations but in the reasoning that motivates the calculations. In a test of significance we focus on a single hypothesis (H_0) and a single probability (the P-value). The goal is to measure the strength of the sample evidence against H_0. Calculations of power are done to check the sensitivity of the test. If we cannot reject H_0, we conclude only that there is not sufficient evidence against H_0, not that H_0 is actually true. If the same inference problem is thought of as a decision problem, we focus on two hypotheses and give a rule for deciding between them based on the sample evidence. We therefore must focus equally on two probabilities, the probabilities of the two types of error. We must choose one or the other hypothesis and cannot abstain on grounds of insufficient evidence.

Hypothesis testing Such a clear distinction between the two ways of thinking is helpful for understanding. In practice, the two approaches often merge. We continued to call one of the hypotheses in a decision problem H_0. The common practice of *testing hypotheses* mixes the reasoning of significance tests and decision rules as follows:

1. State H_0 and H_a just as in a test of significance.
2. Think of the problem as a decision problem, so that the probabilities of Type I and Type II errors are relevant.
3. Because of Step 1, Type I errors are more serious. So choose an α (significance level) and consider only tests with probability of Type I error no greater than α.
4. Among these tests, select one that makes the probability of a Type II error as small as possible (that is, power as large as possible). If this probability is too large, you will have to take a larger sample to reduce the chance of an error.

Testing hypotheses may seem to be a hybrid approach. It was, historically, the effective beginning of decision-oriented ideas in statistics. An impressive mathematical theory of hypothesis testing was developed between 1928 and 1938 by Jerzy Neyman and the English statistician Egon Pearson. The decision-making approach came later (1940s). Because decision theory in its pure form leaves you with two error probabilities and no simple rule on how to balance them, it has been used less often than either tests of significance or tests of hypotheses. Decision ideas have been applied in testing problems mainly by way of the Neyman-Pearson hypothesis-testing theory. That theory asks you first to choose α, and the influence of Fisher has often led users of hypothesis testing comfortably back to $\alpha = 0.05$ or $\alpha = 0.01$. Fisher, who was exceedingly argumentative, violently attacked the Neyman-Pearson decision-oriented ideas, and the argument still continues.

SUMMARY

P-values are more informative than the reject-or-not result of a fixed level α test. Beware of placing too much weight on traditional values of α, such as $\alpha = 0.05$.

Very small effects can be highly significant (small P), especially when a test is based on a large sample. A statistically significant effect need not be practically important. Plot the data to display the effect you are seeking, and use confidence intervals to estimate the actual value of parameters.

On the other hand, lack of significance does not imply that H_0 is true, especially when the test has low power.

Significance tests are not always valid. Faulty data collection, outliers in the data, and testing a hypothesis on the same data that suggested the hypothesis can invalidate a test. Many tests run at once will probably produce some significant results by chance alone, even if all the null hypotheses are true.

The **power** of a significance test measures its ability to detect an alternative hypotheses. Power against a specific alternative is calculated as the probability that the test will reject H_0 when the alternative is true. This calculation requires knowledge of the sampling distribution of the test statistic under the alternative hypothesis. Increasing the size of the sample increases the power when the significance level remains fixed.

An alternative to significance testing regards H_0 and H_a as two statements of equal status that we must decide between. This **decision theory** point of view regards statistical inference in general as giving rules for making decisions in the presence of uncertainty.

In the case of testing H_0 versus H_a, decision analysis chooses a decision rule on the basis of the probabilities of two types of error. A **Type I error** occurs if H_0 is rejected when it is in fact true. A **Type II error** occurs if H_0 is accepted when in fact H_a is true.

In a fixed level α significance test, the significance level α is the probability of a Type I error, and the power against a specific alternative is 1 minus the probability of a Type II error for that alternative.

SECTION 6.3 EXERCISES

6.45 Which of the following questions does a test of significance answer?
(a) Is the sample or experiment properly designed?
(b) Is the observed effect due to chance?
(c) Is the observed effect important?

6.46 In a study of the suggestion that taking vitamin C will prevent colds, 400 subjects are assigned at random to one of two groups. The experimental group takes a vitamin C tablet daily, while the control group takes a placebo. At the end of the experiment, the researchers calculate the difference between the percents of subjects in the two groups who were free of colds. This difference is statistically significant ($P = 0.03$) in favor of the vitamin C group. Can we conclude that vitamin C has a strong effect in preventing colds? Explain your answer.

6.47 Every user of statistics should understand the distinction between statistical significance and practical importance. A sufficiently large sample will declare very small effects statistically significant. Let us suppose that Scholastic Aptitude Test mathematics (SAT-M) scores in the absence of coaching vary normally with mean $\mu = 475$ and $\sigma = 100$. Suppose further that coaching may change μ but does not change σ. An increase in the SAT-M score from 475 to 478 is of no importance in seeking admission to college, but this unimportant change can be statistically very significant. To see this, calculate the *P*-value for the test of

$$H_0: \mu = 475$$
$$H_a: \mu > 475$$

in each of the following situations:

(a) A coaching service coaches 100 students; their SAT-M scores average $\bar{x} = 478$.
(b) By the next year, the service has coached 1000 students; their SAT-M scores average $\bar{x} = 478$.
(c) An advertising campaign brings the number of students coached to 10,000; their average score is still $\bar{x} = 478$.

6.48 Give a 99% confidence interval for the mean SAT-M score μ after coaching in each part of the previous exercise. For large samples, the confidence interval says, "Yes, the mean score is higher after coaching, but only by a small amount."

6.49 As in the previous exercises, suppose that SAT-M scores vary normally with $\sigma = 100$. One hundred students go through a rigorous training program designed to raise their SAT-M scores by improving their mathematics skills. Carry out a test of

$$H_0: \mu = 475$$
$$H_a: \mu > 475$$

in each of the following situations:

(a) The students' average score is $\bar{x} = 491.4$. Is this result significant at the 5% level?
(b) The average score is $\bar{x} = 491.5$. Is this result significant at the 5% level?
The difference between the two outcomes in (a) and (b) is of no importance. Beware attempts to treat $\alpha = 0.05$ as sacred.

6.50 A local television station announces a question for a call-in opinion poll on the six o'clock news and then gives the response on the eleven o'clock news. Today's question concerns a proposed gun-control ordinance. Of the 2372 calls received, 1921 oppose the new law. The station, following standard statistical practice, makes a confidence statement: "81% of the Channel 13 Pulse Poll sample oppose gun control. We can be 95% confident that the proportion of all viewers who oppose the law is within 1.6% of the sample result." Is the station's conclusion justified? Explain your answer.

6.51 A researcher looking for evidence of extrasensory perception (ESP) tests 500 subjects. Four of these subjects do significantly better ($P < 0.01$) than random guessing.

(a) Is it proper to conclude that these four people have ESP? Explain your answer.

(b) What should the researcher now do to test whether any of these four subjects have ESP?

6.52 The text cites an example in which researchers carried out 77 separate significance tests, of which two were significant at the 5% level. Suppose that these tests are independent of each other. (In fact they were not independent, because all involved the same subjects.) If all of the null hypotheses are true, each test has probability 0.05 of being significant at the 5% level.

(a) What is the distribution of the number X of tests that are significant?

(b) Find the probability that two or more of the tests are significant.

The following exercises concern the optional sections beginning on page 477:

6.53 Example 6.12 (page 459) gives a test of a hypothesis about the SAT scores of California high school students based on an SRS of 500 students. The hypotheses are

$$H_0: \mu = 450$$
$$H_a: \mu > 450$$

Assume that the population standard deviation is $\sigma = 100$. The test rejects H_0 at the 1% level of significance when $z \geq 2.326$, where

$$z = \frac{\bar{x} - 450}{100/\sqrt{500}}$$

Is this test sufficiently sensitive to usually detect an increase of 10 points in the population mean SAT score? Answer this question by calculating the power of the test against the alternative $\mu = 460$.

6.54 Example 6.16 (page 463) discusses a test about the mean contents of cola bottles. The hypotheses are

$$H_0: \mu = 300$$
$$H_a: \mu < 300$$

The sample size is $n = 6$, and the population is assumed to have a normal distribution with $\sigma = 3$. A 5% significance test rejects H_0 if $z \le -1.645$, where the test statistic z is

$$z = \frac{\bar{x} - 300}{3/\sqrt{6}}$$

Power calculations help us see how large a shortfall in the bottle contents the test can be expected to detect.

(a) Find the power of this test against the alternative $\mu = 299$.

(b) Find the power against the alternative $\mu = 295$.

(c) Is the power against $\mu = 290$ higher or lower than the value you found in (b)? Explain why this result makes sense.

6.55 Increasing the sample size increases the power of a test when the level α is unchanged. Suppose that in the previous exercise a sample of n bottles had been measured (in that exercise, $n = 6$). The 5% significance test still rejects H_0 when $z \le -1.645$, but the z statistic is now

$$z = \frac{\bar{x} - 300}{3/\sqrt{n}}$$

(a) Find the power of this test against the alternative $\mu = 299$ when $n = 25$.

(b) Find the power against $\mu = 299$ when $n = 100$.

6.56 In Example 6.10 (page 456) a company medical director failed to find significant evidence that the mean blood pressure of a population of executives differed from the national mean $\mu = 128$. The medical director now wonders if the test used would detect an important difference if one were present. For the SRS of size 72 from a population with standard deviation $\sigma = 15$, the z statistic is

$$z = \frac{\bar{x} - 128}{15/\sqrt{72}}$$

The two-sided test rejects

$$H_0\colon \mu = 128$$

at the 5% level of significance when $|z| \ge 1.96$.

(a) Find the power of the test against the alternative $\mu = 134$.

(b) Find the power of the test against $\mu = 122$. Can the test be relied on to detect a mean that differs from 128 by 6?

(c) If the alternative were farther from H_0, say $\mu = 136$, would the power be higher or lower than the values calculated in (a) and (b)?

6.57 You have an SRS of size $n = 9$ from a normal distribution with $\sigma = 1$. You wish to test

$$H_0: \mu = 0$$
$$H_a: \mu > 0$$

You decide to reject H_0 if $\bar{x} > 0$ and to accept H_0 otherwise.

(a) Find the probability of a Type I error, that is, the probability that your test rejects H_0 when in fact $\mu = 0$.

(b) Find the probability of a Type II error when $\mu = 0.3$. This is the probability that your test accepts H_0 when in fact $\mu = 0.3$.

(c) Find the probability of a Type II error when $\mu = 1$.

6.58 Use the result of Exercise 6.54 to give the probabilities of Type I and Type II error for the test discussed there. Take the alternative hypothesis to be $\mu = 295$.

6.59 Use the result of Exercise 6.53 to give the probability of Type I error for the test in that exercise. Then find its probability of Type II error when the alternative is $\mu = 460$.

6.60 You must decide which of two discrete distributions a random variable X has. We will call the distributions p_0 and p_1. Here are the probabilities they assign to the values x of X:

x	0	1	2	3	4	5	6
p_0	.1	.1	.1	.1	.2	.1	.3
p_1	.2	.1	.1	.2	.2	.1	.1

You have a single observation on X and wish to test

$$H_0: p_0 \text{ is correct}$$
$$H_a: p_1 \text{ is correct}$$

One possible decision procedure is to accept H_0 if $X = 4$ or $X = 6$ and reject H_0 otherwise.

(a) Find the probability of a Type I error, that is, the probability that you reject H_0 when p_0 is the correct distribution.

(b) Find the probability of a Type II error.

6.61 You are designing a computerized medical diagnostic program. The program will scan the results of routine medical tests (pulse rate, blood pressure, urinalysis, etc.) and either clear the patient or refer the case to a doctor. The program will be used as part of a preventive medicine system to screen many thousands of persons who do not have specific medical complaints. The program makes a decision about each patient.

(a) What are the two hypotheses and the two types of error that the program can make? Describe the two types of error in terms of "false positive" and "false negative" test results.

(b) The program can be adjusted to decrease one error probability, at the cost of an increase in the other error probability. Which error probability would you choose to make smaller, and why? (This is a matter of judgment. There is no single correct answer.)

6.62 The acceptance sampling test in Example 6.19 (page 484) has probability 0.05 of rejecting a good lot of bearings and probability 0.08 of accepting a bad lot. The consumer of the bearings may imagine that acceptance sampling guarantees that most accepted lots are good. Alas, it is not so. Suppose that 90% of all lots shipped by the producer are bad.

(a) Draw a tree diagram for shipping a lot (the branches are "bad" and "good") and then inspecting it (the branches at this stage are "accept" and "reject").

(b) Write the appropriate probabilities on the branches, and find the probability that a lot shipped is accepted.

(c) Use the definition of conditional probability or Bayes's formula (page 351) to find the probability that a lot is bad given that the lot is accepted. This is the proportion of bad lots among the lots that the sampling plan accepts.

CHAPTER 6 EXERCISES

6.63 Patients with chronic kidney failure may be treated by dialysis, using a machine that removes toxic wastes from the blood, a function normally performed by the kidneys. Kidney failure and dialysis can cause other changes, such as retention of phosphorus, that must be corrected by changes in diet. A study of the nutrition of dialysis patients measured the level of phosphorus in the blood of several patients on six occasions. Here are the data for one patient (milligrams of phosphorus per deciliter of blood):

5.6 5.1 4.6 4.8 5.7 6.4

The measurements are separated in time and can be considered an SRS of the patient's blood phosphorus level. If this level varies normally with $\sigma = 0.9$ mg/dl, give a 90% confidence interval for the mean blood phosphorus level. (Data provided by Joan M. Susic.)

6.64 The normal range of phosphorus in the blood is considered to be 2.6 to 4.8 mg/dl. Is there strong evidence that the patient in the previous exercise has a mean phosphorus level that exceeds 4.8?

6.65 Sulfur compounds cause "off-odors" in wine. Oenologists (wine experts) have determined the odor threshold, the lowest concentration of a compound that the human nose can detect. For example, the odor threshold for dimethyl sulfide (DMS) is given in the oenology literature as 25 micrograms per liter of wine (μg/l). Untrained noses may be less sensitive, however. Here are the DMS odor thresholds for 10 beginning students of oenology:

31 31 43 36 23 34 32 30 20 24

Assume (this is not realistic) that the standard deviation of the odor threshold for untrained noses is known to be $\sigma = 7$ μg/l.

(a) Make a stemplot to verify that the distribution is roughly symmetric with no outliers. (A normal quantile plot confirms that there are no systematic departures from normality.)

(b) Give a 95% confidence interval for the mean DMS odor threshold among all beginning oenology students.

(c) Are you convinced that the mean odor threshold for beginning students is higher than the published threshold, 25 μg/l? Carry out a significance test to justify your answer.

6.66 A government report gives a 99% confidence interval for the 1988 median family income as $\$30,853 \pm \$397$. This result was calculated by advanced methods from the Current Population Survey, a multistage random sample of about 60,000 households.

(a) Would a 95% confidence interval be wider or narrower? Explain your answer.

(b) Would the null hypothesis that the 1988 median family income was \$32,000 be rejected at the 1% significance level in favor of the two-sided alternative?

6.67 An agronomist examines the cellulose content of a variety of alfalfa hay. Suppose that the cellulose content in the population has standard deviation $\sigma = 8$ mg/g. A sample of 15 cuttings has mean cellulose content $\bar{x} = 145$ mg/g.

(a) Give a 90% confidence interval for the mean cellulose content in the population.

(b) A previous study claimed that the mean cellulose content was $\mu = 140$ mg/g, but the agronomist believes that the mean is higher than that figure. State H_0 and H_a and carry out a significance test to see if the new data support this belief.

(c) The statistical procedures used in (a) and (b) are valid when several assumptions are met. What are these assumptions?

6.68 In a study of possible iron deficiency in infants, researchers compared several groups of infants who were following different feeding patterns. One group of 26 infants was being breast-fed. At 6 months of age, these children had a mean hemoglobin level of $\bar{x} = 12.9$ grams per 100 milliliters of blood and a standard deviation of 1.6. Taking the standard deviation to be the population value σ, give a 95% confidence interval for the mean hemoglobin level of breast-fed infants. What assumptions are required for the validity of the method you used to get the confidence interval?

6.69 Statisticians prefer large samples. Describe briefly the effect of increasing the size of a sample (or the number of subjects in an experiment) on each of the following:
(a) The width of a level C confidence interval.
(b) The P-value of a test, when H_0 is false and all facts about the population remain unchanged as n increases.
(c) The power of a fixed level α test, when α, the alternative hypothesis, and all facts about the population remain unchanged.

6.70 A roulette wheel has 18 red slots among its 38 slots. You observe many spins and record the number of times that red occurs. Now you want to use these data to test whether the probability of a red has the value that is correct for a fair roulette wheel. State the hypotheses H_0 and H_a that you will test. (We will describe the test for this situation in Chapter 8.)

6.71 When asked to explain the meaning of "statistically significant at the $\alpha = 0.05$ level," a students says, "This means there is only probability 0.05 that the null hypothesis is true." Is this an essentially correct explanation of statistical significance? Explain your answer.

6.72 Another student, when asked why statistical significance appears so often in research reports, says, "Because saying that results are significant tells us that they cannot easily be explained by chance variation alone." Do you think that this statement is essentially correct? Explain your answer.

6.73 A study compares two groups of mothers with young children who were on welfare 2 years ago. One group attended a voluntary training program offered free of charge at a local vocational school and advertised in the local news media. The other group did not choose to attend the training program. The study finds a significant difference ($P < 0.01$) between the proportions of the mothers in the two groups who are still on welfare. The difference is not only significant but quite large. The report says that with 95% confidence the percent of the nonattending group still on welfare is $21\% \pm 4\%$ higher than that of the group who attended the program. You are on the staff of a member of Congress who is interested in the plight of welfare mothers and who asks you about the report.
(a) Explain briefly and in nontechnical language what "a significant difference ($P < 0.01$)" means.
(b) Explain clearly and briefly what "95% confidence" means.
(c) Is this study good evidence that requiring job training of all welfare

mothers would greatly reduce the percent who remain on welfare for several years?

CHAPTER 6 COMPUTER EXERCISES

6.74 Figure 6.2 (page 432) demonstrates the behavior of a confidence interval in repeated sampling by showing the results of 25 samples from the same population. Now you will do a similar demonstration. Suppose that (unknown to the researcher) the mean SAT-M score of all California high school seniors is $\mu = 460$, and that the standard deviation is known to be $\sigma = 100$. The scores vary normally.

(a) Simulate the drawing of 25 SRSs of size $n = 100$ from this population. In Minitab, for example, you can put an SRS of 100 scores into each of columns 1 to 25 with the command

```
MTB > random 100 into c1-c25;
 SUBC > normal 460 100.
```

(b) The 95% confidence interval for the population mean μ has the form $\bar{x} \pm m$. What is the margin of error m? (Remember that we know $\sigma = 100$.)

(c) Use your software to calculate the 95% confidence interval for μ when $\sigma = 100$ for each of your 25 samples. Minitab does this with the single command

```
MTB > zinterval sigma=100, c1-c25
```

Verify the computer's calculations by checking the interval given for the first sample against your result in (b). Use the $\bar{x}$ reported by the software.

(d) How many of the 25 confidence intervals contained the true mean $\mu = 460$? If you repeated the simulation, would you expect exactly the same number of intervals to contain μ? In a very large number of samples, what percent of the confidence intervals would contain μ?

6.75 In the previous exercise you simulated the SAT-M scores of 25 SRSs of 10 California seniors. Now use these samples to demonstrate the behavior of a significance test. We know that the population of all SAT-M scores is normal with standard deviation $\sigma = 100$.

(a) Use your software to carry out a test of

$$H_0\colon \mu = 460$$
$$H_a\colon \mu \neq 460$$

for each of the 25 samples. You can do this in Minitab with the command

```
MTB > ztest mu=460 sigma=100, c1-c25
```

(b) Verify the computer's calculations by using Table A to find the *P*-value of the test for the first of your samples. Use the $\bar{x}$ reported by your software.

(c) How many of your 25 tests reject the null hypothesis at the $\alpha = 0.05$ significance level? (That is, how many have *P*-values 0.05 or smaller?) Because the simulation was done with $\mu = 460$, samples that lead to rejecting H_0 produce the wrong conclusion. In a very large number of samples, what percent would falsely reject the hypothesis?

6.76 Suppose that in fact the mean SAT-M score of California high school seniors is $\mu = 480$. Would the test in the previous exercise usually detect a mean this far from the hypothesized value? This is a question about the power of the test.

(a) Simulate the drawing of 25 SRSs from a normal population with mean $\mu = 480$ and $\sigma = 100$. These represent the results of sampling when in fact the alternative $\sigma = 480$ is true.

(b) Repeat on these new data the test of

$$H_0: \mu = 460$$
$$H_a: \mu \neq 460$$

that you did in the previous exercise. How many of the 25 tests have *P*-values 0.05 or smaller? These tests reject the null hypothesis at the $\alpha = 0.05$ significance level, which is the correct conclusion.

(c) The power of the test against the alternative $\mu = 480$ is the probability that the test will reject H_0: $\mu = 460$ when in fact $\mu = 480$. Calculate this power. In a very large number of samples from a population with mean 480, what percent would reject H_0?

NOTES

1. The correlation here relates two orderings of men, by birth date and by draft number. We will not study the distribution of *r* in this case, but the probability calculation required is not hard. This and other analyses of the draft lottery data appear in Stephen E. Fienberg, "Randomization and social affairs: The 1970 draft lottery," *Science*, 171 (1971), pp. 255–261.
2. R. A. Fisher, "The arrangement of field experiments," *Journal of the Ministry of Agriculture of Great Britain*, 33 (1926), p. 504, quoted in Leonard J. Savage, "On rereading R. A. Fisher," *Annals of Statistics*, 4 (1976), p. 471.
3. Fisher's work is described in a biography by his daughter: Joan Fisher Box, *R. A. Fisher: The Life of a Scientist*, Wiley, New York, 1978.

4. T. D. Sterling, "Publication decisions and their possible effects on inferences drawn from tests of significance—or vice versa," *Journal of the American Statistical Association*, 54 (1959), pp. 30–34. Related comments appear in J. K. Skipper, A. L. Guenther, and G. Nass, "The sacredness of 0.05: A note concerning the uses of statistical levels of significance in social science," *American Sociologist*, 1 (1967), pp. 16–18.
5. This example is cited by William Feller, "Are life scientists overawed by statistics?" *Scientific Research*, February 3, 1969, p. 26.
6. Robert J. Schiller, "The volatility of stock market prices," *Science*, 235 (1987), pp. 33–36.

CHAPTER

7

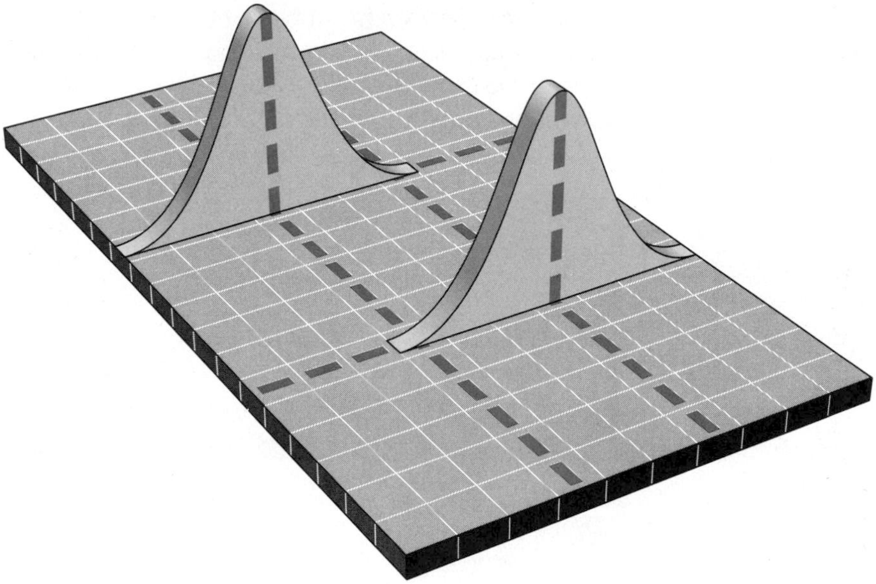

Inference for Distributions

PRELUDE

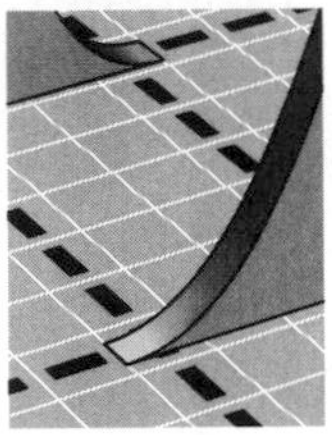

We began our study of data analysis in Chapter 1 by learning graphical and numerical tools for describing the distribution of a single variable and for comparing several distributions. Our study of the practice of statistical inference begins in the same way, with inference about a single distribution and comparison of two distributions. Comparing more than two distributions requires more elaborate methods, which appear in Chapter 10.

Two important aspects of any distribution are its center and spread. If the distribution is normal, we describe its center by the mean μ and its spread by the standard deviation σ. In this chapter, we will meet confidence intervals and significance tests for inference about a population mean μ and for comparing the means of two populations. The previous chapter emphasized the reasoning of tests and confidence intervals; now we emphasize statistical practice, so we no longer assume that population standard deviations are known. The t procedures for inference about means are among the most common statistical methods. Inference about the spread of a population, as we will see, poses some difficult practical problems. The methods of this chapter allow us to answer questions like these:

- Twenty high school French teachers spend 4 weeks at a summer institute that emphasizes spoken French. They gain an average of 2.5 points on a 36-point test of understanding of spoken French. Is this good evidence that the institute improved the teachers' comprehension of spoken French?
- Do male and female college students differ in "social insight," their ability to appraise other people? We have the scores of almost 300 students on a test that measures social insight. What conclusions can we draw from the data?
- Preliminary studies suggest that adding calcium to the diet reduces blood pressure, especially among black males. Now researchers perform a randomized comparative experiment to compare calcium with a placebo. Do the data convince us that calcium works?
- A bank wants to increase the amount its credit card users charge on their cards. Will eliminating the annual fee for cardholders who charge large amounts do this, or will cash rebates that are a percent of the amount charged be more effective?

• • •

With the principles in hand, we proceed to practice. This chapter describes confidence intervals and significance tests for the mean of a single population and for comparing the means of two populations. Optional sections discuss several other inference problems for a single population and for comparing two populations. Later chapters will present procedures for categorical data, for studying relations among variables, and for comparing more than two populations.

7.1 INFERENCE FOR THE MEAN OF A POPULATION

Both confidence intervals and tests of significance for the mean μ of a normal population are based on the sample mean $\bar{x}$, which estimates the unknown μ. The sampling distribution of $\bar{x}$ depends on σ. This fact causes no difficulty when σ is known. When σ is unknown, however, we must estimate σ even though we are primarily interested in μ. The sample standard deviation s is used to estimate the population standard deviation σ.

The one-sample *t* procedures

Suppose that we have a simple random sample (SRS) of size n from a normally distributed population with mean μ and standard deviation σ. The sample mean $\bar{x}$ then has the normal distribution with mean μ and standard deviation $\sigma/\sqrt{n}$. When σ is not known, we estimate the standard deviation of $\bar{x}$ by $s/\sqrt{n}$. This quantity is called the standard error of the sample mean $\bar{x}$.

> **STANDARD ERROR**
>
> When the standard deviation of a statistic is estimated from the data, the result is called the standard error of the statistic.

The term "standard error" is sometimes used for the actual standard deviation of a statistic, $\sigma/\sqrt{n}$ in the case of $\bar{x}$. The estimated value $s/\sqrt{n}$ is then called the "estimated standard error." In this book we will use the term "standard error" only when the standard deviation of a statistic is estimated from the data. The term has this meaning in the output of many statistical computer packages and in reports of research in many fields that apply statistical methods.

The standardized sample mean

$$z = \frac{\bar{x} - \mu}{\sigma/\sqrt{n}}$$

is the basis of the z procedures for inference about μ when σ is known. This statistic has the standard normal distribution $N(0, 1)$. When we substitute the standard error $s/\sqrt{n}$ for the standard deviation $\sigma/\sqrt{n}$ of $\bar{x}$, the statistic does *not* have a normal distribution. It has a distribution that is new to us, called a t distribution.

THE t DISTRIBUTIONS

Suppose that an SRS of size n is drawn from an $N(\mu, \sigma)$ population. Then the one-sample t statistic

$$t = \frac{\bar{x} - \mu}{s/\sqrt{n}} \tag{7.1}$$

has the t distribution with $n - 1$ degrees of freedom.

degrees of freedom

There is a different t distribution for each sample size. A particular t distribution is specified by giving the *degrees of freedom*. The degrees of freedom for this t statistic come from the sample standard deviation s in the denominator of t. We saw in Chapter 1 that s has $n - 1$ degrees of freedom. This is true because the n deviations $x_i - \bar{x}$ that are used to calculate s always have sum zero. Therefore, any $n - 1$ of the deviations determine the remaining deviation. We think of $n - 1$ of the deviations being free to change, and this number is the degrees of freedom. There are other t statistics with different degrees of freedom, some of which we will meet later in this chapter. We will denote the t distribution with k degrees of freedom by $t(k)$ for short.*

The density curves of the $t(k)$ distributions are similar in shape to the standard normal curve. That is, they are symmetric about 0 and are bell-shaped. The spread of the t distributions is a bit greater than that of the standard normal distribution. This is due to the extra variability caused by substituting the random variable s for the fixed parameter σ. As the degrees of freedom k increase, the $t(k)$ density curve approaches

*The t distributions were discovered in 1908 by William S. Gosset. Gosset was a statistician employed by the Guinness brewing company, which required that he not publish his discoveries under his own name. He therefore wrote under the pen name "Student." The t distribution is often called "Student's t" in his honor.

the $N(0, 1)$ curve ever more closely. This reflects the fact that s approaches σ as the sample size increases. Figure 7.1 compares the density curves of the standard normal distribution and the t distribution with 5 degrees of freedom. The similarity in shape is apparent, as is the fact that the t distribution has more probability in the tails and less in the center than does the standard normal distribution.

Table E in the back of the book gives upper p critical values for the t distributions. For convenience, we have labeled the table entries both by p, the upper tail probability needed for significance tests, and by the confidence level C (in percent) required for confidence intervals. The standard normal critical values from Table D are repeated in the bottom row of entries. The degrees of freedom for this row are given as ∞ because the t critical values approach these standard normal critical values as the degrees of freedom increase. As in the case of the normal table, computer software often makes Table E unnecessary.

With the t distributions to help us, we can analyze samples from normal populations with unknown σ by replacing the standard deviation $\sigma/\sqrt{n}$ of $\bar{x}$ by its standard error $s/\sqrt{n}$ in the z procedures of Chapter 6. The z statistic then becomes the one-sample t statistic of Equation 7.1. We must now employ P-values or critical values from t in place of the corresponding normal values.

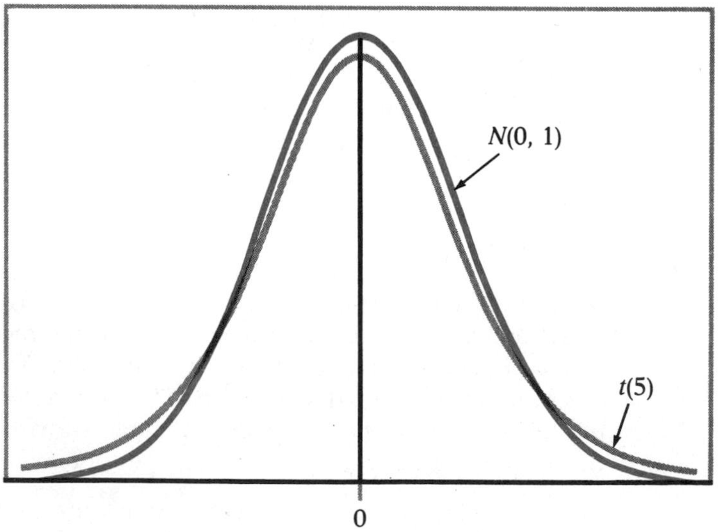

FIGURE 7.1 **Density curves for the standard normal and $t(5)$ distributions. Both are symmetric with center 0. The t distributions have more probability in the tails than the standard normal distribution.**

THE ONE-SAMPLE t PROCEDURES

Suppose that an SRS of size n is drawn from a population having unknown mean μ. A level C confidence interval for μ is

$$\bar{x} \pm t^* \frac{s}{\sqrt{n}}$$

where t^* is the upper $(1 - C)/2$ critical value for the $t(n - 1)$ distribution. This interval is exact when the population distribution is normal and is approximately correct for large n in other cases.

To test the hypothesis H_0: $\mu = \mu_0$ based on an SRS of size n, compute the one-sample t statistic

$$t = \frac{\bar{x} - \mu_0}{s/\sqrt{n}}$$

In terms of a random variable T having the $t(n-1)$ distribution, the P-value for a test of H_0 against

$$\begin{aligned} H_a&: \mu > \mu_0 \quad \text{is} \quad P(T \geq t) \\ H_a&: \mu < \mu_0 \quad \text{is} \quad P(T \leq t) \\ H_a&: \mu \neq \mu_0 \quad \text{is} \quad 2P(T \geq |t|) \end{aligned}$$

These P-values are exact if the population distribution is normal and approximately correct for large n in other cases.

The one-sample t procedures are similar in both reasoning and computational detail to the z procedures of Chapter 6. We can therefore put more emphasis on understanding their effective use in statistical practice.

EXAMPLE 7.1

In an experiment on the metabolism of insects, American cockroaches were fed measured amounts of a sugar solution after being deprived of food for a week and of water for 3 days. After 2, 5, and 10 hours, the researchers dissected some of the cockroaches and measured the amount of sugar in various tissues.[1] Five cockroaches fed the sugar D-glucose and dissected after 10 hours had the following amounts (in micrograms) of D-glucose in their hindguts:

55.95 68.24 52.73 21.50 23.78

The researchers gave a 95% confidence interval for the mean amount of D-glucose in cockroach hindguts under these conditions.

First calculate that

$$\bar{x} = 44.44$$
$$s = 20.741$$

The degrees of freedom are $n - 1 = 4$. From Table E we find that for 95% confidence $t^* = 2.776$. The confidence interval is

$$\begin{aligned}\bar{x} \pm t^* \frac{s}{\sqrt{n}} &= 44.44 \pm 2.776 \frac{20.741}{\sqrt{5}} \\ &= 44.44 \pm 25.75 \\ &= (18.69,\ 70.19)\end{aligned}$$

Comparing this estimate with those for other body tissues and different times before dissection led to new insight into cockroach metabolism and to new ways of eliminating roaches from homes and restaurants. The large margin of error is due to the small sample size and the rather large variation among the cockroaches, reflected in the large value of s. ●

The use of t procedures in Example 7.1 rests on assumptions that cannot easily be checked but are reasonable in this case. The carefully controlled treatment and random assignment of the cockroaches to different sugars and different times before dissection allow the researchers to consider this an SRS of all similarly treated American cockroaches. The assumption that the population distribution is normal cannot be effectively checked with only five observations. Experience with this and similar variables led the researchers to believe that approximate normality holds, despite the wide gap between the two smallest and the three largest observations. In observational data, this might suggest two different species of cockroach. In this case we know that all five cockroaches came from a homogeneous population grown in the laboratory for research purposes.

EXAMPLE 7.2

The first data that we looked at were Simon Newcomb's measurements of the passage time of light. Table 1.1 records his 66 measurements. A normal quantile plot (Figure 7.2) reminds us that when the 2 outliers to the left are omitted, the remaining 64 observations follow a normal distribution quite closely. What result should Newcomb report from these 64 observations?

We want to estimate the mean μ of the distribution of measurements from which Newcomb's 64 are a sample. The sample mean $\bar{x} = 27.750$ estimates μ, but to indicate the precision of this estimate we decide to give a 99% confidence interval. The sample standard deviation is $s = 5.083$. The degrees of freedom are $n - 1 = 63$. Table E contains no entry for 63 degrees of freedom. We therefore use the entry for the next smaller degrees of freedom, which is 60, and find that $t^* = 2.660$. (The next smaller entry gives a slightly larger margin of error than would the exact degrees of freedom, so this is the conservative choice.) The 99% confidence interval is

$$\begin{aligned}\bar{x} \pm t^* \frac{s}{\sqrt{n}} &= 27.750 \pm 2.660 \frac{5.083}{\sqrt{64}} \\ &= 27.75 \pm 1.69 \\ &= (26.06,\ 29.44)\end{aligned}$$

We can report our conclusion in a way that emphasizes the actual estimate: "Newton estimated the passage time of light to be 27.75, with margin of error 1.69 for 99% confidence."

The best modern measurements of the speed of light correspond to a passage time of 33.02 in Newcomb's experiment. Is his result significantly different from this modern value? To answer this question, we test

$$H_0: \mu = 33.02$$
$$H_a: \mu \neq 33.02$$

The t test statistic is

$$t = \frac{\bar{x} - \mu_0}{s/\sqrt{n}} = \frac{27.75 - 33.02}{5.083/\sqrt{64}} = -8.29$$

The P-value is $2P(T \geq 8.29)$, where T has the $t(63)$ distribution. P for a two-sided test is found from Table E by comparing $|t| = 8.29$ to the critical values t^* and *doubling* the corresponding levels p. Table E shows that 8.29 is beyond the 0.0005 critical value. The P-value of the two-sided test is therefore less than 2 times 0.0005, or 0.001. In fact, the P-value is 0 to many decimal places. It is in effect certain that Newcomb's result is not the same as the modern value. That the difference is significant at the 1% level could be seen from the confidence interval. Because 33.02 lies outside the 99% confidence interval, H_0 can be rejected at the 1% level. ●

df = 60

p	.001	.0005
t^*	3.232	3.460

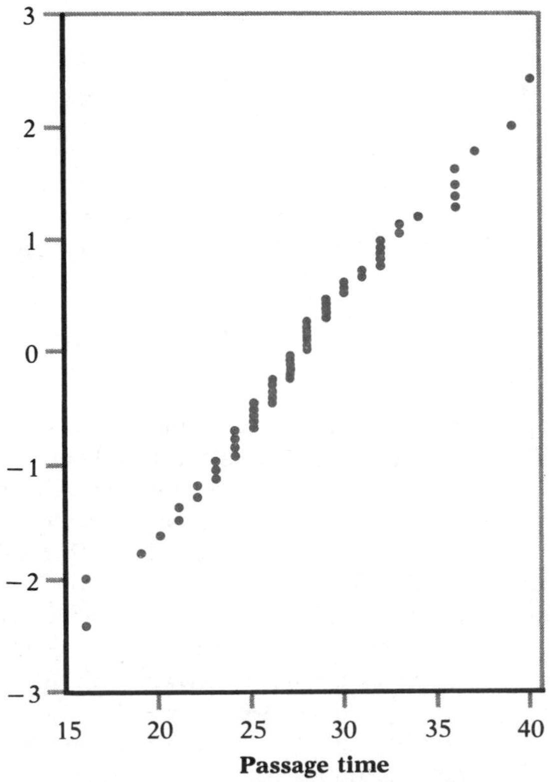

FIGURE 7.2 **Normal quantile plot for Newcomb's passage time data with the outliers omitted (Example 7.2).**

Newcomb's measurements can be regarded as independent observations drawn from the population of all measurements he might make. We verified the normality of the measurements with the normal quantile plot in Figure 7.2. Moreover, the sample mean of 64 observations will have a distribution that is nearly normal even if the population is not normal. Only normality of $\bar{x}$ is required by the t procedures. Use of these procedures in this example is very well justified.

Because the t procedures are so common, all statistical software systems will do the calculations for you. For example, if Newcomb's data are entered into the Minitab system as column C1, the t procedures of Example 7.2 are carried out as follows:

```
MTB> TINTERVAL 99 C1

        N      MEAN     STDEV    SE MEAN       99.0 PERCENT C.I.
C1     64    27.750     5.083      0.635    (  26.062,   29.438)

MTB> TTEST 33.02 C1

TTEST OF MU = 33.020 VS MU N.E. 33.020

        N      MEAN     STDEV    SE MEAN        T      P VALUE
C1     64    27.750     5.083      0.635    -8.29       0.0000
```

Compare these results with those obtained in Example 7.2. The output includes $\bar{x}$, s, and the standard error $s/\sqrt{n}$ of the mean (SE MEAN), as well as the 99% confidence interval and the t statistic with its P-value. Minitab rounds the P-value to four decimal places.

Matched pairs *t* procedures

Newcomb wanted to estimate a constant of nature, so that only a single population was involved. We saw in discussing data production that comparative studies are usually preferred to single-sample investigations because of the protection they offer against confounding. For that reason, inference about a parameter of a single distribution is less common than comparative inference. One common comparative design, however, makes use of single-sample procedures. In a *matched pairs* study, subjects are matched in pairs and the outcomes are compared within each matched pair. The experimenter can toss a coin to assign two treatments to the two subjects in each pair. Matched pairs are also common when randomization is not possible. One situation calling for matched pairs is before-and-after observations on the same subjects, as illustrated in the next example.

matched pairs design

EXAMPLE 7.3

The National Endowment for the Humanities sponsors summer institutes to improve the skills of high school teachers of foreign languages. One such institute hosted 20 French teachers for 4 weeks. At the beginning of the period, the teachers were given the Modern Language Association's listening test of understanding of spoken French. After 4 weeks of immersion in French in and out of class, the listening test was given again. (The actual French spoken in the two tests was different, so that simply taking the first test should not improve the score on the second test.) Table 7.1 gives the pretest and posttest scores. The maximum possible score on the test is 36.[2]

To analyze these data, we first subtract the pretest score from the posttest score to obtain the improvement for each student. These 20 differences form a single sample. They appear in the "Gain" columns in Table 7.1. The first teacher, for example, improved from 32 to 34, so the gain is $34 - 32 = 2$.

To assess whether the institute significantly improved the teachers' comprehension of spoken French, we test

$$H_0\colon \mu = 0$$
$$H_a\colon \mu > 0$$

Here μ is the mean improvement that would be achieved if the entire population of French teachers attended a summer institute. The null hypothesis says that no improvement occurs, and H_a says that posttest scores are higher on the average.

The 20 differences have

$$\bar{x} = 2.5 \quad \text{and} \quad s = 2.893$$

The one-sample t statistic of Equation 7.1 is therefore

$$t = \frac{\bar{x} - 0}{s/\sqrt{n}} = \frac{2.5}{2.893/\sqrt{20}}$$
$$= 3.86$$

df = 19

p	.001	.0005
t^*	3.579	3.883

The P-value is found from the $t(19)$ distribution (remember that the degrees of freedom are 1 less than the sample size). Table E shows that 3.86 lies between the upper 0.001 and 0.0005 critical values of the $t(19)$ distribution. The P-value therefore lies between these values. A computer statistical package gives the value $P = 0.00053$. The improvement in listening scores is very unlikely to be due to chance alone. We have strong evidence that the institute was effective in raising scores. In scholarly publications, the details of routine statistical procedures are omitted; our test would be reported in the form: "The improvement in scores was significant ($t = 3.86$, df = 19, $P = .00053$)."

A 90% confidence interval for the mean improvement in the entire population requires the critical value $t^* = 1.729$ from Table E. The confidence interval is

$$\bar{x} \pm t^* \frac{s}{\sqrt{n}} = 2.5 \pm 1.729 \frac{2.893}{\sqrt{20}}$$
$$= 2.5 \pm 1.12$$
$$= (1.38,\ 3.62)$$

The estimated average improvement is 2.5 points, with margin of error 1.12 for 90% confidence. Though statistically significant, the effect of the institute was rather small. ●

TABLE 7.1 Modern Language Association listening scores for French teachers

Teacher	Pretest	Posttest	Gain	Teacher	Pretest	Posttest	Gain
1	32	34	2	11	30	36	6
2	31	31	0	12	20	26	6
3	29	35	6	13	24	27	3
4	10	16	6	14	24	24	0
5	30	33	3	15	31	32	1
6	33	36	3	16	30	31	1
7	22	24	2	17	15	15	0
8	25	28	3	18	32	34	2
9	32	26	−6	19	23	26	3
10	20	26	6	20	23	26	3

Example 7.3 illustrates how matched pairs data are restated as single-sample data by taking differences within each pair. We are in fact making inferences about a single population, the population of all differences within matched pairs. It is incorrect to ignore the pairs and analyze such data as if we had two samples, one from teachers who had attended an institute and a second from teachers who had not attended. Inference procedures for comparing two samples assume that the samples are selected independently of each other. This assumption does not hold when the same subjects are measured twice. The proper analysis depends on the design used to produce the data.

However, the use of the t procedures in Example 7.3 faces several difficulties. First, the teachers are not an SRS from the population of high school French teachers. There is some selection bias in favor of energetic, committed teachers who are willing to give up 4 weeks of their summer vacation. It is therefore not clear to what population the results apply. Second, a look at the data shows that several of the teachers had pretest scores close to the maximum of 36. They could not improve their scores very much even if their mastery of French increased substantially. This is a weakness in the listening test that is the measuring instrument in this study. The differences in scores may not adequately indicate the effectiveness of the institute. This is one reason why the average increase was small.

A final difficulty facing the t procedures in Example 7.3 is that the data show departures from normality. In a matched pairs analysis, we assume that the population of *differences* has a normal distribution because the t procedures are applied to the differences. In Example 7.3, one teacher actually lost 6 points between the pretest and the posttest. This one subject lowered the sample mean from 2.95 for the other 19 subjects to 2.5 for all 20. A normal quantile plot (Figure 7.3) displays this outlier and also

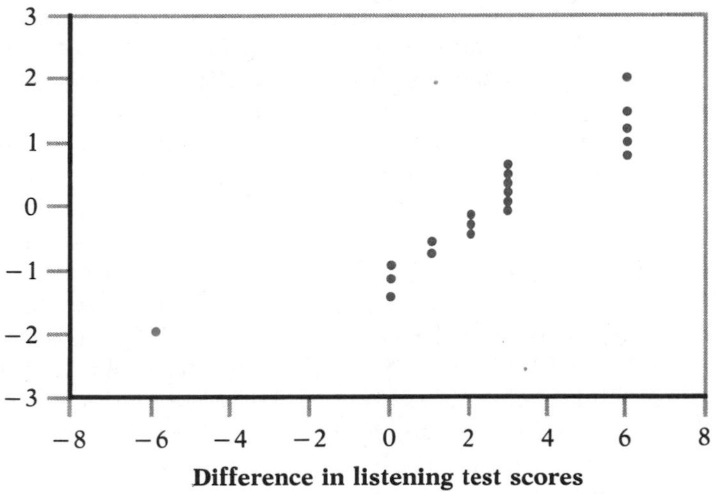

FIGURE 7.3 Normal quantile plot for the change in French listening score (Example 7.3).

granularity due to the fact that only whole-number scores are possible. The overall pattern of the plot is otherwise roughly straight. Does this nonnormality forbid use of the t test? The behavior of the t procedures when the population does not have a normal distribution is one of their most important properties.

Robustness of *t* procedures The results of one-sample t procedures are exactly correct only when the population is normal. Real populations are never exactly normal. The usefulness of the t procedures in practice therefore depends on how strongly they are affected by nonnormality.

ROBUST PROCEDURES

A statistical inference procedure is called robust if the probability calculations required are insensitive to violations of the assumptions made.

The assumption that the population is normal rules out outliers, so the presence of outliers shows that this assumption is not valid. The t procedures are not robust against outliers, because $\bar{x}$ and s are not resistant to outliers. If we dropped the single outlier in Example 7.3, the test statistic would change from $t = 3.86$ to $t = 5.98$, and the P-value would be much smaller. In this case, the outlier makes the test result *less* significant and the margin of error of a confidence interval *larger* than they would otherwise be. The results of the t procedures in Example 7.3 are conservative in the sense that the conclusions show a smaller effect than would be the case if the outlier were not present.

Fortunately, the t procedures are quite robust against nonnormality of the population except in the case of outliers or strong skewness. Larger samples improve the accuracy of P-values and critical values from the t distributions when the population is not normal. This is true for two reasons. First, the sampling distribution of the sample mean $\bar{x}$ from a large sample is close to normal (that's the central limit theorem). We need be less concerned about the normality of the individual observations when the sample is large. Second, as the size n of a sample grows, the sample standard deviation s approaches the population standard deviation σ. This fact is closely related to the law of large numbers. In large samples, s will be an accurate estimate of σ whether or not the population has a normal distribution.

A normal quantile plot or other method to check for skewness and outliers is an important preliminary to the use of t procedures for small samples. For most purposes, the one-sample t procedures can be safely used when $n \geq 15$ unless an outlier or clearly marked skewness is present. Except in the case of small samples, the assumption that the data are an SRS from the population of interest is more crucial than the assumption that the population distribution is normal. Here are practical guidelines for inference on a single mean.[3]

- *Sample size less than 15:* Use t procedures if the data are close to normal. If the data are clearly nonnormal or if outliers are present, do not use t.
- *Sample size at least 15:* The t procedures can be used except in the presence of outliers or strong skewness.
- *Large samples:* The t procedures can be used even for clearly skewed distributions when the sample is large, roughly $n \geq 40$.

Consider, for example, the data whose normal quantile plots appear in Figure 1.21 (pages 75 and 76). Newcomb's data on the passage time of light in Figure 1.21(a) contain two outliers, which make use of t procedures risky even though the sample size is $n = 66$. When the outliers are removed (Figure 1.21(d)), the data are quite normal, and the t procedures are well justified. We would not employ t for inference about the meat hot dog calorie data in Figure 1.21(b), a small sample ($n = 17$) with two clusters and an outlier. On the other hand, the grocery spending data in Figure 1.21(c), although clearly skewed, have no outliers and a moderate sample size ($n = 50$); we would apply the t procedures in this case.

The power of the t test*

The power of a statistical test measures its ability to detect deviations from the null hypothesis. In practice we carry out the test in the hope of showing that the null hypothesis is false, so high power is important. The power of the one-sample t test against a specific alternative value of the population mean μ is the probability that the test will reject the null hypothesis when the alternative value of the mean is true. To calculate the power, we assume a fixed level of significance, usually $\alpha = 0.05$.

Calculation of the exact power of the t test takes into account the estimation of σ by s and is a bit complex. But an approximate calculation that acts as if σ were known is almost always adequate for planning a study. This calculation is very much like that for the z test, presented in Section 6.3. The method is: Write the event that the test rejects H_0 in terms of $\bar{x}$ and then find the probability of this event when the population mean has the alternative value.

EXAMPLE 7.4

It is the winter before the summer language institute of Example 7.3. The director of the institute, thinking ahead to the report he must write, hopes that the planned 20 students will enable him to be quite certain of detecting an average improvement of 2 points in the mean listening score. Is this realistic?

We wish to compute the power of the t test for

$$H_0\colon \mu = 0$$
$$H_a\colon \mu > 0$$

against the alternative $\mu = 2$ when $n = 20$. We must have a rough guess of the size of σ in order to compute the power. In planning a large study, a pilot study is often run for this and other purposes. In this case, listening-score improvements in past summer language institutes have had sample standard deviations of about 3. We therefore take both $\sigma = 3$ and $s = 3$ in our approximate calculation.

The t test with 20 observations rejects H_0 at the 5% significance level if the t statistic

$$t = \frac{\bar{x} - 0}{s/\sqrt{20}}$$

exceeds the upper 5% point of $t(19)$, which is 1.729. Taking $s = 3$, the event that the test rejects H_0 is therefore

$$t = \frac{\bar{x}}{3/\sqrt{20}} \geq 1.729$$
$$\bar{x} \geq 1.729 \frac{3}{\sqrt{20}}$$
$$\bar{x} \geq 1.160$$

*This section can be omitted without loss of continuity.

The power is the probability that $\bar{x} \geq 1.160$ when $\mu = 2$. Taking $\sigma = 3$, this probability is found by standardizing $\bar{x}$,

$$P(\bar{x} \geq 1.160 | \mu = 2) = P\left(\frac{\bar{x} - 2}{3/\sqrt{20}} \geq \frac{1.160 - 2}{3/\sqrt{20}}\right)$$
$$= P(Z \geq -1.252)$$
$$= 1 - .1056 = .8944$$

A true difference of 2 points in the population mean scores will produce significance at the 5% level in 89% of all possible samples. The director can be reasonably confident of detecting a difference this large. ●

Inference for nonnormal populations*

We have not discussed how to do inference about the mean of a clearly nonnormal distribution based on a small sample. If you face this problem, you should consult an expert. Three general strategies are available.

- In some cases a distribution other than a normal distribution will describe the data well. There are many nonnormal models for data, and inference procedures for these models are available.
- Because skewness is the chief barrier to the use of t procedures on data without outliers, you can attempt to transform skewed data so that the distribution is symmetric and as close to normal as possible. Confidence levels and P-values from the t procedures applied to the transformed data will be quite accurate for even moderate sample sizes.
- The third strategy is to use a *distribution-free* inference procedure. Such procedures do not assume that the population distribution has any specific form, such as normal. Distribution-free procedures are often called *nonparametric procedures*.

distribution-free procedure

nonparametric procedure

Each of these strategies can be effective, but each quickly carries us beyond the basic practice of statistics. We emphasize procedures based on normal distributions because they are the most common in practice, because their robustness makes them widely useful, and (most important) because we are first of all concerned with understanding the principles of inference. We will therefore not discuss procedures for nonnormal continuous distributions and will present only one of the many distribution-free procedures that do not require that the population have any specific type of distribution. We will be content with illustrating by example the use of a transformation and of a simple distribution-free procedure.

*This section can be omitted without loss of continuity.

Transforming data When the distribution of a variable is skewed, it often happens that a simple transformation results in a variable whose distribution is symmetric and even close to normal. The most common transformation is the *logarithm*. The logarithm tends to pull in the right tail of a distribution. For example, the data 2, 3, 4, 20 show an outlier in the right tail. Their logarithms 0.30, 0.48, 0.60, 1.30 are much less skewed. Taking logarithms is a possible remedy for right skewness. Instead of analyzing values of the original variable X, we first compute their logarithms and analyze the values of $\log X$. Here is an example of this approach.

logarithm transformation

EXAMPLE 7.5

Table 2.9 (page 214) presents data on the amounts (in grams per mile) of three pollutants in the exhaust of 46 vehicles of the same type, measured under standard conditions prescribed by the Environmental Protection Agency. We will concentrate on emissions of carbon monoxide (CO). We would like to give a confidence interval for the mean emissions μ for this vehicle type.[4]

A normal quantile plot of the CO data from Table 2.9 (Figure 7.4) shows that the distribution is skewed to the right. Because there are no extreme outliers, the sample mean of 46 observations will nonetheless have an approximately normal sampling distribution. The t procedures could be used for approximate inference. For more exact inference, we will seek to transform the data so that the distribution is more nearly normal. Figure 7.5 is a normal quantile plot of the logarithms of the CO measurements. The transformed data are very close to normal, so t procedures will give quite exact results.

The application of the t procedures to the transformed data is straightforward. Call the original CO values from Table 2.9 values of the variable

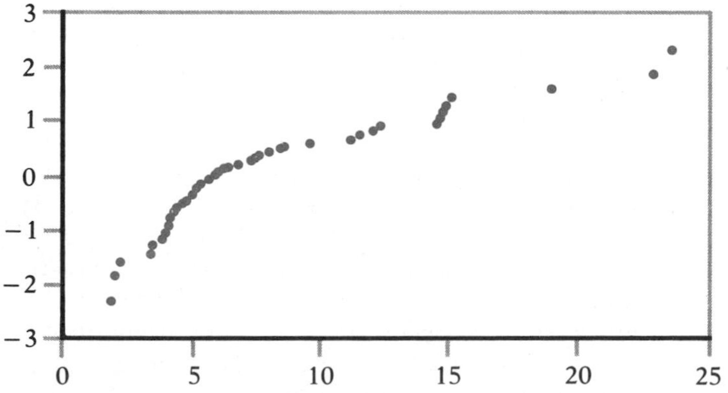

FIGURE 7.4 Normal quantile plot for CO emissions from Example 7.5. The distribution is skewed to the right.

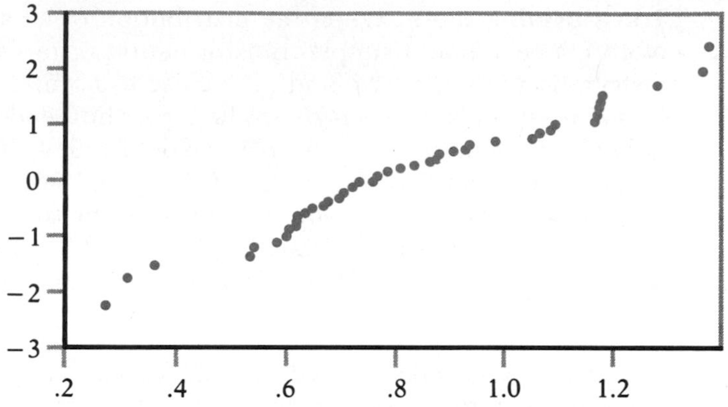

FIGURE 7.5 Normal quantile plot for the logarithms of the CO emissions from Example 7.5. This distribution is close to normal.

X. The transformed data are values of $X^* = \log X$. If the 46 values of X are entered into Minitab as column C1, a 95% confidence interval for the mean μ^* of the transformed values X^* is found as follows:

```
MTB> LET C2 = LOGTEN(C1)

MTB> TINTERVAL 95 C2

      N      MEAN      STDEV      SE MEAN       95.0 PERCENT C.I.
C2   46    0.8198     0.2654       0.0391       ( 0.7409, 0.8986)
```

These commands put the logarithms of the CO measurements in column C2 and then find the 95% t confidence interval for μ^* from the transformed values. For comparison, the 95% t confidence interval for the original mean μ is found from the original data as follows:

```
MTB> TINTERVAL 95 C1

      N      MEAN      STDEV      SE MEAN       95.0 PERCENT C.I.
C1   46     7.960     5.261        0.776       (  6.398,   9.523)
```

The advantage of analyzing transformed data is that use of procedures based on the normal distributions is better justified and the results are more exact. The disadvantage is that a confidence interval for the mean μ in the original scale of emissions measured in grams per mile cannot be recovered from the confidence interval for μ^*. The reason is that the mean μ^* of $\log X$ is *not* the logarithm of the mean μ of X. The Minitab output above illustrates this annoying fact in the case of the sample mean. The mean $\bar{x}$ of the values of X is 7.960. The logarithm of 7.960 is 0.9009, which is not equal to the sample mean 0.8198 of the logarithms of the emissions. So we cannot transform the endpoints of the 95% confidence interval for μ^* back to the original scale and obtain a 95% confidence interval for μ.

In some cases we are content to abandon the original scale and do all our work in the scale that leads to a normal distribution. The use of logarithmic scales in particular is common in science. In Example 7.5, however, we would like to state our conclusions in terms of actual emissions (so many grams of CO per mile driven). If we are interested only in the mean, analysis of the original data is more attractive than working with the transformed data; this analysis is justified by the moderately large sample size. For other purposes that require normality of the individual observations rather than just normality of $\bar{x}$, the transformed data are superior.

The sign test Perhaps the most straightforward way to cope with nonnormal data is to use a *distribution-free* procedure. As the name indicates, these procedures do not require the population distribution to have any specific form, such as normal. Distribution-free significance tests are quite simple and are available in most statistical software systems. Distribution-free tests have two drawbacks. First, they are generally less powerful than tests designed for use with a specific distribution, such as the t test. Second, we must often modify the statement of the hypotheses in order to use a distribution-free test. A distribution-free test concerning the center of a distribution, for example, is usually stated in terms of the median rather than the mean. This is sensible when the distribution may be skewed. But the distribution-free test does not ask the same question (Has the mean changed?) that the t test does. The simplest distribution-free test, and one of the most useful, is the *sign test*. The following example illustrates this test.

sign test

EXAMPLE 7.6

Return to the data of Example 7.3 (page 507) showing the improvement in French listening scores after a summer institute. In that example we used the one-sample t test on these data, despite granularity and an outlier that make the P-value only roughly correct. The sign test is based on the following simple observation: Of the 17 teachers whose scores changed, 16 improved and only 1 did more poorly. This is evidence that the institute improved French listening skills.

To perform a significance test based on the count of teachers whose scores improved, let p be the probability that a randomly chosen teacher would improve if she attended the institute. The null hypothesis of "no effect" says that the posttest is just a repeat of the pretest with no change in ability, so a teacher is equally likely to do better on either test. We therefore want to test

$$H_0: p = 1/2$$
$$H_a: p > 1/2$$

The 17 teachers whose scores changed are 17 independent trials, so the number who improve has the binomial distribution $B(17, 1/2)$ if H_0 is true. The P-value for the observed count 16 is therefore $P(X \geq 16)$, where X has the $B(17, 1/2)$ distribution. You can compute this probability with computer software or from the binomial probability formula, Equation 5.3 (page 378):

$$\begin{aligned} P(X \geq 16) &= P(X = 16) + P(X = 17) \\ &= \binom{17}{16}\left(\frac{1}{2}\right)^{16}\left(\frac{1}{2}\right)^{1} + \binom{17}{17}\left(\frac{1}{2}\right)^{17}\left(\frac{1}{2}\right)^{0} \\ &= (17)\left(\frac{1}{2}\right)^{17} + \left(\frac{1}{2}\right)^{17} \\ &= .00014 \end{aligned}$$

As in Example 7.3, there is very strong evidence that participation in the institute has improved performance on the listening test. ●

There are several varieties of sign test, all based on counts and the binomial distribution. The sign test for matched pairs (Example 7.6) is the most useful. The null hypothesis of "no effect" is then always H_0: $p = 1/2$. The alternative can be one-sided in either direction or two-sided, depending on the type of change we are looking for. The test gets its name from the fact that we only look at the signs of the differences, not their actual values.

THE SIGN TEST FOR MATCHED PAIRS

Ignore pairs with difference 0; the number of trials n is the count of the remaining pairs. The test statistic is the count X of pairs with a positive difference. P-values for X are based on the binomial $B(n, 1/2)$ distribution.

The matched pairs t test in Example 7.3 tested the hypothesis that the mean of the distribution of differences (score after the institute minus score before) is 0. The sign test in Example 7.6 is in fact testing the hypothesis that the *median* of the differences is 0. If p is the probability that a difference is positive, then $p = 1/2$ when the median is 0. This is true because the median of the distribution is the point with probability $1/2$ lying to its right. As Figure 7.6 illustrates, $p > 1/2$ when the median is greater than 0, again because the probability to the right of the median is always $1/2$. Let η (the Greek letter eta) stand for the median of the difference in scores for an entire population of French teachers who might attend a summer institute. Then the sign test of H_0: $p = 1/2$ against

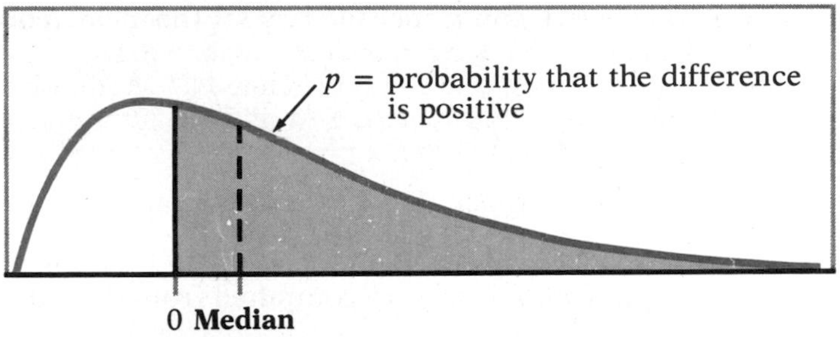

FIGURE 7.6 Why the sign test tests the median difference: When the median is greater than 0, the probability p of a positive difference is greater than $1/2$, and vice-versa.

H_a: $p > 1/2$ is a test of

$$H_0\colon \eta = 0$$
$$H_0\colon \eta > 0$$

The sign test in Example 7.6 makes no use of the actual scores—it just counts how many teachers improved. The teachers whose scores did not change were ignored altogether. Because the sign test uses so little of the available information, it is much less powerful than the t test when the population is close to normal. There are other distribution-free tests that are more powerful than the sign test.[5]

SUMMARY

Significance tests and confidence intervals for the mean μ of a normal population are based on the sample mean $\bar{x}$ of an SRS. Because of the central limit theorem, the resulting procedures are approximately correct for other population distributions when the sample is large.

The standardized sample mean, or **one-sample z statistic,**

$$z = \frac{\bar{x} - \mu}{\sigma/\sqrt{n}}$$

has the $N(0, 1)$ distribution. If the standard deviation $\sigma/\sqrt{n}$ of $\bar{x}$ is replaced by the **standard error** $s/\sqrt{n}$, the **one-sample t statistic**

$$t = \frac{\bar{x} - \mu}{s/\sqrt{n}}$$

has the **t distribution** with $n - 1$ degrees of freedom.

There is a t distribution for every positive **degrees of freedom** k. All are symmetric distributions similar in shape to normal distributions. The $t(k)$ distribution approaches the $N(0, 1)$ distribution as k increases.

An exact level C confidence interval for the mean μ of a normal population is

$$\bar{x} \pm t^* \frac{s}{\sqrt{n}}$$

where t^* is the upper $(1 - C)/2$ critical value of the $t(n - 1)$ distribution.

Significance tests for H_0: $\mu = \mu_0$ are based on the t statistic. P-values or fixed significance levels are computed from the $t(n - 1)$ distribution.

These one-sample procedures are used to analyze **matched pairs** data by first taking the differences within the matched pairs to produce a single sample.

The t procedures are relatively **robust** against nonnormal populations, especially for larger sample sizes. The t procedures are useful for nonnormal data when $n \geq 15$ unless the data show outliers or strong skewness.

The power of the t test is calculated like that of the z test, using an approximate value for both σ and s.

Small samples from skewed populations can sometimes be analyzed by first applying a **transformation** (such as the logarithm) to obtain an approximately normally distributed variable. The t procedures then apply to the transformed data.

The **sign test** is a **distribution-free test** because it uses probability calculations that are correct for a wide range of population distributions.

The sign test for "no treatment effect" in matched pairs counts the number of positive differences. The P-value is computed from the $B(n, 1/2)$ distribution, where n is the number of non-0 differences. The sign test is less powerful than the t test in cases where use of the t test is justified.

SECTION 7.1 EXERCISES

If you are not using computer software, find the P-values in these exercises by using Table E to give two values between which P lies.

7.1 What critical value t^* from Table E should be used for a confidence interval for the mean of the population in each of the following situations?
(a) A 95% confidence interval based on $n = 10$ observations.
(b) A 99% confidence interval from an SRS of 20 observations.
(c) An 80% confidence interval from a sample of size 7.

7.2 The one-sample t statistic for testing

$$H_0: \mu = 0$$
$$H_a: \mu > 0$$

from a sample of $n = 15$ observations has the value $t = 1.82$.

(a) What are the degrees of freedom for this statistic?

(b) Give the two critical values t^* from Table E that bracket t. What are the right-tail probabilities p for these two entries?

(c) Between what two values does the P-value of the test fall?

(d) Is the value $t = 1.82$ significant at the 5% level? Is it significant at the 1% level?

7.3 The one-sample t statistic from a sample of $n = 25$ observations for the two-sided test of

$$H_0: \mu = 64$$
$$H_a: \mu \neq 64$$

has the value $t = 1.12$.

(a) What are the degrees of freedom for t?

(b) Locate the two critical values t^* from Table E that bracket t. What are the right-tail probabilities p for these two values?

(c) Between what two values does the P-value of the test fall? (Note that H_a is two-sided.)

(d) Is the value $t = 1.12$ statistically significant at the 10% level? At the 5% level?

7.4 The one-sample t statistic for a test of

$$H_0: \mu = 10$$
$$H_a: \mu < 10$$

based on $n = 10$ observations has the value $t = -2.25$.

(a) What are the degrees of freedom for this statistic?

(b) Between what two probabilities p from Table E does the P-value of the test fall?

7.5 The scores of four roommates on the Law School Aptitude Test have mean $\bar{x} = 589$ and standard deviation $s = 37$. What is the standard error of the mean?

7.6 A manufacturer of small appliances employs a market research firm to estimate retail sales of its products by gathering information from a sample of retail stores. This month an SRS of 75 stores in the Midwest sales region finds that these stores sold an average of 24 of the manufacturer's hand mixers, with standard deviation 11.

(a) Give a 95% confidence interval for the mean number of mixers sold by all stores in the region.

(b) The distribution of sales is strongly right skewed, because there are many smaller stores and a few very large stores. The use of t in (a) is reasonably safe despite this violation of the normality assumption. Why?

7.7 A bank wonders whether omitting the annual credit card fee for customers who charge at least $2400 in a year would increase the amount charged on its credit card. The bank makes this offer to an SRS of 200 of its existing credit card customers. It then compares how much these customers charge this year with the amount that they charged last year. The mean increase is $332, and the standard deviation is $108.

(a) Is there significant evidence at the 1% level that the mean amount charged increases under the no-fee offer? State H_0 and H_a and carry out a t test.

(b) Give a 99% confidence interval for the mean amount of the increase.

(c) The distribution of the amount charged is skewed to the right, but outliers are prevented by the credit limit that the bank enforces on each card. Use of the t procedures is justified in this case even though the population distribution is not normal. Explain why.

(d) A critic points out that the customers would probably have charged more this year than last even without the new offer, because the economy is more prosperous and interest rates are lower. Briefly describe the design of an experiment to study the effect of the no-fee offer that would avoid this criticism.

7.8 The level of various substances in the blood of kidney dialysis patients is of concern because kidney failure and dialysis can lead to nutritional problems. A researcher performed blood tests on several dialysis patients on six consecutive clinic visits. One variable measured was the level of phosphate in the blood. Phosphate levels for a single person tend to vary normally over time. The data on one patient, in milligrams of phosphate per deciliter (mg/dl) of blood, are given below. (The data are from Joan M. Susic, "Dietary phosphorus intakes, urinary and peritoneal phosphate excretion and clearance in continuous ambulatory peritoneal dialysis patients," M.S. thesis, Purdue University, 1985.)

5.6	5.1	4.6	4.8	5.7	6.4

(a) Calculate the sample mean $\bar{x}$ and its standard error.

(b) Use the t procedures to give a 90% confidence interval for this patient's mean phosphate level.

7.9 Poisoning by the pesticide DDT causes tremors and convulsions. In a study of DDT poisoning, researchers fed several rats a measured amount of DDT. They then measured electrical characteristics of the rats' nervous systems that might explain how DDT poisoning causes tremors. One important variable was the "absolutely refractory period," the time required for a nerve to recover after a stimulus. This period varies normally. Measurements on four rats gave the data below (in milliseconds). (Data from D. L. Shankland, "Involvement of spinal cord and peripheral nerves in DDT-poisoning syndrome in albino rats," *Toxicology and Applied Pharmacology*, 6 (1964), pp. 97–213.)

1.6 1.7 1.8 1.9

(a) Find the mean refractory period $\bar{x}$ and the standard error of the mean.
(b) Give a 90% confidence interval for the mean "absolutely refractory period" for all rats of this strain when subjected to the same treatment.

7.10 The normal range of values for blood phosphate levels is 2.6 to 4.8 mg/dl. The sample mean for the patient in Exercise 7.8 falls above this range. Is this good evidence that the patient's mean level in fact falls above 4.8? State H_0 and H_a and use the data in Exercise 7.8 to carry out a t test. Between which levels from Table E does the P-value lie? Are you convinced that the patient's phosphate level is higher than normal?

7.11 Suppose that the mean "absolutely refractory period" for unpoisoned rats is known to be 1.3 milliseconds. DDT poisoning should slow nerve recovery and so increase this period. Do the data in Exercise 7.9 give good evidence for this supposition? State H_0 and H_a and do a t test. Between what levels from Table E does the P-value lie? What do you conclude from the test?

7.12 In a randomized comparative experiment on the effect of dietary calcium on blood pressure, 54 healthy white males were divided at random into two groups. One group received calcium; the other, a placebo. At the beginning of the study, the researchers measured many variables on the subjects. The paper reporting the study gives $\bar{x} = 114.9$ and $s = 9.3$ for the seated systolic blood pressure of the 27 members of the placebo group.
(a) Give a 95% confidence interval for the mean blood pressure of the population from which the subjects were recruited.
(b) What assumptions about the population and the study design are required by the procedure you used in (a)? Which of these assumptions are important for the validity of the procedure in this case?

7.13 The Acculturation Rating Scale for Mexican Americans (ARSMA) measures the extent to which Mexican Americans have adopted Anglo/English culture. During the development of ARSMA, the test was given to a group of 17 Mexicans. Their scores, from a possible range of 1.00 to 5.00, had $\bar{x} = 1.67$ and $s = 0.25$. Because low scores should indicate a Mexican cultural orientation, these results helped to establish the validity of the test. (Based on I. Cuellar, L. C. Harris, and R. Jasso, "An acculturation scale for Mexican American normal and clinical populations," *Hispanic Journal of Behavioral Sciences*, 2 (1980), pp. 199–217.)
(a) Give a 95% confidence interval for the mean ARSMA score of Mexicans.

(b) What assumptions does your confidence interval require? Which of these assumptions is most important in this case?

7.14 Here are measurements (in millimeters) of a critical dimension for 16 auto engine crankshafts:

224.120	224.001	224.017	223.982	223.989	223.961
223.960	224.089	223.987	223.976	223.902	223.980
224.098	224.057	223.913	223.999		

The mean dimension is supposed to be 224 mm, and the variability of the manufacturing process is unknown. Is there evidence that the mean dimension is not 224 mm?

(a) Check the data graphically for outliers or strong skewness that might threaten the validity of the t procedures.

(b) State H_0 and H_a and carry out a t test. Use Table E to give two levels between which the P-value falls, or give the exact P-value if you use software. What do you conclude?

7.15 How accurate are radon detectors of a type sold to homeowners? To answer this question, university researchers placed 12 detectors in a chamber that exposed them to 105 picocuries per liter (pCi/l) of radon. The detector readings were as follows. (Data provided by Diana Schellenberg, Purdue University School of Health Sciences.)

91.9	97.8	111.4	122.3	105.4	95.0
103.8	99.6	96.6	119.3	104.8	101.7

(a) Make a stemplot of the data. The distribution is somewhat skewed to the right, but not strongly enough to forbid use of the t procedures.

(b) Is there convincing evidence that the mean reading of all detectors of this type differs from the true value 105? Carry out a test in detail and write a brief conclusion.

7.16 Gas chromatography is a sensitive technique used by chemists to measure small amounts of compounds. The response of a gas chromatograph is calibrated by repeatedly testing specimens containing a known amount of the compound to be measured. A calibration study for a specimen containing 1 nanogram (ng) (that's 10^{-9} gram) of a compound gave the following response readings:

21.6 20.0 25.0 21.9

The response is known from experience to vary according to a normal distribution unless an outlier indicates an error in the analysis. Estimate the mean response to 1 ng of this substance, and give the margin of error for your choice of confidence level. Then explain to a chemist who knows no statistics what your margin of error means. (Data from the appendix of D. A. Kurtz (ed.), *Trace Residue Analysis*, American Chemical Society Symposium Series, No. 284, 1985.)

7.17 The embryos of brine shrimp can enter a dormant phase in which metabolic activity drops to a low level. Researchers studying this dormant phase measured the level of several compounds important to normal metabolism. The results were reported in a table, with the note, "Values are means $\pm$ SEM for three independent samples." The table entry for the compound ATP was 0.84 ± 0.01. Biologists reading the article are presumed to be able to decipher this. (From S. C. Hand and E. Gnaiger, "Anaerobic dormancy quantified in *Artemia* embryos," *Science*, 239 (1988), pp. 1425–1427.)

(a) What does the abbreviation "SEM" stand for?

(b) The researchers made three measurements of ATP, which had $\bar{x} = 0.84$. What was the sample standard deviation s for these measurements?

(c) Give a 90% confidence interval for the mean ATP level in dormant brine shrimp embryos.

7.18 The design of controls and instruments has a large effect on how easily people can use them. A student project investigated this effect by asking 25 right-handed students to turn a knob (with their right hands) that moved an indicator by screw action. There were two identical instruments, one with a right-hand thread (the knob turns clockwise) and the other with a left-hand thread (the knob must be turned counterclockwise). The table below gives the times required (in seconds) to move the indicator a fixed distance. (Data provided by Timothy Sturm.)

Subject	Right thread	Left thread	Subject	Right thread	Left thread
1	113	137	14	107	87
2	105	105	15	118	166
3	130	133	16	103	146
4	101	108	17	111	123
5	138	115	18	104	135
6	118	170	19	111	112
7	87	103	20	89	93
8	116	145	21	78	76
9	75	78	22	100	116
10	96	107	23	89	78
11	122	84	24	85	101
12	103	148	25	88	123
13	116	147			

(a) Each of the 25 students used both instruments. Discuss briefly how the experiment should be arranged and how randomization should be used.

(b) The project hoped to show that right-handed people find right-hand threads easier to use. State the appropriate H_0 and H_a about the mean time required to complete the task.
(c) Carry out a test of your hypotheses. Give the P-value and report your conclusions.

7.19 Give a 90% confidence interval for the mean time advantage of right-hand over left-hand threads in the setting of the previous exercise. Do you think that the time saved would be of practical importance if the task were performed many times, for example by an assembly line worker? To help answer this question, find the mean time for right-hand threads as a percent of the mean time for left-hand threads.

7.20 The table below gives the pretest and posttest scores on the MLA listening test in Spanish for 20 high school Spanish teachers who attended an intensive summer course in Spanish. The setting is identical to the one described in Example 7.3. (Data provided by Joseph A. Wipf, Department of Foreign Languages and Literatures, Purdue University.)

Subject	Pretest	Posttest	Subject	Pretest	Posttest
1	30	29	11	30	32
2	28	30	12	29	28
3	31	32	13	31	34
4	26	30	14	29	32
5	20	16	15	34	32
6	30	25	16	20	27
7	34	31	17	26	28
8	15	18	18	25	29
9	28	33	19	31	32
10	20	25	20	29	32

(a) We hope to show that attending the institute improves listening skills. State an appropriate H_0 and H_a. Be sure to identify the parameters appearing in the hypotheses.
(b) Make a graphical check for outliers or strong skewness in the data that you will use in your statistical test, and report your conclusions on the validity of the test.
(c) Carry out a test. Can you reject H_0 at the 5% significance level? At the 1% significance level?
(d) Give a 90% confidence interval for the mean increase in listening score due to attending the summer institute.

7.21 The ARSMA test (Exercise 7.13) was compared with a similar test, the Bicultural Inventory (BI), by administering both tests to 22 Mexican Americans. Both tests have the same range of scores (1.00 to 5.00) and are scaled to have similar means for the groups used to develop them.

There was a high correlation between the two scores, giving evidence that both are measuring the same characteristics. The researchers wanted to know whether the population mean scores for the two tests were the same. The differences in scores (ARSMA − BI) for the 22 subjects had $\bar{x} = 0.2519$ and $s = 0.2767$.

(a) Describe briefly how the administration of the two tests to the subjects should be conducted, including randomization.

(b) Carry out a significance test for the hypothesis that the two tests have the same population mean. Give the *P*-value and state your conclusion.

(c) Give a 95% confidence interval for the difference between the two population mean scores.

7.22 The developer of a new filter for filter-tipped cigarettes claims that it leaves less nicotine in the smoke than does the current filter. Because cigarette brands differ in a number of ways, he tests each filter on one cigarette of each of nine brands and records the difference between the nicotine content for the current filter and the new filter. The mean difference is $\bar{x} = 1.32$ milligrams (mg), and the standard deviation of the differences is $s = 2.35$ mg.

(a) How significant is the observed difference in means? State H_0 and H_a, and give a *P*-value.

(b) Give a 90% confidence interval for the mean amount of additional nicotine removed by the new filter.

7.23 An agricultural field trial compares the yield of two varieties of tomatoes for commercial use. The researchers divide in half each of 10 small plots of land in different locations and plant each tomato variety on one half of each plot. After harvest, they compare the yields in pounds per plant at each location. The 10 differences (variety A − variety B) give the following statistics: $\bar{x} = 0.34$ and $s = 0.83$. Is there convincing evidence that variety A has the higher mean yield? State H_0 and H_a, and give a *P*-value to answer this question.

7.24 The following situations all require inference about a mean or means. Identify each as (1) a single sample, (2) matched pairs, or (3) two independent samples. The procedures of this section apply to cases (1) and (2). We will learn procedures for (3) in the next section.

(a) An education researcher wants to learn whether inserting questions before or after introducing a new concept in an elementary school mathematics text is more effective. He prepares two text segments that teach the concept, one with motivating questions before and the other with review questions after. Each text segment is used to teach a group of children, and their scores on a test over the material are compared.

(b) Another researcher approaches the same problem differently. She prepares text segments on two unrelated topics. Each segment comes in two versions, one with questions before and the other with questions after. Each of a group of children is taught both topics, one (chosen at random) with questions before and the other with questions after. Each child's test scores on the two topics are compared to see which topic he or she learned better.

(c) To evaluate a new analytical method, a chemist obtains a reference specimen of known concentration from the National Institute of Standards and Technology. She then makes 20 measurements of the concentration of this specimen with the new method and checks for bias by comparing the mean result with the known concentration.

(d) Another chemist is evaluating the same new method. He has no reference specimen, but a familiar analytic method is available. He wants to know if the new and old methods agree. He takes a specimen of unknown concentration and measures the concentration 10 times with the new method and 10 times with the old method.

7.25 Exercise 1.18 (page 26) gives a table of the percent of residents 65 years of age and over in each of the 50 states. It does not make sense to use the t procedures (or any other statistical procedures) to give a 95% confidence interval for the mean percent of over-65 residents in the population of the American states. Explain why not.

The following exercises concern the optional material in the sections on the power of the t test and on nonnormal populations.

7.26 The bank in Exercise 7.7 tested a new idea on a sample of 200 customers. Suppose that the bank wanted to be quite certain of detecting a mean increase of $\mu = \$100$ in the amount charged, at the $\alpha = 0.01$ significance level. Perhaps a sample of only $n = 50$ customers would accomplish this. Find the approximate power of the test with $n = 50$ against the alternative $\mu = \$100$ as follows:

(a) What is the t critical value for the one-sided test with $\alpha = 0.01$ and $n = 50$?

(b) Write the criterion for rejecting H_0: $\mu = 0$ in terms of the t statistic. Then take $s = 108$ (an estimate based on the data in Exercise 7.7) and state the rejection criterion in terms of $\bar{x}$.

(c) Assume that $\mu = 100$ (the given alternative) and that $\sigma = 108$ (an estimate from the data in Exercise 7.7). The approximate power is the probability of the event you found in (b), calculated under these assumptions. Find the power. Would you recommend that the bank do a test on 50 customers, or should more customers be included?

7.27 The tomato experts who carried out the field trial described in Exercise 7.23 suspect that the relative lack of significance there is due to low power. They would like to be able to detect a mean difference in yields of 0.5 pound per plant at the 0.05 significance level. Based on the previous study, use 0.83 as an estimate of both the population σ and the value of s in future samples.

(a) What is the power of the test from Exercise 7.23 with $n = 10$ against the alternative $\mu = 0.5$?

(b) If the sample size is increased to $n = 25$ plots of land, what will be the power against the same alternative?

7.28 Exercise 7.21 reports a small study comparing ARSMA and BI, two tests of the acculturation of Mexican Americans. Would this study usually detect a difference in mean scores of 0.2? To answer this question, calculate the approximate power of the test (with $n = 22$ subjects and $\alpha = 0.05$) of

$$H_0: \mu = 0$$
$$H_a: \mu \neq 0$$

against the alternative $\mu = 0.2$. Note that this is a two-sided test.

(a) From Table E, what is the critical value for $\alpha = 0.05$?

(b) Write the criterion for rejecting H_0 at the $\alpha = 0.05$ level. Then take $s = 0.3$, the approximate value observed in Exercise 7.21, and restate the rejection criterion in terms of $\bar{x}$.

(c) Find the probability of this event when $\mu = 0.2$ (the alternative given) and $\sigma = 0.3$ (estimated from the data in Exercise 7.21) by a normal probability calculation. This is the approximate power.

7.29 Apply the sign test to the data in Exercise 7.18 to assess whether the subjects can complete a task with right-hand thread significantly faster than with left-hand thread.

(a) State the hypotheses two ways, in terms of a population median and in terms of the probability of completing the task faster with a right-hand thread.

(b) Carry out the sign test. Find the approximate P-value using the normal approximation to the binomial distributions, and report your conclusion.

7.30 Use the sign test to assess whether the summer institute of Exercise 7.20 improves Spanish listening skills. State the hypotheses, give the P-value using the binomial table, and report your conclusion.

7.31 The paper reporting the results on ARSMA used in Exercise 7.21 does not give the raw data or any discussion of normality. You would like to replace the t procedure used in Exercise 7.21 by a sign test. Can you do this from the available information? Carry out the sign test and state your conclusion, or explain why you are unable to carry out the test.

7.32 In the tomato field trial of Exercise 7.23, variety A had the higher yield in 6 of the 10 locations. Variety B had the higher yield in the other 4 locations. Use the sign test and the binomial table to give a *P*-value for testing the hypothesis that the median difference in yields (A minus B) is positive.

7.33 The data below are the survival times of 72 guinea pigs after they were injected with tubercle bacilli in a medical experiment. Figure 1.26 (page 64) is a normal quantile plot of these data. The distribution is strongly skewed to the right. (Data from T. Bjerkedal, "Acquisition of resistance in guinea pigs infected with different doses of virulent tubercle bacilli," *American Journal of Hygiene*, 72 (1960), pp. 130–148.)

43	45	53	56	56	57	58	66	67	73
74	79	80	80	81	81	81	82	83	83
84	88	89	91	91	92	92	97	99	99
100	100	101	102	102	102	103	104	107	108
109	113	114	118	121	123	126	128	137	138
139	144	145	147	156	162	174	178	179	184
191	198	211	214	243	249	329	380	403	511
522	598								

(a) Give a 95% confidence interval for the mean survival time by applying the *t* procedures to these data.
(b) Transform the data by taking the logarithm of each value. Display the transformed data by either a histogram or a normal quantile plot. The distribution of the logarithms remains somewhat right skewed but is much closer to symmetry than the original distribution. Probability values from the *t* distribution will be more accurate for the transformed data.
(c) Give a 95% confidence interval for the mean of the log survival time by applying the *t* procedures to the transformed data.

7.34 A manufacturer of electric motors tests insulation at a high temperature (250° C) and records the number of hours until the insulation fails. The data for 5 specimens are

300 324 372 372 444

(Data from Wayne Nelson, *Applied Life Data Analysis*, Wiley, New York, 1982, p. 471.) The small sample size makes judgment from the data difficult, but engineering experience suggests that the logarithm of the failure time will have a normal distribution. Take the logarithms of the 5 observations, and use *t* procedures to give a 90% confidence interval for the mean of the log failure time for insulation of this type.

7.2 COMPARING TWO MEANS

A medical researcher is interested in the effect on blood pressure of added calcium in our diet. She conducts a randomized comparative experiment in which one group of subjects receives a calcium supplement and a control group gets a placebo. A psychologist develops a test that measures social insight. He compares the social insight of male college students with that of female college students by giving the test to a large group of students of each sex. A bank wants to know which of two incentive plans will most increase the use of its credit cards. It offers each incentive to a random sample of credit card customers and compares the amount charged during the following 6 months. Two-sample problems such as these are among the most common situations encountered in statistical practice.

> **TWO-SAMPLE PROBLEMS**
>
> - The goal of inference is to compare the responses in two groups.
> - Each group is considered to be a sample from a distinct population.
> - The responses in each group are independent of those in the other group.

A two-sample problem can arise from a randomized comparative experiment that randomly divides the subjects into two groups and exposes each group to a different treatment. Comparing random samples separately selected from two populations is also a two-sample problem. Unlike the matched pairs designs studied earlier, there is no matching of the units in the two samples and the two samples may be of different sizes. Inference procedures for two-sample data differ from those for matched pairs.

We can present two-sample data graphically by a back-to-back stemplot (for small samples) or by side-by-side boxplots (for larger samples). Now we will apply the ideas of formal inference in this setting. When both population distributions are symmetric, and especially when they are at least approximately normal, a comparison of the mean responses in the two populations is most often the goal of inference.

We have two independent samples, from two distinct populations (such as subjects given a treatment and those given a placebo). The same variable is measured for both samples. We will call the variable x_1 in the first population and x_2 in the second because the variable may have different distributions in the two populations. Here is the notation that we will use to describe the two populations:

Population	Variable	Mean	Standard deviation
1	x_1	μ_1	σ_1
2	x_2	μ_2	σ_2

We want to compare the two population means, either by giving a confidence interval for $\mu_1 - \mu_2$ or by testing the hypothesis of no difference, H_0: $\mu_1 = \mu_2$.

Inference is based on two independent SRSs, one from each population. Here is the notation that describes the samples:

Population	Sample size	Sample mean	Sample standard deviation
1	n_1	$\bar{x}_1$	s_1
2	n_2	$\bar{x}_2$	s_2

Throughout this section, the subscripts 1 and 2 show the population to which a parameter or a sample statistic refers.

The two-sample z statistic

The natural estimator of the difference $\mu_1 - \mu_2$ is the difference between the sample means, $\bar{x}_1 - \bar{x}_2$. If we are to base inference on this statistic, we must know its sampling distribution. Our knowledge of probability is equal to the task. First, the mean of the difference $\bar{x}_1 - \bar{x}_2$ is the difference of the means, $\mu_1 - \mu_2$. This follows from the addition rule for means (Equation 4.4 on page 329) and the fact that the mean of any $\bar{x}$ is the same as the mean of the population. Because the samples are independent, their sample means $\bar{x}_1$ and $\bar{x}_2$ are independent random variables. The addition rule for variances (Equation 4.6 on page 332) says that the variance of the difference $\bar{x}_1 - \bar{x}_2$ is the sum of their variances, which is

$$\frac{\sigma_1^2}{n_1} + \frac{\sigma_2^2}{n_2}$$

We now know the mean and variance of the distribution of $\bar{x}_1 - \bar{x}_2$ in terms of the parameters of the two populations. If the two population distributions are both normal, then the distribution of $\bar{x}_1 - \bar{x}_2$ is also normal. This is true because each sample mean alone is normally distributed and because a difference of independent normal random variables is also normal.

EXAMPLE 7.7

The Survey of Study Habits and Attitudes (SSHA) is a psychological test designed to measure the motivation, study habits, and attitudes toward learning of college students. These factors, along with ability, are important in explaining success in school. Scores on the SSHA range from 0 to 200. The mean score for women students on the SSHA is typically somewhat higher than the mean score for men at the same college. Suppose that the scores of all first-year women at Upper Wabash Tech have mean $\mu_1 = 120$ and standard deviation $\sigma_1 = 28$, and the scores for the population of first-year men have mean $\mu_2 = 105$ and standard deviation $\sigma_2 = 35$. A psychologist gives the SSHA test to an SRS of 10 women and an SRS of 12 men from the first-year class and compares the mean scores for these samples.

The difference $\bar{x}_1 - \bar{x}_2$ between the female and male mean scores varies in repeated sampling. The sampling distribution has mean

$$\mu_1 - \mu_2 = 120 - 105 = 15$$

and variance

$$\begin{aligned}\frac{\sigma_1^2}{n_1} + \frac{\sigma_2^2}{n_2} &= \frac{28^2}{10} + \frac{35^2}{12}\\ &= 180.48\end{aligned}$$

The standard deviation of the difference in sample means is therefore $\sqrt{180.48} = 13.43$.

What is the probability that the particular 12 men chosen will have a higher mean SSHA score than the particular 10 women chosen? If scores vary normally, the difference in sample means is also normally distributed. We can standardize $\bar{x}_1 - \bar{x}_2$ by subtracting its mean 15 and dividing by its standard deviation 13.43. Therefore,

$$\begin{aligned}P(\bar{x}_1 - \bar{x}_2 < 0) &= P\left(\frac{(\bar{x}_1 - \bar{x}_2) - 15}{13.43} < \frac{0 - 15}{13.43}\right)\\ &= P(Z < -1.12) = .1314\end{aligned}$$

The men chosen will score higher than the women chosen in about 13% of all samples, despite the superior scores of women in the entire population. ●

As Example 7.7 reminds us, any normal random variable has the $N(0, 1)$ distribution when standardized. We have arrived at a new z statistic.

TWO-SAMPLE z STATISTIC

Suppose that $\bar{x}_1$ is the mean of an SRS of size n_1 drawn from an $N(\mu_1, \sigma_1)$ population and that $\bar{x}_2$ is the mean of an independent SRS of size n_2 drawn from an $N(\mu_2, \sigma_2)$ population. Then the two-sample z statistic

$$z = \frac{(\bar{x}_1 - \bar{x}_2) - (\mu_1 - \mu_2)}{\sqrt{\dfrac{\sigma_1^2}{n_1} + \dfrac{\sigma_2^2}{n_2}}} \tag{7.2}$$

has the standard normal $N(0, 1)$ sampling distribution.

In the unlikely event that both population standard deviations are known, the two-sample z statistic is the basis for inference about $\mu_1 - \mu_2$. Exact z procedures are seldom used, because σ_1 and σ_2 are rarely known. In Chapter 6, we discussed the one-sample z procedures in order to introduce the ideas of inference. Now we pass immediately to the more useful t procedures.

The two-sample *t* procedures

Suppose now that the population standard deviations σ_1 and σ_2 are not known. Following the pattern of the one-sample case, we substitute the standard errors $s_i/\sqrt{n_i}$ for the standard deviations $\sigma_i/\sqrt{n_i}$ in the two-sample z statistic in Equation 7.2. The result is the *two-sample t statistic*

two-sample t statistic

$$t = \frac{(\bar{x}_1 - \bar{x}_2) - (\mu_1 - \mu_2)}{\sqrt{\dfrac{s_1^2}{n_1} + \dfrac{s_2^2}{n_2}}} \tag{7.3}$$

Unfortunately, this statistic does *not* have a t distribution. A t distribution replaces a $N(0, 1)$ distribution only when a single standard deviation in a z statistic is replaced by a standard error. In this case, we replaced two standard deviations by the corresponding standard errors, which does not produce a statistic having a t distribution.

Nonetheless, the statistic in Equation 7.3 is used with t critical values in inference for two-sample problems. There are two ways to do this:

Option 1. The distribution of the statistic t is closely approximated by a t distribution with degrees of freedom computed from the data. The degrees of freedom are generally not a whole number.

Option 2. Procedures based on the statistic t can be used with critical values from the t distribution with degrees of freedom equal to the smaller of n_1-1 and n_2-1. These procedures are always conservative for any two normal populations.

Most statistical software systems use the two-sample t statistic with option 1 for two-sample problems unless the user requests another method. Use of this option without software is a bit complicated. We will therefore present the second, simpler, option first. We recommend that you use option 2 when doing calculations without a computer. If you use a computer package, it should automatically do the calculations for option 1. Here is a statement of the option 2 procedures that includes a statement of just how they are "conservative."

THE TWO-SAMPLE t PROCEDURES

Suppose that an SRS of size n_1 is drawn from a normal population with unknown mean μ_1 and that an independent SRS of size n_2 is drawn from another normal population with unknown mean μ_2. The confidence interval for $\mu_1 - \mu_2$ given by

$$(\bar{x}_1 - \bar{x}_2) \pm t^* \sqrt{\frac{s_1^2}{n_1} + \frac{s_2^2}{n_2}}$$

has confidence level at least C no matter what the population standard deviations may be. Here t^* is the upper $(1 - C)/2$ critical value for the $t(k)$ distribution, with k the smaller of $n_1 - 1$ and $n_2 - 1$.

To test the hypothesis H_0: $\mu_1 = \mu_2$ compute the two-sample t statistic

$$t = \frac{\bar{x}_1 - \bar{x}_2}{\sqrt{\frac{s_1^2}{n_1} + \frac{s_2^2}{n_2}}}$$

and use P-values or critical values for the $t(k)$ distribution. The true P-value or fixed significance level will always be equal to or less than the value calculated from $t(k)$ no matter what values the unknown population standard deviations have.

These two-sample t procedures always err on the safe side, reporting higher P-values and lower confidence than may be true. The gap between what is reported and the truth is quite small unless the sample sizes are both small and unequal. As the sample sizes increase, probability values based on t with the smaller of $n_1 - 1$ and $n_2 - 1$ as the degrees of freedom become more accurate.[6] The following examples illustrate the two-sample t procedures.

EXAMPLE 7.8

An educator believes that new directed reading activities in the classroom will help elementary school pupils improve some aspects of their reading ability. She arranges for a third grade class of 21 students to take part in these activities for an 8-week period. A control classroom of 23 third graders follows the same curriculum without the activities. At the end of the 8 weeks, all students are given a Degree of Reading Power (DRP) test, which measures the aspects of reading ability that the treatment is designed to improve. The data appear in Table 7.2.[7]

First examine the data. A back-to-back stemplot

Control		Treatment
970	1	
860	2	4
773	3	3
8632221	4	3334699
5543	5	23467789
20	6	127
	7	1
5	8	

suggests that there is a mild outlier in the control group but no deviation from normality serious enough to forbid use of t procedures. Separate normal quantile plots for both groups (Figure 7.7) confirm that both are approximately

TABLE 7.2 DRP scores for third graders

Treatment group		Control group	
24	56	42	46
43	59	43	10
58	52	55	17
71	62	26	60
43	54	62	53
49	57	37	42
61	33	33	37
44	46	41	42
67	43	19	55
49	57	54	28
53		20	48
		85	

normal. The scores of the treatment group appear somewhat higher than those of the control group. The summary statistics are

Group	n	$\bar{x}$	s
Treatment	21	51.48	11.01
Control	23	41.52	17.15

Because we hope to show that the treatment (group 1) is better than the control (group 2), the hypotheses are

$$H_0: \mu_1 = \mu_2$$
$$H_a: \mu_1 > \mu_2$$

The two-sample t test statistic is

$$t = \frac{\bar{x}_1 - \bar{x}_2}{\sqrt{\dfrac{s_1^2}{n_1} + \dfrac{s_2^2}{n_2}}} = \frac{51.48 - 41.52}{\sqrt{\dfrac{11.01^2}{21} + \dfrac{17.15^2}{23}}} = 2.31$$

The P-value comes from the $t(k)$ distribution with degrees of freedom k equal to the smaller of

$$n_1 - 1 = 21 - 1 = 20 \quad \text{and} \quad n_2 - 1 = 23 - 1 = 22$$

df = 20

p	.02	.01
t^*	2.197	2.528

The P-value for the one-sided test is $P(T \geq 2.31)$. Comparing 2.31 with the entries in Table E for 20 degrees of freedom, we see that P lies between 0.02 and 0.01. The data strongly support the thesis that directed reading activity improves the DRP score ($t = 2.31$, df $= 20$, $P < 0.02$).

A 95% confidence interval for the mean amount of the improvement in the entire population of third graders uses the critical value $t^* = 2.086$ of the $t(20)$ distribution. The interval is

$$\begin{aligned}(\bar{x}_1 - \bar{x}_2) &\pm t^* \sqrt{\frac{s_1^2}{n_1} + \frac{s_2^2}{n_2}} \\ &= (51.48 - 41.52) \pm 2.086\sqrt{\frac{11.01^2}{21} + \frac{17.15^2}{23}} \\ &= 9.96 \pm 8.99 \\ &= (.97,\ 18.95)\end{aligned}$$

We estimate the mean improvement to be about 10 points, but with a margin of error of almost 9 points. Although we have good evidence of some improvement, the data do not allow a very precise estimate of the size of the average improvement. ●

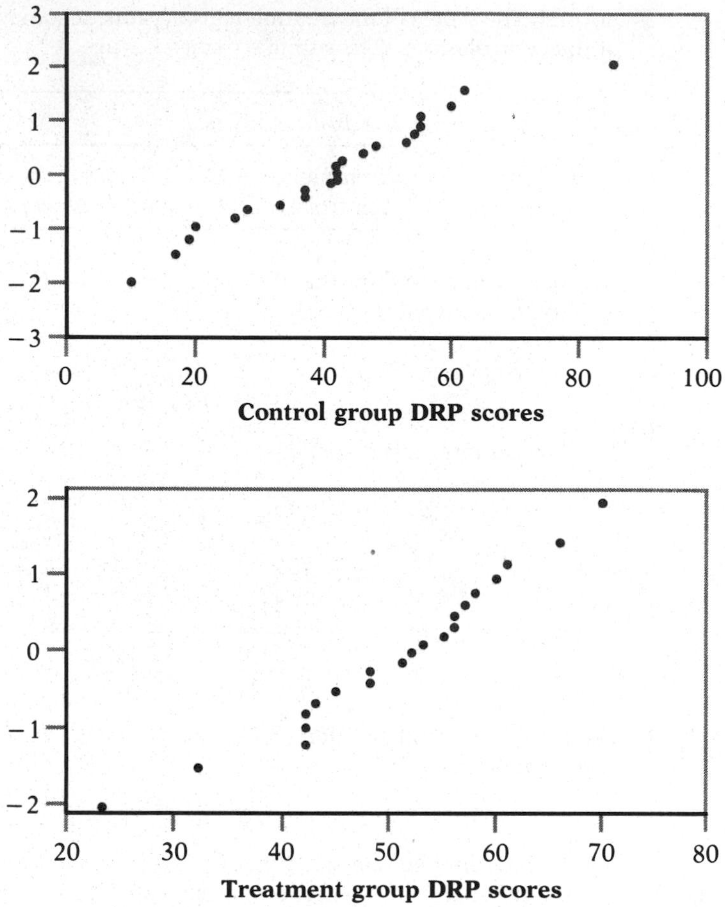

FIGURE 7.7 Normal quantile plots for the DRP scores in Table 7.2.

The design of the study in Example 7.8 is not ideal. Random assignment of students was not possible in a school environment, so existing third-grade classes were used. The effect of the reading programs is therefore confounded with any other differences between the two classes. The classes were chosen to be as similar as possible—for example, in the social and economic status of the students. Extensive pretesting showed that the two classes were on the average quite similar in reading ability at the beginning of the experiment. To avoid the effect of two different teachers, the researcher herself taught reading in both classes during the 8-week period of the experiment. We can therefore be somewhat confident that the two-sample test is detecting the effect of the treatment and not some other difference between the classes. This example is typical of many situations in which an experiment is carried out but randomization is not possible.

EXAMPLE 7.9

The Chapin Social Insight Test is a psychological test designed to measure how accurately the subject appraises other people. The possible scores on the test range from 0 to 41. During the development of the Chapin Test, it was given to several different groups of people. Here are the results for male and female college students majoring in the liberal arts:[8]

Group	Sex	n	$\overline{x}$	s
1	Male	133	25.34	5.05
2	Female	162	24.94	5.44

Do these data support the contention that female and male students differ in average social insight? Because no specific direction for the male/female difference was hypothesized before looking at the data, we choose a two-sided alternative. The hypotheses are

$$H_0: \mu_1 = \mu_2$$
$$H_a: \mu_1 \neq \mu_2$$

The two-sample t statistic is

$$\begin{aligned} t &= \frac{\overline{x}_1 - \overline{x}_2}{\sqrt{\dfrac{s_1^2}{n_1} + \dfrac{s_2^2}{n_2}}} \\ &= \frac{25.34 - 24.94}{\sqrt{\dfrac{5.05^2}{133} + \dfrac{5.44^2}{162}}} \\ &= .654 \end{aligned}$$

df = 100

p	.25	.20
t^*	.677	.845

The P-value is found by comparing 0.654 to critical values for the $t(132)$ distribution and then doubling p because the alternative is two-sided. Table E (for 100 degrees of freedom) shows that 0.654 does not reach the 0.25 critical value, which is the largest upper tail probability in Table E. The P-value is therefore greater than 0.50. The data give no evidence of a male/female difference in mean social insight score ($t = 0.654$, df = 132, $P > 0.50$). ●

The researcher in Example 7.9 did not do an experiment but compared samples from two populations. The large samples imply that the assumption that the populations have normal distributions is of little importance. The sample means will be nearly normal in any case. The major question concerns the population to which the conclusions apply. The student subjects are certainly not an SRS of all liberal arts majors. If they are volunteers from a single college, the sample results may not extend to a wider population.

Comments The two-sample t procedures are more robust than the one-sample t methods. When the sizes of the two samples are equal and the distributions of the two populations being compared have similar shapes, probability values from the t table are quite accurate for a broad range of distributions when the sample sizes are as small as $n_1 = n_2 = 5$.[9] When the two population distributions have different shapes, larger samples are needed. The guidelines given on page 510 for the use of one-sample t procedures can be adapted to two-sample procedures by replacing "sample size" with the "sum of the sample sizes" $n_1 + n_2$. These guidelines are rather conservative, especially when the two samples are of equal size. In planning a two-sample study, you should usually choose equal sample sizes. The two-sample t procedures are most robust against nonnormality in this case, and the conservative probability values are most accurate.

As in the one-sample case, the power of the t test of H_0: $\mu_1 = \mu_2$ can be calculated approximately by a normal probability calculation. The calculation uses approximate values (often guesses based on past data) for both the σ_i and the s_i. Exercise 7.56 leads you through an example.

More accurate levels in the *t* procedures* The two-sample t statistic 7.3 does not have a t distribution. Moreover, the exact distribution changes as the unknown population standard deviations σ_1 and σ_2 change. However, the distribution can be approximated by a t distribution with degrees of freedom given by

$$\text{df} = \frac{\left(\dfrac{s_1^2}{n_1} + \dfrac{s_2^2}{n_2}\right)^2}{\dfrac{1}{n_1 - 1}\left(\dfrac{s_1^2}{n_1}\right)^2 + \dfrac{1}{n_2 - 1}\left(\dfrac{s_2^2}{n_2}\right)^2} \tag{7.4}$$

This approximation is quite accurate when both sample sizes n_1 and n_2 are 5 or larger. The t procedures remain exactly as before except that the t distribution with df degrees of freedom is used to give critical values and P-values.

EXAMPLE 7.10

In the DRP study of Example 7.8 we had for the data in Table 7.2

Group	n	$\bar{x}$	s
1	21	51.48	11.01
2	23	41.52	17.15

*This material can be omitted unless you are using statistical software and wish to understand what the software does.

For greatest accuracy, we will use critical points from the t distribution with degrees of freedom df given by Equation 7.4:

$$\text{df} = \frac{\left(\frac{11.01^2}{21} + \frac{17.15^2}{23}\right)^2}{\frac{1}{20}\left(\frac{11.01^2}{21}\right)^2 + \frac{1}{22}\left(\frac{17.15^2}{23}\right)^2}$$
$$= \frac{344.486}{9.099} = 37.86$$

Notice that the degrees of freedom df is not a whole number.

The conservative 95% confidence interval in Example 7.8 used the critical value $t^* = 2.086$ based on 20 degrees of freedom. A more exact confidence interval replaces this critical value with the critical value for $df = 37.86$ degrees of freedom. We cannot find this critical value exactly without using a computer package. A close approximation can be found by interpolating between the two closest entries (for 30 and 40 degrees of freedom) in Table E. Instead, we just use the 30 degrees of freedom entry in Table E, $t^* = 2.042$. The 95% confidence interval is now

$$(\bar{x}_1 - \bar{x}_2) \pm t^*\sqrt{\frac{s_1^2}{n_1} + \frac{s_2^2}{n_2}} = (51.48 - 41.52) \pm 2.042\sqrt{\frac{11.01^2}{21} + \frac{17.15^2}{23}}$$
$$= 9.96 \pm 8.80$$
$$= (1.16,\ 18.76)$$

This confidence interval is a bit shorter (margin of error 8.80 rather than 8.99) than the conservative interval in Example 7.8. ●

As Example 7.10 illustrates, the two-sample t procedures are exactly as before, except that a t distribution with more degrees of freedom is used. The number df given by Equation 7.4 is always at least as large as the smaller of $n_1 - 1$ and $n_2 - 1$. On the other hand, df is never larger than the sum $n_1 + n_2 - 2$ of the two individual degrees of freedom. The number of degrees of freedom df is generally not a whole number. There is a t distribution with any positive degrees of freedom, even though Table E contains entries only for whole-number degrees of freedom. When df is small and is not a whole number, interpolation between entries in Table E may be needed to obtain an accurate critical value or P-value. Because of this and the need to calculate df, we do not recommend regular use of Equation 7.4 if a computer is not doing the arithmetic. With a computer, the more accurate procedures are painless, as the following example illustrates.

EXAMPLE 7.11

The pesticide DDT causes tremors and convulsions if it is ingested by humans or other mammals. Researchers seek to understand how the convulsions are caused. In a randomized comparative experiment, 6 white rats poisoned with DDT were compared with a control group of 6 unpoisoned rats. Electrical measurements of nerve activity are the main clue to the nature of DDT poisoning. When a nerve is stimulated, its electrical response shows a sharp spike followed by a much smaller second spike. Researchers found that the second spike is larger in rats fed DDT than in normal rats. This observation helps biologists understand how DDT causes tremors.[10]

The researchers measured the amplitude of the second spike as a percentage of the first spike when a nerve in the rat's leg was stimulated. For the poisoned rats the results were

12.207 16.869 25.050 22.429 8.456 20.589

The control group data were

11.074 9.686 12.064 9.351 8.182 6.642

Normal quantile plots (Figure 7.8) show no evidence of outliers or strong skewness. Both populations are reasonably normal, as far as can be judged from six observations. The difference in means is quite large, but in such small samples the sample mean is highly variable. A significance test can help confirm that we are seeing a real effect. Because the researchers did not conjecture in advance that the size of the second spike would increase in rats fed DDT, we test

$$H_0\colon \mu_1 = \mu_2$$
$$H_a\colon \mu_1 \neq \mu_2$$

Here is the output from the SAS statistical software system for these data:[11]

```
                          TTEST PROCEDURE
Variable: SPIKE

GROUP        N           Mean         Std Dev      Std Error
-------------------------------------------------------------
DDT          6     17.60000000      6.34014839     2.58835474
CONTROL      6      9.49983333      1.95005932     0.79610839

Variances         T         DF     Prob>|T|
-------------------------------------------
Unequal      2.9912        5.9       0.0247
Equal        2.9912       10.0       0.0135
```

SAS reports the results of two t procedures, the general two-sample procedure ("Unequal" variances) and a special procedure that assumes the two population variances are equal (see next section). We are interested in the first of these procedures. The two-sample t statistic has the value $t = 2.9912$, the degrees of freedom from Equation 7.4 are df $= 5.9$, and the P-value from the $t(5.9)$ distribution is 0.0247. There is good evidence that the mean size of the secondary spike is larger in rats fed DDT. ●

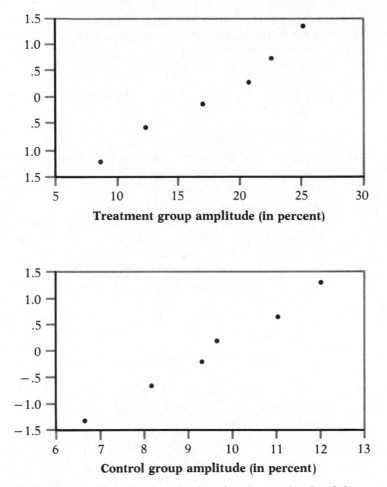

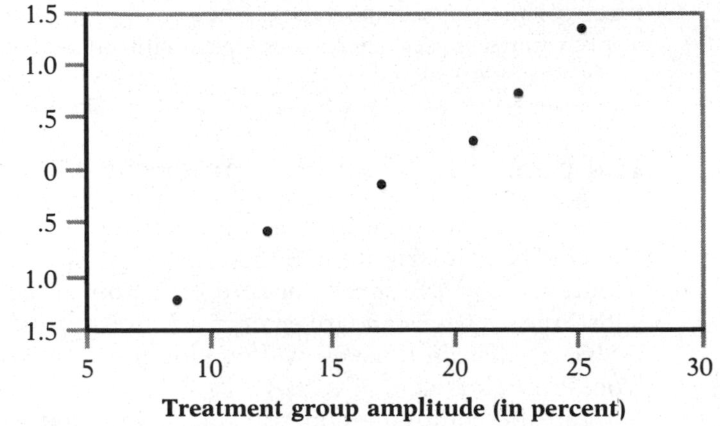

FIGURE 7.8 Normal quantile plots for the amplitude of the second spike as a percent of the first spike in Example 7.11.

Would the conservative test based on 5 degrees of freedom (both $n_1 - 1$ and $n_2 - 1$ are 5) have given a different result in Example 7.11? The statistic is exactly the same, so $t = 2.9912$ as in the example. The conservative P-value is $2P(T \geq 2.9912)$ where T has the $t(5)$ distribution. Table E shows that 2.9912 lies between the 0.02 and 0.01 upper critical values of the $t(5)$ distribution, so P for the two-sided test lies between 0.02 and 0.04. For practical purposes this is the same result as that given by the software. As this example and Example 7.10 suggest, the difference between the t procedures using the conservative and approximately correct distributions is rarely of practical importance. That is why we recommend the simpler approximate procedure for inference without a computer.

The pooled two-sample t procedures*

There is one situation in which a t statistic for comparing two means has exactly a t distribution. Suppose that the two normal population distributions have the *same* standard deviation. In this case we need only substitute a single standard error in a z statistic, and the resulting t statistic has a t distribution. We will develop the z statistic first, as usual, and from it the t statistic.

Call the common—and still unknown—standard deviation of both populations σ. Both sample variances s_1^2 and s_2^2 estimate σ^2. The best way to combine these two estimates is to average them with weights equal to their degrees of freedom. This gives more weight to the information from the larger sample, which is reasonable. The resulting estimator of σ^2 is

$$s_p^2 = \frac{(n_1 - 1)s_1^2 + (n_2 - 1)s_2^2}{n_1 + n_2 - 2} \tag{7.5}$$

pooled estimator of variance

This is called the *pooled estimator* of σ^2 because it combines the information in both samples.

When both populations have variance σ^2, the addition rule for variances says that $\bar{x}_1 - \bar{x}_2$ has variance equal to the *sum* of the individual variances, which is

$$\frac{\sigma^2}{n_1} + \frac{\sigma^2}{n_2} = \sigma^2\left(\frac{1}{n_1} + \frac{1}{n_2}\right)$$

The standardized difference of means in this equal-variance case is therefore

$$z = \frac{(\bar{x}_1 - \bar{x}_2) - (\mu_1 - \mu_2)}{\sigma\sqrt{\dfrac{1}{n_1} + \dfrac{1}{n_2}}}$$

This is a special two-sample z statistic for the case in which the populations have the same σ. Replacing the unknown σ by the estimate s_p gives a t statistic. The degrees of freedom are $n_1 + n_2 - 2$, the sum of the degrees of freedom of the two sample variances. This statistic is the basis of the pooled two-sample t inference procedures.

*This section can be omitted if desired, but it should be read if you plan to read Chapter 10.

THE POOLED TWO-SAMPLE t PROCEDURES

Suppose that an SRS of size n_1 is drawn from a normal population with unknown mean μ_1 and that an independent SRS of size n_2 is drawn from another normal population with unknown mean μ_2. Suppose also that the two populations have the same standard deviation. A level C confidence interval for $\mu_1 - \mu_2$ is

$$(\bar{x}_1 - \bar{x}_2) \pm t^* s_p \sqrt{\frac{1}{n_1} + \frac{1}{n_2}}$$

Here t^* is the upper $(1 - C)/2$ critical value for the $t(n_1 + n_2 - 2)$ distribution.

To test the hypothesis H_0: $\mu_1 = \mu_2$, compute the pooled two-sample t statistic

$$t = \frac{\bar{x}_1 - \bar{x}_2}{s_p \sqrt{\frac{1}{n_1} + \frac{1}{n_2}}} \qquad (7.6)$$

In terms of a random variable T having the $t(n_1+n_2-2)$ distribution, the P-value for a test of H_0 against

$$\begin{aligned} H_a&: \mu_1 > \mu_2 \quad \text{is} \quad P(T \geq t) \\ H_a&: \mu_1 < \mu_2 \quad \text{is} \quad P(T \leq t) \\ H_a&: \mu_1 \neq \mu_2 \quad \text{is} \quad 2P(T \geq |t|) \end{aligned}$$

TABLE 7.3 Seated systolic blood pressure

Calcium group			Placebo group		
Begin	End	Decrease	Begin	End	Decrease
107	100	7	123	124	−1
110	114	−4	109	97	12
123	105	18	112	113	−1
129	112	17	102	105	−3
112	115	−3	98	95	3
111	116	−5	114	119	−5
107	106	1	119	114	5
112	102	10	112	114	2
136	125	11	110	121	−11
102	104	−2	117	118	−1
			130	133	−3

EXAMPLE 7.12 Does increasing the amount of calcium in our diet reduce blood pressure? Examination of a large sample of people revealed a relationship between calcium intake and blood pressure, but such observational studies do not establish causation. Animal experiments then showed that calcium supplements do reduce blood pressure in rats, justifying an experiment with human subjects. A randomized comparative experiment gave one group of 10 black men a calcium supplement for 12 weeks. The control group of 11 black men received a placebo that appeared identical. (In fact, a block design with black and white men as the blocks was used. We will look only at the results for blacks, because the earlier survey suggested that calcium is more effective for blacks.) The experiment was double-blind. Table 7.3 gives the seated systolic (heart contracted) blood pressure for all subjects at the beginning and end of the 12-week period, in millimeters of mercury. Because the researchers were interested in decreasing blood pressure, Table 7.3 also shows the decrease for each subject. An increase appears as a negative entry.[12]

As usual, we first examine the data. To compare the effects of the two treatments, take the response variable to be the amount of the decrease in

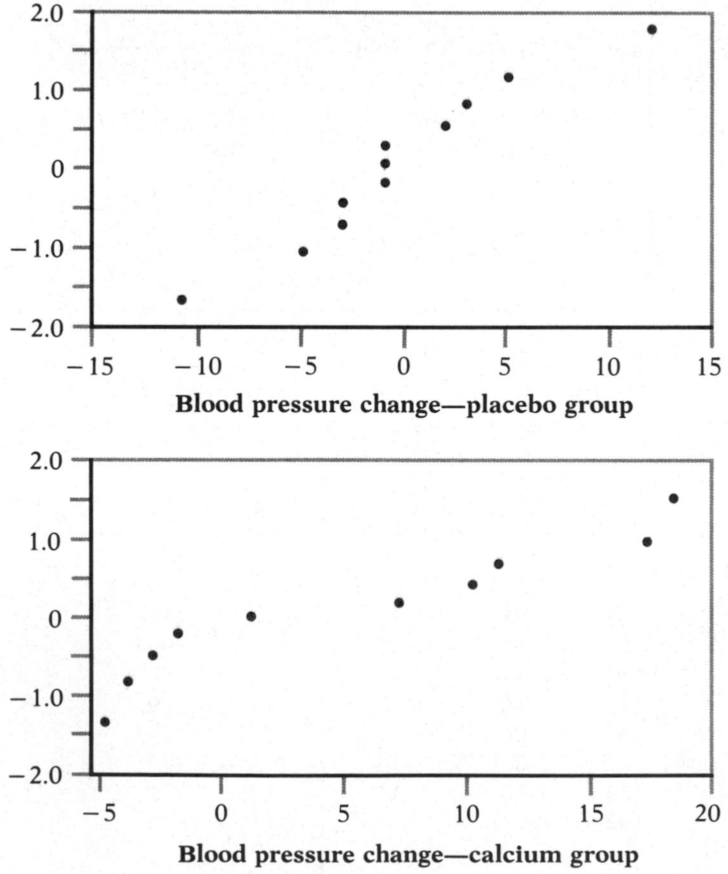

FIGURE 7.9 Normal quantile plots for the decrease in blood pressure from Table 7.3.

blood pressure. Here is a back-to-back stemplot for the responses. (We have split the stems. Notice that negative responses require −0 and 0 to be separate stems, and that the ordering of leaves out from the stems recognizes that −3 is smaller than −1.)

Calcium		Placebo
	−1	1
5	−0	5
234	−0	33111
1	0	23
7	0	5
10	1	2
87	1	

The placebo responses are quite symmetric, but the calcium group has an irregular distribution. This is not unusual when we have only a few observations. There are no outliers. Normal quantile plots (Figure 7.9) give a more detailed picture. The calcium group has a somewhat short left tail, but there are no departures from normality that will prevent use of t procedures.

Take group 1 to be the calcium group and group 2 to be the placebo group. The evidence that calcium lowers blood pressure more than a placebo is assessed by testing

$$H_0\colon \mu_1 = \mu_2$$
$$H_a\colon \mu_1 > \mu_2$$

Here are the summary statistics for the decrease in blood pressure:

Group	Treatment	n	$\overline{x}$	s
1	Calcium	10	5.000	8.743
2	Placebo	11	−.273	5.901

The calcium group shows a drop in blood pressure, and the placebo group has a small increase. The sample standard deviations do not rule out equal population standard deviations. A difference this large will often arise by chance in samples this small. We are willing to assume equal population standard deviations. The pooled sample variance is

$$\begin{aligned} s_p^2 &= \frac{(n_1 - 1)s_1^2 + (n_2 - 1)s_2^2}{n_1 + n_2 - 2} \\ &= \frac{(9)(8.743)^2 + (10)(5.901)^2}{10 + 11 - 2} = 54.536 \end{aligned}$$

so that

$$s_p = \sqrt{54.536} = 7.385$$

The pooled two-sample t statistic is

$$\begin{aligned} t &= \frac{\overline{x}_1 - \overline{x}_2}{s_p\sqrt{\dfrac{1}{n_1} + \dfrac{1}{n_2}}} = \frac{5.000 - (-.273)}{7.385\sqrt{\dfrac{1}{10} + \dfrac{1}{11}}} \\ &= \frac{5.273}{3.227} = 1.634 \end{aligned}$$

df = 19

p	.10	.05
t^*	1.328	1.729

The P-value is $P(T \geq 1.634)$, where T has the $t(19)$ distribution. From Table E we can see that P falls between the $\alpha = 0.10$ and $\alpha = 0.05$ levels. Statistical software gives the exact value $P = 0.059$. The experiment found evidence that calcium reduces blood pressure, but the evidence falls a bit short of the traditional 5% and 1% levels. ●

Sample size strongly influences the P-value of a test. An effect that fails to be significant at a specified level α in a small sample can be significant in a larger sample. In the light of the rather small samples in Example 7.12, the evidence for some effect of calcium on blood pressure is rather good. The published account of the study combined these results for blacks with the results for whites and adjusted for pretest differences among the subjects. Using this more detailed analysis, the researchers were able to report the P-value $P = 0.008$.

We can also estimate the size of the difference in means in Example 7.12. A 90% confidence interval for $\mu_1 - \mu_2$ uses the critical value $t^* = 1.729$ from the $t(19)$ distribution. The interval is

$$\begin{aligned}(\bar{x}_1 - \bar{x}_2) \pm t^* s_p \sqrt{\frac{1}{n_1} + \frac{1}{n_2}} &= [5.000 - (-.273)] \pm (1.729)(7.385)\sqrt{\frac{1}{10} + \frac{1}{11}} \\ &= 5.273 \pm 5.579 \\ &= (-.306, 10.852)\end{aligned}$$

The calcium treatment reduced blood pressure by about 5.3 mm more than a placebo on the average, but the margin of error for this estimate is 5.6 mm.

The pooled two-sample t procedures are anchored in statistical theory and so have long been the standard version of the two-sample t in textbooks. But they require the assumption that the two unknown population standard deviations are equal. As we shall see in Section 7.3, this assumption is hard to verify. The pooled t procedures are therefore a bit risky. They are reasonably robust against both nonnormality and unequal standard deviations when the sample sizes are nearly the same. When the samples are quite different in size, the pooled t procedures become sensitive to unequal standard deviations and should be used with caution unless the samples are large. Unequal standard deviations are quite common. In particular, it is common for the spread of data to increase when the center moves up, as happened in Example 7.11. Statistical software usually calculates both the pooled and the unpooled t statistics, as in Example 7.11. The very unequal sample standard deviations in that example suggest that we not use the pooled procedures.

SUMMARY

Significance tests and confidence intervals for the difference of the means μ_1 and μ_2 of two normal populations are based on the difference $\bar{x}_1 - \bar{x}_2$ of the sample means from two independent SRSs. Because of the central limit theorem, the resulting procedures are approximately correct for other population distributions when the sample sizes are large.

When independent SRSs of sizes n_1 and n_2 are drawn from two normal populations with parameters μ_1, σ_1 and μ_2, σ_2 the **two-sample z statistic**

$$z = \frac{(\bar{x}_1 - \bar{x}_2) - (\mu_1 - \mu_2)}{\sqrt{\frac{\sigma_1^2}{n_1} + \frac{\sigma_2^2}{n_2}}}$$

has the $N(0, 1)$ distribution.

The **two-sample t statistic**

$$t = \frac{(\bar{x}_1 - \bar{x}_2) - (\mu_1 - \mu_2)}{\sqrt{\frac{s_1^2}{n_1} + \frac{s_2^2}{n_2}}}$$

does *not* have exactly a t distribution.

Conservative inference procedures for comparing μ_1 and μ_2 are obtained from the two-sample t statistic by using the $t(k)$ distribution with degrees of freedom k equal to the smaller of $n_1 - 1$ and $n_2 - 1$. More accurate probability values can be obtained by estimating the degrees of freedom from the data. This is the usual procedure in statistical software.

The confidence interval for $\mu_1 - \mu_2$ given by

$$(\bar{x}_1 - \bar{x}_2) \pm t^* \sqrt{\frac{s_1^2}{n_1} + \frac{s_2^2}{n_2}}$$

has confidence level at least C if t^* is the upper $(1 - C)/2$ critical value for $t(k)$, with k the smaller of $n_1 - 1$ and $n_2 - 1$.

Significance tests for H_0: $\mu_1 = \mu_2$ based on

$$t = \frac{\bar{x}_1 - \bar{x}_2}{\sqrt{\frac{s_1^2}{n_1} + \frac{s_2^2}{n_2}}}$$

have a true P-value no higher than that calculated from $t(k)$.

The guidelines for practical use of two-sample t procedures are similar to those for one-sample t procedures. Equal sample sizes are recommended.

If we can assume that the two populations have equal variances, **pooled two-sample t procedures** can be used. These are based on the **pooled estimator**

$$s_p^2 = \frac{(n_1 - 1)s_1^2 + (n_2 - 1)s_2^2}{n_1 + n_2 - 2}$$

of the unknown common variance and the $t(n_1 + n_2 - 2)$ distribution.

SECTION 7.2 EXERCISES

In exercises that call for two-sample t procedures, you may use as the degrees of freedom either the smaller of $n_1 - 1$ and $n_2 - 1$ or the more exact value given by Equation 7.4. We recommend the first choice unless you are using a computer.

7.35 In a study of cereal leaf beetle damage on oats, researchers measured the number of beetle larvae per stem in small plots of oats after randomly applying one of two treatments: no pesticide or Malathion at the rate of 0.25 pound per acre. The data gave the summary statistics below. (Based on M. C. Wilson et al., "Impact of cereal leaf beetle larvae on yields of oats," *Journal of Economic Entomology*, 62 (1969), pp. 699–702.)

Group	Treatment	n	$\bar{x}$	s
1	Control	13	3.47	1.21
2	Malathion	14	1.36	.52

Is there significant evidence at the 1% level that the mean number of larvae per stem is reduced by malathion? Be sure to state H_0 and H_a.

7.36 Physical fitness is related to personality characteristics. In one study of this relationship, middle-aged college faculty who had volunteered for a fitness program were divided into low-fitness and high-fitness groups based on a physical examination. The subjects then took the Cattell Sixteen Personality Factor Questionnaire. Here are the data for the "ego strength" personality factor. (From A. H. Ismail and R. J. Young, "The effect of chronic exercise on the personality of middle-aged men," *Journal of Human Ergology*, 2 (1973), pp. 47–57.)

Group	Fitness	n	$\bar{x}$	s
1	Low	14	4.64	.69
2	High	14	6.43	.43

(a) Is the difference in mean ego strength significant at the 5% level? At the 1% level? Be sure to state H_0 and H_a.

(b) You should be hesitant to generalize these results to the population of all middle-aged men. Explain why.

7.37 A market research firm supplies manufacturers with estimates of the retail sales of their products from samples of retail stores. Marketing managers are prone to look at the estimate and ignore sampling error. Suppose that an SRS of 75 stores this month shows mean sales of 52 units of a small appliance, with standard deviation 13 units. During the same month last year, an SRS of 53 stores gave mean sales of 49 units, with standard deviation 11 units. An increase from 49 to 52 is a rise of 6%. The marketing manager is happy, because sales are up 6%.

(a) Use the two-sample t procedure to give a 95% confidence interval for the difference in mean number of units sold at all retail stores.

(b) Explain in language that the manager can understand why he cannot be certain that sales rose by 6%, and that in fact sales may even have dropped.

7.38 A bank compares two proposals to increase the amount that its credit card customers charge on their cards. (The bank earns a percentage of the amount charged, paid by the stores that accept the card.) Proposal A offers to eliminate the annual fee for customers who charge $2400 or more during the year. Proposal B offers a small percent of the total amount charged as a cash rebate at the end of the year. The bank offers each proposal to an SRS of 150 of its existing credit card customers. At the end of the year, the total amount charged by each customer is recorded. Here are the summary statistics:

Group	n	$\bar{x}$	s
A	150	$1987	$392
B	150	$2056	$413

(a) Do the data show a significant difference between the mean amounts charged by customers offered the two plans? Give the null and alternative hypotheses, and calculate the two-sample t statistic. Obtain the P-value (either approximately from Table E or more accurately from software). State your practical conclusions.

(b) The distributions of amounts charged are skewed to the right, but outliers are prevented by the limits that the bank imposes on credit balances. Do you think that skewness threatens the validity of the test that you used in (a)? Explain your answer.

7.39 A study of iron deficiency among infants compared samples of infants following different feeding regimens. One group contained breast-fed infants, while the children in another group were fed a standard baby formula without any iron supplements. Here are summary results on blood hemoglobin levels at 12 months of age. (From M. F. Picciano and R. H. Deering, "The influence of feeding regimens on iron status during infancy," *American Journal of Clinical Nutrition*, 33 (1980), pp. 746–753.)

Group	n	$\overline{x}$	s
Breast-fed	23	13.3	1.7
Formula	19	12.4	1.8

(a) Is there significant evidence that the mean hemoglobin level is higher among breast-fed babies? State H_0 and H_a and carry out a t test. Give the P-value. What is your conclusion?
(b) Give a 95% confidence interval for the mean difference in hemoglobin level between the two populations of infants.
(c) State the assumptions that your procedures in (a) and (b) require in order to be valid.

7.40 In a study of heart surgery, one issue was the effect of drugs called beta-blockers on the pulse rate of patients during surgery. The available subjects were divided at random into two groups of 30 patients each. One group received a beta-blocker, the other, a placebo. The pulse rate of each patient at a critical point during the operation was recorded. The treatment group had mean 65.2 and standard deviation 7.8. For the control group, the mean was 70.3 and the standard deviation 8.3.
(a) Do beta-blockers reduce the pulse rate? State the hypotheses and do a t test. Is the result significant at the 5% level? At the 1% level?
(b) Give a 99% confidence interval for the difference in mean pulse rates.

7.41 What aspects of rowing technique distinguish between novice and skilled competitive rowers? Researchers compared two groups of female competitive rowers: a group of skilled rowers and a group of novices. The researchers measured many mechanical aspects of rowing style as the subjects rowed on a Stanford Rowing Ergometer. One important variable is the angular velocity of the knee (roughly, the rate at which the knee joint opens as the legs push the body back on the sliding seat). This variable was measured when the oar was at right angles to the machine. (Based on W. N. Nelson and C. J. Widule, "Kinematic analysis and efficiency estimate of intercollegiate female rowers," unpublished manuscript, 1983.) The data show no outliers or strong skewness. Here is the SAS computer output:

TTEST PROCEDURE

Variable: KNEE

GROUP	N	Mean	Std Dev	Std Error
SKILLED	10	4.18283335	0.47905935	0.15149187
NOVICE	8	3.01000000	0.95894830	0.33903942

Variances	T	DF	Prob>|T|
Unequal	3.1583	9.8	0.0104
Equal	3.3918	16.0	0.0037

(a) The researchers believed that the knee velocity would be higher for skilled rowers. State H_0 and H_a.

(b) Give the value of the two-sample t statistic and its P-value (note that SAS provides two-sided P-values). What do you conclude?

(c) Give a 90% confidence interval for the mean difference between the knee velocities of skilled and novice female rowers.

7.42 The novice and skilled rowers in the previous exercise were also compared with respect to several physical variables. Here is the SAS computer output for weight in kilograms:

TTEST PROCEDURE

Variable: WEIGHT

GROUP	N	Mean	Std Dev	Std Error
SKILLED	10	70.3700000	6.10034898	1.92909973
NOVICE	8	68.4500000	9.03999930	3.19612240

Variances	T	DF	Prob>|T|
Unequal	0.5143	9.8	0.6184
Equal	0.5376	16.0	0.5982

Is there significant evidence of a difference in the mean weights of skilled and novice rowers? State H_0 and H_a, report the two-sample t statistic and its P-value, and state your conclusion.

7.43 The Johns Hopkins Regional Talent Searches give the Scholastic Aptitude Tests (intended for high school juniors and seniors) to 13-year-olds. In all, 19,883 males and 19,937 females took the tests between 1980 and 1982. The mean scores of males and females on the verbal test are nearly equal, but there is a clear difference between the sexes on the mathematics test. The reason for this difference is not understood. Here are the data. (From a news article in *Science*, 224 (1983), pp. 1029–1031.)

Group	$\bar{x}$	s
Males	416	87
Females	386	74

Give a 99% confidence interval for the difference between the mean score for males and the mean score for females in the population that Johns Hopkins searches.

7.44 The SSHA is described in Example 7.7 (page 531). A selective private college gives the SSHA to an SRS of both male and female first-year students. The data for the women are as follows:

154	109	137	115	152	140	154	178	101
103	126	126	137	165	165	129	200	148

Here are the scores of the men:

108	140	114	91	180	115	126	92	169	146
109	132	75	88	113	151	70	115	187	104

(a) Examine each sample graphically, with special attention to outliers and skewness. Is use of a t procedure acceptable for these data?

(b) Most studies have found that the mean SSHA score for men is lower than the mean score in a comparable group of women. Test this supposition here. That is, state hypotheses, carry out the test and obtain a P-value, and give your conclusions.

(c) Give a 90% confidence interval for the mean difference between the SSHA scores of male and female first-year students at this college.

7.45 Plant scientists have developed varieties of corn that have increased amounts of the essential amino acid lysine. In a test of the protein quality of this corn, an experimental group of 20 one-day-old male chicks was fed a ration containing the new corn. A control group of another 20 chicks received a ration that was identical except that it contained normal corn. Here are the weight gains (in grams) after 21 days. (Based on G. L. Cromwell et al., "A comparison of the nutritive value of *opaque-2*, *floury-2* and normal corn for the chick," *Poultry Science*, 47 (1968), pp. 840–847.)

Control				Experimental			
380	321	366	356	361	447	401	375
283	349	402	462	434	403	393	426
356	410	329	399	406	318	467	407
350	384	316	272	427	420	477	392
345	455	360	431	430	339	410	326

(a) Present the data graphically. Are there outliers or strong skewness that might prevent the use of t procedures?

(b) State the hypotheses for a statistical test of the claim that chicks fed high-lysine corn gain weight faster. Carry out the test. Is the result significant at the 10% level? At the 5% level? At the 1% level?

(c) Give a 95% confidence interval for the mean extra weight gain in chicks fed high-lysine corn.

7.46 Table 7.3 (page 544) gives data on the blood pressure before and after treatment for two groups of black males. One group took a calcium supplement, and the other group received a placebo. Example 7.12 compares the decrease in blood pressure in the two groups using pooled two-sample t procedures.

(a) Repeat the significance test using a two-sample t test that does not require equal population standard deviations. Compare your P-value with the result $P = 0.059$ for the pooled t test.

(b) Give a 90% confidence interval for the difference in means, again using a procedure that does not require equal standard deviations. How does the margin of error of your interval compare with that in the discussion following Example 7.12?

7.47 Table 1.4 (page 36) gives data on the calories in a sample of brands of each of three kinds of hot dogs. The boxplot in Figure 1.11 shows that poultry hot dogs have fewer calories than either beef or meat hot dogs.

(a) Give a 95% confidence interval for the difference in mean calorie content between beef and poultry hot dogs.

(b) Based on your confidence interval, can the hypothesis that the population means are equal be rejected at the 5% significance level? Explain your answer.

(c) What assumptions does your statistical procedure in (a) require? Which of these assumptions are justified or not important in this case? Are any of the assumptions doubtful in this case?

7.48 Researchers studying the learning of speech often compare measurements made on the recorded speech of adults and children. One variable of interest is called the voice onset time (VOT). Here are the results for 6-year-old children and adults asked to pronounce the word "bees." The VOT is measured in milliseconds and can be either positive or negative. (From M. A. Zlatin and R. A. Koenigsknecht, "Development of the voicing contrast: A comparison of voice onset time in stop perception and production," *Journal of Speech and Hearing Research*, 19 (1976), pp. 93–111.)

Group	n	$\bar{x}$	s
Children	10	−3.67	33.89
Adults	20	−23.17	50.74

(a) What is the standard error of the sample mean VOT for the 20 adult subjects? What is the standard error of the difference $\bar{x}_1 - \bar{x}_2$ between the mean VOT for children and adults?

(b) The researchers were investigating whether VOT distinguishes adults from children. State H_0 and H_a and carry out a two-sample t test. Give a P-value and report your conclusions.

(c) Give a 95% confidence interval for the difference in mean VOTs when pronouncing the word "bees." Explain why you knew from your result in (b) that this interval would contain 0 (no difference).

7.49 The researchers in the study discussed in the previous problem looked at VOTs for adults and children pronouncing several different words. Explain why they should not perform a separate two-sample t test for each word and conclude that the words with a significant difference (say $P < 0.05$) distinguish children from adults. (The researchers did not make this mistake.)

7.50 College financial aid offices expect students to use summer earnings to help pay for college. But how large are these earnings? One college studied this question by asking a sample of students how much they earned. Omitting students who were not employed, 1296 responses were received. Here are the data in summary form. (Data for 1982, provided by Marvin Schlatter, Division of Financial Aid, Purdue University.)

Group	n	$\bar{x}$	s
Males	675	\$1884.52	\$1368.37
Females	621	\$1360.39	\$1037.46

(a) Use the two-sample t procedures to give a 90% confidence interval for the difference between the mean summer earnings of male and female students.

(b) The distribution of earnings is strongly skewed to the right. Nevertheless, use of t procedures is justified. Why?

(c) Once the sample size was decided, the sample was chosen by taking every kth name from an alphabetical list of undergraduates. Is it reasonable to consider the samples as SRSs chosen from the male and female undergraduate populations?

(d) What other information about the study would you request before accepting the results as describing all undergraduates?

The following exercises concern optional material on the pooled two-sample t procedures and on the power of tests.

7.51 Example 7.11 (page 540) reports the analysis of some biological data. The software uses the two-sample t test with degrees of freedom given by Equation 7.4. Starting from the computer's results for $\bar{x}_i$ and s_i, verify the values given for the test statistic $t = 2.99$ and the degrees of freedom $\text{df} = 5.9$.

7.52 Example 7.9 (page 537) analyzed the Chapin Social Insight Test scores of male and female college students. The sample standard deviations s_i for the two samples are almost identical. This suggests that the two populations have similar standard deviations σ_i. The use of the pooled two-sample t procedures is justified in this case. Repeat the analysis in Example 7.9 using the pooled procedures. Compare your results with those obtained in the example.

7.53 The pooled two-sample t procedures can be used to analyze the hemoglobin data in Exercise 7.39, because the very similar s values suggest that the assumption of equal population standard deviations is justified. Repeat (a) and (b) of Exercise 7.39 using the pooled t procedures. Compare your results with those you obtained with the two-sample t procedures in Exercise 7.39.

7.54 The data on weights of skilled and novice rowers in Exercise 7.42 can be analyzed by the pooled t procedures, which assume equal population variances. Report the value of the t statistic, its degrees of freedom, and its P-value, and then state your conclusion. (The pooled procedures should not be used for the more important comparison of knee velocities in Exercise 7.41, because the s-values in the two groups are different enough to cast doubt on the assumption of a common standard deviation.)

7.55 Repeat the comparison of mean VOTs for children and adults in Exercise 7.48 using a pooled t procedure. (In practice, we would not pool in this case, because the data suggest some difference in the population standard deviations.)

(a) Carry out the significance test, and give a P-value.

(b) Give a 95% confidence interval for the difference in population means.

(c) How similar are your results to those you obtained in Exercise 7.48 from the two-sample t procedures?

7.56 In Example 7.12, a small study of black men suggested that a calcium supplement can reduce blood pressure. Now we are planning a larger clinical trial of this effect. We plan to use 100 subjects in each of the two groups. Are these sample sizes large enough to make it very likely that the study will give strong evidence ($\alpha = 0.01$) of the effect of calcium if in fact calcium lowers blood pressure by 5 millimeters more than a placebo? To answer this question, we will compute the power of the two-sample t test of

$$H_0: \mu_1 = \mu_2$$
$$H_a: \mu_1 > \mu_2$$

against the specific alternative $\mu_1 - \mu_2 = 5$. (We plan to use the two-sample t because we are not convinced that the calcium and placebo groups have equal standard deviations.) Based on the pilot study reported in Example 7.12, we take 8, the larger of the two observed s-values, as a rough estimate of both the population σ's and future sample s's.

(a) What is the approximate value of the $\alpha = 0.01$ critical value t^* for the two-sample t statistic when $n_1 = n_2 = 100$?

(b) The test rejects H_0 when

$$\frac{\bar{x}_1 - \bar{x}_2}{\sqrt{\dfrac{s_1^2}{n_1} + \dfrac{s_2^2}{n_2}}} \geq t^*$$

Take both s_1 and s_2 to be 8, and n_1 and n_2 to be 100. What is the number c such that the test rejects H_0 when $\bar{x}_1 - \bar{x}_2 \geq c$?

(c) Suppose that $\mu_1 - \mu_2 = 5$ and that both σ_1 and σ_2 are 8. The power we seek is the probability that $\bar{x}_1 - \bar{x}_2 \geq c$ under these assumptions. Calculate the power.

7.57 You are planning a larger study of VOTs, based on the pilot study reported in Exercise 7.48. Not all words distinguish children (group 1) from adults (group 2) as well as "bees," so you want high power against the alternative that $\mu_1 - \mu_2 = 10$ in a *two-sided* t test. From the pilot study, take 30 as an estimate of σ_1 and s_1, and 50 as an estimate of σ_2 and s_2.

(a) State H_0 and H_a, and write the formula for the test statistic.

(b) Give the $\alpha = 0.05$ critical value for the test when $n_1 = 100$ and $n_2 = 300$. (Although we recommend equal sample sizes to improve the robustness of t procedures against nonnormality, sample sizes that are proportional to the variances give higher power for the same total number of observations.)

(c) Find the approximate power against the given alternative.

7.58 A major bank asks you to design a study of ways to increase the use of their credit cards. One plan (plan A) would offer customers a cash-back rebate of a small percent of their total amount charged each 6 months. Plan B would reduce the interest rate charged by an amount that would cost the bank as much as the rebate. The response variable is the total amount each customer charges during the test period. You decide to offer plan A and plan B each to a separate SRS of the bank's existing charge card customers. In the past, the mean amount charged in a 6-month period has been about \$1100, with a standard deviation of \$400. Will a two-sample t test based on SRSs of 350 customers each detect a difference of \$100 in the mean amounts spent under the two plans?

(a) State H_0 and H_a, and write the formula for the test statistic.

(b) Give the $\alpha = 0.05$ critical value for the test when $n_1 = n_2 = 350$.

(c) Calculate the power of the test with $\alpha = 0.05$, using \$400 as a rough estimate of all standard deviations.

7.3 INFERENCE FOR POPULATION SPREAD*

The two most basic descriptive features of a distribution are its center and spread. In a normal population, these aspects are measured by the mean and the standard deviation. We have described procedures for inference about population means for normal populations and found that these procedures are often useful for nonnormal populations as well. It is natural to turn next to inference about the standard deviations of normal populations. Our advice here is short and clear: Don't do it without expert advice.

There are indeed inference procedures appropriate for the standard deviations of normal populations. We will describe the most common such procedure, the F test for comparing the spread of two normal populations. Unlike the t procedures for means, the F test and other procedures for standard deviations are extremely sensitive to nonnormal distributions. This lack of robustness does not improve in large samples. It is difficult in practice to tell whether a significant F-value is evidence of unequal population spreads or simply evidence that the populations are not normal.

The deeper difficulty that underlies the very poor robustness of normal population procedures for inference about spread already appeared in our work on describing data. The standard deviation is a natural measure of spread for normal distributions, but not for distributions in general. In fact, because skewed distributions have unequally spread tails, no single numerical measure is adequate to describe the spread of a skewed distribution. Thus, the standard deviation is not always a useful parameter, and even when it is (in the normal case), the results of inference are not trustworthy. Consequently, we do not recommend use of inference about population standard deviations in basic statistical practice.[13]

Sometimes equality of standard deviations is tested as a preliminary to performing the pooled two-sample t test for equality of two population means. It is better practice to check the distributions graphically, with special attention to skewness and outliers. The pooled t test is reasonably

*This section can be omitted without loss of continuity, but we recommend reading the introductory paragraphs.

robust against unequal population standard deviations, at least when the population distributions are roughly symmetric and the two sample sizes are similar. On the other hand, the test for equal standard deviations is often misleading because of its extreme sensitivity to departures from normality.

The situation is similar when we wish to compare the means of more than two populations. Chapter 10 discusses the analysis of variance procedures for comparing several means. These procedures are extensions of the pooled t test and require that the populations have a common standard deviation. Like the t test, analysis of variance comparisons of means are quite robust. (Analysis of variance uses F statistics, but these are not the same as the F statistic for comparing two population standard deviations.) Formal tests for the assumption of equal standard deviations are very non robust and will often give misleading results. In the words of one distinguished statistician, "To make a preliminary test on variances is rather like putting to sea in a rowing boat to find out whether conditions are sufficiently calm for an ocean liner to leave port!"[14]

The *F* test

Because of the limited usefulness of procedures for inference about the standard deviations of normal distributions, we will present only one such procedure. Suppose that we have independent SRSs from two normal populations, a sample of size n_1 from $N(\mu_1, \sigma_1)$ and a sample of size n_2 from $N(\mu_2, \sigma_2)$. The population means and standard deviations are all unknown. The hypothesis of equal spread,

$$H_0: \sigma_1 = \sigma_2$$
$$H_a: \sigma_1 \neq \sigma_2$$

is tested by a simple statistic, the ratio of sample variances.

THE *F* STATISTIC AND *F* DISTRIBUTIONS

When s_1^2 and s_2^2 are sample variances from independent SRSs of sizes n_1 and n_2 drawn from normal populations, the F statistic

$$F = \frac{s_1^2}{s_2^2}$$

has the F distribution with $n_1 - 1$ and $n_2 - 1$ degrees of freedom when $H_0: \sigma_1 = \sigma_2$ is true.

F distributions

The *F distributions* are a family of distributions with two parameters, the degrees of freedom of the sample variances in the numerator and denominator of the F statistic.* The numerator degrees of freedom are always mentioned first. Interchanging the degrees of freedom changes the distribution, so the order is important. Our brief notation will be $F(j, k)$ for the F distribution with j degrees of freedom in the numerator and k in the denominator. The F distributions are not symmetric but are right skewed. The density curve in Figure 7.10 illustrates the shape. Because sample variances cannot be negative, the F statistic takes only positive values and the F distribution has no probability below 0. The peak of the F density curve is near 1; values far from 1 in either direction provide evidence against the hypothesis of equal standard deviations.

Tables of F critical values are awkward, because a separate table is needed for every pair of degrees of freedom j and k. Table F in the back of the book gives upper p critical values of the F distributions for $p = 0.10$, 0.05, 0.025, 0.01, and 0.001. For example, these critical points for the $F(9, 10)$ distribution shown in Figure 7.10 are

p	.10	.05	.025	.01	.001
F^*	2.35	3.02	3.78	4.94	8.96

*The F distributions are another of R. A. Fisher's contributions to statistics and are called F in his honor. Fisher introduced F statistics for comparing several means. We will meet these useful statistics in later chapters.

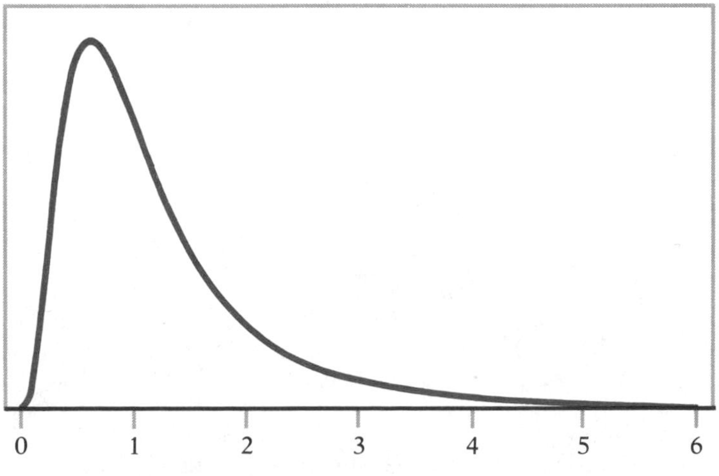

FIGURE 7.10 **The density curve for the $F(9, 10)$ distribution. The F distributions are skewed to the right.**

The skewness of the F distributions causes additional complications. In the symmetric normal and t distributions, the point with probability 0.05 below it is just the negative of the point with probability 0.05 above it. This is not true for F distributions. We therefore need either tables of both the upper and lower tails or means of eliminating the need for lower tail critical values. Statistical software that eliminates the need for tables is plainly very convenient. If you do not use statistical software, arrange the F test as follows:

1. Take the test statistic to be

$$F = \frac{\text{larger } s^2}{\text{smaller } s^2}$$

   This amounts to naming the populations so that s_1^2 is the larger of the observed sample variances. The resulting F is always 1 or greater.

2. Compare the value of F with critical values from Table F. Then *double* the significance levels from the table to obtain the significance level for the two-sided F test.

The idea is that we calculate the probability in the upper tail and double to obtain the probability of all ratios on either side of 1 that are at least as improbable as that observed. Remember that the order of the degrees of freedom is important in using Table F.

EXAMPLE 7.13

Example 7.12 recounts a medical experiment comparing the effects of calcium and a placebo on the blood pressure of black men. The analysis employed the pooled two-sample t procedures. Because these procedures require equal population standard deviations, it is tempting to first test

$$H_0: \sigma_1 = \sigma_2$$
$$H_a: \sigma_1 \neq \sigma_2$$

The larger of the two sample standard deviations is $s = 8.743$ from 10 observations. The other is $s = 5.901$ from 11 observations. The two-sided test statistic is therefore

$$F = \frac{\text{larger } s^2}{\text{smaller } s^2} = \frac{8.743^2}{5.901^2} = 2.195$$

We compare the calculated value $F = 2.20$ with critical points for the $F(9, 10)$ distribution. Table F shows that 2.20 is *less* than the 0.10 critical value of the $F(9, 10)$ distribution, which is $F^* = 2.35$. Doubling 0.10, we know that the observed F falls short of the 0.20 critical value. The results are not significant at the 20% level (or any lower level). Statistical software shows that the exact upper tail probability is 0.118, and hence $P = 0.236$. *If* the populations were normal, the observed standard deviations would give little reason to suspect unequal population standard deviations. Because one of the populations shows some nonnormality, we cannot be fully confident of this conclusion. ●

Robustness of normal inference procedures

We have claimed that

- The t procedures for inference about means are quite robust against nonnormal population distributions. These procedures are particularly robust when the population distributions are symmetric and (for the two-sample case) when the two sample sizes are equal.
- The F test and other procedures for inference about variances are so lacking in robustness as to be of little use in practice.

Figure 7.11 presents evidence for these claims. The paper from which the figure is reproduced contains more information.[15] This figure requires careful attention because it displays a great deal of information in compact form. It presents the results of a number of large simulations. All were carried out with samples of size 25, and all concern significance tests with fixed level $\alpha = 0.05$. The four types of tests studied are the one-sample and pooled two-sample t tests, the F test (called "two-sample F"), and the test for the variance of a single normal population. We have not discussed this last test.

Figure 7.11 reports results for one-sided tests ("above" and "below" the null hypothesis of no difference) and two-sided tests ("beyond"). The 12 small boxes in each half of the figure show the behavior of the 12 combinations of test statistic (four choices) and alternative hypothesis (three choices). Figure 7.11(a) describes performance for symmetric populations, both normal and nonnormal. Figure 7.11(b) depicts the tests when the population is nonnormal and strongly skewed to the right. The horizontal scale in each box is *kurtosis,* a parameter that describes whether the particular distribution concentrates its probability in a central peak or in the tails. Normal populations lie at 3 on this scale in Figure 7.11(a); nonnormal populations lie on either side of 3. All of the skewed populations of Figure 7.11(b) are nonnormal.

kurtosis

All of the tests are carried out at the fixed significance level 5% ($\alpha = 0.05$) using critical values from tables such as t or F that are correct for normal populations. If the population distribution is not normal, the actual significance level (probability of rejecting the hypothesis when it is true) will generally be higher or lower than 5%. A test is robust if the actual level stays close to 5% even when the population is not normal. The colored dots in Figure 7.11 show the actual significance level in each case. The significance levels appear on the vertical scale, with 5% as the target value. The results for normal populations appear in Figure 7.11(a) above 3 in each box within the figure. These points are all close to 5% on the vertical scale, as they should be. The other points show what happens when the tests are used on nonnormal data. The faster and farther the points move from the 5% level, the less robust the test is.

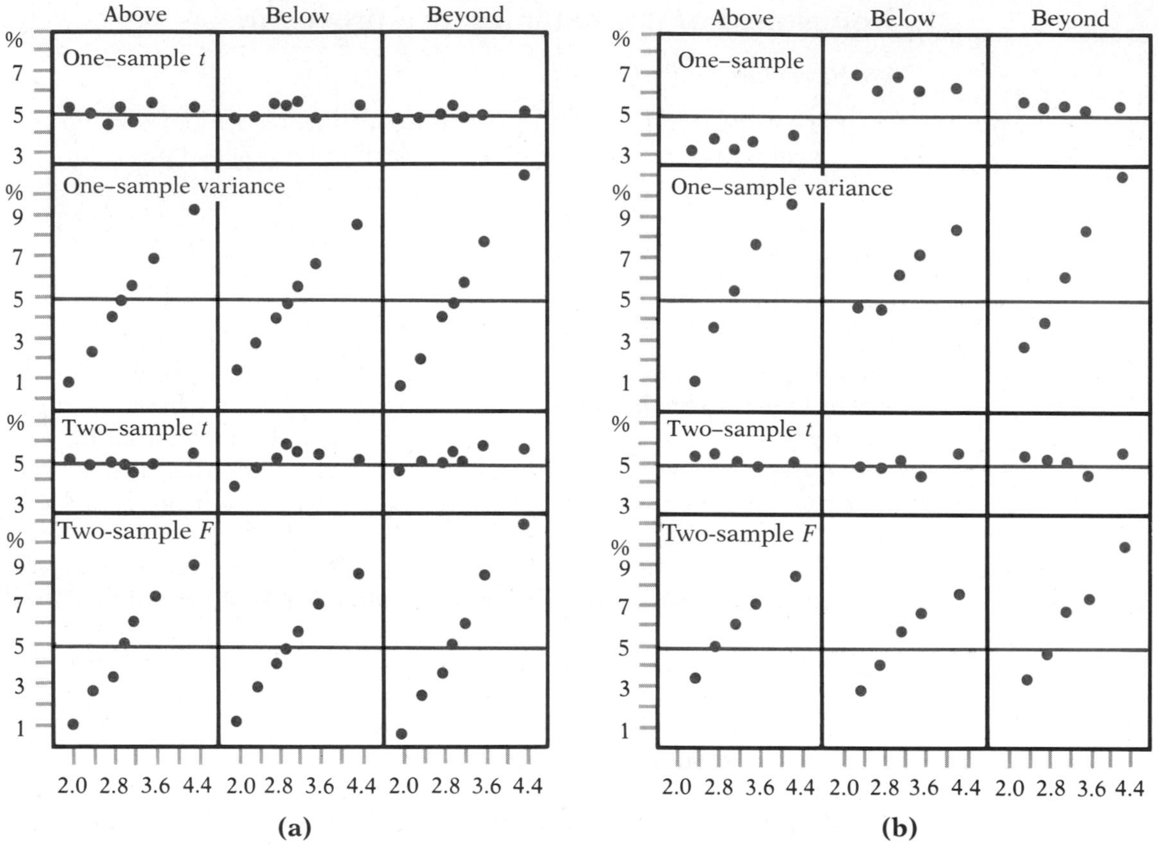

FIGURE 7.11 **The robustness of common tests of significance. The points plotted are the actual significance levels for tests made at the 5% level. The tests assume normality, while the actual distributions are mostly nonnormal. (a) Symmetric distributions; (b) right-skewed distributions. From E. S. Pearson and N. W. Please, "Relation between the shape of population distribution and the robustness of four simple test statistics,"** ***Biometrika*****, 62 (1975), pp. 223–241. Reproduced by permission of the** ***Biometrika*** **trustees.**

EXAMPLE 7.14

A symmetric population has kurtosis equal to 4.4, a quite extreme value. We carry out a two-sample t test for equality of means against a two-sided alternative using 5% critical values. Although the population is not at all normal, the actual significance level of this test is about 5.9%. Look at the rightmost dot in the "two-sample t" box of Figure 7.11(a), above 4.4 on the horizontal scale.

Now consider the F test in the same situation. The actual significance level of what is supposed to be a 5% test is about 11.2%. This is the rightmost dot in the "two-sample F" box in Figure 7.11(a). ●

The robustness of the two-sample t is remarkable. The true significance level remains between about 4% and 6% for the entire range of populations studied. This test and the corresponding confidence interval are among the most reliable tools in the statistician's shop. Do remember, however, that outliers can greatly disturb the t procedures and that they are less robust when the sample sizes are not similar. Note also that the happy results of Figure 7.11 are obtained when both populations have the same shape, even though that shape is not normal. Dissimilar shapes, especially skewness in only one of the populations, can cause the type of results that occur for the one-sample t discussed below.

The lack of robustness of the tests for variances is equally remarkable. The true levels depart rapidly from the target 5% as the population distribution departs from normality. The two-sided F test carried out with 5% critical values can have a true level of less than 1% or greater than 11% even in symmetric populations with no outliers. Look at the lower right box in Figure 7.11(a). Results such as these are the basis for our recommendation that these procedures not be used.

The one-sample t is pleasingly robust in symmetric populations, but the top row of Figure 7.11(b) shows that one-sided tests are systematically disturbed by skewed distributions. For example, the t test of H_0: $\mu = \mu_0$ against H_a: $\mu > \mu_0$ rejects H_0 too seldom (the true significance level is less than 5%) when the population distribution is right skewed.

SUMMARY

Inference procedures for comparing the standard deviations of two normal populations are based on the ***F* statistic,** which is the ratio of sample variances

$$F = \frac{s_1^2}{s_2^2}$$

If an SRS of size n_1 is drawn from the x_1 population and an independent SRS of size n_2 is drawn from the x_2 population, the F statistic has the ***F* distribution** $F(n_1 - 1, n_2 - 1)$ if the two population standard deviations σ_1 and σ_2 are in fact equal.

The two-sided test of H_0: $\sigma_1 = \sigma_2$ uses the statistic

$$F = \frac{\text{larger } s^2}{\text{smaller } s^2}$$

and doubles the upper tail probability to obtain the P-value.

The F tests and other procedures for inference about the spread of one or more normal distributions are so strongly affected by nonnormality that we do not recommend them for regular use.

SECTION 7.3 EXERCISES

In all exercises calling for use of the F test, assume that both population distributions are very close to normal. The actual data are not always sufficiently normal to justify use of the F test.

7.59 The F statistic $F = s_1^2/s_2^2$ is calculated from samples of size $n_1 = 10$ and $n_2 = 8$. (Remember that n_1 is the numerator sample size.)

(a) What is the upper 5% critical value for this F?

(b) In a test of equality of standard deviations against the two-sided alternative, this statistic has the value $F = 3.45$. Is this value significant at the 10% level? Is it significant at the 5% level?

7.60 The F statistic for equality of standard deviations based on samples of sizes $n_1 = 21$ and $n_2 = 16$ takes the value $F = 2.78$.

(a) Is this significant evidence of unequal population standard deviations at the 5% level? At the 1% level?

(b) Between which two values obtained from Table F does the P-value of the test fall?

7.61 The sample variance for the treatment group in the DDT experiment of Example 7.11 (page 540) is more than 10 times as large as the sample variance for the control group. Calculate the F statistic. Can you reject the hypothesis of equal population standard deviations at the 5% significance level? At the 1% level?

7.62 Exercise 7.41 (page 550) records the results of comparing a measure of rowing style for skilled and novice female competitive rowers. Is there significant evidence of inequality between the standard deviations of the two populations?

(a) State H_0 and H_a.

(b) Calculate the F statistic. Between which two levels does the P-value lie?

7.63 Answer the same questions for the weights of the two groups, recorded in Exercise 7.42 (page 551).

7.64 The data for VOTs of children and adults in Exercise 7.48 (page 553) show quite different sample standard deviations. How statistically significant is the observed inequality?

7.65 Return to the SSHA data in Exercise 7.44 (page 552). SSHA scores are generally less variable among women than among men. We want to know whether this is true for this college.

(a) State H_0 and H_a. Note that H_a is one-sided in this case.

(b) Because Table F contains only upper critical values for F, a one-sided test requires that in calculating F the numerator s^2 belong to the group that H_a claims to have the larger σ. Calculate this F.

(c) Compare F to the entries in Table F (no doubling of p) to obtain the P-value. Be sure the degrees of freedom are in the proper order. What do you conclude about the variation in SSHA scores?

7.66 The observed inequality between the sample standard deviations of male and female SAT scores in Exercise 7.43 (page 552) is clearly significant. You can say this without doing any calculations. Find F and look in Table F. Then explain why the significance of F could be seen without arithmetic.

7.67 The paper from which Figure 7.11 was taken also investigated several sets of industrial data. One variable, the failure time of a metal part when repeatedly bent, had the same degree of right skewness as the simulated distributions in Figure 7.11(b). The kurtosis measure that appears on the horizontal scales in Figure 7.11 was about 3.6 for this distribution. Answer the following questions from Figure 7.11.

(a) You are doing a two-sample t test for the equality of mean failure times for two similar types of part. You measure 25 specimens of each type of part, the same sample sizes as in Figure 7.11. The alternative is two-sided. You do the test at fixed significance level 5% by using the 0.025 critical value from Table E of the t distribution. About what will be the actual significance level of your test?

(b) You do an F test for equality of standard deviations against a two-sided alternative for the same data. Again you choose the 5% level and carry out the test using Table F. About what will be the actual significance level?

7.68 In the setting of the previous exercise, suppose that you were doing inference on a single sample. You carry out a one-sample t test at the $\alpha = 0.05$ significance level.

(a) If the alternative is one-sided on the high side, about what will be the actual significance level of your test?

(b) About what will be the actual significance level for the test against a two-sided alternative?

(c) Instead of a t test for the mean, you carry out the test for a specific value of the variance, or standard deviation, of a single distribution, again at the 5% level. About what will be the actual level of this test if the alternative is two-sided?

CHAPTER 7 EXERCISES

7.69 In a study of the effectiveness of weight-loss programs, 47 subjects who were at least 20% overweight took part in a group support program for 10 weeks. Private weighings determined each subject's weight at the beginning of the program and 6 months after the program's end. The matched pairs t test was used to assess the significance of the average weight loss. The paper reporting the study said, "The subjects lost a significant amount of weight over time, $t(46) = 4.68$, $p < .01$." It is common to report the results of statistical tests in this abbreviated style. (Based loosely on D. R. Black et al., "Minimal interventions for weight control: A cost-effective alternative," *Addictive Behaviors*, 9 (1984), pp. 279–285.)

(a) Why was the matched pairs statistic appropriate?

(b) Explain to someone who knows no statistics but is interested in weight-loss programs what the practical conclusion is.

(c) The paper follows the tradition of reporting significance only at fixed levels such as $\alpha = 0.01$. In fact, the results are more significant than "$p < .01$" suggests. What can you say about the P-value of the t test?

7.70 A major study of alternative welfare programs randomly assigned women on welfare to one of two programs, called "WIN" and "Options." The new Options program was designed to give more incentives to work and earn than the existing WIN program. An important question was how much more (on the average) women in Options earned than those in Win. Here is Minitab output for earnings (dollars) over a 3-year period. (Based on D. Friedlander, *Supplemental Report on the Baltimore Options Program*, Manpower Demonstration Research Corporation, New York, 1987.)

```
MTB> TWOSAMPLE T 'OPT' 'WIN'

TWOSAMPLE T FOR 'OPT' VS 'WIN'

          N     MEAN    STDEV    SE MEAN
OPT    1362     7638      289     7.8309
WIN    1395     6595      247     6.6132

95 PCT CI FOR MU OPT - MU WIN: (1022.90, 1063.10)
```

(a) Give a 99% confidence interval for the amount by which the mean earnings of Options participants exceeded the mean earnings of WIN subjects. (Minitab will give a 99% confidence interval if you instruct it to do so. Here we have only the basic output, which includes the 95% confidence interval.)

(b) The distribution of incomes is strongly skewed to the right but includes no extreme outliers because all the subjects were on welfare. What fact about these data allows us to use t procedures despite the strong skewness?

7.71 Nitrites are often added to meat products as preservatives. In a study of the effect of these chemicals on bacteria, the rate of uptake of a radio-labeled amino acid was measured for a number of cultures of bacteria, some growing in a medium to which nitrites had been added. Here are the summary statistics from this study:

Group	n	$\bar{x}$	s
Nitrite	30	7880	1115
Control	30	8112	1250

Carry out a test of the research hypothesis that nitrites decrease amino acid uptake, and report your results.

7.72 The one-hole test is used to test the manipulative skill of job applicants. This test requires subjects to grasp a pin, move it to a hole, insert it, and return for another pin. The score on the test is the number of pins inserted in a fixed time interval. In one study, male college students were compared with experienced female industrial workers. Here are the data for the first minute of the test. (Based on G. Salvendy, "Selection of industrial operators: The one-hole test," *International Journal of Production Research*, 13 (1973), pp. 303–321.)

Group	n	$\bar{x}$	s
Students	750	35.12	4.31
Workers	412	37.32	3.83

(a) It was expected that the experienced workers would outperform the students, at least during the first minute, before learning occurs. State the hypotheses for a statistical test of this expectation and perform the test. Give a P-value and state your conclusions.

(b) The distribution of scores is slightly skewed to the left. Explain why the procedure you used in (a) is nonetheless acceptable.

(c) One purpose of the study was to develop performance norms for job applicants. Based on the data above, what is the range that covers the middle 95% of experienced workers? (Be careful! This is not the same as a 95% confidence interval for the mean score of experienced workers.)

(d) The five-number summary of the distribution of scores among the workers is

23 33.5 37 40.5 46

for the first minute, and

32 39 44 49 59

for the fifteenth minute of the test. Display these facts graphically, and describe briefly the differences between the distributions of scores in the first and fifteenth minute.

7.73 The composition of the earth's atmosphere may have changed over time. One attempt to discover the nature of the atmosphere long ago studies the gas trapped in bubbles inside ancient amber. Amber is tree resin that has hardened and been trapped in rocks. The gas in bubbles within amber should be a sample of the atmosphere at the time the amber was formed. Measurements on specimens of amber from the late Cretaceous era (75 to 95 million years ago) give these percents of nitrogen:

63.4	65.0	64.4	63.3	54.8	64.5	60.8	49.1	51.0

These values are quite different from the present 78.1% of nitrogen in the atmosphere. Assume (this is not yet agreed on by experts) that these observations are an SRS from the late Cretaceous atmosphere. (Data from R. A. Berner and G. P. Landis, "Gas bubbles in fossil amber as possible indicators of the major gas composition of ancient air," *Science*, 239 (1988), pp. 1406–1409.)

(a) Graph the data, and comment on skewness and outliers.

(b) The t procedures will be only approximate in this case. Give a 95% t confidence interval for the mean percent of nitrogen in ancient air.

7.74 Do various occupational groups differ in their diets? A British study of this question compared 98 drivers and 83 conductors of London double-decker buses. The conductors' jobs require more physical activity. The article reporting the study gives the data as "Mean daily consumption ($\pm$ s. e.)." Some of the study results appear below. (From J. W. Marr and J. A. Heady, "Within- and between-person variation in dietary surveys: Number of days needed to classify individuals," *Human Nutrition: Applied Nutrition,* 40A (1986), pp. 347–364.)

	Drivers	Conductors
Total calories	2821 ± 44	2844 ± 48
Alcohol (grams)	$.24 \pm .06$	$.39 \pm .11$

(a) What does "s. e." stand for? Give $\bar{x}$ and s for each of the four sets of measurements.

(b) Is there significant evidence at the 5% level that conductors consume more calories per day than do drivers? Use the two-sample t method to give a P-value, and then assess significance.

(c) How significant is the observed difference in mean alcohol consumption? Use two-sample t methods to obtain the P-value.

(d) Give a 90% confidence interval for the mean daily alcohol consumption of London double-decker bus conductors.

(e) Give an 80% confidence interval for the difference in mean daily alcohol consumption between drivers and conductors.

7.75 The pooled two-sample t test is justified in part (b) of the previous exercise. Explain why. Find the P-value for the pooled t statistic, and compare with your result in the previous exercise.

7.76 The report cited in Exercise 7.74 says that the distribution of alcohol consumption among the individuals studied is "grossly skew."

(a) Do you think that this skewness prevents the use of the two-sample t test for equality of means? Explain your answer.

(b) Do you think that the skewness of the distributions prevents the use of the F test for equality of standard deviations? Explain your answer.

7.77 Do the data in Example 7.9 (page 537) provide evidence of different standard deviations in Chapin Test scores in the populations of female and male college liberal arts majors?

(a) State the hypotheses and carry out the test. Software can assess significance exactly, but inspection of the F table is enough to draw a conclusion.

(b) Do the large sample sizes allow us to ignore the assumption that the population distributions are normal?

7.78 Exercise 1.109 (page 88) gives the populations of all 92 counties in the state of Indiana. Is it proper to apply the one-sample t method to these data to give a 95% confidence interval for the mean population of an Indiana county? Explain your answer.

7.79 A pharmaceutical manufacturer checks the potency of products during manufacture by chemical analysis. The standard release potency for cephalothin crystals is set at 910. An assay of the previous 16 lots gives the following potency data:

897	914	913	906	916	918	905	921
918	906	895	893	908	906	907	901

(a) Check the data for outliers or strong skewness that might threaten the validity of the t procedures.

(b) Give a 95% confidence interval for the mean potency.

(c) Is there significant evidence at the 5% level that the mean potency is not equal to the standard release potency?

7.80 The amount of lead in a certain type of soil, when released by a standard extraction method, averages 86 parts per million (ppm). A new extraction method is tried on 40 specimens of the soil, yielding a mean of 83 ppm lead and a standard deviation of 10 ppm.

(a) Is there significant evidence at the 1% level that the new method frees less lead from the soil?

(b) A critic argues that because of variations in the soil, the effectiveness of the new method is confounded with characteristics of the particular soil specimens used. Briefly describe a better data production design that avoids this criticism.

7.81 High levels of cholesterol in the blood are not healthy in either humans or dogs. Because a diet rich in saturated fats raises the cholesterol level, it is plausible that dogs owned as pets have higher cholesterol levels than dogs owned by a veterinary research clinic. "Normal" levels of cholesterol based on the clinic's dogs would then be misleading. A clinic compared healthy dogs it owned with healthy pets brought to the clinic to be neutered. The summary statistics for blood cholesterol levels (milligrams per deciliter of blood) appear below. (From V. D. Bass, W. E. Hoffmann, and J. L. Dorner, "Normal canine lipid profiles and effects of experimentally induced pancreatitis and hepatic necrosis on lipids," *American Journal of Veterinary Research*, 37 (1976), pp. 1355–1357.)

Group	n	$\overline{x}$	s
Pets	26	193	68
Clinic	23	174	44

(a) Is there strong evidence that pets have a higher mean cholesterol level than clinic dogs? State the H_0 and H_a and carry out an appropriate test. Give the P-value and state your conclusion.
(b) Give a 95% confidence interval for the difference in mean cholesterol levels between pets and clinic dogs.
(c) Give a 95% confidence interval for the mean cholesterol level in pets.
(d) What assumptions must be satisfied to justify the procedures you used in (a), (b), and (c)? Assuming that the cholesterol measurements have no outliers and are not strongly skewed, what is the chief threat to the validity of the results of this study?

7.82 Exercise 2.32 (page 142) gives data on the concentration of airborne particulate matter in a rural area upwind from a small city and in the center of the city. In that exercise, the focus was on using the rural readings to predict the city readings for the same day. Now we want to compare the mean level of particulates in the city and in the rural area. We suspect that pollution is higher in the city and hope to find evidence for this suspicion.
(a) State H_0 and H_a.
(b) Which type of t procedure is appropriate: one-sample, matched pairs, or two-sample?
(c) Make a graph to check for outliers or strong skewness that might prevent the use of t procedures. Your graph should reflect the type of procedure that you will use.
(d) Carry out the appropriate t test. Give the P-value and report your conclusion.
(e) Give a 90% confidence interval for the mean amount by which the city particulate level exceeds the rural level.

7.83 The sign test allows us to assess whether city particulate levels are higher than nearby rural levels on the same day without the use of normal distributions. Carry out a sign test for the data used in the previous exercise. State H_0 and H_a and give the P-value and your conclusion.

7.84 Exercise 1.19 (page 26) gives data on the daily egg production of a population of 25 female and 10 male *Ctenocephalides felis* fleas. Provide the flea experts with an estimate of the mean daily production together with a margin of error.

7.85 Exercise 1.25 (page 28) gives 29 measurements of the density of the earth made in 1798 by Henry Cavendish. Display the data graphically to check for skewness and outliers. Then give an estimate for the density of the earth from Cavendish's data and a margin of error for your estimate.

7.86 Elite distance runners are thinner than the rest of us. Here are data on skinfold thickness, which indirectly measures body fat, for 20 elite runners and 95 ordinary men in the same age group. The data are in millimeters and are given in the form "mean (standard deviation)." (From M. L. Pollock et al., "Body composition of elite class distance runners," in P. Milvey (ed.), *The Marathon: Physiological, Medical, Epidemiological, and Psychological Studies*, New York Academy of Sciences, New York, 1977, p. 366.)

	Runners	Others
Abdomen	7.1 (1.0)	20.6 (9.0)
Thigh	6.1 (1.8)	17.4 (6.6)

Use confidence intervals to describe the difference between runners and typical young men.

CHAPTER 7 COMPUTER EXERCISES

7.87 Table 2.9 (page 214) gives the levels of three pollutants in the exhaust of 46 randomly selected vehicles of the same type. You will investigate emissions of nitrogen oxides (NOX).

(a) Make a stemplot and, if your software allows, a normal quantile plot of the NOX levels. Do the plots suggest that the distribution of NOX emissions is approximately normal? Can you safely employ t procedures to analyze these data?

(b) Give a 99% confidence interval for the mean NOX level in vehicles of this type.

(c) Your supervisor would like the average NOX level to be less than 1 gram per mile. You will have to tell him that it's not so. Carry out a significance test to assess the strength of the evidence that the mean NOX level is greater than 1, and then write a short report to your supervisor based on your work in (b) and (c). (Your supervisor never heard of P-values, so you must use plain language.)

7.88 Is there a difference between the average SAT scores of males and females? The CSDATA data set gives the math (SATM) and verbal (SATV) scores for a group of 224 computer science majors. The variable SEX indicates whether each individual is male or female.

(a) Compare the two distributions graphically, and then use the two-sample t test to compare the average SATM scores of males and females. Is it appropriate to use the pooled t test for this comparison? Write a brief summary of your results and conclusions that refers to both versions of the t test and the F test for equality of standard deviations. Also give a 99% confidence interval for the difference in the means.

(b) Answer part (a) for the SAT verbal score.

(c) The students in the CSDATA data set were all computer science majors who began college during a particular year. To what extent do you think that your results would generalize to (*i*) computer science students entering in different years, (*ii*) computer science majors at other colleges and universities, and (*iii*) college students in general?

7.89 The WOOD data set gives first ($T1$) and second ($T2$) measurements of the modulus of elasticity for 50 strips of wood. We would like to see if the process of taking the first measurement changes the strips so that the second measurement tends to be higher or lower than the first.

(a) Compute the differences $D = T1 - T2$ and describe the distribution of D using graphical and numerical summaries.

(b) Give a 95% confidence interval for the mean of D.

(c) Perform the test of the null hypothesis that the mean of D is 0. Summarize your results and draw a conclusion.

7.90 In the READING data set the response variable POST3 is to be compared for three methods of teaching reading. The Basal method is the standard, or control, method and the two new methods are DRTA and Strat. We can use the methods of this chapter to compare Basal with DRTA and Basal with Strat. Note that to make comparisons among three treatments it is more appropriate to use the procedures that we will learn in Chapter 10.

(a) Is the mean reading score with the DRTA method higher than that for the Basal method? Perform an analysis to answer this question and summarize your results.

(b) Answer part (a) for the Strat method in place of DRTA.

NOTES

1. This example is based on information in D. L. Shankland et al., "The effect of 5-thio-D-glucose on insect development and its absorption by insects," *Journal of Insect Physiology*, 14 (1968), pp. 63–72.
2. Data provided by Joseph A. Wipf, Department of Foreign Languages and Literatures, Purdue University.
3. These recommendations are based on extensive computer work. See, for example, Harry O. Posten, "The robustness of the one-sample t-test over the Pearson system," *Journal of Statistical Computation and Simulation*, 9 (1979), pp. 133–149, and E. S. Pearson and N. W. Please, "Relation between the shape of population distribution and the robustness of four simple test statistics," *Biometrika*, 62 (1975), pp. 223–241.

4. The data and the use of the log transformation come from Thomas J. Lorenzen, "Determining statistical characteristics of a vehicle emissions audit procedure," *Technometrics*, 22 (1980), pp. 483–493.

5. You can find a practical discussion of distribution-free inference in Myles Hollander and Douglas A. Wolfe, *Nonparametric Statistical Methods*, Wiley, New York, 1973.

6. Detailed information about the conservative t procedures can be found in Paul Leaverton and John J. Birch, "Small sample power curves for the two sample location problem," *Technometrics*, 11 (1969), pp. 299–307; in Henry Scheffé, "Practical solutions of the Behrens-Fisher problem," *Journal of the American Statistical Association*, 65 (1970), pp. 1501–1508; and in D. J. Best and J. C. W. Rayner, "Welch's approximate solution for the Behrens-Fisher problem," *Technometrics*, 29 (1987), pp. 205–210.

7. This example is adapted from Maribeth C. Schmitt, "The effects of an elaborated directed reading activity on the metacomprehension skills of third graders," Ph.D. dissertation, Purdue University, 1987.

8. From H. G. Gough, *The Chapin Social Insight Test*, Consulting Psychologists Press, Palo Alto, Calif., 1968.

9. See the extensive simulation studies in Harry O. Posten, "The robustness of the two-sample t test over the Pearson system," *Journal of Statistical Computation and Simulation*, 6 (1978), pp. 295–311.

10. This example is loosely based on D. L. Shankland, "Involvement of spinal cord and peripheral nerves in DDT-poisoning syndrome in albino rats," *Toxicology and Applied Pharmacology*, 6 (1964), pp. 197–213.

11. We did not use Minitab in Example 7.11 because Minitab shortcuts the two-sample t procedure: It calculates the degrees of freedom df using Equation 7.4 but then truncates to the next lower whole-number degrees of freedom to obtain the P-value. The result is slightly less accurate than the P-value from the t(df) distribution.

12. This study is reported in Roseann M. Lyle et al., "Blood pressure and metabolic effects of calcium supplementation in normotensive white and black men," *Journal of the American Medical Association*, 257 (1987), pp. 1772–1776. The individual measurements in Table 7.3 were provided by Dr. Lyle.

13. The problem of comparing spreads is difficult even with advanced methods. Common distribution-free procedures do not offer a satisfactory alternative to the F test, because they are sensitive to unequal shapes when comparing two distributions. A good introduction to the available methods is W. J. Conover, M. E. Johnson, and M. M. Johnson, "A comparative study of tests for homogeneity of variances, with applications to outer continental shelf bidding data," *Technometrics*, 23 (1981), pp. 351–361. Modern resampling procedures often work well. See Dennis D. Boos and Colin Brownie, "Bootstrap methods for testing homogeneity of variances," *Technometrics*, 31 (1989), pp. 69–82.

14. G. E. P. Box, "Non-normality and tests on variances," *Biometrika*, 40 (1953), pp. 318–335. The quote appears on page 333.

15. Figure 7.11 is from the paper of Pearson and Please cited in Note 3.

CHAPTER

8

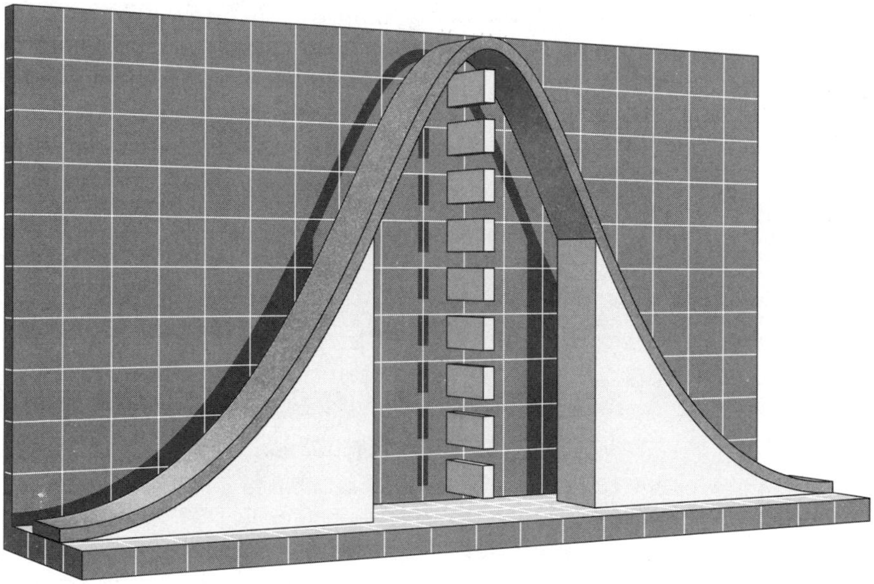

Inference for Count Data

PRELUDE

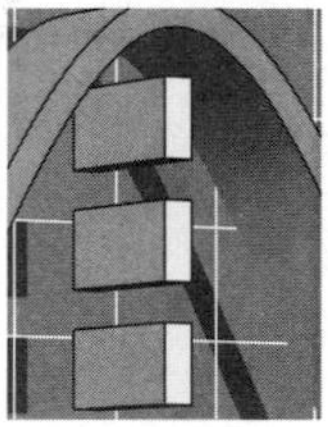

Some statistical studies concern variables measured in a scale of equal units, such as dollars or grams. Other studies examine categorical variables, such as the race or occupation of a person, the make of a car, or the type of complaint received from a customer. When we record categorical variables, our data consist of counts or of percents obtained from counts. This chapter presents confidence intervals and significance tests for use with count data.

The parameters we want to do inference about are population proportions. Just as in the case of inference about population means, we may be concerned with a single population or with comparing two populations. Inference about proportions in these one-sample and two-sample settings is very similar to inference about means, which we met in Chapter 7. The final section of this chapter describes how to compare more than two populations and how to test whether two categorical variables are independent. A single statistical test handles both of these cases. The corresponding procedures for measured variables are more elaborate; they will appear in Chapters 9 and 10. The methods of this chapter answer questions such as these:

- An association of Christmas tree growers sponsored a sample survey to gather data about the use of natural Christmas trees. What percent of households display a Christmas tree? Is there a difference between urban and rural households in their preference for natural or artificial Christmas trees?
- Does taking aspirin regularly help prevent heart attacks? A double-blind randomized comparative experiment assigned 11,000 male doctors to take aspirin and another 11,000 to take a placebo. After 5 years, 104 of the aspirin group and 189 of the control group had died of heart attacks. Is this difference large enough to convince us that aspirin works?
- Does smoking behavior vary according to the socioeconomic status (SES) of adults? To find out, researchers classify several hundred men according to their SES (high, medium, or low) and also according to their smoking behavior (current smoker, former smoker, never smoked). Is there a significant relationship between SES and smoking, and if so, how can we describe it?

• • •

Many statistical studies produce counts rather than measurements. An opinion poll asks a sample of adults whether they approve of the president's conduct of his office; the data are the counts of "yes," "no," and "don't know." An experiment compares the effectiveness of four treatments to prevent the common cold; the data are the number of subjects given each treatment who catch a cold during the next month. A researcher classifies each of a sample of students according to their major field of study and their political preference; the data are the counts of students in each cell of a two-way table. This chapter presents procedures for statistical inference in these settings.

We begin in Section 8.1 with inference about a single population proportion. The statistical model for a count is then the binomial distribution, which we studied in Section 5.1. Section 8.2 concerns methods for comparing two proportions. Binomial distributions again play an important role. Finally, in Section 8.3, we present a general method for analyzing count data when the observations are classified in a two-way table. We have already studied such tables descriptively in Section 2.5.

8.1 INFERENCE FOR A SINGLE PROPORTION

We want to estimate the proportion p of some characteristic, such as approving the president's conduct of his office, among the members of a large population. We select a simple random sample (SRS) of size n from the population and record the count X of "successes" (such as "yes" answers to a question about the president). We will use "success" as shorthand for whatever characteristic we are interested in. The sample proportion of successes $\hat{p} = X/n$ estimates the unknown population proportion p. If the population is much larger than the sample (say, at least 10 times as large), the individual responses are nearly independent and the count X has approximately the binomial distribution $B(n, p)$. In statistical terms, we are concerned with inference about the probability p of a success in the binomial setting.

If the sample size n is small, we must base tests and confidence intervals for p on the binomial distributions. These are awkward to work with because of the discreteness of the binomial distributions.[1] But we know that when the sample is large, both the count X and the sample proportion $\hat{p}$ are approximately normal. We will consider only inference procedures based on the normal approximation. These procedures are similar to those for inference about the mean of a normal distribution.

Confidence intervals and significance tests

The unknown population proportion p is estimated by the sample proportion $\hat{p} = X/n$. We know (Chapter 5, page 382) that if the sample size n is sufficiently large, $\hat{p}$ has approximately the normal distribution with mean $\mu_{\hat{p}} = p$ and standard deviation $\sigma_{\hat{p}} = \sqrt{p(1-p)/n}$. If p were known, standardizing $\hat{p}$ would produce a z statistic

$$z = \frac{\hat{p} - p}{\sqrt{\dfrac{p(1-p)}{n}}}$$

that has approximately the $N(0, 1)$ distribution. In order to test H_0: $p = p_0$, we calculate P-values by acting as though H_0 were true. That is, we substitute p_0 for p in this z statistic.

standard error of $\hat{p}$

To find a confidence interval for p, on the other hand, we must estimate the standard deviation of $\hat{p}$ from the data. To do this, replace p by $\hat{p}$ in the expression for $\sigma_{\hat{p}}$. This gives the *standard error of* $\hat{p}$,

$$s_{\hat{p}} = \sqrt{\frac{\hat{p}(1-\hat{p})}{n}}$$

The z procedure for a confidence interval involves two approximations: the normal approximation to the binomial distribution and the approximation of p by $\hat{p}$ in the standard deviation. Here is a summary of the inference procedures that result.

LARGE-SAMPLE INFERENCE FOR A POPULATION PROPORTION

Draw an SRS of size n from a large population with unknown proportion p of successes. An approximate level C confidence interval for p is

$$\hat{p} \pm z^* \sqrt{\frac{\hat{p}(1-\hat{p})}{n}}$$

where z^* is the upper $(1 - C)/2$ standard normal critical value. To test the hypothesis H_0: $p = p_0$, compute the z statistic

$$z = \frac{\hat{p} - p_0}{\sqrt{\dfrac{p_0(1-p_0)}{n}}}$$

In terms of a standard normal random variable Z, the approximate P-value for a test of H_0 against

$$H_a: p > p_0 \quad \text{is} \quad P(Z \geq z)$$

$$H_a: p < p_0 \quad \text{is} \quad P(Z \leq z)$$

$$H_a: p \neq p_0 \quad \text{is} \quad 2P(Z \geq |z|)$$

Note that for confidence intervals we use the standard error $s_{\hat{p}}$, whereas for significance testing we use the hypothesized value p_0 in the expression for $\sigma_{\hat{p}}$. The confidence interval recipe is quite accurate when n is so large that both $n\hat{p} \geq 10$ and $n(1-\hat{p}) \geq 10$. For tests, we will use the z statistic whenever both $np_0 \geq 10$ and $n(1-p_0) \geq 10$. If these rules of thumb are not met, or if the population is less than 10 times as large as the sample, accurate inference requires more elaborate procedures.

EXAMPLE 8.1

An association of Christmas tree growers in Indiana sponsored a sample survey of Indiana households to help improve the marketing of Christmas trees.[2] An SRS of 500 households was contacted by telephone and asked several questions in a 2-minute interview. One question was "Did you have a Christmas tree this year?" Of the 500 respondents, 421 answered "Yes." The sample proportion who had a tree is therefore

$$\hat{p} = \frac{421}{500} = .842$$

We compute a 95% confidence interval for the true proportion of all Indiana households that displayed a Christmas tree. From Table D we find the value of z^* to be 1.96. The interval is

$$\begin{aligned} \hat{p} \pm z^* \sqrt{\frac{\hat{p}(1-\hat{p})}{n}} &= .842 \pm 1.960\sqrt{\frac{(.842)(.158)}{500}} \\ &= .842 \pm .032 \\ &= (.810,\ .874) \end{aligned}$$

We are 95% confident that between 81% and 87% of Indiana homes displayed Christmas trees. ●

Remember that the margin of error in this confidence interval includes only random sampling error. There are other sources of error that are not accounted for. For example, the sample was chosen from telephone directories and so omits households without telephones or with unlisted numbers. Moreover, the interviewers called households at random until 500 responded. About 700 households were called in all, with no second call if the number dialed was busy or no one answered. These facts of real statistical life introduce some bias into the survey. The survey team checked the bias by comparing demographic data collected from the respondents with census data for the state as a whole.

EXAMPLE 8.2

Of the 500 responding households in the Christmas tree market survey, 38% were from rural areas (including small towns) and the other 62% were from urban areas (including suburbs). According to the census, 36% of Indiana households are in rural areas and the remaining 64% are in urban areas. To examine how well the sample represents the state in regard to rural versus urban residence, we perform a significance test of

$$H_0: p = .36$$

versus the alternative

$$H_a: p \neq .36$$

where p is the probability that a household reached by the survey is rural. That is, p represents the proportion of rural households that would be obtained by the telephone sampling procedure if it were repeated over and over again.

The test statistic is

$$z = \frac{\hat{p} - .36}{\sqrt{\frac{(.36)(.64)}{500}}} = \frac{.38 - .36}{\sqrt{\frac{(.36)(.64)}{500}}} = .93$$

From Table A, we find that the probability that a Z is less than or equal to 0.93 is 0.8238. Therefore,

$$P(Z \geq .93) = 1 - .8238 = .1762$$

and the P-value is $2(0.1762) = 0.35$. Figure 8.1 illustrates the calculation of the P-value. There is a 35% chance of getting a value of Z larger than 0.93 or smaller than -0.93 if the null hypothesis is true. We conclude that the survey households (38% rural) reasonably reflect the census figure (36% rural); the actual difference is small and the null hypothesis of equality is not rejected ($\hat{p} = 0.38$, $z = 0.93$, $P = 0.35$). ●

In this example we arbitrarily chose to state the hypotheses in terms of the proportion of rural respondents. We could as easily have used the proportion of urban respondents. The results of the significance test should not depend on our choice, and the following example shows that they do not.

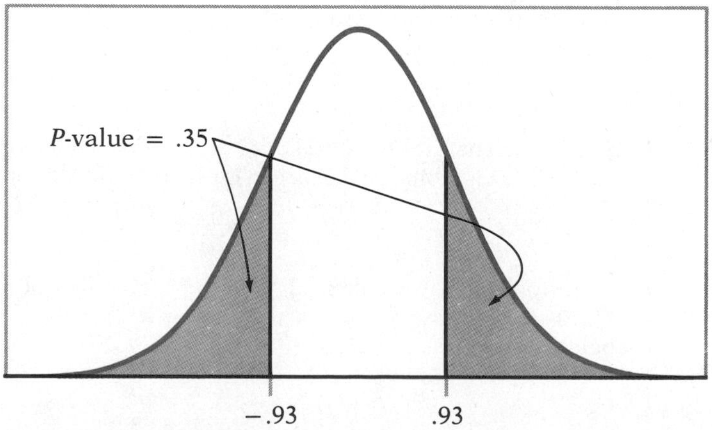

FIGURE 8.1 The *P*-value in Example 8.2.

EXAMPLE 8.3

For the Christmas tree market survey, we let p now represent the proportion of *urban* households that would be obtained by the telephone sampling procedure if it were repeated over and over again. To check agreement with the census results, we test

$$H_0: p = .64$$

versus the alternative

$$H_a: p \neq .64$$

The test statistic is

$$z = \frac{\hat{p} - .64}{\sqrt{\frac{(.64)(.36)}{500}}} = \frac{.62 - .64}{\sqrt{\frac{(.64)(.36)}{500}}} = -.93$$

Because only the sign of z has changed, the P-value remains the same as in the previous example. ●

The fact illustrated in Example 8.3 is true in general. When performing significance tests on a single proportion, we obtain the same value of the z statistic except for a reversed sign when we interchange the labels "success" and "failure" in the binomial model. This corresponds to interchanging the probabilities p and $1 - p$.

For some other demographic variables there were statistically significant differences between the Christmas tree survey respondents and census results. The responding households had somewhat higher incomes and were more likely to live in a house rather than in an apartment or mobile home. These facts lead to some overestimation of Christmas tree usage, because well-off people living in a house are more likely to have a tree. In other words, the survey's estimate is slightly biased upward. More advanced statistical techniques allowed the survey team to assess the bias and to adjust the estimate for the bias.

EXAMPLE 8.4

The French naturalist Buffon once tossed a coin 4040 times and obtained 2048 heads. This is a binomial experiment with $n = 4040$. The sample proportion is

$$\hat{p} = \frac{2048}{4040} = .5069$$

If Buffon's coin was balanced, then the probability of obtaining heads on any toss is 0.5. To assess whether the data provide evidence that the coin was not balanced, we test

$$H_0: p = .5$$
$$H_a: p \neq .5$$

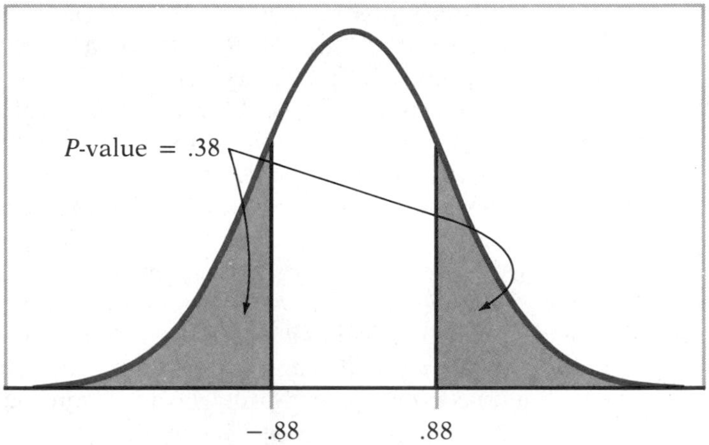

FIGURE 8.2 **The *P*-value in Example 8.4.**

The test statistic is

$$z = \frac{\hat{p} - .5}{\sqrt{\frac{(.5)(.5)}{4040}}} = \frac{.5069 - .5}{\sqrt{\frac{(.5)(.5)}{4040}}} = .88$$

Figure 8.2 illustrates the calculation of the *P*-value. From Table A we find $P(Z \leq 0.88) = 0.8106$. Therefore, the probability in each tail is $1 - 0.8106 = 0.1894$, and the *P*-value is $P = 2(0.1894) = 0.38$. The data are compatible with the balanced-coin hypothesis ($\hat{p} = 0.5069$, $z = 0.88$, $P = 0.38$). ●

Unless a coin is perfectly balanced, its true probability of heads will not be *exactly* 0.5. The significance test in Example 8.4 demonstrates only that Buffon's coin is statistically indistinguishable from a fair coin on the basis of 4040 tosses. To see what other values of p are compatible with the sample results, calculate a confidence interval.

EXAMPLE 8.5

For Buffon's coin-tossing experiment, the 99% confidence interval for the probability of a head is

$$\begin{aligned} \hat{p} \pm z^* \sqrt{\frac{\hat{p}(1-\hat{p})}{n}} &= .5069 \pm 2.576 \sqrt{\frac{(.5069)(.4931)}{4040}} \\ &= .5069 \pm .0203 \\ &= (.4866,\ .5272) \end{aligned}$$

We are 99% confident that the probability of a head is between 0.4866 and 0.5272. ●

The confidence interval of Example 8.5 is much more informative than the significance test of Example 8.4. It indicates the range of values of p that are consistent with the observed results. This is the range of p_0's that would not be rejected by a test of level 0.01. We would not be surprised if the true probability of heads for Buffon's coin were something like 0.51.

Significance tests for a single proportion are relatively rare in statistical practice because it is uncommon to have an exactly specified p_0. For physical experiments such as coin tossing or drawing cards from a well-shuffled deck, probability arguments lead to an ideal p_0, as in Example 8.4. Even here, however, it can be argued that no real coin has a probability of heads *exactly* equal to 0.5. Census results can specify p_0, as in Example 8.2, but populations are constantly changing and the census p_0 is not exactly correct today. Similarly, data from past large samples can determine p_0. In some types of cancer research, for example, "historical controls" from past studies serve as the benchmark for evaluating new treatments. This practice is the subject of much controversy among medical researchers, because the past never quite resembles the present. In general, statisticians prefer comparative studies whenever possible.

Choosing a sample size

In Chapter 6, we showed how to choose the sample size n to obtain a confidence interval with specified margin of error m for a normal mean. Because we are using a normal approximation for inference about a population proportion, sample size selection proceeds in much the same way.

Recall that an approximate level C confidence interval for p is

$$\hat{p} \pm z^* \sqrt{\frac{\hat{p}(1-\hat{p})}{n}}$$

where z^* is the appropriate normal critical value. The margin of error of the confidence interval is therefore

$$z^* \sqrt{\frac{\hat{p}(1-\hat{p})}{n}}$$

Because the value of $\hat{p}$ is not known before the data are gathered, we must guess a value to use in the calculations. Let p^* be the guessed value. The margin of error of the confidence interval is largest when $\hat{p} = 0.5$. So a conservative approach to the problem is to use $p^* = 0.5$ as the guessed value. This ensures that our final confidence interval will have margin of error less than or equal to the specified value. If m is the desired margin of error, set

$$m = z^* \sqrt{\frac{p^*(1-p^*)}{n}}$$

and solve this equation for n to find the sample size required. Here is the result.

SAMPLE SIZE FOR DESIRED MARGIN OF ERROR

The level C confidence interval for a proportion p will have margin of error approximately equal to a specified value m when the sample size is

$$n = \left(\frac{z^*}{m}\right)^2 p^*(1 - p^*)$$

where p^* is a guessed value for the true proportion.

The margin of error will be less than or equal to m if p^* is chosen to be 0.5. This gives

$$n = \left(\frac{z^*}{2m}\right)^2$$

The value of n obtained by this method is not particularly sensitive to the choice of p^* when p^* is fairly close to 0.5. However, if the value of p is smaller than about 0.3 or larger than about 0.7, use of $p^* = 0.5$ may select a sample size that is much larger than needed.

EXAMPLE 8.6

You are doing a sample survey to determine which of two candidates a group of voters prefers. Let p be the true proportion of voters who prefer the first candidate. Because you judge that the value of p is not very different from 0.5, you use $p^* = 0.5$ for planning the sample size.

You want to estimate p with 95% confidence and a margin of error less than or equal to 3%, or 0.03. The sample size required is

$$n = \left[\frac{1.96}{(2)(.03)}\right]^2 = 1067.1$$

which we round up to $n = 1068$. (Rounding down would give a margin of error slightly less than 0.03.)

Similarly for a 2.5% margin of error we have (after rounding up)

$$n = \left[\frac{1.96}{(2)(.025)}\right]^2 = 1537$$

and for a 2% margin of error the required sample size is

$$n = \left[\frac{1.96}{(2)(.02)}\right]^2 = 2401$$

●

News reports frequently describe the results of surveys with sample sizes between 1000 and 1500 and a margin of error of about 3%. These surveys generally use sampling procedures more complicated than simple random sampling, so the calculation of confidence intervals is more involved than we have studied in this section. The calculations in Example 8.6 nonetheless show in principle how such surveys are planned.

In practice, many factors influence the choice of a sample size. In the Christmas tree market survey, the researchers consulted a statistician when planning the study. They estimated that each telephone interview would take about 2 minutes. Nine trained students in agribusiness marketing were to make the phone calls between 1:00 P.M. and 8:00 P.M. on a Sunday. After discussing problems related to people not being at home or being unwilling to answer the questions, the survey team proposed a sample size of 500. To evaluate this proposal, the statistician calculated the margins of error of 95% confidence intervals for various values of $\hat{p}$.

EXAMPLE 8.7

In the Christmas tree market survey, the margin of error of a 95% confidence interval for any value of $\hat{p}$ and $n = 500$ is

$$\begin{aligned} m &= z^* \sqrt{\frac{\hat{p}(1-\hat{p})}{n}} \\ &= 1.96\sqrt{\frac{\hat{p}(1-\hat{p})}{500}} \\ &= .175\sqrt{\hat{p}(1-\hat{p})} \end{aligned}$$

The results for various values of $\hat{p}$ are

$\hat{p}$	m	$\hat{p}$	m
.05	.019	.60	.043
.10	.026	.70	.040
.20	.035	.80	.035
.30	.040	.90	.026
.40	.043	.95	.019
.50	.044		

The survey team judged these margins of error to be acceptable, and they used sample size of 500 in their survey. ●

The table in Example 8.7 illustrates several points. First, the margins of error for $\hat{p} = 0.05$ and $\hat{p} = 0.95$ are the same. The margins of error will always be the same for $\hat{p}$ and $1 - \hat{p}$. This is a direct consequence of the form of the confidence interval. Second, the margin of error varies only between 0.040 and 0.044 as $\hat{p}$ varies from 0.3 to 0.7, and the margin of error is greatest when $\hat{p} = 0.5$, as we claimed earlier. It is true in general that the margin of error will vary relatively little for values of $\hat{p}$ between 0.3 and 0.7. Therefore, when planning a study, it is not necessary to have a very precise guess for p. If $p^* = 0.5$ is used and the observed $\hat{p}$ is between 0.3 and 0.7, the actual interval will be a little shorter than needed but the difference will be small.

Finally, note that the margins of error m for very small and very large values of $\hat{p}$ are not included in the table. Suppose, for example, that we expect p to be about 0.01. Then for $n = 500$, we have $np = 5$. Recall that for the normal approximation to be reasonably accurate, we require $np \geq 10$ and $n(1-p) \geq 10$. We would not be justified in using the normal approximation to the binomial to calculate the confidence interval when p is as small as 0.01. More advanced methods are needed in this case.

SUMMARY

Inference about a population proportion p from an SRS of size n is based on the **sample proportion** $\hat{p} = X/n$. When n is large, $\hat{p}$ has approximately the normal distribution with mean p and standard deviation $\sqrt{p(1-p)/n}$.

The level C confidence interval for p is

$$\hat{p} \pm z^* \sqrt{\frac{\hat{p}(1-\hat{p})}{n}}$$

where z^* is the upper $(1-C)/2$ standard normal critical value.

Tests of $H_0: p = p_0$ are based on the **z statistic**

$$z = \frac{\hat{p} - p_0}{\sqrt{\dfrac{p_0(1-p_0)}{n}}}$$

with P-values calculated from the $N(0,1)$ distribution.

The sample size required to obtain a confidence interval of approximate margin of error m for a proportion is

$$n = \left(\frac{z^*}{m}\right)^2 p^*(1-p^*)$$

where p^* is a guessed value for the proportion, and z^* is the standard normal critical value for the desired level of confidence. To ensure that the margin of error of the interval is less than or equal to m no matter what p may be, use

$$n = \left(\frac{z^*}{2m}\right)^2$$

SECTION 8.1 EXERCISES

8.1 In each of the following cases state whether or not the normal approximation to the binomial should be used for a significance test on the population proportion p.
(a) $n = 10$ and $H_0: p = 0.4$.
(b) $n = 100$ and $H_0: p = 0.6$.
(c) $n = 1000$ and $H_0: p = 0.996$.
(d) $n = 500$ and $H_0: p = 0.3$.

8.2 In each of the following cases state whether or not the normal approximation to the binomial should be used for a confidence interval for the population proportion p.
(a) $n = 30$ and we observe $\hat{p} = 0.9$.
(b) $n = 25$ and we observe $\hat{p} = 0.5$.
(c) $n = 100$ and we observe $\hat{p} = 0.04$.
(d) $n = 600$ and we observe $\hat{p} = 0.6$.

8.3 As part of a quality improvement program, your mail-order company is studying the process of filling customer orders. According to company standards, an order is shipped on time if it is sent within 3 working days of the time it is received. You select an SRS of 100 of the 5000 orders received in the past month for an audit. The audit reveals that 86 of these orders were shipped on time. Find a 95% confidence interval for the true proportion of the month's orders that were shipped on time.

8.4 Large trees growing near power lines can cause power failures during storms when their branches fall on the lines. Power companies spend a great deal of time and money trimming and removing trees to prevent this problem. Researchers are developing hormone and chemical treatments that will stunt or slow tree growth. If the treatment is too severe, however, the tree will die. In one series of laboratory experiments on 216 sycamore trees, 41 trees died. Give a 99% confidence interval for the proportion of sycamore trees that would be expected to die from this particular treatment.

8.5 In recent years over 70% of first-year college students responding to a national survey have identified "being well-off financially" as an important personal goal. A state university finds that 132 of an SRS of 200 of its first-year students say that this goal is important. Give a 95% confidence interval for the proportion of all first-year students at the university who would identify being well-off as an important personal goal.

8.6 The Gallup Poll asked a sample of 1785 U.S. adults, "Did you, yourself, happen to attend church or synagogue in the last 7 days?" Of the respondents, 750 said "Yes." Suppose (it is not, in fact, true) that Gallup's sample was an SRS.

(a) Give a 99% confidence interval for the proportion of all U.S. adults who attended church or synagogue during the week preceding the poll.

(b) Do the results provide good evidence that less than half of the population attended church or synagogue?

(c) How large a sample would be required to obtain a margin of error of ± 0.01 in a 99% confidence interval for the proportion who attend church or synagogue? (Use Gallup's result as the guessed value of p.)

8.7 A national opinion poll found that 44% of all American adults agree that parents should be given vouchers good for education at any public or private school of their choice. The result was based on a small sample. How large an SRS is required to obtain a margin of error of ± 0.03 (that is, $\pm 3\%$) in a 95% confidence interval? (Use the previous poll's result to obtain the guessed value p^*.)

8.8 An entomologist samples a field for egg masses of a harmful insect by placing a yard-square frame at random locations and carefully examining the ground within the frame. An SRS of 75 locations selected from a county's pasture land found egg masses in 13 locations. Give a 95% confidence interval for the proportion of all possible locations that are infested.

8.9 Is there really a home-field advantage in baseball? In the 1991 National League season, the home team won 532 games and lost 438 games.

(a) Is this convincing evidence that the probability p that the home team wins is greater than 0.5? (Assume that the binomial model holds; this is at best a rough approximation because the teams vary in ability.)

(b) What values of p are compatible with the data in the sense that they would not be rejected at the 5% significance level? (Use a confidence interval.) What do you conclude about the home-field advantage?

8.10 Of the 500 respondents in the Christmas tree market survey, 44% had no children at home and 56% had at least one child at home. The corresponding figures for the most recent census are 48% with no children and 52% with at least one child. Test the null hypothesis that the telephone survey technique has a probability of selecting a household with no children that is equal to the value obtained by the census. Give the z statistic and the P-value. What do you conclude?

8.11 The English statistician Karl Pearson once tossed a coin 24,000 times and obtained 12,012 heads.

(a) Find the z statistic for testing the null hypothesis that Pearson's coin had probability 0.5 of coming up heads versus the two-sided alternative. Give the P-value. Do you reject H_0 at the 1% significance level?

(b) Find a 99% confidence interval for the probability of heads for Pearson's coin. This is the range of probabilities that cannot be rejected at the 1% significance level.

8.12 The English mathematician John Kerrich tossed a coin 10,000 times and obtained 5067 heads.

(a) Is this significant evidence at the 5% level that the probability that Kerrich's coin comes up heads is not 0.5?

(b) Use a 95% confidence interval to find the range of probabilities of heads that would not be rejected at the 5% level.

8.13 A matched pairs experiment compares the taste of instant versus fresh-brewed coffee. Each subject tastes two unmarked cups of coffee, one of each type, in random order and states which he or she prefers. Of the 50 subjects who participate in the study, 19 prefer the instant coffee. Let p be the probability that a randomly chosen subject prefers freshly brewed coffee to instant coffee. (In practical terms, p is the proportion of the population who prefer fresh-brewed coffee.)

(a) Test the claim that a majority of people prefer the taste of fresh-brewed coffee. Report the z statistic and its P-value. Is your result significant at the 5% level? What is your practical conclusion?

(b) Find a 90% confidence interval for p.

8.14 LeRoy, a starting player for a major college basketball team, made only 40% of his free throws last season. During the summer he worked on developing a softer shot in the hope of improving his free-throw accuracy. In the first eight games of this season LeRoy made 25 free throws in 40 attempts. Let p be his probability of making each free throw he shoots this season.

(a) State the null hypothesis H_0 that LeRoy's free-throw probability has remained the same as last year and the alternative H_a that his work in the summer resulted in a higher probability of success.

(b) Calculate the z statistic for testing H_0 versus H_a.

(c) Do you accept or reject H_0 for $\alpha = 0.05$? Find the P-value.

(d) Give a 90% confidence interval for LeRoy's free-throw success probability for the new season. Are you convinced that he is now a better free-throw shooter than last season?

(e) What assumptions are needed for the validity of the test and confidence interval calculations that you performed?

8.15 You want to estimate the proportion of students at your college or university who are employed for 10 or more hours per week while classes are in session. You plan to present your results by a 95% confidence interval. Using the guessed value $p^* = 0.3$, find the sample size required if the interval is to have approximate margin of error of $m = 0.05$.

8.16 A magazine publisher would like to know the proportion of subscribers who have annual household incomes in excess of \$75,000. To do this they will survey an SRS of their subscribers. They would like the margin of error of the 99% confidence interval for the proportion to be 0.025 or less. Use the guessed value $p^* = 0.5$ to find the required sample size.

8.17 A student organization wants to start a nightclub for students under the age of 21. To assess support for this proposal, they will select an SRS of students and ask each respondent if he or she would patronize this type of establishment. They expect that about 70% of the student body would respond favorably. What sample size is required to obtain a 90% confidence interval with approximate margin of error of 0.04? Suppose that 50% of the sample respond favorably. Calculate the margin of error of the 90% confidence interval.

8.18 An automobile manufacturer would like to know what proportion of its customers are dissatisfied with the service received from their local dealer. The customer relations department will survey a random sample of customers and compute a 99% confidence interval for the proportion who are dissatisfied. From past studies, they believe that this proportion will be about 0.2. Find the sample size needed if the margin of error of the confidence interval is to be about 0.015. Suppose 10% of the sample say that they are dissatisfied. What is the margin of error of the 99% confidence interval?

8.19 You have been asked to survey students at a large college to determine the proportion who favor an increase in student fees to support an expansion of the student newspaper. Each student will be asked whether he or she is in favor of the proposed increase. Using records provided by the registrar you can select a random sample of students from the college. After careful consideration of your resources, you decide that it is reasonable to conduct a study with a sample of 100 students.

(a) For this sample size, construct a table of the margins of error for 95% confidence intervals when $\hat{p}$ takes the values 0.1, 0.2, 0.3, 0.4, 0.5, 0.6, 0.7, 0.8, and 0.9.

(b) For a sample of size 100, would you use a confidence interval based on a normal approximation if $\hat{p} = 0.04$? Why or why not?

A former editor of the student newspaper agrees to underwrite the study in the previous exercise because she believes the results will demonstrate that most students support an increase in fees. She is willing to provide funds for a sample of size 500. Answer all of the questions posed in the previous exercise. Then write a short summary for your benefactor of why the increased sample size will provide better results.

8.2 COMPARING TWO PROPORTIONS

Because comparative studies are so common, we often want to compare the proportions of two groups (such as men and women) that have some characteristic. We call the two groups being compared population 1 and population 2, and the two population proportions of "successes" p_1 and p_2. The data consist of two independent SRSs, of size n_1 from population 1 and size n_2 from population 2. The proportion of successes in each sample estimates the corresponding population proportion. Here is the notation we will use in this section:

Population	Population proportion	Sample size	Count of successes	Sample proportion
1	p_1	n_1	X_1	$\hat{p}_1 = \dfrac{X_1}{n_1}$
2	p_2	n_2	X_2	$\hat{p}_2 = \dfrac{X_2}{n_2}$

To compare the two populations, we use the difference

$$D = \hat{p}_1 - \hat{p}_2$$

between the two sample proportions. When both sample sizes are sufficiently large, the sampling distribution of the difference D is approximately normal.

Inference procedures for comparing proportions are z procedures based on the normal approximation and on standardizing the difference D. The first step is to obtain the mean and standard deviation of D. By the addition rule for means, the mean of D is the difference of the means,

$$\mu_D = \mu_{\hat{p}_1} - \mu_{\hat{p}_2} = p_1 - p_2$$

That is, the difference $D = \hat{p}_1 - \hat{p}_2$ between the sample proportions is an unbiased estimator of the population difference $p_1 - p_2$. Similarly, the addition rule for variances tells us that the variance of D is the *sum* of the variances,

$$\sigma_D^2 = \sigma_{\hat{p}_1}^2 + \sigma_{\hat{p}_2}^2$$
$$= \frac{p_1(1-p_1)}{n_1} + \frac{p_2(1-p_2)}{n_2}$$

Therefore, when n_1 and n_2 are large, D is approximately normal with mean $\mu_D = p_1 - p_2$ and standard deviation

$$\sigma_D = \sqrt{\frac{p_1(1-p_1)}{n_1} + \frac{p_2(1-p_2)}{n_2}}$$

Confidence intervals

For both confidence intervals and hypothesis tests, we need to estimate the unknown standard deviation σ_D. First, consider a confidence interval for $p_1 - p_2$. Substitute the sample values $\hat{p}_1$ and $\hat{p}_2$ for p_1 and p_2 in the expression for σ_D to obtain

$$s_D = \sqrt{\frac{\hat{p}_1(1-\hat{p}_1)}{n_1} + \frac{\hat{p}_2(1-\hat{p}_2)}{n_2}}$$

This is the standard error of D that we will use in our confidence interval calculations.

CONFIDENCE INTERVALS FOR COMPARING TWO PROPORTIONS

Draw an SRS of size n_1 from a large population having proportion p_1 of successes and an independent SRS of size n_2 from another population having proportion p_2 of successes. When n_1 and n_2 are large, an approximate level C confidence interval for $p_1 - p_2$ is

$$(\hat{p}_1 - \hat{p}_2) \pm z^* s_D$$

where

$$s_D = \sqrt{\frac{\hat{p}_1(1-\hat{p}_1)}{n_1} + \frac{\hat{p}_2(1-\hat{p}_2)}{n_2}}$$

and z^* is the upper $(1-C)/2$ standard normal critical value.

This interval is approximately correct when the sample sizes n_1 and n_2 are large. As a general rule, we will use this method when $n_1\hat{p}_1$, $n_1(1-\hat{p}_1)$, $n_2\hat{p}_2$, and $n_2(1-\hat{p}_2)$ are all 5 or more.

EXAMPLE 8.8

If a respondent to the Christmas tree survey introduced in Example 8.1 did display a tree during the holiday season, the next question asked was whether the tree was natural or artificial. Respondents were also asked if they lived in an urban area or in a rural area. Of the 421 households displaying a Christmas tree, 261 were urban and 160 lived in rural areas.

Take population 1 to be urban tree users and population 2 to be rural tree users. We can consider that we have independent SRSs of size $n_1 = 261$ from population 1 and size $n_2 = 160$ from population 2. The tree growers want to know if there is a difference in preference for natural trees versus artificial trees between urban and rural households. The survey results are tabulated below.

Population	n	X	$\hat{p} = X/n$
1 (urban)	261	89	.341
2 (rural)	160	64	.400

In this table the $\hat{p}$ column gives the sample proportions of households with a natural tree. To compute a 90% confidence interval for the difference between urban and rural tree users in the proportions of households that prefer a natural tree, we first calculate the standard error of the observed difference:

$$\begin{aligned} s_D &= \sqrt{\frac{\hat{p}_1(1-\hat{p}_1)}{n_1} + \frac{\hat{p}_2(1-\hat{p}_2)}{n_2}} \\ &= \sqrt{\frac{(.341)(.659)}{261} + \frac{(.400)(.600)}{160}} \\ &= .04859 \end{aligned}$$

The 90% confidence interval is

$$\begin{aligned} (\hat{p}_1 - \hat{p}_2) \pm z^* s_D &= (.341 - .400) \pm (1.645)(.04859) \\ &= -.059 \pm .080 \\ &= (-.139,\ .021) \end{aligned}$$

With 90% confidence we can say that the difference in the proportions is between −0.14 and +0.02. Because the interval contains 0, we are not confident that either group has a stronger preference for natural trees than the other group. ●

The two samples in this example were not drawn separately and did not have sizes chosen in advance. The sample sizes are in fact random, arising from asking 500 randomly chosen respondents where they lived and whether they had a tree last holiday season. If the SRS were repeated, we would no doubt not have exactly 261 urban and 160 rural tree users among the 500 new respondents. It is in fact true—we do not say that it is obvious—that we can analyze the data just as if the sample sizes had been fixed in the design of the study. This fact extends the usefulness of procedures for comparison of two proportions. The issue of fixed versus random sample sizes will be discussed in Section 8.3, where we consider more elaborate analyses of count data.

Significance tests

Although we prefer to compare two proportions by giving a confidence interval for the difference between the two population proportions, it is sometimes useful to test the null hypothesis that the two population proportions are the same.

For the testing problem, we do not use the estimate s_D of σ_D that we employed for confidence intervals. Although s_D would lead to a valid test, we instead adopt the more common practice of using an estimate of σ_D based on the fact that the null hypothesis states that $p_1 = p_2$. If these two proportions are equal, then we can view all of the data as coming from a single population. Let p denote the common value of p_1 and p_2; then the standard deviation of $D = \hat{p}_1 - \hat{p}_2$ is

$$\begin{aligned} \sigma_D &= \sqrt{\frac{p_1(1-p_1)}{n_1} + \frac{p_2(1-p_2)}{n_2}} \\ &= \sqrt{p(1-p)\left(\frac{1}{n_1} + \frac{1}{n_2}\right)} \end{aligned}$$

We estimate the common value of p by the overall proportion of successes in both samples,

$$\hat{p} = \frac{X_1 + X_2}{n_1 + n_2}$$

pooled estimate of p

This estimate of p is called the *pooled estimate* because it combines, or pools, the information from both samples.

To estimate σ_D under the null hypothesis, we substitute $\hat{p}$ for p in the expression for σ_D. The result is a standard error for D that assumes the truth of H_0: $p_1 = p_2$:

$$s_p = \sqrt{\hat{p}(1-\hat{p})\left(\frac{1}{n_1} + \frac{1}{n_2}\right)}$$

The subscript on s_p reminds us that we pooled data from the two samples to construct the estimate.

SIGNIFICANCE TESTS FOR COMPARING TWO PROPORTIONS

To test the hypothesis

$$H_0: p_1 = p_2$$

compute the z statistic

$$z = \frac{\hat{p}_1 - \hat{p}_2}{s_p}$$

where

$$s_p = \sqrt{\hat{p}(1-\hat{p})\left(\frac{1}{n_1} + \frac{1}{n_2}\right)}$$

and

$$\hat{p} = \frac{X_1 + X_2}{n_1 + n_2}$$

In terms of a standard normal random variable Z, the P-value for a test of H_0 against

$$H_a: p_1 > p_2 \quad \text{is} \quad P(Z \geq z)$$
$$H_a: p_1 < p_2 \quad \text{is} \quad P(Z \leq z)$$
$$H_a: p_1 \neq p_2 \quad \text{is} \quad 2P(Z \geq |z|)$$

This z test is based on the normal approximation to the binomial distribution. As a general rule, we will use it when $n_1\hat{p}$, $n_1(1-\hat{p})$, $n_2\hat{p}$, and $n_2(1-\hat{p})$ are all 5 or more.

EXAMPLE 8.9

Do urban and rural households that display Christmas trees have the same preferences for natural versus artificial trees? Other studies have shown that rural households are more likely to choose natural Christmas trees. A one-sided test examines whether our sample supports this claim. Again, take population 1 to consist of the urban households that use a tree and population 2 to be the rural tree users. We want to test the hypotheses

$$H_0: p_1 = p_2$$
$$H_a: p_1 < p_2$$

The survey responses show that 89 of the 261 urban households and 64 of the 160 rural households that displayed a tree chose a natural tree. So

$$\hat{p}_1 = \frac{89}{261} = .341$$

$$\hat{p}_2 = \frac{64}{160} = .400$$

A higher proportion of the rural households chose natural trees, but it is not yet clear whether the observed difference could plausibly be due to the accidents of random selection.

For the combined sample, the proportion of respondents who chose a natural tree was

$$\hat{p} = \frac{89 + 64}{261 + 160} = .363$$

The test statistic is calculated as follows:

$$s_p = \sqrt{(.363)(.637)\left(\frac{1}{261} + \frac{1}{160}\right)} = .04828$$

$$z = \frac{\hat{p}_1 - \hat{p}_2}{s_p} = \frac{.341 - .400}{.04828}$$

$$= -1.22$$

The calculation of the *P*-value is illustrated in Figure 8.3. From Table A we find

$$P(Z \leq -1.22) = .1112$$

Because we are doing a one-sided test, the *P*-value is 0.11. Even though rural households in the survey chose natural Christmas trees more often than the urban households, there is not strong evidence that this difference in preferences is true in the population of all Indiana tree users. If the preferences of rural and urban households were identical, rural usage in 11% of all random samples of this size would exceed urban usage by an amount at least as large as that observed, simply because of chance variation. ●

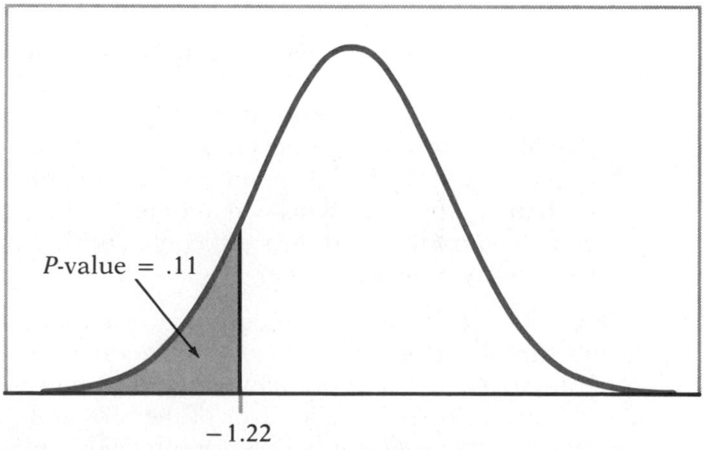

FIGURE 8.3 The *P*-value in Example 8.9.

SUMMARY

Comparison of two population proportions from independent SRSs of sizes n_1 and n_2 is based on the difference of sample proportions $D = \hat{p}_1 - \hat{p}_2$. The level C confidence interval for $p_1 - p_2$ is

$$(\hat{p}_1 - \hat{p}_2) \pm z^* s_D$$

where

$$s_D = \sqrt{\frac{\hat{p}_1(1-\hat{p}_1)}{n_1} + \frac{\hat{p}_2(1-\hat{p}_2)}{n_2}}$$

Significance tests of $H_0: p_1 = p_2$ use the **z statistic**

$$z = \frac{\hat{p}_1 - \hat{p}_2}{s_p}$$

with $N(0, 1)$ probabilities. In this statistic,

$$s_p = \sqrt{\hat{p}(1-\hat{p})\left(\frac{1}{n_1} + \frac{1}{n_2}\right)}$$

and $\hat{p}$ is the **pooled estimate** of the common value of p_1 and p_2,

$$\hat{p} = \frac{X_1 + X_2}{n_1 + n_2}$$

SECTION 8.2 EXERCISES

8.21 In the 1991 regular baseball season, the World Series Champion Minnesota Twins played 81 games at home and 81 games away. They won 51 of their home games and 44 of the games played away. We can consider these games as samples from potentially large populations of games played at home and away. How much advantage does the home field provide?

(a) Find the proportion of wins for the home games. Do the same for the away games.

(b) Find the standard error needed to compute a confidence interval for the difference in the proportions.

(c) Compute a 90% confidence interval for the difference between the probability that the Twins win at home and the probability that they win when on the road. Are you confident that the 1991 Twins were more likely to win at home?

8.22 The state agriculture department asked random samples of Indiana farmers in each county whether they favored a mandatory corn checkoff program to pay for corn product marketing and research. In Tippecanoe County, 263 farmers were in favor of the program and 252 were not. In neighboring Benton County, 260 were in favor and 377 were not.

(a) Find the proportions of farmers in favor of the program in each of the two counties.

(b) Find the standard error needed to compute a confidence interval for the difference in the proportions.
(c) Compute a 99% confidence interval for the difference between the proportions of farmers favoring the program in Tippecanoe County and in Benton County. Do you think opinions differed in the two counties?

8.23 Return to the Minnesota Twins baseball data in Exercise 8.21.
(a) Combining all of the games played, what proportion did the Twins win?
(b) Find the standard error needed for testing that the probability of winning is the same at home and away.
(c) Most people think that it is easier to win at home than away. Formulate null and alternative hypotheses to examine this idea.
(d) Compute the z statistic and its P-value. What conclusion do you draw?

8.24 Return to the survey of farmers described in Exercise 8.22.
(a) Combine the two samples and find the overall proportion of farmers who favor the corn checkoff program.
(b) Find the standard error needed for testing that the population proportions of farmers favoring the program are the same in the two counties.
(c) Formulate null and alternative hypotheses for comparing the two counties.
(d) Compute the z statistic and its P-value. What conclusion do you draw?

8.25 A major court case on the health effects of drinking contaminated water took place in the town of Woburn, Massachusetts. A town well in Woburn was contaminated by industrial chemicals. During the period that residents drank water from this well, there were 16 birth defects among 414 births. In years when the contaminated well was shut off and water was supplied from other wells, there were 3 birth defects among 228 births. The plaintiffs suing the firm responsible for the contamination claimed that these data show that the rate of birth defects was higher when the contaminated well was in use. How statistically significant is the evidence? What assumptions does your analysis require? Do these assumptions seem reasonable in this case?

8.26 A study of chromosome abnormalities and criminality examined data on 4124 Danish males born in Copenhagen. Each man was classified as having a criminal record or not, using the penal registers maintained in the offices of the local police chiefs. Each was also classified as having the normal male XY chromosome pair or one of the abnormalities XYY or XXY. Of the 4096 men with normal chromosomes, 381 had criminal records, while 8 of the 28 men with chromosome abnormalities had criminal records. Some experts believe that chromosome abnormalities are associated with increased criminality. Do these data lend support to this belief? Report your analysis and draw a conclusion. (Data from

H. A. Witkin et al., "Criminality in XYY and XXY men," *Science*, 193 (1976), pp. 547–555.)

8.27 Thc 1958 Detroit Area Study was an important sociological investigation of the influence of religion on everyday life. It is described in Gerhard Lenski, *The Religious Factor*, Doubleday, New York, 1961. The sample "was basically a simple random sample of the population of the metropolitan area." Of the 656 respondents, 267 were white Protestants and 230 were white Catholics. One question asked whether the government was doing enough in areas such as housing, unemployment, and education; 161 of the Protestants and 136 of the Catholics said "No." Is there evidence that white Protestants and white Catholics differed on this issue?

8.28 The respondents in the Detroit Area Study (see the previous exercise) were also asked whether they believed that the right of free speech included the right to make speeches in favor of communism. Of the white Protestants, 104 said "Yes," while 75 of the white Catholics said "Yes." Give a 95% confidence interval for the amount by which the proportion of Protestants who agreed that communist speeches are protected exceeds the proportion of Catholics who held this opinion.

8.29 A university financial aid office polled an SRS of undergraduate students to study their summer employment. Not all students were employed the previous summer. Here are the results for men and women:

	Men	Women
Employed	718	593
Not employed	79	139
Total	797	732

(a) Is there evidence that the proportion of male students employed during the summer differs from the proportion of female students who were employed? State H_0 and H_a, compute the test statistic, and give its P-value.

(b) Give a 99% confidence interval for the difference between the proportions of male and female students who were employed during the summer. Does the difference seem practically important to you?

8.30 The power takeoff driveline on farm tractors is a potentially serious hazard to farmers. A shield covers the driveline on new tractors, but for a variety of reasons, the shield is often missing on older tractors. Two types of shield are the bolt-on and the flip-up. A study initiated by the National Safety Council took a sample of older tractors to examine the proportions of shields removed. The study found that 35 shields had been removed from the 83 tractors having bolt-on shields and that 15 had been removed from the 136 tractors with flip-up shields. (Data from W. E. Sell and W. E.

Field, "Evaluation of PTO master shield usage on John Deere tractors," paper presented at the American Society of Agricultural Engineers 1984 Summer Meeting.)

(a) Test the null hypothesis that there is no difference between the proportions of the two types of shields removed. Give the z statistic and the P-value. State your conclusion in words.

(b) Give a 90% confidence interval for the difference in the proportions of removed shields for the bolt-on and the flip-up types. Based on the data, what recommendation would you make about the type of shield to be used on new tractors?

8.31 A clinical trial examined the effectiveness of aspirin in the treatment of cerebral ischemia (stroke). Patients were randomized into treatment and control groups. The study was double-blind in the sense that neither the patients nor the physicians who evaluated the patients knew which patients received aspirin and which the placebo tablet. After 6 months of treatment, the attending physicians evaluated each patient's progress as either favorable or unfavorable. Of the 78 patients in the aspirin group, 63 had favorable outcomes; 43 of the 77 control patients had favorable outcomes. (From William S. Fields et al., "Controlled trial of aspirin in cerebral ischemia," *Stroke*, 8 (1977), pp. 301–315.)

(a) Compute the sample proportions of patients having favorable outcomes in the two groups.

(b) Give a 95% confidence interval for the difference between the favorable proportions in the treatment and control groups.

(c) The physicians conducting the study had concluded from previous research that aspirin was likely to increase the chance of a favorable outcome. Carry out a significance test to confirm this conclusion. State hypotheses, find the P-value, and write a summary of your results.

8.32 The pesticide diazinon is in common use to treat infestations of the German cockroach, *Blattella germanica*. A study investigated the persistence of this pesticide on various types of surfaces. Researchers applied a 0.5% emulsion of diazinon to glass and plasterboard. After 14 days, they placed 18 cockroaches on each surface and recorded the number that died within 48 hours. On glass, 9 cockroaches died, but on plasterboard, 13 died. (Based on Elray M. Roper and Charles G. Wright, "German cockroach (Orthoptera: Blatellidae) mortality on various surfaces following application of diazinon," *Journal of Economic Entomology*, 78 (1985), pp. 733–737.)

(a) Calculate the mortality rates (sample proportion that died) for the two surfaces.

(b) Find a 90% confidence interval for the difference in the two population proportions.

(c) Chemical analysis of the residues of diazinon suggests that it may persist longer on plasterboard than on glass because it binds to the paper covering on the plasterboard. The researchers therefore expected the mortality rate to be greater on plasterboard than on glass. Conduct a significance test to assess the evidence that this is true.

8.33 Refer to the study of undergraduate student summer employment described in Exercise 8.29. Similar results from a smaller number of students may not have the same statistical significance. Specifically, suppose that 72 of 80 men surveyed were employed and 59 of 73 women surveyed were employed. The sample proportions are essentially the same as in the earlier exercise.

(a) Compute the z statistic for these data and report the P-value. What do you conclude?

(b) Compare the results of this significance test with your results in Exercise 8.29. What do you observe about the effect of the sample size on the results of these significance tests?

8.34 Suppose that the experiment of Exercise 8.32 placed more cockroaches on each surface and observed similar mortality rates. Specifically, suppose that 36 cockroaches were placed on each surface and that 26 died on the plasterboard while 18 died on the glass.

(a) Compute the z statistic for these data and report its P-value. What do you conclude?

(b) Compare the results of this significance test with those you gave in Exercise 8.32. What do you observe about the effect of the sample size on the results of these significance tests?

8.3 INFERENCE FOR TWO-WAY TABLES

Comparison of the proportion of "successes" in two populations begins with the count of successes in two samples, one from each population. We could of course choose to count failures rather than successes. One clear way to present the data is in a *two-way table* that gives both the counts of successes and the counts of failures. To construct a two-way table, classify each observation first by which population it represents and then by whether it is a success or a failure. Here is a two-way table of the Christmas tree survey data analyzed in Example 8.8 (page 592):

two-way table

	Urban	Rural
Natural tree	89	64
Artificial tree	172	96

We call this particular two-way table a 2×2 table because there are two rows (natural tree, artificial tree) and two columns (urban, rural). The advantage of the two-way table is that it can present data for variables having more than two categories by simply increasing the number of rows or columns. In this section we study statistical inference for count data that are classified according to two variables. The question of interest is whether there is a relation between the row variable and the column variable. For example, is there a relation between whether a household is rural or urban and its preferred type of Christmas tree? Example 8.8 found that there was a relation, that rural households were more likely to prefer natural trees. Now we will extend our methods to handle larger tables. Here is an example of a 4×2 table (two categorical variables, one with four categories and one with two categories).

EXAMPLE 8.10

Do men and women participate in sports for the same reasons? One goal for sports participants is social comparison—the desire to win or to do better than other people. Another is mastery—the desire to improve one's skills or to try one's best. A study on why students participate in sports collected data from 67 male and 67 female undergraduates at a large university.[3] Each student was classified into one of four categories based on his or her responses to a questionnaire about sports goals. The four categories were high social comparison–high mastery (HSC-HM), high social comparison–low mastery (HSC-LM), low social comparison–high mastery (LSC-HM), and low social comparison–low mastery (LSC-LM). One purpose of the study was to compare the goals of male and female students. Here are the data displayed in a two-way table:

Observed counts for sports goals

	Sex		
Goal	Female	Male	Total
HSC-HM	14	31	45
HSC-LM	7	18	25
LSC-HM	21	5	26
LSC-LM	25	13	38
Total	67	67	134

The entries in this table are the observed, or sample, counts. For example, there are 14 females in the high social comparison–high mastery group. Note that the marginal totals are given with the table. These are not part of the raw data but are calculated by summing over the rows or columns. The column totals are the numbers of observations sampled in the two populations. The grand total, 134, can be obtained by summing the row totals or the column totals. It is the total number of observations in the study. ●

The rows and columns of a two-way table represent values of two categorical variables. These are goal and sex in Example 8.10. Each com-

cell

bination of values for these two variables defines a *cell*. A two-way table with r rows and c columns contains $r \times c$ cells. The 4×2 table in Example 8.10 has 8 cells. The objective in this example is to compare men and women. That is, the column variable describes which population an observation comes from. The row variable is a categorical response variable, type of sports goal. It is not always the case that one direction of the table identifies populations to be compared. Two-way tables can display observations on any two categorical variables.

Analysis of two-way tables is best done using statistical software to carry out the considerable arithmetic required. We will use Example 8.10 and output from a typical statistical computing package to describe inference for two-way tables. Later in this section we will show how to do the work with a calculator if software is not available.

Describing relations in two-way tables

To describe relations between the variables, we compute and compare percents. The count in each cell can be viewed as a percent of the grand total, the row total, or the column total. You must decide which percents are most appropriate. Software usually prints out all three, but not all are of interest in a specific problem.

EXAMPLE 8.11

Figure 8.4 shows the output of the procedure FREQ in the SAS statistical software package for the data of Example 8.10. The two-way table appears in the output in expanded form. Each cell contains five entries, which are labeled in the upper left-hand corner of the output. FREQUENCY is simply the cell count. The row and column totals appear in the margins, just as in Example 8.10. The last three entries in each cell give the cell count as a percent of the overall total, the row total, and the column total. These are labeled PERCENT, ROW PCT, and COL PCT. For the HSC-HM females, the three percents are 10.45%, 31.11%, and 20.90%. You can verify these percents by dividing the cell count 14 by the total count 134, by the row total 45, and by the column total 67, in that order.

In this example, we are interested in the effect of sex on the distribution of sports goals. To compare the sexes we examine the column percents.

Column percents for sports goals

	Sex	
Goal	Female	Male
HSC-HM	21	46
HSC-LM	10	27
LSC-HM	31	7
LSC-LM	37	19
Total	100	100

We rounded the percents from the output to get a clearer summary of the data. The total row reminds us that these groups account for 100% of both men and women. (In fact, the sums differ slightly from 100% because of roundoff error.) The bar graph in Figure 8.5 compares the male and female percents. The data reveal something interesting: It appears that females and males have different goals when they participate in recreational sports. ●

TABLE OF GOAL BY SEX

GOAL / FREQUENCY, EXPECTED, PERCENT, ROW PCT, COL PCT	SEX: FEMALE	MALE	TOTAL
HSC-HM	14	31	45
	22.5	22.5	
	10.45	23.13	33.58
	31.11	68.89	
	20.90	46.27	
HSC-LM	7	18	25
	12.5	12.5	
	5.22	13.43	18.66
	28.00	72.00	
	10.45	26.87	
LSC-HM	21	5	26
	13.0	13.0	
	15.67	3.73	19.40
	80.77	19.23	
	31.34	7.46	
LSC-LM	25	13	38
	19.0	19.0	
	18.66	9.70	28.36
	65.79	34.21	
	37.31	19.40	
TOTAL	67	67	134
	50.00	50.00	100.00

STATISTICS FOR TABLE OF GOAL BY SEX

STATISTIC	DF	VALUE	PROB
CHI-SQUARE	3	24.898	0.000

SAMPLE SIZE=134

FIGURE 8.4 Computer output for the sports goals study (Example 8.11).

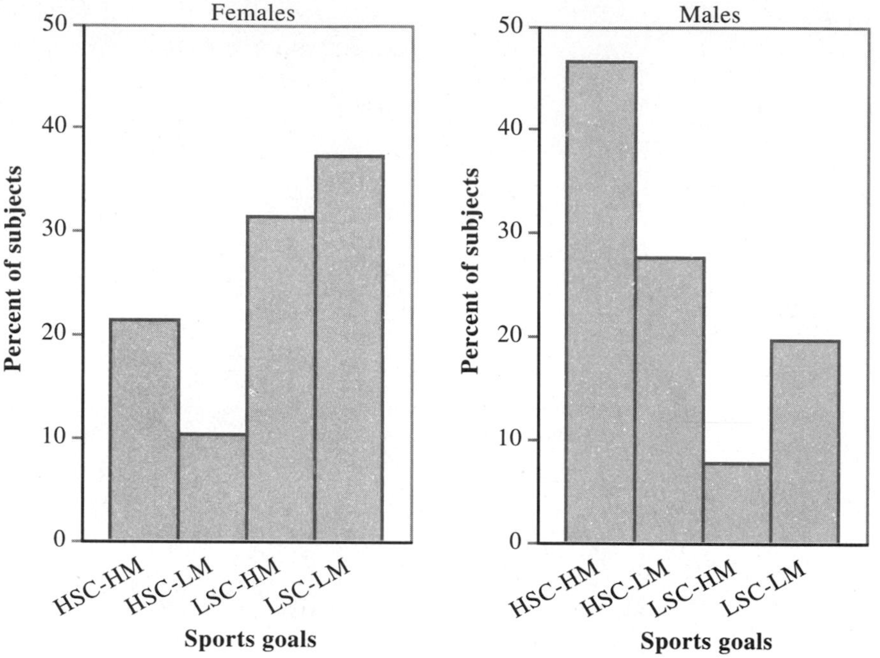

FIGURE 8.5 Comparison of sports goal distributions for females and males (Example 8.11).

The differences between the distributions of male and female sports goals in the sample appear to be large. A statistical test will tell us whether or not these differences can be plausibly attributed to chance. Specifically, if the two population distributions were the same, how likely is it that a sample would show differences as large or larger than those displayed in Figure 8.5?

The chi-square test

A null hypothesis H_0 of interest in a two-way table is: There is *no association* between the row variable and the column variable. In the sports goals example, this null hypothesis says that sex and sports goals are not related. The alternative hypothesis H_a is that there is an association between these two variables. The alternative H_a does not specify any particular direction for the association, such as "men rate social comparison higher as a goal than do women." Moreover, we cannot describe H_a as either one-sided or two-sided, because it includes all of the many kinds of association that are possible.

All statistical hypotheses can be stated in terms of population parameters. In our example, the hypothesis H_0 that there is no association between sex and sports goal is equivalent to the statement that the distributions of sports goals in the male and female populations are the same. Because a precise statement of H_0 and H_a involves some fine points, we leave it to a later, optional subsection. For $r \times c$ tables like that in Example 8.10, where the columns correspond to independent samples from distinct populations, there are c distributions for the row variable, one for each population. The null hypothesis then says that the c distributions of the row variable are identical. The alternative hypothesis is that the distributions are not all the same.

expected counts

To test the null hypothesis in $r \times c$ tables, we compare the actually observed cell counts with *expected* cell counts calculated under the assumption that the null hypothesis is true.

EXAMPLE 8.12

The expected counts for the sports goals example appear in the computer output shown in Figure 8.4. They are labeled EXPECTED and are the second entry in each cell. For example, the expected number of high social comparison–high mastery females is 22.5.

How is this expected count obtained? Look at the percents in the right margin of the table in Figure 8.4. We see that 33.58% of all respondents (female and male together) are in the HSC-HM group. If the null hypothesis of no sex difference in sports goals is true, we expect this overall percentage to apply to both men and women. So we expect 33.58% of the 67 females in the study to be in this group. The expected count is therefore 33.58% of 67, which is 22.5. The other expected counts are calculated similarly. Because the number of males is the same as the number of females in this study, the expected counts for males are the same as the expected counts for females. ●

To test the H_0 that there is no association between the row and column classifications, we use a statistic that compares the entire set of observed counts with the set of expected counts. First, take the difference between each observed count and its corresponding expected count, and square these values so that they are all 0 or positive. A large difference means less if it comes from a cell expected to have a large count, so divide each squared difference by the expected count, a kind of standardization. Finally, sum over all cells. The result is called the chi-square statistic.*

*The chi-square statistic was invented by the English statistician Karl Pearson (1857–1936) in 1900, for purposes slightly different from ours. It is the oldest inference procedure still used in its original form. With the work of Pearson and his contemporaries at the beginning of this century, statistics first emerges as a separate discipline.

CHI-SQUARE STATISTIC

The chi-square statistic is a measure of how much the observed cell counts in a two-way table diverge from the expected cell counts. The recipe for the statistic is

$$X^2 = \sum \frac{(\text{observed} - \text{expected})^2}{\text{expected}} \qquad (8.1)$$

where "observed" represents an observed sample count, "expected" represents the expected count for the same cell, and the sum is over all $r \times c$ cells in the table.

If the expected counts and the observed counts are very different, a large value of X^2 will result. So large values of X^2 provide evidence against the null hypothesis. To obtain a P-value for the test, we need the sampling distribution of X^2 under the assumption that H_0 (no association between the row and column variables) is true. We once again use an approximation, related to the normal approximation for binomial distributions. The result is a new distribution, the *chi-square distribution*, which we denote by χ^2. (χ is the Greek letter chi.)

chi-square distribution

Like the t distributions, the χ^2 distributions form a family described by a single parameter, the degrees of freedom. We use $\chi^2(\text{df})$ to indicate a particular member of this family. Figure 8.6 displays the density curves of the $\chi^2(2)$ and $\chi^2(4)$ distributions. As the figure suggests, chi-square distributions take only positive values and are skewed to the right. Table G in the back of the book gives upper critical values for the χ^2 distributions.

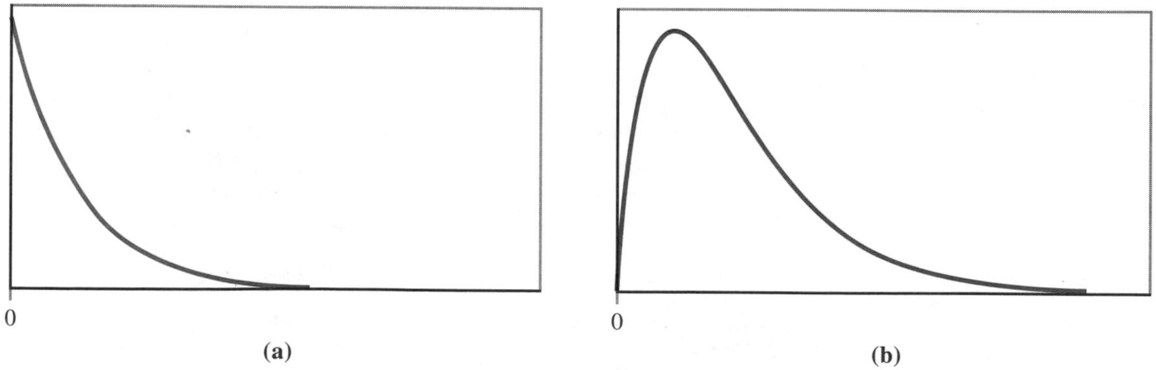

FIGURE 8.6 (a) The $\chi^2(2)$ density curve. (b) The $\chi^2(4)$ density curve.

CHI-SQUARE TEST FOR $r \times c$ TABLES

The null hypothesis H_0 for a two-way table is that there is no association between the row and column variables. The alternative is that these variables are related.

If H_0 is true, the statistic X^2 has approximately a χ^2 distribution with $(r-1)(c-1)$ degrees of freedom. The *P*-value for the chi-square test is

$$P(\chi^2 \geq X^2)$$

where χ^2 is a random variable having the $\chi^2(\text{df})$ distribution with $\text{df} = (r-1)(c-1)$.

The chi-square test always uses the upper tail of the χ^2 distribution, because any deviation from the null hypothesis makes the statistic larger. The approximation of the distribution of X^2 by χ^2 becomes more accurate as the cell counts increase. Moreover, it is more accurate for tables larger than 2×2 tables. For tables larger than 2×2, we will use this approximation whenever the average of the expected counts is 5 or more and the smallest expected count is 1 or more. For 2×2 tables, we require that all expected cell counts be 5 or more.[4] The 2×2 table is special in another way. A comparison of the proportions of "successes" in two populations leads to a 2×2 table. We can compare two population proportions either by the chi-square test or by the two-sample z test from Section 8.2. In fact, these tests always give exactly the same result, because the chi-square statistic is equal to the square of the z statistic and $\chi^2(1)$ critical values are equal to the squares of the corresponding $N(0, 1)$ critical values. The advantage of the z test is that we can test either one-sided or two-sided alternatives. The chi-square test always tests the two-sided alternative. Of course, the chi-square test can compare more than two populations, whereas the z test compares only two.

EXAMPLE 8.13

The results of the chi-square significance test for the sports goals example appear in the lower part of the computer output in Figure 8.4. Because all of the expected cell counts are moderately large, the χ^2 distribution provides accurate *P*-values. We see that $X^2 = 24.898$, $\text{df} = 3$, and the *P*-value is given as 0.000 under the heading PROB. Because the *P*-value is rounded to the nearest 0.001, we know that it is less than 0.0005. As a check we verify that the df are correct for a 4×2 table:

$$(r-1)(c-1) = (4-1)(2-1) = 3$$

The chi-square test confirms that the data contain clear evidence against the null hypothesis that female and male students have the same distributions of sports goals. Under H_0, the chance of obtaining a value of X^2 greater than or equal to the calculated value of 24.898 is very small—less than 0.0005. ●

The test indicates that the male and female distributions are not the same, but it does not say how they differ. Always combine the test with a description that shows what kind of relationship is present. From the percents in Figure 8.4 and the graph in Figure 8.5, we see that the percent of males in each of the HSC goal classes is more than twice the percent of females. The HSC-HM group contains 46.27% of the males but only 20.90% of the females, and the HSC-LM group contains 26.87% of the males and 10.45% of the females. The pattern is reversed for the LSC goal classes. We conclude that males are more likely to be motivated by social comparison goals and females are more likely to be motivated by mastery goals.

Computations

The calculations required to analyze a two-way table are straightforward but tedious. Though we recommend turning them over to software, it is possible to do the work with a calculator and patience. Here is an outline of the steps required.

COMPUTATIONS FOR $r \times c$ TABLES

1. Calculate descriptive statistics that convey the important information in the table. Usually these will be column or row percents.
2. Find the expected counts and use these to compute the X^2 statistic.
3. Use Table G to find the approximate P-value.
4. Draw a conclusion about the association between the row and column variables.

The following example illustrates these steps. The two-way table in this example does not compare several populations. Instead, it arises by classifying observations on a single population in two ways.

EXAMPLE 8.14

In a study of heart disease in male federal employees, researchers classified 356 volunteer subjects according to their socioeconomic status (SES) and their smoking habits.[5] There were three categories of SES: high, middle, and low. Individuals were asked whether they were current smokers, former smokers, or had never smoked, producing three categories for smoking habits as well. Here is the two-way table that summarizes the data:

Observed counts for smoking and SES

	SES			
Smoking	High	Middle	Low	Total
Current	51	22	43	116
Former	92	21	28	141
Never	68	9	22	99
Total	211	52	93	356

This is a 3×3 table, to which we have added the marginal totals obtained by summing across rows and columns. For example, the first row total is $51 + 22 + 43 = 116$. The grand total, the number of subjects in the study, can be computed by summing the row totals, $116 + 141 + 99 = 356$, or the column totals, $211 + 52 + 93 = 356$. It is good statistical practice to do both as a check on your arithmetic.

We start our analysis by computing descriptive statistics that summarize the observed relation between SES and smoking. The researchers suspected that SES helps explain smoking, so in this situation SES is the explanatory variable and smoking is the response variable. In general, the clearest description of the relationship is provided by comparing the conditional distributions of the response variable for each value of the explanatory variable. That is, compare the column percents that give the distribution of smoking for each SES category.

EXAMPLE 8.15

We must calculate the column percents. For the high-SES group, there are 51 current smokers out of a total of 211 people. The column proportion for this cell is

$$\frac{51}{211} = .242$$

That is, 24.2% of the high-SES group are current smokers. Similarly, 92 of the 211 people in this group are former smokers. The column proportion is

$$\frac{92}{211} = .436$$

or 43.6%. In all, we must calculate nine percents. Here are the results:

Column percents for smoking and SES

	SES			
Smoking	High	Middle	Low	All
Current	24.2	42.3	46.2	32.6
Former	43.6	40.4	30.1	39.6
Never	32.2	17.3	23.7	27.8
Total	100.0	100.0	100.0	100.0

In addition to the distributions of smoking behavior in each of the three SES categories, the table gives the overall breakdown of smoking among all 356 subjects. These percents appear in the rightmost column, labeled All. They make up the marginal distribution of the row variable smoking. ●

The sum of the percents in each column should be 100, except for possible small roundoff errors. It is good practice to calculate each percent separately and then sum each column as a check. In this way we can find arithmetic errors that would not be uncovered if, for example, we calculated the column percent for the Never row by subtracting the sum of the percents for Current and Former from 100.

Figure 8.7 compares the distributions of smoking behavior in the three SES groups. The percent of current smokers decreases as SES increases from low to middle to high; in particular, relatively few high-SES subjects smoke. The percent of former smokers increases as SES increases, suggesting that higher-SES people who once smoked are more likely to quit. The percent of people who never smoked is highest in the high-SES group, but the middle-SES group has a somewhat lower percentage than the low-SES group. Overall, the column percents suggest that there is a negative association between smoking and SES: Higher-SES people tend to smoke less.

The chi-square test assesses whether this observed association is statistically significant. That is, is the SES-smoking relationship in the sample sufficiently strong for us to conclude that it is due to a relationship between these two variables in the underlying population and not merely to chance? Note that the test only asks whether there is evidence of some relationship. To explore the direction or nature of the relationship we must examine the column or row percents. Note also that in using the chi-square test we are acting as if the subjects were a simple random sample from the population of interest. If the volunteers are a biased sample—for example, if smokers are reluctant to volunteer for a study of employee health—then conclusions about the entire population of employees are not justified.

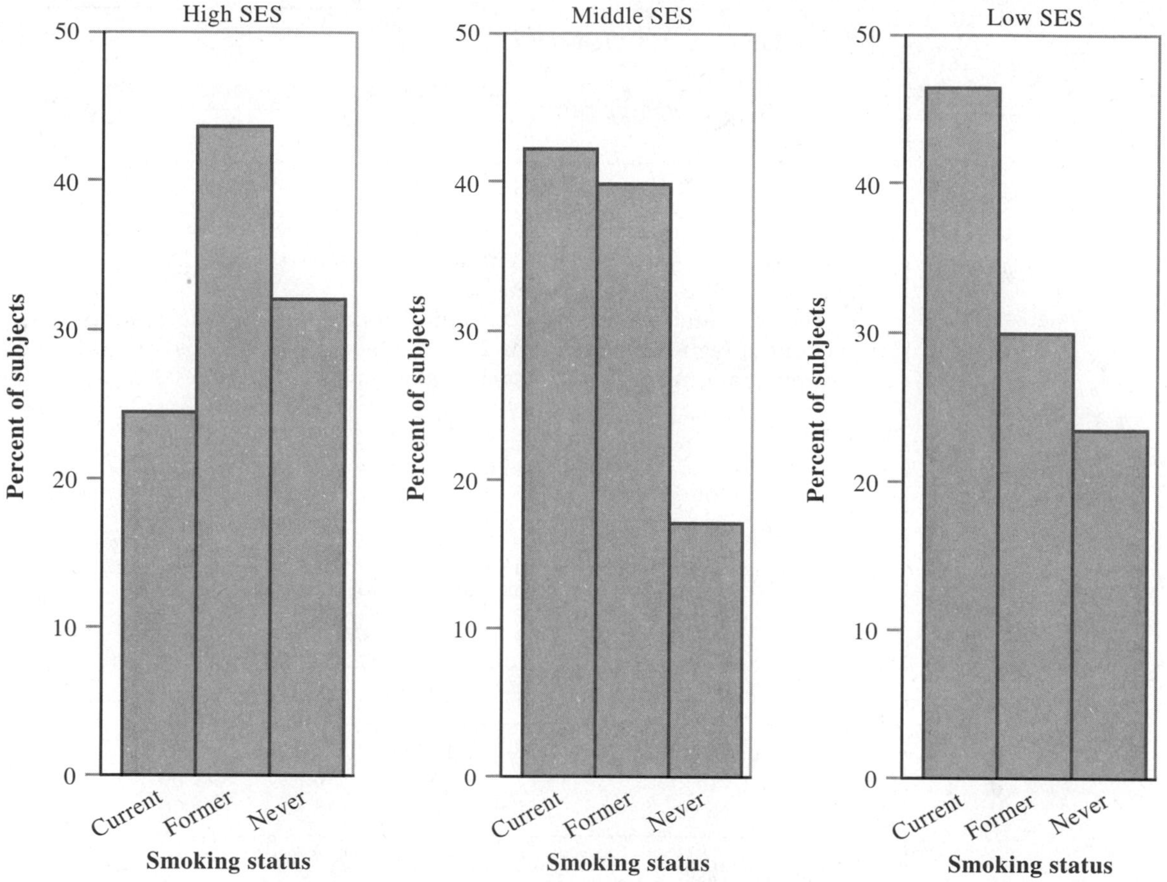

FIGURE 8.7 Comparison of smoking behavior distributions for high, middle, and low SES (Example 8.15).

The null hypothesis is that there is no relationship between SES and smoking in the population. The alternative is that these two variables are related. More exactly, H_0 in this example states that the two variables smoking and SES are *independent*. The precise statement of H_0 in Example 8.10 was that a single categorical variable had the same distribution in several populations. The precise statement of H_0 here is that two categorical variables are independent. Fortunately, expected counts and the chi-square statistic are found in exactly the same way in both cases. We need only keep in mind that we are testing whether the two directions in the table are related. Here is the recipe for the expected cell counts under the hypothesis of no relationship.

EXPECTED CELL COUNTS

$$\text{expected count} = \frac{\text{row total} \times \text{column total}}{n} \quad (8.2)$$

EXAMPLE 8.16

What is the expected count in the upper left cell in the table of Example 8.14, corresponding to high-SES current smokers, under the null hypothesis that smoking and SES are independent?

The row total, the count of current smokers, is 116. The column total, the count of high-SES subjects, is 211. The total sample size is $n = 356$. The expected number of high-SES current smokers is therefore

$$\frac{(116)(211)}{356} = 68.75$$

You can find the expected counts for the other cells in the same way. We summarize these calculations in a table of expected counts.

Expected counts for smoking and SES

	SES			
Smoking	High	Middle	Low	Total
Current	68.75	16.94	30.30	115.99
Former	83.57	20.60	36.83	141.00
Never	58.68	14.46	25.86	99.00
Total	211.00	52.00	92.99	355.99

We can check our work by adding the expected counts to obtain the row and column totals, as in the table. These should be the same as those in the table of observed counts, except for small roundoff errors, such as 115.99 rather than 116 for the first row total.

The expected cell counts are all large, so we proceed with the chi-square test. We compare the table of observed counts with the table of expected counts using the X^2 statistic given by Equation 8.1.[6] We must calculate the term for each cell, then sum over all nine cells. For the high-SES current smokers, the observed count is 51 and the expected count is 68.75. The contribution to the X^2 statistic for this cell is therefore

$$\frac{(51 - 68.75)^2}{68.75} = 4.583$$

Similarly, the calculation for the middle-SES current smokers is

$$\frac{(22 - 16.94)^2}{16.94} = 1.511$$

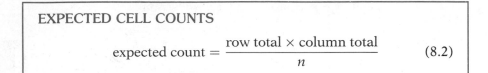

The X^2 statistic is the sum of nine such terms.

$$\begin{aligned} X^2 &= \sum \frac{(\text{observed} - \text{expected})^2}{\text{expected}} \\ &= \frac{(51 - 68.75)^2}{68.75} + \frac{(22 - 16.94)^2}{16.94} + \frac{(43 - 30.30)^2}{30.30} \\ &\quad + \frac{(92 - 83.57)^2}{83.57} + \frac{(21 - 20.60)^2}{20.60} + \frac{(28 - 36.83)^2}{36.83} \\ &\quad + \frac{(68 - 58.68)^2}{58.68} + \frac{(9 - 14.46)^2}{14.46} + \frac{(22 - 25.86)^2}{25.86} \\ &= 4.583 + 1.511 + 5.323 + .850 + .008 + 2.117 \\ &\quad + 1.480 + 2.062 + .576 \\ &= 18.51 \end{aligned}$$

Because there are $r = 3$ smoking categories and $c = 3$ SES groups, the degrees of freedom for this statistic are

$$(r - 1)(c - 1) = (3 - 1)(3 - 1) = 4$$

Under the null hypothesis that smoking and SES are independent, the test statistic X^2 has a $\chi^2(4)$ distribution. To obtain the P-value, refer to the row in Table G corresponding to 4 df. The calculated value $X^2 = 18.52$ lies between upper critical points corresponding to probabilities 0.001 and 0.0005. The P-value is therefore between 0.001 and 0.0005. Because the expected cell counts are all large, the P-value from Table G will be quite accurate. There is strong evidence ($X^2 = 18.52$, df $= 4$, $P < 0.001$) of an association between smoking and SES in the population of federal employees. The size and nature of this association are described by the table of percents examined in Example 8.15 and the display of these percents in Figure 8.7. The chi-square test assesses only the significance of the association, so the descriptive percents are essential to an understanding of the data.

df = 4

p	.001	.0005
χ^2	18.47	20.00

Models for two-way tables*

The chi-square test for the presence of a relationship between the two directions in a two-way table is valid for data produced from several different study designs. The precise statement of the null hypothesis "no relationship" in terms of population parameters is different for different designs. We now describe two of these settings in detail. The essential requirement is that *each experimental unit or subject is counted only once in the data table*. In the sports goals example, each student is either male

*The analysis of two-way tables is based on statistical models that require a fair amount of notation to describe. This optional subsection gives the details of these models.

or female and is classified as having only one sports goal. In the SES-smoking example, each federal employee fits into one SES-smoking cell. If, for example, the same individual is observed at several different times and so may appear in several cells of the table, more advanced statistical procedures are needed.

Comparing several populations The first model for a two-way table is illustrated by the comparison of two proportions discussed in Section 8.2 and also by the comparison of the sports goals of men and women in Example 8.10. We have *separate and independent random samples* from each of c populations. Each individual falls into one of r categories, and we wish to compare the distributions of these categories in the populations. In Section 8.2, we compared just two populations ($c = 2$) and we categorized each observation as a success or a failure ($r = 2$). The data could have been displayed in a two-way 2×2 table. Because the counts of failures are known once we know the counts of successes, we simply compared the two proportions of successes and did not need the two-way table format. The $r \times c$ table allows us to compare more than two populations or to use more than two categories, or both.

EXAMPLE 8.17

In the sports goals example, we compare two populations (men and women) and assign each subject to one of four categories. This yields a 4×2 table. The parameters of the model are the population proportions of each category, one set of proportions for each population.

The population proportions for females (column 1) are

$$p_{1(1)},\ p_{2(1)},\ p_{3(1)},\ \text{and } p_{4(1)}$$

Thus, $p_{i(1)}$ is the probability that a randomly selected female student falls into goals category i. Each female student must fall into exactly one of these categories, so the sum of the four population proportions is 1. Similarly, the population proportions for the males (column 2) are

$$p_{1(2)},\ p_{2(2)},\ p_{3(2)},\ \text{and } p_{4(2)}$$

These four proportions also sum to 1. Here is a table of these parameters, arranged in the same form as the tables of observed counts and expected counts:

	Sex	
Goal	Female	Male
HSC-HM	$p_{1(1)}$	$p_{1(2)}$
HSC-LM	$p_{2(1)}$	$p_{2(2)}$
LSC-HM	$p_{3(1)}$	$p_{3(2)}$
LSC-LM	$p_{4(1)}$	$p_{4(2)}$
Total	1	1

Just as the column totals in this table must be 1, the column totals (67 females and 67 males) in the table of observed cell counts are fixed in advance. They are the sample sizes for the two groups, determined by the researchers as part of the study design before the data were collected. ●

More generally, if we take an independent SRS from each of c populations and classify the outcomes into one of r categories, we have an $r \times c$ table of population proportions.

Rows (outcomes)	Columns (populations) 1	2	...	j	...	c
1	$p_{1(1)}$	$p_{1(2)}$	...	$p_{1(j)}$	...	$p_{1(c)}$
2	$p_{2(1)}$	$p_{2(2)}$	...	$p_{2(j)}$	...	$p_{2(c)}$
⋮	⋮	⋮	⋮	⋮	⋮	⋮
i	$p_{i(1)}$	$p_{i(2)}$	...	$p_{i(j)}$	...	$p_{i(c)}$
⋮	⋮	⋮	⋮	⋮	⋮	⋮
r	$p_{r(1)}$	$p_{r(2)}$	...	$p_{r(j)}$	...	$p_{r(c)}$
Total	1	1	1	1	1	1

Because the sum of the probabilities for each of the c populations is 1, the column sums in this table are all 1. The row sums are not equal to 1 and are usually not meaningful.

EXAMPLE 8.18

The null hypothesis for the sports goals example is that there is no relationship between sex and goals. That is, the sports goals distributions for females and males are the same. Therefore, the probability $p_{1(1)}$ that a randomly selected female is in the high social comparison high mastery group is equal to $p_{1(2)}$, the corresponding probability for a randomly selected male. In other words, the two values in the first row of the population probability table are equal. The same is true for each of the other three goals. The null hypothesis H_0, expressed in terms of the population parameters, says that the two probabilities in each row of the table of parameters are equal. ●

Testing independence A second model for which our analysis of $r \times c$ tables is valid is illustrated by the SES-smoking study, Example 8.14. There, a *single* sample from a *single* population was classified according to two categorical variables.

MODEL FOR COMPARING SEVERAL POPULATIONS USING $r \times c$ TABLES

Select independent SRSs from each of c populations, of sizes $n_1, n_2, \ldots, n_c$. Classify each individual in a sample according to a categorical response variable with r possible values. The probability that an individual from the jth population falls into category i is $p_{i(j)}$.

The null hypothesis is that the distributions of the response variable are the same in all c populations:

$$\begin{aligned} H_0\colon p_{1(1)} &= p_{1(2)} = \ldots = p_{1(c)} \\ p_{2(1)} &= p_{2(2)} = \ldots = p_{2(c)} \\ &\vdots \\ p_{r(1)} &= p_{r(2)} = \ldots = p_{r(c)} \end{aligned}$$

The alternative hypothesis is

H_a: at least one of the equalities in H_0 does not hold

EXAMPLE 8.19

In this study, researchers classified 356 federal employees according to two categorical variables—SES and smoking. Because there are nine possible categories in the 3 × 3 table that results, we view the process of selecting a federal employee at random and then classifying the individual according to the two categorical variables as a random phenomenon with nine possible outcomes.

The parameters are the probabilities of these nine outcomes. These probabilities are the proportions of each outcome in the population from which the sample is drawn. To keep track of both SES and smoking, we again display the parameters in the same two-way arrangement as the observed and expected cell counts. Let p_{ij} be the population proportion in row i and column j of the table. In this case, $i = 1, 2, 3$ describes the three smoking categories and $j = 1, 2, 3$ describes the three SES groups. Here is the two-way table of parameters:

	SES			
Smoking	High	Middle	Low	Total
Current	p_{11}	p_{12}	p_{13}	r_1
Former	p_{21}	p_{22}	p_{23}	r_2
Never	p_{31}	p_{32}	p_{33}	r_3
Total	c_1	c_2	c_3	1

Because each individual falls into exactly one of the nine cells, the sum of the nine probabilities must be 1. The row and column sums r_i and c_j differ in different populations.

In the table of observed cell counts, only the table total is fixed before the data are observed. This total (356) is the number of subjects in the study. The row and column sums are random and would no doubt be different if we took another sample. In both the models we are examining, the sums that are fixed are the same for the table of parameters and for the table of observed counts. ●

More generally, take a single SRS from a single population and classify each observation into one cell of an $r \times c$ table. Here is the two-way table of population proportions:

	Columns						
Rows	1	2	...	j	...	c	Total
1	p_{11}	p_{12}	...	p_{1j}	...	p_{1c}	r_1
2	p_{21}	p_{22}	...	p_{2j}	...	p_{2c}	r_2
⋮	⋮	⋮	⋮	⋮	⋮	⋮	⋮
i	p_{i1}	p_{i2}	...	p_{ij}	...	p_{ic}	r_i
⋮	⋮	⋮	⋮	⋮	⋮	⋮	⋮
r	p_{r1}	p_{r2}	...	p_{rj}	...	p_{rc}	r_r
Total	c_1	c_2	...	c_j	...	c_c	1

The marginal proportions in this table are the sums of the proportions in the rows and columns. The row sum r_i is the sum of the proportions across row i, and c_j is the sum down column j. These marginal proportions are also probabilities of interest. Each r_i is the probability that a randomly selected member of the population falls in the ith row category. Similarly, each c_j is the probability of the jth column category. These are the marginal distributions of the row and column variables.

The null hypothesis "no relationship" now states that the row and column variables are independent. The multiplication rule for independent events (page 295) then expresses the joint probability p_{ij} as the product of the marginal probabilities r_i and c_j.

EXAMPLE 8.20

The probability that a member of the population is a current smoker is r_1, and the probability that the person has high SES is c_1. If smoking and SES are independent, then the probability that a federal employee is *both* a current smoker and high SES is $p_{11} = r_1 c_1$. The hypothesis that smoking and SES are independent says that the multiplication rule applies to all outcomes, so that $p_{ij} = r_i c_j$. ●

MODEL FOR EXAMINING INDEPENDENCE IN $r \times c$ TABLES

Select an SRS of size n from a population. Classify each individual in the sample according to two categorical variables. The probabilities for the row classification are r_i, and the probabilities for the column classification are c_j.

The null hypothesis is that the row and column classifications are independent. If p_{ij} is the probability of an observation being classified in row i and column j, the null hypothesis is

$$H_0\colon p_{ij} = r_i c_j \quad \text{for all } i \text{ and } j$$

The alternative hypothesis is that the row and column classifications are not independent; that is,

$$H_a\colon p_{ij} \neq r_i c_j \quad \text{for some } i \text{ and } j$$

The independence model explains the recipe for expected counts given in Equation 8.2 (page 612). The count in any cell follows the binomial distribution with n trials and probability of success p_{ij}. The expected count is therefore np_{ij}. If the hypothesis of independence is true, this expected count is nr_ic_j. We cannot observe r_i, so we estimate it by the sample proportion in row i,

$$\frac{\text{count in row } i}{n}$$

Similarly, we estimate c_j by the sample proportion in column j,

$$\frac{\text{count in column } j}{n}$$

The estimated value of the expected count is therefore

$$\begin{aligned} nr_ic_j &= n \times \frac{\text{row total}}{n} \times \frac{\text{column total}}{n} \\ &= \frac{\text{row total} \times \text{column total}}{n} \end{aligned}$$

Concluding remarks You can distinguish between the two models by examining the design of the study. In the independence model, there is a single sample. The column totals are random variables. The total sample size n is set by the researcher and the column sums are only known when the data are analyzed. For the comparison-of-populations model, on the other hand, there is a sample from each of two or more populations. The column sums are the sample sizes selected at the design phase of the research. The null hypothesis in both models says that there is no relationship between the column variable and the row variable. The precise statement of the hypothesis differs, depending on the sampling design. Fortunately, *the test of the hypothesis of no relationship is the same for both models;* it is the chi-square test.

There are yet other statistical models for two-way tables that justify the chi-square test of the null hypothesis "no relation," made precise in ways suitable for these models. Statistical methods related to the chi-square test also allow the analysis of three-way and higher-way tables of count data. You can find a discussion of these topics in advanced texts on categorical data.[7]

SUMMARY

The **null hypothesis** for $r \times c$ tables of count data is that there is no relationship between the row variable and column variable.

Expected cell counts under the null hypothesis are computed using the formula

$$\text{expected count} = \frac{\text{row total} \times \text{column total}}{n}$$

The null hypothesis is tested by the **chi-square statistic**

$$X^2 = \sum \frac{(\text{observed} - \text{expected})^2}{\text{expected}}$$

Under the null hypothesis, X^2 has approximately the χ^2 **distribution** with $(r-1)(c-1)$ degrees of freedom. The P-value for the test is

$$P(\chi^2 \geq X^2)$$

where χ^2 is a random variable having the $\chi^2(\text{df})$ distribution with $\text{df} = (r-1)(c-1)$.

The chi-square approximation is adequate for practical use when the average expected cell count is 5 or greater and all individual expected counts are 1 or greater, except in the case of 2×2 tables. All four expected counts in a 2×2 table should be 5 or greater.

Two different models for generating $r \times c$ tables lead to the chi-square test. In the first model, independent SRSs are drawn from each of c populations, and each observation is classified according to a categorical variable with r possible values. The null hypothesis is that the distributions of the row categorical variable are the same for all c populations. In the second model, a single SRS is drawn from a population, and observations are classified according to two categorical variables having r and c possible values. In this model, H_0 states that the row and column variables are independent.

SECTION 8.3 EXERCISES

8.35 Psychological and social factors can influence the survival of patients with serious diseases. One study examined the relationship between survival of patients with coronary heart disease (CHD) and pet ownership. Each of 92 patients was classified as having a pet or not and by whether they survived for 1 year. Here are the data. (Data from Erika Friedmann et al., "Animal companions and one-year survival of patients after discharge from a coronary care unit," *Public Health Reports*, 96 (1980), pp. 307–312.)

	Pet ownership	
Patient status	No	Yes
Alive	28	50
Dead	11	3

(a) Was this study an experiment? Why or why not?
(b) The researchers thought that having a pet might improve survival, so pet ownership is the explanatory variable. Compute appropriate percentages to describe the data and state your preliminary findings.
(c) State in words the null hypothesis for this problem. What is the alternative hypothesis?
(d) Find the chi-square statistic, its degrees of freedom, and the *P*-value.
(e) What do you conclude? Do the data give convincing evidence that owning a pet is an effective treatment for increasing the survival of CHD patients?

8.36 Investors use many "indicators" in their attempts to predict the behavior of the stock market. One of these is the "January indicator." Some investors believe that if the market is up in January, then it will be up for the rest of the year. On the other hand, if it is down in January, then it will be down for the rest of the year. The following table gives the data for the Standard and Poor's 500 stock index for the 75 years from 1916 to 1990:

	January	
Rest of year	Up	Down
Up	35	13
Down	13	14

These data do not strictly conform to either of the sampling models described in this section. However, the chi-square analysis is valid for this problem if we assume that the yearly data are independent observations of a process that generates either an "up" or a "down" both in January and for the rest of the year.

(a) Calculate the column percents for this table. Explain briefly what they express.
(b) Do the same for the row percents.
(c) State appropriate null and alternative hypotheses for this problem. Use words rather than symbols.
(d) Find the table of expected counts under the null hypothesis. In which cells do the expected counts exceed the observed counts? In what cells are they less than the observed counts? Explain why the pattern suggests that the January indicator is valid.
(e) Give the value of the chi-square statistic, its degrees of freedom, and the *P*-value. What do you conclude?
(f) Write a short discussion of the evidence for the January indicator, referring to your analysis for substantiation.

8.37 The baseball player Reggie Jackson had a reputation for hitting better in the World Series than during the regular season. In his 21-year career, Jackson was at bat 9864 times in regular-season play and had 2584 hits. During World Series games, he was at bat 98 times and had 35 hits. We can view Jackson's regular-season at bats as a random sample from a population of potential at bats (he might have batted many more times if the season were longer, for example), and his World Series at bats as a sample from a second population.

(a) Display the data in a 2×2 table of counts with regular season and World Series as the column headings, and fill in the marginal sums.
(b) Calculate appropriate percents to compare Jackson's regular season and World Series performances. Did he hit better in World Series games?
(c) Is there a significant difference between Jackson's regular season and World series performances? State hypotheses (in words), and then calculate the chi-square statistic, its degrees of freedom, and its *P*-value. What is your conclusion?

8.38 A survey on the severity of rodent problems in commercial poultry houses studied a random sample of poultry operations. Each operation were classified by type (egg or turkey production) and by the extent of the rodent problems. Here are the results:

	Type	
Rodent problem	Egg	Turkey
Mild	34	22
Moderate	33	22
Severe	7	4

(a) The type of poultry operation is a natural explanatory variable. Calculate a table of percents that describes how rodent problems vary with the type of operation. Summarize the results in words.
(b) State H_0 and H_a for this problem.
(c) Conduct a significance test for your hypotheses and give the results. What do you conclude?

8.39 Do businesses of different sizes respond more or less readily to questionnaires sent out by business schools? A study sent questionnaires to 200 randomly selected businesses of each of three sizes. Here are data on the responses. Note that the column sums are fixed by the design of the survey.

	Size		
	Small	Medium	Large
Response	125	81	40
No response	75	119	160
Total	200	200	200

(a) For each size of business find the percent that responded and the percent that did not respond. Do the same for all of the data combined and present your results in a table. In what way do the response rates appear to vary with the size of the business?
(b) State in words an appropriate H_0 for this problem. What is H_a?
(c) Test your hypothesis and give a full report of your conclusions.

8.40 In January 1975, the Committee on Drugs of the American Academy of Pediatrics recommended that tetracycline drugs not be given to children under the age of 8. A 2-year study conducted in Tennessee investigated the extent to which physicians had prescribed these drugs between 1973 and 1975. The study categorized family practice physicians according to whether the county of their practice was urban, intermediate, or rural. The researchers examined how many doctors in each of these categories prescribed tetracycline to at least one patient under the age of 8. Here is the table of observed counts. (Data from Wayne A. Ray et al., "Prescribing of tetracycline to children less than 8 years old," *Journal of the American Medical Association*, 237 (1977), pp. 2069–2074.)

	County type		
	Urban	Intermediate	Rural
Tetracycline	65	90	172
No tetracycline	149	136	158

(a) Find the row and column sums and put them in the margins of the table.
(b) For each type of county find the percent of physicians who prescribed tetracycline and the percent of those who did not. Do the same for the combined sample. Display the percents in a table and describe briefly what they show.
(c) Write null and alternative hypotheses to assess whether county type and prescription practices are unrelated.
(d) Carry out a significance test, give a full report of the results, and interpret them in plain language.

8.41 Another part of the medical study described in the previous exercise compared tetracycline prescribing by physicians engaged in different types of practices. The practices were classified as family practice, pediatrics, and other. Here are the data:

	Family practice	Pediatrics	Other
Tetracycline	327	32	159
No tetracycline	443	122	808

(a) Find the row and column sums and put them in the margins of the table.
(b) For each type of practice find the percent of physicians who prescribed tetracycline and the percent of those who did not. Do the same for the combined sample. Display the results in a table and describe them briefly.
(c) Test whether type of practice is related to tetracycline prescribing. Report your results in detail and state your overall conclusions.

8.42 Alcohol and nicotine consumption during pregnancy may harm children. Because drinking and smoking behaviors may be related, it is important to understand the nature of this relationship when assessing the possible effects on children. One study classified 452 mothers according to their alcohol intake prior to pregnancy recognition and their nicotine intake during pregnancy. The data are summarized in the following table. (Data from Ann P. Streissguth et al., "Intrauterine alcohol and nicotine exposure: Attention and reaction time in 4-year-old children," *Developmental Psychology*, 20 (1984), pp. 533–541.)

	Nicotine (milligrams/day)		
Alcohol (ounces/day)	None	1–15	16 or more
None	105	7	11
.01–.10	58	5	13
.11–.99	84	37	42
1.00 or more	57	16	17

Carry out a complete analysis of the association between alcohol and nicotine consumption. That is, describe the nature and strength of this association and assess its statistical significance.

8.43 Nutrition and illness are related in a complex way. If the diet is inadequate, the ability to resist infections can be impaired and illness results. On the other hand, some illnesses cause lack of appetite, so that poor nutrition can be the result of illness. In a study of morbidity and nutritional status in 1165 preschool children living in poor conditions in Delhi, India, data were obtained on nutrition and illness. Nutrition was described by a standard method as normal or as one of four levels of inadequate: I, II, III, and IV. For the purpose of analysis, the two most severely undernourished groups, III and IV, were combined. One part of the study examined four categories of illness during the past year: upper respiratory infection (URI), diarrhea, URI and diarrhea, and none. The following table gives the data. (Data from Vimlesh Seth et al., "Profile of morbidity and nutritional status and their effect on the growth potentials in preschool children in Delhi, India," *Tropical Pediatrics and Environmental Health*, 25 (1979), pp. 23–29.)

	Nutritional status			
Illness	Normal	I	II	III and IV
URI	95	143	144	70
Diarrhea	53	94	101	48
URI and diarrhea	27	60	76	27
None	113	48	44	22
Total	288	345	365	167

Carry out a complete analysis of the association between degree of undernourishment and type of illness. That is, describe the association numerically, assess its significance, and write a brief summary of your findings that refers to your analysis for substantiation.

CHAPTER 8 EXERCISES

8.44 A television news program conducts a call-in poll about a proposed city ban on handgun ownership. Of the 2372 calls, 1921 oppose the ban. The station, following recommended practice, makes a confidence statement: "81% of the Channel 13 Pulse Poll sample opposed the ban. We can be 95% confident that the true proportion of citizens opposing a handgun ban is within 1.6% of the sample result." Is this conclusion justified?

8.45 Eleven percent of the products produced by an industrial process over the past several months fail to conform to the specifications. The company modifies the process in an attempt to reduce the rate of nonconformities.

In a trial run, the modified process produces 16 nonconforming items out of a total of 300 produced. Do these results demonstrate that the modification is effective? Support your conclusion with a clear statement of your assumptions and the results of your statistical calculations.

8.46 In the setting of the previous exercise, give a 95% confidence interval for the proportion of nonconforming items for the modified process. Then, taking $p_0 = 0.11$ to be the old proportion and p the proportion for the modified process, give a 95% confidence interval for $p - p_0$.

8.47 In a random sample of 950 students from a large public university, it was found that 444 of the students changed majors sometime during their college years.

(a) Give a 99% confidence interval for the proportion of students at this university who change majors.
(b) Express your results from (a) in terms of the *percent* of students who change majors.
(c) University officials concerned with counseling students are interested in the number of students who change majors rather than the proportion. The university has 35,000 undergraduate students. Convert the confidence interval you found in (a) to a confidence interval for the *number* of students who change majors during their college years.

8.48 Many colleges that once enrolled only male or only female students have became coeducational. Some administrators and alumni were concerned that the academic standards of the institutions would decrease with the change. One formerly all-male college undertook a study of the first class to contain women. The class consisted of 851 students, 214 of whom were women. An examination of first-semester grades revealed that 15 of the top 30 students were female.

(a) What is the proportion of women in the class? Call this value p_0.
(b) Assume that the number of females in the top 30 is approximately a binomial random variable with $n = 30$ and unknown probability p of success. In this case success corresponds to the student being female. What is the value of $\hat{p}$?
(c) Are women more likely to be top students than their proportion in the class would suggest? State hypotheses that ask this question, carry out a significance test, and report your conclusion.

8.49 In a recent year, 75% of the 20 fatal accidents in Tippecanoe County, Indiana, were alcohol related. The national average is 50%. Is there evidence in these data to conclude that alcohol is involved in a different proportion of fatal accidents in Tippecanoe County than in the nation as a whole? Give a summary of your analysis and reasons for your conclusion. It would be tempting to use a one-sided alternative for this problem. Why is this choice not appropriate?

8.50 In a study on blood pressure and diet, a random sample of Seventh Day Adventists were interviewed at a national meeting. Because many people who belong to this denomination are vegetarians, they are a very useful group for studying the effects of a meatless diet. Blacks in the population as a whole have a higher average blood pressure than whites. A study of this type should therefore take race into account in the analysis. The 312 people in the sample were categorized by race and whether or not they were vegetarians. The data are given in the following table. (Data provided by Chris Melby and David Goldflies, Department of Physical Education, Health, and Recreation Studies, Purdue University.)

	Black	White
Vegetarian	42	135
Not vegetarian	47	88

Are the proportions of vegetarians the same among all black and white Seventh Day Adventists who attended this meeting? Analyze the data, paying particular attention to this question. Summarize your analysis and conclusions. What can you infer about the proportions of vegetarians among black and white Seventh Day Adventists in general? What about blacks and whites in general?

8.51 There is much evidence that high blood pressure is associated with increased risk of death from cardiovascular disease. A major study of this association examined 2676 men with low blood pressure and 3338 men with high blood pressure. During the period of the study, 21 men in the low blood pressure and 55 in the high blood pressure group died from cardiovascular disease. (See Exercise 2.75 on page 194 for a more detailed description of this study.)

(a) What is the explanatory variable? Describe the association in these data numerically and in words.

(b) Do the study data confirm that death rates are higher among men with high blood pressure? State hypotheses, carry out a significance test, and give your conclusions.

(c) Present the data in a two-way table. Is the chi-square test appropriate for the hypotheses you stated in (b)?

(d) Give a 95% confidence interval for the difference between the death rates for the low and high blood pressure groups.

8.52 It is traditional practice in Egypt to withhold food from children with diarrhea. Because it is known that feeding children with this illness reduces mortality, medical authorities undertook a nationwide program designed to promote feeding sick children. To evaluate the impact of the program, surveys were taken before and after the program was implemented. In the first survey 457 of 1003 surveyed mothers followed the practice of feeding children with diarrhea. For the second survey,

437 of 620 surveyed followed this practice. (Data taken from O. M. Galal et al., "Feeding the child with diarrhea: A strategy for testing a health education message within the primary health care system in Egypt," *Socio-Economic Planning Sciences*, 21 (1987), pp. 139–147.)

(a) Assume that the data come from two independent samples. Test the hypothesis that the program was effective, that is, that the practice of feeding children with diarrhea increased between the time of the first study and the time of the second. State H_0 and H_a, give the test statistic and its P-value, and summarize your conclusion.

(b) Present the data in a two-way table. Can the chi-square statistic test your hypotheses?

(c) Describe the results using a 95% confidence interval for the difference in proportions.

8.53 Aluminum is suspected as a factor in the development of Alzheimer's disease (AD). In one study, researchers compared a group of AD patients with a carefully selected control group of people who did not have AD but were similar in other ways. (Selection of a matching control group is a difficult task. In epidemiological studies such as this, however, experiments are not possible.) The focus of the study was on the use of antacids that contain aluminum. Each subject was classified according to the use of these antacids. The two-way table below gives the data. (Data from Amy Borenstein Graves et al., "The association between aluminum-containing products and Alzheimer's disease," *Journal of Clinical Epidemiology*, 43 (1990), pp. 35–44.)

	Aluminum-containing antacid use			
	None	Low	Medium	High
Alzheimer's patients	112	3	5	8
Control group	114	9	3	2

Analyze the data and summarize your results. Does the use of aluminum-containing antacids appear to be associated with Alzheimer's disease?

8.54 Are there sex differences in the progress of students in doctoral programs? A major university classified all students entering Ph.D. programs in a given year by their status 6 years later. The categories used were completed the degree, still enrolled, and dropped out. Here are the data:

Status	Men	Women
Completed	423	98
Still enrolled	134	33
Dropped out	238	98

Assume that these data can be viewed as a random sample giving us information on student progress. Describe the data using whatever percents are appropriate. State and test a null hypothesis and alternative that address the question of sex differences. Summarize your conclusions. What factors not given might be relevant to this study?

8.55 PTC is a compound that has a strong bitter taste for some people and is tasteless for others. The ability to taste this compound is an inherited trait. Many studies have assessed the proportions of people in different populations who can taste PTC. The following table gives results for samples from several countries. (Data from A. E. Mourant et al., *The Distribution of Human Blood Groups and Other Polymorphisms*, Oxford University Press, London, 1976.)

	Ireland	Portugal	Norway	Italy
Tasters	558	345	185	402
Nontasters	225	109	81	134

Complete the table and describe the data. Do they provide evidence that the proportion of PTC tasters varies among the four countries? Give a complete summary of your analysis.

8.56 There are four major blood types in humans: O, A, B, and AB. In a study conducted using blood specimens from the Blood Bank of Hawaii, individuals were classified according to blood type and ethnic group. The ethnic groups were Hawaiian, Hawaiian-white, Hawaiian-Chinese, and white. Assume that the blood bank specimens are random samples from the Hawaiian populations of these ethnic groups. (Data are from the book cited in the previous exercise.)

Blood type	Hawaiians	Hawaiian-White	Hawaiian-Chinese	White
O	1903	4469	2206	53,759
A	2490	4671	2368	50,008
B	178	606	568	16,252
AB	99	236	243	5001

Summarize the data. Is there evidence to conclude that blood type and ethnic group are related? Explain how you arrived at your conclusion.

8.57 An article in *the New York Times* of January 30, 1988, described the results of an experiment on the effects of aspirin on cardiovascular disease. The subjects were 5139 male British medical doctors. The doctors were randomly assigned to two groups. One group of 3429 doctors took one aspirin daily, and the other group did not take aspirin. After 6 years, there were 148 deaths from heart attack or stroke in the first group and 79 in the second group. A similar experiment using male American medical doctors as subjects was reported in the *New York Times* on January 27, 1988. These doctors were also randomly assigned to one of two groups. The 11,037 doctors in the first group took one aspirin every other day, and the 11,034 doctors in the second group took no aspirin. After nearly 5 years there were 104 deaths from heart attacks in the first group and 189 in the second. Analyze the data from these two studies and summarize the results. How do the conclusions of the two studies differ, and why?

8.58 Gastric freezing was once a recommended treatment for ulcers in the upper intestine. A randomized comparative experiment found that 28 of the 82 patients who were subjected to gastric freezing improved, while 30 of the 78 patients in the control group improved. The hypothesis of "no difference" between the two groups can be tested in two ways: using a z statistic or using the chi-square statistic.

(a) State the appropriate hypothesis and a two-sided alternative and carry out a z test. What is the P-value?

(b) Present the data in a 2×2 table. State the appropriate hypothesis and carry out the chi-square test. What is the P-value? Verify that the X^2 statistic is the square of the z statistic.

(c) What do you conclude about the effectiveness of gastric freezing as a treatment for ulcers? (See Example 3.5 on page 231 for a discussion of gastric freezing.)

8.59 Example 2.25 (page 189) presents artificial data that illustrate Simpson's paradox. The data concern the survival rates of surgery patients at two hospitals.

(a) Apply the chi-square test to the data for all patients combined and summarize the results.

(b) Run separate chi-square analyses for the patients in good condition and for those in poor condition. Summarize these results.

(c) Are the effects that illustrate Simpson's paradox in this example statistically significant?

8.60 In this exercise we examine the effect of the sample size on the significance test for comparing two proportions. In each case suppose that $\hat{p}_1 = 0.6$, $\hat{p}_2 = 0.4$, and take n to be the common value of n_1 and n_2. Use the z statistic to test $H_0\colon p_1 = p_2$ versus the alternative $H_a\colon p_1 \neq p_2$. Compute the statistic and the associated P-value for the following values of n: 15, 25, 50, 75, 100, and 500. Summarize the results in a table. Explain what you observe about the effect of the sample size on statistical significance when the same sample proportions $\hat{p}_1$ and $\hat{p}_2$ are unchanged.

8.61 In the first section of this chapter, we studied the effect of the sample size on the margin of error of the confidence interval for a single proportion. In this exercise we perform some calculations to observe this effect for the two-sample problem. As in the exercise above, suppose that $\hat{p}_1 = 0.6$, $\hat{p}_2 = 0.4$, and n represents the common value of n_1 and n_2. Compute the

95% confidence intervals for the differences in the two proportions for $n = 15, 25, 50, 75, 100$, and 500. For each interval calculate the margin of error. Summarize and explain your results.

8.62 For a single proportion the margin of error of a confidence interval is largest for any given sample size n and confidence level C when $\hat{p} = 0.5$. This led us to use $p^* = 0.5$ for planning purposes. The same kind of result is true for the two-sample problem. The margin of error of the confidence interval for the difference between two proportions is largest when $\hat{p}_1 = \hat{p}_2 = 0.5$. Use these conservative values in the following calculations, and assume that the sample sizes n_1 and n_2 have the common value n. Calculate the margins of error of the 99% confidence intervals for the difference in two proportions for the following choices of n: 10, 30, 50, 100, 200, and 500. Present the results in a table or with a graph. Summarize your conclusions.

8.63 As the previous problem noted, using the guessed value 0.5 for both $\hat{p}_1$ and $\hat{p}_2$ gives a conservative margin of error in confidence intervals for the difference between two population proportions. You are planning a survey and will calculate a 95% confidence interval for the difference in two proportions when the data are collected. You would like the margin of error of the interval to be less than or equal to 0.05. You will use the same sample size n for both populations.

(a) How large a value of n is needed?

(b) Give a general formula for n in terms of the desired margin of error m and the critical value z^*.

8.64 You are planning a survey in which a 90% confidence interval for the difference between two proportions will present the results. You will use the conservative guessed value 0.5 for $\hat{p}_1$ and $\hat{p}_2$ in your planning. You would like the margin of error of the confidence interval to be less than or equal to 0.1. It is very difficult to sample from the first population, so that it will be impossible for you to obtain more than 20 observations from this population. Taking $n_1 = 20$, can you find a value of n_2 that will guarantee the desired margin of error? If so, report the value; if not, explain why not.

8.65 A study of alcohol and nicotine consumption among 452 pregnant women produced these counts:

	Nicotine (milligrams/day)		
Alcohol (ounces/day)	None	1–15	16 or more
None	105	7	11
.01–.10	58	5	13
.11–.99	84	37	42
1.00 or more	57	16	17

In Exercise 8.42 (page 623) you found that the association was statistically significant. Present one or more charts or figures that clearly display this association.

8.66 In Exercise 8.56 you found significant differences in the distribution of blood types among four ethnic groups. Display the data using charts or figures in a way that clearly presents the differences among the four ethnic groups.

8.67 You are asked to evaluate a proposal for an experiment on the effects of aspirin on cardiovascular disease similar to the experiments described in Exercise 8.57. The researchers will randomly assign subjects to a treatment group or to a control group. The proposed sample sizes are 200 for each group. Write a short evaluation of this proposal, using any relevant information from Exercise 8.57.

8.68 The sports goals study described in Example 8.10 (page 601) actually involved three categorical variables: sex, social comparison, and mastery. In that example, we combined the social comparison and mastery categories to obtain a two-way table. There are effective statistical methods for analyzing three-way tables, but these methods are beyond the scope of this text. We can, however, look at the several two-way tables formed by rearranging the data.

The analyses and discussion of this example presented earlier suggest that the major sex difference is in the social comparison variable. To investigate this idea, rearrange the data into a 2 × 2 table that classifies the students by sex and social comparison. Analyze this table and summarize the results. Do the same using sex and mastery as the classification variables. Considering these analyses and the analysis presented in Example 8.10, what conclusions do you draw?

CHAPTER 8 COMPUTER EXERCISES

8.69 A recent study of 865 college students found that 42.5% had student loans. The students were randomly selected from the approximately 30,000 undergraduates enrolled in a large public university. The overall purpose of the study was to examine the effects of student loan burdens on the choice of a career. A student with a large debt may be more likely to choose a field where starting salaries are high so that the loan can more easily be repaid. The following table classifies the students by field of study and whether or not they have a loan. (Data provided by Susan Prohofsky, from her Ph.D. dissertation, "Selection of undergraduate major: The influence of expected costs and expected benefits," Purdue University, 1991.)

	Student loan	
Field of study	Yes	No
Agriculture	32	35
Child development and family studies	37	50
Engineering	98	137
Liberal arts and education	89	124
Management	24	51
Science	31	29
Technology	57	71

Carry out a complete analysis of the association between having a loan and field of study, including a description of the association and an assessment of its statistical significance.

8.70 In the study described in the previous exercise, students were asked to respond to some questions regarding their interests and attitudes. Some of these questions form a scale called PEOPLE that measures altruism or an interest in the welfare of others. Each student was classified as low, medium, or high on this scale. Is there an association between PEOPLE score and field of study? Here are the data:

	PEOPLE score		
Field of study	Low	Medium	High
Agriculture	5	27	35
Child development and family studies	1	32	54
Engineering	12	129	94
Liberal arts and education	7	77	129
Management	3	44	28
Science	7	29	24
Technology	2	62	64

Analyze the data and summarize your results. Are there some fields of study that have very large or very small proportions of students in the high PEOPLE category?

8.71 An article in the *New York Times* of April 24, 1991, discussed data from the Centers for Disease Control that showed an increase in cases of measles in the United States. Of particular concern are complications from measles that can lead to death. The article noted that young children, who do not have fully developed immune systems, face an increased risk of death from complications of measles. Here are data on the 23,067 cases of measles reported in 1990. For each age group, the probability of death from measles is a parameter of interest. A comparison of the estimates of these parameters across age groups will provide information about the relationship between age and survival of an attack of measles.

	Survival	
Age group	Dead	Survived
Under 1 year	17	3806
1–4 years	37	7113
5–9 years	3	2208
10–14 years	3	1888
15–19 years	8	2715
20–24 years	6	2209
25–29 years	9	1492
30 years and over	14	1636

Summarize the death rates by age group. Prepare a plot to illustrate the pattern. Test the hypothesis that survival and age are related, report the results, and summarize your conclusion. From the data given, is it possible to study the association between catching measles and age? Explain why or why not.

8.72 In healthy individuals the concentration of various substances in the blood remains within relatively narrow bounds. One such substance is potassium. A person is said to be hypokalemic if the potassium level is too low (less than 3.5 milliequivalents per liter) and hyperkalemic if the level is too high (above 5.5 meq/l). Hypokalemia is associated with a variety of symptoms such as excessive tiredness, while hyperkalemia is generally an indication of a serious problem. Patients being treated with diuretics (pharmaceuticals that help the body to eliminate water) sometimes have abnormal potassium concentrations. In a large study of patients on chronic diuretic therapy, several risk factors were studied to see if they were associated with abnormal potassium levels. Of the 5180 patients studied, 1094 were hypokalemic, 4689 had normal potassium levels, and 27 were hyperkalemic. The following table gives the percentages of patients having each of four risk factors in the three potassium groups. (Data taken from William M. Tierney, Clement J. McDonald, and George P. McCabe, "Serum potassium testing in diuretic-treated outpatients," *Medical Decision Making*, 5 (1985), pp. 91–104.)

	Potassium group		
	Hypokalemic	Normal	Hyperkalemic
n	1094	4689	27
Hypertension	88.3%	78.1%	40.7%
Heart failure	16.5%	24.7%	55.6%
Diabetes	20.6%	25.5%	29.6%
Sex (% female)	72.5%	68.0%	48.1%

For example, 88.3% of the 1094 hypokalemic patients had hypertension, and 78.1% of the 4689 normal patients had hypertension. For each of the four risk factors, use the percents and n's given to compute the counts for the 2×3 table needed to study the association between the factor and potassium. Then analyze each table using the methods presented in this chapter. Note that there are very few patients in the hyperkalemic group. Therefore, reanalyze the data dropping this category from the tables. Write a short summary explaining what you have found.

8.73 The proportion of women entering many professions has undergone considerable change in recent years. A study of students enrolled in pharmacy programs describes the changes in this field. A random sample of 700 students in their third or higher year of study at colleges of pharmacy was taken in each of nine years. The following table gives the numbers of women in each of these samples. (Data based on *Seventh Report to the President and Congress on the Status of Health Personnel in the United States*, Public Health Service, Washington, D.C., 1990.)

Year	1970	1972	1974	1976	1978	1980	1982	1984	1986
Women	164	195	226	283	302	343	369	384	412

Use the chi-square test to assess the change in the percentage of women pharmacy students over time and summarize your results. (You will need to calculate the number of male students for each year using the fact that the sample size each year is 700.) Plot the percentage of women versus year. Describe the plot. Is it roughly linear? Find the least-squares line that summarizes the relation between time and the percentage of women pharmacy students. Would you be willing to use this line to predict the percentage of women pharmacy students in the year 2000? Explain why or why not.

8.74 *Castaneda v. Partida* is an important court case in which statistical methods were used as part of a legal argument. When reviewing this case, the Supreme Court used the phrase "two or three standard deviations" as a criterion for statistical significance. This Supreme Court review has served as the basis for many subsequent applications of statistical methods in legal settings. (The two or three standard deviations referred to by the Court are values of the z statistic and correspond to P-values of approximately 0.05 and 0.0026.) In *Castaneda* the plaintiffs alleged that the method for selecting juries in a county in Texas was biased against Mexican Americans. For the period of time at issue, there were 181,535 persons eligible for jury duty, of whom 143,611 were Mexican Americans. Of the 870 people selected for jury duty, 339 were Mexican Americans.

(a) Compute the proportion of Mexican Americans eligible for jury duty. Let this value be p_0.

(b) Let p be the probability that a randomly selected juror is a Mexican American. The null hypothesis to be tested is $H_0\colon p = p_0$. Find the value of $\hat{p}$ for this problem, compute the z statistic, and find the P-value. What do you conclude? (A finding of statistical significance in this circumstance does not constitute a proof of discrimination. It can be used, however, to establish a prima facie case. The burden of proof then shifts to the defense.)

(c) We can reformulate this exercise as a two-sample problem. Here we wish to compare the proportion of Mexican Americans who are selected as jurors with the proportion of Mexican Americans who were not selected as jurors. Let p_1 be the probability that a randomly selected juror is a Mexican American, and let p_2 be the probability that a randomly selected nonjuror is a Mexican American. Find the z statistic and its P-value. How do your answers compare with the results in (a)?

(d) This problem can also be formulated using a two-way table of counts. We are then interested in finding out if there is an association between being a Mexican American and being selected as a juror. Construct the 2×2 table using the variables Mexican American or not and juror or not. Find the chi-square statistic and its P-value. Square the z statistic that you obtained in (c) and verify that the result is equal to the X^2 statistic. (For a further discussion of this case see D. H. Kaye and M. Aickin (eds.), *Statistical Methods in Discrimination Litigation*, Marcel Dekker, New York, 1986.)

NOTES

1. Details of exact binomial procedures can be found in Chapter 2 of Myles Hollander and Douglas Wolfe, *Nonparametric Statistical Methods*, Wiley, New York, 1973.
2. This example is adapted from a survey directed by Professor Joseph N. Uhl of the Department of Agricultural Economics, Purdue University. The survey was sponsored by the Indiana Christmas Tree Growers Association and was conducted in April 1987.
3. This study is reported in Joan L. Duda, "The relationship between goal perspectives, persistence and behavioral intensity among male and female recreational sport participants," *Leisure Sciences*, 10 (1988), pp. 95–106.
4. When the expected cell counts are small, it is best to use a test based on the exact distribution rather than the chi-square approximation, particularly for 2×2 tables. Many statistical software systems offer an "exact" test as well as the chi-square test for 2×2 tables.
5. These data were taken from Ray H. Rosenman et al., "A 4-year prospective study of the relationship of different habitual vocational physical activity to risk and incidence of ischemic heart disease in volunteer male federal employees," in P. Milvey (ed.), *The Marathon: Physiological, Medical, Epidemiological and Psychological Studies*, New York Academy of Sciences, 301 (1977), pp. 627–641.
6. An alternative formula that can be used for hand or calculator computations is
$$X^2 = \sum \frac{(\text{observed})^2}{\text{expected}} - n$$
7. See, for example, Alan Agresti, *Categorical Data Analysis*, Wiley, New York, 1990.

CHAPTER

9

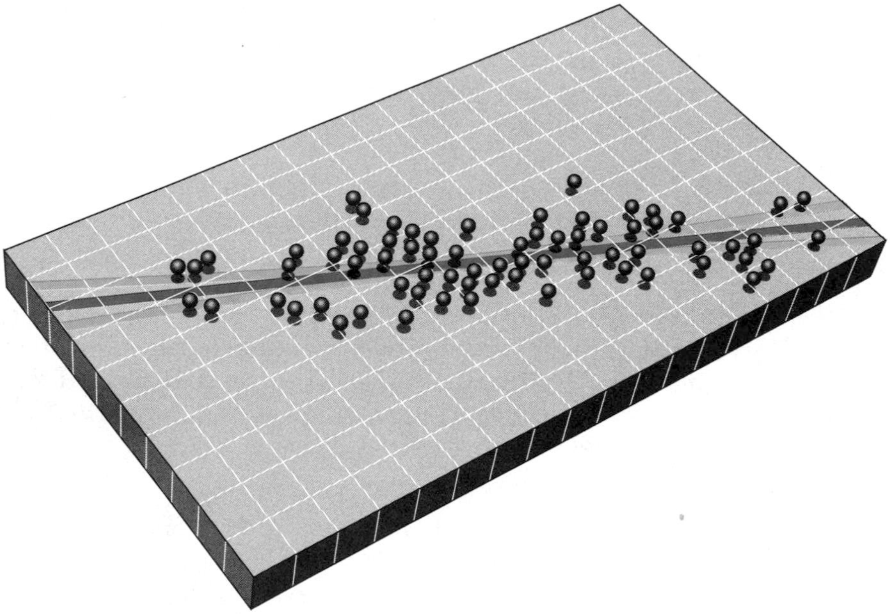

Inference for Regression

PRELUDE

Throughout this book we have followed several strategies for working with data. Two of these strategies are (1) descriptive analysis and exploration of data precede formal inference and (2) examination of each variable separately precedes study of relations among several variables. In this chapter we reach the second part of both strategies in the study of formal inference for relations among quantitative variables.

We will describe methods for inference when there is a single response variable and one or several explanatory variables, and all of these are quantitative variables. Section 9.1 concerns the case of a single explanatory variable, and Section 9.2 briefly introduces the more complex setting in which several explanatory variables work together to explain the response. The descriptive tools we learned in Chapter 2—scatterplots, least-squares regression, and correlation—are essential preliminaries to inference and also provide a foundation for confidence intervals and significance tests. Here are some examples of questions we can answer from data using the methods in this chapter:

- A scatterplot shows a straight-line relationship between how much natural gas a household consumes and how cold the outside temperature is. How accurately can we predict gas consumption from temperature?
- How are ticket prices for professional basketball games related to attendance at the games? Is there a statistically significant relationship? Can we say whether raising prices would hurt attendance?
- We want to predict the college grade index of newly admitted students. We have data on their high school grades in several subjects and their scores on the two parts of the Scholastic Aptitude Test (SAT). How well can we predict college grades from this information? Do high school grades or SAT scores predict college grades more accurately?

• • •

We first met the sample mean $\bar{x}$ in Chapter 1 as a measure of the center of a collection of observations. Later we learned that when the data are a random sample from a population, the sample mean is an estimate of the population mean μ. In Chapters 6 and 7, we used $\bar{x}$ as the basis for confidence intervals and significance tests for inference about μ.

Now we will follow the same program for the problem of fitting straight lines to data. In Chapter 2 we met the least-squares regression line $\hat{y} = a + bx$ as a description of a straight-line relationship between a response variable y and an explanatory variable x. At that point we did not distinguish between sample and population. Now we will think of the least-squares line computed from a sample as an estimate of a *true* regression line for the population, just as $\bar{x}$ is an estimate of the true population mean μ. To emphasize the change in thinking, we introduce new notation. Following the common practice of using Greek letters for population parameters, we will write the population line as $\beta_0 + \beta_1 x$. The least-squares line fitted to sample data is now $b_0 + b_1 x$. This notation reminds us that the intercept b_0 of the fitted line estimates the intercept β_0 of the population line, and the slope b_1 estimates the slope β_1.

simple linear regression

Section 9.1 concerns inference in the setting of *simple linear regression*, the familiar case of a single explanatory variable x. By extending the ideas in Chapters 6 and 7, we can give confidence intervals and significance tests for inference about the slope β_1 and the intercept β_0. Because regression lines are often used for prediction, we also consider inference about either the mean response or an individual future observation on y for a given value of the explanatory variable x. Finally, Section 9.1 also discusses statistical inference about the correlation between two variables x and y that need not have an explanatory-response relation.

multiple linear regression

Section 9.2 presents some basic ideas about *multiple linear regression* through examination of a case study. Multiple regression problems involve one response variable and *more than one* explanatory variable. The presence of several explanatory variables introduces a number of new concepts as well as more elaborate computations.

9.1 SIMPLE LINEAR REGRESSION

Statistical model for linear regression

Simple linear regression studies the relationship between a response variable y and a single explanatory variable x. We expect that different values of x will produce different mean responses. We encountered a similar but simpler situation in Chapter 7 when we discussed methods for comparing two population means. Figure 9.1 illustrates the statistical model for a comparison of the blood pressure in two groups of experimental sub-

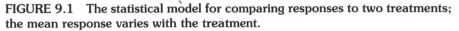

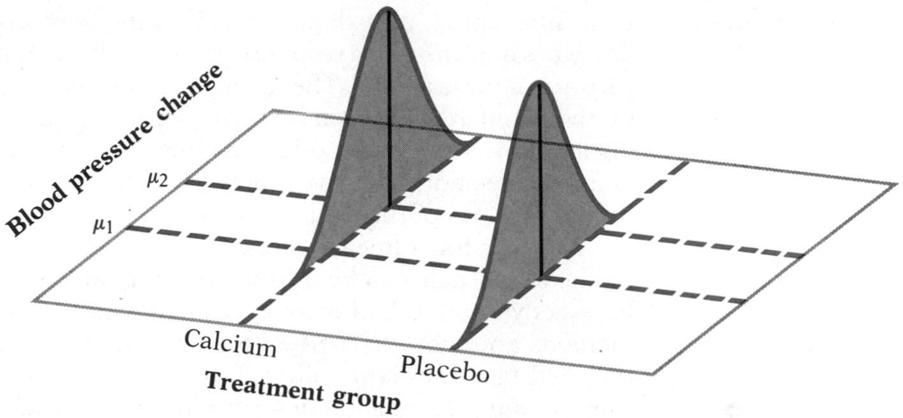

FIGURE 9.1 The statistical model for comparing responses to two treatments; the mean response varies with the treatment.

jects, one taking a calcium supplement and the other a placebo. In each population the observed change in blood pressure varies according to a normal distribution. The mean change will generally be different for the two populations. In the figure, μ_1 and μ_2 represent the two population means. The two normal curves in Figure 9.1 have the same spread, indicating that the population standard deviations are assumed to be equal. We can think of the treatment (placebo or calcium) as the explanatory variable in this example.

In linear regression the explanatory variable x can have many different values. Imagine, for example, giving different amounts x of calcium to different groups of subjects. We can think of the values of x as defining different subpopulations, one for each possible value of x. Each subpopulation consists of all individuals in the population having the same value of x. For example, we can imagine giving $x = 1500$ milligrams of calcium per day to all possible experimental subjects; that forms a subpopulation. The subjects who actually received 1500 milligrams of calcium are a sample from this subpopulation.

The statistical model for simple linear regression assumes that for each value of x the observed values of the response variable y are normally distributed about a mean that depends on x. We use μ_y to represent these means. Rather than just two means μ_1 and μ_2, we are interested in how the many means μ_y change as x changes. In general the means μ_y can change according to any sort of pattern as x changes. In *linear* regression we assume that they all lie on a line when plotted against x. The equation of the line is given by

$$\mu_y = \beta_0 + \beta_1 x$$

population regression line

with intercept β_0 and slope β_1. This is the *population regression line;* it describes how the mean response changes with x. Actually observed y's will vary about these means. The model assumes that this variation, measured by the standard deviation σ, is the same for all values of x. Figure 9.2 displays this statistical model. The line is the population regression line, and the three normal curves show how the response y will vary for three different values of the explanatory variable x.

The data for a linear regression are observed values of y and x. The model takes each x to be a fixed known quantity. In practice, x may not be exactly known. If the error in measuring x is large, more advanced methods are needed. The response y to a given x is a random variable that will take different values if we have several observations with the same x-value. The model describes the mean and standard deviation of the random variable y. These unknown parameters must be estimated from the data.

We will use the following example to explain the fundamentals of simple linear regression. Because regression calculations in practice are always done by statistical software, we will rely on computer output for the arithmetic in this example. In the last part of this section, we give another example that illustrates how to do the work with a calculator if software is unavailable

EXAMPLE 9.1

"The fat content of the human body has physiological and medical importance. It may influence morbidity and mortality, it may alter the effectiveness of drugs and anesthetics, and it may affect the ability to withstand exposure to cold and starvation."[1]

In practice, fat content is found by measuring body density, the weight per unit volume of the body. High fat content corresponds to low body density. Body density is hard to measure directly—the standard method requires that subjects be weighed under water. For this reason, scientists have sought variables that are easier to measure and that can be used to predict body density. Research suggests that *skinfold thickness* can accurately predict body density. To measure skinfold thickness, pinch a fold of skin between calipers at four body locations to determine the thickness, and add the four thicknesses. There is a linear relationship between body density and the logarithm of the skinfold thickness measure. The explanatory variable x is the log of the sum of the skinfold measures, and the response variable y is body density. ●

In the statistical model for predicting body density from skinfold thickness, subpopulations are defined by the x variable. All individuals

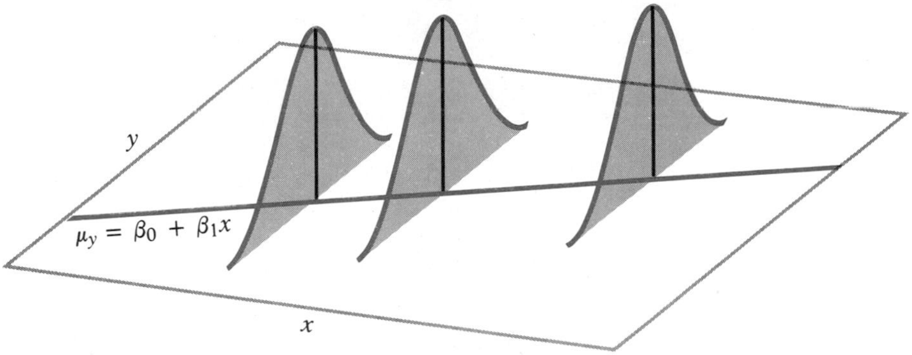

FIGURE 9.2 The statistical model for linear regression; the mean response is a straight-line function of the explanatory variable.

with the same skinfold thickness are in the same subpopulation. Their body densities are assumed to be normally distributed about a mean density that has a straight-line relation to skinfold thickness. The mean density for people with log skinfold thickness x is

$$\mu_y = \beta_0 + \beta_1 x$$

This population regression line gives the mean body density for all values of x. We cannot observe this line, because the observed responses y vary about their means. The statistical model for linear regression consists of the population regression line and a description of the variation of y about the line. Figure 9.2 displays that variation in a picture. Now we want an algebraic statement of the model. In Chapter 2 the principle of seeking an overall pattern and deviations from it when examining data was expressed by the equation

$$\text{DATA} = \text{FIT} + \text{RESIDUAL}$$

The statistical model for linear regression has a similar form. The FIT part of the model consists of the subpopulation means, given by the expression $\beta_0 + \beta_1 x$. The RESIDUAL part represents deviations of the data from the line of population means. We assume that these deviations are normally distributed with standard deviation σ. We use ϵ, the Greek letter epsilon, to stand for the RESIDUAL part of the statistical model. A response y is the sum of its mean and a chance deviation ϵ from the mean. The deviations ϵ represent "noise," that is, variation in y due to other causes that prevent the observed (x, y) values from forming a perfectly straight line on the scatterplot.

> **SIMPLE LINEAR REGRESSION MODEL**
>
> Given n observations on the explanatory variable x and the response variable y,
>
> $$(x_1, y_1), \ (x_2, y_2), \ \ldots, \ (x_n, y_n)$$
>
> the statistical model for simple linear regression states that for each i from 1 to n the observed response is
>
> $$y_i = \beta_0 + \beta_1 x_i + \epsilon_i$$
>
> Here $\beta_0 + \beta_1 x_i$ is the mean response when $x = x_i$. The deviations ϵ_i are assumed to be independent and normally distributed with mean 0 and standard deviation σ. In other words, they are a simple random sample (SRS) from the $N(0, \sigma)$ distribution. The parameters of the model are β_0, β_1, and σ.

Because the means μ_y lie on the line $\mu_y = \beta_0 + \beta_1 x$, they are all determined by β_0 and β_1. Once we have estimates of β_0 and β_1, the linear relationship determines the estimates of μ_y for all values of x. Linear regression allows us to do inference not only for subpopulations for which we have data but also for those corresponding to x's not present in the data. We will learn how to do inference about

- the slope β_0 and the intercept β_1 of the population regression line;
- the mean response μ_y for a given value of x, and
- an individual future response y for a given value of x.

Estimating the regression parameters

The method of least squares presented in Chapter 2 summarizes a straight-line relationship between the observed values of an explanatory variable and a response variable. Now we want to use the least-squares line as a basis for inference about a population from which our observations are a sample. We can do this only when the statistical model just presented holds. In that setting, the slope b_1 and intercept b_0 of the least-squares line

$$\hat{y} = b_0 + b_1 x$$

estimate the slope β_1 and the intercept β_0 of the population regression line.

Using the formulas from Chapter 2 and our new notation, the slope of the least-squares line is

$$b_1 = \frac{\sum xy - \frac{1}{n}(\sum x)(\sum y)}{\sum x^2 - \frac{1}{n}(\sum x)^2}$$

and the intercept is

$$b_0 = \bar{y} - b_1\bar{x}$$

Some algebra based on the rules for means of random variables (Section 4.3) shows that b_0 and b_1 are unbiased estimators of β_0 and β_1. Furthermore, b_0 and b_1 are normally distributed with means β_0 and β_1 and standard deviations that can be estimated from the data. Normality of these sampling distributions is a consequence of the assumption that the ϵ_i are normal. A general form of the central limit theorem tells us that the distributions of b_0 and b_1 will still be approximately normal even if the ϵ_i are not. On the other hand, outliers and influential observations can invalidate the results of inference for regression.

The predicted value of y for a given value x^* of x is the point on the least-squares line $\hat{y} = b_0 + b_1x^*$. This is an unbiased estimator of the mean response μ_y when $x = x^*$. The *residuals* are

residuals

$$\begin{aligned} e_i &= \text{observed response} - \text{predicted response} \\ &= y_i - \hat{y}_i \\ &= y_i - b_0 - b_1x_i \end{aligned}$$

The residuals e_i correspond to the model deviations ϵ_i. The e_i sum to 0, and the ϵ_i come from a population with mean 0.

The remaining parameter to be estimated is σ, which measures the variation of y about the population regression line. Because this parameter is the standard deviation of the model deviations, it should come as no surprise that we use the residuals to estimate it. As usual, we work first with the variance and take the square root to obtain the standard deviation. For simple linear regression the estimate of σ^2 is the average squared residual

$$\begin{aligned} s^2 &= \frac{\sum e_i^2}{n-2} \\ &= \frac{\sum (y_i - \hat{y}_i)^2}{n-2} \end{aligned}$$

We average by dividing the sum by $n - 2$ in order to make s^2 an unbiased estimator of σ^2. Recall that in finding the sample variance of n observations we divided the sum of squared deviations from the mean by $n - 1$ for this same reason. The quantity $n - 2$ is called the *degrees of freedom* for s^2. The estimate of σ is given by

degrees of freedom

$$s = \sqrt{s^2}$$

In practice, we use software to do the unpleasant arithmetic needed to obtain b_1, b_0, and s. Here are the results for the example of body density and skinfold thickness.

EXAMPLE 9.2

We use the SAS statistical software package to calculate the regression of body density on log of skinfolds for Example 9.1. The output of other software packages is quite similar. In entering the data, we chose the names LSKIN and DEN for the explanatory and response variables. It is good practice to use names, rather than just x and y, to remind yourself which data the output describes. Because the relationship between skinfold and body density varies slightly with age and sex, we look only at data from a sample of 92 males aged 20 to 29. In the discussion that follows, we often round off the numbers from the computer output. Software often reports many more digits than are meaningful or useful. ●

First, we examine the variables and their relationship. Figure 9.3 contains the output from the SAS procedure MEANS. We see that

$$\bar{x} = 1.568 \quad s_x = .2159$$
$$\bar{y} = 1.064 \quad s_y = .0160$$

Figure 9.4 is a scatterplot of the data. The relationship appears to be linear, and no unusual observations are evident.

We then proceed to fit the least-squares line. Figure 9.5 gives part of the output of the SAS regression procedure REG for this example. The results of the least-squares fit appear at the bottom of the output. The values of b_0 and b_1 are given in the column labeled PARAMETER ESTIMATE. The column labeled VARIABLE tells us that the entry in the INTERCEP row is b_0 and the entry for LSKIN is b_1. We see that $b_0 = 1.1631$ and $b_1 = -0.0632$, so the least-squares line is

$$\hat{y} = 1.1631 - .0632x$$

That line appears on the scatterplot of Figure 9.4. Moving to the upper section of the output, the estimated standard deviation s is labeled ROOT

VARIABLE	N	MEAN	STANDARD DEVIATION	VARIANCE
LSKIN	92	1.56800000	0.21590000	0.04661281
DEN	92	1.06400240	0.01602319	0.00025674

FIGURE 9.3 Descriptive statistics output for LSKIN and DEN.

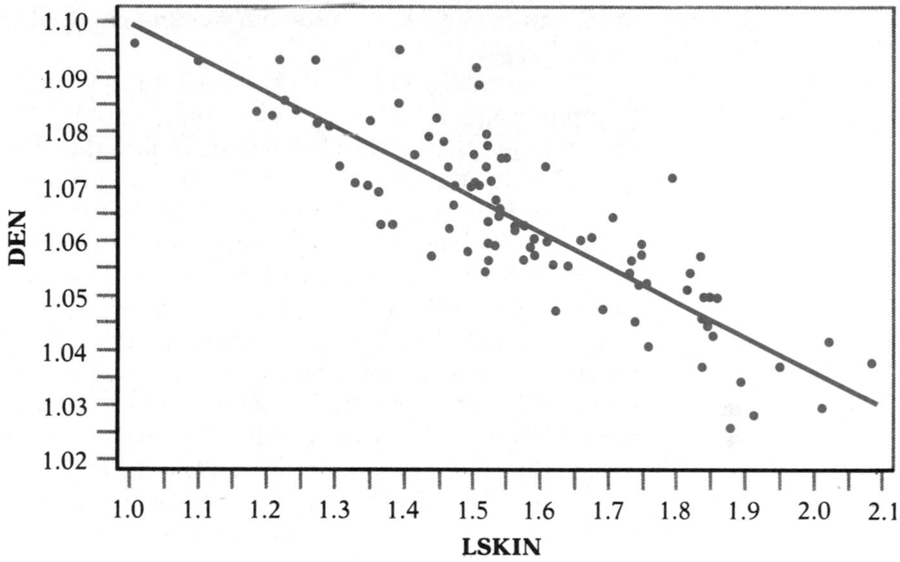

FIGURE 9.4 Scatterplot of DEN versus LSKIN with the least-squares regression line.

```
DEP VARIABLE:  DEN
                          ANALYSIS OF VARIANCE

                          SUM OF           MEAN
    SOURCE      DF       SQUARES         SQUARE    F VALUE     PROB>F

    MODEL        1    0.01694263     0.01694263    237.478     0.0001
    ERROR       90   0.006420960    0.000071344
    C TOTAL     91    0.02336359

         ROOT MSE    0.008446538      R-SQUARE      0.7252
         DEP MEAN       1.064002      ADJ R-SQ      0.7221
         C.V.          0.7938457

                          PARAMETER ESTIMATES

                   PARAMETER       STANDARD      T FOR HO:
  VARIABLE          ESTIMATE          ERROR    PARAMETER=0     PROB > |T|

  INTERCEP        1.16310000    0.006490615        179.197         0.0001
  LSKIN          -0.06320000    0.004101147        -15.410         0.0001
```

FIGURE 9.5 Regression output for the body density example.

MSE and is 0.0084. We now have estimates of all of the parameters β_0, β_1, and σ.

The output contains much other information that we ignore for now. Computer outputs often give more information than we want or need. On the other hand, it is very frustrating to find that a software package does not print out the particular statistics that we want for an analysis. The experienced user of statistical software learns to ignore the parts of the output that are not needed for the problem at hand.

Now that we have fitted a line, we should examine the residuals. Plot the residuals both against the case number (especially if this reflects the order in which the observations were taken) and against the explanatory variable. Figure 9.6 is a plot of the residuals versus case number, and Figure 9.7 plots the residuals versus LSKIN. No unusual values or patterns are evident in either plot. Finally, Figure 9.8 is a normal quantile plot of the residuals. The plot looks fairly straight, so the assumption of normally distributed deviations appears to be reasonable. That is important for the inference that will follow. We need not make these plots by hand; the software will produce all of them on demand.

For the 92 males aged 20 to 29 in the study, the mean body density was 1.064, and the standard deviation was 0.016. The regression output shows that the standard deviation of body density about the regression line is 0.0084, only about half as large. This decrease in the standard deviation is due to the ability of skinfold thickness to explain body density.

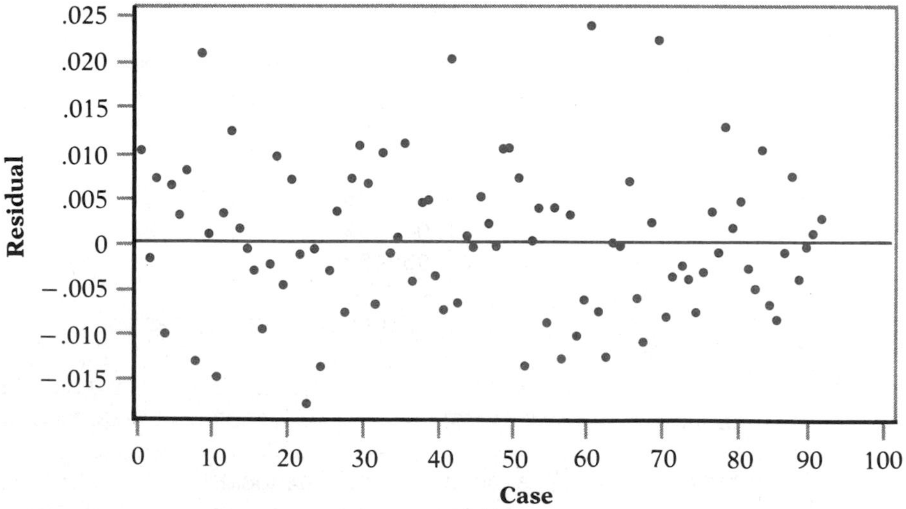

FIGURE 9.6 **Plot of residuals versus case number for the body density example.**

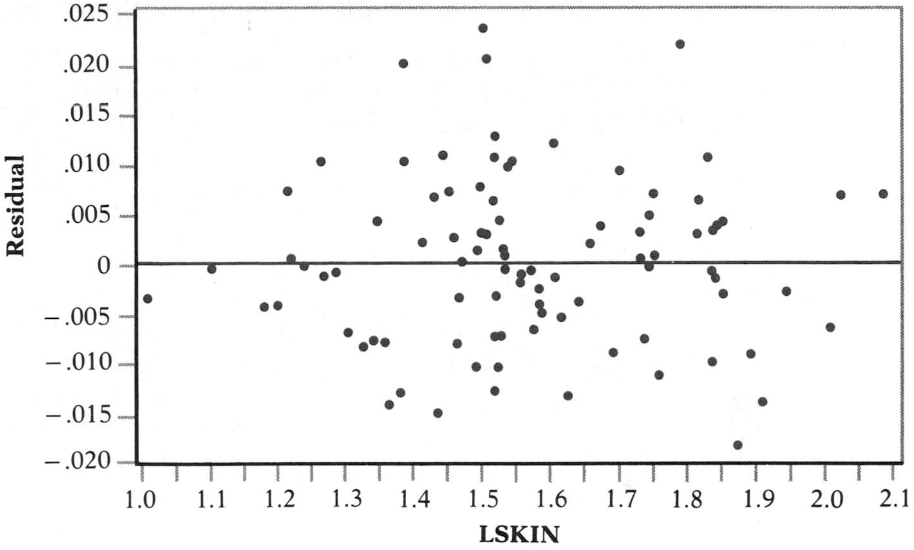

FIGURE 9.7 Plot of residuals versus the explanatory variable for the body density example.

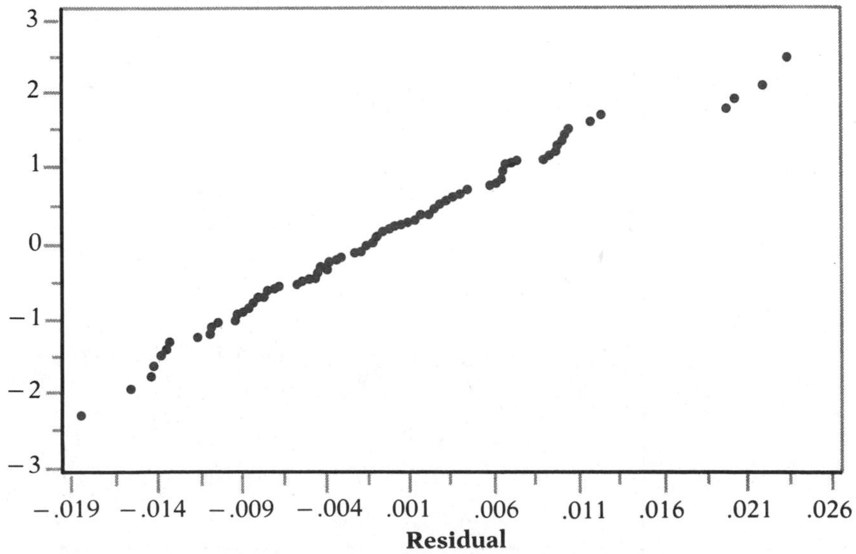

FIGURE 9.8 Normal quantile plot of the residuals for the body density example.

Confidence intervals and significance tests

Chapter 7 presented confidence intervals and significance tests for means and differences in means. In each case, inference rested on the standard errors of estimates and on t distributions. Inference for the intercept and slope in a linear regression is similar in principle, although the recipes are more complicated. All of the confidence intervals, for example, have the form

$$\text{estimate} \pm t^* s_{\text{estimate}}$$

where t^* is a critical point of a t distribution.

Confidence intervals and tests for the slope and intercept are based on the normal sampling distributions of the estimates b_1 and b_0. Standardizing these estimates gives standard normal z statistics. Because the standard deviations are not known, a t distribution replaces the standard normal distribution, a pattern we have seen many times. The degrees of freedom in the t distribution are $n-2$, the same as in the estimate s of the unknown σ. Formulas for the various standard errors s_{estimate} appear in the subsection on calculations. For now we will concentrate on the basic ideas and let the computer do the computations.

CONFIDENCE INTERVALS AND SIGNIFICANCE TESTS FOR REGRESSION SLOPE AND INTERCEPT

A level C confidence interval for the intercept β_0 is

$$b_0 \pm t^* s_{b_0}$$

A level C confidence interval for the slope β_1 is

$$b_1 \pm t^* s_{b_1}$$

In these expressions t^* is the upper $(1-C)/2$ critical value for the $t(n-2)$ distribution. To test the hypothesis H_0: $\beta_1 = 0$, compute

$$t = \frac{b_1}{s_{b_1}}$$

In terms of a random variable T having the $t(n-2)$ distribution, the P-value for a test of H_0 against

$$\begin{array}{lll} H_a\colon \beta_1 > 0 & \text{is} & P(T \geq t) \\ H_a\colon \beta_1 < 0 & \text{is} & P(T \leq t) \\ H_a\colon \beta_1 \neq 0 & \text{is} & 2P(T \geq |t|) \end{array}$$

There are similar formulas for significance tests about the intercept β_0, using s_{b_0} and the $t(n-2)$ distribution. Although computer outputs often include a test of H_0: $\beta_0 = 0$, this information usually has little practical value. From the equation for the population regression line, $\mu_y = \beta_0 + \beta_1 x$, we see that β_0 is the mean response corresponding to $x = 0$. In many practical situations, this subpopulation does not exist or is not interesting.

On the other hand, the test of H_0: $\beta_1 = 0$ is quite useful. When we substitute $\beta_1 = 0$ in the model, the x term drops out and we are left with

$$\mu_y = \beta_0$$

This model says that the mean of y does not vary with x. All of the y's come from a single population with mean β_0, which we would estimate by $\bar{y}$. The hypothesis H_0: $\beta_1 = 0$ therefore says that there is no straight-line relationship between y and x and that linear regression of y on x is of no value for predicting y.

EXAMPLE 9.3

The computer output in Figure 9.5 for the body density problem contains the information needed for inference about the regression coefficients. The column labeled STANDARD ERROR gives the standard errors of the estimated coefficients. The value of s_{b_1} appears to the right of the estimated slope $b_1 = -0.0632$; it is 0.0041. (As usual, we have rounded the values from the output.)

The t statistic and P-value for the test of H_0: $\beta_1 = 0$ against the two-sided alternative H_a: $\beta_1 \neq 0$ appear in the columns labeled T FOR H0: PARAMETER = 0 and PROB >|T|. We can verify the t calculation from the formula for the standardized estimate:

$$t = \frac{b_1}{s_{b_1}} = \frac{-.0632}{.0041} = -15.41$$

The P-value is 0.0001. There is strong evidence against the null hypothesis. We conclude that linear regression on the logarithm of the skinfold measures is useful for predicting body density.

Higher body fat reduces body density and increases skinfold thickness, so we expect a negative association. We might therefore choose the one-sided alternative H_a: $\beta_1 < 0$. This choice does not affect our conclusion for this example, but the P-value for the one-sided alternative is half of the value for the two-sided alternative.

A confidence interval for β_1 requires a critical value t^* from the $t(n - 2) = t(90)$ distribution. In Table E there are entries for 80 and 100 degrees of freedom. The values for these rows are very similar. To be conservative, we will use the larger critical value, for 80 degrees of freedom. For a 95% confidence interval we use the value in the 95% column, which is $t^* = 1.99$.

The 95% confidence interval for β_1 is

$$\begin{aligned} b_1 \pm t^* s_{b_1} &= -.0632 \pm (1.99)(.0041) \\ &= -.0632 \pm .0082 \\ &= (-.0714, -.0550) \end{aligned}$$

We estimate that an increase of 1 in the logarithm of skinfold thickness is associated with a decrease of between 0.0714 and 0.0550 in body density. ●

Note that the intercept in this example is not of practical interest. It estimates the mean body density when the logarithm of the sum of the four skinfold thicknesses (that's x) is 0, a value with no special meaning. The values of LSKIN in the data range from 1.20 to 2.26. For this reason, we do not compute a confidence interval for β_0.

Confidence intervals for mean response

For any specific value of x, say x^*, the mean of the response y in this subpopulation is given by

$$\mu_y = \beta_0 + \beta_1 x^*$$

To estimate this mean from the sample, we substitute the estimates b_0 and b_1 for β_0 and β_1:

$$\hat{\mu}_y = b_0 + b_1 x^*$$

A confidence interval for μ_y adds to this estimate a margin of error based on the standard error $s_{\hat{\mu}}$. (The formula for the standard error again appears in the subsection on calculations.)

CONFIDENCE INTERVAL FOR A MEAN RESPONSE

A level C confidence interval for the mean response μ_y when x takes the value x^* is

$$\hat{\mu}_y \pm t^* s_{\hat{\mu}}$$

where t^* is the upper $(1 - C)/2$ critical value for the $t(n - 2)$ distribution.

Many computer programs calculate confidence intervals for the mean response corresponding to each of the x-values in the data. Some can calculate an interval for any value x^* of the explanatory variable.[2]

EXAMPLE 9.4

Figure 9.9 gives more of the output generated by the SAS procedure REG for the data on body density and skinfold measures. We show the output for only the first 24 of the 92 cases.

The observed data values y_i appear in the column labeled ACTUAL, the predicted values $\hat{\mu}_y$ for the given x are given in the column labeled PREDICT VALUE, and the standard errors $s_{\hat{\mu}}$ of the predicted values appear in the column labeled STD ERR PREDICT. The lower and upper limits for the 95% confidence intervals are in the columns labeled LOWER95% MEAN and UPPER95% MEAN.

Consider the first line of the output. The value of LSKIN for this individual is 1.266. The predicted mean DEN for all individuals with LSKIN = 1.266 is

$$\hat{\mu}_{\text{DEN}} = 1.0831$$

with a standard error of

$$s_{\hat{\mu}} = .0015$$

The 95% confidence interval is (1.0801, 1.0861). We conclude with 95% confidence that the mean body density for men aged 20 to 29 whose logarithm of skinfold measures is 1.266 lies between 1.0801 and 1.0861. ●

We can gain more insight by plotting the upper and lower confidence limits for a range of values of x on a graph with the data and the least-

OBS	LSKIN	ACTUAL	PREDICT VALUE	STD ERR PREDICT	LOWER95% MEAN	UPPER95% MEAN
1	1.26594	1.0930	1.0831	.00152	1.0801	1.0861
2	1.55720	1.0627	1.0647	.00088	1.0629	1.0664
3	1.45025	1.0783	1.0714	.00100	1.0694	1.0734
4	1.52207	1.0564	1.0669	.00090	1.0651	1.0687
5	1.51477	1.0734	1.0674	.00091	1.0656	1.0692
6	1.50624	1.0706	1.0679	.00092	1.0661	1.0697
7	1.49688	1.0761	1.0685	.00093	1.0667	1.0703
8	1.61624	1.0474	1.0610	.00090	1.0592	1.0627
9	1.50313	1.0886	1.0681	.00092	1.0663	1.0699
10	1.75201	1.0531	1.0524	.00116	1.0501	1.0547
11	1.43343	1.0573	1.0725	.00104	1.0704	1.0746
12	1.81344	1.0513	1.0485	.00134	1.0458	1.0511
13	1.60096	1.0738	1.0619	.00089	1.0601	1.0637
14	1.49381	1.0699	1.0687	.00093	1.0668	1.0705
15	1.28574	1.0808	1.0818	.00145	1.0790	1.0847
16	1.51957	1.0636	1.0671	.00090	1.0653	1.0689
17	1.83410	1.0371	1.0472	.00140	1.0444	1.0500
18	1.58400	1.0603	1.0630	.00088	1.0612	1.0647
19	1.69952	1.0649	1.0557	.00103	1.0536	1.0577
20	1.58701	1.0577	1.0628	.00088	1.0610	1.0646
21	2.02173	1.0421	1.0353	.00206	1.0312	1.0394
22	1.84178	1.0451	1.0467	.00143	1.0439	1.0495
23	1.87409	1.0265	1.0447	.00153	1.0416	1.0477
24	1.83495	1.0460	1.0471	.00140	1.0443	1.0499

FIGURE 9.9 Computer output giving confidence intervals for subpopulation means for the body density example.

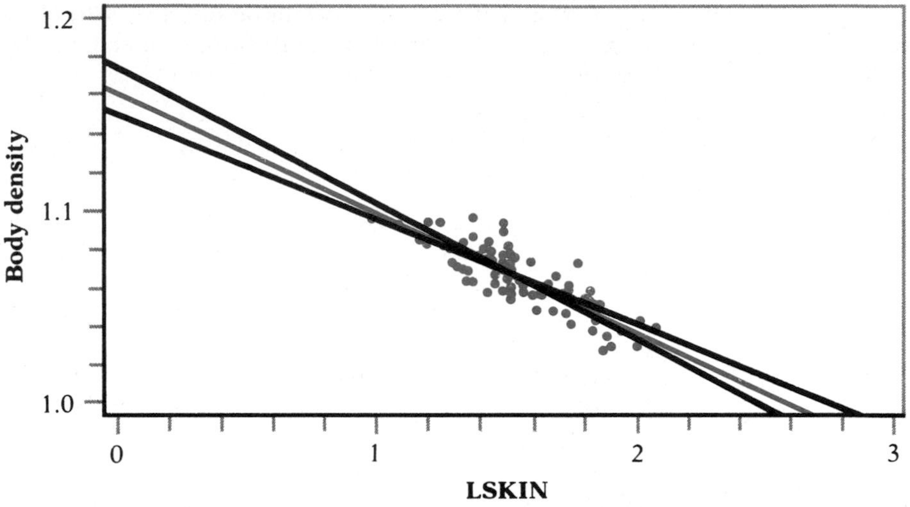

FIGURE 9.10 The 95% confidence limits (black curves) for the mean response for the body density example. The range of LSKIN is extended to show that the intervals grow wider as x^* moves away from the mean of the x observations.

squares line. The 95% confidence limits appear as curved black lines in Figure 9.10. For any x^*, the confidence interval for the mean response extends from the lower black line to the upper black line. The intervals are narrowest for values of x^* near the mean of the observed x's and increase as x^* moves away from $\bar{x}$. To show this effect clearly, we have extended the x axis in the plot far beyond the range of the actual data. In practice, you would not do prediction for such extreme values.

Prediction intervals

In the last example, we predicted the mean body density for all young men with LSKIN equal to 1.266. Suppose that John has LSKIN = 1.266. We can also predict John's individual body density. The predicted value is the same as if we were predicting the mean of the entire subpopulation, but the margin of error is larger because it is harder to predict one individual value than to predict the mean.

The predicted response y for an individual case with a specific value x^* of the explanatory variable x is

$$\hat{y} = b_0 + b_1 x^*$$

This is the same as the expression for $\hat{\mu}_y$. That is, the fitted line is used both to estimate the mean response when $x = x^*$ and to predict a single future response. We use the two notations $\hat{\mu}_y$ and $\hat{y}$ to remind ourselves of these two distinct uses.

A useful prediction should include a margin of error to indicate its accuracy. The interval used to predict a future observation is called a *prediction interval*. Although the response y that is being predicted is a random variable, the interpretation of a prediction interval is similar to that for a confidence interval. Consider drawing a sample of n observations (x_i, y_i) and then one additional observation (x^*, y). Do this many times, and each time calculate the 95% prediction interval for y based on the sample. Then 95% of the prediction intervals will contain the y-value of the additional observation. In other words, the probability that this method produces an interval that contains the value of a future observation is 0.95.

prediction interval

The form of the prediction interval is very similar to that of the confidence interval for the mean response. The difference is that the standard error $s_{\hat{y}}$ used in the prediction interval includes both the variability due to the fact that the least-squares line is not exactly equal to the true regression line *and* the variability of the future response variable y around the subpopulation mean. (The formula for $s_{\hat{y}}$ appears in the subsection on regression calculations.)

PREDICTION INTERVAL FOR A FUTURE OBSERVATION

A level C prediction interval for a future observation on the response variable y from the subpopulation corresponding to x^* is

$$\hat{y} \pm t^* s_{\hat{y}}$$

where t^* is the upper $(1 - C)/2$ critical value for the $t(n - 2)$ distribution.

EXAMPLE 9.5

Figure 9.11 gives prediction output from the SAS regression procedure REG for the first 24 of the 92 cases in the body density versus skinfold data set. The format is similar to Figure 9.9, but the two rightmost columns now contain the upper and lower limits of the 95% prediction interval for the body density of a single young man with LSKIN equal to the value in the first column. These limits are wider than the corresponding 95% confidence limits for the mean body density that appear in Figure 9.9.

Consider observation 1, which has LSKIN = 1.266. The 95% prediction interval is 1.066 to 1.100. For a male aged 20 to 29 with LSKIN = 1.266, we predict with 95% confidence that his body density will be between 1.066 and 1.100. ●

Figure 9.12 shows the upper and lower prediction limits for a range of values of x, along with the data and the least-squares line. The 95% prediction limits are indicated by the black curves. Compare this figure

OBS	LSKIN	ACTUAL	PREDICT VALUE	STD ERR PREDICT	LOWER95% PREDICT	UPPER95% PREDICT
1	1.26594	1.0930	1.0831	.00152	1.0660	1.1001
2	1.55720	1.0627	1.0647	.00088	1.0478	1.0816
3	1.45025	1.0783	1.0714	.00100	1.0545	1.0883
4	1.52207	1.0564	1.0669	.00090	1.0500	1.0838
5	1.51477	1.0734	1.0674	.00091	1.0505	1.0842
6	1.50624	1.0706	1.0679	.00092	1.0510	1.0848
7	1.49688	1.0761	1.0685	.00093	1.0516	1.0854
8	1.61624	1.0474	1.0610	.00090	1.0441	1.0778
9	1.50313	1.0886	1.0681	.00092	1.0512	1.0850
10	1.75201	1.0531	1.0524	.00116	1.0354	1.0693
11	1.43343	1.0573	1.0725	.00104	1.0556	1.0894
12	1.81344	1.0513	1.0485	.00134	1.0315	1.0655
13	1.60096	1.0738	1.0619	.00089	1.0450	1.0788
14	1.49381	1.0699	1.0687	.00093	1.0518	1.0856
15	1.28574	1.0808	1.0818	.00145	1.0648	1.0989
16	1.51957	1.0636	1.0671	.00090	1.0502	1.0839
17	1.83410	1.0371	1.0472	.00140	1.0302	1.0642
18	1.58400	1.0603	1.0630	.00088	1.0461	1.0799
19	1.69952	1.0649	1.0557	.00103	1.0388	1.0726
20	1.58701	1.0577	1.0628	.00088	1.0459	1.0797
21	2.02173	1.0421	1.0353	.00206	1.0181	1.0526
22	1.84178	1.0451	1.0467	.00143	1.0297	1.0637
23	1.87409	1.0265	1.0447	.00153	1.0276	1.0617
24	1.83495	1.0460	1.0471	.00140	1.0301	1.0641

FIGURE 9.11 Computer output giving prediction intervals for the body density example.

with Figure 9.10, which shows the 95% confidence limits drawn to the same scale. The upper and lower limits of the prediction intervals are farther from the least-squares line than are the confidence limits. The figures remind us that the interval for a single future observation must be larger than an interval for the mean of its subpopulation.

Analysis of variance for regression*

The usual computer output for regression includes additional calculations called *analysis of variance*. Analysis of variance, often abbreviated ANOVA,

*This optional section presents material that should be studied if you plan to read Section 9.2 on multiple regression.

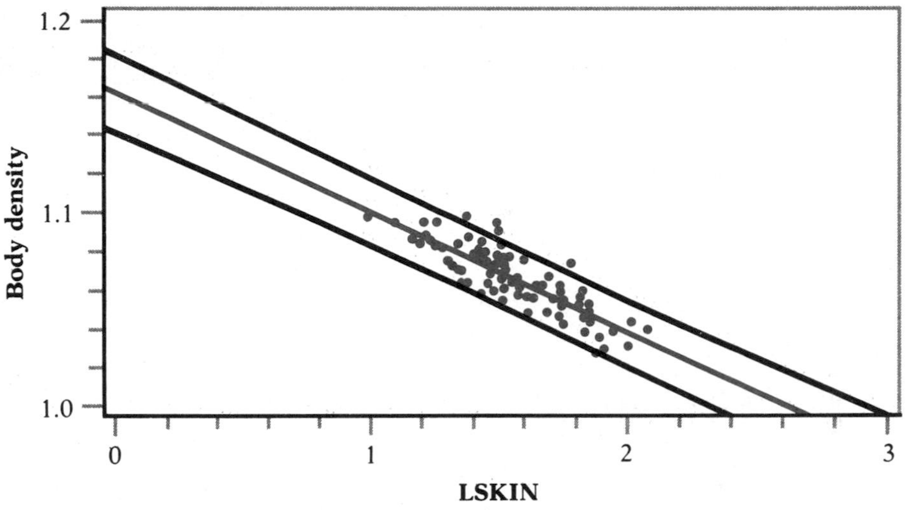

FIGURE 9.12 The 95% prediction limits (black curves) for individual responses for the body density example. Compare with Figure 9.10.

is essential for multiple regression (Section 9.2) and for comparing several means (Chapter 10). Analysis of variance summarizes information about the sources of variation in the data. It is based on the

$$\text{DATA} = \text{FIT} + \text{RESIDUAL}$$

framework.

The total variation in the response y is expressed by the deviations $y_i - \bar{y}$. If these deviations were all 0, all observations would be equal and there would be no variation in the response. There are two reasons why the individual observations y_i are not all equal to their mean $\bar{y}$. First, the responses y_i correspond to different values of the explanatory variable x and will differ because of that. The fitted value $\hat{y}_i$ estimates the mean response for the specific x_i. The differences $\hat{y}_i - \bar{y}$ reflect the variation in mean response due to differences in the x_i. This variation is accounted for by the regression line, because the $\hat{y}$'s lie exactly on the line. Second, individual observations will vary about their mean because of variation within the subpopulation of responses to a fixed x_i. This variation is represented by the residuals $y_i - \hat{y}_i$ that record the scatter of the actual observations about the fitted line. The overall deviation of any y observation from the mean of the y's is the sum of these two deviations:

$$(y_i - \bar{y}) = (\hat{y}_i - \bar{y}) + (y_i - \hat{y}_i)$$

For deviations, this equation expresses the idea that DATA = FIT + RESIDUAL.

We have several times measured variation by an average of squared deviations. If we square each of the three deviations above and then sum over all n observations, it is an algebraic fact that the sums of squares add:

$$\sum(y_i - \bar{y})^2 = \sum(\hat{y}_i - \bar{y})^2 + \sum(y_i - \hat{y}_i)^2$$

We rewrite this equation as

$$\text{SST} = \text{SSM} + \text{SSE}$$

where

$$\text{SST} = \sum(y_i - \bar{y})^2$$
$$\text{SSM} = \sum(\hat{y}_i - \bar{y})^2$$
$$\text{SSE} = \sum(y_i - \hat{y}_i)^2$$

sum of squares

The SS in each abbreviation stands for *sum of squares,* and the T, M, and E stand for total, model, and error, respectively. ("Error" here stands for deviations from the line, which might better be called "residual" or "unexplained variation.") Thus the total variation, as expressed by SST, is composed of the variation due to the straight-line model (SSM) and the variation due to deviations from this model (SSE). This partition of the variation in the data among several sources is the heart of analysis of variance.

If H_0: $\beta_1 = 0$ were true, there would be no subpopulations and all of the y's should be viewed as coming from a single population with mean μ_y. The variation of the y's would then be described by the sample variance

$$s_y^2 = \frac{\sum(y_i - \bar{y})^2}{n - 1}$$

degrees of freedom

The numerator in this expression is SST. The denominator is the total *degrees of freedom,* or simply DFT.

Just as the total sum of squares SST is the sum of SSM and SSE, the total degrees of freedom DFT is the sum of DFM and DFE, the degrees of freedom for the model and for the error.

$$\text{DFT} = \text{DFM} + \text{DFE}$$

mean square

The model has one explanatory variable x, so the degrees of freedom for this source is DFM $= 1$. Because DFT $= n - 1$, this leaves DFE $= n - 2$ as the degrees of freedom for error. For each source, the ratio of the sum of squares to the degrees of freedom is called the *mean square,* or simply MS. The general formula for a mean square is

$$\text{MS} = \frac{\text{sum of squares}}{\text{degrees of freedom}}$$

Each mean square is an average squared deviation. MST is just s_y^2, the sample variance that we would calculate if all of the data came from a single population. MSE is familiar to us:

$$\text{MSE} = s^2 = \frac{\sum(y_i - \hat{y}_i)^2}{n - 2}$$

It is our estimate of σ^2, the variance about the population regression line.

SUMS OF SQUARES, DEGREES OF FREEDOM, AND MEAN SQUARES

Sums of squares represent variation present in the responses. They are calculated by summing squared deviations. Analysis of variance seeks to partition the total variation among several sources.

The sums of squares are related by the formula

$$\text{SST} = \text{SSM} + \text{SSE}$$

That is, the total variation is partitioned into two parts, one due to the model and one due to deviations from the model.

Degrees of freedom are associated with each sum of squares. They are related in the same way:

$$\text{DFT} = \text{DFM} + \text{DFE}$$

To calculate mean squares, use the formula

$$\text{MS} = \frac{\text{sum of squares}}{\text{degrees of freedom}}$$

The null hypothesis H_0: $\beta_1 = 0$ that y is not linearly related to x can be tested by comparing MSM with MSE. The ANOVA test statistic is an F statistic,

$$F = \frac{\text{MSM}}{\text{MSE}}$$

When H_0 is true, this statistic has an F distribution with 1 degree of freedom in the numerator and $n - 2$ degrees of freedom in the denominator. These degrees of freedom are those of MSM and MSE. The F distributions were introduced in Section 7.3. Figure 7.10 page 560 shows a typical F density curve. Just as there are many t statistics, there are many F statistics. The ANOVA F statistic is not the same as the F statistic of Section 7.3.

When $\beta_1 \neq 0$, MSM tends to be large relative to MSE. So large values of F are evidence against H_0 in favor of the two-sided alternative.

ANALYSIS OF VARIANCE F TEST

In the simple linear regression model, the hypotheses

$$H_0: \beta_1 = 0$$
$$H_a: \beta_1 \neq 0$$

are tested by the F statistic

$$F = \frac{\text{MSM}}{\text{MSE}}$$

The P-value is the probability that a random variable having the $F(1,\ n-2)$ distribution is greater than or equal to the calculated value of the F statistic.

The F statistic tests the same null hypothesis as the t statistic that we encountered earlier in this chapter, so it is not surprising that the two are related. It is an algebraic fact that $t^2 = F$ in this case. For linear regression with one explanatory variable, we prefer the t form of the test because it more easily allows us to test one-sided alternatives and is closely related to the confidence interval for β_1.

analysis of variance table

The ANOVA calculations are displayed in an *analysis of variance table*, often abbreviated ANOVA table. Here is the format of the table for simple linear regression.

Source	Degrees of freedom	Sum of squares	Mean square	F
Model	1	$\sum(\hat{y}_i - \bar{y})^2$	SSM/DFM	MSM/MSE
Error	$n-2$	$\sum(y_i - \hat{y}_i)^2$	SSE/DFE	
Total	$n-1$	$\sum(y_i - \bar{y})^2$	SST/DFT	

EXAMPLE 9.6

For the problem of predicting body density from skinfold measures, Figure 9.13 gives the output of the SAS regression procedure REG, including the ANOVA table. The output corresponds to the ANOVA table format just given. The only difference is that the label C TOTAL is used in the output rather than TOTAL.

The calculated value of the F statistic is 237.478 and the P-value is 0.0001. There is strong evidence against the null hypothesis. Note that the t statistic for LSKIN is -15.410. The square of this t is 237.468, which agrees reasonably well with the F statistic given on the output. Whenever numbers are rounded off in calculations such as this, there will be roundoff error. ●

```
DEP VARIABLE: DEN
                      ANALYSIS OF VARIANCE

                         SUM OF            MEAN
    SOURCE     DF       SQUARES          SQUARE    F VALUE    PROB>F

    MODEL       1    0.01694263      0.01694263    237.478    0.0001
    ERROR      90   0.006420960     0.000071344
    C TOTAL    91    0.02336359

        ROOT MSE   0.008446538        R-SQUARE      0.7252
        DEP MEAN      1.064002        ADJ R-SQ      0.7221
        C.V.         0.7938457

                      PARAMETER ESTIMATES

                 PARAMETER        STANDARD       T FOR HO:
VARIABLE          ESTIMATE           ERROR     PARAMETER=0     PROB > |T|

INTERCEP        1.16310000     0.006490615         179.197         0.0001
LSKIN          -0.06320000     0.004101147         -15.410         0.0001
```

FIGURE 9.13 Regression output for the body density example, with ANOVA table.

Calculations for regression inference*

We recommend using statistical software for regression calculations. With time and care, however, the work is feasible with a calculator. We will use the following example to illustrate how to perform regression analysis using a calculator.

EXAMPLE 9.7

The amount of natural gas required to heat a home depends on the outdoor temperature. When the weather is cold, more gas will be consumed. A study[3] of one home recorded the average daily gas consumption y (in hundreds of cubic feet) for each month during one heating season. The explanatory variable x is the average number of heating degree days per day during the month. (One heating degree day is accumulated for each degree a day's average temperature

*The material in this optional section is needed if regression calculations are to be performed with a calculator rather than a computer.

falls below 65° F. An average temperature of 50°, for example, corresponds to 15 degree days.) Here are the data:

	Oct.	Nov.	Dec.	Jan.	Feb.	Mar.	Apr.	May	June
x	15.6	26.8	37.8	36.4	35.5	18.6	15.3	7.9	.0
y	5.2	6.1	8.7	8.5	8.8	4.9	4.5	2.5	1.1

We will refer to these two variables as degree days and gas consumption. The data and the least-squares line are plotted in Figure 9.14. The strong straight-line pattern suggests that we can use linear regression to predict gas consumption from degree days. ●

To do inference for this problem, we assume that the statistical model for simple linear regression applies. The model makes two important assumptions, both of which can to some extent be verified from data. First, for each fixed value of heating degree days x, gas consumption y has a normal distribution with a standard deviation σ that is the same for all values of x. If we observed many months with $x = 20$, for example, gas consumption y in these months would follow the normal distribution with mean μ_y and standard deviation σ. Second, the mean gas consumption is a straight-line function of x, given by the population regression line $\mu_y = \beta_0 + \beta_1 x$.

The value $x = 20$ is not present in the data collected. However, because this value is within the range of the data available, we may want to

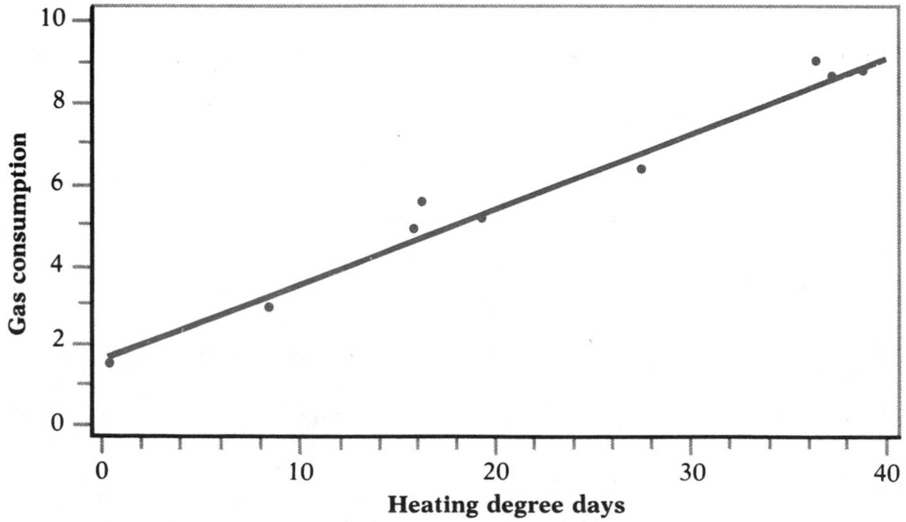

FIGURE 9.14 Data and regression line for the gas consumption example.

predict mean gas consumption for a month with 20 heating degree days per day. For this reason, our model assumes a linear relationship for the means over the range of values of degree days spanned by the data. In our example, it would be reasonable to make inferences for values of degree days from 0 to about 40. Negative values of degree days do not make sense, and for very large values we would face the dangers of extrapolation.

We begin our regression calculations by fitting the least-squares line. Fitting the line gives estimates b_1 and b_0 of the model parameters β_1 and β_0. Next we examine the residuals from the fitted line and obtain an estimate s of the remaining parameter σ. These calculations are preliminary to inference. Finally, we use s to obtain the standard errors needed for the various interval estimates and significance tests. Roundoff errors that accumulate during these calculations can ruin the final results. Be sure to carry many significant digits and check your work carefully.

Preliminary calculations After examining the scatterplot (Figure 9.14) to verify that the data show a straight-line pattern with no unusual points, we fit the least-squares line. The calculations are identical to those presented in Chapter 2 when we first met least-squares regression.

EXAMPLE 9.8

Table 9.1 contains the data x_i and y_i from Example 9.7 and the building block sums needed for the least-squares line. The slope and intercept of the least-squares line are found from the column sums in the usual way:

$$b_1 = \frac{\sum xy - \frac{1}{n}(\sum x)(\sum y)}{\sum x^2 - \frac{1}{n}(\sum x)^2}$$

$$= \frac{1375.00 - \frac{1}{9}(193.9)(50.3)}{5618.11 - \frac{1}{9}(193.9)^2} = .20221$$

and

$$b_0 = \bar{y} - b_1\bar{x}$$
$$= 5.589 - (.20221)(21.54) = 1.233$$

The equation of the least-squares regression line is therefore

$$\hat{y} = 1.233 + .20221x$$

This is the line shown in Figure 9.14. ●

For inference we will need two other quantities that it is convenient to obtain here. The mean and standard deviation of the x (degree days) observations are

TABLE 9.1 Basic regression calculations for the gas consumption example

Case	x_i	x_i^2	y_i	x_iy_i
1	15.6	243.36	5.2	81.12
2	26.8	718.24	6.1	163.48
3	37.8	1428.84	8.7	328.86
4	36.4	1324.96	8.5	309.40
5	35.5	1260.25	8.8	312.40
6	18.6	345.96	4.9	91.14
7	15.3	234.09	4.5	68.85
8	7.9	62.41	2.5	19.75
9	.0	.00	1.1	.00
Sum	193.9	5618.11	50.3	1375.00

$$\bar{x} = 21.54$$
$$s_x = 13.4194$$

(Your calculator should give s_x directly from keyed-in data. If not, see the computing formula 1.3 on page 48.) The quantities we need are $\bar{x}$ and the sum of squared deviations of the x observations from their mean, $\sum(x_i - \bar{x})^2$. From the definition of the sample variance,

$$s_x^2 = \frac{1}{n-1}\sum(x_i - \bar{x})^2$$

we can find the sum of squared deviations:

$$\begin{aligned}\sum(x_i - \bar{x})^2 &= (n-1)s_x^2 \\ &= (9-1)(13.4194)^2 = 1440.642\end{aligned}$$

To find the standard deviation s about the fitted line we first compute the residuals. There are simpler computational formulas for s but, as we learned in Chapter 2, a careful examination of the residuals is an essential step in a regression analysis. Table 9.2 organizes the calculations. The first three columns give the case and values of the explanatory and response variables, just as in Table 9.1. The next three columns give the predicted values, the residuals, and the squares of the residuals. For case 1, for example,

$$\begin{aligned}\hat{y}_1 &= b_0 + b_1x_1 \\ &= 1.233 + (.20221)(15.6) = 4.387\end{aligned}$$

The residual for the first case is

$$e_1 = y_1 - \hat{y}_1 = 5.2 - 4.387 = .813$$

TABLE 9.2 Residual calculations for the gas consumption example

Case	x_i	y_i	$\hat{y}_i$	e_i	e_i^2
1	15.6	5.2	4.387	.813	.661
2	26.8	6.1	6.652	−.552	.305
3	37.8	8.7	8.876	−.176	.031
4	36.4	8.5	8.593	−.093	.009
5	35.5	8.8	8.411	.389	.151
6	18.6	4.9	4.994	−.094	.009
7	15.3	4.5	4.327	.173	.030
8	7.9	2.5	2.830	−.330	.109
9	.0	1.1	1.233	−.133	.018
Sum	193.9	50.3	50.303	−.003	1.323

and the squared residual is

$$e_1^2 = (.813)^2 = .661$$

Examination of the residuals confirms the impressions formed from the scatterplot. The first case has the largest residual, but it is not particularly large relative to the other values given. The residuals are plotted versus case in Figure 9.15 and versus degree days in Figure 9.16. No suspicious patterns or unusual observations are evident.

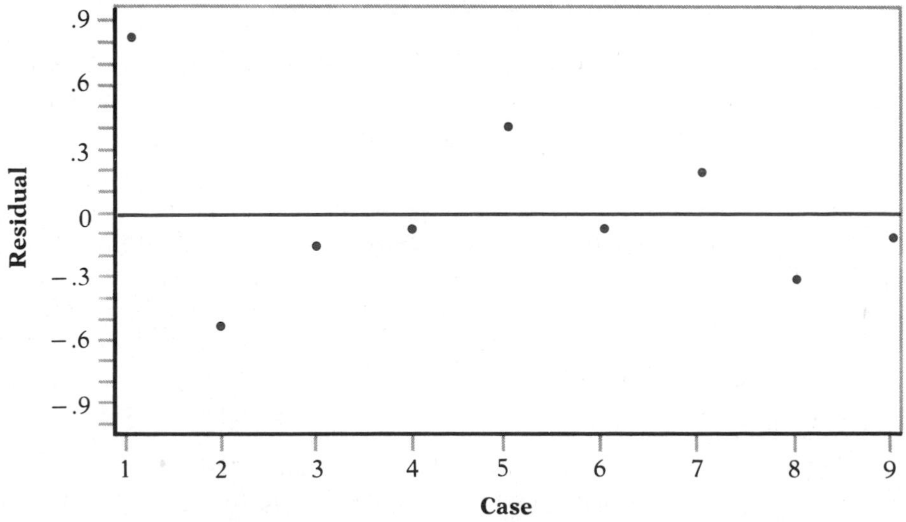

FIGURE 9.15 Plot of residuals versus case for the gas consumption example.

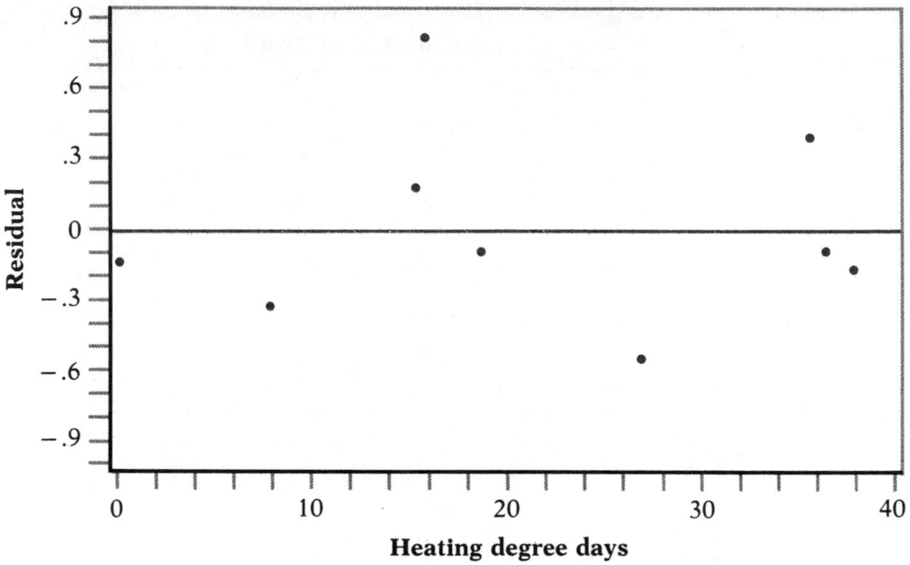

FIGURE 9.16 Plot of residuals versus the explanatory variable for the gas consumption example.

The sum of the residuals from a least-squares line is always 0, a fact that helps us check our calculations. The sum of the residuals in Table 9.2 is −0.003 because each residual was rounded to three places after the decimal point. If the sum were far from 0, we would need to compute $\hat{y}$ more accurately. Another useful check on the accuracy of our calculations is based on the fact that the fitted regression line goes through the point $(\bar{x}, \bar{y})$ on the scatterplot. So $b_0 + b_1\bar{x}$, the point on the line when $x = \bar{x}$, should equal $\bar{y}$ up to roundoff error. Try this for yourself in our example.

To find s^2, the sample estimate of the population σ^2, we obtain the sum of squares of the residuals

$$\sum e_i^2 = 1.323$$

from the bottom row of Table 9.2. Therefore,

$$s^2 = \frac{\sum e_i^2}{n-2} = \frac{1.323}{9-2} = .189$$

The estimate of σ is

$$s = \sqrt{.189} = .435$$

Calculating the least-squares line and s is tedious even with only nine observations. We must carry many digits to obtain accurate results, and it is wise to check that the sum of the residuals is close to 0. Software is both faster and more accurate, and it is essential for regression problems with many observations.

Inference for slope and intercept Confidence intervals and significance tests for the slope β_1 and intercept β_0 of the population regression line make use of the estimates b_1 and b_0 and their standard errors.

Some algebra using the rules for variances (Section 4.3) establishes that the standard deviation of b_1 is

$$\sigma_{b_1} = \frac{\sigma}{\sqrt{\sum(x_i - \bar{x})^2}}$$

Similarly, the standard deviation of b_0 is

$$\sigma_{b_0} = \sigma\sqrt{\frac{1}{n} + \frac{\bar{x}^2}{\sum(x_i - \bar{x})^2}}$$

To estimate these standard deviations, we need only replace σ by its estimate s.

STANDARD ERRORS FOR ESTIMATED REGRESSION COEFFICIENTS

The standard error of the slope b_1 of the least-squares regression line is

$$s_{b_1} = \frac{s}{\sqrt{\sum(x_i - \bar{x})^2}}$$

The standard error of the intercept b_0 is

$$s_{b_0} = s\sqrt{\frac{1}{n} + \frac{\bar{x}^2}{\sum(x_i - \bar{x})^2}}$$

EXAMPLE 9.9

The least-squares line for the regression of gas consumption on degree days has slope and intercept

$$b_1 = .20221$$
$$b_0 = 1.233$$

The standard deviation of the residuals about the fitted line and the sum of squared deviations of the x observations are

$$s = .435$$
$$\sum(x_i - \bar{x})^2 = 1440.642$$

From these preliminary results and the recipes for the standard errors of b_1 and b_0 we now obtain confidence intervals for β_1 and β_0 and test H_0: $\beta_1 = 0$ versus H_a: $\beta_1 > 0$. The one-sided alternative is appropriate because we expect gas consumption to be positively related to degree days. ●

The standard error of b_1 is

$$s_{b_1} = \frac{s}{\sqrt{\sum(x_i - \bar{x})^2}} = \frac{.435}{\sqrt{1440.642}} = .01146$$

To test

$$H_0: \beta_1 = 0$$
$$H_a: \beta_1 > 0$$

calculate the t statistic:

$$t = \frac{b_1}{s_{b_1}} = \frac{.20221}{.01146} = 17.64$$

The degrees of freedom for s are $n - 2 = 7$, so the value of t is compared with the critical values for the $t(7)$ distribution in Table E. The largest entry for 7 degrees of freedom is 5.408, corresponding to $p = 0.0005$. Because $t = 17.64$ is greater than 5.408, we have strong evidence ($P < 0.0005$) that the slope β_1 is positive.

For a confidence interval for β_1 we again need a critical value from the $t(7)$ distribution. For 95% confidence, Table E gives $t^* = 2.365$. The 95% confidence interval for β_1 is

$$\begin{aligned} b_1 \pm t^* s_{b_1} &= .20221 \pm (2.365)(.01146) \\ &= .20221 \pm .02710 \\ &= (.17511, \ .22931) \end{aligned}$$

We can be 95% confident that each additional degree day results in an increased consumption of between 0.175 and 0.229 hundreds of cubic feet of gas per day.

In this example, the intercept β_0 also has a meaningful interpretation. Because β_0 is the mean of y when $x = 0$, β_0 represents average gas consumption for cooking, hot water, and other uses that are present when there is no demand for heating. To construct a confidence interval for β_0, first calculate the standard error

$$s_{b_0} = s\sqrt{\frac{1}{n} + \frac{\bar{x}^2}{\sum(x_i - \bar{x})^2}} = .435\sqrt{\frac{1}{9} + \frac{(21.54)^2}{1440.642}} = .2863$$

The 95% confidence interval for β_0 is then

$$\begin{aligned} b_0 \pm t^* s_{b_0} &= 1.233 \pm (2.36)(.2863) \\ &= 1.233 \pm .6757 \end{aligned}$$

We estimate that average gas consumption for purposes other than heating is between 0.56 and 1.91 hundreds of cubic feet of gas per day.

Confidence intervals for the mean response and prediction intervals for a future observation When we substitute a particular value x^* of the explanatory variable into the regression equation and obtain a value of $\hat{y}$, we can view the result in two ways. We have estimated the mean response μ_y, and we have also predicted a future value of the response y. The margins of error for these two uses are quite different. Prediction intervals for an individual response are wider than confidence intervals for estimating a mean response. We now proceed with the details of these calculations. Once again, standard errors are the essential quantities.

STANDARD ERRORS FOR $\hat{\mu}$ AND $\hat{y}$

The standard error of $\hat{\mu}$ is

$$s_{\hat{\mu}} = s\sqrt{\frac{1}{n} + \frac{(x^* - \bar{x})^2}{\sum(x_i - \bar{x})^2}}$$

The standard error for predicting an individual response is[4]

$$s_{\hat{y}} = s\sqrt{1 + \frac{1}{n} + \frac{(x^* - \bar{x})^2}{\sum(x_i - \bar{x})^2}}$$

Note that the only difference between the recipes for these two standard errors is the extra 1 under the square root sign in the standard error for prediction. This standard error is larger due to the additional variation of individual responses about the mean response. It produces prediction intervals that are wider than the confidence intervals for the mean response.

EXAMPLE 9.10

We will give a 95% confidence interval for the mean gas consumption in months with an average of 20 degree days per day. The estimate of this mean from the least-squares line is

$$\begin{aligned} \hat{\mu}_y &= b_0 + b_1 x \\ &= 1.233 + (.20221)(20) = 5.277 \end{aligned}$$

The standard error of the estimate $\hat{\mu}$ is

$$\begin{aligned} s_{\hat{\mu}} &= s\sqrt{\frac{1}{n} + \frac{(x^* - \bar{x})^2}{\sum(x_i - \bar{x})^2}} \\ &= .435\sqrt{\frac{1}{9} + \frac{(20 - 21.54)^2}{1440.642}} \\ &= .435\sqrt{\frac{1}{9} + \frac{2.372}{1440.642}} = .146 \end{aligned}$$

The numbers in this computation are taken from the previous calculations performed for the gas consumption example.

Finally, we need a value for t^*. For a 95% confidence interval and 7 degrees of freedom, Table E gives $t^* = 2.365$. The 95% confidence interval is

$$\begin{aligned} \hat{\mu}_y \pm t^* s_{\hat{\mu}} &= 5.277 \pm (2.365)(.146) \\ &= 5.277 \pm .345 \\ &= (4.932,\ 5.622) \end{aligned}$$

For months with 20 degree days per day, we estimate the mean consumption to be between 4.93 and 5.62 hundreds of cubic feet of gas per day, with 95% confidence. ●

The calculations for prediction intervals are very similar to those in Example 9.10. The major change is that we use a different expression for the standard error.

EXAMPLE 9.11

If next March averages 20 degree days per day, how much natural gas will this household consume? We want a 95% prediction interval for the gas consumption during a future month with 20 degree days. For $x^* = 20$, the predicted value of y is

$$\begin{aligned} \hat{y} &= b_0 + b_1 x^* \\ &= 1.233 + (.20221)(20) = 5.277 \end{aligned}$$

This is the same value that we found for $\hat{\mu}_y$ in the previous example. The computation of the standard error is also very similar—but don't forget the additional 1 in the recipe:

$$\begin{aligned} s_{\hat{y}} &= s\sqrt{1 + \frac{1}{n} + \frac{(x^* - \bar{x})^2}{\sum(x_i - \bar{x})^2}} \\ &= .435\sqrt{1 + \frac{1}{9} + \frac{(20 - 21.54)^2}{1440.646}} \\ &= .435\sqrt{1 + \frac{1}{9} + \frac{2.372}{1440.646}} = .459 \end{aligned}$$

The standard error $s_{\hat{y}}$ for prediction is considerably larger than the standard error $s_{\hat{\mu}}$ for estimating the mean gas consumption for the same number of degree days. The prediction interval is correspondingly wider than the confidence interval of Example 9.10.

The 95% prediction interval is

$$\begin{aligned}\hat{y} \pm t^* s_{\hat{y}} &= 5.277 \pm (2.36)(.459)\\ &= 5.277 \pm 1.083\\ &= (4.194,\ 6.360)\end{aligned}$$

If next March averages 20 degree days per day, we predict with 95% confidence that this household will consume between 4.19 and 6.36 hundreds of cubic feet of gas per day. ●

Inference for correlation*

The correlation coefficient is a measure of the strength and direction of the linear association between two variables. Correlation does not require an explanatory-response relationship between the variables. We can consider the sample correlation r as an estimate of the correlation in the population and base inference about the population correlation on r.

population correlation

The correlation between the variables x and y when they are measured for every member of a population is the *population correlation*. As usual, we use Greek letters to represent population parameters. In this case ρ, the Greek letter rho, is the population correlation. When $\rho = 0$, there is no linear association in the population. In the important case where the two variables x and y are both normally distributed, the condition $\rho = 0$ is equivalent to the statement that x and y are independent. That is, there is no association of any kind between x and y. (Technically, the condition required is that x and y be *jointly normal*. This means that the distribution of x is normal and also that the conditional distribution of y given any fixed value of x is normal.) We therefore may wish to test the null hypothesis that a population correlation is 0.

jointly normal variables

TEST FOR A ZERO POPULATION CORRELATION

To test the hypothesis H_0: $\rho = 0$, compute the t statistic:

$$t = \frac{r\sqrt{n-2}}{\sqrt{1-r^2}}$$

where n is the sample size and r is the sample correlation.

In terms of a random variable T having the $t(n-2)$ distribution, the P-value for a test of H_0 against

$$\begin{aligned} H_a&: \rho > 0 \quad \text{is} \quad P(T \geq t)\\ H_a&: \rho < 0 \quad \text{is} \quad P(T \leq t)\\ H_a&: \rho \neq 0 \quad \text{is} \quad 2P(T \geq |t|)\end{aligned}$$

*This material is optional and can be omitted without loss of continuity.

```
PEARSON CORRELATION COEFFICIENTS / PROB > |R| UNDER HO:RHO=0 / N = 92

                    LSKIN         DEN

LSKIN             1.00000    -0.85157
                   0.0000      0.0001

DEN              -0.85157     1.00000
                   0.0001      0.0000
```

FIGURE 9.17 Correlation output for the body density example.

We illustrate the calculation with our two examples. Most computer packages have routines for calculating and testing correlations.

EXAMPLE 9.12

For the body density and skinfold measures problem, the output of the SAS procedure CORR appears in Figure 9.17. The sample correlation between body density and log of skinfolds is $r = -0.85157$ and is called a Pearson correlation coefficient in this output. The P-value for a two-sided test of H_0: $\rho = 0$ is given below the correlation; it is 0.0001. To test the one-sided alternative that the population correlation is negative, we divide the P-value in the output by 2, after checking that the sample coefficient is in fact negative. We conclude that there is a negative correlation between body density and log of skinfolds. ●

EXAMPLE 9.13

For the gas consumption problem we test

$$H_0\colon \rho = 0$$
$$H_a\colon \rho > 0$$

We expect gas consumption to be positively correlated with degree days, so the one-sided alternative is appropriate.

Formulas for calculating the sample correlation appear in Section 2.4. If we use the computational form given in Equation 2.3 (page 164), we can compute the correlation from the building block sums in Table 9.1 and the standard deviations of the x and y observations:

$$r = \frac{\sum x_i y_i - \frac{1}{n}(\sum x_i)(\sum y_i)}{(n-1)s_x s_y}$$
$$= \frac{1375.00 - \frac{1}{9}(193.9)(50.3)}{(9-1)(13.42)(2.744)} = .989$$

Because the sample size is $n = 9$, the t statistic is

$$t = \frac{.989\sqrt{9-2}}{\sqrt{1-.989^2}} = \frac{2.6166}{.1479} = 17.69$$

Compare $t = 17.69$ with the critical values of the $t(7)$ distribution given in Table E. The critical value for 7 degrees of freedom and $p = 0.0005$ is 5.408. We therefore have strong evidence that gas consumption is positively correlated with degree days. ●

When the variables x and y can be viewed as explanatory and response variables, there is a close connection between correlation and regression. In Section 2.4, we found that the sample correlation and the slope of the least-squares regression line are related by the equation

$$b_1 = r\frac{s_y}{s_x}$$

From this fact we see that if the slope is 0, so is the correlation, and vice versa. It should come as no surprise to learn that the procedures for testing H_0: $\beta_1 = 0$ and H_0: $\rho = 0$ are also closely related. In fact, the t statistics for testing these hypotheses are numerically equal. That is,

$$\frac{b_1}{s_{b_1}} = \frac{r\sqrt{n-2}}{\sqrt{1-r^2}}$$

Check that this holds in both of our examples.

In our examples, the conclusion that there is a significant correlation between the two variables would not come as a surprise to anyone familiar with the meaning of these variables. The significance test simply tells us whether or not there is evidence in the data to conclude that the population correlation is different from 0. The actual size of the correlation is of considerably more interest. We would therefore like to give a confidence interval for the population correlation. Unfortunately, most software packages do not perform this calculation. Because hand calculation of the confidence interval is very tedious, we do not give the method here.[5]

In Section 2.4 we also noted that r^2 is the fraction of variation in the values of y that is explained by the least-squares regression of y on x. The ANOVA table makes this interpretation precise. Recall that SST = SSM + SSE. It is an algebraic fact that

$$r^2 = \frac{\text{SSM}}{\text{SST}} = \frac{\sum(\hat{y}_i - \bar{y})^2}{\sum(y_i - \bar{y})^2}$$

Because SST is the total variation in y and SSM is the variation due to the regression of y on x, this equation is the precise statement of the fact that r^2 is the fraction of variation explained by the regression.

SUMMARY

The statistical model for **simple linear regression** is

$$y_i = \beta_0 + \beta_1 x_i + \epsilon_i$$

where $i = 1, 2, \ldots, n$. The ϵ_i are assumed to be independent and normally distributed with mean 0 and standard deviation σ. The **parameters** of the model are β_0, β_1, and σ. The intercept and slope β_0 and β_1 are estimated by the intercept and slope of the **least-squares regression line,** b_0 and b_1. The parameter σ is estimated by

$$s = \sqrt{\frac{\sum e_i^2}{n-2}}$$

where the e_i are the **residuals**

$$e_i = y_i - \hat{y}_i$$

The **standard errors** for b_0 and b_1 are

$$s_{b_0} = s\sqrt{\frac{1}{n} + \frac{\bar{x}^2}{\sum(x_i - \bar{x})^2}}$$

$$s_{b_1} = \frac{s}{\sqrt{\sum(x_i - \bar{x})^2}}$$

A level C **confidence interval for** β_1 is

$$b_1 \pm t^* s_{b_1}$$

where t^* is the upper $(1 - C)/2$ critical value for the $t(n-2)$ distribution.

The **test of the hypothesis** H_0: $\beta_1 = 0$ is based on the statistic

$$t = \frac{b_1}{s_{b_1}}$$

and the $t(n-2)$ distribution. There are similar formulas for confidence intervals and tests for β_0, but these are meaningful only in special cases.

The **estimated mean response** for the subpopulation corresponding to the value x^* of the explanatory variable is

$$\hat{\mu}_y = b_0 + b_1 x^*$$

A level C **confidence interval for the mean response** is

$$\hat{\mu}_y \pm t^* s_{\hat{\mu}}$$

where

$$s_{\hat{\mu}} = s\sqrt{\frac{1}{n} + \frac{(x^* - \bar{x})^2}{\sum(x_i - \bar{x})^2}}$$

is the standard error and t^* is the upper $(1 - C)/2$ critical value for the $t(n - 2)$ distribution.

The **estimated value of the response variable** y for a future observation from the subpopulation corresponding to the value x^* of the explanatory variable is

$$\hat{y} = b_0 + b_1 x^*$$

A level C **prediction interval** for the estimated response is

$$\hat{y} \pm t^* s_{\hat{y}}$$

where

$$s_{\hat{y}} = s\sqrt{1 + \frac{1}{n} + \frac{(x^* - \bar{x})^2}{\sum(x_i - \bar{x})^2}}$$

and t^* is the upper $(1 - C)/2$ critical value for the $t(n - 2)$ distribution.

The **ANOVA table** for a linear regression gives the degrees of freedom, sum of squares, and mean squares for the model, error, and total sources of variation.

The **ANOVA *F* statistic** is the ratio MSM/MSE. Under H_0: $\beta_1 = 0$, this statistic has an $F(1, n-2)$ distribution and is used to test H_0 versus the two-sided alternative.

When the variables y and x are jointly normal, the sample correlation is an estimate of the population correlation ρ. The test of H_0: $\rho = 0$ is based on the statistic

$$t = \frac{r\sqrt{n - 2}}{\sqrt{1 - r^2}}$$

which has a $t(n-2)$ distribution under H_0. This test statistic is numerically identical to the t statistic used to test H_0: $\beta_1 = 0$.

The **square of the sample correlation** can be expressed as

$$r^2 = \frac{\text{SSM}}{\text{SST}}$$

and is interpreted as the proportion of the variability in the response variable y that is explained by the explanatory variable x in the linear regression.

SECTION 9.1 EXERCISES

9.1 Manatees are large sea creatures that live in the shallow water along the coast of Florida. Many manatees are injured or killed each year by power boats. Here are data on manatees killed and powerboat registrations (in thousands of boats) in Florida for the period 1977 to 1990:

Year	Powerboat registrations	Manatees killed
1977	447	13
1978	460	21
1979	481	24
1980	498	16
1981	513	24
1982	512	20
1983	526	15
1984	559	34
1985	585	33
1986	614	33
1987	645	39
1988	675	43
1989	711	50
1990	719	47

Here is the output from the Minitab regression command for these data, with boat registrations ("Boats" in the output) as the explanatory variable and manatees killed ("Killed") as the response variable:

```
The regression equation is
Killed = - 41.4 + 0.125 Boats

Predictor        Coef        Stdev     t-ratio        p
Constant      -41.430        7.412       -5.59    0.000
Boats         0.12486      0.01290        9.68    0.000

s = 4.276        R-sq = 88.6%     R-sq(adj) = 87.7%

Analysis of Variance

SOURCE          DF         SS          MS         F         p
Regression       1     1712.0      1712.0     93.61     0.000
Error           12      219.4        18.3
Total           13     1931.4
```

```
Unusual Observations
Obs.  Boats  Killed    Fit  Stdev.Fit  Residual  St.Resid
  7     526   15.00  24.25       1.26     -9.25    -2.26R

R denotes an obs. with a large st. resid.

     Fit  Stdev.Fit         95% C.I.          95% P.I.
   45.97       2.06  (  41.49,  50.46)  (  35.63,  56.31)
```

(a) Make a scatterplot of boats registered and manatees killed. Is there a strong straight-line pattern? What is r^2 for these data? Minitab calls attention to one observation that has a moderately large residual. Circle that observation on your plot. We have no reason to remove this observation, so we retain it in our analysis.

(b) What is the equation of the least-squares regression line? Draw this line on your scatterplot. Is there strong evidence that the mean number of manatees killed increases as the number of powerboats increases? State this question as null and alternative hypotheses about the slope of the population regression line, obtain the t statistic, and give your conclusion.

9.2 The previous exercise gives data and Minitab output for the regression of manatees killed on powerboat registrations in Florida.

(a) In recent years Florida has established Manatee refuge zones with reduced speed limits and other restrictions on powerboats. Circle the points for the years 1988 to 1990 on your scatterplot of the data. Do the data suggest that the state's actions have reduced the numbers of manatees killed in these years?

(b) Suppose that Florida were to restrict the number of powerboats to 700,000. How many manatees per year would be killed on the average if only 700,000 powerboats were allowed? Give both a point prediction and a suitable 95% confidence or prediction interval. (The last line of the output, produced by the Minitab subcommand PREDICT 700, contains the information you need.)

9.3 The Leaning Tower of Pisa is an architectural wonder. Engineers concerned about the tower's stability have done extensive studies of its increasing tilt. Measurements of the lean of the tower over time provide much useful information. The following table gives measurements for the years 1975 to 1987. The variable "lean" represents the difference between where a point on the tower would be if the tower were straight and where it actually is. The data are coded as tenths of a millimeter in excess of 2.9 meters, so that the 1975 lean, which was 2.9642 meters, appears in the table as 642. Only the last 2 digits of the year were entered into the computer. (Data from G. Geri and B. Palla, "Considerazioni sulle più recenti osservazioni ottiche alla Torre Pendente di Pisa," *Estratto dal Bollettino della Società Italiana di Topografia e Fotogrammetria*, 2 (1988), pp. 121–135. Professor Julia Mortera of the University of Rome provided valuable assistance with the translation.)

Year	75	76	77	78	79	80	81	82	83	84	85	86	87
Lean	642	644	656	667	673	688	696	698	713	717	725	742	757

The following output was produced by the SAS regression procedure with year as the explanatory variable and lean as the response variable:

```
                      Sum of              Mean
Source     DF        Squares            Square   F Value   Prob>F

Model       1    15804.48352   15804.48352   904.120   0.0001
Error      11      192.28571      17.48052
C Total    12    15996.76923

Root MSE          4.18097      R-square      0.9880
Dep Mean        693.69231      Adj R-sq      0.9869
C.V.              0.60271

                        Parameter Estimates

              Parameter       Standard    T for H0:
Variable       Estimate          Error  Parameter=0   Prob>|T|

INTERCEP   -61.120879    25.12981850       -2.432     0.0333
YEAR         9.318681     0.30991420       30.069     0.0001
```

(a) Plot the data. Does the trend in lean over time appear to be linear?
(b) What is the equation of the least-squares line? What percentage of the variation in lean is explained by this line?
(c) Give a 95% confidence interval for the average rate of change (tenths of a millimeter per year) of the lean.

9.4 The previous exercise gives regression output for the lean of the Leaning Tower of Pisa.
(a) In 1918 the lean was 2.9071 meters. (The coded value is 71.) Using the least-squares equation for the years 1975 to 1987, calculate a predicted value for the lean in 1918. (Note that you must use the coded value 18 for year.)
(b) Although the least-squares line gives an excellent fit to the data for 1975 to 1987, this pattern did not extend back to 1918. Write a short statement explaining why this conclusion follows from the

information available. Use numerical and graphical summaries to support your explanation.

9.5 Refer once more to the data and regression output in Exercise 9.3.

(a) The engineers studying the Leaning Tower of Pisa are most interested in what will happen to the tower in the future. Use the least-squares equation to predict the tower's lean in the year 1997.

(b) To give a margin of error for the lean in 1997, would you use a confidence interval for a mean response or a prediction interval? Explain your choice.

9.6 Are ticket prices for professional basketball games related to attendance at the games? One way to address this question is to look at data on ticket prices and attendance for the 27 teams in the National Basketball Association. Here is the SAS output for the regression of average ticket price (in dollars) on average paid attendance for the 1989–1990 NBA season. (Data from an article in the *Lafayette Journal and Courier*, June 26, 1990.)

```
                          Analysis of Variance

                         Sum of            Mean
Source     DF           Squares          Square   F Value   Prob>F

Model       1       27899.12992     27899.12992     0.002   0.9637
Error      25       330803772.5      13232150.9
C Total    26      330831671.63

Root MSE      3637.60236     R-square      0.0001
Dep Mean     15710.70370     Adj R-sq     -0.0399
C.V.            23.15366

                          Parameter Estimates

             Parameter         Standard    T for H0:
Variable      Estimate            Error   Parameter=0   Prob>|T|

INTERCEP         15553     3495.7571607         4.449     0.0002
PRICE         8.470316     184.46733775         0.046     0.9637
```

(a) What percentage of the variation in attendance is explained by ticket price?

(b) Is linear regression on price of any value in explaining attendance? State hypotheses, report the test statistic and its *P*-value from the output, and state your conclusion.

(c) The regression output suggests that ticket prices do not influence attendance. Explain carefully why it is not sound statistical practice to base such a conclusion on the computer analysis alone. What other analysis should be done?

9.7 Table 9.3 gives the raw data for the analysis in the previous exercise.

(a) Plot the data. In recent years, four new teams were added to the NBA. They are Charlotte, Miami, Minnesota, and Orlando. Circle the points corresponding to these teams on your plot, and write the name of the team next to each of the circled points.

(b) Are any of the expansion teams outliers or influential points? Which ones?

(c) Redo the regression analysis omitting the data for Charlotte and Minnesota. How do the results of this analysis differ from those in the computer output of Exercise 9.6?

(d) The analysis with all 27 teams suggests that there is no linear relation

TABLE 9.3 NBA attendance and ticket prices

Team	Attendance	Price
Atlanta	13,993	20.06
Boston	14,916	22.54
Charlotte	23,901	17.00
Chicago	18,404	21.98
Cleveland	16,969	19.63
Dallas	16,868	17.05
Denver	12,668	17.40
Detroit	21,454	24.42
Golden State	15,025	17.04
Houston	15,846	17.56
Indiana	12,885	13.77
LA Clippers	11,869	21.95
LA Lakers	17,378	29.18
Miami	15,008	17.60
Milwaukee	16,088	14.08
Minnesota	26,160	10.92
New Jersey	12,160	13.31
New York	17,815	22.70
Orlando	15,606	20.47
Philadelphia	14,017	19.04
Phoenix	14,114	16.59
Portland	12,884	22.19
Sacramento	17,014	16.96
San Antonio	14,722	16.79
Seattle	12,244	18.11
Utah	12,616	18.41
Washington	11,565	14.55

between price and attendance, while the analysis with Charlotte and Minnesota excluded suggests that higher prices are associated with larger attendance. Suppose that you are the general manager of an NBA team. You are considering raising the price of your tickets. On the basis of the analyses in this and the previous exercise can you conclude that there will be no effect on the attendance if you increase the price of your tickets? Can you conclude that attendance will increase? Suggest some lurking variables that might influence the average attendance at NBA basketball games for different teams.

9.8 Example 9.7 (page 659) demonstrates that there is a strong linear relationship between household consumption of natural gas and outdoor temperature, measured by heating degree days. The slope and intercept depend on the particular house and on the habits of the household living there. Data for two heating seasons (18 months) for another household produce the least-squares line $\hat{y} = 2.405 + 0.26896x$ for predicting average daily gas consumption y from average degree days per day x. The standard error of the slope is $s_{b_1} = 0.00815$.

(a) Explain briefly what the slope β_1 of the population regression line represents. Then give a 95% confidence interval for β_1.

(b) This interval is based on twice as many observations as the one calculated in Example 9.9 for the household of Example 9.7, and the two standard errors are of similar size. How would you expect the margins of error of the two intervals to be related? Check your answer by comparing the two margins of error.

9.9 The standard error of the intercept in the regression of gas consumption on degree days for the household in Exercise 9.8 is $s_{b_0} = 0.20351$.

(a) Explain briefly what the intercept represents in this setting. Find a 95% confidence interval for the intercept.

(b) Compare the width of your interval with the one calculated for a different household in Example 9.9. Explain why it is shorter.

9.10 Exercise 9.8 gives information about the regression of natural gas consumption on degree days for a particular household.

(a) Calculate the t statistic for testing H_0: $\beta_1 = 0$.

(b) For the alternative H_a: $\beta_1 > 0$, what critical value would you use for a test at the $\alpha = 0.05$ significance level? Do you reject H_0 at this level?

(c) How would you report the P-value for this test?

9.11 Can a pretest on mathematics skills predict success in a statistics course? The 55 students in an introductory statistics class took a pretest at the beginning of the semester. The least-squares regression line for predicting the score y on the final exam from the pretest score x was $\hat{y} = 10.5 + 0.82x$. The standard error of b_1 was 0.38. Test the null hypothesis that there is no linear relationship between the pretest and the score on the final exam against the two-sided alternative.

9.12 Use the preliminary calculations given in the text for the gas consumption example and the procedure illustrated in Example 9.10 (page 667) to answer the following questions:

(a) Find a 99% confidence interval for the mean gas consumption for months with 30 degree days per day.

(b) Repeat the calculation for months with 80 degree days per day.

(c) Which interval has the larger margin of error? Explain why.

(d) Do either of these estimates involve extrapolation? Explain your answer.

9.13 Use the preliminary calculations given in the text for the regression of natural gas consumption on degree days to obtain both a 90% confidence interval for the mean gas consumption in all future months with 31.54 degree days per day and a 90% prediction interval for the gas consumption in an individual future month with 31.54 degree days per day. Compare the margins of error of these intervals. Would 95% intervals be longer or shorter?

9.14 Example 9.11 (page 668) illustrates a prediction interval for the gas consumption example. Use the method of that example and the preliminary calculations in the text to answer the following questions:

(a) Find a 99% prediction interval for gas consumption in a future month having 30 degree days per day.

(b) Repeat the calculation for a month with 80 degree days per day.

(c) Compare the margins of error of the intervals found in (a) and (b) with those of the confidence intervals in (a) and (b) of Exercise 9.12. Explain the results of these comparisons.

9.15 Economists know that the consumption of a product increases when the price drops. In a study of this relationship, researchers looked at the price and consumption of textiles in the Netherlands in the 17 years from 1923 to 1939. During this period, the price of textiles fell and consumption rose. The explanatory variable x is the price of textiles, adjusted for changes in the overall cost of living, and the response variable y is per person consumption of textile goods. Both x and y are reported as index numbers with 1925 = 100. That is, they are given as percents of their values in the year 1925. A least-squares regression routine gives the following output:

```
                Coeff          Std Err          t Value
Intercept     235.4897        9.079104          25.93755
Price          -1.323306      0.1163298        -11.37547
Residual Standard Error = 7.84818
R-Square = 0.896123   N = 17
```

"Residual Standard Error" is the standard deviation s about the fitted line. A scatterplot of the data shows a negative linear association, with no outliers or influential observations. (From a larger set of data in Henry Theil, *Principles of Econometrics*, Wiley, New York, 1971, p. 102.)

(a) Explain in plain language what the slope β_1 of the true regression line means. This is an important economic quantity. At your request, the computer tells you that the mean and variance of the explanatory variable are

$$\bar{x} = 76.3118 \quad \text{and} \quad s_x^2 = 284.47$$

Give a 90% confidence interval for β_1. (Hint: Remember that $\sum(x - \bar{x})^2 = (n-1)s_x^2$.)

(b) Explain why the intercept β_0 of the true regression line is of no interest in this example.

(c) In the year 1925, $x = 100$ and $y = 100$ because of the way the two variables are defined. Predict the mean consumption in years when the price of textiles is the same as the 1925 price. Then give a 90% confidence interval for this mean.

(d) Suppose you are told that the 1940 price of textiles was 75. Give a 90% prediction interval for textile consumption based on the regression line fitted to the 1923 to 1939 data. (In fact this prediction is very inaccurate. In 1940 Germany invaded and conquered the Netherlands. Extrapolation by even one year is risky in economics because outside events can change economic conditions quickly.)

9.16 Exercise 2.28 (page 140) gives the following data from a study of two methods for measuring the blood flow in the stomachs of dogs:

Spheres	4.0	4.7	6.3	8.2	12.0	15.9	17.4	18.1	20.2	23.9
Vein	3.3	8.3	4.5	9.3	10.7	16.4	15.4	17.6	21.0	21.7

"Spheres" is an experimental method that the researchers hope will predict "Vein," the standard but difficult method. Examination of the data gives no reason to doubt the validity of the simple linear regression model. The estimated regression line is $\hat{y} = 1.031 + 0.902x$, where y is the response variable vein and x is the explanatory variable spheres. The estimate of σ is $s = 1.757$.

(a) Find $\bar{x}$ and $\sum(x_i - \bar{x})^2$ from the data.

(b) We expect x and y to be positively associated. State hypotheses in terms of the slope of the population regression line that express this expectation, and carry out a significance test. What conclusion do you draw?

(c) Find a 99% confidence interval for the slope.

(d) Suppose that we observe a value of spheres equal to 15.0 for one dog. Give a 90% interval for predicting the variable vein for that dog.

9.17 Exercise 2.32 (page 142) describes data on air pollution measurements in two locations, rural and city. Readings for both locations are available for 26 cases. Examination of these data suggests that the simple linear regression model fits. The regression equation for predicting the city reading from the rural one is $\hat{y} = -2.580 + 1.0935x$, and the estimated standard deviation about the line is $s = 4.4792$.

(a) Calculate $\bar{x}$ and $\sum(x_i - \bar{x})^2$ for the 26 cases.

(b) State appropriate null and alternative hypotheses for assessing whether or not there is a linear relationship between the city and rural readings. Give the test statistic and report the P-value for testing your null hypothesis. Summarize your conclusion.

(c) The rural reading is 43 for a period during which the city equipment is out of service. Give a 95% interval for the missing city reading.

9.18 Ohm's law $I = V/R$ states that the current I in a metal wire is proportional to the voltage V applied to its ends and is inversely proportional to the resistance R in the wire. Students in a physics lab performed experiments to study Ohm's law. They varied the voltage and measured the current at each voltage with an ammeter. The goal was to determine the resistance R of the wire. We can rewrite Ohm's law in the form of a linear regression as $I = \beta_0 + \beta_1 V$, where $\beta_0 = 0$ and $\beta_1 = 1/R$. Because voltage is set by the experimenter, we think of V as the explanatory variable. The current I is the response. Here are the data for one experiment. (Data provided by Sara McCabe.)

V	.5	1.0	1.5	1.8	2.0
I	.52	1.19	1.62	2.00	2.4

(a) Plot the data. Are there any outliers or unusual points?

(b) Find the least-squares fit to the data, and estimate $1/R$ for this wire. Then give a 95% confidence interval for $1/R$.

(c) If b_1 estimates $1/R$, then $1/b_1$ estimates R. Estimate the resistance R. Similarly, if L and U represent the lower and upper confidence limits for $1/R$, then the corresponding limits for R are given by $1/U$ and $1/L$, as long as L and U are positive. Use this fact and your answer to (b) to find a 95% confidence interval for R.

(d) Ohm's law states that β_0 in the model is 0. Calculate the test statistic for this hypothesis and give an approximate P-value.

9.19 The data on another wire for the Ohm's law experiment described in the previous exercise are as follows:

V	.5	1.0	2.0	3.0	4.0
I	.12	.31	.69	1.10	1.30

Answer the questions in the previous exercise for this wire.

9.20 Most statistical software systems have an option for doing regressions in which the intercept is set in advance to 0. In SAS this is the NOINT option, and in Minitab it is the NOCONSTANT subcommand. If you have access to such software, reanalyze the Ohm's law data given in Exercise 9.18 with this option and report the estimate of R. The output should also include an estimated standard error for $1/R$. Use this to calculate the 95% confidence interval for R. Note: With this option the degrees of freedom for t^* will be 1 greater than for the model with the intercept.

9.21 Answer the questions posed in the previous exercise for the Ohm's law data given in Exercise 9.19.

9.22 The human body takes in more oxygen when exercising than when it is at rest. To deliver the oxygen to the muscles, the heart must beat faster. Heart rate is easy to measure, but measuring oxygen uptake requires elaborate equipment. If oxygen uptake (VO2) can be accurately predicted from heart rate (HR), the predicted values can replace actually measured values for various research purposes. Unfortunately, not all human bodies are the same, so no single prediction equation works for all people. Researchers can, however, measure both HR and VO2 for one person under varying sets of exercise conditions and calculate a regression equation for predicting that person's oxygen uptake from heart rate. They can then use predicted oxygen uptakes in place of measured uptakes for this individual in later experiments. Here are data for one individual. (Data provided by Paul Waldsmith from experiments conducted in Don Corrigan's laboratory at Purdue University.)

HR	94	96	95	95	94	95	94	104	104	106
VO2	.473	.753	.929	.939	.832	.983	1.049	1.178	1.176	1.292

HR	108	110	113	113	118	115	121	127	131
VO2	1.403	1.499	1.529	1.599	1.749	1.746	1.897	2.040	2.231

(a) Plot the data. Are there any outliers or unusual points?
(b) Compute the least-squares regression line for predicting oxygen uptake from heart rate for this individual.
(c) Test the null hypothesis that the slope of the regression line is 0. Explain in words the meaning of your conclusion from this test.
(d) Calculate a 95% interval for the oxygen uptake of this individual on a future occasion when his heart rate is 95. Repeat the calculation for heart rate 110.
(e) From what you have learned in (a), (b), (c), and (d) of this exercise, do you think that the researchers should use predicted VO2 in place of measured VO2 for this individual under similar experimental conditions? Explain your answer.

9.23 The data in the previous exercise were collected as part of a study designed to assess the effects of wearing different protective masks on worker productivity. When those data were obtained, the individual was wearing a half-mask, which may have affected oxygen intake. Here are data for the same person wearing no mask:

HR	103	112	113	110	111	112	115	120	118	126
VO2	.768	.953	1.090	1.167	1.025	1.130	1.173	1.296	1.358	1.507

HR	123	128	132	136	137	138	140	145	147
VO2	1.565	1.740	1.807	1.891	1.958	2.023	2.148	2.395	2.507

Analyze these data, using the questions in the previous exercise as a guide.

9.24 The data given in Exercise 9.22 were taken while the individual performed progressively more vigorous exercise. The observations are therefore in time order. In fact, the observations correspond to consecutive 1-minute periods. The statistical model for inference in linear regression assumes that the errors are independent and all have the same variance. Plot the residuals versus the order in which the data were taken. Do you see any patterns that would lead you to question the validity of the regression methodology that we used in Exercise 9.22?

9.25 Answer the questions of the previous exercise for the data on the individual with no mask given in Exercise 9.23.

9.26 Premature infants are often kept in intensive care nurseries after they are born. It is common practice to measure their blood pressure frequently. The oscillometric method of measuring blood pressure is noninvasive and easy to use. The traditional procedure, called the direct intra-arterial method, is believed to be more accurate but is invasive and more difficult to perform. Several studies have reported high correlations between measurements made by the two methods, ranging from $r = 0.49$ to $r = 0.98$. These correlations are statistically significant. One study that investigated the relation between the two methods reported the regression equation $\hat{y} = 15 + 0.83x$. Here x represents the easy method and y represents the difficult one. (From John A. Wareham et al., "Prediction of arterial blood pressure on the premature neonate using the oscillometric method," *American Journal of Diseases of Children*, 141 (1987), pp. 1108–1110.)

(a) The standard error of the slope is 0.065 and the sample size is 81. Calculate the t statistic for testing H_0: $\beta_1 = 0$. Specify an appropriate alternative hypothesis for this problem, and give an approximate P-value for the test. Then explain your conclusion in words a physician can understand.

(b) Give a 99% prediction interval for y for an infant with blood pressure 130 as measured by the oscillometric method. (The authors of the study calculated and plotted prediction intervals. They found the widths to be unacceptably large and concluded that statistical significance does not imply that results are clinically useful.)

9.27 Soil aeration and soil water evaporation involve the exchange of gases between the soil and the atmosphere. Experimenters have investigated the effect of the airflow above the soil on this process. One such experiment varied the speed of the air x and measured the rate of evaporation y. The fitted regression equation based on 18 observations was $\hat{y} = 5.0 + 0.00665x$. The standard error of the slope was reported to be 0.00182.

(a) It is reasonable to suppose that greater airflow will cause more evaporation. State hypotheses to test this belief and calculate the test statistic. Find an approximate P-value for the significance test and report your conclusion.

(b) Construct a 95% confidence interval for the additional evaporation experienced when airflow increases by 1 unit.

The following exercises concern the optional sections on analysis of variance and inference for correlation.

9.28 Return to the data on current versus voltage given in the Ohm's law experiment of Exercise 9.18.

(a) Compute all values for the ANOVA table.

(b) State the null hypothesis tested by the ANOVA F statistic, and explain in plain language what this hypothesis says.

(c) What is the distribution of this F statistic when H_0 is true? Find an approximate P-value for the test of H_0.

9.29 A second set of data for an Ohm's law experiment appears in Exercise 9.19. Answer the questions in the previous exercise for these data.

9.30 Return to the oxygen uptake and heart rate data given in Exercise 9.22.

(a) Construct the ANOVA table.

(b) What null hypothesis is tested by the ANOVA F statistic? What does this hypothesis say in practical terms?

(c) Give the degrees of freedom for the F statistic and an approximate P-value for the test of H_0.

(d) Verify that the square of the t statistic that you calculated in Exercise 9.22 is equal to the F statistic in your ANOVA table. (Any difference found is due to roundoff error.)

(e) What proportion of the variation in oxygen uptake is explained by heart rate for this set of data?

9.31 A second set of data on oxygen uptake and heart rate appears in Exercise 9.23. Answer the questions in the previous exercise for these data.

9.32 A study conducted in the Egyptian village of Kalama examined the relationship between the birth weights of 40 infants and various socioeconomic variables. (The study is reported in M. El-Kholy, F. Shaheen, and W. Mahmoud, "Relationship between socioeconomic status and birth weight, a field study in a rural community in Egypt," *Journal of the Egyptian Public Health Association*, 61 (1986), pp. 349–358.)

(a) The correlation between monthly income and birth weight was $r = 0.39$. Calculate the t statistic for testing the null hypothesis that the correlation is 0 in the entire population of infants.

(b) The researchers expected that higher birth weights would be associated with higher incomes. Express this expectation as an alternative hypothesis for the population correlation.

(c) Determine a P-value for H_0 versus the alternative that you specified in (b). What conclusion does your test suggest?

9.33 Chinese third-grade students from public schools in Hong Kong were the subjects of a study designed to investigate the relationship between various measures of parental behavior and other variables. The sample size was 713. The data were obtained from questionnaires filled in by the students. One of the variables examined was parental control, an indication of the amount of control that the parents exercised over the behavior of the students. Another was the self-esteem of the students. (This study is reported in S. Lau and P. C. Cheung, "Relations between Chinese adolescents' perception of parental control and organization and their perception of parental warmth," *Developmental Psychology*, 23 (1987), pp. 726–729.)

(a) The correlation between parental control and self-esteem was $r = -0.19$. Calculate the t statistic for testing the null hypothesis that the population correlation is 0.

(b) Find an approximate P-value for testing H_0 versus the two-sided alternative, and report your conclusion.

9.2 MULTIPLE LINEAR REGRESSION*

In Section 9.1 we presented methods for examining a linear relationship between a response variable y and a *single* explanatory variable x. With multiple linear regression we use *more than one* explanatory variable to explain or predict a single response variable. Many of the ideas that we encountered in our study of simple linear regression carry over to this more general case. However, the introduction of several explanatory variables leads to many additional considerations. In this short section we cannot explore all of these issues. Rather, we will outline some basic facts about inference in the multiple regression setting and then illustrate multiple regression analysis with a case study.

*This section presents advanced material that requires statistical software.

Statistical model for multiple regression

The simple linear regression model assumes that the mean of the response variable y depends on the explanatory variable x according to a linear equation

$$\mu_y = \beta_0 + \beta_1 x$$

For any fixed value of x, the response y varies normally around this mean and has a standard deviation σ that is the same for all values of x.

In the multiple regression setting, the response variable y depends on not one but p explanatory variables. We will denote these explanatory variables by $x_1, x_2, \ldots, x_p$. The mean response is a linear function of the explanatory variables,

$$\mu_y = \beta_0 + \beta_1 x_1 + \beta_2 x_2 + \cdots + \beta_p x_p$$

population regression equation

This expression is the *population regression equation.* We cannot directly observe this equation because the observed values of y vary about their means. We can think of subpopulations of responses, each corresponding to a particular set of values for *all* of the explanatory variables $x_1, x_2, \ldots, x_p$. In each subpopulation, y varies normally with mean given by the population regression equation. The regression model assumes that the standard deviation σ of the responses is the same in all subpopulations.

EXAMPLE 9.14

Our case study will concern data collected at a large university on all first-year computer science majors in a particular year.[6] The purpose of the study was to attempt to predict success in the early university years. One measure of success was the cumulative grade point average (GPA) after three semesters. Among the explanatory variables recorded at the time the students enrolled in the university were high school grades in mathematics (HSM), science (HSS), and English (HSE).

We will use high school grades to predict the response variable GPA. There are $p = 3$ explanatory variables: $x_1 = \text{HSM}$, $x_2 = \text{HSS}$, and $x_3 = \text{HSE}$. The high school grades are coded on a scale from 1 to 10, with 10 corresponding to A, 9 to A−, 8 to B+, and so on. These grades define the subpopulations. For example, the straight-C students are the subpopulation defined by HSM = 4, HSS = 4, and HSE = 4.

One possible multiple regression model for the subpopulation mean GPAs is

$$\mu_{\text{GPA}} = \beta_0 + \beta_1 \text{HSM} + \beta_2 \text{HSS} + \beta_3 \text{HSE}$$

For the straight-C subpopulation of students, the model gives the subpopulation mean as

$$\mu_{\text{GPA}} = \beta_0 + 4\beta_1 + 4\beta_2 + 4\beta_3$$

●

The subpopulation means describe the FIT part of our statistical model. The RESIDUAL part represents the variation of observations about the means. We will use the same notation for the RESIDUAL that we used in the simple linear regression model. The symbol ϵ represents the deviation of an individual observation from its subpopulation mean. We assume that these deviations are normally distributed with mean 0 and an unknown standard deviation σ that does not depend on the values of the x variables.

The data for a simple linear regression problem consist of observations (x_i, y_i) on the two variables. Because there are several explanatory variables in multiple regression, the notation needed to describe the data is more elaborate. Each observation or case consists of a value for the response variable and for each of the explanatory variables. Call x_{ij} the value of the jth explanatory variable for the ith case. The data are then

$$\begin{array}{ll} \text{Case 1:} & (x_{11}, x_{12}, \ldots, x_{1p}, y_1) \\ \text{Case 2:} & (x_{21}, x_{22}, \ldots, x_{2p}, y_2) \\ & \vdots \\ \text{Case } n\text{:} & (x_{n1}, x_{n2}, \ldots, x_{np}, y_n) \end{array}$$

Here, n is the number of cases and p is the number of explanatory variables. Data are often entered into computer regression programs in this format. Each row is a case and each column corresponds to a different variable. The data for Example 9.14, with several additional explanatory variables, appear in this format in the CSDATA data set described in the data appendix.

MULTIPLE LINEAR REGRESSION MODEL

The statistical model for a multiple linear regression is

$$y_i = \beta_0 + \beta_1 x_{i1} + \beta_2 x_{i2} + \cdots + \beta_p x_{ip} + \epsilon_i$$

for $i = 1, 2, \ldots, n$.

The mean response is a linear function of the explanatory variables. The deviations ϵ_i are independent and normally distributed with mean 0 and standard deviation σ. In other words, they are an SRS from the $N(0, \sigma)$ distribution.

The parameters of the model are β_0, β_1, β_2, ..., β_p, and σ.

The assumption that the subpopulation means are related to the regression coefficients β by the equation

$$\mu_y = \beta_0 + \beta_1 x_1 + \beta_2 x_2 + \cdots + \beta_p x_p$$

implies that we can estimate all subpopulation means from estimates of the β's. To the extent that this equation is accuratc, we have a useful tool for studying the relation of y to the x's.

Estimation, confidence intervals, and significance tests

For simple linear regression we used the principle of least squares first described in Section 2.2 to obtain estimates of the intercept and slope of the regression line. For multiple regression the principle is the same but the details are more complicated. Let

$$b_0,\ b_1,\ b_2,\ \ldots,\ b_p$$

denote the estimators of the parameters

$$\beta_0,\ \beta_1,\ \beta_2,\ \ldots,\ \beta_p$$

For the ith observation the predicted response is

$$\hat{y}_i = b_0 + b_1 x_{i1} + b_2 x_{i2} + \cdots + b_p x_{ip}$$

residual

The ith *residual,* the difference between the observed and predicted response, is therefore

$$\begin{aligned} e_i &= \text{observed response} - \text{predicted response} \\ &= y_i - \hat{y}_i \\ &= y_i - b_0 - b_1 x_{i1} - b_2 x_{i2} - \cdots - b_p x_{ip} \end{aligned}$$

method of least squares

The *method of least squares* chooses the values of the b's that make the sum of the squares of the residuals as small as possible. In other words, the parameter estimates $b_0, b_1, b_2, \ldots, b_p$ minimize the quantity

$$\sum (y_i - b_0 - b_1 x_{i1} - b_2 x_{i2} - \cdots - b_p x_{ip})^2$$

The exact recipe for the least-squares estimates is complicated. We will be content to understand the principle on which they are based and to let software do the computations.

The parameter σ^2 measures the variability of the responses about the population regression equation. As in the case of simple linear regression, we estimate σ^2 by an average of the squared residuals. The estimator is

$$\begin{aligned} s^2 &= \frac{\sum e_i^2}{n-p-1} \\ &= \frac{\sum (y_i - \hat{y}_i)^2}{n-p-1} \end{aligned}$$

degrees of freedom

The quantity $n - p - 1$ is the *degrees of freedom* associated with s^2. The degrees of freedom equal the sample size n minus $p + 1$, the number of β's we must estimate to fit the model. In the simple linear regression case, $p = 1$ and the degrees of freedom are $n - 2$. To estimate σ we use

$$s = \sqrt{s^2}$$

We can obtain confidence intervals and significance tests for each of the regression coefficients β_j much as in simple linear regression. The recipes for the standard errors of the b's are complicated, though s is the key component. We again rely on statistical software for the calculations.

CONFIDENCE INTERVALS AND SIGNIFICANCE TESTS FOR β_J

A level C confidence interval for β_j is

$$b_j \pm t^* s_{b_j}$$

where s_{b_j} is the standard error of b_j and t^* is the upper $(1 - C)/2$ critical value for the $t(n - p - 1)$ distribution.

To test the hypothesis H_0: $\beta_j = 0$, compute

$$t = \frac{b_j}{s_{b_j}}$$

In terms of a random variable T having the $t(n - p - 1)$ distribution, the P-value for a test of H_0 against

$$H_a\colon \beta_j > 0 \quad \text{is} \quad P(T \geq t)$$
$$H_a\colon \beta_j < 0 \quad \text{is} \quad P(T \leq t)$$
$$H_a\colon \beta_j \neq 0 \quad \text{is} \quad 2P(T \geq |t|)$$

Because regression is often used for prediction, we may wish to construct confidence intervals for a mean response and prediction intervals for a future observation from multiple regression models. The basic ideas are once more the same as in the simple linear regression case. In most statistical software systems, the same commands that give confidence and prediction intervals for simple linear regression work for multiple regression. The only difference is that we specify a list of explanatory variables rather than a single variable. Modern software allows us to perform these rather complex calculations without an intimate knowledge of all of the computational details. This frees us to concentrate on the meaning and appropriate use of the results.

In simple linear regression the F test from the ANOVA table is equivalent to the two-sided t test of the hypothesis that the slope of the regression line is 0. For multiple regression there is a corresponding ANOVA F test, but it tests the hypothesis that *all* of the regression coefficients (with the exception of the intercept) are 0. Here is the general form of the ANOVA table for multiple regression:

Source	Sum of squares	Degrees of freedom	Mean square	F
Model	$\sum(\hat{y}_i - \bar{y})^2$	p	SSM/DFM	MSM/MSE
Error	$\sum(y_i - \hat{y}_i)^2$	$n - p - 1$	SSE/DFE	
Total	$\sum(y_i - \bar{y})^2$	$n - 1$	SST/DFT	

The ANOVA table is similar to that for simple linear regression. The degrees of freedom for the model increase from 1 to p to reflect the fact that we now have p explanatory variables rather than just one. As a consequence, the degrees of freedom for error decrease by the same amount. The sums of squares represent sources of variation. Once again, both sums of squares and their degrees of freedom add:

$$\text{SST} = \text{SSM} + \text{SSE}$$
$$\text{DFT} = \text{DFM} + \text{DFE}$$

The estimate of the variance σ^2 for our model is again given by the MSE in the ANOVA table. That is, s^2 = MSE.

The ratio MSM/MSE is an F statistic for testing the null hypothesis

$$H_0\colon \beta_1 = \beta_2 = \cdots = \beta_p = 0$$

against the alternative hypothesis

$$H_a\colon \beta_j \neq 0 \text{ for at least one } j = 1, 2, \ldots, p$$

The null hypothesis says that none of the explanatory variables are predictors of the response variable when used in the form expressed by the multiple regression equation. The alternative states that at least one of them is linearly related to the response. As in simple linear regression, large values of F give evidence against H_0. When H_0 is true, F has the $F(p,\ n-p-1)$ distribution. The degrees of freedom for the F distribution are those associated with the model and error in the ANOVA table.

> **ANALYSIS OF VARIANCE F TEST**
>
> In the multiple regression model, the hypothesis
>
> $$H_0: \beta_1 = \beta_2 = \cdots = \beta_p = 0$$
>
> is tested by the analysis of variance F statistic
>
> $$F = \frac{\text{MSM}}{\text{MSE}}$$
>
> The P-value is the probability that a random variable having the $F(p,\ n-p-1)$ distribution is greater than or equal to the calculated value of the F statistic.

For simple linear regression we noted that the square of the sample correlation could be written as the ratio of SSM to SST and could be interpreted as the proportion of variation in y explained by x. A similar statistic is routinely calculated for multiple regression.

> **THE SQUARED MULTIPLE CORRELATION**
>
> The statistic
>
> $$R^2 = \frac{\text{SSM}}{\text{SST}} = \frac{\sum(\hat{y}_i - \bar{y})^2}{\sum(y_i - \bar{y})^2}$$
>
> is the proportion of the variation of the response variable y that is explained by the explanatory variables $x_1, x_2, \ldots, x_p$ in a multiple linear regression.

multiple correlation coefficient

Often, R^2 is multiplied by 100 and expressed as a percent. The square root of R^2, called the *multiple correlation coefficient*, is the correlation between the observations y_i and the predicted values $\hat{y}_i$.

A case study

In this section we illustrate multiple regression by analyzing the data from the study described in Example 9.14. The response variable is the cumulative GPA after three semesters for a group of computer science majors at a large university. The explanatory variables previously mentioned are average high school grades, represented by HSM, HSS, and HSE. We also examine the SAT mathematics and SAT verbal scores as explanatory variables. We have data for $n = 224$ students in the study.

Preliminary analysis The first step in the analysis is to carefully examine each of the variables. Means, standard deviations, minimum and maximum values, as given by the SAS procedure MEANS, appear in Figure 9.18.

The minimum value for the SAT mathematics (SATM) variable appears to be rather extreme; it is $(595 - 300)/86 = 3.43$ standard deviations below the mean. We do not discard this case at this time but will take care in our subsequent analyses to see if it has an excessive influence on our results. The mean for the SATM score is higher than the mean for the verbal score (SATV), as we might expect for a group of computer science majors. The two standard deviations are about the same. The means of the three high school grade variables are similar, with the mathematics grades being a bit higher. The standard deviations for the high school grade variables are very close to each other. The mean GPA is 4.635 on a 6-point scale, with standard deviation 0.779.

Because the variables GPA, SATM, and SATV have many possible values, we could use stemplots or histograms to examine the shapes of their distributions. Normal quantile plots indicate whether or not the distributions look normal. It is important to note that the multiple regression model *does not* require any of these distributions to be normal. Only the deviations of the responses y from their means are assumed to be normal. The purpose of examining these plots is to understand something about each variable alone before attempting to use it in a complicated model. Extreme values of any variable should be noted and checked for accuracy. If found to be correct, the cases with these values should be carefully examined to see if they are truly exceptional and perhaps do not belong in

VARIABLE	N	MEAN	STANDARD DEVIATION	VARIANCE	MINIMUM VALUE	MAXIMUM VALUE
GPA	224	4.6352232	0.77939493	0.607456	2.1200000	6.000000
SATM	224	595.2857143	86.40144374	7465.209481	300.0000000	800.000000
SATV	224	504.5491071	92.61045907	8576.697129	285.0000000	760.000000
HSM	224	8.3214286	1.63873672	2.685458	2.0000000	10.000000
HSS	224	8.0892857	1.69966270	2.888853	3.0000000	10.000000
HSE	224	8.0937500	1.50787358	2.273683	3.0000000	10.000000

FIGURE 9.18 Descriptive statistics for the computer science student case study.

HSM	FREQUENCY	PERCENT	CUMULATIVE FREQUENCY	CUMULATIVE PERCENT
2	1	0.4	1	0.4
3	1	0.4	2	0.9
4	4	1.8	6	2.7
5	6	2.7	12	5.4
6	23	10.3	35	15.6
7	28	12.5	63	28.1
8	36	16.1	99	44.2
9	59	26.3	158	70.5
10	66	29.5	224	100.0

HSS	FREQUENCY	PERCENT	CUMULATIVE FREQUENCY	CUMULATIVE PERCENT
3	1	0.4	1	0.4
4	7	3.1	8	3.6
5	9	4.0	17	7.6
6	24	10.7	41	18.3
7	42	18.8	83	37.1
8	31	13.8	114	50.9
9	50	22.3	164	73.2
10	60	26.8	224	100.0

HSE	FREQUENCY	PERCENT	CUMULATIVE FREQUENCY	CUMULATIVE PERCENT
3	1	0.4	1	0.4
4	4	1.8	5	2.2
5	5	2.2	10	4.5
6	23	10.3	33	14.7
7	43	19.2	76	33.9
8	49	21.9	125	55.8
9	52	23.2	177	79.0
10	47	21.0	224	100.0

FIGURE 9.19 The distributions of the high school grade variables.

the same analysis with the other cases. When these data are examined in this way, no obvious problems are evident.

The high school grade variables HSM, HSS, and HSE have relatively few values and are best summarized by giving the relative frequencies for each possible value. The output of the SAS procedure FREQ, given in Figure 9.19, provides these summaries. The distributions are all skewed, with a large proportion of high grades (10 = A and 9 = A−). Again we emphasize that these distributions need not be normal.

Relationships between pairs of variables The second step in our analysis is to examine the relationships between all pairs of variables. Scatterplots and correlations are our tools for studying two-variable relationships. The correlations, given by the SAS procedure CORR, are displayed in Figure 9.20. The output includes the *P*-value for the test of the null hypothesis that the population correlation is 0 versus the two-sided alternative for each pair. Thus, we see that the correlation between GPA and HSM is 0.44, with a *P*-value of 0.0001, whereas the correlation between GPA and SATV is 0.11, with a *P*-value of 0.087. The first is statistically significant by any reasonable standard, while the second is rather marginal.

PEARSON CORRELATION COFFICIENTS / PROB > |R| UNDER HO: RHO=0 / N = 224

	GPA	SATM	SATV	HSM	HSS	HSE
GPA	1.00000	0.25171	0.11449	0.43650	0.32943	0.28900
	0.0000	0.0001	0.0873	0.0001	0.0001	0.0001
SATM	0.25171	1.00000	0.46394	0.45351	0.24048	0.10828
	0.0001	0.0000	0.0001	0.0001	0.0003	0.1060
SATV	0.11449	0.46394	1.00000	0.22112	0.26170	0.24371
	0.0873	0.0001	0.0000	0.0009	0.0001	0.0002
HSM	0.43650	0.45351	0.22112	1.00000	0.57569	0.44689
	0.0001	0.0001	0.0009	0.0000	0.0001	0.0001
HSS	0.32943	0.24048	0.26170	0.57569	1.00000	0.57937
	0.0001	0.0003	0.0001	0.0001	0.0000	0.0001
HSE	0.28900	0.10828	0.24371	0.44689	0.57937	1.00000
	0.0001	0.1060	0.0002	0.0001	0.0001	0.0000

FIGURE 9.20 **Correlations among the case study variables.**

The high school grades all have higher correlations with GPA than do the SAT scores. As we might expect, math grades have the highest correlation ($r = 0.44$), followed by science grades (0.33) and then English grades (0.29). The two SAT scores have a rather high correlation with each other (0.46), and the high school grades also correlate well with each other (0.45 to 0.58). The SAT mathematics score correlates well with HSM (0.45), less well with HSS (0.24), and rather poorly with HSE (0.11). The correlations of the SAT verbal score with the three high school grades are about equal, ranging from 0.22 to 0.26.

It is important to keep in mind that by examining pairs of variables we are seeking a better understanding of the data. The fact that the correlation of a particular explanatory variable with the response variable does not achieve statistical significance does not *necessarily* imply that it will not be a useful (and significant) predictor in a multiple regression.

Numerical summaries such as correlations are useful, but plots are generally more informative when seeking to understand data. Plots tell us whether the numerical summary gives a fair representation of the data. For a multiple regression, each pair of variables should be plotted. For the six variables in our case study, this means that we should examine 15 plots. In general there are $p + 1$ variables in a multiple regression analysis with p explanatory variables, so that $p(p + 1)/2$ plots are required. Multiple regression is a complicated procedure. If we do not do the necessary preliminary work, we are in serious danger of producing useless or misleading results. We leave the task of making these plots as an exercise.

Regression on high school grades To explore the relationship between the explanatory variables and our response variable GPA, we run several multiple regressions. The explanatory variables fall into two classes. High school grades are represented by HSM, HSS, and HSE, and standardized tests are represented by the two SAT scores. We begin our analysis by using the high school grades to predict GPA. Figure 9.21 gives the output generated by the SAS procedure REG for this problem.

The output contains an ANOVA table, some additional descriptive statistics, and information about the parameter estimates. When examining any ANOVA table, it is a good idea to first verify the degrees of freedom. This ensures that we have not made some serious error in specifying the model for the program or in entering the data. Because there are $n = 224$ cases, we have DFT $= n - 1 = 223$. The three explanatory variables give DFM $= p = 3$ and DFE $= n - p - 1 = 223 - 3 = 220$.

The ANOVA F statistic is 18.86, with a P-value of 0.0001. Under the null hypothesis

$$H_0\colon \beta_1 = \beta_2 = \beta_3 = 0$$

the F statistic has an $F(3, 220)$ distribution. According to this distribution, the chance of obtaining an F statistic of 18.86 or larger is 0.0001. We

DEP VARIABLE: GPA

ANALYSIS OF VARIANCE

SOURCE	DF	SUM OF SQUARES	MEAN SQUARE	F VALUE	PROB>F
MODEL	3	27.71233132	9.23744377	18.861	0.0001
ERROR	220	107.75046	0.48977481		
C TOTAL	223	135.46279			

ROOT MSE	0.6998391	R-SQUARE	0.2046
DEP MEAN	4.635223	ADJ R-SQ	0.1937
C.V.	15.09828		

PARAMETER ESTIMATES

VARIABLE	PARAMETER ESTIMATE	STANDARD ERROR	T FOR H0: PARAMETER=0	PROB > |T|
INTERCEP	2.58987662	0.29424324	8.802	0.0001
HSM	0.16856664	0.03549214	4.749	0.0001
HSS	0.03431557	0.03755888	0.914	0.3619
HSE	0.04510182	0.03869585	1.166	0.2451

FIGURE 9.21 Multiple regression output for regression using high school grades to predict GPA.

therefore conclude that at least one of the three regression coefficients for the high school grades is different from 0 in the population regression equation.

In the descriptive statistics that follow the ANOVA table we find that ROOT MSE is 0.6998. This value is the square root of the MSE given in the ANOVA table and is the estimate s of the parameter σ of our model. The value of R^2 is 0.20. That is, 20% of the observed variation in the GPA scores is explained by linear regression on high school grades.

From the PARAMETER ESTIMATES section we obtain the fitted regression equation

$$\widehat{\text{GPA}} = 2.590 + .169\text{HSM} + .034\text{HSS} + .045\text{HSE}$$

Recall that the t statistics for testing the regression coefficients are obtained by dividing the estimates by their standard errors. Thus, for the coefficient of HSM we obtain the t-value given in the output by calculating

$$t = \frac{.16856664}{.3549214} = 4.749$$

The *P*-values appear in the last column. HSM has a *P*-value of 0.0001, and we conclude that the regression coefficient for this explanatory variable is significantly different from 0. The *P*-values for the other explanatory variables (0.36 for HSS and 0.25 for HSE) do not achieve statistical significance. This result seems to contradict the impression obtained by examining the correlations in Figure 9.20. In that display we see that the correlation between GPA and HSS is 0.33 and the correlation between GPA and HSE is 0.29. The *P*-values for both of these correlations are 0.0001. In other words, if we used HSS alone in a regression to predict GPA, or if we used HSE alone, we would obtain statistically significant regression coefficients.

This phenomenon is not unusual in multiple regression analysis. Part of the explanation lies in the correlations between HSM and the other two explanatory variables. These are rather high (at least compared with the other correlations in Figure 9.20). The correlation between HSM and HSS is 0.58, and that between HSM and HSE is 0.45. Thus, when we have a regression model that contains all three high school grades as explanatory variables, there is considerable overlap of the predictive information contained in these variables. *The significance tests for individual regression coefficients assess the significance of each predictor variable assuming that all other predictors are included in the regression equation*. Given that we use a model with HSM and HSS as predictors, the coefficient of HSE is not statistically significant. Similarly, given that we have HSM and HSE in the model, HSS does not have a significant regression coefficient. HSM, however, adds significantly to our ability to predict GPA even after HSS and HSE are already in the model.

Unfortunately, we cannot conclude from this analysis that the *pair* of explanatory variables HSS and HSE contribute nothing significant to our model for predicting GPA once HSM is in the model. The impact of relations among the several explanatory variables on fitting models for the response is the most important new phenomenon encountered in moving from simple linear regression to multiple regression. We can only hint at the many complicated problems that arise.

As in simple linear regression, we should always examine the residuals as an aid to determining whether the multiple regression model is appropriate for the data. Because there are several explanatory variables, we must examine several residual plots. It is usual to plot the residuals versus the predicted values $\hat{y}$ and also versus each of the explanatory variables. Look for outliers, influential observations, evidence of a curved (rather than linear) relation, and anything else unusual. Again, we leave the task of making these plots as an exercise. The plots all appear to show more or less random noise around the center value of 0.

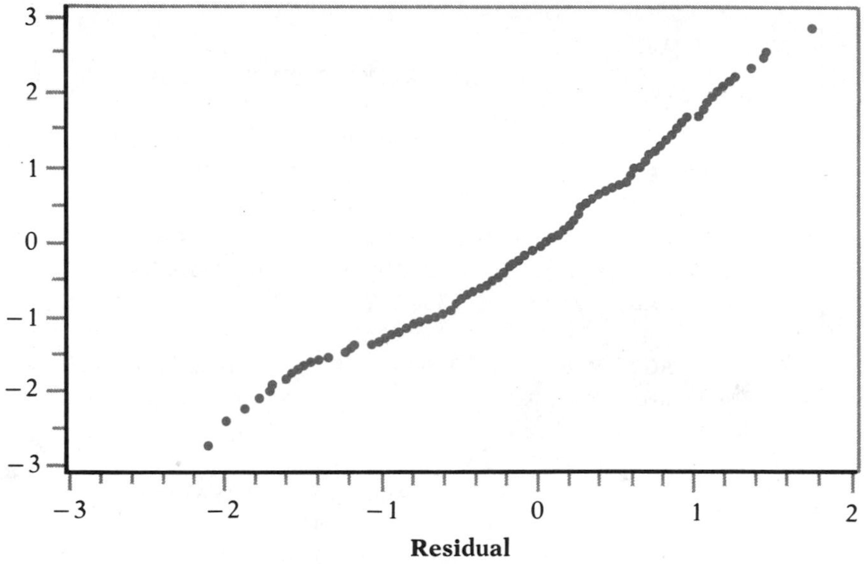

FIGURE 9.22 **Normal quantile plot of the residuals from the high school grade model. There are no important deviations from normality.**

If the deviations ϵ in the model are normally distributed, the residuals should be normally distributed. Figure 9.22 presents a normal quantile plot of the residuals. The distribution appears to be approximately normal. There are many other specialized plots that help detect departures from the multiple regression model. Discussion of these, however, is more than we can undertake in this chapter.

Because the variable HSS has the largest *P*-value of the three explanatory variables (see Figure 9.21) and therefore appears to contribute the least to our explanation of GPA, we rerun the regression using only HSM and HSE as explanatory variables. The SAS output appears in Figure 9.23. The *F* statistic indicates that we can reject the null hypothesis that the regression coefficients for the two explanatory variables are both 0. The *P*-value is still 0.0001. The value of R^2 has dropped very slightly compared with our previous run, from 0.2046 to 0.2016. Dropping HSS from the model resulted in the loss of very little explanatory power. The measure s of variation about the fitted equation (ROOT MSE in the printout) is nearly identical for the two regressions, another indication that dropping HSS loses very little. The t statistics for the individual regression coefficients indicate that HSM is still clearly significant ($P = 0.0001$), while the coefficient of HSE is larger than before (1.747 versus 1.166) and approaches the traditional 0.05 level of significance ($P = 0.082$).

Comparison of the fitted equations for the two multiple regression analyses tells us something more about the intricacies of this procedure.

```
DEP VARIABLE:  GPA
                          ANALYSIS OF VARIANCE

                           SUM OF            MEAN
    SOURCE      DF        SQUARES          SQUARE    F VALUE    PROB>F

    MODEL        2    27.30349117     13.65174558     27.894    0.0001
    ERROR      221      108.15930      0.48940859
    C TOTAL    223      135.46279

        ROOT MSE        0.6995774        R-SQUARE     0.2016
        DEP MEAN         4.635223        ADJ R-SQ     0.1943
        C.V.             15.09264

                          PARAMETER ESTIMATES

                  PARAMETER        STANDARD      T FOR HO:
  VARIABLE         ESTIMATE           ERROR    PARAMETER=0     PROB > |T|

  INTERCEP       2.62422848      0.29172204          8.996         0.0001
  HSM            0.18265442      0.03195581          5.716         0.0001
  HSE            0.06067015      0.03472914          1.747         0.0820
```

FIGURE 9.23 Multiple regression output for regression using HSM and HSE to predict GPA.

For the first run we have

$$\widehat{\text{GPA}} = 2.590 + .169\text{HSM} + .034\text{HSS} + .045\text{HSE}$$

whereas the second gives us

$$\widehat{\text{GPA}} = 2.624 + .183\text{HSM} + .061\text{HSE}$$

Eliminating HSS from the model changes the regression coefficients for all of the remaining variables and the intercept. This phenomenon occurs quite generally in multiple regression. *Individual regression coefficients, their standard errors, and significance tests are meaningful only when interpreted in the context of the other explanatory variables in the model.*

Regression on SAT scores We now turn to the problem of predicting GPA using the two SAT scores. Figure 9.24 gives the output from the SAS procedure REG. The degrees of freedom are as expected: 2, 221, and 223. The F statistic is 7.476, with a P-value of 0.0007. We conclude that the

DEP VARIABLE: GPA

ANALYSIS OF VARIANCE

SOURCE	DF	SUM OF SQUARES	MEAN SQUARE	F VALUE	PROB>F
MODEL	2	8.58383910	4.29191955	7.476	0.0007
ERROR	221	126.87895	0.57411289		
C TOTAL	223	135.46279			

ROOT MSE	0.7577024	R-SQUARE	0.0634
DEP MEAN	4.635223	ADJ R-SQ	0.0549
C.V.	16.34662		

PARAMETER ESTIMATES

VARIABLE	PARAMETER ESTIMATE	STANDARD ERROR	T FOR H0: PARAMETER=0	PROB > |T|
INTERCEP	3.28867737	0.37603684	8.746	0.0001
SATM	0.002282834	0.000662914	3.444	0.0007
SATV	-0.000024562	0.000618470	-0.040	0.9684

FIGURE 9.24 Multiple regression output for regression using SAT scores to predict GPA.

regression coefficients for SATM and SATV are not both 0. Recall that we obtained the *P*-value 0.0001 when we used high school grades to predict GPA. Both multiple regression equations are highly significant, but this obscures the fact that the two models have quite different explanatory power. For the SAT regression, $R^2 = 0.0634$, whereas for the high school grades model even with only HSM and HSE (Figure 9.23), we have $R^2 = 0.2016$, a value more than three times as large. Stating that we have a statistically significant result is quite different from saying that an effect is large or important.

Further examination of the output in Figure 9.24 reveals that the coefficient of SATM is significant ($t = 3.44$, $P = 0.0007$), and that for SATV is not ($t = -0.04$, $P = 0.9684$). For a complete analysis we should carefully examine the residuals. Also, we might want to run the analysis with SATM as the only explanatory variable.

Regression using all variables We have seen that either the high school grades or the SAT scores give a highly significant regression equation. The mathematics component of each of these groups of explanatory variables appears to be the key predictor. Comparing the values of R^2 for the two models indicates that high school grades are better predictors than SAT scores. Can we get a better prediction equation using all of the explanatory variables together in one multiple regression?

To address this question we run the regression with all five explanatory variables. The output appears in Figure 9.25. The F statistic is 11.69, with a P-value of 0.0001, so at least one of our explanatory variables has a nonzero regression coefficient. This result is not surprising given that we

```
DEP VARIABLE: GPA
                          ANALYSIS OF VARIANCE

                             SUM OF           MEAN
     SOURCE       DF        SQUARES         SQUARE    F VALUE     PROB>F

     MODEL         5    28.64364489     5.72872898     11.691     0.0001
     ERROR       218      106.81914     0.48999607
     C TOTAL     223      135.46279

         ROOT MSE         0.6999972       R-SQUARE     0.2115
         DEP MEAN          4.635223       ADJ R-SQ     0.1934
         C.V.              15.10169

                          PARAMETER ESTIMATES

                 PARAMETER          STANDARD       T FOR HO:
  VARIABLE        ESTIMATE             ERROR     PARAMETER=0     PROB > |T|

  INTERCEP      2.32671874        0.39999643           5.817         0.0001
  SATM         0.000943593       0.000685657           1.376         0.1702
  SATV        -0.000407850       0.000591893          -0.689         0.4915
  HSM           0.14596108        0.03926097           3.718         0.0003
  HSS           0.03590532        0.03779841           0.950         0.3432
  HSE           0.05529258        0.03956869           1.397         0.1637

TEST: SAT       NUMERATOR:      0.465657    DF:      2    F VALUE:      0.9503
                DENOMINATOR:    0.489996    DF:    218    PROB >F:      0.3882

TEST: HS        NUMERATOR:        6.6866    DF:      3    F VALUE:     13.6462
                DENOMINATOR:    0.489996    DF:    218    PROB >F:      0.0001
```

FIGURE 9.25 Multiple regression output for regression using all variables to predict GPA.

have already seen that HSM and SATM are strong predictors of GPA. The value of R^2 is 0.2115, not much higher than the value of 0.2046 that we found for the high school grades regression.

Examination of the t statistics and the associated P-values for the individual regression coefficients reveals that HSM is the only one that is significant ($P = 0.0003$). That is, only HSM makes a significant contribution when it is added to a model that already has the other four explanatory variables. Once again it is important to understand that this result does not necessarily mean that the regression coefficients for the four other explanatory variables are *all* 0. Many statistical software packages provide the capability for testing whether a collection of regression coefficients in a multiple regression model are *all* 0. We use this approach to address two interesting questions about this set of data. We did not discuss such tests in the outline that opened this section, but the basic idea is quite simple.

In the context of the multiple regression model with all five predictors, we ask first whether or not the coefficients for the two SAT scores are both 0. In other words, do the SAT scores add any significant predictive information to that already contained in the high school grades? To be fair, we also ask the complementary question—do the high school grades add any significant predictive information to that already contained in the SAT scores?

The answers are given in the last two parts of the output in Figure 9.25. For the first test we see that $F = 0.9503$. Under the null hypothesis that the two SAT coefficients are 0, this statistic has an $F(2, 218)$ distribution and the P-value is 0.39. We conclude that the SAT scores are not significant predictors of GPA in a regression that already contains the high school scores as predictor variables. Recall that the model with just SAT scores has a highly significant F statistic. We now see that whatever predictive information is in the SAT scores can also be found in the high school grades. In this sense, the SAT scores are superfluous.

The test statistic for the three high school grade variables is $F = 13.6462$. Under the null hypothesis that these three regression coefficients are 0, the statistic has an $F(3, 218)$ distribution and the P-value is 0.0001. We conclude that high school grades contain useful information for predicting GPA that is not contained in SAT scores.

Of course, our statistical analysis of these data does not imply that SAT scores are less useful than high school grades for predicting college grades for all groups of students. We have studied a select group of students—computer science majors—from a specific university. Generalizations to other situations are beyond the scope of inference based on these data alone.

SUMMARY

The statistical model for **multiple linear regression** with response variable y and p explanatory variables $x_1, x_2, \ldots, x_p$ is

$$y_i = \beta_0 + \beta_1 x_{i1} + \beta_2 x_{i2} + \cdots + \beta_p x_{ip} + \epsilon_i$$

where $i = 1, 2, \ldots, n$. The ϵ_i are assumed to be independent and normally distributed with mean 0 and standard deviation σ. The parameters of the model are β_0, β_1, β_2, ..., β_p, and σ.

The β's are estimated by b_0, b_1, b_2, ..., b_p, which are obtained by the **principle of least squares.** The parameter σ is estimated by

$$s = \sqrt{\text{MSE}} = \sqrt{\frac{\sum e_i^2}{n - p - 1}}$$

where the e_i are the **residuals**

$$e_i = y_i - \hat{y}_i$$

A level C **confidence interval** for β_j is

$$b_j \pm t^* s_{b_j}$$

where t^* is the upper $(1 - C)/2$ critical value for the $t(n - p - 1)$ distribution.

The **test of the hypothesis** H_0: $\beta_j = 0$ is based on the statistic

$$t = \frac{b_j}{s_{b_j}}$$

and the $t(n - p - 1)$ distribution.

The estimate b_j of β_j and the test and confidence interval for β_j are all based on a specific multiple linear regression model. The results of all of these procedures change if other explanatory variables are added to or deleted from the model.

The **ANOVA table** for a multiple linear regression gives the degrees of freedom, sum of squares, and mean squares for the model, error, and total sources of variation. The **ANOVA *F* statistic** is the ratio MSM/MSE and is used to test the null hypothesis

$$H_0\text{: } \beta_1 = \beta_2 = \cdots = \beta_p = 0$$

If H_0 is true, this statistic has an $F(p, n-p-1)$ distribution. The **squared multiple correlation** is given by the expression

$$R^2 = \frac{\text{SSM}}{\text{SST}}$$

and is interpreted as the proportion of the variability in the response variable y that is explained by the explanatory variables $x_1, x_2, \ldots, x_p$ in the multiple linear regression.

SECTION 9.2 EXERCISES

Because multiple regression calculations require software, most exercises on this material appear in the Computer Exercises section.

9.34 One model for subpopulation means for the computer science study is described in Example 9.14 (page 687) as

$$\mu_{\text{GPA}} = \beta_0 + \beta_1 \text{HSM} + \beta_2 \text{HSS} + \beta_3 \text{HSE}$$

(a) Give the model for the subpopulation mean GPA for students having high school grade scores HSM = 9 (A−), HSS = 8 (B+), and HSE = 7 (B).

(b) Using the parameter estimates given in Figure 9.21 (page 697), calculate the estimate of this subpopulation mean. Then briefly explain in words what your numerical answer means.

9.35 Use the model given in the previous exercise to do the following:

(a) For students having high school grade scores HSM = 6 (B−), HSS = 7 (B), and HSE = 8 (B+), express the subpopulation mean in terms of the parameters β_j.

(b) Calculate the estimate of this subpopulation mean using the b_j given in Figure 9.21 (page 697). Briefly explain the meaning of the number you obtain.

9.36 Consider the regression problem of predicting GPA from the three high school grade variables. The computer output appears in Figure 9.21 (page 697).

(a) Use information given in the output to calculate a 95% confidence interval for the regression coefficient of HSM. Explain in words what this regression coefficient means in the context of this particular model.

(b) Do the same for the explanatory variable HSE.

9.37 Computer output for the regression run for predicting GPA from the high school grade variables HSM and HSE appears in Figure 9.23 (page 700). Answer (a) and (b) of the previous exercise for this regression. Why are the results for this exercise different from those that you calculated for the previous exercise?

9.38 Consider the regression problem of predicting GPA from the three high school grade variables. The computer output appears in Figure 9.21 (page 697.)

(a) Write the estimated regression equation.

(b) What is the value of s, the estimate of σ?

(c) State the H_0 and H_a tested by the ANOVA F statistic for this problem. After stating the hypotheses in symbols, explain them in words.

(d) What is the distribution of the F statistic under H_0? What conclusion do you draw from the F test?

(e) What percent of the variation in GPA is explained by these two high school grade variables?

9.39 Figure 9.24 (page 701) gives the output for the regression analysis in which the two SAT scores are used to predict GPA. Answer the questions in the previous exercise for this analysis.

9.40 Multiple regressions are sometimes used in litigation. In the case of *Cargill, Inc. v. Hardin*, the prosecution charged that the cash price of wheat was manipulated in violation of the Commodity Exchange Act. In a statistical study conducted for this case, a multiple regression model was constructed to predict the price of wheat using three supply-and-demand explanatory variables. Data for 14 years were used to construct the regression equation, and a prediction for the suspect period was computed from this equation. The value of R^2 was 0.989. The predicted value was reported as \$2.136 with a standard error of \$0.013. Express the prediction as an interval. (The degrees of freedom were large for this analysis, so use 100 as the df to determine t^*.) The actual price for the period in question was \$2.13. The judge in this case decided that the analysis provided evidence that the price was not artificially depressed, and the opinion was sustained by the Court of Appeals. Write a short summary of the results of the analysis that relate to the decision and explain why you agree or disagree with it. (A description of this case as well as other examples of the use of statistics in legal settings is given in Michael O. Finkelstein, *Quantitative Methods in Law*, Free Press, New York, 1978.)

CHAPTER 9 COMPUTER EXERCISES

Because regression calculations in practice are always done by statistical software, the chapter exercises for this chapter are all computer exercises. The first six exercises below use the CSDATA data set described in the data appendix. This is the same data set to which Example 9.14 and the exposition in Section 9.2 refer.

9.41 Use software to make a plot of GPA versus SATM. Do the same for GPA versus SATV. Describe the general patterns. Are there any unusual values?

9.42 Make a plot of GPA versus HSM. Do the same for the other two high school grade variables. Describe the three plots. Are there any outliers or influential points?

9.43 Regress GPA on the three high school grade variables. Calculate and store the residuals from this regression. Plot the residuals versus each of the three predictors and versus the predicted value of GPA. Are there any unusual points or patterns in these four plots?

9.44 Use the two SAT scores in a multiple regression to predict GPA. Calculate and store the residuals. Plot the residuals versus each of the explanatory variables and versus the predicted GPA. Describe the plots.

9.45 It appears that the mathematics explanatory variables are strong predictors of GPA in the computer science study. Run a multiple regression using HSM and SATM to predict GPA.

(a) Give the fitted regression equation.

(b) State the H_0 and H_a tested by the ANOVA F statistic, and explain their meaning in plain language. Report the value of the F statistic, its P-value, and your conclusion.

(c) Give 95% confidence intervals for the regression coefficients of HSM and SATM. Do either of these include the point 0?

(d) Report the t statistics and P-values for the tests of the regression coefficients of HSM and SATM. What conclusions do you draw from these tests?

(e) What is the value of s, the estimate of σ?

(f) What percent of the variation in GPA is explained by HSM and SATM in your model?

9.46 How well do verbal variables predict the performance of computer science students? Perform a multiple regression analysis to predict GPA from HSE and SATV. Summarize the results and compare them with those obtained in the previous exercise. In what ways do the regression results indicate that the mathematics variables are better predictors?

9.47 The WOOD data set described in the data appendix contains two measurements of the strength of each of 50 strips of yellow poplar wood. In this problem we look at this experiment as one in which the process of measuring strength is to be evaluated. Specifically, we ask how well a repeat measurement T2 can be predicted from the first measurement T1 for strips of wood of this type.

(a) Plot T2 versus T1 and describe the pattern you see.

(b) Carry out a linear regression with T1 as the explanatory variable and T2 as the response variable. Write the fitted equation and give s, the estimate of the standard deviation σ of the model.

(c) What is the correlation between T1 and T2? This quantity is sometimes called the *reliability* of the strength-measuring procedure. What proportion of the variability in T2 is explained by T1?

(d) Give the t statistic for testing the null hypothesis that the slope is 0. State an appropriate alternative hypothesis for this problem and state the P-value for the significance test. Report your conclusions from parts (b), (c), and (d) in plain language.

(e) Verify that the square of the t statistic for the slope hypothesis is equal to the F statistic given in the ANOVA table.

9.48 Refer to the previous exercise. Rerun the regression for the 25 observations with odd-numbered strips; that is, $S = 1, 3, \ldots, 49$. We want to compare the results of this run with those for the full data set given in the previous exercise. In particular, we want to see which quantities remain approximately the same and which change.

(a) Construct a table giving b_0, b_1, s, and s_{b_1} for the two runs. Which values are approximately the same and which have changed?

(b) Summarize the differences between the two ANOVA tables.

(c) Compare the two correlations.

(d) Summarize your results in plain language. In particular, did your results in the previous exercise depend strongly on the quite large sample size?

(e) Suppose that we had available another 50 wood strips from the same population measured under the same conditions. If all 100 observations were included in a single analysis, approximately what values would you expect to see for b_0, b_1, s, and r?

9.49 This and the following exercise use the Gesell Adaptive Score data given in Table 2.4 (page 132). Run a linear regression to predict the Gesell score from age at first word. Use all 21 cases for this analysis and assume that the simple linear regression model holds.

(a) Write the fitted regression equation.

(b) What is the value of the t statistic for testing the null hypothesis that the slope is 0? Give an approximate P-value for the test of H_0 versus the alternative that the slope is negative.

(c) Give a 95% confidence interval for the slope.

(d) What percentage of the variation in Gesell scores is explained by the age at first word?

(e) What is s, the estimate of the model standard deviation σ?

9.50 In Chapter 2 we noted that cases 18 and 19 in the Gesell data set were somewhat unusual. The simple linear regression model is therefore suspect when all cases are included. Rerun the analysis described in the previous exercise for this set of data with these two cases excluded. Compare your results with those obtained in the previous exercise. In particular, to what extent did the conclusions from the first regression run depend on these two cases?

9.51 Refer again to the CSDATA data set. The variable SEX has the value 1 for males and 2 for females. Create a data set containing the values for males only. Run a multiple regression analysis for predicting GPA from the three high school grade variables for this group. Using the case study in the text as a guide, interpret the results and state what conclusions can be drawn from this analysis. In what way (if any) do the results for males alone differ from those for all students?

9.52 Refer to the previous exercise. Perform the analysis using the data for females only. Are there any important differences between female and male students in predicting GPA?

9.53 The data on gas consumption and degree days from Example 9.7 (page 659) are

	Oct.	Nov.	Dec.	Jan.	Feb.	Mar.	Apr.	May	June
x	15.6	26.8	37.8	36.4	35.5	18.6	15.3	7.9	.0
y	5.2	6.1	8.7	8.5	8.8	4.9	4.5	2.5	1.1

Suppose that the gas consumption for December was incorrectly recorded as 87 instead of 8.7.

(a) Calculate the least-squares regression line for the incorrect set of data.

(b) Find the standard error of b_1.

(c) Compute the test statistic for H_0: $\beta_1 = 0$ and find the P-value. How do the results compare with those for the correct set of data?

Exercises 9.54 to 9.62 refer to the CHEESE data set described in the data appendix.

9.54 For each of the four variables in the CHEESE data set, find the mean, median, standard deviation, and interquartile range. Display each distribution by means of a stemplot and use a normal quantile plot to assess normality of the data. Summarize your findings. Note that when doing regressions with these data, we do not assume that these distributions are normal. Only the residuals from our model need be (approximately) normal. The careful study of each variable to be analyzed is nonetheless an important first step in any statistical analysis.

9.55 Make a scatterplot for each pair of variables in the CHEESE data set (you will have six plots). Describe the relationships. Calculate the correlation for each pair of variables and report the P-value for the test of zero population correlation in each case.

9.56 Perform a simple linear regression analysis using taste as the response variable and acetic as the explanatory variable. Be sure to examine the residuals carefully. Summarize your results. Include a plot of the data with the least-squares regression line. Plot the residuals versus each of the other two chemicals. Are any patterns evident? (The concentrations of the other chemicals are lurking variables for the simple linear regression.)

9.57 Repeat the analysis of Exercise 9.56 using taste as the response variable and H2S as the explanatory variable.

9.58 Repeat the analysis of Exercise 9.56 using taste as the response variable and lactic as the explanatory variable.

9.59 Compare the results of the regressions performed in the three previous exercises. Construct a table with values of the F statistic, its P-value, R^2, and the estimate s of the standard deviation for each model. Report the three regression equations. Why are the intercepts in these three equations different?

9.60 Carry out a multiple regression using acetic and H2S to predict taste. Summarize the results of your analysis. Compare the statistical significance of acetic in this model with its significance in the model with acetic alone as a predictor (Exercise 9.56). Which model do you prefer? Give a simple explanation for the fact that acetic alone appears to be a good predictor of taste but with H2S in the model, it is not.

9.61 Carry out a multiple regression using H2S and lactic to predict taste. Comparing the results of this analysis with the simple linear regressions using each of these explanatory variables alone, it is evident that a better result is obtained by using both predictors in a model. Support this statement with explicit information obtained from your analysis.

9.62 Use the three explanatory variables acetic, H2S, and lactic in a multiple regression to predict taste. Write a short summary of your results, including an examination of the residuals. Based on all of the regression analyses you have carried out on these data, which model do you prefer and why?

NOTES

1. The quote and the example are taken from J. V. G. A. Durnin and J. Womersley, "Body fat assessed from total body density and its estimation from skinfold thicknesses: Measurements on 481 men and women aged 16 to 72 years," *British Journal of Nutrition*, 32 (1974), pp. 77–97. The data analyzed in the examples that follow have been generated by computer simulation in such a way that they give the same statistical results reported in the article.
2. Minitab does this directly in response to the PREDICT subcommand of the REGRESS command. Some software packages, such as SAS, use an indirect method: Define an additional observation with LSKIN equal to the value of x^* and DEN missing. The software ignores the additional case in computing the

least-squares line but computes and prints the predicted value and associated statistics.

3. These data were provided by Robert Dale of Purdue University.

4. This quantity is the estimated standard deviation of $\hat{y} - y$, not the estimated standard deviation of $\hat{y}$ alone.

5. The method is described in Chapter 10 of G. W. Snedecor and W. G. Cochran, *Statistical Methods*, 8th ed., Iowa State University Press, Ames, Iowa, 1989.

6. Results of the study are reported in P. F. Campbell and G. P. McCabe, "Predicting the success of freshmen in a computer science major," *Communications of the ACM*, 27 (1984), pp. 1108–1113.

CHAPTER

10

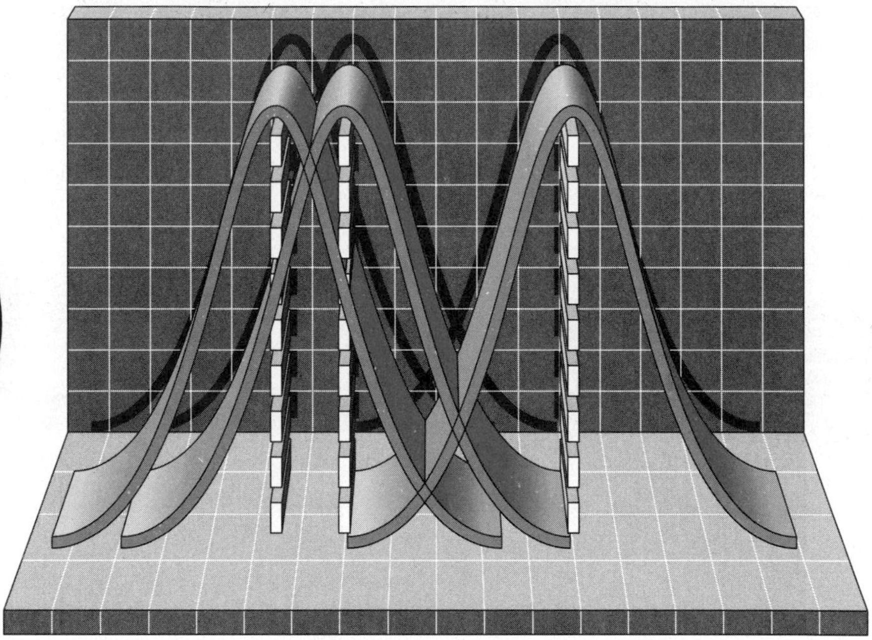

Analysis of Variance

PRELUDE

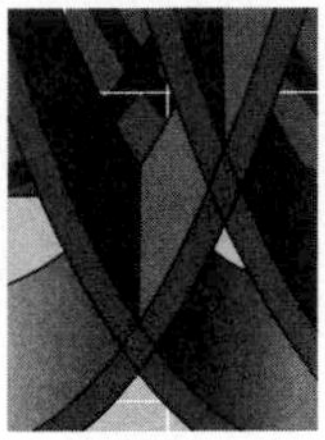

Many of the most effective statistical studies are comparative, whether we compare the salaries of women and men in a sample of firms or the responses to two treatments in a clinical trial. We display comparisons by back-to-back stemplots or side-by-side boxplots, and measure them by comparing five-number summaries or means and standard deviations. Now we ask: Is the difference between groups statistically significant?

When only two groups are compared, Chapter 7 provides the tools we need. Two-sample t procedures compare the means of two normal populations, and we saw that these procedures, unlike comparisons of spread, are sufficiently robust to be widely useful. Now we will compare any number of means by techniques that generalize the two-sample t and share its robustness and usefulness.

- Do three methods for teaching reading in second grade differ in effectiveness? Do children taught by any of these methods learn significantly faster than children taught by the other methods?
- Careful observation of small samples of young children in Egypt, Kenya, and Mexico shows that the average food intake in the samples is different. Egyptian children average 1217 kilocalories daily, while Mexican children receive 1119 and Kenyan children only 844 kilocalories. Are these just chance differences, or are they statistically significant? Like the comparison of methods of teaching reading, this is a one-way design; countries or teaching methods are just lined up for comparison.
- A study of the salaries earned by scientists classifies a sample of scientists by field (biology, medicine, physics, and so on) and by sex. Are there systematic differences in salary across fields? Do male scientists earn more than female scientists? In all fields, or only in some? This two-way design classifies scientists both by field and by sex, leading to more complex questions about the relation between field and sex in their influence on salaries.

• • •

Which of four advertising offers mailed to sample households produces the highest sales in dollars? Which of ten types of automobile tires wears longest? How long do cancer patients live under each of three therapies for their cancer? In each of these settings we wish to compare several treatments. In each case the data are subject to sampling variability—if we mailed the advertising offers to another set of households, we would get different data. We therefore pose the question for inference in terms of the *mean* response. We compare, for example, the mean tread lifetime of different types of tires. In Chapter 7 we met procedures for comparing the means of two populations. We are now ready to extend those methods to problems involving more than two populations. The statistical methodology for comparing several means is called *analysis of variance*, or simply *ANOVA*.

ANOVA

We will consider two ANOVA techniques. When there is only one way to classify the populations of interest, we use *one-way ANOVA* to analyze the data. Comparing the tread lifetimes of ten specific types of tires is a job for one-way ANOVA. Section 10.1 presents this technique in detail. However, in many practical situations there is more than one way to classify the populations. A mail-order firm might want to compare mailings that offer different discounts and also have different layouts. Will a lower price offered in a plain format draw more sales on the average than a higher price offered in a fancy brochure? Analyzing the effect of price and layout together requires *two-way ANOVA*, and adding yet more factors necessitates higher-way ANOVA techniques. Most of the new ideas in ANOVA with more than one factor already appear in two-way ANOVA, which we discuss in Section 10.2.

one-way ANOVA

two-way ANOVA

10.1 ONE-WAY ANALYSIS OF VARIANCE

Do two population means differ? If we have random samples from the two populations, we compute a two-sample t statistic and its P-value to assess the statistical significance of the difference in the sample means. We compare several means in much the same way. Instead of a t statistic, ANOVA uses an F statistic and its P-value to evaluate the null hypothesis that all of several population means are equal.

In the sections that follow, we will examine the basic ideas and assumptions that are needed for the analysis of variance. Although the details differ, many of the concepts are similar to those discussed in the two-sample case.

Comparing means

One-way analysis of variance is a statistical method for comparing several population means. We draw a simple random sample (SRS) from each population and use the data to test the null hypothesis that the population means are all equal.

EXAMPLE 10.1

A medical researcher wants to compare the effectiveness of three different treatments to lower the cholesterol of patients with high blood cholesterol levels. He assigns 60 individuals at random to the three treatments (20 to each) and records the reduction in cholesterol for each patient. ●

EXAMPLE 10.2

An ecologist is interested in comparing the concentration of the pollutant cadmium in five streams. She collects 50 water specimens from each stream and measures the concentration of cadmium in each specimen. ●

These two examples have several similarities. In both there is a single response variable measured on several units; the units are people in the first example and water specimens in the second. We expect the data to be approximately normal and will consider a transformation if they are not. We wish to compare several populations, three treatments in the first example and five streams in the second. But there is also an important difference between the examples. The first is an experiment in which patients are randomly assigned to treatments. The second example is an observational study. By taking water specimens at randomly chosen locations, the ecologist tries to ensure that her specimens from each stream are a random sample of all water specimens that could be obtained from that stream. In both cases we can use ANOVA to compare the mean responses. The same ANOVA methods apply to data from random samples and randomized experiments. Do keep the data-production method in mind when interpreting the results, however. A strong case for causation is best made by a randomized experiment. We will often use the term *groups* for the populations to be compared in a one-way ANOVA.

To assess whether several populations all have the same mean, we compare the means of samples drawn from each population. Figure 10.1 displays the sample means for Example 10.1. Treatment 2 had the highest average cholesterol reduction. But is the observed difference among the groups just the result of chance variation? We do not expect sample means to be equal even if the population means are all identical. The purpose of ANOVA is to assess whether the observed differences among sample means are *statistically significant*. In other words, could a variation this large be plausibly due to chance, or is it good evidence for a difference among the population means? This question can't be answered from the sample means alone. Because the standard deviation of a sample mean $\bar{x}$ is the population standard deviation σ divided by $\sqrt{n}$, the answer depends upon both the variation within the groups of observations and the sizes of the samples.

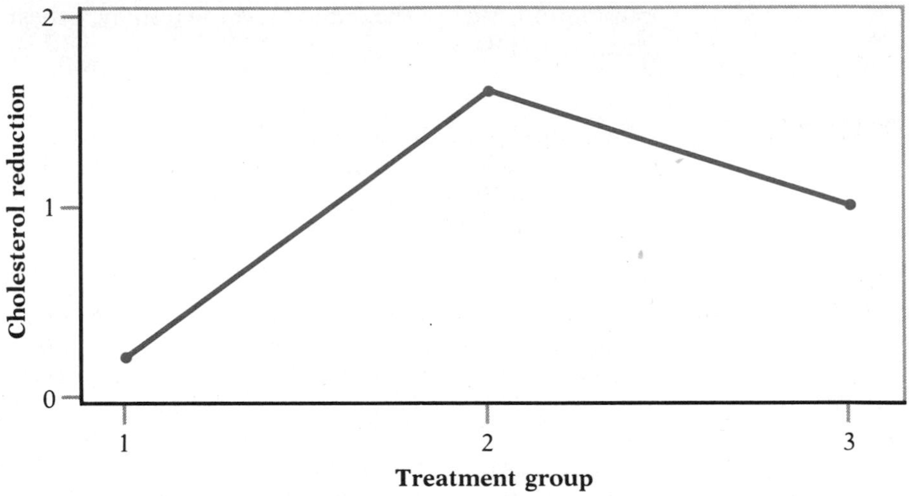

FIGURE 10.1 Mean serum cholesterol reduction in three groups.

Side-by-side boxplots help us see the within-group variation. Compare Figures 10.2(a) and 10.2(b). The sample medians are the same in both figures, but the large variation within the groups in Figure 10.2(a) suggests that the differences among the sample medians could be due simply to chance variation. The data in Figure 10.2(b) are much more convincing evidence that the populations differ. Even the boxplots omit essential information, however. To assess the observed differences we must also know how large the samples are. Nonetheless, boxplots are a good preliminary display of ANOVA data. (ANOVA compares means, and boxplots display medians. If the distributions are nearly symmetric, these two measures of center will be close together.)

Two-sample t statistics compare the means of two populations. If the two populations are assumed to have equal but unknown standard deviations and the sample sizes are both equal to n, the t statistic is

$$t = \frac{\bar{x} - \bar{y}}{s_p\sqrt{\frac{1}{n} + \frac{1}{n}}} = \frac{\sqrt{\frac{n}{2}}(\bar{x} - \bar{y})}{s_p}$$

The square of this t statistic is

$$t^2 = \frac{\frac{n}{2}(\bar{x} - \bar{y})^2}{s_p^2}$$

If we use ANOVA to compare two populations, the ANOVA F statistic

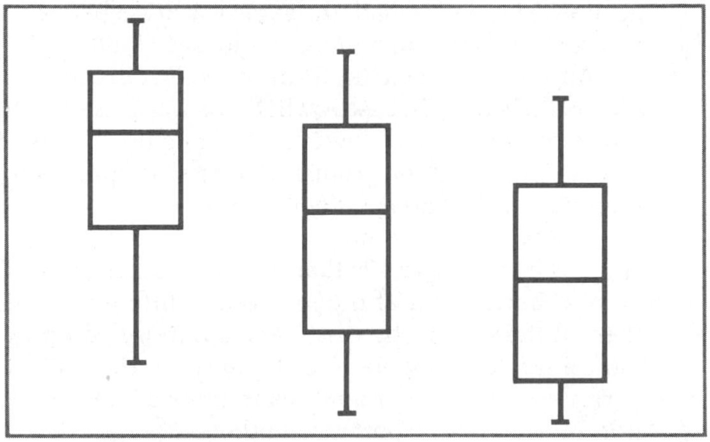

FIGURE 10.2(a) Side-by-side boxplots for three groups with large within-group variation. The differences among centers may be just chance variation.

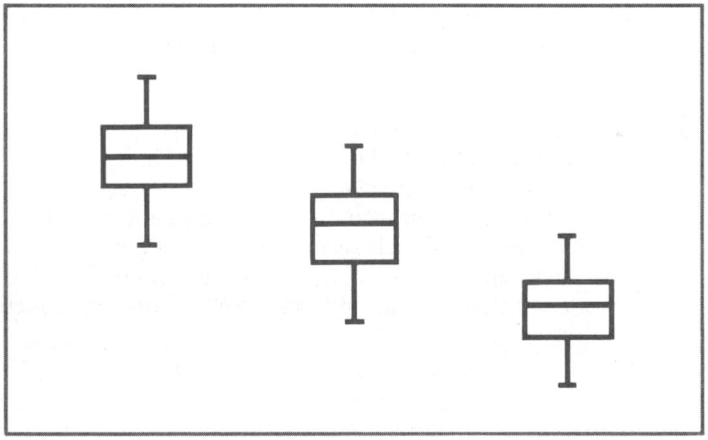

FIGURE 10.2(b) Side-by-side boxplots for three groups with the same centers as in Figure 10.2(a) but with small within-group variation. The differences among centers are more likely to be significant.

is exactly equal to this t^2. We can therefore learn something about how ANOVA works by looking carefully at the statistic in this form.

The numerator in the t^2 statistic measures the variation *between* the groups in terms of the difference between their sample means $\bar{x}$ and $\bar{y}$. It includes a factor for the common sample size n. The numerator can be large because of a large difference between the sample means or because the sample sizes are large. The denominator measures the variation *within* groups by s_p^2, the pooled estimator of the common variance. If the within-group variation is small, the same variation between the groups produces a larger statistic and a more significant result.

Although the general form of the F statistic is more complicated, the idea is the same. To assess whether several populations all have the same mean, we compare the variation *among* the means of several groups with the variation *within* groups. Because we are comparing variation, the method is called *analysis of variance*.

ANOVA tests the null hypothesis that the population means are *all* equal. The alternative is that they are not all equal. This alternative could be true because all of the means are different, or simply because one of them differs from the rest. This is a more complex situation than comparing just two populations. If we reject the null hypothesis, we need to perform some further analysis to draw conclusions about which population means differ from which others.

The computations needed for an ANOVA are more lengthy than those for the t test. For this reason we generally use computer programs to perform the calculations. Automating the calculations frees us from the burden of arithmetic and allows us to concentrate on interpretation. The following example illustrates the practical use of ANOVA in analyzing data. Later we will explore the technical details.

EXAMPLE 10.3

In the computer science department of a large university, many students change their major after the first year. A detailed study[1] of the 256 students enrolled as first-year computer science majors in one year was undertaken to help understand this phenomenon. Students were classified on the basis of their status at the beginning of their second year, and several variables measured at the time of their entrance to the university were obtained. Here are summary statistics for the Scholastic Aptitude Test (SAT) mathematics scores:

Second-year major	n	$\bar{x}$	s
Computer science	103	619	86
Engineering and other sciences	31	629	67
Other	122	575	83

Figure 10.3 gives side-by-side boxplots of the SAT data. Compare Figure 10.3 with the plot of the mean scores in Figure 10.4. The means appear to be different but there is a large amount of overlap in the three distributions. When we perform an ANOVA on these data, we ask a question about the group means. The null hypothesis is that the population mean SAT scores for the three groups are equal, and the alternative is that they are not all equal. The report on the study states that the ANOVA F statistic is 10.35 with $P < 0.001$. ●

When P-values are very small, it is common practice to report them as simply being less than some small number, as was done in the example. If

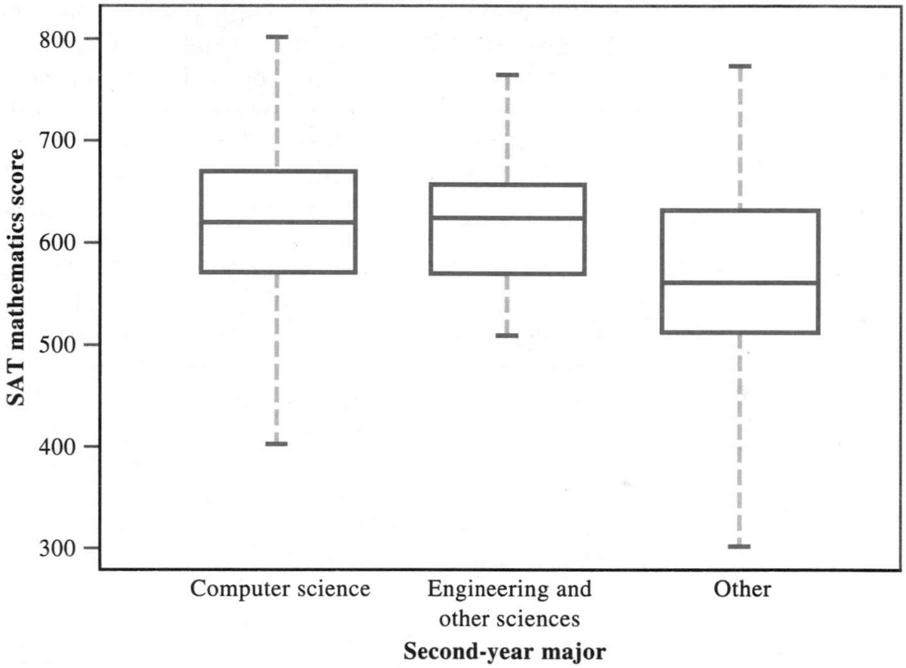

FIGURE 10.3 Side-by-side boxplots of SAT mathematics scores for the change-of-majors study.

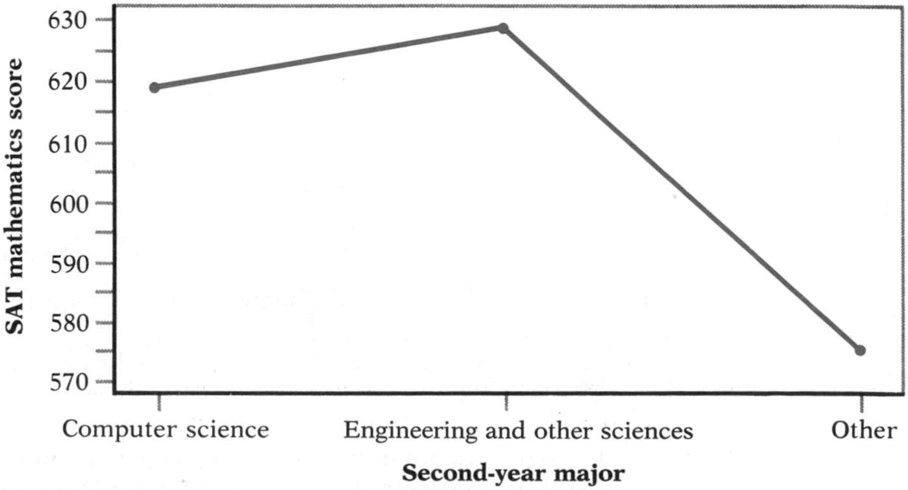

FIGURE 10.4 SAT mathematics score means for the change-of-majors study.

the null hypothesis is true, we would obtain an F at least as large as 10.35 less than 0.1% of the time. The variation in the sample means is much larger than we would expect to occur by chance, so we conclude that the mean SAT scores in the three populations are not all identical.

Although we rejected the null hypothesis, we have not demonstrated that all three population means are different. Inspection of Figures 10.3 and 10.4 suggests that the means of the first two groups are nearly equal and that the third group differs from these two. We need to do some additional analysis to clarify the situation. The researchers expected the mathematics scores of the engineering and other sciences group to be similar to those of the computer science group. They suspected that the students in the third (other major) group would have lower mathematics scores, because these students chose to transfer to a program of study that requires less mathematics. The researchers had this specific comparison in mind before seeing the data, so they used *contrasts* to assess the significance of this specific relation among the means. If no specific relations among the means are in mind before looking at the data, we instead use a *multiple comparison* procedure to determine which pairs of population means differ significantly. In later sections we will explore both contrasts and multiple comparisons in detail.

A purist might argue that ANOVA is inappropriate in Example 10.3. *All* first-year computer science majors in this university for the year studied were included in the analysis. There was no random sampling from larger populations. On the other hand, we can think of these three groups of students as samples from three populations of students in similar circumstances. They may be representative of students in the next few years at the same university or other universities with similar programs. Judgments such as this are very important and must be made not by statisticians but by people who are knowledgeable about the subject of the study.

The ANOVA model

In Chapter 2 we learned that in examining data we seek an overall pattern and deviations from it. We expressed this idea through the equation

$$\text{DATA} = \text{FIT} + \text{RESIDUAL}$$

In the regression model of Chapter 9, the FIT was the population regression line and the RESIDUAL represented the deviations of the data from this line. We now apply this framework to describe the statistical models used in ANOVA. These models provide a convenient way to summarize the assumptions which are the foundation for our analysis. They also give us the necessary notation to describe the calculations needed.

First, recall the statistical model for a random sample of observations from a single normal population with mean μ and standard deviation σ. If the observations are

$$x_1, x_2, \ldots, x_n$$

we can describe this model by saying that the x_j are an SRS from the $N(\mu, \sigma)$ distribution. Another way to describe the same model is to think of the x's varying about their population mean. To do this, write each observation x_j as

$$x_j = \mu + \epsilon_j$$

The ϵ_j are then an SRS from the $N(0, \sigma)$ distribution. Because μ is unknown, the ϵ's cannot actually be observed. This form more closely corresponds to our DATA = FIT + RESIDUAL way of thinking. The FIT part of the model is represented by μ. It is the systematic part of the model, like the line in a regression. The RESIDUAL part is represented by ϵ_j. It represents the deviations of the data from the fit and is due to random, or chance, variation.

There are two unknown parameters in this statistical model: μ and σ. We estimate μ by $\overline{x}$, the sample mean, and σ by s, the sample standard deviation. The differences $e_j = x_j - \overline{x}$ are the sample residuals and correspond to the ϵ_j in the statistical model.

The model for one-way ANOVA is very similar. We take random samples from each of I different populations. The sample size is n_i for the ith population. Let x_{ij} represent the jth observation from the ith population. The I population means are the FIT part of the model and are represented by μ_i. The random variation, or RESIDUAL, part of the model is represented by the deviations ϵ_{ij} of the observations from the means. The model is

$$x_{ij} = \mu_i + \epsilon_{ij}$$

for $i = 1, \ldots, I$ and $j = 1, \ldots, n_i$. The ϵ_{ij} are assumed to be an SRS from an $N(0, \sigma)$ distribution. The sample sizes n_i may differ, but the standard deviation σ is the same in all of the populations. Figure 10.5 pictures

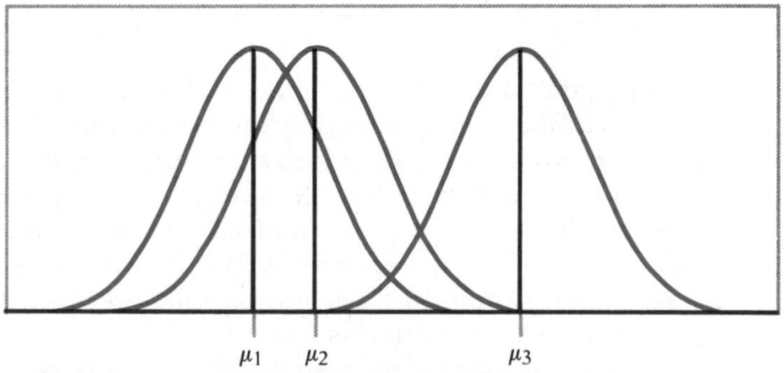

FIGURE 10.5 Model for one-way ANOVA with three groups. The three populations have normal distributions with the same standard deviation.

this model for $I = 3$. The three population means μ_i are different, but the shapes of the three normal distributions are the same, reflecting the assumption that all three populations have the same standard deviation.

EXAMPLE 10.4

A survey of college students attempted to determine how much money they spend per year on textbooks and to compare students by year of study. From lists of students provided by the registrar, SRSs of size 50 were chosen from each of the four classes (freshmen, sophomores, juniors, and seniors). The students selected were asked how much they spent on textbooks during the current semester.

There are $I = 4$ populations. The population means μ_1, μ_2, μ_3, and μ_4 are the average amounts spent on textbooks by *all* freshmen, sophomores, juniors, and seniors at this college for this semester. The sample sizes n_i are 50, 50, 50, and 50.

Suppose the first freshman sampled is Eve Brogden. Then $x_{1,1}$ represents the amount spent by Eve. The data for the other freshmen sampled are denoted by $x_{1,2}, x_{1,3}, \ldots, x_{1,50}$. Similarly, the data for the other groups have a first subscript indicating the group and a second subscript indicating the student in the group.

According to our model, Eve's spending is $x_{1,1} = \mu_1 + \epsilon_{1,1}$, where μ_1 is the average for *all* of the students in the freshman class and $\epsilon_{1,1}$ is the chance variation due to Eve's specific needs. We are assuming that the ϵ_{ij} are independent and normally distributed (at least approximately) with mean 0 and standard deviation σ. ●

The unknown parameters in the statistical model for ANOVA are the population means μ_i and the common population standard deviation σ. To estimate μ_i we use the sample mean for the ith group

$$\bar{x}_i = \frac{1}{n_i}\sum_{j=1}^{n_i} x_{ij}$$

The sample residuals are $e_{ij} = x_{ij} - \bar{x}_i$. Thus, for the *sample* data we represent the FIT by $\bar{x}_i$ and the RESIDUAL by e_{ij}.

The ANOVA model assumes that the population standard deviations are all equal. If we have unequal standard deviations, we generally try to transform the data so that they are approximately equal. We might, for example, work with $\sqrt{x_{ij}}$ or $\log x_{ij}$. Fortunately, we can often find a transformation that *both* makes the group standard deviations more nearly equal and also makes the distributions of observations in each group more nearly normal. If the standard deviations are markedly different and cannot be made similar by a transformation, inference requires methods that are beyond the scope of this text.

Unfortunately, formal tests for the equality of standard deviations in several groups share the lack of robustness against nonnormality that we noted in Chapter 7 for the case of two groups. Because ANOVA procedures are not extremely sensitive to unequal standard deviations, we do not recommend a formal test of equality of standard deviations as a preliminary to the ANOVA. Instead, we will use the following rule of thumb:

> **RULE FOR EXAMINING STANDARD DEVIATIONS IN ANOVA**
>
> If the ratio of the largest sample standard deviation to the smallest sample standard deviation is less than 2, we can use methods based on the assumption of equal standard deviations and our results will still be approximately correct.[2]

When we assume that the population standard deviations are equal, each sample standard deviation is an estimate of σ. To combine these into a single estimate, we use a generalization of the pooling method introduced in Chapter 7.

> **POOLED ESTIMATOR OF VARIANCE**
>
> Suppose we have sample variances $s_1^2, s_2^2, \ldots, s_I^2$ from I independent SRSs of sizes $n_1, n_2, \ldots, n_I$ from populations with common variance σ^2. The pooled sample variance
>
> $$s_p^2 = \frac{(n_1 - 1)s_1^2 + (n_2 - 1)s_2^2 + \cdots + (n_I - 1)s_I^2}{(n_1 - 1) + (n_2 - 1) + \cdots + (n_I - 1)}$$
>
> is an unbiased estimator of σ^2.

Pooling gives more weight to groups with larger sample sizes. If the sample sizes are equal, s_p^2 is just the average of the I sample variances. To compute the pooled standard deviation s_p, we take the square root of the pooled variance s_p^2. Note that this is *not* the same as averaging the standard deviations.

EXAMPLE 10.5

In the change-of-majors study of Example 10.3 there are $I = 3$ groups and the sample sizes are $n_1 = 103$, $n_2 = 31$, and $n_3 = 122$. The sample means for the SAT mathematics scores are $\bar{x}_1 = 619$, $\bar{x}_2 = 629$, and $\bar{x}_3 = 575$. The sample standard deviations are $s_1 = 86$, $s_2 = 67$, and $s_3 = 83$.

Because the ratio of the largest to the smallest standard deviation is

$$\frac{86}{67} = 1.28$$

which is less than 2, our rule of thumb indicates that we can use the assumption of equal population standard deviations. The pooled variance estimate is

$$\begin{aligned} s_p^2 &= \frac{(n_1-1)s_1^2+(n_2-1)s_2^2+(n_3-1)s_3^2}{(n_1-1)+(n_2-1)+(n_3-1)} \\ &= \frac{(102)(86)^2+(30)(67)^2+(121)(83)^2}{102+30+121} \\ &= \frac{1,722,631}{253} = 6809 \end{aligned}$$

The pooled standard deviation is $s_p = \sqrt{6809} = 82.5$. This is our estimate of the common standard deviation σ of the SAT mathematics scores in the three populations of students. ●

The ANOVA table and the *F* test

Comparison of several means is accomplished by using an F statistic to compare the variation among groups with the variation within groups. We now show how to calculate the F statistic. The calculations are organized in an ANOVA table, which contains numerical measures of the variation among groups and within groups.

First we will summarize our assumptions and hypotheses for one-way ANOVA. As usual, I represents the number of populations to be compared and x_{ij} represents the jth observation from the ith population.

ASSUMPTIONS AND HYPOTHESES FOR ONE-WAY ANOVA

We have independent SRSs of size n_i from each of I normal populations. The population means μ_i may differ, but all populations have the same standard deviation σ. The μ_i and σ are unknown parameters.

The statistical model is

$$x_{ij} = \mu_i + \epsilon_{ij}$$

for $i = 1, \ldots, I$ and $j = 1, \ldots, n_i$, where the deviations ϵ_{ij} are an SRS from an $N(0, \sigma)$ distribution.

The null and alternative hypotheses are

$$H_0\colon \mu_1 = \mu_2 = \ldots = \mu_I$$
$$H_a\colon \text{not all of the } \mu_i \text{ are equal}$$

The example that follows illustrates how to do a one-way ANOVA. The calculations are generally performed using statistical software on a computer, so we focus on interpretation of the output.

TABLE 10.1 Pretest reading scores

Group	Subject	Score	Group	Subject	Score	Group	Subject	Score
Basal	1	4	DRTA	23	7	Strat	45	11
Basal	2	6	DRTA	24	7	Strat	46	7
Basal	3	9	DRTA	25	12	Strat	47	4
Basal	4	12	DRTA	26	10	Strat	48	7
Basal	5	16	DRTA	27	16	Strat	49	7
Basal	6	15	DRTA	28	15	Strat	50	6
Basal	7	14	DRTA	29	9	Strat	51	11
Basal	8	12	DRTA	30	8	Strat	52	14
Basal	9	12	DRTA	31	13	Strat	53	13
Basal	10	8	DRTA	32	12	Strat	54	9
Basal	11	13	DRTA	33	7	Strat	55	12
Basal	12	9	DRTA	34	6	Strat	56	13
Basal	13	12	DRTA	35	8	Strat	57	4
Basal	14	12	DRTA	36	9	Strat	58	13
Basal	15	12	DRTA	37	9	Strat	59	6
Basal	16	10	DRTA	38	8	Strat	60	12
Basal	17	8	DRTA	39	9	Strat	61	6
Basal	18	12	DRTA	40	13	Strat	62	11
Basal	19	11	DRTA	41	10	Strat	63	14
Basal	20	8	DRTA	42	8	Strat	64	8
Basal	21	7	DRTA	43	8	Strat	65	5
Basal	22	9	DRTA	44	10	Strat	66	8

EXAMPLE 10.6

A study of reading comprehension in children compared three methods of instruction.[3] As is common in such studies, several pretest variables were measured before any instruction was given. One purpose of the pretest was to see if the three groups of children were similar in their comprehension skills. One of the pretest variables was an "intruded sentences" measure, which measures one type of reading comprehension skill. The data for the 22 subjects in each group are given in Table 10.1. The three methods of instruction are called basal, DRTA, and strategies. We use Basal, DRTA, and Strat as values for the categorical variable indicating which method each student received. ●

Before proceeding with ANOVA, several important preliminaries need attention. Normal quantile plots for the three groups appear in Figures 10.6(a), 10.6(b), and 10.6(c). The data look reasonably normal. Although there are a small number of possible values for the score (the values are all integers between 4 and 17), this should not cause difficulties in using ANOVA. Side-by-side boxplots do not give a particularly good picture of these data. This is a consequence of the relatively small number of observations in each group (22) combined with the fact that the variable has only a few possible values.

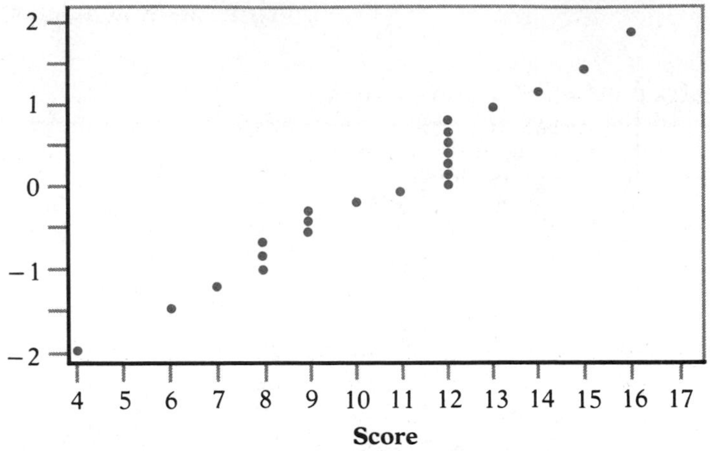

FIGURE 10.6(a) Normal quantile plot of the pretest scores in the Basal group of the reading comprehension study.

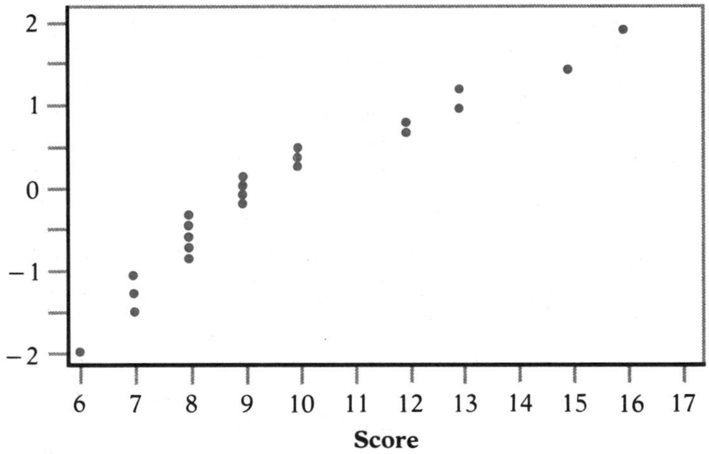

FIGURE 10.6(b) Normal quantile plot of the pretest scores in the DRTA group of the reading comprehension study.

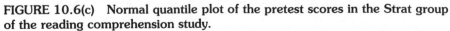

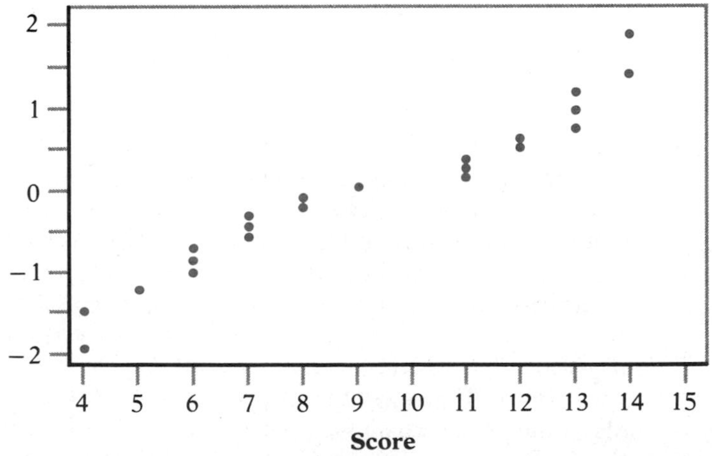

FIGURE 10.6(c) Normal quantile plot of the pretest scores in the Strat group of the reading comprehension study.

VARIABLE	N	MEAN	STANDARD DEVIATION
----------GROUP=Basal----------			
SCORE	22	10.50000000	2.97209242
----------GROUP=DRTA----------			
SCORE	22	9.72727273	2.69358706
----------GROUP=Strat----------			
SCORE	22	9.13636364	3.34230395

FIGURE 10.7 Summary statistics for the pretest scores in the three groups of the reading comprehension study.

Figure 10.7 gives the summary statistics generated by the SAS procedure MEANS. The ratio of the largest standard deviation to the smallest standard deviation is $3.34/2.69 = 1.24$, which is less than 2. Our rule of thumb tells us that we can assume that the three populations have the same standard deviation. The pooled variance is

$$\begin{aligned} s_p^2 &= \frac{(n_1-1)s_1^2 + (n_2-1)s_2^2 + (n_3-1)s_3^2}{(n_1-1)+(n_2-1)+(n_3-1)} \\ &= \frac{(21)(2.972)^2 + (21)(2.694)^2 + (21)(3.342)^2}{21+21+21} \\ &= \frac{572.45}{63} = 9.09 \end{aligned}$$

Because the sample sizes are equal, this s_p^2 is simply the average of the three variances. The pooled standard deviation is $s_p = \sqrt{9.09} = 3.01$. Note that we would *not* get the same answer if we averaged the three standard deviations.

The ANOVA model assumes that the ϵ_{ij} are an SRS from an $N(0, \sigma)$ distribution. The sample versions of these deviations are the residuals $e_{ij} = x_{ij} - \bar{x}_i$, the differences between each observation and its group mean. Subtracting the mean from each observation lets us examine the normality of the combined data from the three groups. Figure 10.8 is a normal quantile plot for the residuals.

The data look reasonably normal and the assumption of equal variances can be used, so we proceed with ANOVA. The results produced by the SAS General Linear Models (GLM) procedure are shown in Figure

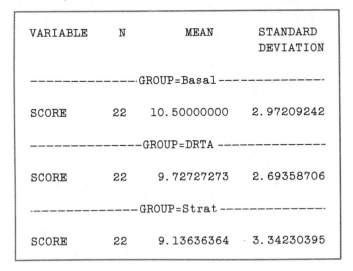

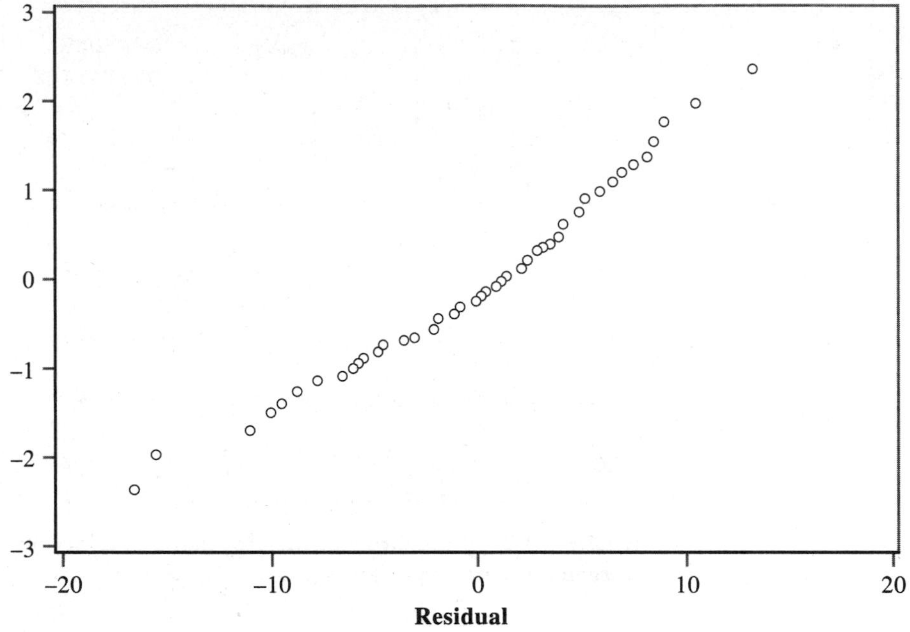

FIGURE 10.8 Normal quantile plot of the residuals for the pretest scores in the reading comprehension study.

10.9. The calculated value of the F statistic appears under the heading F VALUE and its P-value is given under the heading PR > F. The value of F is 1.13, with a P-value of 0.3288. That is, an F of 1.13 or larger would occur about 33% of the time by chance, when the population means are equal. Because the P-value is large, the observed variation in the sample means can be attributed to chance, and we have no reason to reject the null hypothesis that the three populations have equal means.

ANOVA table

The ANOVA table The information in an analysis of variance is organized in an *ANOVA table*. In the computer output in Figure 10.9, the columns of this table are labeled SOURCE, DF, SUM OF SQUARES, MEAN SQUARE, and F VALUE. The rows are labeled MODEL, ERROR, and CORRECTED TOTAL. These are the three sources of variation in the one-way ANOVA.

The MODEL row in the table corresponds to the FIT term in our DATA = FIT + RESIDUAL way of thinking. It gives information related to the variation *among* group means. In writing ANOVA tables we will use the generic label *groups* or some other term which describes the factor being studied for this row.

The ERROR row in the table corresponds to the RESIDUAL term in our DATA = FIT + RESIDUAL. It gives information related to the variation

GENERAL LINEAR MODELS PROCEDURE

DEPENDENT VARIABLE: SCORE

SOURCE	DF	SUM OF SQUARES	MEAN SQUARE	F VALUE
MODEL	2	20.57575758	10.28787879	1.13
ERROR	63	572.45454545	9.08658009	PR > F
CORRECTED TOTAL	65	593.03030303		0.3288

R-SQUARE	C.V.	ROOT MSE	SCORE MEAN
0.034696	30.7972	3.01439548	9.78787879

SOURCE	DF	TYPE I SS	F VALUE	PR > F
GROUP	2	20.57575758	1.13	0.3288

FIGURE 10.9 Analysis of variance output for the pretest scores in the reading comprehension study.

within groups. The term "error" is most appropriate for experiments in the physical sciences where the observations within a group differ because of measurement error. In business and the biological and social sciences, on the other hand, the within-group variation is often due to the fact that not all firms or plants or people are the same. This sort of variation is not due to errors and is better described as "residual." Finally, the CORRECTED TOTAL row in the table corresponds to the DATA term in our DATA = FIT + RESIDUAL framework. Often this row is simply labeled TOTAL.

sum of squares

As you might expect, each *sum of squares* is a sum of squared deviations. We use SSG, SSE, and SST for the entries in this column corresponding to groups, error, and total. To see how the sums of squares are calculated, it is helpful to write the data in the following form:

$$(x_{ij} - \overline{x}) = (\overline{x}_i - \overline{x}) + (x_{ij} - \overline{x}_i)$$

Here

$$\overline{x} = \frac{\sum_{\text{obs}} x_{ij}}{N}$$

is the overall mean of all of the observations and

$$N = \sum n_i$$

is the total number of observations in all of our samples.

Each of SST, SSG, and SSE measures a different type of deviation. The first is the deviation of an observation from the overall mean, $x_{ij} - \bar{x}$. These deviations reflect the overall variation among the observations. If we calculate this deviation for each observation, square, and sum, we obtain SST. The sum of squares SST divided by $N - 1$ is just the ordinary sample variance of all the observations. SST therefore represents the total variation in all of the data. Some of this variation is due to the fact that the group means are not all the same, and the rest is due to variation among the individual observations within each group.

The deviations of the group means from the overall mean, $\bar{x}_i - \bar{x}$, reflect the variation among the groups. If we calculate this deviation for each observation, square, and sum, we obtain SSG. Because this deviation is the same for all observations in a group, we can calculate the square once for each group and multiply by n_i to get the sum of squares. SSG represents that part of SST that is due to the fact that the groups have different means. If all groups had the same $\bar{x}_i$, SSG would be 0.

The deviation of an individual observation from its group mean, $x_{ij} - \bar{x}_i$, reflects the error or residual variation within the groups. Again we calculate this deviation for each observation, square, and sum to obtain SSE. The sample residuals are $e_{ij} = x_{ij} - \bar{x}_i$, so SSE is the sum of the squared residuals. It represents the part of SST that is due to the within-group variation.

The SUM OF SQUARES column in Figure 10.9 gives the values for these three sums of squares. The output gives many more digits than we need. Rounded-off values are SSG = 21, SSE = 572, and SST = 593. In this example it appears that most of the variation is coming from ERROR, that is, from within groups. You can verify that SST = SSG + SSE for this example. This fact is true in general. The total variation is always equal to the among-group variation plus the within-group variation.

degrees of freedom

The entries in the DF column are called *degrees of freedom*. They are a bit more difficult to understand than the sums of squares but are much easier to calculate. Each entry in the DF column is associated with the sum of squares in the SUM OF SQUARES column and can be found by examining the deviations that make up the sum of squares. Degrees of freedom represent the number of terms that are free to vary independently of the other terms in the sum of squares. The sums of squares in general all grow larger as the number of terms in the sums increases. The degrees of freedom supply a number by which to divide a sum of squares to yield a measure of the "average" variation.

First, SST is based on the deviations $x_{ij} - \bar{x}$. When all but one of these deviations are given, the last one can be calculated. Another way to put this is that the N observations are compared with one mean. Therefore, DFT $= N - 1$. This is the same as the degrees of freedom for the ordinary sample variance. Next, consider SSG. It is based on the deviations $\bar{x}_i - \bar{x}$. Now I sample means are compared with one overall mean. So the degrees of freedom for groups are DFG $= I - 1$. Finally, SSE is the sum of squares of the deviations $x_{ij} - \bar{x}_i$. Here we have N observations being compared with I sample means. There are $N - I$ degrees of freedom for error; that is, DFE $= N - I$.

In our example, we have $I = 3$ and $N = 66$. Therefore,

$$\text{DFT} = N - 1 = 66 - 1 = 65$$
$$\text{DFG} = I - 1 = 3 - 1 = 2$$
$$\text{DFE} = N - I = 66 - 3 = 63$$

These are the entries in the DF column of Figure 10.9. Note that the degrees of freedom add in the same way that the sums of squares add. That is, DFT = DFG + DFE.

mean square

For each source of variation, the *mean square* is the sum of squares divided by the degrees of freedom. You can verify this by doing the divisions for the values given on the output in Figure 10.9.

> **SUMS OF SQUARES, DEGREES OF FREEDOM, AND MEAN SQUARES**
>
> Sums of squares represent variation present in the data. They are calculated by summing squared deviations. In the one-way ANOVA there are three sources of variation: groups, error, and total. The sums of squares are related by the formula
>
> $$\text{SST} = \text{SSG} + \text{SSE}$$
>
> Thus the total variation is composed of two parts, one due to groups and one due to error.
>
> Degrees of freedom are related to the deviations that are used in the sums of squares. The degrees of freedom are related in the same way as the sums of squares,
>
> $$\text{DFT} = \text{DFG} + \text{DFE}$$
>
> To calculate each mean square, divide the corresponding sum of squares by its degrees of freedom.

At the beginning of this section we found the pooled variance for Example 10.6 to be $572.45/63 = 9.09$. Now look at the ERROR row on the computer output in Figure 10.9. The mean square in the error row is MSE = 9.09. It is true in general that

$$s_p^2 = \text{MSE} = \frac{\text{SSE}}{\text{DFE}}$$

That is, the error mean square is an estimate of the within-group variance, σ^2. The output gives s_p under the heading ROOT MSE.

There is no MEAN SQUARE entry in the CORRECTED TOTAL row of the SAS output because this mean square (MST) is not needed in later calculations. To supply it, we calculate

$$\text{MST} = \frac{593}{65} = 9.12$$

MST is simply the usual sample variance that we would calculate if all of the observations came from a single population.

The *F* Test If H_0 is true, there are no differences among the group means. Both MSE and MSG are then estimates of σ^2. On the other hand, if H_a is true, MSG tends to overestimate σ^2 because it includes group-to-group variation as well as within-group variation. Because MSE is based only on the within-group variation, it is a valid estimate of σ^2 both when H_0 is true *and* when H_a is true. The ratio MSG/MSE is a statistic that is approximately 1 if H_0 is true and tends to be larger if H_a is true. This is the ANOVA F statistic. In our example, MSG = 10.29 and MSE = 9.09, so the ANOVA F statistic is

$$F = \frac{\text{MSG}}{\text{MSE}} = \frac{10.29}{9.09} = 1.13$$

When H_0 is true, the F statistic has an F distribution that depends upon two numbers: the *degrees of freedom for the numerator* and the *degrees of freedom for the denominator*. These degrees of freedom are those associated with the mean squares in the numerator and denominator of the F statistic. For one-way ANOVA, the degrees of freedom for the numerator are DFG $= I - 1$ and the degrees of freedom for the denominator are DFE $= N - I$. We use the notation $F(I-1, N-I)$ for this distribution.

THE ANOVA *F* TEST

To test the null hypothesis in a one-way ANOVA, calculate the statistic

$$F = \frac{\text{MSG}}{\text{MSE}}$$

When H_0 is true, the F statistic has the $F(I-1, N-I)$ distribution. When H_a is true, the F statistic tends to be large. We reject H_0 in favor of H_a if the F statistic is sufficiently large.

The P-value of the F test is the probability that a random variable having the $F(I-1, N-I)$ distribution is greater than or equal to the calculated value of the F statistic.

Tables of F critical values are available for use when software does not give the P-value. Table F in the back of the book contains the F critical values for tail probabilities $p = 0.100, 0.050, 0.025, 0.010$, and 0.001. For one-way ANOVA we use critical values from the table corresponding to $I - 1$ degrees of freedom in the numerator and $N - I$ degrees of freedom in the denominator. We have already seen several examples where the F statistic and its P-value were used to choose between H_0 and H_a.

EXAMPLE 10.7

p	Critical value
0.100	2.33
0.050	3.04
0.025	3.76
0.010	4.71
0.001	7.15

In the study of computer science students in Example 10.3, $F = 10.35$. There were three populations, so the degrees of freedom in the numerator are DFG $= I - 1 = 2$. For this example the degrees of freedom in the denominator are $N - I = 256 - 3 = 253$. In Table F we first find the column corresponding to 2 degrees of freedom in the numerator. For the degrees of freedom in the denominator (DFD in the table), we see that there are entries for 200 and 1000. These entries are very close. To be conservative we use critical values corresponding to 200 degrees of freedom in the denominator since these are slightly larger. Because 10.35 is larger than all of the tabulated values, we reject H_0 and conclude that the differences in means are statistically significant with $P < 0.001$. ●

EXAMPLE 10.8

p	Critical value
0.100	2.39
0.050	3.15
0.025	3.93
0.010	4.98
0.001	7.77

In the comparison of teaching methods of Example 10.6, $F = 1.13$. Because we are comparing three populations, the degrees of freedom in the numerator are DFG $= I - 1 = 2$ and the degrees of freedom in the denominator are DFE $= N - I = 66 - 3 = 63$. Table F has entries for 2 degrees of freedom in the numerator. We do not find entries for 63 degrees of freedom in the denominator, so we use the slightly larger values corresponding to 60. Because 1.13 is less than any of the tabulated values, we do not have strong evidence in the data against the hypothesis of equal means. We conclude that the differences in sample means are not significant at the 10% level. Statistical software (Figure 10.9) gives the exact value $P = 0.3288$. ●

Remember that the F test is always one-sided because any differences among the group means tend to make F large. The ANOVA F test shares the robustness of the two-sample t test. It is relatively insensitive to moderate nonnormality and unequal variances, especially when the sample sizes are similar.

The following display shows the general form of a one-way ANOVA table with the F statistic. The formulas in the sum of squares column can be used for calculations in small problems. There are other formulas that are more efficient for hand or calculator use, but ANOVA calculations are usually done by computer software.

Source	Degrees of freedom	Sum of squares	Mean square	F
Groups	$I-1$	$\sum_{\text{groups}} n_i(\overline{x_i}-\overline{x})^2$	SSG/DFG	MSG/MSE
Error	$N-I$	$\sum_{\text{groups}}(n_i-1)s_i^2$	SSE/DFE	
Total	$N-1$	$\sum_{\text{obs}}(x_{ij}-\overline{x})^2$	SST/DFT	

EXAMPLE 10.9

In Example 10.8 we could not reject H_0 that the three groups of students in Example 10.6 had equal mean scores on a pretest of reading comprehension. We now change the data in a way that favors H_a. Add one point to the score of each student in the Basal group, and subtract one point from each of the observations in the Strat group. Leave the observations in the DRTA group as they were. Call the new variable SCOREX. The summary statistics for SCOREX appear in Figure 10.10. Compared with the original means, given in Figure 10.7, the SCOREX means are farther apart. Note that we have not changed any of the standard deviations.

The ANOVA results appear in Figure 10.11. Compare this table with the ANOVA for the original variable SCORE given in Figure 10.9. The degrees of freedom are the same. The values of SSG and SST have increased, but SSE has not changed. In constructing SCOREX, we have added more total variation to the data, and all of this added variation went into the variation between groups. The increase in SSG is reflected in MSG. The new value is about six times the old. Because MSE has not changed, the F statistic is also about six times as large. The P-value has decreased dramatically and we now have evidence to reject H_0 in favor of H_a. ●

```
VARIABLE     N         MEAN        STANDARD
                                   DEVIATION

----------GROUP=Basal----------

SCOREX      22    11.50000000    2.97209242

---------- GROUP=DRTA ----------

SCOREX      22     9.72727273    2.69358706

----------GROUP=Strat----------

SCOREX      22     8.13636364    3.34230395
```

FIGURE 10.10 **Summary statistics for the adjusted scores SCOREX in the reading comprehension study.**

```
                    GENERAL LINEAR MODELS PROCEDURE

  DEPENDENT VARIABLE: SCOREX

SOURCE                DF    SUM OF SQUARES          MEAN SQUARE    F VALUE

MODEL                  2      124.57575758          62.28787879       6.85

ERROR                 63      572.45454545           9.08658009     PR > F

CORRECTED TOTAL       65      697.03030303                          0.0020

R-SQUARE            C.V.          ROOT MSE          SCOREX MEAN

0.178724         30.7972        3.01439548           9.78787879


SOURCE                DF         TYPE I SS    F VALUE    PR > F

GROUP                  2      124.57575758       6.85    0.0020
```

FIGURE 10.11 Analysis of variance output for SCOREX in the reading comprehension study.

One other item in the computer output for ANOVA is worth noting. For an analysis of variance, we define the *coefficient of determination* as

$$R^2 = \frac{\text{SSG}}{\text{SST}}$$

The coefficient of determination plays the same role as the squared multiple correlation R^2 in a multiple regression. In Figure 10.9, R^2 is listed under the heading R-SQUARE. The value is 0.035. This result says that the FIT part of the model (that is, differences among the means of the groups) accounts for 3.5% of the total variation in the data. Compare this value with that for the changed data described in Example 10.9 and given in Figure 10.11. The change has increased the coefficient of determination to 17.9%.

Contrasts

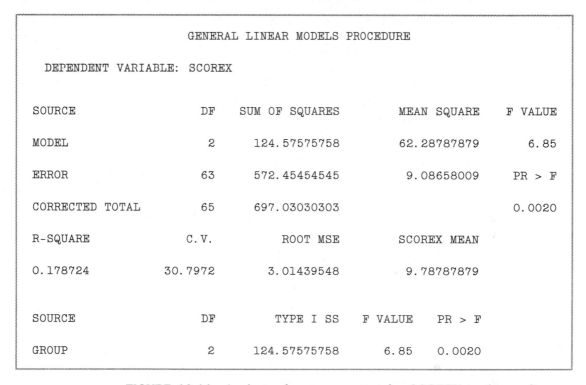

The ANOVA F test gives a general answer to a general question: Are the differences among observed group means significant? Unfortunately, a small P-value simply tells us that the group means are different; it does not tell us specifically which means differ from each other. Plotting and inspecting the means give us some indication of where the differences lie, but we would like to supplement inspection with formal inference.

In the ideal situation, specific questions regarding comparisons among the means are posed before the data are collected. We can answer specific questions of this kind and attach a level of confidence to the answers we give. We now explore these ideas through an example.

EXAMPLE 10.10

Example 10.6 describes a randomized comparative experiment to compare three methods for teaching reading. We analyzed the pretest scores and found no reason to reject H_0 that the three groups had similar population means on this measure. This was the desired outcome. We now turn to the response variable, a measure of reading comprehension called COMP that was measured by a test taken after the instruction was complete.

Figure 10.12 gives the summary statistics computed by the SAS procedure MEANS for COMP. Side-by-side boxplots appear in Figure 10.13 and Figure 10.14 plots the group means. The ANOVA results generated by the SAS procedure GLM are given in Figure 10.15.

The ANOVA null hypothesis is

$$H_0\colon \mu_B = \mu_D = \mu_S$$

where the subscripts correspond to the group labels Basal, DRTA, and Strat. Figure 10.15 shows that $F = 4.48$ with degrees of freedom 2 and 63. The P-value is 0.0152 and R^2 is 0.12. We have good evidence against H_0.

What can the researchers conclude from this analysis? The alternative hypothesis is true if $\mu_B \neq \mu_D$ or if $\mu_B \neq \mu_S$ or if $\mu_D \neq \mu_S$ or if any combination of these statements is true. We would like to be more specific. ●

```
VARIABLE     N         MEAN          STANDARD
                                     DEVIATION

----------GROUP=Basal----------

COMP        22    41.04545455     5.63557808

---------- GROUP=DRTA ----------

COMP        22    46.72727273     7.38841963

----------GROUP=Strat           ---

COMP        22    44.27272727     5.76675049
```

FIGURE 10.12 Summary statistics for the comprehension scores in the three groups of the reading comprehension study.

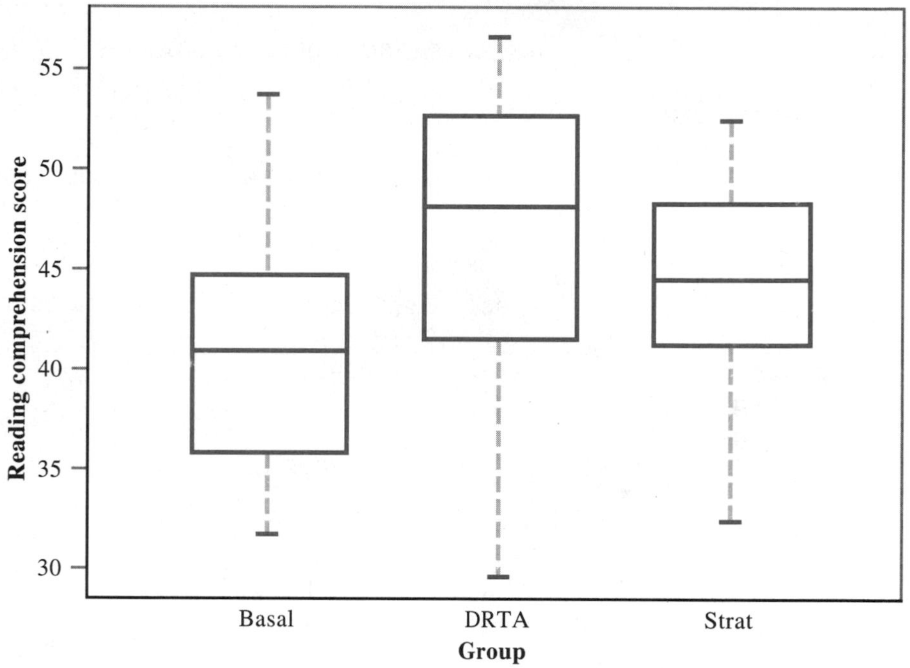

FIGURE 10.13 Side-by-side boxplots of the comprehension scores for the reading comprehension study.

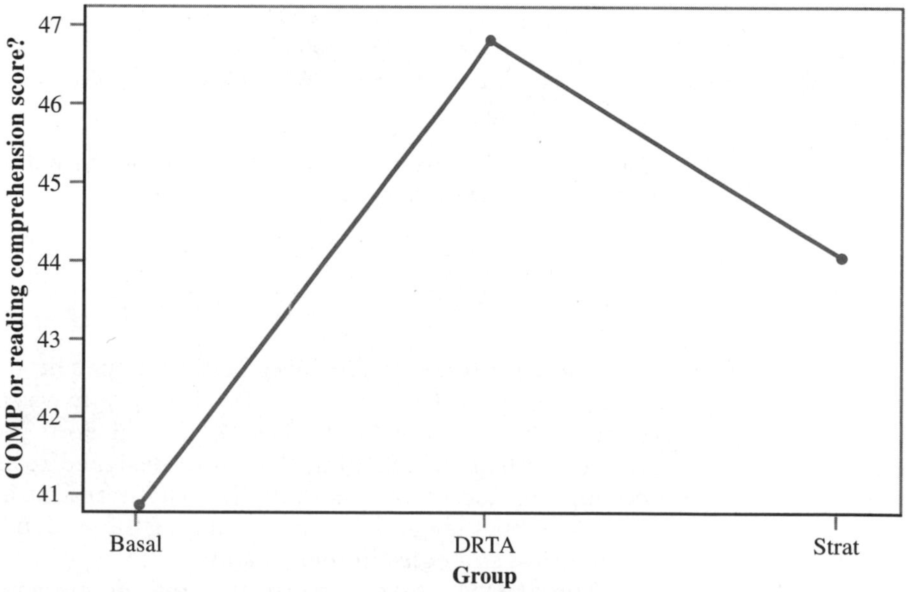

FIGURE 10.14 Comprehension score group means for the reading comprehension study.

GENERAL LINEAR MODELS PROCEDURE

DEPENDENT VARIABLE: COMP

SOURCE	DF	SUM OF SQUARES	MEAN SQUARE	F VALUE
MODEL	2	357.30303030	178.65151515	4.48
ERROR	63	2511.68181818	39.86796537	PR > F
CORRECTED TOTAL	65	2868.98484848		0.0152

R-SQUARE	C.V.	ROOT MSE	COMP MEAN
0.124540	14.3453	6.31410844	44.01515152

SOURCE	DF	TYPE I SS	F VALUE	PR > F
GROUP	2	357.30303030	4.48	0.0152

CONTRAST	DF	SS	F VALUE	PR > F
B VS D AND S	1	291.03030303	7.30	0.0088
D VS S	1	66.27272727	1.66	0.2020

FIGURE 10.15 Analysis of variance and contrasts output for the comprehension scores in the reading comprehension study.

The researchers were investigating a specific theory about reading comprehension. The instruction for Basal group was the standard method commonly used in schools. The DRTA and Strat groups received innovative methods of teaching that were designed to increase the reading comprehension of the children. The DRTA and Strat methods were not identical, but they both involved teaching the students to use similar comprehension strategies in their reading.

The ANOVA F test demonstrates that there is some difference among the mean responses, but it does not directly answer the researchers' questions. The primary question can be formulated as a null hypothesis:

$$H_{01}: \frac{1}{2}(\mu_D + \mu_S) = \mu_B$$

with the alternative

$$H_{a1}: \frac{1}{2}(\mu_D + \mu_S) > \mu_B$$

The hypothesis H_{01} compares the average of the two innovative methods (DRTA and Strat) with the standard method (Basal). The alternative is one-sided because the researchers are interested in demonstrating that the new methods are better than the old.

A secondary question involves a comparison of the two new methods. We formulate this as the hypothesis that the methods DRTA and Strat are equally effective:

$$H_{02}: \mu_D = \mu_S$$

versus the alternative

$$H_{a2}: \mu_D \neq \mu_S$$

The subscripts 1 and 2 distinguish the two sets of hypotheses.

Each of H_{01} and H_{02} says that a combination of population means is 0. These combinations of means are called *contrasts*. We use ψ, the Greek letter psi, for contrasts among population means. The two contrasts that arise from our two null hypotheses are

$$\begin{aligned}\psi_1 &= -\mu_B + \frac{1}{2}(\mu_D + \mu_S) \\ &= (-1)\mu_B + (.5)\mu_D + (.5)\mu_S\end{aligned}$$

and

$$\psi_2 = \mu_D - \mu_S$$

In each case, the value of the contrast is 0 when H_0 is true.

A contrast expresses an effect in the population as a combination of population means. To estimate the contrast, form the corresponding *sample contrast* by using sample means in place of population means. Under the ANOVA assumptions, a sample contrast is a linear combination of independent normal variables and therefore has a normal distribution. We can obtain the standard error of a contrast by using the rules for variances given in Section 4.3. Inference is based on t statistics. Here are the details.

CONTRASTS

A contrast is a combination of population means of the form

$$\psi = \sum a_i \mu_i$$

where the coefficients a_i have sum 0. The corresponding sample contrast is

$$c = \sum a_i \bar{x}_i$$

The standard error of c is

$$s_c = s_p \sqrt{\sum \frac{a_i^2}{n_i}}$$

To test the null hypothesis

$$H_0\colon \psi = 0$$

use the t statistic

$$t = \frac{c}{s_c}$$

with degrees of freedom DFE that are associated with s_p. The alternative hypothesis may be one-sided or two-sided.

A level C confidence interval for ψ is

$$c \pm t^* s_c$$

where t^* is the upper $(1 - C)/2$ critical value for the $t(\text{DFE})$ distribution.

Because each $\bar{x}_i$ estimates the corresponding μ_i, the addition rule for means tells us that the mean μ_c of the sample contrast c is ψ. In other words, c is an unbiased estimator of ψ. Testing the hypothesis that a contrast is 0 assesses the significance of the effect measured by the contrast. It is often more informative to estimate the size of the effect using a confidence interval for the population contrast.

In our example the coefficients in the contrasts are $a_1 = -1, a_2 = 0.5$, $a_3 = 0.5$ for ψ_1 and $a_1 = 0, a_2 = 1, a_3 = -1$ for ψ_2, where the subscripts 1, 2, and 3 correspond to B, D, and S. In each case the sum of the a_i is 0. We look at inference for each of these contrasts in turn.

The sample contrast that estimates ψ_1 is

$$\begin{aligned} c_1 &= \bar{x}_B + \frac{1}{2}(\bar{x}_D + \bar{x}_S) \\ &= -41.05 + \frac{1}{2}(46.73 + 44.27) = 4.45 \end{aligned}$$

with standard error

$$\begin{aligned} s_{c_1} &= 6.314\sqrt{\frac{(-1)^2}{22} + \frac{(.5)^2}{22} + \frac{(.5)^2}{22}} \\ &= 1.65 \end{aligned}$$

The t statistic for testing $H_{01}\colon \psi_1 = 0$ versus $H_{a1}\colon \psi_1 > 0$ is

$$\begin{aligned} t &= \frac{c_1}{s_{c_1}} \\ &= \frac{4.45}{1.65} = 2.70 \end{aligned}$$

Because s_p has 63 degrees of freedom, software using the $t(63)$ distribution gives the one-sided P-value as 0.0044. If we used Table E, we would conclude that $P < 0.005$. The P-value is small, so there is strong evidence against H_{01}. The researchers have shown that the new methods produce higher mean scores than the old. The size of the improvement can be described by a confidence interval. The 95% confidence interval for ψ_1 is

$$\begin{aligned} c_1 \pm t^* s_{c_1} &= 4.45 \pm (2.00)(1.65) \\ &= 4.45 \pm 3.30 \\ &= (1.15,\ 7.75) \end{aligned}$$

We are 95% confident that the mean improvement obtained by using the innovative methods as compared with the old method is between 1.15 and 7.75 points.

The second sample contrast, which compares the two new methods, is

$$\begin{aligned} c_2 &= 46.73 - 44.27 \\ &= 2.46 \end{aligned}$$

with standard error

$$\begin{aligned} s_{c_2} &= 6.314\sqrt{\frac{(1)^2}{22} + \frac{(-1)^2}{22}} \\ &= 1.90 \end{aligned}$$

The t statistic for assessing the significance of this contrast is

$$t = \frac{2.46}{1.90} = 1.29$$

The P-value for the two-sided alternative is 0.2020. We conclude that either the two new methods have the same population means or the sample sizes are not sufficiently large to distinguish them. A confidence interval helps clarify this statement. The 95% confidence interval for ψ_2 is

$$\begin{aligned} c_2 \pm t^* s_{c_2} &= 2.46 \pm (2.00)(1.90) \\ &= 2.46 \pm 3.80 \\ &= (-1.34,\ 6.26) \end{aligned}$$

With 95% confidence we state that the difference between the population means for the two new methods is between -1.34 and 6.26.

Many statistical software packages report the test statistics associated with contrasts as F statistics rather than t statistics. These F statistics are the squares of the t statistics described above. The SAS output from the GLM procedure is given at the bottom of Figure 10.15. The F statistics 7.30 and 1.66 for our two contrasts appear in the F VALUE column. These are the squares of the t-values 2.70 and 1.29 (up to roundoff error). The degrees of freedom for each F are 1 and 63. The P-values are correct for two-sided alternative hypotheses. They are 0.0088 and 0.2020. To convert the computer-generated results to apply to our one-sided hypothesis concerning ψ_1, simply divide the reported P-value by 2 after checking that the value of c is in the direction of H_a (that is, that c is positive).

Questions about population means are expressed as hypotheses about contrasts. A contrast should express a specific question that we have in mind when designing the study. When contrasts are formulated before seeing the data, *inference about contrasts is valid whether or not the ANOVA H_0 of equality of means is rejected.* Because the F test answers a very general question, it is less powerful than tests for contrasts designed to answer specific questions. Specifying the important questions before the analysis is undertaken enables us to use this powerful statistical technique.

Multiple comparisons

In many studies, specific questions cannot be specified in advance of the analysis. If H_0 is not rejected, we conclude that the population means are indistinguishable on the basis of the data given. On the other hand, if H_0 is rejected, we would like to know which pairs of means differ. *Multiple comparisons* methods address this issue. It is important to keep in mind that multiple comparisons methods are used only *after rejecting* the ANOVA H_0.

Return once more to the reading comprehension data given in Example 10.10. There are three pairs of population means, comparing groups 1 and 2, 1 and 3, and 2 and 3. We can write a t statistic for each of these pairs. For example, the statistic

$$\begin{aligned} t_{12} &= \frac{\bar{x}_1 - \bar{x}_2}{s_p\sqrt{\dfrac{1}{n_1} + \dfrac{1}{n_2}}} \\ &= \frac{41.05 - 46.73}{6.31\sqrt{\dfrac{1}{22} + \dfrac{1}{22}}} \\ &= -2.99 \end{aligned}$$

compares populations 1 and 2. The subscripts on t specify which groups are compared. The t statistics for the other two pairs are

$$\begin{aligned} t_{13} &= \frac{\bar{x}_1 - \bar{x}_3}{s_p\sqrt{\dfrac{1}{n_1} + \dfrac{1}{n_3}}} \\ &= \frac{41.05 - 44.27}{6.31\sqrt{\dfrac{1}{22} + \dfrac{1}{22}}} \\ &= -1.69 \end{aligned}$$

and

$$\begin{aligned} t_{23} &= \frac{\bar{x}_2 - \bar{x}_3}{s_p\sqrt{\dfrac{1}{n_2} + \dfrac{1}{n_3}}} \\ &= \frac{46.73 - 44.27}{6.31\sqrt{\dfrac{1}{22} + \dfrac{1}{22}}} \\ &= 1.29 \end{aligned}$$

We performed the last calculation when we analyzed the contrast ψ_2 in the previous section, because that contrast was $\mu_2 - \mu_3$. These t statistics are very similar to the pooled two-sample t statistic for comparing two population means, Equation 7.6 on page 543. The difference is that we now have more than two populations, so each statistic uses the pooled estimator s_p from all groups rather than the pooled estimator from just the two groups being compared. This additional information about the common σ increases the power of the tests. The degrees of freedom for all of these statistics are DFE $= 63$, those associated with s_p.

Because we do not have any specific ordering of the means in mind as an alternative to equality, we must use a two-sided approach to the problem of deciding which pairs of means are significantly different.

MULTIPLE COMPARISONS

To perform a multiple comparisons procedure, compute t statistics for all pairs of means using the formula

$$t_{ij} = \frac{\bar{x}_i - \bar{x}_j}{s_p\sqrt{\frac{1}{n_i} + \frac{1}{n_j}}}$$

If

$$|t_{ij}| \geq t^{**}$$

we declare that the population means μ_i and μ_j are different. Otherwise, we conclude that the data do not distinguish between them. The value of t^{**} depends upon which multiple comparisons procedure we choose.

One obvious choice for t^{**} is the upper $\alpha/2$ critical value for the $t(\text{DFE})$ distribution. This choice simply carries out as many separate significance tests of fixed level α as there are pairs of means to be compared. The procedure based on this choice is called the *least-significant differences* method, or simply LSD. LSD has some undesirable properties, particularly if the number of means being compared is large. Suppose, for example, that there are $I = 20$ groups and we use LSD with $\alpha = 0.05$. There are 190 different pairs of means. If we perform 190 t tests, each with a Type I error rate of 5%, our overall error rate will be unacceptably large. We expect about 5% of the 190 to be significant even if the corresponding population means are the same. Since 5% of 190 is 9.5, we expect 9 or 10 false rejections.

LSD method

The LSD procedure fixes the probability of a false rejection for each single pair of means being compared. It does not control the overall probability of *some* false rejection among all pairs. Other choices of t^{**} control possible errors in other ways. The choice of t^{**} is therefore a complex problem and a detailed discussion of it is beyond the scope of this text. Many choices for t^{**} are used in practice. One major statistical package allows selection from a list of over a dozen choices.

Bonferroni method

We will discuss only one of these, called the *Bonferroni method*. Use of this procedure with $\alpha = 0.05$, for example, guarantees that the probability of *any* false rejection among all comparisons made is no greater than 0.05. This is much stronger protection than controlling the probability of a false rejection at 0.05 for *each separate* comparison.

Many computer programs display the output for multiple comparisons procedures as a table of the group means along with the smallest minimum significant difference, MSD

difference between two means that is statistically significant, called the *minimum significant difference,* or MSD. The MSD depends on the type of multiple comparison we choose. Any two sample means that differ by more than the MSD are significantly different. That is, we conclude that the corresponding population means are different. In our notation, the MSD is

$$MSD = t^{**} s_p \sqrt{\frac{1}{n_i} + \frac{1}{n_j}}$$

EXAMPLE 10.11

We apply the Bonferroni multiple comparisons procedure with $\alpha = 0.05$ to the data from the reading comprehension study in Example 10.10. The value of t^{**} for this procedure (from software or special tables) is 2.46. Of the statistics $t_{12} = -2.99$, $t_{13} = -1.69$, and $t_{23} = 1.29$ calculated in the beginning of this section, only t_{12} is significant.

The output generated by the MEANS statement in the SAS procedure GLM appears in Figure 10.16(a). The critical value t^{**} is given on the output as CRITICAL VALUE OF T. The MSD is 4.6825. The output shows which group means *are not* significantly different. The letters A under the heading GROUPING extend from the DRTA group to the Strat group. This means that these two groups are not distinguishable. Similarly, the letters B extend from the Strat group to the Basal group, indicating that these two groups also cannot be distinguished. There is no collection of letters that connects the DRTA and Basal groups. This means that we can declare these two means to be significantly different. ●

This conclusion appears to be illogical. If μ_1 is the same as μ_3, and μ_2 is the same as μ_3, doesn't it follow that μ_1 is the same as μ_2? Logically, the answer must be yes.

This apparent contradiction points out dramatically the nature of the conclusions of statistical tests of significance. Some of the difficulty can be resolved by noting the choice of words used above. In describing the inferences, we talk about being able or unable to distinguish between pairs of means. In making the logical statements, we say, "is the same as." There is a big difference between the two modes of thought. Statistical tests ask, "Do we have adequate evidence to distinguish two means?" It is not illogical to conclude that we have evidence to distinguish μ_1 from μ_3, but not μ_1 from μ_2 or μ_2 from μ_3. It is very unlikely that any two methods of teaching reading comprehension would give *exactly* the same population means, but the data can fail to provide good evidence of a difference.

One way to deal with these difficulties of interpretation is to give confidence intervals for the differences. The intervals remind us that the differences are not known exactly. We want to give *simultaneous* confidence intervals, that is, intervals for *all* differences among the population means at once. Again, we must face the problem that there are many compet-

```
                    GENERAL LINEAR MODELS PROCEDURE

BONFERRONI (DUNN) T TESTS FOR VARIABLE: COMP
NOTE: THIS TEST CONTROLS THE TYPE I EXPERIMENTWISE ERROR RATE
      BUT GENERALLY HAS A HIGHER TYPE II ERROR RATE THAN REGWQ

           ALPHA=0.05 DF=63 MSE=39.868
           CRITICAL VALUE OF T=2.45958
           MINIMUM SIGNIFICANT DIFFERENCE=4.6825

 MEANS WITH THE SAME LETTER ARE NOT SIGNIFICANTLY DIFFERENT.

     BON        GROUPING              MEAN       N  GROUP

                      A             46.727      22  DRTA
                      A
               B      A             44.273      22  Strat
               B
               B                    41.045      22  Basal
```

FIGURE 10.16(a) Bonferroni multiple comparisons output for the comprehension scores in the reading comprehension study.

ing procedures—in this case, many methods of obtaining simultaneous intervals.

SIMULTANEOUS CONFIDENCE INTERVALS FOR DIFFERENCES BETWEEN MEANS

Simultaneous confidence intervals for all differences $\mu_i - \mu_j$ between population means have the form

$$(\bar{x}_i - \bar{x}_j) \pm t^{**} s_p \sqrt{\frac{1}{n_i} + \frac{1}{n_j}}$$

The critical values t^{**} are the same as those used for the multiple comparisons procedure chosen.

The confidence intervals generated by a particular choice of t^{**} are closely related to the multiple comparisons results for that same method. If one of the confidence intervals includes the value 0, then that pair of means will not be declared significantly different, and vice versa.

```
                  GENERAL LINEAR MODELS PROCEDURE

BONFERRONI (DUNN) T TESTS FOR VARIABLE: COMP
NOTE: THIS TEST CONTROLS THE TYPE I EXPERIMENTWISE ERROR RATE
      BUT GENERALLY HAS A HIGHER TYPE II ERROR RATE THAN TUKEY'S

      ALPHA=0.05  CONFIDENCE=0.95  DF=63  MSE=39.868
      CRITICAL VALUE OF T=2.45958
      MINIMUM SIGNIFICANT DIFFERENCE=4.6825

COMPARISONS SIGNIFICANT AT THE 0.05 LEVEL ARE INDICATED BY '***'

                       SIMULTANEOUS                   SIMULTANEOUS
                          LOWER        DIFFERENCE        UPPER
       GROUP           CONFIDENCE       BETWEEN       CONFIDENCE
     COMPARISON           LIMIT          MEANS           LIMIT

  DRTA   - Strat         -2.228          2.455           7.137
  DRTA   - Basal          0.999          5.682          10.364     ***

  Strat  - DRTA          -7.137         -2.455           2.228
  Strat  - Basal         -1.455          3.227           7.910

  Basal  - DRTA         -10.364         -5.682          -0.999     ***
  Basal  - Strat         -7.910         -3.227           1.455
```

FIGURE 10.16(b) Bonferroni simultaneous confidence intervals output for the comprehension scores in the reading comprehension study.

EXAMPLE 10.12 Adding the CLDIFF option to the MEANS statement in the SAS GLM procedure gives the output in Figure 10.16(b) for the data in Example 10.5. These are Bonferroni simultaneous confidence intervals for 95% confidence. We are 95% confident that *all three* intervals simultaneously contain the true values of the population mean differences. ●

Power*

Recall that the power of a test is the probability of rejecting H_0 when H_a is in fact true. Power measures how likely a test is to detect a specific alternative. When planning a study in which ANOVA will be used for the analysis, it is important to perform power calculations to check that the sample sizes are adequate to detect differences among means that are

*This section is optional.

judged to be important. Power calculations also help evaluate and interpret the results of studies in which H_0 was not rejected. We sometimes find that the power of the test was so low against reasonable alternatives that there was little chance of obtaining a significant F.

We want to find the power of the ANOVA F test when the true population means are $\mu_1, \mu_2, \ldots, \mu_I$. As usual, σ is the common population standard deviation. We proceed in three steps. First, calculate a quantity called the *noncentrality parameter*, represented by λ, the Greek letter lambda. It measures the variation among population means relative to the within-group variance σ^2. Specifically,[4]

noncentrality parameter

$$\lambda = \frac{\sum n_i(\mu_i - \overline{\mu})^2}{\sigma^2}$$

where $\overline{\mu}$ is a weighted average of the group means,

$$\overline{\mu} = \sum w_i \mu_i$$

and the weights are proportional to the sample sizes,

$$w_i = \frac{n_i}{\sum n_i} = \frac{n_i}{N}$$

If the n_i are all equal with common value n, then $\overline{\mu}$ is the ordinary average of the μ_i and

$$\lambda = \frac{n \sum (\mu_i - \overline{\mu})^2}{\sigma^2}$$

If the means are all equal (the ANOVA H_0), then $\lambda = 0$. The noncentrality parameter measures how unequal the given set of means is. Large λ points to an alternative far from H_0, and we expect the ANOVA F test to have high power.

The second step in the calculation of power is to find the critical value that would be used to determine whether H_0 is rejected. This value, which we denote by F^*, is the upper α critical value for the $F(\text{DFG}, \text{DFE})$ distribution. Power calculations use a standard value of α, such as 0.05, even though the actual analysis will report a P-value.

The final step is to find the power, which is the probability of rejecting H_0 when the given set of means is true. Software makes this calculation quite easy, but tables and charts are available. Under H_a, the F statistic has a distribution known as the *noncentral F distribution.* Many software packages have routines that can find noncentral F probabilities. In SAS the function PROBF(x, DFN, DFD, λ) gives the probability of falling below x for the noncentral F distribution with noncentrality parameter λ and degrees of freedom DFN and DFD for the numerator and denominator. Using this routine, we find the power as

noncentral F distribution

$$\text{Power} = 1 - \text{PROBF}(F^*, \text{DFG}, \text{DFE}, \lambda)$$

EXAMPLE 10.13

Suppose that a study on reading comprehension similar to the one described in Example 10.10 has 10 students in each group. How likely is this study to detect differences in the mean responses similar to those observed in the actual study? Based on the results of the actual study, we will use $\mu_1 = 41$, $\mu_2 = 47$, $\mu_3 = 44$ and $\sigma = 7$ in a calculation of power. The n_i are equal, so $\overline{\mu}$ is simply the average of the μ_i,

$$\overline{\mu} = \frac{41 + 47 + 44}{3} = 44$$

The noncentrality parameter is therefore

$$\begin{aligned}\lambda &= \frac{n \sum(\mu_i - \overline{\mu})^2}{\sigma^2} \\ &= \frac{(10)[(41-44)^2 + (47-44)^2 + (44-44)^2]}{49} \\ &= \frac{(10)(18)}{49} = 3.67\end{aligned}$$

Because there are three groups with 10 observations per group, DFG = 2 and DFE = 27. The critical value for $\alpha = 0.05$ is $F^* = 3.35$. The power is therefore

$$1 - \text{PROBF}(3.35, 2, 27, 3.67) = .3486$$

The chance that we reject the ANOVA H_0 at the 5% significance level is only about 35%. ●

If the assumed values of the μ_i in this example describe differences among the groups that the experimenter wants to detect, then there is little point in conducting the experiment with only 10 subjects per group. Although H_0 is assumed to be false, the chance of rejecting it is only about 35%. This chance can be increased to acceptable levels by increasing the sample sizes.

EXAMPLE 10.14

To decide on an appropriate sample size for the experiment described in the previous example, we repeat the power calculation for different values of n, the number of subjects in each group. Here are the results:

n	DFG	DFE	F^*	λ	Power
20	2	57	3.16	7.35	.65
30	2	87	3.10	10.02	.80
40	2	117	3.07	14.69	.93
50	2	147	3.06	18.37	.97
100	2	297	3.03	36.73	≈ 1

●

With $n = 40$ the experimenters have a 93% chance of rejecting H_0 with $\alpha = 0.05$ and thereby demonstrating that the groups have different means.

In the long run, 93 out of every 100 such experiments would reject H_0 at the $\alpha = 0.05$ level of significance. Using 50 subjects per group increases their chance of finding significance to 97%. With 100 subjects per group, they are virtually certain to reject H_0. The exact power for $n = 100$ is 0.99989. In most cases the additional cost of increasing the sample size from 50 to 100 subjects per group would not be justified by the relatively small increase in the chance of obtaining statistically significant results.

SUMMARY

One-way analysis of variance (ANOVA) is used to compare several population means based on independent SRSs from each population. The populations are assumed to be normal with possibly different means and the same standard deviation. To do an analysis of variance, first compute sample means and standard deviations for all groups. Side-by-side boxplots give an overview of the data. Examine normal quantile plots (either for each group separately or for the residuals) to detect outliers or extreme deviations from normality. Compute the ratio of the largest to the smallest sample standard deviation. If this ratio is less than 2 and the normal quantile plots are satisfactory, ANOVA can be performed.

The **null hypothesis** is that the population means are *all* equal. The **alternative hypothesis** is true if there are *any* differences among the population means.

ANOVA is based on separating the total variation observed in the data into two parts, variation *among* group means and variation *within* groups. If the variation among groups is large relative to the variation within groups, we have evidence against the null hypothesis.

An **analysis of variance table** organizes the ANOVA calculations. **Sums of squares**, **degrees of freedom**, and **mean squares** appear in the table. The **F statistic** and its *P*-value are used to test the null hypothesis.

Specific questions formulated before examination of the data can be expressed as **contrasts**. Tests and confidence intervals for contrasts provide answers to these questions.

If no specific questions are formulated before examination of the data and the null hypothesis of equality of population means is rejected, **multiple comparisons** methods are used to assess the statistical significance of the differences between pairs of means.

The **power** of the F test depends upon the sample sizes, the variation among population means, and the within-group standard deviation.

SECTION 10.1 EXERCISES

10.1 For each of the following situations, identify the response variable and the populations to be compared, and give I, the n_i, and N.

(a) To compare four varieties of tomato plants, 10 plants of each variety are grown and the yield in pounds of tomatoes is recorded.
(b) A marketing experiment compares six different types of packaging for a laundry detergent. Each package is shown to 120 different potential consumers, who rate the attractiveness of the product on a 1 to 10 scale.
(c) To compare the effectiveness of three different weight-loss programs, 10 people are randomly assigned to each. At the end of the program, the weight loss for each of the participants is recorded.

10.2 For each of the following situations, identify the response variable and the populations to be compared, and give I, the n_i, and N.
(a) In a study on smoking, subjects are classified as nonsmokers, moderate smokers, or heavy smokers. A sample of size 200 is drawn from each group. Each person is asked to report the number of hours of sleep he or she gets on a typical night.
(b) The strength of concrete depends upon the formula used to prepare it. One study compared five different mixtures. Six batches of each mixture were prepared, and the strength of the concrete made from each batch was measured.
(c) Which of four methods of teaching sign language is most effective? Ten students are randomly assigned to each of the methods, and their scores on a final exam are recorded.

10.3 A randomized comparative experiment compares three programs designed to help people lose weight. There are 20 subjects in each program. The sample standard deviations for the amount of weight lost (in pounds) are 5.2, 8.9, and 10.1. Can you use the assumption of equal standard deviations to analyze these data? Compute the pooled variance and find s_p.

10.4 A study of physical fitness collected data on the weight (in kilograms) of men in four different age groups. The sample sizes for the groups were 92, 34, 35, and 24. The sample standard deviations for the groups were 12.2, 10.4, 9.2, and 11.7. Can you use the assumption of equal standard deviations to analyze these data? Compute the pooled variance and find s_p.

10.5 For each part of Exercise 10.1, outline the ANOVA table, giving the sources of variation and the degrees of freedom. (Do not compute the numerical values for the sums of squares and mean squares.)

10.6 For each part of Exercise 10.2, outline the ANOVA table, giving the sources of variation and the degrees of freedom. (Do not compute the numerical values for the sums of squares and mean squares.)

10.7 Return to the change-of-majors study described in Example 10.3 (page 718).
(a) State H_0 and H_a for ANOVA.

(b) Outline the ANOVA table, giving the sources of variation and the degrees of freedom.
(c) What is the distribution of the F statistic under the assumption that H_0 is true?
(d) Using Table F, find the critical value for an $\alpha = 0.05$ test.

10.8 Return to the survey of college students described in Example 10.4 (page 722).
(a) State H_0 and H_a for ANOVA.
(b) Outline the ANOVA table, giving the sources of variation and the degrees of freedom.
(c) What is the distribution of the F statistic under the assumption that H_0 is true?
(d) Using Table F, find the critical value for an $\alpha = 0.05$ test.

10.9 A study of the effects of exercise on physiological and psychological variables compared four groups of male subjects. The treatment group (T) consisted of 10 participants in an exercise program. A control group (C) of 5 subjects volunteered for the program but were unable to attend for various reasons. Subjects in the other two groups were selected to be similar to those in the first two groups in age and other characteristics. These were 11 joggers (J) and 10 sedentary people (S) who did not regularly exercise. (Data provided by Dennis Lobstein, from his Ph.D. dissertation, "A multivariate study of exercise training effects on beta-endorphin and emotionality in psychologically normal, medically healthy men," Purdue University, 1983.) One of the variables measured at the end of the program was a physical fitness score. Part of the ANOVA table used to analyze these data is given below.

Source	Degrees of freedom	Sum of squares	Mean square	F
Groups	3	104,855.87		
Error	32	70,500.59		
Total				

(a) Fill in the missing entries in the ANOVA table.
(b) State H_0 and H_a for this experiment.
(c) What is the distribution of the F statistic under the assumption that H_0 is true? Using Table F, give an approximate P-value for the ANOVA test. Write a brief conclusion.
(d) What is s_p^2, the estimate of the within-group variance? What is s_p?

10.10 Another variable measured in the experiment described in the previous exercise was a depression score. Higher values of this score indicate more depression. Part of the ANOVA table for these data appears below.

Source	Degrees of freedom	Sum of squares	Mean square	F
Groups	3	476.87		
Error	32	2009.88		
Total				

(a) Fill in the missing entries in the ANOVA table.
(b) State H_0 and H_a for this experiment.
(c) What is the distribution of the F statistic under the assumption that H_0 is true? Using Table F, give an approximate P-value for the ANOVA test. What do you conclude?
(d) What is s_p^2, the estimate of the within-group variance? What is s_p?

10.11 The weight gain of women during pregnancy has an important effect on the birth weight of their children. If the weight gain is not adequate, the infant is more likely to be small and will tend to be less healthy. In a study conducted in three countries, weight gains (in kilograms) of women during the third trimester of pregnancy were measured. The results are summarized in the following table. (Data taken from Collaborative Research Support Program in Food Intake and Human Function, *Management Entity Final Report*, University of California, Berkeley, 1988.)

Country	n	$\bar{x}$	s
Egypt	46	3.7	2.5
Kenya	111	3.1	1.8
Mexico	52	2.9	1.8

(a) Find the pooled estimate of the within-country variance s_p^2. What entry in the ANOVA table gives this quantity?
(b) The sum of squares for countries (groups) is 17.22. Use this information and that given above to complete all entries in an ANOVA table.
(c) State H_0 and H_a for this study.
(d) What is the distribution of the F statistic under the assumption that H_0 is true? Use Table F to find an approximate P-value for the significance test. Report your conclusion.
(e) Calculate R^2, the coefficient of determination.

10.12 In another part of the study described in the previous exercise, measurements of food intake in kilocalories were taken on many individuals several times during the period of a year. From these data, average daily food intake values were computed for each individual. The results for toddlers aged 18 to 30 months are summarized in the following table:

Country	n	$\overline{x}$	s
Egypt	88	1217	327
Kenya	91	844	184
Mexico	54	1119	285

(a) Find the pooled estimate of the within-country variance s_p^2. What entry in the ANOVA table gives this quantity?
(b) The sum of squares for countries (groups) is 6,572,551. Use this information and that given above to complete all entries in an ANOVA table.
(c) State H_0 and H_a for this study.
(d) What is the distribution of the F statistic under the assumption that H_0 is true? Use Table F to find an approximate P-value for the significance test. Report your conclusion.
(e) Calculate R^2, the coefficient of determination.

10.13 Refer to the change-of-majors study described in Example 10.3 (page 718). Let μ_1, μ_2, and μ_3 represent the population mean SAT scores for the three groups.
(a) Because the first two groups (computer science, engineering and other sciences) are majoring in areas that require mathematics skills, it is natural to compare the average of these two with the third group. Write an expression in terms of the μ_i for a contrast ψ_1 that represents this comparison.
(b) Let ψ_2 be a contrast which compares the first two groups. Write an expression in terms of the μ_i for ψ_2 that represents this comparison.

10.14 Return to the survey of college students described in Example 10.4 (page 722). Let μ_1, μ_2, μ_3, and μ_4 represent the population mean expenditures on textbooks for the freshmen, sophomores, juniors, and seniors.
(a) Because juniors and seniors take higher-level courses, which might use more expensive textbooks, we want to compare the average of the freshmen and sophomores with the average of the juniors and seniors. Write a contrast that expresses this comparison.
(b) Write a contrast for comparing the freshmen with the sophomores.
(c) Write a contrast for comparing the juniors with the seniors.

10.15 Return to the SAT mathematics scores for the change-of-majors study in Example 10.3 (page 718). Answer the following questions for the two contrasts that you defined in Exercise 10.13.
(a) For each contrast give H_0 and an appropriate H_a. In choosing the alternatives you should use information given in the description of

the problem, but you may not consider any impressions obtained by inspection of the sample means.

(b) Find the values of the corresponding sample contrasts.

(c) Using the value $s_p = 82.5$, calculate the standard errors for the sample contrasts.

(d) Give the test statistics and approximate P-values for the two significance tests. What do you conclude?

(e) Compute 95% confidence intervals for the two contrasts.

10.16 The following table presents data for the high school mathematics grades of the students in the change-of-majors study described in Example 10.3 (page 718). The sample sizes in this exercise are different from those in Example 10.3 because complete information was not available on all students. The values have been coded so that 10 = A, 9 = A−, and so on.

Second-year major	n	$\overline{x}$	s
Computer science	90	8.77	1.41
Engineering and other sciences	28	8.75	1.46
Other	106	7.83	1.74

Answer the following questions for the two contrasts that you defined in (a) and (b) of Exercise 10.13.

(a) For each contrast state H_0 and an appropriate H_a. In choosing the alternatives you should use information given in the description of the problem, but you may not consider any impressions obtained by inspection of the sample means.

(b) Find the value of the corresponding sample contrasts c_1 and c_2.

(c) Using the value $s_p = 1.581$, calculate the standard errors s_{c_1} and s_{c_2} for the sample contrasts.

(d) Give the test statistics and approximate P-values for the two significance tests. What do you conclude?

(e) Compute 95% confidence intervals for the two contrasts.

10.17 In the exercise program study described in Exercise 10.9, the summary statistics for physical fitness scores are as follows:

Group	n	$\overline{x}$	s
Treatment (T)	10	291.91	38.17
Control (C)	5	308.97	32.07
Joggers (J)	11	366.87	41.19
Sedentary (S)	10	226.07	63.53

The researchers wanted to address the following questions for the physical fitness scores. In these questions "better" means a higher fitness score.

(1) Is T better than C? (2) Is T better than the average of C and S? (3) Is J better than the average of the other three groups?

(a) For each of the three questions, define an appropriate contrast. Translate the questions into null and alternative hypotheses about these contrasts.

(b) Test your hypotheses and give approximate *P*-values. Summarize your conclusions. Do you think that the use of contrasts in this way gives an adequate summary of the results?

(c) You found that the groups differ significantly in the physical fitness scores. Does this study allow conclusions about causation—for example, that a sedentary lifestyle causes people to be less physically fit? Explain your answer.

10.18 Exercise 10.10 gives the ANOVA table for depression scores from the exercise program study described in Exercise 10.9. Here are the summary statistics for the depression scores:

Group	n	$\bar{x}$	s
Treatment (T)	10	51.90	6.42
Control (C)	5	57.40	10.46
Joggers (J)	11	49.73	6.27
Sedentary (S)	10	58.20	9.49

In planning the experiment, the researchers wanted to address the following questions for the depression scores. In these questions "better" means a lower depression score. (1) Is T better than C? (2) Is T better than the average of C and S? (3) Is J better than the average of the other three groups?

(a) For each of the three questions, define an appropriate contrast. Translate the questions into null and alternative hypotheses about these contrasts.

(b) Test your hypotheses and give approximate *P*-values. Summarize your conclusions. Do you think that the use of contrasts in this way gives an adequate summary of the results?

(c) You found that the groups differ significantly in the depression scores. Does this study allow conclusions about causation—for example, that a sedentary lifestyle causes people to be more depressed? Explain your answer.

10.19 Exercise 10.11 gives data on the weight gains of pregnant women in Egypt, Kenya, and Mexico. Computer software gives the critical value for the Bonferroni multiple comparisons procedure with $\alpha = 0.05$ as $t^{**} = 2.41$. Explain in plain language what $\alpha = 0.05$ means in the Bonferroni procedure. Use this procedure to compare the mean weight gains for the three countries. Summarize your conclusions.

10.20 Exercise 10.12 gives summary statistics for the food intake values for toddlers in Egypt, Kenya, and Mexico. Computer software gives the critical value for the Bonferroni multiple comparisons procedure with $\alpha = 0.05$ as $t^{**} = 2.41$. Explain in plain language what $\alpha = 0.05$ means in the Bonferroni procedure. Use this procedure to compare the toddler food intake means for the three countries. What do you conclude?

10.21 Refer to the physical fitness scores for the four groups in the exercise program study discussed in Exercises 10.9 and 10.17. Computer software gives the critical value for the Bonferroni multiple comparisons procedure with $\alpha = 0.05$ as $t^{**} = 2.53$. Use this procedure to compare the mean fitness scores for the four groups. Summarize your conclusions.

10.22 Refer to the depression scores for the four groups in the exercise program study discussed in Exercises 10.10 and 10.18. Computer software gives the critical value for the Bonferroni multiple comparisons procedure with $\alpha = 0.05$ as $t^{**} = 2.53$. Use this procedure to compare the mean depression scores for the four groups. Summarize your conclusions.

10.23 You are planning a study of the weight gains of pregnant women during the third trimester of pregnancy similar to that described in Exercise 10.11. The standard deviations given in that exercise range from 1.8 to 2.5. To perform power calculations, assume that the standard deviation is $\sigma = 2.3$. You have three groups, each with n subjects, and you would like to reject the ANOVA H_0 when the alternative $\mu_1 = 2.5$, $\mu_2 = 3.0$, and $\mu_3 = 3.5$ is true. Use software to make a table of powers against this alternative (similar to the table in Example 10.14) for the following numbers of women in each group: $n = 50$, 100, 150, 175, and 200. What sample size would you choose for your study?

10.24 Repeat the previous exercise for the alternative $\mu_1 = 2.7$, $\mu_2 = 3.0$, and $\mu_3 = 3.3$.

10.25 How do nematodes (microscopic worms) affect plant growth? A botanist prepares 16 identical planting pots and then introduces different numbers of nematodes into the pots. A tomato seedling is transplanted into each plot. Here are data on the increase in height of the seedlings (in centimeters) 16 days after planting. (Data provided by Matthew Moore.)

Nematodes	Seedling growth			
0	10.8	9.1	13.5	9.2
1000	11.1	11.1	8.2	11.3
5000	5.4	4.6	7.4	5.0
10,000	5.8	5.3	3.2	7.5

(a) Make a table of means and standard deviations for the four treatments, and plot the means.

(b) State H_0 and H_a for an ANOVA on these data, and explain in words what ANOVA tests in this setting.
(c) Using computer software, run the ANOVA. What are the F statistic and its P-value? Give the values of s_p and R^2. Report your conclusion.

10.26 The presence of harmful insects in farm fields is detected by erecting boards covered with a sticky material and then examining the insects trapped on the board. To investigate which colors are most attractive to cereal leaf beetles, researchers placed six boards of each of four colors in a field of oats in July. The table below gives data on the number of cereal leaf beetles trapped. (Modified from M. C. Wilson and R. E. Shade, "Relative attractiveness of various luminescent colors to the cereal leaf beetle and the meadow spittlebug," *Journal of Economic Entomology*, 60 (1967), pp. 578–580.)

Color	Insects trapped					
Lemon yellow	45	59	48	46	38	47
White	21	12	14	17	13	17
Green	37	32	15	25	39	41
Blue	16	11	20	21	14	7

(a) Make a table of means and standard deviations for the four colors, and plot the means.
(b) State H_0 and H_a for an ANOVA on these data, and explain in words what ANOVA tests in this setting.
(c) Using computer software, run the ANOVA. What are the F statistic and its P-value? Give the values of s_p and R^2. What do you conclude?

10.27 Return to the nematode problem in Exercise 10.25.
(a) Define the contrast that compares the 0 treatment (the control group) with the average of the other three.
(b) State H_0 and H_a for using this contrast to test whether or not the presence of nematodes causes decreased growth in tomato seedlings.
(c) Perform the significance test and give the P-value. Do you reject H_0?
(d) Define the contrast that compares the 0 treatment with the treatment with 10,000 nematodes. This contrast is a measure of the decrease in growth due to having a very large nematode infestation. Give a 95% confidence interval for this decrease in growth.

10.28 Return to the color attractiveness problem in Exercise 10.26. For the Bonferroni procedure with $\alpha = 0.05$, the value of t^{**} is 2.61. Use this multiple comparisons procedure to decide which pairs of colors are significantly different. Summarize your results. Which color would you recommend for attracting cereal leaf beetles?

10.29 In large classes instructors sometimes use different forms of an examination. When average scores for the different forms are calculated, students who received the form with the lowest average score may complain that their examination was more difficult than the others. Analysis of variance can help determine whether the variation in mean scores is larger than would be expected by chance. One such class used three forms. Summary statistics were as follows. (Data provided by Peter Georgeoff of the Purdue University Department of Educational Studies.)

Form	n	$\bar{x}$	s	min	Q_1	Median	Q_3	max
1	79	31.78	4.45	18	29	32	35	42
2	81	32.88	4.40	20	30	33	36	42
3	81	34.47	4.29	24	32	35	38	46

Here is the SAS output for a one-way ANOVA run on the exam scores:

```
                                  SUM OF          MEAN
SOURCE                    DF     SQUARES        SQUARE  F VALUE  PR > F
MODEL                      2   292.01871     146.00936     7.61  0.0006
ERROR                    238  4566.28004      19.18605
CORRECTED TOTAL          240  4858.29876

                R-SQUARE        C.V.      ROOT MSE          SCORE MEAN
                0.060107    13.25164        4.3802              33.054

           BONFERRONI (DUNN) T TESTS FOR VARIABLE: SCORE

NOTE: THIS TEST CONTROLS THE TYPE I EXPERIMENTWISE ERROR RATE BUT
      GENERALLY HAS A HIGHER TYPE II ERROR RATE THAN TUKEY'S FOR
      ALL PAIRWISE COMPARISONS.

        ALPHA= 0.05  CONFIDENCE= 0.95  DF= 238  MSE= 19.18605
                     CRITICAL VALUE OF T= 2.41102

 COMPARISONS SIGNIFICANT AT THE 0.05 LEVEL ARE INDICATED BY '***'.

                     SIMULTANEOUS              SIMULTANEOUS
                         LOWER    DIFFERENCE      UPPER
          FORM        CONFIDENCE    BETWEEN    CONFIDENCE
        COMPARISON       LIMIT       MEANS        LIMIT

        3    - 2        -0.0669      1.5926      3.2521
        3    - 1         1.0144      2.6843      4.3543     ***

        2    - 3        -3.2521     -1.5926      0.0669
        2    - 1        -0.5782      1.0917      2.7617

        1    - 3        -4.3543     -2.6843     -1.0144     ***
        1    - 2        -2.7617     -1.0917      0.5782
```

(a) Compare the distributions of exam scores for the three forms with side-by-side boxplots. Give a short summary of the information contained in these plots.
(b) Summarize and interpret the results of the ANOVA, including the multiple comparisons procedure.

10.30 The presence of lead in the soil of forests is an important ecological concern. One source of lead contamination is the exhaust from automobiles. In recent years this source has been greatly reduced by the elimination of lead from gasoline. Can the effects be seen in our forests? The Hubbard Brook Experimental Forest in West Thornton, New Hampshire, is the site of an ongoing study of the forest floor. Lead measurements of samples taken from this forest are available for several years. The variable of interest is lead concentration recorded as milligrams per square meter. Because the data are strongly skewed to the right, logarithms of the concentrations were analyzed. Here are some summary statistics for 5 years. (Data provided by Tom Siccama of the Yale University School of Forestry and Environmental Studies.)

Year	n	$\overline{x}$	s	min	Q_1	Median	Q_3	max
76	59	6.80	.58	5.74	6.33	6.73	7.32	8.05
77	58	6.75	.68	3.95	6.39	6.80	7.23	8.10
78	58	6.76	.50	5.01	6.50	6.78	7.10	7.66
82	68	6.50	.55	5.15	6.11	6.53	6.83	7.82
87	70	6.40	.68	4.38	6.09	6.46	6.85	8.15

Here is the SAS output for a one-way ANOVA run on the logs of the lead concentrations:

```
                                     SUM OF          MEAN
SOURCE                     DF       SQUARES        SQUARE  F VALUE   PR > F
MODEL                       4     8.4437799     2.1109450     5.75   0.0002
ERROR                     308   113.1440666     0.3673509
CORRECTED TOTAL           312   121.5878465

                  R-SQUARE          C.V.     ROOT MSE          LLEAD MEAN
                  0.069446      9.143762       0.6061              6.6285

             BONFERRONI (DUNN) T TESTS FOR VARIABLE: LLEAD

 NOTE: THIS TEST CONTROLS THE TYPE I EXPERIMENTWISE ERROR RATE BUT
       GENERALLY HAS A HIGHER TYPE II ERROR RATE THAN TUKEY'S FOR
       ALL PAIRWISE COMPARISONS.

     ALPHA= 0.05  CONFIDENCE= 0.95  DF= 308  MSE= 0.367351
                  CRITICAL VALUE OF T= 2.82740
```

COMPARISONS SIGNIFICANT AT THE 0.05 LEVEL ARE INDICATED BY '***'.

YEAR COMPARISON		SIMULTANEOUS LOWER CONFIDENCE LIMIT	DIFFERENCE BETWEEN MEANS	SIMULTANEOUS UPPER CONFIDENCE LIMIT	
76	- 78	-0.2745	0.0424	0.3592	
76	- 77	-0.2687	0.0482	0.3650	
76	- 82	-0.0046	0.3003	0.6052	
76	- 87	0.0995	0.4024	0.7052	***
78	- 76	-0.3592	-0.0424	0.2745	
78	- 77	-0.3124	0.0058	0.3240	
78	- 82	-0.0484	0.2579	0.5642	
78	- 87	0.0557	0.3600	0.6643	***
77	- 76	-0.3650	-0.0482	0.2687	
77	- 78	-0.3240	-0.0058	0.3124	
77	- 82	-0.0542	0.2521	0.5584	
77	- 87	0.0499	0.3542	0.6585	***
82	- 76	-0.6052	-0.3003	0.0046	
82	- 78	-0.5642	-0.2579	0.0484	
82	- 77	-0.5584	-0.2521	0.0542	
82	- 87	-0.1897	0.1021	0.3938	
87	- 76	-0.7052	-0.4024	-0.0995	***
87	- 78	-0.6643	-0.3600	-0.0557	***
87	- 77	-0.6585	-0.3542	-0.0499	***
87	- 82	-0.3938	-0.1021	0.1897	

(a) Display the data with side-by-side boxplots. Describe the major features of the data.
(b) Summarize the ANOVA results. Do the data suggest that the (log) concentration of lead in the Hubbard Forest floor is decreasing?

10.2 TWO-WAY ANALYSIS OF VARIANCE

Two-way ANOVA compares the means of populations that are classified in two ways or the mean responses in two-factor experiments. Many of the key concepts are similar to those of one-way ANOVA, but the presence of more than one factor also introduces some new ideas. We once more assume that the data are approximately normal and that groups have possibly different means but the same standard deviation; we again pool to estimate the variance; and we again use F statistics for significance tests. The major difference between one-way and two-way ANOVA is in the FIT part of the model. We will carefully study this term, and we will find much that is both new and useful.

Advantages of two-way ANOVA

In one-way ANOVA populations are classified according to one categorical variable, or factor. In the two-way ANOVA model there are two factors, each with its own number of levels. When we are interested in the effects of two factors, a two-way design offers great advantages over several single-factor studies. Several examples will illustrate these advantages.

EXAMPLE 10.15 In an experiment on the influence of dietary minerals on blood pressure, rats receive diets prepared with varying amounts of calcium and varying amounts of magnesium, but with all other ingredients of the diets the same. Calcium and magnesium are the two factors in this experiment.

As is common in such experiments, high, normal, and low values for each of the two minerals were selected for study. So there are three levels for each of the factors and a total of nine diets, or treatment combinations. The following diagram summarizes the factors and levels for this experiment:

	Calcium		
Magnesium	L	M	H
L	1	2	3
M	4	5	6
H	7	8	9

For example, diet 2 contains magnesium at its low level combined with the normal level of calcium. Each diet is fed to 9 rats, giving a total of 81 rats used in this experiment. The response variable is the blood pressure of a rat after some time on the diet. (This experiment was conducted by Connie Weaver, Department of Foods and Nutrition, Purdue University.) ●

The researcher in this example was primarily interested in the effect of calcium on the blood pressure. She could have conducted a single-factor experiment with magnesium set at its normal value and three levels of calcium, using one-way ANOVA to analyze the responses. Had she then decided to study the effect of magnesium, she could have performed another experiment with calcium set at its normal value and three levels of magnesium. Suppose that the budget allowed for only 80 to 90 rats in all. For the first experiment, the researcher might assign 15 rats to each of the three calcium diets. Similarly, she could assign 15 rats to each of the three magnesium diets in the second experiment. This would require a total of 90 rats, 9 more than the two-factor design in Example 10.15.

Let us compare the two-way experiment with the two one-way experiments. In the two-way experiment 27 rats are assigned to each of the three calcium diets and 27 rats to each of the three magnesium diets. In the one-way experiments there are only 15 rats per diet. Therefore, the two-way experiment provides more information to estimate the effects of both calcium and magnesium. This fact can be expressed in terms of standard deviations. If σ is the standard deviation of blood pressure for rats fed the same diet, then the standard deviation of a sample mean for n rats is $\sigma/\sqrt{n}$. For the two-way experiment this is $\sigma/\sqrt{27} = 0.19\sigma$, whereas for the two one-way experiments we have $\sigma/\sqrt{15} = 0.26\sigma$. The sample mean responses for any one level of either calcium or magnesium are less variable in the two-way design.

When the actual experiment was analyzed using two-way ANOVA, no effect of magnesium on blood pressure was found. On the other hand, the data was clearly demonstrated that the high-calcium diet had the beneficial effect of lowering the average blood pressure. Means for the three calcium levels were reported with standard errors based on 27 rats per group.

The conclusions about the effect of calcium on blood pressure are also more general in the two-way experiment. In the one-way experiments, calcium is varied only for the normal level of magnesium. All levels of calcium are combined with all levels of magnesium in the two-way design, so that conclusions about calcium are not confined to one level of magnesium. The same increased generality also applies to conclusions about magnesium. So the conclusions are both more precise and more general, yet the two-way design requires fewer rats. Experiments with several factors are an efficient use of resources.

EXAMPLE 10.16

A group of students and administrators concerned about abuse of alcohol is planning an education and awareness program for the student body at a large university. To aid in the design of their program they decide to gather information on the attitudes of undergraduate students toward alcohol.

Initially, they consider taking an SRS of all the students on the campus. They soon realize that the attitudes of freshman, sophomores, juniors, and seniors may be quite different. A stratified sample is therefore more appropriate. The study planners decide to use year in school as a factor with four levels and to take separate random samples from each class of students.

Further discussion suggests that men and women may have different attitudes toward alcohol. The team therefore decides to add sex as a second factor. The final sampling design has two factors: year with four levels, and sex with two levels. An SRS of size 50 is drawn from each of the eight populations. The eight groups for this study are shown in the following diagram:

	Year			
Sex	Fr	So	Jr	Sr
Male	1	2	3	4
Female	5	6	7	8

Example 10.16 illustrates another advantage of two-way designs analyzed by two-way ANOVA. In the one-way design with year as the only factor, all four random samples would include both men and women. If a sex difference exists, the one-way ANOVA would assign this variation to the RESIDUAL (within groups) part of the model. In the two-way ANOVA, sex is included as a factor, and therefore this variation is included in the FIT part of the model. Whenever we can move variation from RESIDUAL to FIT, we reduce the σ of our model and increase the power of our tests.

EXAMPLE 10.17

A textile researcher is interested in how four different colors of dye affect the durability of fabrics. Because the effects of the dyes may be different for different types of cloth, he applies each dye to five different kinds of cloth. The two factors in this experiment are dyes with four levels and cloth types with five levels. Six fabric specimens are dyed for each of the 20 dye-cloth combinations, and all 120 specimens are tested for durability.

The researcher in Example 10.17 could have used only one type of cloth and compared the four dyes using a one-way analysis. However, based on knowledge of the chemistry of textiles and dyes, he suspected that some of the dyes might have a chemical reaction with some types of cloth. Such a reaction could have a strong negative effect on the durability of the cloth. He wanted his results to be useful to manufacturers who would use the dyes on a variety of cloths. He chose several cloth types for the study to represent the different kinds that would be used with these dyes.

If one or more specific dye-cloth combinations produced exceptionally bad or exceptionally good durability measures, the experiment should discover this combined effect. Effects of the dyes that may be different for different types of cloth are represented in the FIT part of a two-way model as *interactions*. In contrast, the average values for dyes and cloths are represented as *main effects*. The two-way model represents FIT as the sum of a main effect for each of the two factors *and* an interaction. One-way designs that vary a single factor and hold other factors fixed cannot

interaction
main effect

discover interactions. We will discuss interactions more fully in a later section.

These examples illustrate several reasons why two-way designs are preferable to one-way designs:

> ADVANTAGES OF TWO-WAY ANOVA
>
> 1. Valuable resources can be spent more efficiently by studying two factors simultaneously rather than separately.
> 2. The residual variation in a model can be reduced by including a second factor thought to influence the response.
> 3. We can investigate interactions between factors.

These considerations also apply to study designs with more than two factors. We will be content to explore only the two-way case. The choice of sampling or experimental design is fundamental to any statistical study. Factors and levels must be carefully selected by an individual or team who understands both the statistical models and the issues that the study will address.

The two-way ANOVA model

When discussing two-way models in general, we will use the labels A and B for the two factors. For particular examples and when using statistical software, it is better to use names for these categorical variables that suggest their meaning. Thus, in Example 10.15 we would say that the factors are calcium and magnesium. The numbers of levels of the factors are often used to describe the model. Again using Example 10.15 as an illustration, we would call this a 3 × 3 ANOVA. Similarly, Example 10.16 illustrates a 4 × 2 ANOVA. In general, factor A will have I levels and factor B will have J levels. Therefore, we call the general two-way problem an $I \times J$ ANOVA.

In a two-way design every level of A appears in combination with every level of B, so that $I \times J$ groups are compared. The sample size[5] for level i of factor A and level j of factor B is n_{ij}. The total number of observations is

$$N = \sum n_{ij}$$

> **ASSUMPTIONS FOR TWO-WAY ANOVA**
>
> We have independent SRSs of size n_{ij} from each of $I \times J$ normal populations. The population means μ_{ij} may differ, but all populations have the same standard deviation σ. The μ_{ij} and σ are unknown parameters.
>
> Let x_{ijk} represent the kth observation from the population having factor A at level i and factor B at level j. The statistical model is
>
> $$x_{ijk} = \mu_{ij} + \epsilon_{ijk}$$
>
> for $i = 1, \ldots, I$ and $j = 1, \ldots, J$ and $k = 1, \ldots, n_{ij}$, where the deviations ϵ_{ijk} are an SRS from an $N(0, \sigma)$ distribution.

The FIT part of the model is the means μ_{ij}, and the RESIDUAL part is the deviations ϵ_{ijk} of the individual observations from their group means. To estimate the μ_{ij} we use the sample averages of the observations in each sample,

$$\bar{x}_{ij} = \frac{1}{n_{ij}} \sum_k x_{ijk}$$

The k below the $\sum$ means that we sum the n_{ij} observations that belong to the (i, j)th sample.

The RESIDUAL part of the model contains the unknown σ. We calculate the sample variances for each SRS and pool these to estimate σ^2:

$$s_p^2 = \frac{\sum(n_{ij} - 1)s_{ij}^2}{\sum(n_{ij} - 1)}$$

Just as in one-way ANOVA, the numerator in this fraction is SSE and the denominator is DFE. Also as in the one-way analysis, DFE is the total number of observations minus the number of groups. That is, DFE $=$ $N - IJ$. The estimator of σ is s_p.

Main effects and interactions

In this section we will explore in detail the FIT part of the two-way ANOVA, which is represented in the model by the population means μ_{ij}. The two-way design gives some structure to the set of means μ_{ij}.

Because we have independent samples from each of $I \times J$ groups, we can think of the problem initially as a one-way ANOVA with IJ groups. Each population mean μ_{ij} is estimated by the corresponding sample mean $\bar{x}_{ij}$. We can calculate sums of squares and degrees of freedom as in one-way ANOVA. In accordance with the conventions used by many computer software packages, we use the term *model* when discussing the sums of squares and degrees of freedom calculated as in one-way ANOVA with IJ groups. Thus, SSM is a model sum of squares constructed from deviations of the form $\bar{x}_{ij} - \bar{x}$, where $\bar{x}$ is the average of all of the observations and $\bar{x}_{ij}$ is the mean of the (i, j)th group. Similarly, DFM is simply $IJ - 1$.

In two-way ANOVA, the terms SSM and DFM are broken down into terms corresponding to a main effect for A, a main effect for B, and an AB interaction. Each of SSM and DFM is then a sum of terms,

$$\text{SSM} = \text{SSA} + \text{SSB} + \text{SSAB}$$

and

$$\text{DFM} = \text{DFA} + \text{DFB} + \text{DFAB}$$

The term SSA represents variation among the means for the different levels of the factor A. Because there are I such means, DFA $= I - 1$. Similarly, SSB represents variation among the means for the different levels of the factor B, with DFB $= J - 1$ degrees of freedom.

Interactions are a bit more involved. We can see that SSAB, which is SSM $-$ SSA $-$ SSB, represents the variation in the model that is not accounted for by the main effects. By subtraction we see that its degrees of freedom are

$$\text{DFAB} = (IJ - 1) - (I - 1) - (J - 1) = (I - 1)(J - 1)$$

There are many kinds of interactions. The easiest way to study them is through examples.

EXAMPLE 10.18 A National Science Foundation survey of 1985 salaries of men and women scientists in the United States[6] reported summary statistics for scientists in different fields. Some of these results appear in the following table, which gives mean yearly salaries in thousands of dollars:

Field	Women	Men	Mean
Physics	41.2	48.6	44.90
Math	34.7	42.3	38.50
Biology	34.5	42.0	38.25
Mean	36.8	44.3	40.55

The table includes averages of the m :ans in the rows and columns. For example, the first entry in the far right margin is the average of the salaries for women and men in physics:

$$\frac{41.2 + 48.6}{2} = 44.90$$

Similarly, the average of women's salaries in the three fields is

$$\frac{41.2 + 34.7 + 34.5}{3} = 36.8$$

marginal mean

These averages are called *marginal means* (because of their location at the *margins* of such tabulations).

Figure 10.17 is a plot of the group means. From the plot it is clear that men earn more on the average than women. In statistical language, there is a main effect for sex. We also see that physicists earn more than mathematicians and biologists, whose salaries are approximately the same. There is a main effect for field that is due to the higher salaries of the physicists. These main effects are described by differences among the marginal means. For example, the mean for men is 44.3, which is 7.5 higher than the female mean. The body of the table does not enter into the main effects.

What about the interaction between sex and field? An interaction is present if the main effects provide an incomplete description of the data. The marginal means for sex show that the salary difference for men versus women is 7.5. This is a main effect of sex. The body of the table shows that the male-female salary differences for the three fields are 7.4, 7.6, and 7.5. These values are remarkably similar. We do not need to consider field in describing salary differences between the sexes—the difference is about the same for all three fields, so that the marginal means are an adequate description. Similarly, if we want to discuss the differences among the fields, the marginal means provide an adequate summary. For each of the two sexes, the differences in salary among fields are very much alike. In this example, there are no important interactions.

The absence of interaction can be seen in Figure 10.17. The lines connecting the means for women and the lines connecting the means for

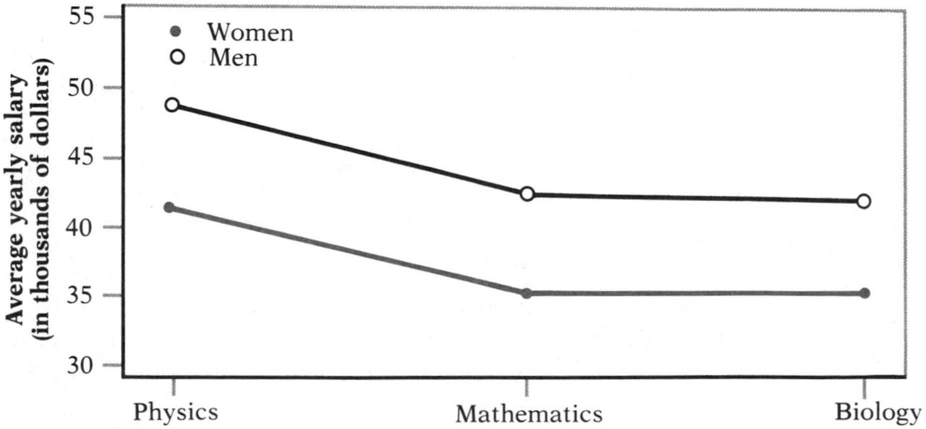

FIGURE 10.17 Plot of mean salaries of women and men scientists in three fields.

men are roughly parallel. This happens because the difference between the lines is the difference between the men's and women's salaries for each field. As we have seen, this difference is approximately the same for all three fields.

In the next example, we add another field to our data. We will see that in this case an interaction is present.

EXAMPLE 10.19 The survey described in the previous example also gave mean salaries for medical scientists. Adding these gives the following table:

Field	Women	Men	Mean
Physics	41.2	48.6	44.90
Math	34.7	42.3	38.50
Biology	34.5	42.0	38.25
Medicine	36.2	50.4	43.30
Mean	36.65	45.83	41.24

Including a fourth field changes the marginal means for sex and the overall mean of all groups. Figure 10.18 is a plot of the group means. ●

The two lines in Figure 10.18 are not parallel. The male-female difference in salaries depends on which field we examine. This is an interaction between sex and field. The sex difference in salary for the medical scientists is 14.2, nearly double the amount for the other fields. The male-female difference in the marginal means is 9.18, but this overall difference hides the fact that the difference is about 7.5 for three fields and 14.2 for medicine. The marginal means alone are not an adequate summary of the table.

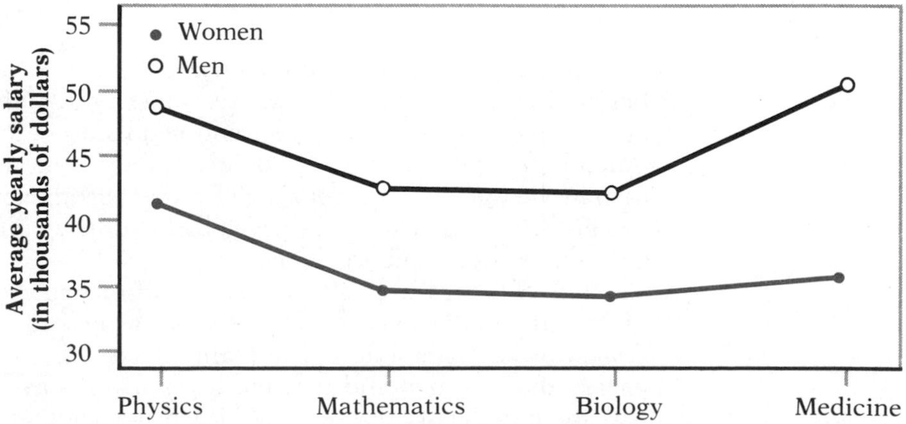

FIGURE 10.18 Plot of mean salaries of women and men scientists in four fields.

The presence of an interaction does not necessarily mean that the main effects are uninformative. It is still clear that men earn more than women, and this is evident in the marginal means for the two sexes. On the other hand, the marginal means for fields are less informative, as we see when we compare medicine with the other fields. The marginal means for the physicists and the medical scientists are approximately the same. But women physicists earn more than women in medicine, whereas male medical scientists earn more than male physicists. The conclusion that medical scientists and physicists earn about the same is incomplete because of this interaction.

Interactions come in many shapes and forms. When we find them, a careful examination of the means is needed to properly interpret the data. Simply stating that interactions are significant tells us little. Plots of the group means are very helpful.

EXAMPLE 10.20 A study of the energy expenditure of farmers in Burkina Faso (previously known as Upper Volta) collected data during the wet and dry seasons.[7] The farmers grow millet during the wet season. In the dry season, there is relatively little activity because the ground is too hard to grow crops. The mean energy expended (in kilocalories) by men and women in Burkina Faso during the wet and dry seasons is given in the following table:

Season	Men	Women	Mean
Dry	2310	2320	2315
Wet	3460	2890	3175
Mean	2885	2605	2745

Figure 10.19 is the plot of the group means. ●

During the dry season both men and women use about the same number of kilocalories. When the wet season arrives, both sexes use more energy. The men, who by social convention do most of the field work, expend much more energy than the women. The amount of energy used by the men in the wet season is very high by any reasonable standard. Such values are sometimes found in developed countries for people engaged in coal mining and lumberjacking.

In a statistical analysis, the pattern of means shown in Figure 10.19 produced significant main effects for season and sex in addition to a season-by-sex interaction. The main effects record that men use more energy than women and that the wet season is associated with greater activity than the dry season. This clearly does not tell the whole story. We need to discuss the men and women in each of the two seasons to fully understand how energy is used in Burkina Faso.

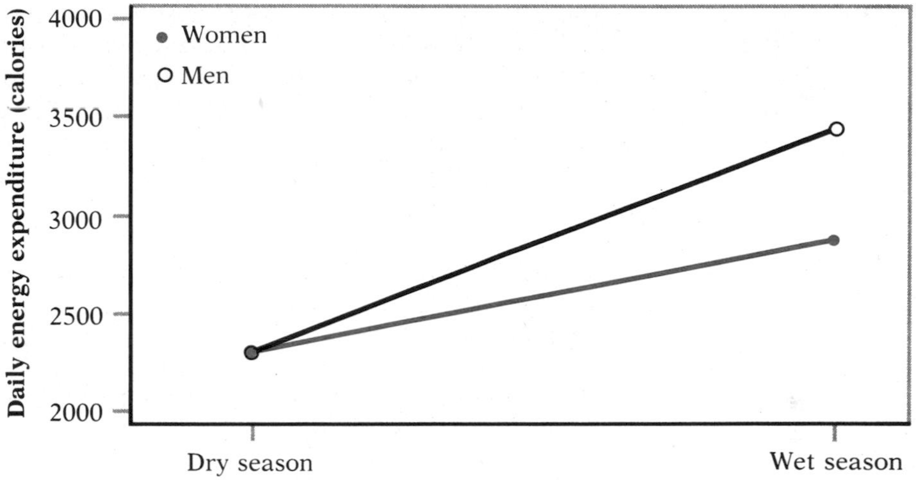

FIGURE 10.19 Plot of mean energy expenditures of farmers in Burkina Faso by sex and season.

A different kind of interaction is present in the next example. Here, we must be very cautious in our interpretation of the main effects since one of them can lead to a distorted conclusion.

EXAMPLE 10.21

An experiment to study how noise affects the performance of children tested second-grade hyperactive children and a control group of second graders who were not hyperactive.[8] One of the tasks involved solving math problems. The children solved problems under both high-noise and low-noise conditions. Here are the mean scores:

Group	High noise	Low noise	Mean
Control	214	170	192
Hyperactive	120	140	130
Mean	167	155	161

The means are plotted in Figure 10.20. In the analysis of this experiment, both of the main effects and the interaction were statistically significant. How are we to interpret these results? ●

What catches our eye in the plot is that the lines cross. The hyperactive children did better in low noise than in high, but the control children performed better in high noise than in low. This interaction between noise and type of children is an important result. Which noise level is preferable depends on which type of children are in the class. The main effect for type of children is also straightforward: The normal children did better than the hyperactive children at both levels of noise. However, the main

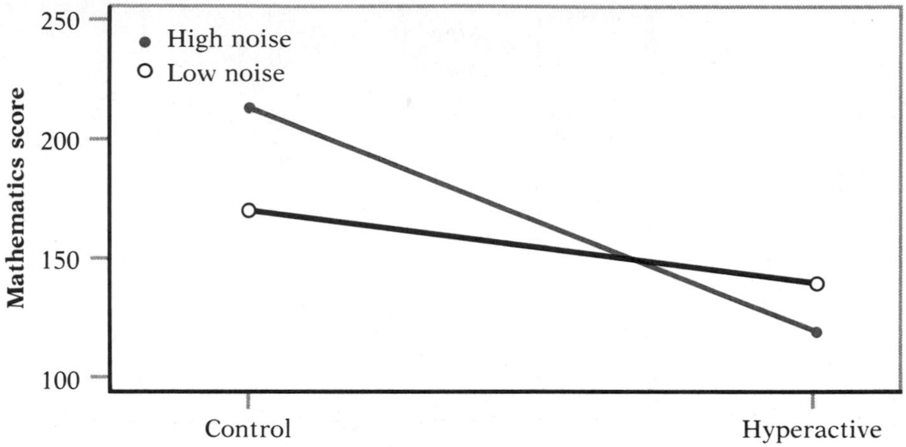

FIGURE 10.20 Plot of mean math scores of control and hyperactive children under low- and high-noise conditions.

effect for noise could easily be misinterpreted if we do not look at the means carefully. Under high-noise conditions the marginal mean was 167, compared with a low-noise mean of 155. This suggests that high noise is, on the average, a more favorable condition than low noise. In fact, high noise is more favorable only for the control group. The hyperactive children performed better under the low-noise conditions. Because of the interaction, we must know which type of children we have in order to say which noise level is better for working math problems.

ANOVA table for two-way ANOVA

Two-way ANOVA is the statistical analysis for a two-way design with a quantitative response variable. The results of a two-way ANOVA are summarized in an ANOVA table based on splitting the total variation SST and the total degrees of freedom DFT among the two main effects and the interaction. Both the sums of squares (which measure variation) and the degrees of freedom add:

$$\text{SST} = \text{SSA} + \text{SSB} + \text{SSAB} + \text{SSE}$$
$$\text{DFT} = \text{DFA} + \text{DFB} + \text{DFAB} + \text{DFE}$$

The sums of squares are always calculated in practice by statistical software. From a sum of squares and its degrees of freedom we find a mean square in the usual way,

$$\text{mean square} = \frac{\text{sum of squares}}{\text{degrees of freedom}}$$

The significance of each of the main effects and the interaction is assessed by an F statistic that compares the variation due to the effect of interest with the within-group variation. Each F statistic is the mean square for the source of interest divided by MSE. Here is the general form of the two-way ANOVA table:

Source	Degrees of freedom	Sum of squares	Mean square	F
A	$I - 1$	SSA	SSA/DFA	MSA/MSE
B	$J - 1$	SSB	SSB/DFB	MSB/MSE
AB	$(I - 1)(J - 1)$	SSAB	SSAB/DFAB	MSAB/MSE
Error	$N - IJ$	SSE	SSE/DFE	
Total	$N - 1$	SST	SST/DFT	

There are three null hypotheses in two-way ANOVA, with an F test for each. We can test for significance of the main effect of A, the main effect of B, and the AB interaction. It is generally good practice to examine the test for interaction first, since the presence of a strong interaction may influence the interpretation of the main effects. Be sure to plot the means as an aid to interpreting the results of the significance tests.

SIGNIFICANCE TESTS IN TWO-WAY ANOVA

To test the main effect of A, use the F statistic

$$F_{\text{A}} = \frac{\text{MSA}}{\text{MSE}}$$

If there is no main effect of A, F_{A} has the $F(\text{DFA}, \text{DFE})$ distribution. The P-value for the null hypothesis of no main effect of A is the probability that a random variable having the $F(\text{DFA}, \text{DFE})$ distribution is greater than or equal to the calculated value of F_{A}.

To test the main effect of B, use the F statistic

$$F_{\text{B}} = \frac{\text{MSB}}{\text{MSE}}$$

If there is no main effect of B, F_{B} has the $F(\text{DFB}, \text{DFE})$ distribution. The P-value for the null hypothesis of no main effect of B is the probability that a random variable having the $F(\text{DFB}, \text{DFE})$ distribution is greater than or equal to the calculated value of F_{B}.

To test the interaction of A and B, use the F statistic

$$F_{\text{AB}} = \frac{\text{MSAB}}{\text{MSE}}$$

If there is no interaction, F_{AB} has the $F(\text{DFAB}, \text{DFE})$ distribution. The P-value for the null hypothesis of no interaction is the probability that a random variable having the $F(\text{DFAB}, \text{DFE})$ distribution is greater than or equal to the calculated value of F_{AB}.

The following example illustrates how to do a two-way ANOVA. As with the one-way ANOVA, we focus our attention on interpretation of the computer output.

EXAMPLE 10.22

A study of cardiovascular risk factors compared runners who averaged at least 15 miles per week with a control group described as "generally sedentary." Both men and women were included in the study.[9] The design is a 2×2 ANOVA with the factors group and sex. There were 200 subjects in each of the four combinations. One of the variables measured was the heart rate after 6 minutes of exercise on a treadmill. A computer analysis produced the outputs in Figure 10.21 (from the SAS procedure MEANS) and Figure 10.22 (from the SAS procedure GLM). ●

We begin with the usual preliminary examination. From Figure 10.21 we see that the ratio of the largest to the smallest standard deviation is less than 2. Therefore, we are not concerned about violating the assumption of equal population standard deviations. Normal quantile plots (not shown) do not reveal any outliers, and the data appear to be reasonably normal.

The ANOVA table at the top of the output in Figure 10.22 is in effect a one-way ANOVA with the four groups female control, female runner,

```
VARIABLE      N          MEAN          STANDARD
                                       DEVIATION

----------GROUP=CONTROL SEX=FEMALE----------

HR          200    150.16768826    15.90882113

-----------GROUP=CONTROL SEX=MALE-----------

HR          200    129.40308425    16.91896949

----------GROUP=RUNNERS SEX=FEMALE----------

HR          200    117.86423502    15.21759686

-----------GROUP=RUNNERS SEX=MALE-----------

HR          200    103.09755344    12.66789092
```

FIGURE 10.21 Summary statistics for heart rates in the four groups of a 2 × 2 ANOVA.

GENERAL LINEAR MODELS PROCEDURE

DEPENDENT VARIABLE: HR

SOURCE	DF	SUM OF SQUARES	MEAN SQUARE	F VALUE
MODEL	3	236673.01696169	78891.00565390	338.81
ERROR	796	185347.17372171	232.84820819	PR > F
CORRECTED TOTAL	799	422020.19068340		0.0001

R-SQUARE	C.V.	ROOT MSE	HR MEAN
0.560810	12.1945	15.25936461	125.13314024

SOURCE	DF	TYPE I SS	F VALUE	PR > F
GROUP	1	171750.65050977	737.61	0.0001
SEX	1	63123.61278194	271.09	0.0001
GROUP*SEX	1	1798.75366998	7.73	0.0056

FIGURE 10.22 **Two-way analysis of variance output for heart rates.**

male control, and male runner. In this analysis MODEL has 3 degrees of freedom and there are 796 degrees of freedom for ERROR. The F test and its associated P-value for this analysis refer to the hypothesis that all four groups have the same population mean. We are interested in the main effects and interaction, so we ignore this test.

The sums of squares for the group and sex main effects and the group-by-sex interaction appear at the bottom of Figure 10.22 under the heading TYPE I SS. These sum to the sum of squares for MODEL. Similarly, the degrees of freedom for these sums of squares sum to the degrees of freedom for MODEL. Two-way ANOVA splits the variation among the means (expressed by the MODEL sum of squares) into three parts that reflect the two-way layout.

This output omits the mean squares for the main effects and interaction. If they were needed, we would divide the sum of squares by the degrees of freedom. Because the degrees of freedom are all 1 in this example, the mean squares are the same as the sums of squares. The F statistics for the three effects appear in the column labeled F VALUE and

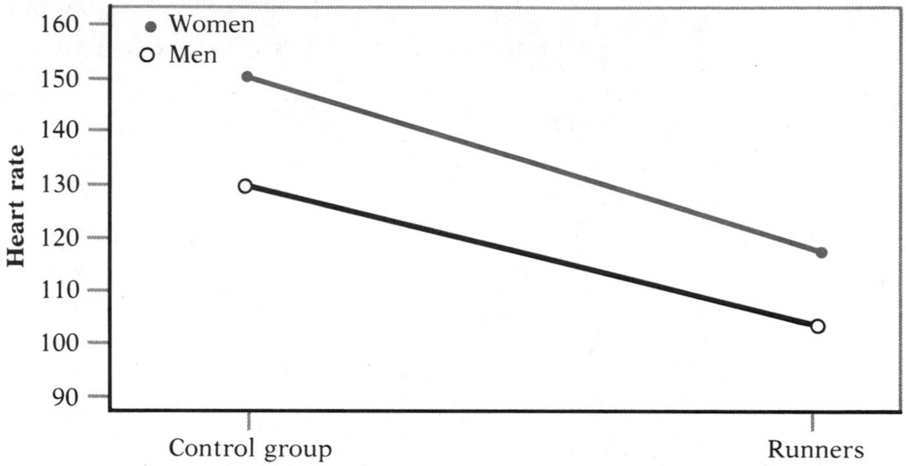

FIGURE 10.23 Plot of the group means for heart rates in a 2 × 2 ANOVA.

the *P*-values are under the heading PR > F. For the group main effect, we verify the calculation of *F* as follows:

$$F = \frac{\text{MSG}}{\text{MSE}} = \frac{171,751}{232.848} = 737.61$$

All three effects are statistically significant. The group effect has the largest *F*, followed by the sex effect and then the group-by-sex interaction. To interpret these results, we examine the plot of means in Figure 10.23. The significance of the main effect for group is due to the fact that the controls have higher average heart rates than the runners for both sexes. This is the largest effect evident in the plot.

The significance of the main effect for sex is due to the fact that the females have higher heart rates than the men in both groups. The differences are not as large as those for the group effect, and this is reflected in the smaller value of the *F* statistic.

The analysis indicates that a complete description of the average heart rates requires consideration of the interaction in addition to the main effects. The two lines in the plot are not parallel. This interaction can be described in two ways. The female-male difference in average heart rates is greater for the controls than for the runners. Alternatively, the difference in average heart rates between controls and runners is greater for women than for men. As the plot suggests, the interaction is not large. It is statistically significant because there were 800 subjects in the study.

SUMMARY

Two-way analysis of variance is used to compare population means when populations are classified according to two factors.

ANOVA assumes that the populations are normal with possibly different means and the same standard deviation and that independent SRSs are drawn from each population.

As with one-way ANOVA, preliminary analysis includes examination of means, standard deviations, and normal quantile plots. **Marginal means** are calculated by taking averages of the cell means across rows and columns. Pooling is used to estimate the within-group variance.

ANOVA separates the total variation into parts for the **model** and **error.** The model variation is separated into parts for each of the **main effects** and the **interaction.**

The calculations are organized into an **ANOVA table.** F statistics and P-values are used to test hypotheses about the main effects and the interaction.

Careful inspection of the means is necessary to interpret significant main effects and interactions. Plots are a useful aid.

SECTION 10.2 EXERCISES

10.31 Each of the following situations is a two-way study design. For each case, identify the response variable and both factors, and state the number of levels for each factor (I and J) and the total number of observations (N).

(a) A study of the productivity of tomato plants compares four varieties of tomatoes and two types of fertilizer. Five plants of each variety are grown with each type of fertilizer. The yield in pounds of tomatoes is recorded for each plant.

(b) A marketing experiment compares six different types of packaging for a laundry detergent. A survey is conducted to determine the attractiveness of the packaging in four different parts of the country. Each type of packaging is shown to 30 different consumers in each part of the country, who rate the attractiveness of the product on a 1 to 10 scale.

(c) To compare the effectiveness of three different weight-loss programs, five men and five women are randomly assigned to each. At the end of the program, the weight loss for each of the participants is recorded.

10.32 Each of the following situations is a two-way study design. For each case, identify the response variable and both factors, and state the number of levels for each factor (I and J) and the total number of observations (N).

(a) A study of smoking classifies subjects as nonsmokers, moderate smokers, or heavy smokers. Samples of 100 men and 100 women are drawn from each group. Each person reports the number of hours of sleep he or she gets on a typical night.

(b) The strength of concrete depends upon the formula used to prepare it. An experiment compares five different mixtures. Six specimens of concrete are poured from each mixture. Two of these specimens are subjected to 0 cycles of freezing and thawing, two are subjected to 100 cycles, and two specimens are subjected to 500 cycles. The strength of each specimen is then measured.

(c) Four methods for teaching sign language are to be compared. Five students in special education and five students in linguistics are randomly assigned to each of the methods, and the scores on a final exam are recorded.

10.33 For each part of Exercise 10.31, outline the ANOVA table, giving the sources of variation and the degrees of freedom. (Do not compute the numerical values for the sums of squares and mean squares.)

10.34 For each part of Exercise 10.32, outline the ANOVA table, giving the sources of variation and the degrees of freedom. (Do not compute the numerical values for the sums of squares and mean squares.)

10.35 In the course of a clinical trial of measures to prevent coronary heart disease, blood pressure measurements were taken on 12,866 men. Individuals were classified by age group and race. The means for systolic blood pressure are given in the following table. (Data from W. M. Smith et al., "The multiple risk factor intervention trial," in H. M. Perry, Jr., and W. M. Smith (eds.), *Mild Hypertension: To Treat or Not to Treat*, New York Academy of Sciences, 1978, pp. 293–308.)

	35–39	40–44	45–49	50–54	55–59
White	131.0	132.3	135.2	139.4	142.0
Nonwhite	132.3	134.2	137.2	141.3	144.1

(a) Plot the group means, with age on the x axis and blood pressure on the y axis. For each racial group connect the points for the different ages.

(b) Describe the patterns you see. Does there appear to be a difference between the two racial groups? Does systolic blood pressure appear to vary with age? If so, how does it vary? Is there an interaction?

(c) Compute the marginal means. Then find the differences between the white and nonwhite mean blood pressures for each age group. Use this information to summarize numerically the patterns in the plot.

10.36 The means for diastolic blood pressure recorded in the clinical trial described in the previous exercise are:

	35–39	40–44	45–49	50–54	55–59
White	89.4	90.2	90.9	91.6	91.4
Nonwhite	91.2	93.1	93.3	94.5	93.5

(a) Plot the group means with age on the x axis and blood pressure on the y axis. For each racial group connect the points for the different ages.
(b) Describe the patterns you see. Does there appear to be a difference between the two racial groups? Does diastolic blood pressure appear to vary with age? If so, how does it vary? Is there an interaction between race and age?
(c) Compute the marginal means. Find the differences between the white and nonwhite mean blood pressures for each age group. Use this information to summarize numerically the patterns in the plot.

10.37 The amount of chromium in the diet has an effect on the way the body processes insulin. In an experiment designed to study this phenomenon, four diets were fed to male rats. There were two factors. Chromium had two levels: low (L) and normal (N). The rats were allowed to eat as much as they wanted (M) or the total amount that they could eat was restricted (R). We call the second factor Eat. One of the variables measured was the amount of an enzyme called GITH. The means for this response variable appear in the following table. (Data provided by Julie Hendricks and V. J. K. Liu of the Department of Foods and Nutrition, Purdue University.)

	Eat	
Chromium	M	R
L	4.545	5.175
N	4.425	5.317

(a) Make a plot of the mean GITH for these diets, with the factor Chromium on the x axis and GITH on the y axis. For each Eat group connect the points for the two Chromium means.
(b) Describe the patterns you see. Does the amount of chromium in the diet appear to affect the GITH mean? Does restricting the diet as compared with letting the rats eat as much as they want appear to have an effect? Is there an interaction?
(c) Compute the marginal means. Compute the differences between the M and R diets for each level of Chromium. Use this information to summarize numerically the patterns in the plot.

10.38 The Chapin Social Insight Test measures how well people can appraise others and predict what they may say or do. A study administered this test to different groups of people and compared the mean scores. Some of the results are given in the table below. Means for males and females who were psychology graduate students (PG) and liberal arts undergraduates (LU) are presented. The two factors are labeled Sex and Group. (This example is based on results reported in H. G. Gough, *The Chapin Social Insight Test*, Consulting Psychologists Press, Palo Alto, Calif., 1968.)

	Group	
Sex	PG	LU
Males	27.56	25.34
Females	29.25	24.94

Plot the means and describe the essential features of the data in terms of main effects and interactions.

10.39 The change-of-majors study described in Example 10.3 (page 718) classified students into one of three groups depending upon their major in the sophomore year. In this exercise we also consider the sex of the students. There are now two factors: major with three levels, and sex with two levels. The mean SAT mathematics scores for the six groups appear in the following table. For convenience we use the labels CS for computer science majors, EO for engineering and other science majors, and O for other majors.

	Major		
Sex	CS	EO	O
Males	628	618	589
Females	582	631	543

Describe the main effects and interaction using appropriate graphs and calculations.

10.40 The mean high school mathematics grades for the students in the previous exercise are summarized in the following table. The grades have been coded so that 10 = A, 9 = A−, etc.

	Major		
Sex	CS	EO	O
Males	8.68	8.35	7.65
Females	9.11	9.36	8.04

Summarize the results of this study using appropriate plots and calculations to describe the main effects and interaction.

10.41 Return to the chromium-insulin experiment described in Exercise 10.37. Here is part of the ANOVA table for these data:

Source	Degrees of freedom	Sum of squares	Mean square	F
A(Chromium)		.00121		
B(Eat)		5.79121		
AB		.17161		
Error		1.08084		
Total				

(a) In all, 40 rats were used in this experiment. Fill in the missing values in the ANOVA table.
(b) What is the value of the F statistic to test the null hypothesis that there is no interaction? What is its distribution when the null hypothesis is true? Using Table F, find an approximate P-value for this test.
(c) Answer the questions in part (b) for the main effect of Chromium and the main effect of Eat.
(d) What is s_p^2, the within-group variance? What is s_p?
(e) Using what you have learned in this exercise and your answer to Exercise 10.37, summarize the results of this experiment.

10.42 Return to the Chapin Social Insight Test study described in Exercise 10.38. Part of the ANOVA table for these data is given below:

Source	Degrees of freedom	Sum of squares	Mean square	F
A(Sex)		62.40		
B(Group)		1599.03		
AB				
Error		13633.29		
Total		15458.52		

(a) There were 150 individuals tested in each of the groups. Fill in the missing values in the ANOVA table.
(b) What is the value of the F statistic to test the null hypothesis that there is no interaction? What is its distribution when the null hypothesis is true? Using Table F, find an approximate P-value for this test.
(c) Answer the questions in part (b) for the main effect of Sex and the main effect of Group.

(d) What is s_p^2, the within-group variance? What is s_p?
(e) Using what you have learned in this exercise and your answer to Exercise 10.38, summarize the results of this study.

10.43 Do left-handed people live shorter lives than right-handed people? A study of this question examined a sample of 949 death records and contacted next of kin to determine handedness. Note that there are many possible definitions of "left-handed." The researchers examined the effects of different definitions on the results of their analysis and found that their conclusions were not sensitive to the exact definition used. For the results presented here, people were defined to be right-handed if they wrote, drew, and threw a ball with the right hand. All others were defined to be left-handed. People were classified by sex (female or male) and handedness (left or right), and a 2×2 ANOVA was run with the age at death as the response variable. The F statistics were 22.36 (handedness), 37.44 (sex), and 2.10 (interaction). The following marginal mean ages at death (in years) were reported: 77.39 (females), 71.32 (males), 75.00 (right-handed), and 66.03 (left-handed). (For a summary of this study and other research in this area see Stanley Coren and Diane F. Halpern, "Left-handedness: A marker for decreased survival fitness," *Psychological Bulletin*, 109 (1991), pp. 90–106.)
(a) For each of the F statistics given above find the degrees of freedom and an approximate P-value. Summarize the results of these tests.
(b) Using the information given, write a short summary of the results of the study.

10.44 Scientists believe that exposure to the radioactive gas radon is associated with some types of cancers in the respiratory system. Radon from natural sources is present in many homes in the United States. A group of researchers decided to study the problem in dogs because dogs get similar types of cancers and are exposed to environments similar to their owners. Radon detectors are available for home monitoring but the researchers wanted to obtain actual measures of the exposure of a sample of dogs. To do this they placed the detectors in holders and attached them to the collars of the dogs. One problem was that the holders might in some way affect the radon readings. The researchers therefore devised a laboratory experiment to study the effects of the holders. Detectors from four series of production were available, so they used a two-way ANOVA design (Series with 4 levels and Holder with 2, representing the presence or absence of a holder). All detectors were exposed to the same radon source and the radon measure in picocuries per liter was recorded. (Data provided by Neil Zimmerman of the Purdue University School of Health Sciences.) Here is the SAS output:

SOURCE	DF	SUM OF SQUARES	MEAN SQUARE	F VALUE	PR > F
MODEL	7	15612.073	2230.296	3.82	0.0017
ERROR	61	35608.165	583.740		
CORRECTED TOTAL	68	51220.238			

R-SQUARE	C.V.	ROOT MSE	RADON MEAN
0.304803	7.837457	24.161	308.27

SOURCE	DF	TYPE I SS	MEAN SQUARE	F VALUE	PR > F
SERIES	3	12293.636	4097.879	7.02	0.0004
HOLDER	1	1144.666	1144.666	1.96	0.1665
SERIES*HOLDER	3	2173.771	724.590	1.24	0.3026

(a) Summarize the results of the ANOVA. Is there evidence to suggest that the holders make a difference in the radon readings?

(b) The mean radon readings for the four series were 330, 303, 302, and 295. The results of the significance test for series were of great concern to the researchers. Explain why.

CHAPTER 10 COMPUTER EXERCISES

10.45 Example 10.6 (page 725) describes an experiment to compare three methods of teaching reading. One of the pretest measures is analyzed in that example and one of the posttest measures is analyzed in Example 10.10 (page 736). In the actual study there were two pretest and three posttest measures. The data are given in the READING data set described in the data appendix. The pretest variable analyzed in Example 10.6 is PRE1 and the posttest variable analyzed in Example 10.10 is POST3. For this exercise, use the data for the pretest variable PRE2.

(a) Summarize the data with a table giving the sample sizes, means, and standard deviations. Plot the means.

(b) Examine the distribution of the scores in the three groups by constructing normal quantile plots. Do the data look normal?

(c) What is the ratio of the largest to the smallest standard deviation? Is it reasonable to proceed with an ANOVA?

(d) Run a one-way ANOVA on these data. State H_0 and H_a for the ANOVA significance test. Report the value of the F statistic and its P-value. Do you have strong evidence against H_0?

(e) Summarize your conclusions from this analysis.

10.46 Refer to the previous exercise. To examine the effect on the ANOVA of changing the means, you will do an analysis similar to that described in Example 10.9. Use the pretest variable PRE2 for this exercise. For each of the observations in the Basal group add one point to the PRE2. For each of the observations in the Strat group, subtract one point. Leave the observations in the DRTA group unchanged. Call the new variable PRE2X.

(a) Give the sample means and standard deviations for PRE2X in each group. (You can find these from the means and standard deviations for PRE2.)
(b) Run the ANOVA on the variable PRE2X.
(c) Compare the results of this analysis with the ANOVA for PRE2.
(d) Summarize your conclusions.

10.47 Example 10.10 analyzed the scores for the third posttest variable in the reading comprehension study. For this exercise you will analyze the first posttest variable, given as POST1 in the READING data set.

(a) Summarize the data in a table giving the sample sizes, means, and standard deviations. Plot the means.
(b) Examine the distribution of the scores in the three groups by constructing normal quantile plots. Do the data look normal?
(c) What is the ratio of the largest to the smallest standard deviation? Is it reasonable to proceed with the ANOVA?
(d) Run the ANOVA and summarize the results.
(e) Following the procedure given in the text in the section on contrasts, analyze the contrast for comparing the Basal group with the average of the other two groups. Test the one-sided H_a and give a 95% confidence interval.
(f) Analyze the contrast which compares the DRTA and the Strat groups. Use a two-sided test and give a 95% confidence interval.
(g) Based on your work in (a), (d), (e), and (f), write a brief nontechnical summary of your conclusions.

10.48 Example 10.10 analyzed the scores for the third posttest variable in the reading comprehension study. For this exercise you will analyze the second posttest variable, given as POST2 in the READING data set.

(a) Summarize the data in a table of the sample sizes, means, and standard deviations. Plot the means.
(b) Examine the distribution of the scores in the three groups by constructing normal quantile plots. Do the data look normal?
(c) What is the ratio of the largest to the smallest standard deviation? Is it reasonable to proceed with the ANOVA?
(d) Run the ANOVA and summarize the results.
(e) Following the procedure given in the text in the section on contrasts, analyze the contrast for comparing the Basal group with the average of the other two groups. Test the one-sided H_a and give a 95% confidence interval.
(f) Analyze the contrast which compares the DRTA and the Strat groups. Use a two-sided test and give a 95% confidence interval.

(g) Based on your work in (a), (d), (e), and (f), write a brief nontechnical summary of your conclusions.

10.49 Return to the nematode experiment described in Exercise 10.25 (page 757). Suppose that when entering the data into the computer, you accidentally entered the first observation as 108 rather than 10.8.

(a) Run the ANOVA with the incorrect observation. Summarize the results.

(b) Compare this run with the results obtained with the correct data set. What does this illustrate about the effect of outliers in an ANOVA?

(c) Compute a table of means and standard deviations for each of the four treatments using the incorrect data. How would this table have helped you to detect the incorrect observation?

10.50 Refer to the color attractiveness experiment described in Exercise 10.26 (page 758). Suppose that when entering the data into the computer, you accidentally entered the first observation as 450 rather than 45.

(a) Run the ANOVA with the incorrect observation. Summarize the results.

(b) Compare this run with the results obtained with the correct data set. What does this illustrate about the effect of outliers in an ANOVA?

(c) Compute a table of means and standard deviations for each of the four treatments using the incorrect data. How would this table have helped you to detect the incorrect observation?

10.51 With small numbers of observations in each group, it can be difficult to detect deviations from normality and violations of the equal standard deviations assumption for ANOVA. Return to the nematode experiment described in Exercise 10.25 (page 757). The log transformation is often used for variables such as the growth of plants. In many cases this will tend to make the standard deviations more similar across groups and to make the data within each group look more normal. Rerun the ANOVA using the logarithms of the recorded values. Answer the questions given in Exercise 10.25. Compare these results with those obtained by analyzing the raw data.

10.52 Refer to the color attractiveness experiment described in Exercise 10.26 (page 758). The square root transformation is often used for variables that are counts, such as the number of insects trapped in this example. In many cases data transformed in this way will conform more closely to the assumptions of normality and equal standard deviations. Rerun the ANOVA using the square roots of the original counts of insects. Answer the questions given in Exercise 10.26. Compare these results with those obtained by analyzing the raw data.

10.53 You are planning a study of the SAT mathematics scores of four groups of students. From Example 10.3, we know that the standard deviations of the three groups considered in that study were 86, 67, and 83. In Example 10.5, we found the pooled standard deviation to be 82.5. Since the power of the F test decreases as the standard deviation increases, use $\sigma = 90$ for the calculations in this exercise. This choice will lead to sample sizes that are perhaps a little larger than we need but will prevent us from choosing sample sizes that are too small to detect the effects of interest. You would like to conclude that the population means are different when $\mu_1 = 620$, $\mu_2 = 600$, $\mu_3 = 580$, and $\mu_4 = 560$.

(a) Pick several values for n (the number of students that you will select from each group) and calculate the power of the ANOVA F test for each of your choices.

(b) Plot the power versus the sample size. Describe the general shape of the plot.

(c) What choice of n would you choose for your study? Give reasons for your answer.

10.54 Refer to the previous exercise. Repeat all parts for the alternative $\mu_1 = 610$, $\mu_2 = 600$, $\mu_3 = 590$, and $\mu_4 = 580$.

10.55 A large research project studied the physical properties of wood materials constructed by bonding together small flakes of wood. Different species of trees were used, and the flakes were made in different sizes. One of the physical properties measured was the tension modulus of elasticity in the direction perpendicular to the alignment of the flakes, in pounds per square inch (psi). Some of the data are given in the following table. The size of the flakes are S1 = 0.015 inches by 2 inches and S2 = 0.025 inches by 2 inches. (Data provided by Michael Hunt and Bob Lattanzi of the Purdue University Forestry Department.)

	Size of flakes	
Species	S1	S2
Aspen	308	278
	428	398
	426	331
Birch	214	534
	433	512
	231	320
Maple	272	158
	376	503
	322	220

(a) Compute means and standard deviations for the three observations in each species-size group. Find the marginal mean for each species and for each size of flakes. Display the means and marginal means in a table.

(b) Plot the means of the six groups. Put species on the x axis and modulus of elasticity on the y axis. For each size connect the three points corresponding to the different species. Describe the patterns you see. Do the species appear to be different? What about the sizes? Does there appear to be an interaction?

(c) Run a two-way ANOVA on these data. Summarize the results of the significance tests. What do these results say about the impressions that you described in part (b) of this exercise?

10.56 Refer to the previous exercise. Another of the physical properties measured was the strength, in kilopounds per square inch (ksi), in the direction perpendicular to the alignment of the flakes. Some of the data are given in the following table. The sizes of the flakes are S1 = 0.015 inches by 2 inches and S2 = 0.025 inches by 2 inches.

	Size of flakes	
Species	S1	S2
Aspen	1296	1472
	1997	1441
	1686	1051
Birch	903	1422
	1246	1376
	1355	1238
Maple	1211	1440
	1827	1238
	1541	748

(a) Compute means and standard deviations for the three observations in each species-size group. Find the marginal means for the species and for the flake sizes. Display the means and marginal means in a table.

(b) Plot the means of the six groups. Put species on the x axis and strength on the y axis. For each size connect the three points corresponding to the different species. Describe the patterns you see. Do the species appear to be different? What about the sizes? Does there appear to be an interaction?

(c) Run a two-way ANOVA on these data. Summarize the results of the significance tests. What do these results say about the impressions that you described in (b) of this exercise?

10.57 Refer to the data given for the change-of-majors study described in the data set MAJORS in the data set appendix. Analyze the data for SAT-V, the SAT verbal score. Your analysis should include a table of sample sizes, means, and standard deviations; normal quantile plots; a plot of the means; and a two-way ANOVA using sex and major as the factors. Write a short summary of your conclusions.

10.58 Refer to the data for the change-of-majors study in the data set MAJORS in the data set appendix. Analyze the data for HSS, the high school science grades. Your analysis should include a table of sample sizes, means, and standard deviations; normal quantile plots; a plot of the means; and a two-way ANOVA using sex and major as the factors. Write a short summary of your conclusions.

10.59 Refer to the data given for the change-of-majors study in the data set MAJORS in the data set appendix. Analyze the data for HSE, the high school English grades. Your analysis should include a table of sample sizes, means, and standard deviations; normal quantile plots; a plot of the means; and a two-way ANOVA using sex and major as the factors. Write a short summary of your conclusions.

10.60 Refer to the data given for the change-of-majors study in the data set MAJORS in the data appendix. Analyze the data for GPA, the college grade point average. Your analysis should include a table of sample sizes, means, and standard deviations; normal quantile plots; a plot of the means; and a two-way ANOVA using sex and major as the factors. Write a short summary of your conclusions.

NOTES

1. Results of the study are reported in P. F. Campbell and G. P. McCabe, "Predicting the success of freshmen in a computer science major," *Communications of the ACM*, 27 (1984), pp. 1108–1113.
2. This rule is intended to provide a general guideline for deciding when serious errors may result by applying ANOVA procedures. When the sample sizes in each group are very small, the sample variances will tend to vary much more than when the sample sizes are large. In this case, the rule may be a little too conservative. For unequal sample sizes, particular difficulties can arise when a relatively small sample size is associated with a population having a relatively large standard deviation. Careful judgment is needed in all cases. By considering P-values rather than fixed level α testing, judgments in ambiguous cases can more easily be made; for example, if the P-value is very small, say 0.001, then it is probably safe to reject the H_0 even if there is a fair amount of variation in the sample standard deviations.
3. This example is based on data provided from a study conducted by Jim Baumann and Leah Jones of the Purdue University School of Education.
4. Several different definitions for the noncentrality parameter of the noncentral F distribution are in use. Many authors prefer $\phi = \sqrt{\lambda/I}$. We have chosen to use λ because it is the form needed for the SAS function PROBF.

5. We present the two-way ANOVA model and analysis for the general case in which the sample sizes may be unequal. If the sample sizes vary a great deal, serious complications can arise. There is no longer a single standard ANOVA analysis. Most computer packages offer several options for the computation of the ANOVA table when cell counts are unequal. When the counts are approximately equal, all methods give essentially the same results.
6. The data in Examples 10.18 and 10.19 were taken from an article on the results of the National Science Foundation study: M. Scott, "Women scientists earn 79% of men's salaries," *The Scientist*, November 16, 1987.
7. This example is based on a study described in P. Payne, "Nutrition adaptation in man: Social adjustments and their nutritional implications," in K. Blaxter and J. C. Waterlow, (eds.), *Nutrition Adaptation in Man*, Libbey, London, 1985.
8. Example 10.21 is based on a study described in S. S. Zentall and J. H. Shaw, "Effects of classroom noise on performance and activity of second-grade hyperactive and control children," *Journal of Educational Psychology*, 72 (1980), pp. 830–840.
9. This example is based on a study described in P. D. Wood et al., "Plasma lipoprotein distributions in male and female runners," in P. Milvey (ed.), *The Marathon: Physiological, Medical, Epidemiological, and Psychological Studies*, New York Academy of Sciences, 1977.

Data Appendix

Some of the computer exercises in the text refer to five relatively large data sets that are available from the publisher on a floppy disk. This disk also contains data from many other exercises and examples.

Background information for each of the five data sets is presented below. For the CHEESE, WOOD and READING data sets a complete listing is given. Since the CS-DATA and MAJORS data sets are very large, only a sample of cases are given here.

Text Reference: Computer Exercises 1.123 and 9.54–9.62

1 CHEESE

As cheddar cheese matures a variety of chemical processes take place. The taste of matured cheese is related to the concentration of several chemicals in the final product. In a study of cheddar cheese from the LaTrobe Valley of Victoria, Australia, samples of cheese were analyzed for their chemical composition and were subjected to taste tests.

Data for one type of cheese manufacturing process appear below. The variable "Case" is used to number the observations from 1 to 30. "Taste" is the response variable of interest. The taste scores were obtained by combining the scores from several tasters.

Three of the chemicals whose concentrations were measured were acetic acid, hydrogen sulfide and lactic acid. For acetic acid and hydrogen sulfide (natural) log transformations were taken. Thus the explanatory variables are the transformed concentrations of acetic acid ("Acetic") and hydrogen sulfide ("H2S") and the untransformed concentration of lactic acid ("Lactic").

These data are based on experiments performed by G. T. Lloyd and E. H. Ramshaw of the CSIRO Division of Food Research, Victoria, Australia. Some results of the statistical analyses of these data are given in G. P. McCabe, L. McCabe and A. Miller, "Analysis of taste and chemical composition of cheddar cheese 1982–83 experiments," CSIRO Division of Mathematics and Statistics Consulting Report VT85/6 and I. Barlow et al., "Correlations and changes in flavour and chemical parameters of cheddar cheeses during maturation," *Australian Journal of Dairy Technology*, 44 (1989), pp. 7–18.

Case	Taste	Acetic	H2S	Lactic	Case	Taste	Acetic	H2S	Lactic
1	12.3	4.543	3.135	0.86	16	40.9	6.365	9.588	1.74
2	20.9	5.159	5.043	1.53	17	15.9	4.787	3.912	1.16
3	39.0	5.366	5.438	1.57	18	6.4	5.412	4.700	1.49
4	47.9	5.759	7.496	1.81	19	18.0	5.247	6.174	1.63
5	5.6	4.663	3.807	0.99	20	38.9	5.438	9.064	1.99
6	25.9	5.697	7.601	1.09	21	14.0	4.564	4.949	1.15
7	37.3	5.892	8.726	1.29	22	15.2	5.298	5.220	1.33
8	21.9	6.078	7.966	1.78	23	32.0	5.455	9.242	1.44
9	18.1	4.898	3.850	1.29	24	56.7	5.855	10.199	2.01
10	21.0	5.242	4.174	1.58	25	16.8	5.366	3.664	1.31
11	34.9	5.740	6.142	1.68	26	11.6	6.043	3.219	1.46
12	57.2	6.446	7.908	1.90	27	26.5	6.458	6.962	1.72
13	0.7	4.477	2.996	1.06	28	0.7	5.328	3.912	1.25
14	25.9	5.236	4.942	1.30	29	13.4	5.802	6.685	1.08
15	54.9	6.151	6.752	1.52	30	5.5	6.176	4.787	1.25

2 CSDATA

Text Reference: Section 9.2; Computer Exercises 1.124, 2.107, 7.88, and 9.41–9.46.

The Computer Science department of a large university was interested in understanding why a large proportion of their first-year students failed to graduate as computer science majors. An examination of records from the registrar indicated that most of the attrition occurred during the first three semesters. Therefore, they decided to study all first-year students entering their program in a particular year and to follow their progress for the first three semesters.

The variables studied included the grade point average after three semesters and a collection of variables that would be available as students entered their program. These included standardized tests such as the Scholastic Aptitude Tests (SATs) and high school grades in various subjects. The individuals who conducted the study were also interested in examining differences between men and women in this program. Therefore, sex was included as a variable.

Data on 224 students who began study as computer science majors in a particular year were analyzed. A few exceptional cases were excluded, such as students who did not have complete data available on the variables of interest (a few students were admitted who did not take the SATs). Data for the first 20 students appear in the table below. There are eight variables for each student. OBS is a variable used to identify the student. The data files kept by the registrar identified students by social security number but for this study they were simply given a number from 1 to 224. The grade point average after three semesters is the variable GPA. This university uses a six-point scale with A corresponding to 6, B to 5, C to 4, etc. A straight A student has a 6.00 GPA. If you would like to convert to the more conventional four-point scale, use your software to subtract 2 from the GPA.

The high school grades included in the data set are the variables HSM, HSS, and HSE. These correspond to average high school grades in math, science and English. High schools use different grading systems (some high schools have a grade higher than A for honors courses), so the university's task in constructing these variables is not easy. The researchers were willing to accept the university's judgment and used its values. High school grades were recorded on a scale from 1 to 10, with 10 corresponding to A, 9 to A−, 8 to B+, etc.

The SAT scores are SATM and SATV, corresponding to the mathematics and verbal parts of the SAT. Gender was recorded as 1 for men and 2 for women. This is an arbitrary code. For software packages that are capable of dealing with alphanumeric variables (values do not have to be numbers), it is more convenient to use M and F or Men and Women as values for the sex variable. With this kind of user-friendly capability, you do not have to remember who are the 1's and who are the 2's.

Results of the study are reported in P. F. Campbell and G. P. McCabe, "Predicting the success of freshmen in a computer science major," *Communications of the ACM,* 27 (1984), pp. 1108–1113.

OBS	GPA	HSM	HSS	HSE	SATM	SATV	SEX
1	5.32	10	10	10	670	600	1
2	4.26	6	8	5	700	640	1
3	4.35	8	6	8	640	530	1
4	4.08	9	10	7	670	600	1
5	5.38	8	9	8	540	580	1
6	5.29	10	8	8	760	630	1
7	5.21	8	8	7	600	400	1
8	4.00	3	7	6	460	530	1
9	5.18	9	10	8	670	450	1
10	4.34	7	7	6	570	480	1
11	5.08	9	10	6	491	488	1
12	5.34	5	9	7	600	600	1
13	3.40	6	8	8	510	530	1
14	3.43	10	9	9	750	610	1
15	4.48	8	9	6	650	460	1
16	5.73	10	10	9	720	630	1
17	5.80	10	10	9	760	500	1
18	6.00	9	9	8	800	610	1
19	4.00	9	6	5	640	670	1
20	5.74	9	10	9	750	700	1

Text Reference: Computer Exercises 2.108–2.110, 7.89, 9.47, and 9.48.

3 WOOD

The WOOD data set comes from a study of the strength of wood products conducted by Mike Hunt and Mike Triche of the Purdue University Department of Forestry. These researchers measured the modulus of elasticity for each of 50 strips of yellow poplar wood two times. The units are millions of pounds per square inch.

The complete data set appears below. The variable S identifies the strip. T1 and T2 are the measures of the modulus of elasticity at the two times.

S	T1	T2	S	T1	T2	S	T1	T2	S	T1	T2
1	1.58	1.55	14	1.33	1.33	27	1.65	1.65	39	1.89	1.91
2	1.53	1.54	15	1.72	1.70	28	1.76	1.75	40	1.73	1.75
3	1.38	1.37	16	1.63	1.62	29	1.58	1.56	41	1.80	1.81
4	1.24	1.25	17	1.51	1.50	30	1.69	1.67	42	1.67	1.68
5	1.59	1.57	18	1.42	1.42	31	1.60	1.58	43	1.63	1.64
6	1.52	1.51	19	1.90	1.87	32	1.57	1.56	44	1.87	1.88
7	1.47	1.45	20	1.81	1.79	33	1.70	1.69	45	2.02	2.04
8	1.45	1.44	21	1.88	1.86	34	1.41	1.40	46	1.79	1.81
9	1.54	1.53	22	1.68	1.65	35	1.33	1.33	47	1.86	1.88
10	1.49	1.49	23	1.56	1.54	36	1.59	1.59	48	1.72	1.73
11	1.51	1.51	24	1.68	1.67	37	1.64	1.66	49	1.83	1.83
12	1.60	1.58	25	1.75	1.74	38	1.74	1.74	50	1.68	1.70
13	1.34	1.32	26	1.63	1.62						

Text Reference: Section 10.1, Computer Exercises 7.90 and 10.45–10.48.

4 READING

Jim Baumann and Leah Jones of the Purdue University School of Education conducted a study to compare three methods of teaching reading comprehension. The 66 students who participated in the study were randomly assigned to the methods (22 to each). The standard practice of comparing new methods with a traditional one was used in this study. The traditional method is called Basal and the two innovative methods are called DRTA and Strat.

In the data set the variable Subject is used to identify the individual students. The values are 1 to 66. The method of instruction is indicated by the variable Group with values B, D and S, corresponding to Basal, DRTA and Strat. Two pretests and three post tests were given to all students. These are the variables Pre1, Pre2, Post1, Post2, and Post3.

Subject	Group	Pre1	Pre2	Post1	Post2	Post3	Subject	Group	Pre1	Pre2	Post1	Post2	Post3
1	B	4	3	5	4	41	34	D	6	2	7	0	55
2	B	6	5	9	5	41	35	D	8	4	10	6	57
3	B	9	4	5	3	43	36	D	9	6	8	6	53
4	B	12	6	8	5	46	37	D	9	4	8	7	37
5	B	16	5	10	9	46	38	D	8	4	10	11	50
6	B	15	13	9	8	45	39	D	9	5	12	6	54
7	B	14	8	12	5	45	40	D	13	6	10	6	41
8	B	12	7	5	5	32	41	D	10	2	11	6	49
9	B	12	3	8	7	33	42	D	8	6	7	8	47
10	B	8	8	7	7	39	43	D	8	5	8	8	49
11	B	13	7	12	4	42	44	D	10	6	12	6	49
12	B	9	2	4	4	45	45	S	11	7	11	12	53
13	B	12	5	4	6	39	46	S	7	6	4	8	47
14	B	12	2	8	8	44	47	S	4	6	4	10	41
15	B	12	2	6	4	36	48	S	7	2	4	4	49
16	B	10	10	9	10	49	49	S	7	6	3	9	43
17	B	8	5	3	3	40	50	S	6	5	8	5	45
18	B	12	5	5	5	35	51	S	11	5	12	8	50
19	B	11	3	4	5	36	52	S	14	6	14	12	48
20	B	8	4	2	3	40	53	S	13	6	12	11	49
21	B	7	3	5	4	54	54	S	9	5	7	11	42
22	B	9	6	7	8	32	55	S	12	3	5	10	38
23	D	7	2	7	6	31	56	S	13	9	9	9	42
24	D	7	6	5	6	40	57	S	4	6	1	10	34
25	D	12	4	13	3	48	58	S	13	8	13	1	48
26	D	10	1	5	7	30	59	S	6	4	7	9	51
27	D	16	8	14	7	42	60	S	12	3	5	13	33
28	D	15	7	14	6	48	61	S	6	6	7	9	44
29	D	9	6	10	9	49	62	S	11	4	11	7	48
30	D	8	7	13	5	53	63	S	14	4	15	7	49
31	D	13	7	12	7	48	64	S	8	2	9	5	33
32	D	12	8	11	6	43	65	S	5	3	6	8	45
33	D	7	6	8	5	55	66	S	8	3	4	6	42

Text Reference: Computer Exercises 10.51–10.60.

5 MAJORS

See the the description of the CSDATA data set for background information on the study for this data set. In this data file, the variables described for CSDATA are given with an additional variable "Maj" that specifies the student's major field of study at the end of three semesters. The codes 1, 2, and 3 correspond to Computer Science, Engineering and Other Sciences, and Other. All available data were used in the analysis performed, which resulted in sample sizes that were unequal in the six sex-by-major groups.

For a one-way ANOVA this causes no particular problems. However, for a two-way ANOVA several complications arise when the sample sizes are unequal. A detailed discussion of these complications is beyond the scope of this text. To avoid these difficulties and still use these interesting data, simulated data based on the results of this study are given on the data disk. ANOVA based on these simulated data will result in the same qualitative conclusions as those obtained with the original data.

OBS	SEX	Maj	SATM	SATV	HSM	HSE	HSS	GPA
1	1	1	640	530	8	8	6	4.35
2	1	1	670	600	9	7	10	4.08
3	1	1	600	400	8	7	8	5.21
4	1	1	570	480	7	6	7	4.34
5	1	1	510	530	6	8	8	3.40
6	1	1	750	610	10	9	9	3.43
7	1	1	650	460	8	6	9	4.48
8	1	1	720	630	10	9	10	5.73
9	1	1	760	500	10	9	10	5.80
10	1	1	640	670	9	5	6	4.00
11	1	1	640	490	10	8	9	5.16
12	1	1	520	360	9	7	8	4.73
13	1	1	700	520	7	6	8	3.07
14	1	1	490	550	6	6	8	3.82
15	1	1	640	520	10	7	10	5.12
16	1	1	550	290	9	4	7	4.25
17	1	1	600	520	10	10	10	4.93
18	1	1	710	530	10	9	9	4.83
19	1	1	750	670	9	9	10	5.10
20	1	1	620	480	9	9	9	4.87

Tables

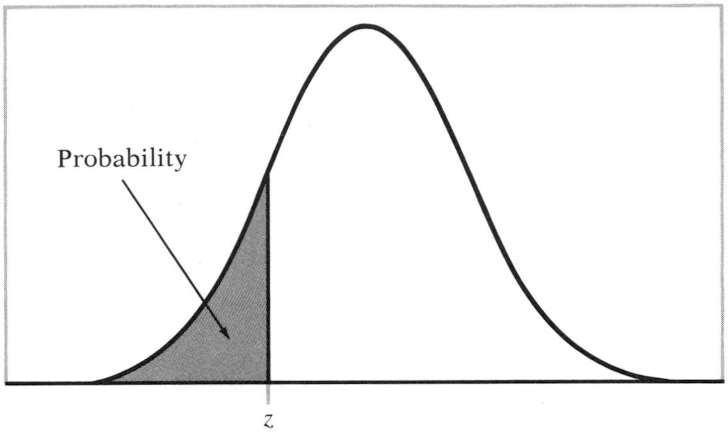

Table entry for *z* is the probability lying below *z*.

Table A Standard normal probabilities

z	.00	.01	.02	.03	.04	.05	.06	.07	.08	.09
−3.4	.0003	.0003	.0003	.0003	.0003	.0003	.0003	.0003	.0003	.0002
−3.3	.0005	.0005	.0005	.0004	.0004	.0004	.0004	.0004	.0004	.0003
−3.2	.0007	.0007	.0006	.0006	.0006	.0006	.0006	.0005	.0005	.0005
−3.1	.0010	.0009	.0009	.0009	.0008	.0008	.0008	.0008	.0007	.0007
−3.0	.0013	.0013	.0013	.0012	.0012	.0011	.0011	.0011	.0010	.0010
−2.9	.0019	.0018	.0018	.0017	.0016	.0016	.0015	.0015	.0014	.0014
−2.8	.0026	.0025	.0024	.0023	.0023	.0022	.0021	.0021	.0020	.0019
−2.7	.0035	.0034	.0033	.0032	.0031	.0030	.0029	.0028	.0027	.0026
−2.6	.0047	.0045	.0044	.0043	.0041	.0040	.0039	.0038	.0037	.0036
−2.5	.0062	.0060	.0059	.0057	.0055	.0054	.0052	.0051	.0049	.0048
−2.4	.0082	.0080	.0078	.0075	.0073	.0071	.0069	.0068	.0066	.0064
−2.3	.0107	.0104	.0102	.0099	.0096	.0094	.0091	.0089	.0087	.0084
−2.2	.0139	.0136	.0132	.0129	.0125	.0122	.0119	.0116	.0113	.0110
−2.1	.0179	.0174	.0170	.0166	.0162	.0158	.0154	.0150	.0146	.0143
−2.0	.0228	.0222	.0217	.0212	.0207	.0202	.0197	.0192	.0188	.0183
−1.9	.0287	.0281	.0274	.0268	.0262	.0256	.0250	.0244	.0239	.0233
−1.8	.0359	.0351	.0344	.0336	.0329	.0322	.0314	.0307	.0301	.0294
−1.7	.0446	.0436	.0427	.0418	.0409	.0401	.0392	.0384	.0375	.0367
−1.6	.0548	.0537	.0526	.0516	.0505	.0495	.0485	.0475	.0465	.0455
−1.5	.0668	.0655	.0643	.0630	.0618	.0606	.0594	.0582	.0571	.0559
−1.4	.0808	.0793	.0778	.0764	.0749	.0735	.0721	.0708	.0694	.0681
−1.3	.0968	.0951	.0934	.0918	.0901	.0885	.0869	.0853	.0838	.0823
−1.2	.1151	.1131	.1112	.1093	.1075	.1056	.1038	.1020	.1003	.0985
−1.1	.1357	.1335	.1314	.1292	.1271	.1251	.1230	.1210	.1190	.1170
−1.0	.1587	.1562	.1539	.1515	.1492	.1469	.1446	.1423	.1401	.1379
−0.9	.1841	.1814	.1788	.1762	.1736	.1711	.1685	.1660	.1635	.1611
−0.8	.2119	.2090	.2061	.2033	.2005	.1977	.1949	.1922	.1894	.1867
−0.7	.2420	.2389	.2358	.2327	.2296	.2266	.2236	.2206	.2177	.2148
−0.6	.2743	.2709	.2676	.2643	.2611	.2578	.2546	.2514	.2483	.2451
−0.5	.3085	.3050	.3015	.2981	.2946	.2912	.2877	.2843	.2810	.2776
−0.4	.3446	.3409	.3372	.3336	.3300	.3264	.3228	.3192	.3156	.3121
−0.3	.3821	.3783	.3745	.3707	.3669	.3632	.3594	.3557	.3520	.3483
−0.2	.4207	.4168	.4129	.4090	.4052	.4013	.3974	.3936	.3897	.3859
−0.1	.4602	.4562	.4522	.4483	.4443	.4404	.4364	.4325	.4286	.4247
−0.0	.5000	.4960	.4920	.4880	.4840	.4801	.4761	.4721	.4681	.4641

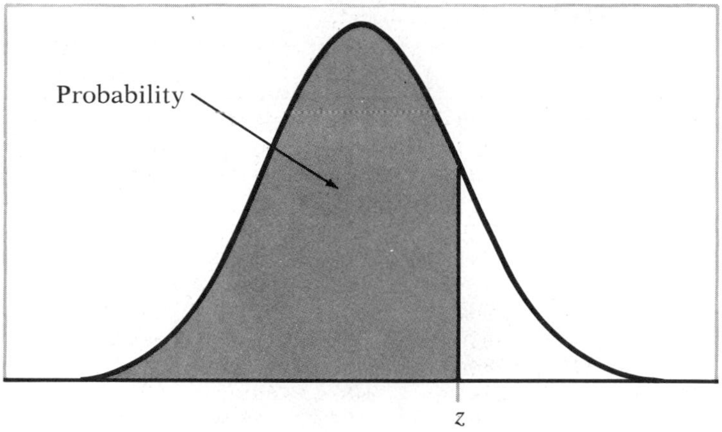

Table entry for *z* is the probability lying below *z*.

Table A *(Continued)*

z	.00	.01	.02	.03	.04	.05	.06	.07	.08	.09
0.0	.5000	.5040	.5080	.5120	.5160	.5199	.5239	.5279	.5319	.5359
0.1	.5398	.5438	.5478	.5517	.5557	.5596	.5636	.5675	.5714	.5753
0.2	.5793	.5832	.5871	.5910	.5948	.5987	.6026	.6064	.6103	.6141
0.3	.6179	.6217	.6255	.6293	.6331	.6368	.6406	.6443	.6480	.6517
0.4	.6554	.6591	.6628	.6664	.6700	.6736	.6772	.6808	.6844	.6879
0.5	.6915	.6950	.6985	.7019	.7054	.7088	.7123	.7157	.7190	.7224
0.6	.7257	.7291	.7324	.7357	.7389	.7422	.7454	.7486	.7517	.7549
0.7	.7580	.7611	.7642	.7673	.7704	.7734	.7764	.7794	.7823	.7852
0.8	.7881	.7910	.7939	.7967	.7995	.8023	.8051	.8078	.8106	.8133
0.9	.8159	.8186	.8212	.8238	.8264	.8289	.8315	.8340	.8365	.8389
1.0	.8413	.8438	.8461	.8485	.8508	.8531	.8554	.8577	.8599	.8621
1.1	.8643	.8665	.8686	.8708	.8729	.8749	.8770	.8790	.8810	.8830
1.2	.8849	.8869	.8888	.8907	.8925	.8944	.8962	.8980	.8997	.9015
1.3	.9032	.9049	.9066	.9082	.9099	.9115	.9131	.9147	.9162	.9177
1.4	.9192	.9207	.9222	.9236	.9251	.9265	.9279	.9292	.9306	.9319
1.5	.9332	.9345	.9357	.9370	.9382	.9394	.9406	.9418	.9429	.9441
1.6	.9452	.9463	.9474	.9484	.9495	.9505	.9515	.9525	.9535	.9545
1.7	.9554	.9564	.9573	.9582	.9591	.9599	.9608	.9616	.9625	.9633
1.8	.9641	.9649	.9656	.9664	.9671	.9678	.9686	.9693	.9699	.9706
1.9	.9713	.9719	.9726	.9732	.9738	.9744	.9750	.9756	.9761	.9767
2.0	.9772	.9778	.9783	.9788	.9793	.9798	.9803	.9808	.9812	.9817
2.1	.9821	.9826	.9830	.9834	.9838	.9842	.9846	.9850	.9854	.9857
2.2	.9861	.9864	.9868	.9871	.9875	.9878	.9881	.9884	.9887	.9890
2.3	.9893	.9896	.9898	.9901	.9904	.9906	.9909	.9911	.9913	.9916
2.4	.9918	.9920	.9922	.9925	.9927	.9929	.9931	.9932	.9934	.9936
2.5	.9938	.9940	.9941	.9943	.9945	.9946	.9948	.9949	.9951	.9952
2.6	.9953	.9955	.9956	.9957	.9959	.9960	.9961	.9962	.9963	.9964
2.7	.9965	.9966	.9967	.9968	.9969	.9970	.9971	.9972	.9973	.9974
2.8	.9974	.9975	.9976	.9977	.9977	.9978	.9979	.9979	.9980	.9981
2.9	.9981	.9982	.9982	.9983	.9984	.9984	.9985	.9985	.9986	.9986
3.0	.9987	.9987	.9987	.9988	.9988	.9989	.9989	.9989	.9990	.9990
3.1	.9990	.9991	.9991	.9991	.9992	.9992	.9992	.9992	.9993	.9993
3.2	.9993	.9993	.9994	.9994	.9994	.9994	.9994	.9995	.9995	.9995
3.3	.9995	.9995	.9995	.9996	.9996	.9996	.9996	.9996	.9996	.9997
3.4	.9997	.9997	.9997	.9997	.9997	.9997	.9997	.9997	.9997	.9998

Table B **Random digits**

Line								
101	19223	95034	05756	28713	96409	12531	42544	82853
102	73676	47150	99400	01927	27754	42648	82425	36290
103	45467	71709	77558	00095	32863	29485	82226	90056
104	52711	38889	93074	60227	40011	85848	48767	52573
105	95592	94007	69971	91481	60779	53791	17297	59335
106	68417	35013	15529	72765	85089	57067	50211	47487
107	82739	57890	20807	47511	81676	55300	94383	14893
108	60940	72024	17868	24943	61790	90656	87964	18883
109	36009	19365	15412	39638	85453	46816	83485	41979
110	38448	48789	18338	24697	39364	42006	76688	08708
111	81486	69487	60513	09297	00412	71238	27649	39950
112	59636	88804	04634	71197	19352	73089	84898	45785
113	62568	70206	40325	03699	71080	22553	11486	11776
114	45149	32992	75730	66280	03819	56202	02938	70915
115	61041	77684	94322	24709	73698	14526	31893	32592
116	14459	26056	31424	80371	65103	62253	50490	61181
117	38167	98532	62183	70632	23417	26185	41448	75532
118	73190	32533	04470	29669	84407	90785	65956	86382
119	95857	07118	87664	92099	58806	66979	98624	84826
120	35476	55972	39421	65850	04266	35435	43742	11937
121	71487	09984	29077	14863	61683	47052	62224	51025
122	13873	81598	95052	90908	73592	75186	87136	95761
123	54580	81507	27102	56027	55892	33063	41842	81868
124	71035	09001	43367	49497	72719	96758	27611	91596
125	96746	12149	37823	71868	18442	35119	62103	39244
126	96927	19931	36089	74192	77567	88741	48409	41903
127	43909	99477	25330	64359	40085	16925	85117	36071
128	15689	14227	06565	14374	13352	49367	81982	87209
129	36759	58984	68288	22913	18638	54303	00795	08727
130	69051	64817	87174	09517	84534	06489	87201	97245
131	05007	16632	81194	14873	04197	85576	45195	96565
132	68732	55259	84292	08796	43165	93739	31685	97150
133	45740	41807	65561	33302	07051	93623	18132	09547
134	27816	78416	18329	21337	35213	37741	04312	68508
135	66925	55658	39100	78458	11206	19876	87151	31260
136	08421	44753	77377	28744	75592	08563	79140	92454
137	53645	66812	61421	47836	12609	15373	98481	14592
138	66831	68908	40772	21558	47781	33586	79177	06928
139	55588	99404	70708	41098	43563	56934	48394	51719
140	12975	13258	13048	45144	72321	81940	00360	02428
141	96767	35964	23822	96012	94591	65194	50842	53372
142	72829	50232	97892	63408	77919	44575	24870	04178
143	88565	42628	17797	49376	61762	16953	88604	12724
144	62964	88145	83083	69453	46109	59505	69680	00900
145	19687	12633	57857	95806	09931	02150	43163	58636
146	37609	59057	66967	83401	60705	02384	90597	93600
147	54973	86278	88737	74351	47500	84552	19909	67181
148	00694	05977	19664	65441	20903	62371	22725	53340
149	71546	05233	53946	68743	72460	27601	45403	88692
150	07511	88915	41267	16853	84569	79367	32337	03316

Table B (*Continued*)

Line								
151	03802	29341	29264	80198	12371	13121	54969	43912
152	77320	35030	77519	41109	98296	18984	60869	12349
153	07886	56866	39648	69290	03600	05376	58958	22720
154	87065	74133	21117	70595	22791	67306	28420	52067
155	42090	09628	54035	93879	98441	04606	27381	82637
156	55494	67690	88131	81800	11188	28552	25752	21953
157	16698	30406	96587	65985	07165	50148	16201	86792
158	16297	07626	68683	45335	34377	72941	41764	77038
159	22897	17467	17638	70043	36243	13008	83993	22869
160	98163	45944	34210	64158	76971	27689	82926	75957
161	43400	25831	06283	22138	16043	15706	73345	26238
162	97341	46254	88153	62336	21112	35574	99271	45297
163	64578	67197	28310	90341	37531	63890	52630	76315
164	11022	79124	49525	63078	17229	32165	01343	21394
165	81232	43939	23840	05995	84589	06788	76358	26622
166	36843	84798	51167	44728	20554	55538	27647	32708
167	84329	80081	69516	78934	14293	92478	16479	26974
168	27788	85789	41592	74472	96773	27090	24954	41474
169	99224	00850	43737	75202	44753	63236	14260	73686
170	38075	73239	52555	46342	13365	02182	30443	53229
171	87368	49451	55771	48343	51236	18522	73670	23212
172	40512	00681	44282	47178	08139	78693	34715	75606
173	81636	57578	54286	27216	58758	80358	84115	84568
174	26411	94292	06340	97762	37033	85968	94165	46514
175	80011	09937	57195	33906	94831	10056	42211	65491
176	92813	87503	63494	71379	76550	45984	05481	50830
177	70348	72871	63419	57363	29685	43090	18763	31714
178	24005	52114	26224	39078	80798	15220	43186	00976
179	85063	55810	10470	08029	30025	29734	61181	72090
180	11532	73186	92541	06915	72954	10167	12142	26492
181	59618	03914	05208	84088	20426	39004	84582	87317
182	92965	50837	39921	84661	82514	81899	24565	60874
183	85116	27684	14597	85747	01596	25889	41998	15635
184	15106	10411	90221	49377	44369	28185	80959	76355
185	03638	31589	07871	25792	85823	55400	56026	12193
186	97971	48932	45792	63993	95635	28753	46069	84635
187	49345	18305	76213	82390	77412	97401	50650	71755
188	87370	88099	89695	87633	76987	85503	26257	51736
189	88296	95670	74932	65317	93848	43988	47597	83044
190	79485	92200	99401	54473	34336	82786	05457	60343
191	40830	24979	23333	37619	56227	95941	59494	86539
192	32006	76302	81221	00693	95197	75044	46596	11628
193	37569	85187	44692	50706	53161	69027	88389	60313
194	56680	79003	23361	67094	15019	63261	24543	52884
195	05172	08100	22316	54495	60005	29532	18433	18057
196	74782	27005	03894	98038	20627	40307	47317	92759
197	85228	93264	61409	03404	09649	55937	60843	66167
198	68309	12060	14762	58002	03716	81968	57934	32624
199	26461	88346	52430	60906	74216	96263	69296	90107
200	42672	67680	42376	95023	82744	03971	96560	55148

Entry is $P(X = k) = \binom{n}{k} p^k (1 - p)^{n-k}$

		p								
n	*k*	.01	.02	.03	.04	.05	.06	.07	.08	.09
2	0	.9801	.9604	.9409	.9216	.9025	.8836	.8649	.8464	.8281
	1	.0198	.0392	.0582	.0768	.0950	.1128	.1302	.1472	.1638
	2	.0001	.0004	.0009	.0016	.0025	.0036	.0049	.0064	.0081
3	0	.9703	.9412	.9127	.8847	.8574	.8306	.8044	.7787	.7536
	1	.0294	.0576	.0847	.1106	.1354	.1590	.1816	.2031	.2236
	2	.0003	.0012	.0026	.0046	.0071	.0102	.0137	.0177	.0221
	3				.0001	.0001	.0002	.0003	.0005	.0007
4	0	.9606	.9224	.8853	.8493	.8145	.7807	.7481	.7164	.6857
	1	.0388	.0753	.1095	.1416	.1715	.1993	.2252	.2492	.2713
	2	.0006	.0023	.0051	.0088	.0135	.0191	.0254	.0325	.0402
	3			.0001	.0002	.0005	.0008	.0013	.0019	.0027
	4									.0001
5	0	.9510	.9039	.8587	.8154	.7738	.7339	.6957	.6591	.6240
	1	.0480	.0922	.1328	.1699	.2036	.2342	.2618	.2866	.3086
	2	.0010	.0038	.0082	.0142	.0214	.0299	.0394	.0498	.0610
	3		.0001	.0003	.0006	.0011	.0019	.0030	.0043	.0060
	4						.0001	.0001	.0002	.0003
	5									
6	0	.9415	.8858	.8330	.7828	.7351	.6899	.6470	.6064	.5679
	1	.0571	.1085	.1546	.1957	.2321	.2642	.2922	.3164	.3370
	2	.0014	.0055	.0120	.0204	.0305	.0422	.0550	.0688	.0833
	3		.0002	.0005	.0011	.0021	.0036	.0055	.0080	.0110
	4					.0001	.0002	.0003	.0005	.0008
	5									
	6									
7	0	.9321	.8681	.8080	.7514	.6983	.6485	.6017	.5578	.5168
	1	.0659	.1240	.1749	.2192	.2573	.2897	.3170	.3396	.3578
	2	.0020	.0076	.0162	.0274	.0406	.0555	.0716	.0886	.1061
	3		.0003	.0008	.0019	.0036	.0059	.0090	.0128	.0175
	4				.0001	.0002	.0004	.0007	.0011	.0017
	5								.0001	.0001
	6									
	7									
8	0	.9227	.8508	.7837	.7214	.6634	.6096	.5596	.5132	.4703
	1	.0746	.1389	.1939	.2405	.2793	.3113	.3370	.3570	.3721
	2	.0026	.0099	.0210	.0351	.0515	.0695	.0888	.1087	.1288
	3	.0001	.0004	.0013	.0029	.0054	.0089	.0134	.0189	.0255
	4			.0001	.0002	.0004	.0007	.0013	.0021	.0031
	5							.0001	.0001	.0002
	6									
	7									
	8									

Entry is $P(X = k) = \binom{n}{k} p^k (1 - p)^{n-k}$

n	k	p = .10	.15	.20	.25	.30	.35	.40	.45	.50
2	0	.8100	.7225	.6400	.5625	.4900	.4225	.3600	.3025	.2500
	1	.1800	.2550	.3200	.3750	.4200	.4550	.4800	.4950	.5000
	2	.0100	.0225	.0400	.0625	.0900	.1225	.1600	.2025	.2500
3	0	.7290	.6141	.5120	.4219	.3430	.2746	.2160	.1664	.1250
	1	.2430	.3251	.3840	.4219	.4410	.4436	.4320	.4084	.3750
	2	.0270	.0574	.0960	.1406	.1890	.2389	.2880	.3341	.3750
	3	.0010	.0034	.0080	.0156	.0270	.0429	.0640	.0911	.1250
4	0	.6561	.5220	.4096	.3164	.2401	.1785	.1296	.0915	.0625
	1	.2916	.3685	.4096	.4219	.4116	.3845	.3456	.2995	.2500
	2	.0486	.0975	.1536	.2109	.2646	.3105	.3456	.3675	.3750
	3	.0036	.0115	.0256	.0469	.0756	.1115	.1536	.2005	.2500
	4	.0001	.0005	.0016	.0039	.0081	.0150	.0256	.0410	.0625
5	0	.5905	.4437	.3277	.2373	.1681	.1160	.0778	.0503	.0313
	1	.3280	.3915	.4096	.3955	.3602	.3124	.2592	.2059	.1563
	2	.0729	.1382	.2048	.2637	.3087	.3364	.3456	.3369	.3125
	3	.0081	.0244	.0512	.0879	.1323	.1811	.2304	.2757	.3125
	4	.0004	.0022	.0064	.0146	.0284	.0488	.0768	.1128	.1562
	5		.0001	.0003	.0010	.0024	.0053	.0102	.0185	.0312
6	0	.5314	.3771	.2621	.1780	.1176	.0754	.0467	.0277	.0156
	1	.3543	.3993	.3932	.3560	.3025	.2437	.1866	.1359	.0938
	2	.0984	.1762	.2458	.2966	.3241	.3280	.3110	.2780	.2344
	3	.0146	.0415	.0819	.1318	.1852	.2355	.2765	.3032	.3125
	4	.0012	.0055	.0154	.0330	.0595	.0951	.1382	.1861	.2344
	5	.0001	.0004	.0015	.0044	.0102	.0205	.0369	.0609	.0937
	6			.0001	.0002	.0007	.0018	.0041	.0083	.0156
7	0	.4783	.3206	.2097	.1335	.0824	.0490	.0280	.0152	.0078
	1	.3720	.3960	.3670	.3115	.2471	.1848	.1306	.0872	.0547
	2	.1240	.2097	.2753	.3115	.3177	.2985	.2613	.2140	.1641
	3	.0230	.0617	.1147	.1730	.2269	.2679	.2903	.2918	.2734
	4	.0026	.0109	.0287	.0577	.0972	.1442	.1935	.2388	.2734
	5	.0002	.0012	.0043	.0115	.0250	.0466	.0774	.1172	.1641
	6		.0001	.0004	.0013	.0036	.0084	.0172	.0320	.0547
	7				.0001	.0002	.0006	.0016	.0037	.0078
8	0	.4305	.2725	.1678	.1001	.0576	.0319	.0168	.0084	.0039
	1	.3826	.3847	.3355	.2670	.1977	.1373	.0896	.0548	.0313
	2	.1488	.2376	.2936	.3115	.2965	.2587	.2090	.1569	.1094
	3	.0331	.0839	.1468	.2076	.2541	.2786	.2787	.2568	.2188
	4	.0046	.0185	.0459	.0865	.1361	.1875	.2322	.2627	.2734
	5	.0004	.0026	.0092	.0231	.0467	.0808	.1239	.1719	.2188
	6		.0002	.0011	.0038	.0100	.0217	.0413	.0703	.1094
	7			.0001	.0004	.0012	.0033	.0079	.0164	.0312
	8					.0001	.0002	.0007	.0017	.0039

		p								
n	k	.01	.02	.03	.04	.05	.06	.07	.08	.09
9	0	.9135	.8337	.7602	.6925	.6302	.5730	.5204	.4722	.4279
	1	.0830	.1531	.2116	.2597	.2985	.3292	.3525	.3695	.3809
	2	.0034	.0125	.0262	.0433	.0629	.0840	.1061	.1285	.1507
	3	.0001	.0006	.0019	.0042	.0077	.0125	.0186	.0261	.0348
	4			.0001	.0003	.0006	.0012	.0021	.0034	.0052
	5						.0001	.0002	.0003	.0005
	6									
	7									
	8									
	9									
10	0	.9044	.8171	.7374	.6648	.5987	.5386	.4840	.4344	.3894
	1	.0914	.1667	.2281	.2770	.3151	.3438	.3643	.3777	.3851
	2	.0042	.0153	.0317	.0519	.0746	.0988	.1234	.1478	.1714
	3	.0001	.0008	.0026	.0058	.0105	.0168	.0248	.0343	.0452
	4			.0001	.0004	.0010	.0019	.0033	.0052	.0078
	5					.0001	.0001	.0003	.0005	.0009
	6									.0001
	7									
	8									
	9									
	10									
12	0	.8864	.7847	.6938	.6127	.5404	.4759	.4186	.3677	.3225
	1	.1074	.1922	.2575	.3064	.3413	.3645	.3781	.3837	.3827
	2	.0060	.0216	.0438	.0702	.0988	.1280	.1565	.1835	.2082
	3	.0002	.0015	.0045	.0098	.0173	.0272	.0393	.0532	.0686
	4		.0001	.0003	.0009	.0021	.0039	.0067	.0104	.0153
	5				.0001	.0002	.0004	.0008	.0014	.0024
	6							.0001	.0001	.0003
	7									
	8									
	9									
	10									
	11									
	12									
15	0	.8601	.7386	.6333	.5421	.4633	.3953	.3367	.2863	.2430
	1	.1303	.2261	.2938	.3388	.3658	.3785	.3801	.3734	.3605
	2	.0092	.0323	.0636	.0988	.1348	.1691	.2003	.2273	.2496
	3	.0004	.0029	.0085	.0178	.0307	.0468	.0653	.0857	.1070
	4		.0002	.0008	.0022	.0049	.0090	.0148	.0223	.0317
	5			.0001	.0002	.0006	.0013	.0024	.0043	.0069
	6						.0001	.0003	.0006	.0011
	7								.0001	.0001
	8									
	9									
	10									
	11									
	12									
	13									
	14									
	15									

		p								
n	*k*	.10	.15	.20	.25	.30	.35	.40	.45	.50
9	0	.3874	.2316	.1342	.0751	.0404	.0207	.0101	.0046	.0020
	1	.3874	.3679	.3020	.2253	.1556	.1004	.0605	.0339	.0176
	2	.1722	.2597	.3020	.3003	.2668	.2162	.1612	.1110	.0703
	3	.0446	.1069	.1762	.2336	.2668	.2716	.2508	.2119	.1641
	4	.0074	.0283	.0661	.1168	.1715	.2194	.2508	.2600	.2461
	5	.0008	.0050	.0165	.0389	.0735	.1181	.1672	.2128	.2461
	6	.0001	.0006	.0028	.0087	.0210	.0424	.0743	.1160	.1641
	7			.0003	.0012	.0039	.0098	.0212	.0407	.0703
	8				.0001	.0004	.0013	.0035	.0083	.0176
	9						.0001	.0003	.0008	.0020
10	0	.3487	.1969	.1074	.0563	.0282	.0135	.0060	.0025	.0010
	1	.3874	.3474	.2684	.1877	.1211	.0725	.0403	.0207	.0098
	2	.1937	.2759	.3020	.2816	.2335	.1757	.1209	.0763	.0439
	3	.0574	.1298	.2013	.2503	.2668	.2522	.2150	.1665	.1172
	4	.0112	.0401	.0881	.1460	.2001	.2377	.2508	.2384	.2051
	5	.0015	.0085	.0264	.0584	.1029	.1536	.2007	.2340	.2461
	6	.0001	.0012	.0055	.0162	.0368	.0689	.1115	.1596	.2051
	7		.0001	.0008	.0031	.0090	.0212	.0425	.0746	.1172
	8			.0001	.0004	.0014	.0043	.0106	.0229	.0439
	9					.0001	.0005	.0016	.0042	.0098
	10							.0001	.0003	.0010
12	0	.2824	.1422	.0687	.0317	.0138	.0057	.0022	.0008	.0002
	1	.3766	.3012	.2062	.1267	.0712	.0368	.0174	.0075	.0029
	2	.2301	.2924	.2835	.2323	.1678	.1088	.0639	.0339	.0161
	3	.0852	.1720	.2362	.2581	.2397	.1954	.1419	.0923	.0537
	4	.0213	.0683	.1329	.1936	.2311	.2367	.2128	.1700	.1208
	5	.0038	.0193	.0532	.1032	.1585	.2039	.2270	.2225	.1934
	6	.0005	.0040	.0155	.0401	.0792	.1281	.1766	.2124	.2256
	7		.0006	.0033	.0115	.0291	.0591	.1009	.1489	.1934
	8		.0001	.0005	.0024	.0078	.0199	.0420	.0762	.1208
	9			.0001	.0004	.0015	.0048	.0125	.0277	.0537
	10					.0002	.0008	.0025	.0068	.0161
	11						.0001	.0003	.0010	.0029
	12								.0001	.0002
15	0	.2059	.0874	.0352	.0134	.0047	.0016	.0005	.0001	.0000
	1	.3432	.2312	.1319	.0668	.0305	.0126	.0047	.0016	.0005
	2	.2669	.2856	.2309	.1559	.0916	.0476	.0219	.0090	.0032
	3	.1285	.2184	.2501	.2252	.1700	.1110	.0634	.0318	.0139
	4	.0428	.1156	.1876	.2252	.2186	.1792	.1268	.0780	.0417
	5	.0105	.0449	.1032	.1651	.2061	.2123	.1859	.1404	.0916
	6	.0019	.0132	.0430	.0917	.1472	.1906	.2066	.1914	.1527
	7	.0003	.0030	.0138	.0393	.0811	.1319	.1771	.2013	.1964
	8		.0005	.0035	.0131	.0348	.0710	.1181	.1647	.1964
	9		.0001	.0007	.0034	.0116	.0298	.0612	.1048	.1527
	10			.0001	.0007	.0030	.0096	.0245	.0515	.0916
	11				.0001	.0006	.0024	.0074	.0191	.0417
	12					.0001	.0004	.0016	.0052	.0139
	13						.0001	.0003	.0010	.0032
	14								.0001	.0005
	15									

Table C (Continued)

		p								
n	k	.01	.02	.03	.04	.05	.06	.07	.08	.09
20	0	.8179	.6676	.5438	.4420	.3585	.2901	.2342	.1887	.1516
	1	.1652	.2725	.3364	.3683	.3774	.3703	.3526	.3282	.3000
	2	.0159	.0528	.0988	.1458	.1887	.2246	.2521	.2711	.2818
	3	.0010	.0065	.0183	.0364	.0596	.0860	.1139	.1414	.1672
	4		.0006	.0024	.0065	.0133	.0233	.0364	.0523	.0703
	5			.0002	.0009	.0022	.0048	.0088	.0145	.0222
	6				.0001	.0003	.0008	.0017	.0032	.0055
	7						.0001	.0002	.0005	.0011
	8								.0001	.0002
	9									
	10									
	11									
	12									
	13									
	14									
	15									
	16									
	17									
	18									
	19									
	20									

		p								
n	k	.10	.15	.20	.25	.30	.35	.40	.45	.50
20	0	.1216	.0388	.0115	.0032	.0008	.0002	.0000	.0000	.0000
	1	.2702	.1368	.0576	.0211	.0068	.0020	.0005	.0001	.0000
	2	.2852	.2293	.1369	.0669	.0278	.0100	.0031	.0008	.0002
	3	.1901	.2428	.2054	.1339	.0716	.0323	.0123	.0040	.0011
	4	.0898	.1821	.2182	.1897	.1304	.0738	.0350	.0139	.0046
	5	.0319	.1028	.1746	.2023	.1789	.1272	.0746	.0365	.0148
	6	.0089	.0454	.1091	.1686	.1916	.1712	.1244	.0746	.0370
	7	.0020	.0160	.0545	.1124	.1643	.1844	.1659	.1221	.0739
	8	.0004	.0046	.0222	.0609	.1144	.1614	.1797	.1623	.1201
	9	.0001	.0011	.0074	.0271	.0654	.1158	.1597	.1771	.1602
	10		.0002	.0020	.0099	.0308	.0686	.1171	.1593	.1762
	11			.0005	.0030	.0120	.0336	.0710	.1185	.1602
	12			.0001	.0008	.0039	.0136	.0355	.0727	.1201
	13				.0002	.0010	.0045	.0146	.0366	.0739
	14					.0002	.0012	.0049	.0150	.0370
	15						.0003	.0013	.0049	.0148
	16							.0003	.0013	.0046
	17								.0002	.0011
	18									.0002
	19									
	20									

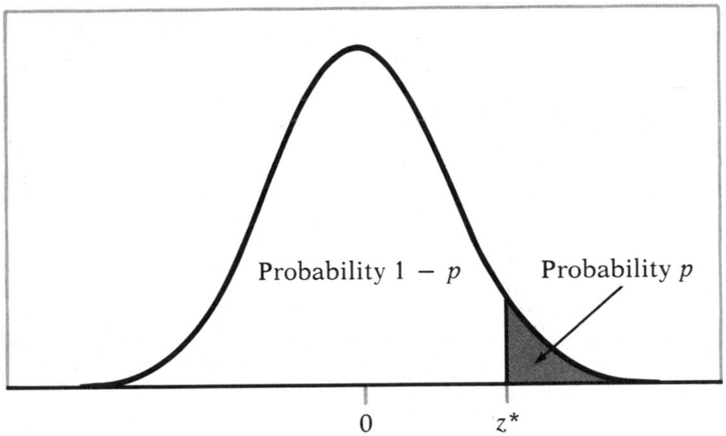

Table entry for *p* and *C* is the point z^* with probability *p* lying above it and probability *C* lying between $-z^*$ and z^*

Table D Standard normal critical values

C	p	z^*	C	p	z^*
50%	.25	.674	96%	.02	2.054
60%	.20	.841	98%	.01	2.326
70%	.15	1.036	99%	.005	2.576
80%	.10	1.282	99.5%	.0025	2.807
90%	.05	1.645	99.8%	.001	3.091
95%	.025	1.960	99.9%	.0005	3.291

Table entry for p and C is the point t^* with probability p lying above it and probability C lying between $-t^*$ and t^*

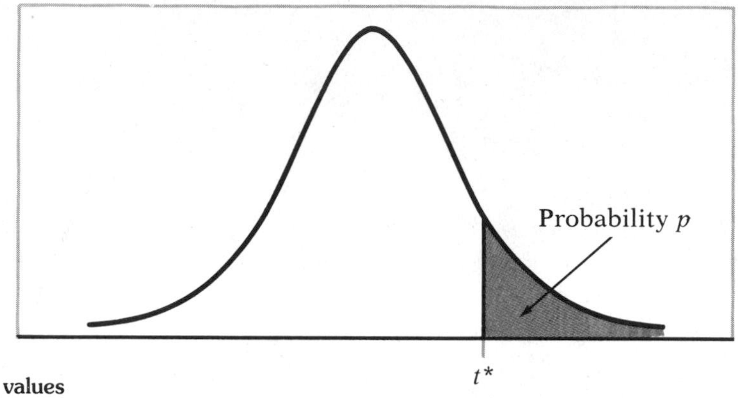

Table E *t* **distribution critical values**

	Tail probability p											
df	.25	.20	.15	.10	.05	.025	.02	.01	.005	.0025	.001	.0005
1	1.000	1.376	1.963	3.078	6.314	12.71	15.89	31.82	63.66	127.3	318.3	636.6
2	.816	1.061	1.386	1.886	2.920	4.303	4.849	6.965	9.925	14.09	22.33	31.60
3	.765	.978	1.250	1.638	2.353	3.182	3.482	4.541	5.841	7.453	10.21	12.92
4	.741	.941	1.190	1.533	2.132	2.776	2.999	3.747	4.604	5.598	7.173	8.610
5	.727	.920	1.156	1.476	2.015	2.571	2.757	3.365	4.032	4.773	5.893	6.869
6	.718	.906	1.134	1.440	1.943	2.447	2.612	3.143	3.707	4.317	5.208	5.959
7	.711	.896	1.119	1.415	1.895	2.365	2.517	2.998	3.499	4.029	4.785	5.408
8	.706	.889	1.108	1.397	1.860	2.306	2.449	2.896	3.355	3.833	4.501	5.041
9	.703	.883	1.100	1.383	1.833	2.262	2.398	2.821	3.250	3.690	4.297	4.781
10	.700	.879	1.093	1.372	1.812	2.228	2.359	2.764	3.169	3.581	4.144	4.587
11	.697	.876	1.088	1.363	1.796	2.201	2.328	2.718	3.106	3.497	4.025	4.437
12	.695	.873	1.083	1.356	1.782	2.179	2.303	2.681	3.055	3.428	3.930	4.318
13	.694	.870	1.079	1.350	1.771	2.160	2.282	2.650	3.012	3.372	3.852	4.221
14	.692	.868	1.076	1.345	1.761	2.145	2.264	2.624	2.977	3.326	3.787	4.140
15	.691	.866	1.074	1.341	1.753	2.131	2.249	2.602	2.947	3.286	3.733	4.073
16	.690	.865	1.071	1.337	1.746	2.120	2.235	2.583	2.921	3.252	3.686	4.015
17	.689	.863	1.069	1.333	1.740	2.110	2.224	2.567	2.898	3.222	3.646	3.965
18	.688	.862	1.067	1.330	1.734	2.101	2.214	2.552	2.878	3.197	3.611	3.922
19	.688	.861	1.066	1.328	1.729	2.093	2.205	2.539	2.861	3.174	3.579	3.883
20	.687	.860	1.064	1.325	1.725	2.086	2.197	2.528	2.845	3.153	3.552	3.850
21	.686	.859	1.063	1.323	1.721	2.080	2.189	2.518	2.831	3.135	3.527	3.819
22	.686	.858	1.061	1.321	1.717	2.074	2.183	2.508	2.819	3.119	3.505	3.792
23	.685	.858	1.060	1.319	1.714	2.069	2.177	2.500	2.807	3.104	3.485	3.768
24	.685	.857	1.059	1.318	1.711	2.064	2.172	2.492	2.797	3.091	3.467	3.745
25	.684	.856	1.058	1.316	1.708	2.060	2.167	2.485	2.787	3.078	3.450	3.725
26	.684	.856	1.058	1.315	1.706	2.056	2.162	2.479	2.779	3.067	3.435	3.707
27	.684	.855	1.057	1.314	1.703	2.052	2.158	2.473	2.771	3.057	3.421	3.690
28	.683	.855	1.056	1.313	1.701	2.048	2.154	2.467	2.763	3.047	3.408	3.674
29	.683	.854	1.055	1.311	1.699	2.045	2.150	2.462	2.756	3.038	3.396	3.659
30	.683	.854	1.055	1.310	1.697	2.042	2.147	2.457	2.750	3.030	3.385	3.646
40	.681	.851	1.050	1.303	1.684	2.021	2.123	2.423	2.704	2.971	3.307	3.551
50	.679	.849	1.047	1.299	1.676	2.009	2.109	2.403	2.678	2.937	3.261	3.496
60	.679	.848	1.045	1.296	1.671	2.000	2.099	2.390	2.660	2.915	3.232	3.460
80	.678	.846	1.043	1.292	1.664	1.990	2.088	2.374	2.639	2.887	3.195	3.416
100	.677	.845	1.042	1.290	1.660	1.984	2.081	2.364	2.626	2.871	3.174	3.390
1000	.675	.842	1.037	1.282	1.646	1.962	2.056	2.330	2.581	2.813	3.098	3.300
∞	.674	.841	1.036	1.282	1.645	1.960	2.054	2.326	2.576	2.807	3.091	3.291
	50%	60%	70%	80%	90%	95%	96%	98%	99%	99.5%	99.8%	99.9%
	Confidence level											

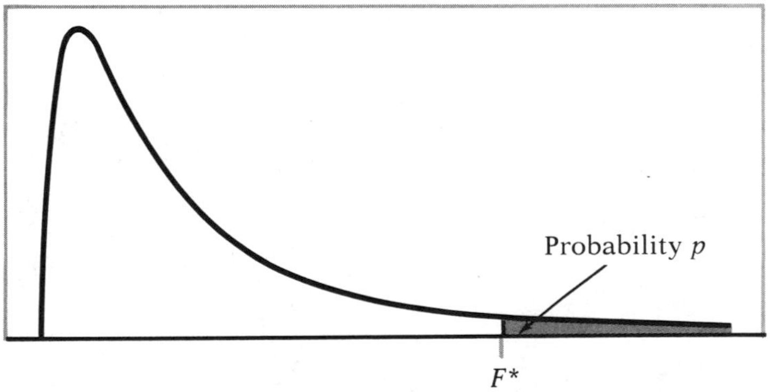

Figure for Table F
Table entry for p is the point F^* with probability p lying above it.

Degrees of freedom in the denominator

		Degrees of freedom in the numerator								
DFD	p	1	2	3	4	5	6	7	8	9
	.100	39.86	49.50	53.59	55.83	57.24	58.20	58.91	59.44	59.86
	.050	161.45	199.50	215.71	224.58	230.16	233.99	236.77	238.88	240.54
1	.025	647.79	799.50	864.16	899.58	921.85	937.11	948.22	956.66	963.28
	.010	4052.2	4999.5	5403.4	5624.6	5763.6	5859.0	5928.4	5981.1	6022.5
	.001	405284	500000	540379	562500	576405	585937	592873	598144	602284
	.100	8.53	9.00	9.16	9.24	9.29	9.33	9.35	9.37	9.38
	.050	18.51	19.00	19.16	19.25	19.30	19.33	19.35	19.37	19.38
2	.025	38.51	39.00	39.17	39.25	39.30	39.33	39.36	39.37	39.39
	.010	98.50	99.00	99.17	99.25	99.30	99.33	99.36	99.37	99.39
	.001	998.50	999.00	999.17	999.25	999.30	999.33	999.36	999.37	999.39
	.100	5.54	5.46	5.39	5.34	5.31	5.28	5.27	5.25	5.24
	.050	10.13	9.55	9.28	9.12	9.01	8.94	8.89	8.85	8.81
3	.025	17.44	16.04	15.44	15.10	14.88	14.73	14.62	14.54	14.47
	.010	34.12	30.82	29.46	28.71	28.24	27.91	27.67	27.49	27.35
	.001	167.03	148.50	141.11	137.10	134.58	132.85	131.58	130.62	129.86
	.100	4.54	4.32	4.19	4.11	4.05	4.01	3.98	3.95	3.94
	.050	7.71	6.94	6.59	6.39	6.26	6.16	6.09	6.04	6.00
4	.025	12.22	10.65	9.98	9.60	9.36	9.20	9.07	8.98	8.90
	.010	21.20	18.00	16.69	15.98	15.52	15.21	14.98	14.80	14.66
	.001	74.14	61.25	56.18	53.44	51.71	50.53	49.66	49.00	48.47
	.100	4.06	3.78	3.62	3.52	3.45	3.40	3.37	3.34	3.32
	.050	6.61	5.79	5.41	5.19	5.05	4.95	4.88	4.82	4.77
5	.025	10.01	8.43	7.76	7.39	7.15	6.98	6.85	6.76	6.68
	.010	16.26	13.27	12.06	11.39	10.97	10.67	10.46	10.29	10.16
	.001	47.18	37.12	33.20	31.09	29.75	28.83	28.16	27.65	27.24
	.100	3.78	3.46	3.29	3.18	3.11	3.05	3.01	2.98	2.96
	.050	5.99	5.14	4.76	4.53	4.39	4.28	4.21	4.15	4.10
6	.025	8.81	7.26	6.60	6.23	5.99	5.82	5.70	5.60	5.52
	.010	13.75	10.92	9.78	9.15	8.75	8.47	8.26	8.10	7.98
	.001	35.51	27.00	23.70	21.92	20.80	20.03	19.46	19.03	18.69
	.100	3.59	3.26	3.07	2.96	2.88	2.83	2.78	2.75	2.72
	.050	5.59	4.74	4.35	4.12	3.97	3.87	3.79	3.73	3.68
7	.025	8.07	6.54	5.89	5.52	5.29	5.12	4.99	4.90	4.82
	.010	12.25	9.55	8.45	7.85	7.46	7.19	6.99	6.84	6.72
	.001	29.25	21.69	18.77	17.20	16.21	15.52	15.02	14.63	14.33
	.100	3.46	3.11	2.92	2.81	2.73	2.67	2.62	2.59	2.56
	.050	5.32	4.46	4.07	3.84	3.69	3.58	3.50	3.44	3.39
8	.025	7.57	6.06	5.42	5.05	4.82	4.65	4.53	4.43	4.36
	.010	11.26	8.65	7.59	7.01	6.63	6.37	6.18	6.03	5.91
	.001	25.41	18.49	15.83	14.39	13.48	12.86	12.40	12.05	11.77
	.100	3.36	3.01	2.81	2.69	2.61	2.55	2.51	2.47	2.44
	.050	5.12	4.26	3.86	3.63	3.48	3.37	3.29	3.23	3.18
9	.025	7.21	5.71	5.08	4.72	4.48	4.32	4.20	4.10	4.03
	.010	10.56	8.02	6.99	6.42	6.06	5.80	5.61	5.47	5.35
	.001	22.86	16.39	13.90	12.56	11.71	11.13	10.70	10.37	10.11
	.100	3.29	2.92	2.73	2.61	2.52	2.46	2.41	2.38	2.35
	.050	4.96	4.10	3.71	3.48	3.33	3.22	3.14	3.07	3.02
10	.025	6.94	5.46	4.83	4.47	4.24	4.07	3.95	3.85	3.78
	.010	10.04	7.56	6.55	5.99	5.64	5.39	5.20	5.06	4.94
	.001	21.04	14.91	12.55	11.28	10.48	9.93	9.52	9.20	8.96
	.100	3.23	2.86	2.66	2.54	2.45	2.39	2.34	2.30	2.27
	.050	4.84	3.98	3.59	3.36	3.20	3.09	3.01	2.95	2.90
11	.025	6.72	5.26	4.63	4.28	4.04	3.88	3.76	3.66	3.59
	.010	9.65	7.21	6.22	5.67	5.32	5.07	4.89	4.74	4.63
	.001	19.69	13.81	11.56	10.35	9.58	9.05	8.66	8.35	8.12
	.100	3.18	2.81	2.61	2.48	2.39	2.33	2.28	2.24	2.21
	.050	4.75	3.89	3.49	3.26	3.11	3.00	2.91	2.85	2.80
12	.025	6.55	5.10	4.47	4.12	3.89	3.73	3.61	3.51	3.44
	.010	9.33	6.93	5.95	5.41	5.06	4.82	4.64	4.50	4.39
	.001	18.64	12.97	10.80	9.63	8.89	8.38	8.00	7.71	7.48

Degrees of freedom in the numerator

10	12	15	20	25	30	40	50	60	120	1000
60.19	60.71	61.22	61.74	62.05	62.26	62.53	62.69	62.79	63.06	63.30
241.88	243.91	245.95	248.01	249.26	250.10	251.14	251.77	252.20	253.25	254.19
968.63	976.71	984.87	993.10	998.08	1001.4	1005.6	1008.1	1009.8	1014.0	1017.7
6055.8	6106.3	6157.3	6208.7	6239.8	6260.6	6286.8	6302.5	6313.0	6339.4	6362.7
605621	610668	615764	620908	624017	626099	628712	630285	631337	633972	636301
9.39	9.41	9.42	9.44	9.45	9.46	9.47	9.47	9.47	9.48	9.49
19.40	19.41	19.43	19.45	19.46	19.46	19.47	19.48	19.48	19.49	19.49
39.40	39.41	39.43	39.45	39.46	39.46	39.47	39.48	39.48	39.49	39.50
99.40	99.42	99.43	99.45	99.46	99.47	99.47	99.48	99.48	99.49	99.50
999.40	999.42	999.43	999.45	999.46	999.47	999.47	999.48	999.48	999.49	999.50
5.23	5.22	5.20	5.18	5.17	5.17	5.16	5.15	5.15	5.14	5.13
8.79	8.74	8.70	8.66	8.63	8.62	8.59	8.58	8.57	8.55	8.53
14.42	14.34	14.25	14.17	14.12	14.08	14.04	14.01	13.99	13.95	13.91
27.23	27.05	26.87	26.69	26.58	26.50	26.41	26.35	26.32	26.22	26.14
129.25	128.32	127.37	126.42	125.84	125.45	124.96	124.66	124.47	123.97	123.53
3.92	3.90	3.87	3.84	3.83	3.82	3.80	3.80	3.79	3.78	3.76
5.96	5.91	5.86	5.80	5.77	5.75	5.72	5.70	5.69	5.66	5.63
8.84	8.75	8.66	8.56	8.50	8.46	8.41	8.38	8.36	8.31	8.26
14.55	14.37	14.20	14.02	13.91	13.84	13.75	13.69	13.65	13.56	13.47
48.05	47.41	46.76	46.10	45.70	45.43	45.09	44.88	44.75	44.40	44.09
3.30	3.27	3.24	3.21	3.19	3.17	3.16	3.15	3.14	3.12	3.11
4.74	4.68	4.62	4.56	4.52	4.50	4.46	4.44	4.43	4.40	4.37
6.62	6.52	6.43	6.33	6.27	6.23	6.18	6.14	6.12	6.07	6.02
10.05	9.89	9.72	9.55	9.45	9.38	9.29	9.24	9.20	9.11	9.03
26.92	26.42	25.91	25.39	25.08	24.87	24.60	24.44	24.33	24.06	23.82
2.94	2.90	2.87	2.84	2.81	2.80	2.78	2.77	2.76	2.74	2.72
4.06	4.00	3.94	3.87	3.83	3.81	3.77	3.75	3.74	3.70	3.67
5.46	5.37	5.27	5.17	5.11	5.07	5.01	4.98	4.96	4.90	4.86
7.87	7.72	7.56	7.40	7.30	7.23	7.14	7.09	7.06	6.97	6.89
18.41	17.99	17.56	17.12	16.85	16.67	16.44	16.31	16.21	15.98	15.77
2.70	2.67	2.63	2.59	2.57	2.56	2.54	2.52	2.51	2.49	2.47
3.64	3.57	3.51	3.44	3.40	3.38	3.34	3.32	3.30	3.27	3.23
4.76	4.67	4.57	4.47	4.40	4.36	4.31	4.28	4.25	4.20	4.15
6.62	6.47	6.31	6.16	6.06	5.99	5.91	5.86	5.82	5.74	5.66
14.08	13.71	13.32	12.93	12.69	12.53	12.33	12.20	12.12	11.91	11.72
2.54	2.50	2.46	2.42	2.40	2.38	2.36	2.35	2.34	2.32	2.30
3.35	3.28	3.22	3.15	3.11	3.08	3.04	3.02	3.01	2.97	2.93
4.30	4.20	4.10	4.00	3.94	3.89	3.84	3.81	3.78	3.73	3.68
5.81	5.67	5.52	5.36	5.26	5.20	5.12	5.07	5.03	4.95	4.87
11.54	11.19	10.84	10.48	10.26	10.11	9.92	9.80	9.73	9.53	9.36
2.42	2.38	2.34	2.30	2.27	2.25	2.23	2.22	2.21	2.18	2.16
3.14	3.07	3.01	2.94	2.89	2.86	2.83	2.80	2.79	2.75	2.71
3.96	3.87	3.77	3.67	3.60	3.56	3.51	3.47	3.45	3.39	3.34
5.26	5.11	4.96	4.81	4.71	4.65	4.57	4.52	4.48	4.40	4.32
9.89	9.57	9.24	8.90	8.69	8.55	8.37	8.26	8.19	8.00	7.84
2.32	2.28	2.24	2.20	2.17	2.16	2.13	2.12	2.11	2.08	2.06
2.98	2.91	2.85	2.77	2.73	2.70	2.66	2.64	2.62	2.58	2.54
3.72	3.62	3.52	3.42	3.35	3.31	3.26	3.22	3.20	3.14	3.09
4.85	4.71	4.56	4.41	4.31	4.25	4.17	4.12	4.08	4.00	3.92
8.75	8.45	8.13	7.80	7.60	7.47	7.30	7.19	7.12	6.94	6.78
2.25	2.21	2.17	2.12	2.10	2.08	2.05	2.04	2.03	2.00	1.98
2.85	2.79	2.72	2.65	2.60	2.57	2.53	2.51	2.49	2.45	2.41
3.53	3.43	3.33	3.23	3.16	3.12	3.06	3.03	3.00	2.94	2.89
4.54	4.40	4.25	4.10	4.01	3.94	3.86	3.81	3.78	3.69	3.61
7.92	7.63	7.32	7.01	6.81	6.68	6.52	6.42	6.35	6.18	6.02
2.19	2.15	2.10	2.06	2.03	2.01	1.99	1.97	1.96	1.93	1.91
2.75	2.69	2.62	2.54	2.50	2.47	2.43	2.40	2.38	2.34	2.30
3.37	3.28	3.18	3.07	3.01	2.96	2.91	2.87	2.85	2.79	2.73
4.30	4.16	4.01	3.86	3.76	3.70	3.62	3.57	3.54	3.45	3.37
7.29	7.00	6.71	6.40	6.22	6.09	5.93	5.83	5.76	5.59	5.44

DFD	p	Degrees of freedom in the numerator: 1	2	3	4	5	6	7	8	9
13	0.100	3.14	2.76	2.56	2.43	2.35	2.28	2.23	2.20	2.16
	0.050	4.67	3.81	3.41	3.18	3.03	2.92	2.83	2.77	2.71
	0.025	6.41	4.97	4.35	4.00	3.77	3.60	3.48	3.39	3.31
	0.010	9.07	6.70	5.74	5.21	4.86	4.62	4.44	4.30	4.19
	0.001	17.82	12.31	10.21	9.07	8.35	7.86	7.49	7.21	6.98
14	0.100	3.10	2.73	2.52	2.39	2.31	2.24	2.19	2.15	2.12
	0.050	4.60	3.74	3.34	3.11	2.96	2.85	2.76	2.70	2.65
	0.025	6.30	4.86	4.24	3.89	3.66	3.50	3.38	3.29	3.21
	0.010	8.86	6.51	5.56	5.04	4.69	4.46	4.28	4.14	4.03
	0.001	17.14	11.78	9.73	8.62	7.92	7.44	7.08	6.80	6.58
15	0.100	3.07	2.70	2.49	2.36	2.27	2.21	2.16	2.12	2.09
	0.050	4.54	3.68	3.29	3.06	2.90	2.79	2.71	2.64	2.59
	0.025	6.20	4.77	4.15	3.80	3.58	3.41	3.29	3.20	3.12
	0.010	8.68	6.36	5.42	4.89	4.56	4.32	4.14	4.00	3.89
	0.001	16.59	11.34	9.34	8.25	7.57	7.09	6.74	6.47	6.26
16	0.100	3.05	2.67	2.46	2.33	2.24	2.18	2.13	2.09	2.06
	0.050	4.49	3.63	3.24	3.01	2.85	2.74	2.66	2.59	2.54
	0.025	6.12	4.69	4.08	3.73	3.50	3.34	3.22	3.12	3.05
	0.010	8.53	6.23	5.29	4.77	4.44	4.20	4.03	3.89	3.78
	0.001	16.12	10.97	9.01	7.94	7.27	6.80	6.46	6.19	5.98
17	0.100	3.03	2.64	2.44	2.31	2.22	2.15	2.10	2.06	2.03
	0.050	4.45	3.59	3.20	2.96	2.81	2.70	2.61	2.55	2.49
	0.025	6.04	4.62	4.01	3.66	3.44	3.28	3.16	3.06	2.98
	0.010	8.40	6.11	5.19	4.67	4.34	4.10	3.93	3.79	3.68
	0.001	15.72	10.66	8.73	7.68	7.02	6.56	6.22	5.96	5.75
18	0.100	3.01	2.62	2.42	2.29	2.20	2.13	2.08	2.04	2.00
	0.050	4.41	3.55	3.16	2.93	2.77	2.66	2.58	2.51	2.46
	0.025	5.98	4.56	3.95	3.61	3.38	3.22	3.10	3.01	2.93
	0.010	8.29	6.01	5.09	4.58	4.25	4.01	3.84	3.71	3.60
	0.001	15.38	10.39	8.49	7.46	6.81	6.35	6.02	5.76	5.56
19	0.100	2.99	2.61	2.40	2.27	2.18	2.11	2.06	2.02	1.98
	0.050	4.38	3.52	3.13	2.90	2.74	2.63	2.54	2.48	2.42
	0.025	5.92	4.51	3.90	3.56	3.33	3.17	3.05	2.96	2.88
	0.010	8.18	5.93	5.01	4.50	4.17	3.94	3.77	3.63	3.52
	0.001	15.08	10.16	8.28	7.27	6.62	6.18	5.85	5.59	5.39
20	0.100	2.97	2.59	2.38	2.25	2.16	2.09	2.04	2.00	1.96
	0.050	4.35	3.49	3.10	2.87	2.71	2.60	2.51	2.45	2.39
	0.025	5.87	4.46	3.86	3.51	3.29	3.13	3.01	2.91	2.84
	0.010	8.10	5.85	4.94	4.43	4.10	3.87	3.70	3.56	3.46
	0.001	14.82	9.95	8.10	7.10	6.46	6.02	5.69	5.44	5.24
21	0.100	2.96	2.57	2.36	2.23	2.14	2.08	2.02	1.98	1.95
	0.050	4.32	3.47	3.07	2.84	2.68	2.57	2.49	2.42	2.37
	0.025	5.83	4.42	3.82	3.48	3.25	3.09	2.97	2.87	2.80
	0.010	8.02	5.78	4.87	4.37	4.04	3.81	3.64	3.51	3.40
	0.001	14.59	9.77	7.94	6.95	6.32	5.88	5.56	5.31	5.11
22	0.100	2.95	2.56	2.35	2.22	2.13	2.06	2.01	1.97	1.93
	0.050	4.30	3.44	3.05	2.82	2.66	2.55	2.46	2.40	2.34
	0.025	5.79	4.38	3.78	3.44	3.22	3.05	2.93	2.84	2.76
	0.010	7.95	5.72	4.82	4.31	3.99	3.76	3.59	3.45	3.35
	0.001	14.38	9.61	7.80	6.81	6.19	5.76	5.44	5.19	4.99
23	0.100	2.94	2.55	2.34	2.21	2.11	2.05	1.99	1.95	1.92
	0.050	4.28	3.42	3.03	2.80	2.64	2.53	2.44	2.37	2.32
	0.025	5.75	4.35	3.75	3.41	3.18	3.02	2.90	2.81	2.73
	0.010	7.88	5.66	4.76	4.26	3.94	3.71	3.54	3.41	3.30
	0.001	14.20	9.47	7.67	6.70	6.08	5.65	5.33	5.09	4.89
24	0.100	2.93	2.54	2.33	2.19	2.10	2.04	1.98	1.94	1.91
	0.050	4.26	3.40	3.01	2.78	2.62	2.51	2.42	2.36	2.30
	0.025	5.72	4.32	3.72	3.38	3.15	2.99	2.87	2.78	2.70
	0.010	7.82	5.61	4.72	4.22	3.90	3.67	3.50	3.36	3.26
	0.001	14.03	9.34	7.55	6.59	5.98	5.55	5.23	4.99	4.80

Degrees of freedom in the numerator										
10	12	15	20	25	30	40	50	60	120	1000
2.14	2.10	2.05	2.01	1.98	1.96	1.93	1.92	1.90	1.88	1.85
2.67	2.60	2.53	2.46	2.41	2.38	2.34	2.31	2.30	2.25	2.21
3.25	3.15	3.05	2.95	2.88	2.84	2.78	2.74	2.72	2.66	2.60
4.10	3.96	3.82	3.66	3.57	3.51	3.43	3.38	3.34	3.25	3.18
6.80	6.52	6.23	5.93	5.75	5.63	5.47	5.37	5.30	5.14	4.99
2.10	2.05	2.01	1.96	1.93	1.91	1.89	1.87	1.86	1.83	1.80
2.60	2.53	2.46	2.39	2.34	2.31	2.27	2.24	2.22	2.18	2.14
3.15	3.05	2.95	2.84	2.78	2.73	2.67	2.64	2.61	2.55	2.50
3.94	3.80	3.66	3.51	3.41	3.35	3.27	3.22	3.18	3.09	3.02
6.40	6.13	5.85	5.56	5.38	5.25	5.10	5.00	4.94	4.77	4.62
2.06	2.02	1.97	1.92	1.89	1.87	1.85	1.83	1.82	1.79	1.76
2.54	2.48	2.40	2.33	2.28	2.25	2.20	2.18	2.16	2.11	2.07
3.06	2.96	2.86	2.76	2.69	2.64	2.59	2.55	2.52	2.46	2.40
3.80	3.67	3.52	3.37	3.28	3.21	3.13	3.08	3.05	2.96	2.88
6.08	5.81	5.54	5.25	5.07	4.95	4.80	4.70	4.64	4.47	4.33
2.03	1.99	1.94	1.89	1.86	1.84	1.81	1.79	1.78	1.75	1.72
2.49	2.42	2.35	2.28	2.23	2.19	2.15	2.12	2.11	2.06	2.02
2.99	2.89	2.79	2.68	2.61	2.57	2.51	2.47	2.45	2.38	2.32
3.69	3.55	3.41	3.26	3.16	3.10	3.02	2.97	2.93	2.84	2.76
5.81	5.55	5.27	4.99	4.82	4.70	4.54	4.45	4.39	4.23	4.08
2.00	1.96	1.91	1.86	1.83	1.81	1.78	1.76	1.75	1.72	1.69
2.45	2.38	2.31	2.23	2.18	2.15	2.10	2.08	2.06	2.01	1.97
2.92	2.82	2.72	2.62	2.55	2.50	2.44	2.41	2.38	2.32	2.26
3.59	3.46	3.31	3.16	3.07	3.00	2.92	2.87	2.83	2.75	2.66
5.58	5.32	5.05	4.78	4.60	4.48	4.33	4.24	4.18	4.02	3.87
1.98	1.93	1.89	1.84	1.80	1.78	1.75	1.74	1.72	1.69	1.66
2.41	2.34	2.27	2.19	2.14	2.11	2.06	2.04	2.02	1.97	1.92
2.87	2.77	2.67	2.56	2.49	2.44	2.38	2.35	2.32	2.26	2.20
3.51	3.37	3.23	3.08	2.98	2.92	2.84	2.78	2.75	2.66	2.58
5.39	5.13	4.87	4.59	4.42	4.30	4.15	4.06	4.00	3.84	3.69
1.96	1.91	1.86	1.81	1.78	1.76	1.73	1.71	1.70	1.67	1.64
2.38	2.31	2.23	2.16	2.11	2.07	2.03	2.00	1.98	1.93	1.88
2.82	2.72	2.62	2.51	2.44	2.39	2.33	2.30	2.27	2.20	2.14
3.43	3.30	3.15	3.00	2.91	2.84	2.76	2.71	2.67	2.58	2.50
5.22	4.97	4.70	4.43	4.26	4.14	3.99	3.90	3.84	3.68	3.53
1.94	1.89	1.84	1.79	1.76	1.74	1.71	1.69	1.68	1.64	1.61
2.35	2.28	2.20	2.12	2.07	2.04	1.99	1.97	1.95	1.90	1.85
2.77	2.68	2.57	2.46	2.40	2.35	2.29	2.25	2.22	2.16	2.09
3.37	3.23	3.09	2.94	2.84	2.78	2.69	2.64	2.61	2.52	2.43
5.08	4.82	4.56	4.29	4.12	4.00	3.86	3.77	3.70	3.54	3.40
1.92	1.87	1.83	1.78	1.74	1.72	1.69	1.67	1.66	1.62	1.59
2.32	2.25	2.18	2.10	2.05	2.01	1.96	1.94	1.92	1.87	1.82
2.73	2.64	2.53	2.42	2.36	2.31	2.25	2.21	2.18	2.11	2.05
3.31	3.17	3.03	2.88	2.79	2.72	2.64	2.58	2.55	2.46	2.37
4.95	4.70	4.44	4.17	4.00	3.88	3.74	3.64	3.58	3.42	3.28
1.90	1.86	1.81	1.76	1.73	1.70	1.67	1.65	1.64	1.60	1.57
2.30	2.23	2.15	2.07	2.02	1.98	1.94	1.91	1.89	1.84	1.79
2.70	2.60	2.50	2.39	2.32	2.27	2.21	2.17	2.14	2.08	2.01
3.26	3.12	2.98	2.83	2.73	2.67	2.58	2.53	2.50	2.40	2.32
4.83	4.58	4.33	4.06	3.89	3.78	3.63	3.54	3.48	3.32	3.17
1.89	1.84	1.80	1.74	1.71	1.69	1.66	1.64	1.62	1.59	1.55
2.27	2.20	2.13	2.05	2.00	1.96	1.91	1.88	1.86	1.81	1.76
2.67	2.57	2.47	2.36	2.29	2.24	2.18	2.14	2.11	2.04	1.98
3.21	3.07	2.93	2.78	2.69	2.62	2.54	2.48	2.45	2.35	2.27
4.73	4.48	4.23	3.96	3.79	3.68	3.53	3.44	3.38	3.22	3.08
1.88	1.83	1.78	1.73	1.70	1.67	1.64	1.62	1.61	1.57	1.54
2.25	2.18	2.11	2.03	1.97	1.94	1.89	1.86	1.84	1.79	1.74
2.64	2.54	2.44	2.33	2.26	2.21	2.15	2.11	2.08	2.01	1.94
3.17	3.03	2.89	2.74	2.64	2.58	2.49	2.44	2.40	2.31	2.22
4.64	4.39	4.14	3.87	3.71	3.59	3.45	3.36	3.29	3.14	2.99

		Degrees of freedom in the numerator								
DFD	p	1	2	3	4	5	6	7	8	9
	0.100	2.92	2.53	2.32	2.18	2.09	2.02	1.97	1.93	1.89
	0.050	4.24	3.39	2.99	2.76	2.60	2.49	2.40	2.34	2.28
25	0.025	5.69	4.29	3.69	3.35	3.13	2.97	2.85	2.75	2.68
	0.010	7.77	5.57	4.68	4.18	3.85	3.63	3.46	3.32	3.22
	0.001	13.88	9.22	7.45	6.49	5.89	5.46	5.15	4.91	4.71
	0.100	2.91	2.52	2.31	2.17	2.08	2.01	1.96	1.92	1.88
	0.050	4.23	3.37	2.98	2.74	2.59	2.47	2.39	2.32	2.27
26	0.025	5.66	4.27	3.67	3.33	3.10	2.94	2.82	2.73	2.65
	0.010	7.72	5.53	4.64	4.14	3.82	3.59	3.42	3.29	3.18
	0.001	13.74	9.12	7.36	6.41	5.80	5.38	5.07	4.83	4.64
	0.100	2.90	2.51	2.30	2.17	2.07	2.00	1.95	1.91	1.87
	0.050	4.21	3.35	2.96	2.73	2.57	2.46	2.37	2.31	2.25
27	0.025	5.63	4.24	3.65	3.31	3.08	2.92	2.80	2.71	2.63
	0.010	7.68	5.49	4.60	4.11	3.78	3.56	3.39	3.26	3.15
	0.001	13.61	9.02	7.27	6.33	5.73	5.31	5.00	4.76	4.57
	0.100	2.89	2.50	2.29	2.16	2.06	2.00	1.94	1.90	1.87
	0.050	4.20	3.34	2.95	2.71	2.56	2.45	2.36	2.29	2.24
28	0.025	5.61	4.22	3.63	3.29	3.06	2.90	2.78	2.69	2.61
	0.010	7.64	5.45	4.57	4.07	3.75	3.53	3.36	3.23	3.12
	0.001	13.50	8.93	7.19	6.25	5.66	5.24	4.93	4.69	4.50
	0.100	2.89	2.50	2.28	2.15	2.06	1.99	1.93	1.89	1.86
	0.050	4.18	3.33	2.93	2.70	2.55	2.43	2.35	2.28	2.22
29	0.025	5.59	4.20	3.61	3.27	3.04	2.88	2.76	2.67	2.59
	0.010	7.60	5.42	4.54	4.04	3.73	3.50	3.33	3.20	3.09
	0.001	13.39	8.85	7.12	6.19	5.59	5.18	4.87	4.64	4.45
	0.100	2.88	2.49	2.28	2.14	2.05	1.98	1.93	1.88	1.85
	0.050	4.17	3.32	2.92	2.69	2.53	2.42	2.33	2.27	2.21
30	0.025	5.57	4.18	3.59	3.25	3.03	2.87	2.75	2.65	2.57
	0.010	7.56	5.39	4.51	4.02	3.70	3.47	3.30	3.17	3.07
	0.001	13.29	8.77	7.05	6.12	5.53	5.12	4.82	4.58	4.39
	0.100	2.84	2.44	2.23	2.09	2.00	1.93	1.87	1.83	1.79
	0.050	4.08	3.23	2.84	2.61	2.45	2.34	2.25	2.18	2.12
40	0.025	5.42	4.05	3.46	3.13	2.90	2.74	2.62	2.53	2.45
	0.010	7.31	5.18	4.31	3.83	3.51	3.29	3.12	2.99	2.89
	0.001	12.61	8.25	6.59	5.70	5.13	4.73	4.44	4.21	4.02
	0.100	2.81	2.41	2.20	2.06	1.97	1.90	1.84	1.80	1.76
	0.050	4.03	3.18	2.79	2.56	2.40	2.29	2.20	2.13	2.07
50	0.025	5.34	3.97	3.39	3.05	2.83	2.67	2.55	2.46	2.38
	0.010	7.17	5.06	4.20	3.72	3.41	3.19	3.02	2.89	2.78
	0.001	12.22	7.96	6.34	5.46	4.90	4.51	4.22	4.00	3.82
	0.100	2.79	2.39	2.18	2.04	1.95	1.87	1.82	1.77	1.74
	0.050	4.00	3.15	2.76	2.53	2.37	2.25	2.17	2.10	2.04
60	0.025	5.29	3.93	3.34	3.01	2.79	2.63	2.51	2.41	2.33
	0.010	7.08	4.98	4.13	3.65	3.34	3.12	2.95	2.82	2.72
	0.001	11.97	7.77	6.17	5.31	4.76	4.37	4.09	3.86	3.69
	0.100	2.76	2.36	2.14	2.00	1.91	1.83	1.78	1.73	1.69
	0.050	3.94	3.09	2.70	2.46	2.31	2.19	2.10	2.03	1.97
100	0.025	5.18	3.83	3.25	2.92	2.70	2.54	2.42	2.32	2.24
	0.010	6.90	4.82	3.98	3.51	3.21	2.99	2.82	2.69	2.59
	0.001	11.50	7.41	5.86	5.02	4.48	4.11	3.83	3.61	3.44
	0.100	2.73	2.33	2.11	1.97	1.88	1.80	1.75	1.70	1.66
	0.050	3.89	3.04	2.65	2.42	2.26	2.14	2.06	1.98	1.93
200	0.025	5.10	3.76	3.18	2.85	2.63	2.47	2.35	2.26	2.18
	0.010	6.76	4.71	3.88	3.41	3.11	2.89	2.73	2.60	2.50
	0.001	11.15	7.15	5.63	4.81	4.29	3.92	3.65	3.43	3.26
	0.100	2.71	2.31	2.09	1.95	1.85	1.78	1.72	1.68	1.64
	0.050	3.85	3.00	2.61	2.38	2.22	2.11	2.02	1.95	1.89
1000	0.025	5.04	3.70	3.13	2.80	2.58	2.42	2.30	2.20	2.13
	0.010	6.66	4.63	3.80	3.34	3.04	2.82	2.66	2.53	2.43
	0.001	10.89	6.96	5.46	4.65	4.14	3.78	3.51	3.30	3.13

Degrees of freedom in the numerator										
10	12	15	20	25	30	40	50	60	120	1000
1.87	1.82	1.77	1.72	1.68	1.66	1.63	1.61	1.59	1.56	1.52
2.24	2.16	2.09	2.01	1.96	1.92	1.87	1.84	1.82	1.77	1.72
2.61	2.51	2.41	2.30	2.23	2.18	2.12	2.08	2.05	1.98	1.91
3.13	2.99	2.85	2.70	2.60	2.54	2.45	2.40	2.36	2.27	2.18
4.56	4.31	4.06	3.79	3.63	3.52	3.37	3.28	3.22	3.06	2.91
1.86	1.81	1.76	1.71	1.67	1.65	1.61	1.59	1.58	1.54	1.51
2.22	2.15	2.07	1.99	1.94	1.90	1.85	1.82	1.80	1.75	1.70
2.59	2.49	2.39	2.28	2.21	2.16	2.09	2.05	2.03	1.95	1.89
3.09	2.96	2.81	2.66	2.57	2.50	2.42	2.36	2.33	2.23	2.14
4.48	4.24	3.99	3.72	3.56	3.44	3.30	3.21	3.15	2.99	2.84
1.85	1.80	1.75	1.70	1.66	1.64	1.60	1.58	1.57	1.53	1.50
2.20	2.13	2.06	1.97	1.92	1.88	1.84	1.81	1.79	1.73	1.68
2.57	2.47	2.36	2.25	2.18	2.13	2.07	2.03	2.00	1.93	1.86
3.06	2.93	2.78	2.63	2.54	2.47	2.38	2.33	2.29	2.20	2.11
4.41	4.17	3.92	3.66	3.49	3.38	3.23	3.14	3.08	2.92	2.78
1.84	1.79	1.74	1.69	1.65	1.63	1.59	1.57	1.56	1.52	1.48
2.19	2.12	2.04	1.96	1.91	1.87	1.82	1.79	1.77	1.71	1.66
2.55	2.45	2.34	2.23	2.16	2.11	2.05	2.01	1.98	1.91	1.84
3.03	2.90	2.75	2.60	2.51	2.44	2.35	2.30	2.26	2.17	2.08
4.35	4.11	3.86	3.60	3.43	3.32	3.18	3.09	3.02	2.86	2.72
1.83	1.78	1.73	1.68	1.64	1.62	1.58	1.56	1.55	1.51	1.47
2.18	2.10	2.03	1.94	1.89	1.85	1.81	1.77	1.75	1.70	1.65
2.53	2.43	2.32	2.21	2.14	2.09	2.03	1.99	1.96	1.89	1.82
3.00	2.87	2.73	2.57	2.48	2.41	2.33	2.27	2.23	2.14	2.05
4.29	4.05	3.80	3.54	3.38	3.27	3.12	3.03	2.97	2.81	2.66
1.82	1.77	1.72	1.67	1.63	1.61	1.57	1.55	1.54	1.50	1.46
2.16	2.09	2.01	1.93	1.88	1.84	1.79	1.76	1.74	1.68	1.63
2.51	2.41	2.31	2.20	2.12	2.07	2.01	1.97	1.94	1.87	1.80
2.98	2.84	2.70	2.55	2.45	2.39	2.30	2.25	2.21	2.11	2.02
4.24	4.00	3.75	3.49	3.33	3.22	3.07	2.98	2.92	2.76	2.61
1.76	1.71	1.66	1.61	1.57	1.54	1.51	1.48	1.47	1.42	1.38
2.08	2.00	1.92	1.84	1.78	1.74	1.69	1.66	1.64	1.58	1.52
2.39	2.29	2.18	2.07	1.99	1.94	1.88	1.83	1.80	1.72	1.65
2.80	2.66	2.52	2.37	2.27	2.20	2.11	2.06	2.02	1.92	1.82
3.87	3.64	3.40	3.14	2.98	2.87	2.73	2.64	2.57	2.41	2.25
1.73	1.68	1.63	1.57	1.53	1.50	1.46	1.44	1.42	1.38	1.33
2.03	1.95	1.87	1.78	1.73	1.69	1.63	1.60	1.58	1.51	1.45
2.32	2.22	2.11	1.99	1.92	1.87	1.80	1.75	1.72	1.64	1.56
2.70	2.56	2.42	2.27	2.17	2.10	2.01	1.95	1.91	1.80	1.70
3.67	3.44	3.20	2.95	2.79	2.68	2.53	2.44	2.38	2.21	2.05
1.71	1.66	1.60	1.54	1.50	1.48	1.44	1.41	1.40	1.35	1.30
1.99	1.92	1.84	1.75	1.69	1.65	1.59	1.56	1.53	1.47	1.40
2.27	2.17	2.06	1.94	1.87	1.82	1.74	1.70	1.67	1.58	1.49
2.63	2.50	2.35	2.20	2.10	2.03	1.94	1.88	1.84	1.73	1.62
3.54	3.32	3.08	2.83	2.67	2.55	2.41	2.32	2.25	2.08	1.92
1.66	1.61	1.56	1.49	1.45	1.42	1.38	1.35	1.34	1.28	1.22
1.93	1.85	1.77	1.68	1.62	1.57	1.52	1.48	1.45	1.38	1.30
2.18	2.08	1.97	1.85	1.77	1.71	1.64	1.59	1.56	1.46	1.36
2.50	2.37	2.22	2.07	1.97	1.89	1.80	1.74	1.69	1.57	1.45
3.30	3.07	2.84	2.59	2.43	2.32	2.17	2.08	2.01	1.83	1.64
1.63	1.58	1.52	1.46	1.41	1.38	1.34	1.31	1.29	1.23	1.16
1.88	1.80	1.72	1.62	1.56	1.52	1.46	1.41	1.39	1.30	1.21
2.11	2.01	1.90	1.78	1.70	1.64	1.56	1.51	1.47	1.37	1.25
2.41	2.27	2.13	1.97	1.87	1.79	1.69	1.63	1.58	1.45	1.30
3.12	2.90	2.67	2.42	2.26	2.15	2.00	1.90	1.83	1.64	1.43
1.61	1.55	1.49	1.43	1.38	1.35	1.30	1.27	1.25	1.18	1.08
1.84	1.76	1.68	1.58	1.52	1.47	1.41	1.36	1.33	1.24	1.11
2.06	1.96	1.85	1.72	1.64	1.58	1.50	1.45	1.41	1.29	1.13
2.34	2.20	2.06	1.90	1.79	1.72	1.61	1.54	1.50	1.35	1.16
2.99	2.77	2.54	2.30	2.14	2.02	1.87	1.77	1.69	1.49	1.22

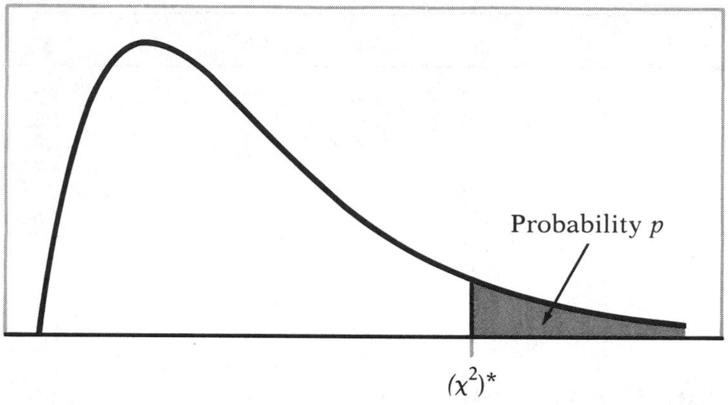

Table entry for p is the point $(X^2)^*$ with probability p lying above it.

Table G χ^2 **critical values**

	Tail probability p											
df	.25	.20	.15	.10	.05	.025	.02	.01	.005	.0025	.001	.0005
1	1.32	1.64	2.07	2.71	3.84	5.02	5.41	6.63	7.88	9.14	10.83	12.12
2	2.77	3.22	3.79	4.61	5.99	7.38	7.82	9.21	10.60	11.98	13.82	15.20
3	4.11	4.64	5.32	6.25	7.81	9.35	9.84	11.34	12.84	14.32	16.27	17.73
4	5.39	5.99	6.74	7.78	9.49	11.14	11.67	13.28	14.86	16.42	18.47	20.00
5	6.63	7.29	8.12	9.24	11.07	12.83	13.39	15.09	16.75	18.39	20.51	22.11
6	7.84	8.56	9.45	10.64	12.59	14.45	15.03	16.81	18.55	20.25	22.46	24.10
7	9.04	9.80	10.75	12.02	14.07	16.01	16.62	18.48	20.28	22.04	24.32	26.02
8	10.22	11.03	12.03	13.36	15.51	17.53	18.17	20.09	21.95	23.77	26.12	27.87
9	11.39	12.24	13.29	14.68	16.92	19.02	19.68	21.67	23.59	25.46	27.88	29.67
10	12.55	13.44	14.53	15.99	18.31	20.48	21.16	23.21	25.19	27.11	29.59	31.42
11	13.70	14.63	15.77	17.28	19.68	21.92	22.62	24.72	26.76	28.73	31.26	33.14
12	14.85	15.81	16.99	18.55	21.03	23.34	24.05	26.22	28.30	30.32	32.91	34.82
13	15.98	16.98	18.20	19.81	22.36	24.74	25.47	27.69	29.82	31.88	34.53	36.48
14	17.12	18.15	19.41	21.06	23.68	26.12	26.87	29.14	31.32	33.43	36.12	38.11
15	18.25	19.31	20.60	22.31	25.00	27.49	28.26	30.58	32.80	34.95	37.70	39.72
16	19.37	20.47	21.79	23.54	26.30	28.85	29.63	32.00	34.27	36.46	39.25	41.31
17	20.49	21.61	22.98	24.77	27.59	30.19	31.00	33.41	35.72	37.95	40.79	42.88
18	21.60	22.76	24.16	25.99	28.87	31.53	32.35	34.81	37.16	39.42	42.31	44.43
19	22.72	23.90	25.33	27.20	30.14	32.85	33.69	36.19	38.58	40.88	43.82	45.97
20	23.83	25.04	26.50	28.41	31.41	34.17	35.02	37.57	40.00	42.34	45.31	47.50
21	24.93	26.17	27.66	29.62	32.67	35.48	36.34	38.93	41.40	43.78	46.80	49.01
22	26.04	27.30	28.82	30.81	33.92	36.78	37.66	40.29	42.80	45.20	48.27	50.51
23	27.14	28.43	29.98	32.01	35.17	38.08	38.97	41.64	44.18	46.62	49.73	52.00
24	28.24	29.55	31.13	33.20	36.42	39.36	40.27	42.98	45.56	48.03	51.18	53.48
25	29.34	30.68	32.28	34.38	37.65	40.65	41.57	44.31	46.93	49.44	52.62	54.95
26	30.43	31.79	33.43	35.56	38.89	41.92	42.86	45.64	48.29	50.83	54.05	56.41
27	31.53	32.91	34.57	36.74	40.11	43.19	44.14	46.96	49.64	52.22	55.48	57.86
28	32.62	34.03	35.71	37.92	41.34	44.46	45.42	48.28	50.99	53.59	56.89	59.30
29	33.71	35.14	36.85	39.09	42.56	45.72	46.69	49.59	52.34	54.97	58.30	60.73
30	34.80	36.25	37.99	40.26	43.77	46.98	47.96	50.89	53.67	56.33	59.70	62.16
40	45.62	47.27	49.24	51.81	55.76	59.34	60.44	63.69	66.77	69.70	73.40	76.09
50	56.33	58.16	60.35	63.17	67.50	71.42	72.61	76.15	79.49	82.66	86.66	89.56
60	66.98	68.97	71.34	74.40	79.08	83.30	84.58	88.38	91.95	95.34	99.61	102.7
80	88.13	90.41	93.11	96.58	101.9	106.6	108.1	112.3	116.3	120.1	124.8	128.3
100	109.1	111.7	114.7	118.5	124.3	129.6	131.1	135.8	140.2	144.3	149.4	153.2

Solutions to Odd-Numbered Exercises

CHAPTER 1

1.1 Height, width or depth measured in inches with a ruler; volume in cubic inches; weight in ounces measured with a scale; number of pages, lines or words. Number of words would relate to reading time and size measures would tell how easily it would fit into your bookbag.

1.3 Population density, number of automobiles, various measures of air quality, commuting times, parking availability, cultural activities, cost of housing taxes, quality of schools. Reasons include cost of living and quality of life.

1.5 Weight because it is more sensitive.

1.7 The one minute period should give the best results because it is based on more information. If the pulse is changing over time, for example, after exercise, the one minute period might not be the best.

1.9 Total profits, number of employees, total value of stock, net sales, and total assets.

1.11 1120, 1001, 1017, 982, 989, 961, 960, 1089, 987, 976, 902, 980, 1098, 1057, 913, 999

1.13 Figure 1.6(a) is strongly skewed to the right with a peak at 0; Figure 1.6(b) is somewhat symmetric with a central peak at 4. The peak is at the lowest value in Figure 1.6(a) and is at a central value in Figure 1.6(b).

1.15 There are two peaks. The ACT states are located in the upper portion of the distribution.

1.17 200 appears to be an outlier. The median is approximately 140.

1.19
(a) The distribution is somewhat skewed to the right. There is one peak. The center is 523 eggs. 915 and 945 appear to be unusual values.
(b) The egg production increases up to day 7; then falls off to day 12. From day 13 to day 27, there is a small decreasing trend.

1.21 Yes. Slightly skewed to the right. One peak.

1.23 Stemplot and histogram give similar impression. Slightly skewed to the left.

1.25 The stemplot gives more information than the histogram but both give the same impression. The distribution is roughly symmetric with one value that is somewhat low. Between 5.4 and 5.5.

1.27 No. The centers are close together.

1.29 Both distributions are skewed to the right. Men tend to have higher salaries than women.

1.31 Shape is roughly symmetric. Center is around 68 inches.

1.33 **(a)** There is a general decreasing trend since 1966 with a relatively flat period from 1974 to 1980. Since 1982 the rate has been decreasing.
(b) The decrease starting in 1974 is evident but there is no apparent increase following the mid-80's changes.

1.35 There is a general increasing trend with clear seasonal variation. There are more passengers in the summer months and fewer in the winter months.

1.37 **(b)** Times are decreasing.

1.39 **(a)** APR, MAY, JUN, JUL are high; SEP and DEC are low.
(b) Yes, the secondary peak is between OCT and NOV.
(c) Months with high diarrhea are followed by months with low weight. Height isn't as sensitive to the short term effects of illness.

1.41 Mean = 141. Median = 138.5. Distribution has a single outlier that causes the mean to be larger than the median.

1.43 Mean = 224.002. Median = 223.988. Distribution is slightly skewed to the right and the mean is a little larger than the median.

1.45 $x^* = -223000 + 1000x$. Mean = 1002. Median = 988.

1.47 Suppose the incomes for five people are 0, 0, \$10,000, \$20,000, \$30,000 and each income increases by \$500. If the zeros are excluded from all calculations, the median decreases from \$20,000 to \$10,500 and the mean decreases from \$20,000 to \$12,500.

1.49 **(a)** 1.12, 1.88, 2.23, 2.86, 4.69.
(b) IQR = .98.
(c) Mean should be greater than the median since the distribution is skewed to the right.

1.51 **(a)** Skewed to the right.
(b) 89.67, 61.33, 7.83.
(c) 88.5, 84.5, 93, 8.5. No outliers.
(d) Quartiles, since distribution is skewed to the right.

1.53 60, 4.97.

1.55 It is not unusual to obtain a standard deviation of 0 when there are four or more zeros in the center of the number.

1.57 61.8, .0618 mm. No.

1.59 Prefer median and quartiles. 435, 523, 590.

1.61 **(a)** 65.5. Yes.

1.63 **(a)** $x* = 0 + (1/7)x$.
(b) 29.09, 7.41, 27.43.

1.65 160, 181, 203, 248.

1.67 Calcium: 102, 107, 111.5, 123, 136. Placebo: 98, 109, 112, 119, 130. No.

1.69 **(a)** $x* = 0 + (1/28.35)x$.
(b) Control: 9.59, 11.89, 12.63, 14.13, 16.30. Experimental: 11.22, 13.53, 14.34, 15.11, 16.83.

1.71 10% trimmed mean = 2.32. 20% trimmed mean = 2.32. Median = 2.23. Mean = 2.37.

1.73 **(a)** .2
(b) .6.
(c) .5.

1.75 The normal curve is centered at 1.45; plus and minus one, two and three standard deviation marks are at 1.85, 1.05, 2.25, 0.65, 2.65 and 0.25.

1.77 0.65 to 2.25.

1.79 Eleanor: 1.8, Gerald: 1.5. Eleanor.

1.81 **(a)** .9978.
(b) .0022.
(c) .9515.
(d) .9493.

1.83 **(a)** $-.68$.
(b) .25.

1.85 **(a)** .7257.
(b) .4514.

1.87 .0668.

1.89 3.4, 1.0 to 2.6.

1.91 .0301, .1995, .4295, .2815, .0594.

1.93 124.6, 134.9.

1.95 50%, 69%, 81%. Yes.

1.97 Approximately normal but the tails are less extreme than expected.

1.99 Approximately normal with one larger outlier.

1.101 83%, 97%, 97%. Yes, there are outliers.

1.103 Distribution is not normal. Both tails have some extreme observations.

1.105 The 1940 distribution has a marked skew to the left while the 1980 distribution is less spread out and is roughly symmetric. In 1980 few states have low percentages reflecting the increased voting by blacks.

1.107 The mean and standard deviation are .9281 and .2141. The small sample size makes it difficult to assess the exact shape, but there are no clear outliers. There may be two or three clusters but no clear deviation from normality is evident in the normal quantile plot.

1.109 Distribution is skewed to the right. The mean and standard deviation are 60.28 and 102.08. The five number summary is 5, 19, 30, 60, 797. The IQR is 41. Small with less than 50,000; medium with between 50,000 and 200,000 and four large counties.

1.111 **(a)** $b = 2, a = -50$.
(b) $a = -49.24, b = 1.82$.

(c) 106, 92.72.
(d) 61.79%, 35.94%.

1.113 (a) $a = 0, b = 4$.
(b) 120.
(c) 74.4, 83.2, 89.6, 95.0, 100.0, 105.0, 110,4, 116.8, 125.6.

1.115 $x* = 68.6393 + .8937x$.

1.117 (a) The distribution is strongly skewed to the right.
(b) 60.28, 30. They differ because the distribution is skewed. The total can be recovered only from the mean; it is 5546.
(c) The counts are 87, 89 and 90. These counts correspond to 95%, 97% and 98%. The rule does not apply.

1.119 (b) It is not symmetric, not normal. There are three peaks that probably correspond to different positions of the players. No outliers.
(c) 5 Number Summary – 145, 183, 210, 235, 295. Peaks – approximately 185, 220, 260.
(d) 5 Number Summary – 65.9, 83.2, 95.5, 106.8, 134.1. Peaks – 84.1, 100.0, 118.2.
(e) Many digits end in zero or five.

1.121 The distribution of the means looks normal with mean 20.07 and standard deviation 1.12. The standard deviations have a high peak in the middle of the distribution; the shape is approximately symmetric; the mean and median are 4.98 and 5.01. Results will vary with different simulations.

1.123 For H2S, $\bar{x} = 5.942$, $s = 2.127$, the median is 5.329 and the IQR is 3.689. The distribution is skewed toward high values. For LACTIC, $\bar{x} = 1.442$, $s = 0.303$, the median is 1.45 and the IQR is 0.43. The distribution is approximately normal.

CHAPTER 2

2.1 (a) Categorical.
(b) Quantitative.
(c) Quantitative.
(d) Categorical.
(e) Quantitative.
(f) Quantitative.

2.3 (a) Negative. Nearly curved. One observation is high on nitrogen oxides.
(b) No. Low nitrogen oxide is associated with high carbon monoxide.

2.5 (a) Flow rate.
(b) As the flow rate increases the amount of eroded soil increases. Yes. Positive.

2.7 (a) Heavier cars cost more. The association is weak and positive.
(b) Foreign cars generally cost more than the domestic cars of similar weight. The five lightest cars are foreign.

2.9 (a) Alaska is an outlier only in median teacher salary. Educators may be more highly valued by the citizens of this state.
(b) There is no apparent relationship between teachers' salaries and students' SAT scores.

2.11 The relationship is negative and not linear. It is fairly flat for high values of CO.

2.13 **(a)** Means = 10.65, 10.43, 5.60, 5.45.
(b) The introduction of 1000 nematodes per pot has no effect on seedling growth. With 5000 nematodes there is a substantial reduction in seedling growth. Introduction of 10,000 nematodes causes essentially the same growth reduction as 5000.

2.15 **(a)** Means = 1520, 1707, 1540, 1816.
(b) Against. Pecking order 1 has the lowest mean weight, and pecking order 4 has the highest mean weight.

2.17 $y = 1500x$.

2.19 $y = 30 + 33x$, 525, yes.

2.21 **(c)** 9.32. The insulation reduced gas consumption.

2.23 **(b)** $\hat{y} = 71.95 + .38x$.
(c) .38 cm per month, .5 cm per month. Sara's growth is less rapid than normal.
(d) 87.15, 96.65.

2.25 Predicted Height = 85.75, 90.35, 91.50, 92.65, 93.80, 94.95. Residuals = 0.25, −0.35, −0.50, 0.35, 0.20, 0.05. There is no clear pattern in the residuals.

2.27 **(a)** Yes.
(b) b = .080, a = 1.766.
(c) The predicted values are 3.039, 3.121, 3.171, 3.261, 3.369, 3.451, 3.541. The residuals are .011, −.001, −.001, −.011, −.009, .003, .009. They sum to .001. The difference is due to roundoff error.
(d) The residuals are all very small, indicating that the line fits the data well. Positive residuals are associated with high and low speeds and negative residuals are associated with intermediate speeds. We cannot plot the residuals versus the time at which the observations were made because this information is not given.

2.29 **(a)** The last observation is the most influential.
(b) The line does not fit the pattern.
(c) No.

2.31 **(b)** No.
(c) Residuals are increasing from 1 to 5 months and decreasing from 6 to 10 months.

2.33 **(a)** The relationship is linear with two clusters and one observation that is low in calories and sodium.
(b) Colored. It is closer to the influential observation.
(c) 407.05.

2.35 **(b)** $\hat{y} = -14.4107 + 46.6287x$.
(c) The residuals for $x = .25$ are all positive; residuals for the next two amounts are all negative. The residuals for $x = 20.0$ show large variation.

2.37 **(a)** 1, 2, 4, 8, 16, 32, 64, 128, 256, 512.
(c) Approximately 9,000,000,000,000,000,000.
(d) 0.00, 0.30, 0.60, 0.90, 1.20, 1.51, 1.81, 2.11, 2.41, 2.71.
(e) .3, −.3, 18.9.

2.39 Alice has $3049.17; Fred has $3000.00. Alice has more.

2.41 **(a)** The pattern appears to be exponential.
(b) Yes.
(d) 5.74112, $550,959.91.
(e) The point is somewhat below the line. The rate of growth appears to be slower.

2.43 **(b)** The log plot looks linear from 1790 to 1880 and also from 1880 to 1990. The slope is less in the second period.
(c) 294.8

2.45 **(a)** Predicted logs are: 1.620, 2.349. Predicted motor vehicles are: 41.7, 223.4.
(b) No.
(c) The actual value for 1989 (188.7) is given in 2.44 (e). The residuals for 1945 and 1989 are both very low.

2.47 **(b)** There is a positive association; it is approximately linear. Bowdin produces an unusually high percent of female doctorates relative to the percent of male doctorates in science.
(c) .69542, 48%.

2.49 **(a)** All points fall very close to a line.
(b) $r = .99438$.
(c) The value of r does not change.

2.51 The means are 91.5 and 51.0. The standard deviations are 3.27 and 8.49. The slopes are .38, 2.58.

2.53 **(b)** The relationship is positive and linear with one outlier.
(c) $r = .784$. The outlier almost destroys the linear relationship.

2.55 The paper suggests a negative relationship. The psychologist is saying there is no linear relationship.

2.57 $r = .4$.

2.59 **(b)** $r = .25$.
(c) No. It will be 100 times as large.

2.61 It is the percentage of variation in the number of students enrolled in 100 level math courses explained by the linear relationship with the number of students in the freshman class.

2.63 .54, 69.31.

2.65 **(b)** There is a positive linear relationship. No species are outliers or influential observations. Sea scallops are high in 1970 and 1980 but fit with the overall pattern.
(c) .96704.
(d) .9352.
(e) .93996 (without sea scallops). No, there is no strong effect on the correlation; it fits the pattern of the other data.
(f) The correlation indicates that the linear relationship is strong.

2.67 Let the slopes be the same for both groups indicating a strong positive relation in each group. The academic group should be shifted to the right (indicating more education) and lower (indicating less income).

2.69 **(a)** .41.
(b) .625, .405, .200.
(d) .5081, .3293, .1626.

(e) Small businesses are overrepresented and large businesses are underrepresented.

2.71 (b) The proportion of people voting increases with age.
(c) No, you would need the total number of registered voters in each age group.

2.73 .4866, .3373, .1095, .0666.

2.75 (a) .78%, 1.65%.
(b) The high blood pressure group has a higher mortality rate. The data suggest that there may be a link.

2.77 (a) Male: 490, 210; female: 280, 220.
(b) 70%, 56%.
(c) 80%, 90%, 10%, 33%.
(d) The admission rate is the highest in the business school where the majority of the applicants are male.

2.79 More men are employed in departments where the salaries are higher.

2.81 (a) White: 19, 141; black: 17, 149.
(b) Overall 11.87% of the whites and 10.24% of the blacks are sentenced to death. Whites are sentenced to death at a higher rate than blacks. For white victims, 13% of whites and 17% of blacks are sentenced to death. For black victims, 0% of whites and 6% of blacks are sentenced to death. For each group of victims, blacks are sentenced to death at a higher rate. This is the paradox.
(c) The percent of death penalties when the victim is white is higher (30/214=14%) than when the victim is black (6/112=5%). For white defendants, 94% of their victims were white. For black defendants, 62% of their victims were black.

2.83 Many things could cause an individual to receive more education and more income. For example, people who can easily afford more education are more likely to be in situations where there are greater opportunities for higher income. Similarly, higher innate abilities or drive to succeed would lead to both more education and higher income.

2.85 There may be some confounding between the class of the secretaries and factors such as seniority and experience.

2.87 Explanatory variable is serving herbal tea or not. Response variable is measure of cheerfulness. The confounding variables are the time and the attention of the students.

2.89 In addition to the diet information, you would want the family history and any other information that relates to blood cholesterol levels.

2.91 (a) Correlation measures the strength and direction of the linear association between two quantitative variables. In this case we are quantifying the association between recalled and historical consumption. The second aim involves prediction, so regression is appropriate.
(b) A correlation of .217 indicates a rather weak association. See plots in Figure 2.34 on page 167.
(c) The value of r^2 is the fraction of the variation in intake at age 30 predicted by each of the explanatory variables.

2.93 The plot shows no clear association. The correlation is $-.23$.

2.95 (a) Yes, the two lines appear to fit the data well. There do not appear to be any outliers or influential observations.

(b) .189, .157.
(c) 7.704, 6.348, $31.53.

2.97 (b) Phase one (0–6 hours): no increase in growth; phase two (6–24 hours): exponential growth; phase three (36 hours): growth is slower than exponential.
(c) 2.65.

2.99 (a) The percentage of people voting decreases over time.
(b) The proportion of younger people eligible to vote has increased from 1960 to 1984 and younger people are less likely to vote than older people.

2.101 $-.33$, $\hat{y} = 105.63 - .78x$; $-.76$, $\hat{y} = 109.30 - 1.19x$; $-.52$, $\hat{y} = 107.59 - 1.05x$; When all cases are included, the correlation is $-.64$ and the equation is $\hat{y} = 109.87 - 1.13x$. Case 18 has a stronger influence on the regression line. Both cases influence the value of the correlation.

2.103 (b) Three high residuals are evident but there are no clear patterns.
(c) The distribution is not symmetric because of the three large residuals. If these are removed, the remaining residuals are approximately normal.

2.105 (a) Both distributions are skewed to the right. No, Alaska is an outlier for pay. There are five states with high values for spending (Alaska, Connecticut, DC, New Jersey and New York), but no clear outliers.
(b) There is a positive linear association. You would expect high spending to be associated with high salaries.
(c) None of the points are outliers from the overall relationship.
(d) $\hat{y} = 14.19 + 3.22x$, $r^2 = .70$.
(e) $\hat{y} = 14.90 + 3.04x$, no.

2.107 $r = .252$, $r^2 = .063$, $\hat{y} = 3.28 + .002x$. SATM is a poor predictor of GPA.

2.109 (a) The new point has a very large residual.
(b) The distribution of T1 and T2 do not appear to have any outliers.
(c) With the new point the equation is $\hat{y} = -.017 + 1.011x$; without the new point it is $\hat{y} = -.033 + 1.018x$. The two regression lines are very similar, so the point is not influential.
(d) With the new point the correlation is .966; without the new point it is .996.

CHAPTER 3

3.1 This was not an experiment. The okra may have been planted in an area of the garden which was not attractive to stinkbugs.

3.3 (a) Treatments were not actively imposed.
(b) This was a survey and no treatment was imposed.

3.5 No. No treatment was imposed. The explanatory variable was whether or not they live in public housing. The response variable was family stability and other variables.

3.7 Yes. The treatment was walking briskly on the treadmill. The fact that eating was not recorded limits the conclusions that can be drawn. The explanatory variable was time after exercise, and the response variable was the metabolic rate.

3.9 Experimental units: pairs of pieces of package liner. Explanatory variable: temperature of jaws. Response variable: peel strength of the seal.

3.11 Experimental units: one day old male chicks. Explanatory variables: corn variety and protein level. Response variable: weight gain.

3.13 250°, 275°, 300°, 325°.

3.15
(a) Patients could be randomized to the two treatments.
(b) No, there is no treatment that can be applied.

3.17 Assume the field has four rows and five plots per row. First ten numbers between 1 and 20 correspond to plots receiving treatment A. Row 1: AABBB, Row 2: AAAAA, Row 3: BBBBB, Row 4: ABBAA.

3.19 250°: Units 4, 7, 10, 16, 19; 275°: Units 5, 8, 9, 13, 15; 300°: Units 1, 3, 6, 11, 18; 325°: Units 2, 12, 14, 17, 20.

3.21 First two numbers between 1 and 20 correspond to plots receiving treatment A and genetic line one, next two numbers correspond to plots receiving treatment A and genetic line two, etc. Row 1: A3, A5, B2, B4, B3; Row 2: A1, A5, A4, A2, A2; Row 3: B3, B2, B4, B5, B1; Row 4: A3, B5, B1, A1, A4.

3.23 The experimenter was aware of which subjects were assigned to the meditation group, and might rate these subjects as having lower anxiety.

3.25 For each person, flip the coin. If heads, measure the right hand and then the left hand. If tails, measure in reverse order.

3.27 First, assign the numbers 1 to 6 to the schools in district one. From line 125 of Table B we use the following digits for the randomization: 6, 4, 1, 2, 3, 5. Assign school number 6 to Fact-Before, 4 to Compute-Before, 1 to Word-Before, 2 to Fact-After, 3 to Compute-After, and 5 to Word-After. For district two the randomization is 1, 6, 2, 3, 4, 5.

3.29
(a) The basic unit is a person. The population is all adult U.S. residents.
(b) The basic unit is a household. The population is all U.S. households.
(c) The basic unit is a voltage regulator. The population is all voltage regulators from the last shipment.

3.31 The sample was not representative of women in general because the questionnaires were distributed through women's groups and the response was voluntary. It is known that voluntary respondents are more likely to have stronger feelings about their opinions that nonrespondents. It is likely that these facts would cause the percents to be higher.

3.33 Beginning with A1096 and going across rows, label the control numbers with the numbers from 1 to 25. From line 111 of Table B we select the following: 12-B0986, 04-A1101, 11-A2220.

3.35 Beginning with Agarwal and going down the columns, label the people with the numbers 1 to 28. From line 139 of Table B we select the following: 04-Bowman, 10-Frank, 17-Mihalko, 19-Naber, 12-Goel, 13-Gupta.

3.37
(a) We will choose one of the first 40 at random and then the addresses 40, 80, 120, and 160 places down the list from it. The addresses selected are 35, 75, 115, 155, 195.
(b) Because we choose 1 of the first 25 at random, each of the first 25 has the same chance to be chosen. But each of the second 25 is chosen exactly when the corresponding address in the first 25 is chosen, and so on. So each address has the same chance of being chosen. This is not an SRS because the only possible samples have exactly 1 address from the first 25, 1 address from the

second 25, and so on. An SRS could contain any 5 of the 200 addresses in the population. Note that this view of systematic sampling assumes that the number in the population is a multiple of the sample size.

3.39 Give each name on the alphabetized lists a number; 001 to 500 for females and 0001 and 2000 for males. From line 122 of Table B, the first five females selected are 138, 159, 052, 087, and 359. If we start at the beginning of line 122 of Table B, the first five males selected are 1387, 0529, 0908, 1369, 0815.

3.41
(a) Households without telephones and those with unlisted numbers are omitted. These households would include people who cannot afford a telephone, people who choose not to have a telephone, and people who prefer not to list their telephone number.
(b) The random digit method includes unlisted numbers in the sampling frame.

3.43
(a) Many people think that "food stamps" refer to the prize stamps that some grocery stores give away. So the question is not clear.
(b) The question is clear but it is slanted because it asks to choose between an extreme position (confiscate) and a general statement taken from the Constitution.
(c) The question is clear but it is slanted toward agreement because it gives reasons to support a freeze.
(d) The question is unclear. It uses fancy words and technical phrases that many people will not understand. It is slanted toward a positive response because it gives reasons to favor recycling.

3.45 Parameter, statistic.

3.47 Statistic, statistic.

3.49
(a) Starting at line 120, the sample is 3, 5, 4, 7. The mean is 67.25.
(b) The means are 67.25, 66.50, 68.00, 70.00, 75.25, 69.00, 67.25, 69.00, 64.50, 71.00. The mean of the means is 68.775. It is quite close to 69.4.

3.51
(a) Use digits 0, 1, 2, 3 for "Yes" and 4 to 9 for "No". Starting at the beginning of line 110, we obtain $\hat{p} = .25$.
(b) For lines 111 to 119, the values of $\hat{p}$ are .30, .30, .45, .40, .40, .45, .50, .50, .25. The mean is .38. Yes.

3.53
(a) Large bias, large variability.
(b) Small bias, small variability.
(c) Small bias, large variability.
(d) Large bias, small variability.

3.55 For this exercise we assume the population proportions in all states are about the same. The effect of the population proportion on the variability will be studied further in Chapter 6.
(a) No. The variability is controlled by the size of the sample.
(b) Yes. The sample sizes will vary from 2100 in Wyoming to 120,000 in California and larger samples are less variable.

3.57 For each taster flip a coin. If heads taste Pepsi first, then Coke. If tails taste Coke first, then Pepsi.

3.59 You want to compare how long it takes to walk to class by two different routes. The experiment will take 20 days. The days are labeled from 01 to 20. Using Table B, the first 10 numbers between 01 and 20 will be assigned to route A; the others will be assigned to route B. Take the designated route on each day and record the time to get to class. Note that this experiment is not blind; you know the route that you take on each day.

3.61 **(a)** Fifteen patients are randomly assigned to receive β-blockers. The other 15 patients will receive a placebo.
(b) The patients are numbered from 01 to 30. Those receiving the β-blockers are 21, 18, 23, 19, 10, 08, 03, 25, 06, 11, 15, 13, 24, 09 and 28.

3.63 **(a)** Assign the numbers 0001 to 3478 to the alphabetized list of students.
(b) 2940, 0769, 1481, 2975, 1315.

3.65 **(a)** The population is all students classified as full-time undergraduates at the beginning of the Fall semester on a list provided by the Registrar.
(b) Stratify on class and randomly select 125 students in each class.
(c) If questionnaires are mailed some students will not respond. If telephones are used, the students without a telephone will not be contacted.

3.67 **(a)** There are two factors and six treatments. The treatments are 50°–60 rpm, 50°–90 rpm, 50°–120 rpm, 60°–60 rpm, 60°–90 rpm, 60°–120 rpm. 12 experimental units.
(c) Label the batches with the numbers 01 to 12. The first two numbers which appear in the table will be given treatment 50°–60 rpm, the next two will be given 50°–90 rpm, etc. The assignment is 06 and 09 to 50°–60 rpm, 03 and 05 to 50°–90 rpm, 04 and 07 to 50°–120 rpm, 02 and 08 to 60°–60 rpm, 10 and 11 to 60°–90 rpm, 12 and 01 to 60°–120 rpm.

3.69 If the patients who choose not to participate in this study are in some way different from those who do choose to participate, then including them in the control group would bias the estimation for that group.

3.71 This is an observational study and no treatment is imposed. There is no information to support the claim that you can alter your moods if you start to run.

3.75 Increasing the size of the sample decreases the variability of the statistic. The mean remains the same.

CHAPTER 4

4.3 **(a)** $S = \{germinates,\ does\ not\ germinate\}$.
(b) If measured in weeks, for example, $S = \{0,\ 1,\ 2, \ldots\}$.
(c) $S = \{A,\ B,\ C,\ D,\ F\}$.
(d) $S = \{makes\ the\ shot,\ misses\ the\ shot\}$.
(e) $S = \{1,\ 2,\ 3,\ 4,\ 5,\ 6,\ 7\}$.

4.5 $S = \{all\ numbers\ between\ 0.00\ and\ 5.00\}$.

4.7 .04

4.9 Model 1: Legitimate. Model 2: Legitimate. Model 3: Sum is less than one. Model 4: Probabilities cannot be negative.

4.11 No. The sum of the probabilities is greater than one.

4.13 .54.

4.15 .67, .33.

4.17 (a) .64, .07.
(b) The event that the student selected ranked in the lower 60% in high school. .36.
(c) .71.

4.19 (a) Each probability is positive and the sum is 1.
(b) .445.
(c) .97.
(d) .28.
(e) .72.

4.21 (a) (Abby, Deborah), (Abby, Jim), (Abby, Julie), (Abby, Sam), (Deborah, Jim), (Deborah, Julie), (Deborah, Sam), (Jim, Julie), (Jim, Sam), (Julie, Sam).
(b) .10.
(c) .40.
(d) .30.

4.23 .54.

4.25 .668.

4.27 .25, .25, .50.

4.29 Dem: .055, .294, .265; Rep: .035, .184, .167.

4.31 (a) All probabilities are between 0 and 1; the sum is 1.
(b) .94.
(c) .86.
(d) $X \geq 4$, .06.
(e) $X > 1$, .52.

4.33 (a) All probabilities are between 0 and 1; the sum is 1.
(b) .983.
(c) .976.
(d) $X \geq 9$, .931.
(e) $X < 12$, .248.

4.35 (a) (1,1), (1,2), (1,3), (1,4), (1,5), (1,6), (2,1), (2,2), (2,3), (2,4), (2,5), (2,6), (3,1), (3,2), (3,3), (3,4), (3,5), (3,6), (4,1), (4,2), (4,3), (4,4), (4,5), (4,6), (5,1), (5,2), (5,3), (5,4), (5,5), (5,6), (6,1), (6,2), (6,3), (6,4), (6,5), (6,6).
(b) 1/36.
(c) (x,prob): (2,1/36), (3,2/36), (4,3/36), (5,4/36), (6,5/36), (7,6/36), (8,5/36), (9,4/36), (10,3/36), (11,2/36), (12,1/36).
(d) 2/9.
(e) 5/6.

4.37 (a) .4.
(b) .6.
(c) .2.
(d) .2.
(e) .487.

4.39 (a) .5.
(b) .5.
(c) .4.
(d) .6.

(e) .3.
(f) 0.

4.41 (a) .6826.
(b) .1359.
(c) .1359.

4.43 (a) 0.
(b) .1379.
(c) .7242.

4.45 2.619.

4.47 2.25.

4.49 The company will earn \$309.75 per person in the long run because of the law of large numbers.

4.51 (a) Independent.
(b) Not independent.
(c) Not independent.

4.53 (a) The spins are independent.
(b) Wrong. The probabilities are not equally likely because the deck now contains more black cards than red cards.

4.55 70.

4.57 2.68, 1.3106.

4.59 2.2361.

4.61 (a) 19,225, 138.65.
(b) 2796.
(c) 5,606,738.

4.63 (a) 550, 5.70.
(b) 0, 5.70.
(c) 1022, 10.26.

4.65 8763.85.

4.67 .321.

4.69 (a) .092, the household is prosperous and educated.
(b) .100, the household is prosperous and not educated.
(c) .129, the household is not prosperous and educated.
(d) .679, the household is not prosperous and not educated.

4.71 (a) .53.
(b) .6923.
(c) .57.

4.73 (a) .1925.
(b) .4150.
(c) .4775.
(d) Not independent, probability that a household earns over 50 thousand given the householder completed at least four years of college is .4150; this is not equal to .1925, the probability that a household earns over 50 thousand.

4.75 .1170.

4.77 (a) $P(A) = .859$, $P(B|A) = .953$, $P(B|A^c) = .899$.
(c) .8186, .1268, .9454.

4.79 .8659.

4.81 .58.

4.83 .25.

4.85 .719; the surgery offers him a slightly better chance of achieving his goal.

4.87 .00409, .17.

4.89 **(a)** .9389.
(b) 1.3.
(c) 12.97.

4.91 $a = -70, b = .05$.

4.93 **(a)** .065, .14.
(b) If $\alpha = .75$, $\mu = .0875$ and $\sigma = .21$.

4.95 **(a)** The 16 outcomes in the sample space are all possible combinations of the 4 choices for each player.
(b) (Ann's choice, Bob's choice, X) = (A, A, 0), (A, B, 2), (A, C, −3), (A, D, 0), (B, A, −2), (B, B, 0), (B, C, 0), (B, D, 3), (C, A, 3), (C, B, 0), (C, C, 0), (C, D, −4), (D, A, 0), (D, B, −3), (D, C, 4), (D, D, 0).
(c) (X, probability) = (−4, .0625), (−3, .125), (−2, .0625), (0, .5), (2, .0625), (3, .125), (4, .0625).
(d) $\mu = 0$, yes, $\sigma = 2.18$.

4.97 .84.

4.99 **(a)** .44.
(b) .1786.

4.101 **(b)** .0249.
(c) .4016.

4.103 **(a)** For 100 students there will be 5000 repetitions. The approximate probability distribution is (x,p) = (0, .002), (1, .007), (2, .043), (3, .117), (4, .203), (5, .250), (6, .208), (7, .120), (8, .041), (9, .008), (10, .001).
(b) .96.
(c) 5.01, 5.16. These values are close. Note that your answer will vary from these results.

CHAPTER 5

5.1 **(a)** Yes.
(b) No. n is random.
(c) No. Trials are not independent.

5.3 **(a)** Yes.
(b) No. Trials are not independent.
(c) No. Each trial is performed under different conditions.

5.5 **(a)** .0020.
(b) .0026.

5.7 6, .65, $P(X = 0) = .0018, P(X = 1) = .0205, P(X = 2) = .0951, P(X = 3) = .2355,$ $P(X = 4) = .3280$, $P(X = 5) = .2437$, $P(X = 6) = .0754$.

5.9 (a) There is only one way in which n successes can be chosen from among n trials.
(b) There are n ways in which $n - 1$ successes can be chosen from among n trials.
(c) The number of ways in which k successes can be chosen from among n trials is the same as the number of ways in which $n - k$ successes can be chosen from among n trials.

5.11 3.9, 1.17, .8072.

5.13 (a) .3174.
(b) .2131.

5.15 (a) 66.00, 7.17.
(b) .0256, (with continuity correction, .0301).

5.17 (a) 1050, 17.75.
(b) .9976, (with continuity correction, .9978).
(c) .0000.
(d) .2981, (with continuity correction, .2877).

5.19 (a) .1251, (with continuity correction, .1492).
(b) .0336, (with continuity correction, .0401).
(c) 400.
(d) Yes.

5.21 (a) .8179.
(b) .0169.
(c) $\mu = 19.8$, $\sigma = .445$.

5.23 $\sigma = .0462$.

5.25 $\mu = 18.6$, $\sigma = .6768$.

5.27 (a) .3409.
(b) .0039.

5.29 (a) .0668.
(b) .0013.

5.31 (a) $N(55000, 1591)$.
(b) .0222.

5.33 (a) $N(.001, .0005)$.
(b) .0228.

5.35 (a) $N(34, 2.3534)$.
(b) $N(37, 2.2454)$.
(c) $N(3, 3.2527)$.
(d) .3783.

5.37 (a) .004, .002.
(b) .0024, .0048, no.

5.39 .0768.

5.41 (a) Yes, by the rules for means.
(b) No, by the rules for variances. Variances add only if the two variables are uncorrelated.

5.43 (a) $N(2.2, .1941)$.
(b) .1515.
(c) .0764.

5.45 1.4623.

5.47 Center line = 75°. Control limits are 75.75 and 74.25.

5.49 **(a)** Center line = 11.5. Control limits are 11.80 and 11.20.
(b) B. C at time 11 or 12. A.

5.51 **(a)** Center line = 10. Control limits are 12.08 and 7.92.
(b) Process is out of control but no remedial action is needed.

5.53 .0058.

5.55 From the table the following would be correct: 3.08, 3.09, and 3.10.

5.57 **(a)** 8.40, .62.
(b) Center line is at 8.40, the lower and upper control limits are at 7.15 and 9.65.
(c) Three points are out of control. The high values on Oct. 27 and Dec. 5 are explained by the truck and the ice as described in the problem. There is a low value on November 28. This was the day after Thanksgiving and there was little traffic. The last ten points are all above the center line, probably due to bad winter weather.

5.59 Center line = p. Control limits are $p \pm 3\sqrt{p(1-p)/n}$.

5.61 .0704, yes

5.63 **(a)** 3525, 55.
(b) .6736.

5.65 **(a)** .3115.
(b) 60.
(c) .9951 (with continuity correction, .9966).

5.67 **(a)** $B(500, .52)$.
(b) .8159 (with continuity correction, .8264).

5.69 .0122 (with continuity correction, .0139).

5.71 Center line = 10. Control limits are 11.47 and 8.53.

5.73 **(a)** $N(32, 1.25)$, $N(29, 1.04)$.
(b) $N(-3, 1.63)$.
(c) .0329.

5.75 Answer will vary with the software used.

CHAPTER 6

6.1 **(a)** Ninety-five percent of all samples we could select would give intervals that capture the true percentage of people who intend to vote for Carter.
(b) The interval is 49% to 53%. This includes the value 50%, so a clear winner is not evident.
(c) Confidence intervals do not give this kind of result. Either half of the voters favor Carter or not. There is no probability associated with the truth of falsehood of this unknown fact.

6.3 No. The confidence interval attempts to capture the true mean, not a proportion of the population.

6.5 **(a)** 1.02.
(b) (58.00, 62.00), yes.

6.7 (57.37, 62.63), wider, more confidence requires a larger interval.

6.9 **(a)** (2.87, 3.53).
(b) (3.01, 3.39).

6.11 (223.973, 224.031).

6.13 **(a)** 2.35.
(b) 7.44.
(c) 23, yes.

6.15 11.

6.17 (23015, 23891).

6.19 (657.14, 670.86).

6.21 **(a)** $H_0 : \mu = 1250$, $H_a : \mu < 1250$.
(b) $H_0 : \mu = 32$, $H_a : \mu > 32$.
(c) $H_0 : \mu = 5$, $H_a : \mu \neq 5$.

6.23 **(a)** $H_0 : p_1 = p_2$, $H_a : p_1 > p_2$, where p_1 is the proportion of males who name mathematics as their favorite and p_2 is the corresponding proportion of females.
(b) $H_0 : \mu_A = \mu_B$, $H_a : \mu_A > \mu_B$, where μ_A is the mean score for group A and μ_B is the mean score for group B.
(c) $H_0 : \rho = 0$, $H_a : \rho > 0$, where ρ is the correlation between income and the percent of disposable income that is saved.

6.25 If we assume that the percentage of church attenders who are ethnocentric is equal to the percentage of nonattenders who are ethnocentric, then the probability of observing a difference as extreme as that actually observed is less than .05.

6.27 1.07, .1423. There is no evidence to conclude that the new sonnets are not by the poet.

6.29 .0108, yes. The normal assumption is not critical because the sample size is 50.

6.31 **(a)** $H_0 : \mu = 224$, $H_a : \mu \neq 224$.
(b) .8966, no.

6.33 **(a)** $H_0 : \mu = 32$, $H_a : \mu > 32$.
(b) $z = 1.86$, $P = .0314$. There is evidence to conclude that the mean score is higher than the national mean.

6.35 **(a)** Yes.
(b) Yes.

6.37 If a signficance test is significant at the 1% level, the probability of observing a result as or more extreme than the result actually observed is less than 1%. Therefore, this probability is also less than 5%.

6.39 .005, .01.

6.41 **(a)** (99.86, 108.41).
(b) No. $H_0 : \mu = 105$, $H_a : \mu \neq 105$. Since the confidence interval includes 105, we do not reject H_0 at the 10% level.

6.43 **(a)** Yes. $\mu = 7$, $\mu \neq 7$.
(b) No.

6.45 b.

6.47 **(a)** .3821.
(b) .1711.
(c) .0013.

6.49 **(a)** $z = 1.64$, no.
(b) $z = 1.65$, yes.

6.51 **(a)** No, you would expect five results to be significant at the .01 level by chance alone.
(b) They should be further tested.

6.53 .4641. The test is not very sensitive to detect an increase of ten points.

6.55 **(a)** .5080.
(b) .9545.

6.57 **(a)** .50.
(b) .1841.
(c) .0013.

6.59 .01, .5359.

6.61 **(a)** Patient does not need medical attention, patient needs medical attention. Referring a patient who does not need medical attention; clearing a patient who needs medical attention.
(b) The second error probability should be made small. Failing to provide medical attention to a patient needing it could have very serious consequences.

6.63 (4.76, 5.98).

6.65 **(b)** (26.06, 34.74).
(c) Yes. $z = 2.44$, $P = .007$.

6.67 **(a)** (141.6, 148.4).
(b) $H_0 : \mu = 140$, $H_a : \mu > 140$, $z = 2.42$, $P = .0078$. There is strong evidence that the mean cellulose content is greater than 140.
(c) The data are an SRS from a normal population with known standard deviation.

6.69 **(a)** $2z * \sigma/\sqrt{n}$. As n increases, the width decreases because the $\sqrt{n}$ is in the denominator.
(b) The P-value decreases. Roughly speaking, the z statistic will increase.
(c) The power will increase.

6.71 The statement is false and misleading. The .05 is a probability statement about the test statistic, not a statement about the null hypothesis.

6.73 **(a)** Under the assumption that the two populations have the same proportions, the chance of seeing a difference as extreme as that observed is less than .01.
(b) The probability that the interval will cover the true difference in population proportions is .95.
(c) No. The participants in the program were volunteers.

6.75 Results will be different for each simulation.
(a) The P values range from .003 to .493.
(b) For the first sample the mean is 455.60, $z = -.44$, $P = .33$.
(c) 1, 5%.

CHAPTER 7

7.1 (a) 2.262.
(b) 2.861.
(c) 1.440.

7.3 (a) 24.
(b) .10, .15.
(c) .20, .30.
(d) No, no.

7.5 18.5.

7.7 (a) Yes. $\mu = 0$, $\mu > 0$. $t = 43.47$.
(b) (312.139, 351.861).
(c) When the sample size is large the t procedure can be used for skewed distributions.
(d) Randomly assign the customers to two groups. One group will receive the no-fee offer, and the other will not receive the offer.

7.9 (a) 1.75, .065.
(b) (1.59, 1.90).

7.11 Yes. $\mu = 1.3$, $\mu > 1.3$. $t = 6.971$. .0025 and .005. The data give evidence that the mean is larger in rats poisoned with DDT.

7.13 (a) (1.54, 1.80).
(b) An SRS is selected from a population with a normal distribution. SRS.

7.15 (b) No, $t = -.32$, df $= 11$, $P = .76$.

7.17 (a) Standard error of the mean.
(b) .0173.
(c) (.811, .869).

7.19 (−21.17, −5.47). No. .8866.

7.21 (a) Toss a coin to decide which test is given first for each person.
(b) $t = 4.27$. $P = .0003$. The two tests do not have the same means.
(c) (.129, .375).

7.23 No. $\mu_A = \mu_B$, $\mu_A \neq \mu_B$. $P = .227$.

7.25 The results are based on a census and there is no sample.

7.27 (a) .5279.
(b) .9032.

7.29 (a) $\eta = 0$, $\eta > 0$, $p = .5$, $p > .5$.
(b) .0033. The right hand thread can be completed faster than the left hand thread.

7.31 The test cannot be done because the number of people with improved scores is not given.

7.33 (a) (116.18, 167.51).
(c) (2.015, 2.129).

7.35 $\mu_1 = \mu_2$, $\mu_1 > \mu_2$. $t = 5.81$. Yes, reject null hypothesis.

7.37 (a) (−1.274, 7.274).
(b) We are 95% confident that the true change in sales is covered by our interval.

Since the interval includes 0 and negative values, we cannot be certain that the sales have increased.

7.39 **(a)** $\mu_1 = \mu_2$, $\mu_1 > \mu_2$. $t = 1.65$, df $= 18$, $P = .057$ (from Table E P-value is between .05 and .10). The difference between the hemoglobin levels is not significant at the .05 level.
(b) $(-.24,\ 2.04)$.
(c) The two means are calculated from independent SRSs from normal populations. The standard deviations are not assumed to be equal.

7.41 **(a)** $\mu_1 = \mu_2$, $\mu_1 > \mu_2$.
(b) Using the Unequal line, $t = 3.1583$, df $= 9.8$, $P = .0052$. Note that the P-value on the output is divided by 2 because we are using a one-sided alternative. The skilled rowers have a higher angular velocity. The conclusion is the same if we use the Equal line, assuming that the standard deviations are the same.
(c) (.47, 1.87).

7.43 (27.91, 32.09).

7.45 **(a)** No.
(b) $\mu_1 = \mu_2$, $\mu_1 < \mu_2$. $t = -2.46$, df $= 19$, $P = .012$, yes, yes, no.
(c) (5.58, 67.72).

7.47 **(a)** (17.38, 51.18).
(b) Yes. The confidence interval does not include 0.
(c) The two means are calculated from independent SRSs from normal populations. The standard deviations are not assumed to be equal. No clear deviation from the normal assumption is evident.

7.49 If there are no differences between the children and adults, approximately 5% of the tests would be significant.

7.53 **(a)** $\mu_1 = \mu_2$, $\mu_1 > \mu_2$. $t = 1.66$, df $= 40$, $P = .052$ (from Table E P-value is between .05 and .10). The difference between the hemoglobin levels is not significant at the .05 level.
(b) $(-.19,\ 1.99)$. The results are essentialy the same.

7.55 **(a)** $t = 1.09$, df $= 28$, $P = .28$.
(b) $(-17.00,\ 56.00)$.
(c) The results are very similar.

7.57 **(a)** $\mu_1 = \mu_2$, $\mu_1 \neq \mu_2$. $t = (\bar{x}_1 - \bar{x}_2)/\sqrt{(s_1^2/n_1) + (s_2^2/n_2)}$.
(b) 1.984, -1.984.
(c) .6619.

7.59 **(a)** 3.68.
(b) No, no.

7.61 10.58. From Table F, the P-value is beween .02 and .05. Yes. No.

7.63 **(a)** $\sigma_1 = \sigma_2$, $\sigma_1 \neq \sigma_2$.
(b) $F = 2.20$. The P-value is greater than .2.

7.65 **(a)** $\sigma_1 = \sigma_2$, $\sigma_1 < \sigma_2$, where group 1 is the females.
(b) $F = 1.54$.
(c) Since there is no entry for 19 df in the numerator, we use the entries for $F(20,\ 17)$. $P > .1$. There is no clear evidence that the men are more variable than the women.

7.67 **(a)** .045.
(b) .075. Thigh (9.65, 12.95).

7.69 **(a)** The change in weight is the variable of interest. Therefore, the pairs are the before and after measurements.
(b) The mean weight change for these subjects was a weight loss. We are not given information on the amount lost.
(c) From Table E we can conclude $P < .0005$.

7.71 $t = .7586$, df $= 29$, $P = .23$. There is no evidence to conclude that nitrites decrease amino acid uptake.

7.73 **(a)** No clear skewness or outliers are evident.
(b) (54.78, 64.40).

7.75 The two standard deviations are quite close. $P = 36$. The conclusion is the same.

7.77 **(a)** $H_0 : \sigma_1 = \sigma_2, H_a : \sigma_1 \neq \sigma_2, F = 1.16, P = .375$ (exact).
(b) If the distributions are not normal, the test for equality of standard deviations is not recommended.

7.79 **(a)** There are no obvious outliers; there is no strong skewness.
(b) (903, 912).
(c) No, the exact P is .3054.

7.81 $H_0 : \mu_1 = \mu_2, H_a : \mu_1 > \mu_2$. $t = 1.17$, $P = .13$ (exact). There is no evidence to conclude that the mean levels are different.
(a) (−14.6, 52.6).
(b) (166, 220).
(c) Samples are SRSs and the distributions are normal. The chief threat is that the data are not SRSs.

7.83 $H_0 : \eta = 0, H_a : \eta > 0, P = .038$. The particulate levels are higher in the city than in the rural area.

7.85 The mean is 5.42 and the margin of error is .13. If we exclude the two lowest values, the mean is 5.49 and the margin of error is .08.

7.87 **(a)** The data are approximately normal. There appears to be one high outlier.
(b) (1.14, 1.52), using the exact $t* = 2.690$.
(c) $t = 4.60$, df $= 45$, $P < .0005$. We are 99% confident that the mean NOX level is between 1.14 and 1.52. There is strong evidence that the mean is greater than 1.

7.89 **(a)** The distribution is very discrete (with values −.02, −.01, 0, .01, .02, and .03). There are no outliers and no skewness is evident. The mean is .0046 and the standard deviation is .0149.
(b) (.00037, .00883).
(c) $t = 2.187$, df $= 49$, $P = .0336$. There is evidence to suggest that the second measure tends to be lower than the first measure.

CHAPTER 8

8.1 **(a)** No.
(b) Yes.
(c) No.
(d) Yes.

8.3 (.792, .928).

8.5 (.595, .725).

8.7 1052.

8.9 **(a)** Yes, $z = 3.01$, $P = .0013$.
(b) (.517, .580). There is a slight advantage.

8.11 **(a)** $z = .155$, H_0 is not rejected at $\alpha = .05$, $P = .88$.
(b) (.492, .509).

8.13 **(a)** $z = 1.70$, $P = .045$, reject H_0 for $\alpha = .05$. More than half of the population prefers freshly brewed coffee.
(b) (.507, .733).

8.15 323.

8.17 356. .0436.

8.19 **(a)** .059, .079, .090, .096, .098, .096, .090, .079, .059.
(b) No. np is too small.

8.21 **(a)** .63, .54.
(b) .077.
(c) $(-.040, .213)$. No, because the interval includes zero.

8.23 **(a)** .586.
(b) .077.
(c) $H_0 : p_1 = p_2$, $H_1 : p_1 > p_2$.
(d) $z = 1.12$, $P = .13$. There is not sufficient evidence to conclude that it is easier to win at home.

8.25 $z = 1.81$, $P = .035$. We assume that the data come from two independent binomial populations. The pooled $\hat{p}$ is .03. Therefore, the values of n_1 and n_2 in this problem are sufficiently large for the normal approximation to be reasonably accurate.

8.27 $\hat{p}_1 = .603$, $\hat{p}_2 = .591$, $z = .265$, $P = .395$. There is no evidence that the Protestants and Catholics differ on this issue.

8.29 **(a)** $p_1 = p_2$, $p_1 \neq p_2$. $\hat{p}_1 = .901$, $\hat{p}_2 = .810$, $z = 5.07$, $P < .001$.
(b) (.045, .137).

8.31 **(a)** .808, .558.
(b) (.108, .391).
(c) $p_1 = p_2$, $p_1 > p_2$, $z = 3.34$, $P < .001$. Use of aspirin increases the probability of a favorable outcome for patients with cerebral ischemia.

8.33 **(a)** $z = 1.62$, $P = .0526$. There is not clear evidence to conclude that the proportions are different.
(b) In Exercise 8.29 the sample proportions were the same but the results were significant because of the larger sample sizes. As the sample sizes increase the standard error decreases and the z statistic increases.

8.35 **(a)** No, since no treatments were imposed.
(b) The survival proportions are .72 for no pets and .94 for pets. It appears that pet owners have a higher chance of survival.
(c) The survival proportion of patients with pets is the same as the survival proportion without pets. The alternative is that the two proportions are not equal.

(d) $X^2 = 8.85$, df $= 1$, $P = .003$.
(e) We conclude that the pet owners have a higher chance of survival. We cannot conclude that pet ownership is an effective treatment because there was no random assignment of patients to the pet and no pet groups

8.37 (a) Hits: 2584, reg; 35, WS; No hits: 7280, reg; 63 WS totals 2619, 7343, 9864, 98.
(b) Jackson got a hit 26.2% of the time in regular season at bats and 35.7% of the time in World Series play. He hit much better in World Series games.
(c) The null hypothesis is that Jackson's hitting percentage is the same for regular season games and World Series games. The alternative is that the percentages are not equal. $X^2 = 4.50$, df $= 1$, $P = .033$. Reggie Jackson was a better hitter in World Series play than in the regular season.

8.39 (a) Small: 62.5, 37.5; medium: 40.5, 59.5; large: 20.0, 80.0; combined 41.0, 59.0. As the size of the company increases, the response rate decreases.
(b) The null hypothesis is that the response rate does not depend on the size of the company. The alternative is that there is a relationship between the size of the company and the response rate.
(c) $X^2 = 74.7$, df $= 2$, $P < .005$. The response rate decreases as the size of the company increases.

8.41 (a) Row sums: 518, 1373; column sums: 770, 154, 967; total sample: 1891.
(b) Family practice: 42.47, 57.53; pediatrics: 20.78, 79.22; other: 16.44, 83.56; combined: 27.39, 72.61. The family practitioners prescribe tetracycline at a higher rate than the two other groups.
(c) $X^2 = 149.66$, df $= 2$, $P < .0005$. The proportion of physicians prescribing tetracycline varies with the type of practice. The family practitioners have a very high rate compared to the others.

8.43 The three nutritionally inadequate groups have similar profiles. The normal group has lower rates for each of the three categories of disease. $X^2 = 101.29$, df $= 9$, $P < .0005$. There is a relationship between nutritional status and illness.

8.45 The modification is effective in reducing the defective rate from 11% to 5%. $z = -3.13$, $P = .0009$. We assume that the old rate is known to be 11% and that the number of defectives in the new sample is binomial.

8.47 (a) (.428, .512).
(b) 42.8% to 51.2%.
(c) 14,980 to 17,920.

8.49 $z = 2.24$, $P = .025$. There is evidence to conclude that the proportion in Tippecanoe County is higher than the national average. It is inappropriate to use a one-sided alternative because we have no reason to believe, before looking at the data, that the rate in Tippecanoe County is higher.

8.51 (a) High or low blood pressure. For low blood pressure the death rate is .78%, for high blood pressure it is 1.65%. It appears that the death rate is higher for the high blood pressure group.
(b) Yes. $H_0 : p_1 = p_2$, $H_a : p_1 < p_2$, where p_1 is the low blood pressure death rate. $z = 2.98$, $P = .0014$. We conclude that high blood pressure is associated with an increased risk of death from cardiovascular disease.
(c) Dead: low 21, high 55; Not dead: low 2655, high 3283. No, the chi-square test is appropriate only for a two-sided alternative.
(d) (.0032, .0142).

8.53 There is no evidence to conclude that there is an association beween the use of aluminum-containing antacids and Alzheimer's disease ($X^2 = 7.118$, df $= 3$, $P = .068$).

8.55 Column sums: 783, 454, 266, 536; Row sums: 1490, 549; Total: 2039. The data do not support the hypothesis that PTC tasting varies with country ($X^2 = 5.96$, df $= 3$, $P = .114$).

8.57 For the British study the death rates are 4.32 (aspirin) and 4.62 (no aspirin). The difference is not significant ($X^2 = .25$, df $= 1$, $P = .618$). For the American study the death rates are .94 (aspirin) and 1.71 (no aspirin). The aspirin group has a lower death rate ($X^2 = 25.01$, df $= 1$, $P < .0005$). The American study found a beneficial effect of aspirin while the British study did not. There is no clear reason why the two studies should give different results. The groups of physicians may have had different characteristics because the death rates in the British study are much higher than those in the American study. The American study used a much larger sample size, the duration was one year shorter and the aspirin was taken every other day rather than every day.

8.59 **(a)** $X^2 = 2.19$, df $= 1$, $P = .136$. There is no evidence to conclude that the death rates in the two hospitals are different.
(b) Good conditions: $X^2 = .29$, df $= 1$, $P = .591$. There is no evidence to conclude that the hospitals differ for good condition patients. Poor condition: $X^2 = .02$, df $= 1$, $P = .89$. No difference between hospitals is evident for the poor condition patients.
(c) Although the sample proportions illustrate Simpson's paradox as described in Example 2.25, none of the differences are statistically significant.

8.61 .35, .27, .19, .16, .14, .06. The margin of error decreases as the sample size increases.

8.63 **(a)** 769.
(b) $n = .5(z*/m)^2$.

8.65 A segmented bar chart giving the percents in the alcohol categories for each nicotine level shows the differences among the groups.

8.67 The death rates are so low that these small sample sizes are very unlikely to show any difference between the groups.

8.69 The percentages of students having loans in the different fields are 47.76, 42.53, 41.70, 41.78, 32.00, 51.67, 44.53. The differences are not statistically significant ($X^2 = 6.525$, df $= 6$, $P = .367$).

8.71 The death percentages are .44, .52, .14, .16, .29, .27, .60, .85. A plot of these percentages shows the relationship between age and death rate. $X^2 = 19.715$, df $= 7$, $P = .006$. Survival and age are related. The death rates are relatively high for children under 4 and for those over 30. The data do not provide any information regarding the chances of catching measles.

8.73 The percentages of women by year are 23.43, 27.8, 32.29, 40.43, 43.14, 49.00, 52.71, 54.86, 58.86. The proportion of women enrolled in pharmacy programs has increased over time ($X^2 = 359.186$, df $= 8$, $P < .0005$). The plot of the percentages is roughly linear. The least squares line is $\hat{y} = -4443 + 2.27x$. No, this would require extrapolation far beyond the range of the data available. The predicted value from the least-squares line is 97%.

CHAPTER 9

9.1 **(a)** The plot indicates a fairly strong straight-line pattern. The value of r^2 is .88 or 88.6%. The observation with the moderately large residual is 1983.
(b) $\hat{y} = -41.4 + 0.125x$. $H_0 : \beta_1 = 0$, $H_a : \beta_1 > 0$, $t = 9.68$, df $= 12$, $P < .0005$. There is strong evidence to conclude that there is a positive linear association between powerboat registrations and manatees killed.

9.3 **(a)** There appears to be a strong linear pattern.
(b) $\hat{y} = -61.12 + 9.31x$, 98.8%.
(c) (8.64, 10.00).

9.5 **(a)** The predicted value from the equation is 842.79. Therefore, the lean is 2.9 plus this value or .0843 or 2.984.
(b) A prediction interval because you want to predict the value of the lean in 1997.

9.7 **(b)** Minnesota and Charlotte have high values for attendance. Both are outliers and influential. Minnesota is much more influential than Charlotte. Miami and Orlando fit within the general linear pattern of the rest of the data.
(c) $\hat{y} = 8962 + 317x$, $r^2 = .23$, $F = 6.945$, $P = .0148$. The least squares line has changed substantially. The value of r^2 is much larger and the linear relationship is now statistically significant.
(d) The results of these analyses show an association with teams as the unit of analysis. Teams with higher priced tickets tend to have greater attendance. The two outliers have very high attendance. The results do not indicate how attendance might change if ticket prices would change. Lurking variables would include the team performance and cost of living as reflected in prices for entertainment.

9.9 **(a)** The intercept represents gas consumption for uses other than heating. (1.92, 2.89).
(b) The margin of error here is shorter because the standard error is smaller.

9.11 $t = 2.16$, df $= 53$, $P = .035$.

9.13 (7.26, 7.96), (6.72, 8.51). The margin of error for the prediction interval is larger (.896 versus .351). The 95% intervals would be longer.

9.15 **(a)** The slope is the change in the textile consumption index associated with a unit change in the price index. (−1.53, −1.12).
(b) The intercept corresponds to textile consumption when the price is zero. This has no practical meaning.
(c) 103.49, (97.6, 109.4).
(d) (122.1, 150.4).

9.17 **(a)** 51.04, 7836.9625.
(b) $H_0 : \beta_1 = 0$, $H_a : \beta_1 \neq 0$. (Note that a one-sided alternative could be justified here.) $t = 21.63$, df $= 24$, $P < .0005$. There is a positive linear relationship between the city and rural readings.
(c) (34.98, 53.90).

9.19 **(a)** There are no outliers or unusual points.
(b) $\hat{y} = -.027927 + .3485x$, .3485, (.2771, .4199).
(c) 2.87, (2.38, 3.61).
(d) $t = -.506$, df $= 3$, $P = .65$.

9.21 2.95, (2.7, 3.2).

9.23 There are no outliers or unusual points. $\hat{y} = -3.151 + .03777$. $t = 26.73$, df $= 17$, $P < .0001$, for the one-sided alternative. The 95% prediction interval for a heart rate of 95 is (.2442, .6306). For a heart rate of 110 the interval is (.8266, 1.1814). The line fits the data very well and can be used in subsequent experiments.

9.25 Observations two and four are rather large. There appears to be a pattern in the last six residuals. The regression methodology is questionable in this case and more information about the data is needed.

9.27 **(a)** $H_0 : \beta_1 = 0$, $H_a : \beta_1 > 0$. $t = 3.65$, df $= 16$, $P < .0025$. There is a positive linear association between the rate of evaporation and the speed of the air.
(b) (.0028, .0105).

9.29 **(a)** Source, df, SS, MS: Model, 1, .9961, .9961; Error, 3, .0124, .0041; Total, 4, 1.0085; $F = 240.95$; $P = .0006$.
(b) $H_0 : \beta_1 = 0$. The null hypothesis is that the slope of the line relating current to voltage is zero.
(c) $F(1, 3)$, $P < .001$.

9.31 **(a)** Source, df, SS, MS: Model, 1, 4.51245, 4.51245; Error, 17, .10735, .00631; Total, 18, 4.61979; $F = 714.618$; $P = .001$.
(b) $H_0 : \beta_1 = 0$. The null hypothesis is that the slope of the line relating oxygen uptake to heart rate is zero.
(c) $F(1, 17)$, $P < .0001$.
(d) $t = 26.813$. The difference between the square of this value and F is due to roundoff error.
(e) .9768.

9.33 **(a)** $t = -5.16$.
(b) $P < .001$. We conclude that there is a negative relation between parental control and self-esteem.

9.35 **(a)** $\mu_{GPA} = \beta_0 + 6\beta_1 + 7\beta_2 + 8\beta_3$.
(b) 4.202. This is the estimated mean GPA for all students having B$-$ in high school math, B in high school science, and B+ in high school English.

9.37 **(a)** (.120, .246). Assuming that the other high school grades remain constant, an increase of one unit in high school math grades is associated with an increase of $\beta_1 = .183$ units in GPA.
(b) ($-$.008, .129). Assuming that the other high school grades remain constant, an increase of one unit in high school English grades is associated with an increase of $\beta_3 = .061$ units in GPA. The confidence interval indicates that this increase is not distinguishable from zero.

9.39 **(a)** $\hat{y} = 3.2887 + .002283x_1 - .00002456x_2$.
(b) .7577.
(c) $H_0 : \beta_1 = \beta_2 = 0$, $H_a : \beta_j \neq 0$ for at least one $j = 1, 2$. The null hypothesis is that neither of the SAT scores predict GPA in a linear model. The alternative is that at least one of them is linearly related to GPA.
(d) $F(2,221)$. Because $P = .0007$, we conclude that there is a linear relation between GPA and SATM, SATV, or both.
(e) 6.34%.

9.41 For SATM and GPA a hint of a positive relationship can be seen. This corresponds to the statistically significant (P = .0001) correlation of .251 reported in Figure

9.20. No clear relationship is evident between SATV and GPA. There are no outliers or unusual points.

9.43 No unusual points or patterns are evident in the residual plots.

9.45 **(a)** $\hat{y} = 2.666 + .193HSM + .0006SATM$.
(b) $\beta_1 = \beta_2 = 0$, $\beta_j \neq 0$ for at least one $j = 1, 2$. The null hypothesis states that there is no linear relation between GPA and either of the variables HSM or SATM. The alternative is that there is a linear relation between GPA and at least one of these variables. $F = 26.63$. $P < .0001$. We conclude that at least one of HSM and SATM have a linear relation with GPA.
(c) (.1295, .2565), (−.0006, .0018). The second interval includes 0.
(d) $t = 5.99$, $P < .001$. $t = 1.00$, $P = .319$. For this model we reject $H_0 : \beta_1 = 0$ and we do not reject $H_0 : \beta_2 = 0$. Given the HSM is in the model, SATM does add statistically significant information for predicting GPA.
(e) .7028.
(f) 19.42%.

9.47 **(a)** There appears to be a strong positive linear relationship between the two measurements. There are no outliers or unusual points.
(b) $\widehat{T2} = .0048 + .9920T1$, .0265.
(c) .988, .977.
(d) $t = 45.22$, $\beta_1 > 0$, $P < .0001$. There is a very strong statistically significant linear relationship between the two measurements.
(e) $(45.22)^2 = 2044.85$, $F = 2044.66$. Difference is due to round-off.

9.49 **(a)** $\hat{y} = 109.87 - 1.127x$.
(b) $t = -3.63$, df $= 19$, $P = .0009$.
(c) $(-1.776, -.478)$.
(d) 41.00%.
(e) 11.023.

9.51 Plots of GPA versus each predictor do not indicate any very strong relationships. The F statistic for the model is 10.63 with 3 and 141 degrees of freedom and $P < .0001$. The proportion of variation explained is .1844. The model with the three high school variables as predictors is statistically significant. For the tests on the individual regression coefficients, only HSM is significant ($t = 3.46$, df $=$ 141, $P = .0007$). No unusual patterns or observations are evident in the residual plots. The model should be rerun deleting either HSS or HSE. The results for males alone are similar to those obtained for all students in the case study. The value of R^2 is a little smaller (.1844 for males versus .2046 for all students). The overall model and the HSM coefficient are slightly less significant, partly because of the reduced sample size.

9.53 **(a)** $\hat{y} = -9.102 + 1.086x$.
(b) .6529.
(c) $t = 1.66$, $P = .14$. All of the results are quite different. The intercept is now negative. The value of t has changed from 17.64 (see Example 9.5) to 1.66 and is no longer statistically significant. The incorrect observation has caused the results of the regression analysis to be meaningless.

9.55 The relationships are weak and positive. The correlations and their P-values are: taste with acetic .55, .0017; taste with H2S .76, .0001; taste with lactic .70, .0001; acetic with H2S .62, .0003; acetic with lactic .60, .0004; H2S with lactic .64, .0001.

9.57 $\hat{y} = -9.79 + 5.78x$, $F = 37.29$, $P = .0001$, $r^2 = .57$. The residuals are smaller for low values of H2S. The residual plots versus acetic and lactic show no clear patterns. There is some tendency for smaller residuals to be associated with smaller values of these variables. The normal quantile plot of the residuals indicates that they are approximately normal with a small amount of skewness to the right.

9.59 $F, P, R^2, \sqrt{MSE}, b_0, b_1$: Acetic 12.114, .0017, .30, 13.82, −61.50, 15.65. H2S 37.293, .0001, .57, 10.83, −9.79, 5.78. Lactic 27.55, .0001, .50, 11.75, −29.86, 37.72. The intercepts are not the same because the models are different.

9.61 For the multiple regression $\hat{y} = -27.59 + 3.95H2S + 19.89Lactic$. The coefficient for H2S is significant ($t = 3.475$, $P = .0018$). The same is true for the coefficient for Lactic ($t = 2.499$, $P = .0188$). For this model $F = 25.26$, $P = .0001$, $R^2 = .652$. The R^2s for the simple regressions are .5712 for H2S and .4959 for Lactic. We prefer the multiple regression because both coefficients are significant and the R^2 is considerably higher.

CHAPTER 10

10.1 **(a)** Yield of tomatoes in pounds, the varieties of tomatoes, 4; 10, 10, 10, 10; 40.
(b) Attractiveness rating, type of packaging, 6; 120, 120, 120, 120, 120, 120; 720.
(c) Weight loss, type of weight loss program, 3; 10, 10, 10; 30.

10.3 Yes. 69.42, 8.33.

10.5 **(a)** Source, df: Variety, 3; Error, 36; Total, 39.
(b) Source, df: Type, 5, Error, 714; Total, 719.
(c) Source, df: Type, 2; Error, 27; Total, 29.

10.7 **(a)** $\mu_1 = \mu_2 = \mu_3$. Not all of the μ_i are equal.
(b) Source, df: Major, 2; Error, 253; Total, 255.
(c) $F(2, 253)$.
(d) 3.04 for $F(2, 200)$; 3.00 for $F(2, 1000)$. From computer, 3.03 for $F(2, 253)$.

10.9 **(a)** SST = 175356.46; DFT = 35; MSG = 34951.96; MSE = 2203.14; F = 15.86.
(b) $\mu_1 = \mu_2 = \mu_3 = \mu_4$. Not all of the μ_i are equal.
(c) $F(3, 32)$. $P < .001$. The mean fitness scores for the four groups are not all the same.
(d) 2203.14, 46.94.

10.11 **(a)** 3.90, MSE.
(b) Source, SS, df, MS: Country, 17.22, 2, 8.61; Error, 802.89, 206, 3.90; Total, 820.11, 208. $F = 2.21$.
(c) $\mu_1 = \mu_2 = \mu_3$. Not all of the μ_i are equal.
(d) $F(2, 206)$. $P > .1$. From computer, $P = .112$. The data do not provide evidence to conclude that the mean birth weights are different in the three countries.
(e) .0210.

10.13 **(a)** $\psi_1 = \frac{1}{2}(\mu_1 + \mu_2) - \mu_3$.
(b) $\psi_2 = \mu_1 - \mu_2$.

10.15 **(a)** $H_{01} : \frac{1}{2}(\mu_1 + \mu_2) = \mu_3$, $H_{a1} : \frac{1}{2}(\mu_1 + \mu_2) > \mu_3$. $H_{02} : \mu_1 = \mu_2$, $H_{a2} : \mu_1 \neq \mu_2$.
(b) 49, −10.

(c) 11.28, 16.90.
(d) $t_1 = 4.34$, df $= 253$, $P < .0005$. $t_2 = -.59$, df $= 253$, $P > .50$. From computer $P = .555$. The average of the means for the first two groups is greater than the mean for the third group. There is no evidence to conclude that the first two groups have different means.
(e) (26.78, 71.22), (−43.29, 23.29).

10.17 (a) $\psi_1 = \mu_1 - \mu_2$, $H_{01} : \mu_1 = \mu_2$, $H_{a1} : \mu_1 > \mu_2$. $\psi_2 = \mu_1 - \frac{1}{2}(\mu_2 + \mu_4)$, $H_{02} : \mu_1 = \frac{1}{2}(\mu_2 + \mu_4)$, $H_{a2} : \mu_1 > \frac{1}{2}(\mu_2 + \mu_4)$. $\psi_3 = \mu_3 - \frac{1}{3}(\mu_1 + \mu_2 + \mu_4)$, $H_{03} : \mu_3 = \frac{1}{3}(\mu_1 + \mu_2 + \mu_4)$, $H_{a3} : \mu_3 > \frac{1}{3}(\mu_1 + \mu_2 + \mu_4)$.
(b) $c_1 = -17.06$, $s_{c1} = 25.71$. $c_2 = 24.39$, $s_{c2} = 19.64$. $c_3 = 91.22$, $s_{c3} = 17.27$. $t_1 = -.66$, df $= 32$, $P = .256$. $t_2 = 1.24$, df $= 32$, $P = .112$. $t_3 = 5.28$, df $= 32$, $P < .0005$. There is no evidence to conclude that the treatment and control grups have different means. There is no evidence to conclude that the mean of the treatment group is greater than the average of the means of the control and sedentary groups. The mean for the joggers is greater than the average of the means for the other three groups. Contrasts are very useful for summarizing the data in this problem.
(c) No conclusion about causation can be drawn because subjects were not randomly assigned to groups.

10.19 The Egypt and Kenya means are not distinguishable ($t = 1.73$). The Egypt and Mexico means are not distinguishable ($t = 2.00$). The Kenya and Mexico means are not distinguishable ($t = .60$).

10.21 $t_{12} = -.66$, not distinguishable. $t_{13} = -3.65$, different. $t_{14} = 3.14$, different. $t_{23} = -2.29$, not distinguishable. $t_{24} = 3.22$, different. $t_{34} = 6.86$, different.

10.23

n	DFG	DFE	$F*$	λ	Power
50	2	147	3.06	4.73	.47
100	2	297	3.03	9.45	.79
150	2	447	3.02	14.18	.93
175	2	522	3.01	16.54	.96
200	2	597	3.01	18.90	.98

A sample size of 175 appears to be adequate.

10.25 (a) Mean, standard deviation: 10.65, 2.05; 10.43; 1.49; 5.60, 1.24; 5.45, 1.77.
(b) $\mu_1 = \mu_2 = \mu_3 = \mu_4$. Not all of the μ_i are equal.
(c) $F = 12.08$, $P = .0006$. There is evidence to conclude that not all of the means are equal. $s_p = 1.67$, $R^2 = .75$.

10.27 (a) $\psi = \mu_1 - \frac{1}{3}(\mu_2 + \mu_3 + \mu_4)$.
(b) $H_0 : \mu_1 = \frac{1}{3}(\mu_2 + \mu_3 + \mu_4)$. $H_a : \mu_1 > \frac{1}{3}(\mu_2 + \mu_3 + \mu_4)$.
(c) $t = 3.63$, df $= 12$, $P < .0025$. From computer $P = .002$. Yes.
(d) $\psi = \mu_1 - \mu_4$. (2.63, 7.77).

10.29 (a) The ranges and IQRs for the three groups are very similar. Group 3 appears to have slightly higher scores.
(b) The group means are significantly different ($F = 7.61$); df $=$ 2,238; $P = .0006$. The Bonferroni procedure indicates that the group 3 average is higher than that of group 1. No other pairs of groups differ according to this procedure, although the confidence interval for the comparison of groups 2 and 3 barely includes zero.

10.31 (a) Yield of tomatoes in pounds; variety and fertilizer; 4, 2; 40.
(b) Attractiveness rating; type of packaging and parts of the country; 6, 4; 720.
(c) Weight loss; type of weight loss program, gender; 3, 2; 30.

10.33 (a) Source, df: Variety, 3; Fertilizer, 1; Variety x Fertilizer, 3; Error, 32; Total, 39.
(b) Source, df: Packaging, 5; Part, 3; Packaging x Part, 15; Error, 696, Total, 719.
(c) Source, df: Program, 2; Sex, 1; Program x Sex, 2; Error, 24, Total, 29.

10.35 (b) Blood pressure for nonwhites is higher than for whites in all age groups. Yes, blood pressure increases with age. There does not appear to be an interaction.
(c) Race: 135.98, 137.82. Age Group: 131.65, 133.25, 136.20, 140.35, 143.05. Differences: 1.3, 1.9, 2.0, 1.9, 2.1.

10.37 (b) The mean amount of GITH for the low chromium diet is about the same as the amount for the normal chromium diet. The GITH for the R diet is greater than the M diet. Yes, there is an increase in GITH for the R diet as the level of chromium is increased; but for the M diet there is a decrease.
(c) Diet: 4.48, 5.25. Chromium: 4.86, 4.87. Differences: .63, .89. There is a larger difference for the normal chromium level.

10.39 The average score for the males is higher than for the females. The scores in engineering are higher than the scores in computer science and the other group has the lowest scores. The marginal means are 611.7, 585.3 for sex and 605.0, 624.5, 566.0 for majors. The male versus female differences for the three majors are 46, −13, 46. Male scores are higher than female scores in CS and O, while the reverse is true in EO. The O group has lower scores than the other two groups.

10.41 (a) SST = 7.04487, DFB = 1, DFAB = 1, DFE = 36, DFT = 39, MSA = .00121, MSB = 5.79121, MSAB = .17161, MSE = .03002, $F_A = .04$, $F_B = 192.89$, $F_{AB} = 5.72$.
(b) 5.72, $F(1, 36)$, $P < .025$.
(c) .04, $F(1, 36)$, $P > .10$; 192.89, $F(1, 36)$, $P < .001$.
(d) .03, .17.
(e) The interpretation given in Exercise 10.37 is supported by the significance tests.

10.43 (a) For each F statistic the degrees of freedom are 1 and 945. Both main effects are highly significant ($P < .001$) while the interaction is not ($P > .1$).
(b) Women live longer than men and right-handed people live longer than left-handed people.

10.45 (a) $n, \bar{x}, s$: 22, 5.27, 2.76; 22, 5.09, 2.00; 22, 4.95, 1.86.
(b) Yes.
(c) 1.48, yes.
(d) $\mu_1 = \mu_2 = \mu_3$, not all of the μ_i are equal. $F = .11$, $P = .8948$. H_0 is not rejected.
(e) There is no evidence to conclude that the three means are different.

10.47 (a) $n, \bar{x}, s$: 22, 6.68, 2.77; 22, 9.77, 2.72; 22, 7.77, 3.93.
(b) Yes.
(c) 1.44, yes.
(d) $F = 5.32$, $P = .0073$. H_0 is rejected.
(e) $\psi = -\mu_1 + \frac{1}{2}(\mu_2 + \mu_3)$. $H_0 : \mu_1 = \frac{1}{2}(\mu_2 + \mu_3)$, $H_a : \mu_1 < \frac{1}{2}(\mu_2 + \mu_3)$, $C = 209$, $s(C) = .833$, $t = 2.51$, df = 63, $P < .01$. From computer $P = .007$. (.43, 3.75).

(f) $\psi = \mu_2 - \mu_3$. $H_0 : \mu_2 = \mu_3$, $H_a : \mu_2 \neq \mu_3$. $C = 2.00$, $s(C) = .96$, $t = 2.08$, df = 63, $P < .05$. From computer $P = .0415$. (.079, 3.921).

(g) Not all of the means are equal. The Basal group mean is smaller than the average of the DRTA and Strat group means, and the DRTA group mean is larger than the Strat group mean.

10.49 **(a)** $F = 1.33$, $P = .31$. H_0 is not rejected.

(b) $F = 12.08$, $P = .0006$, from Exercise 10.25. The outlier has caused a statistically significant result to disappear.

(c) $\bar{x}$, s: 34.95, 48.74; 10.43, 1.49; 5.60, 1.24; 5.45, 1.77. The large standard deviation and mean for the first group relative to the other groups should lead us to check the data in this group.

10.51 **(a)** $\bar{x}$, s: 1.02, .08; 1.01, .07; .74, .09; .72, .15.

(b) $\mu_1 = \mu_2 = \mu_3 = \mu_4$. Not all of the μ_i are equal.

(c) $F = 10.39$, $P = .0012$. The number of nematodes affects the growth of the seedlings. $s_p = .10$, $R^2 = .72$. The results are qualitatively the same.

10.53 **(a)** n, power: 20, .42, 40, .74; 60, .91; 80, .97; 100, .99.

(b) The power increases rapidly from 20 to 60 and levels off afterward.

(c) A sample size between 80 and 100 will give good power.

10.55 **(a)** $\bar{x}$, s: Aspen 387, 69; 336, 60. Birch 293, 122; 455, 118. Maple 323, 52; 294, 184. Species 362, 374, 308. Size 334, 362.

(b) Yes, yes. The interaction is so strong that statements about the marginal means are meaningless.

(c) No statistically significant effects are present. The impressions obtained from the plot are not substantiated by the analysis and are probably due to chance variation. Note that there is a large range of standard deviations and the sample sizes are small.

10.57 n, $\bar{x}$, s: Males 39, 526.95, 100.94; 39, 507.85, 57.21; 39, 487.56, 108.78. Females 39, 543.38, 77.65; 39, 538.21, 102.21; 39, 465.03, 82.18. The interaction is not significant ($F = 1.81$, $P = .17$). The sex effect is not significant ($F = .47$, $P = .49$). The majors differ significantly ($F = 9.32$, $P = .0001$).

10.59 n, $\bar{x}$, s: Males 39, 7.79, 1.51; 39, 7.49, 2.15; 39, 7.41, 1.57. Females 39, 8.85, 1.14; 39, 9.26, .75; 39, 8.62, 1.16. The high school grade variables are clearly not normal. We rely upon the robustness of the ANOVA procedures for nonnormal data for this analysis. The interaction is not significant ($F = 1.33$, $P = .27$). There is a significant sex difference ($F = 50.32$, $P = .0001$). The majors are not significantly different ($F = 1.40$, $P = .25$).

Index

Table B Random digits

Line								
101	19223	95034	05756	28713	96409	12531	42544	82853
102	73676	47150	99400	01927	27754	42648	82425	36290
103	45467	71709	77558	00095	32863	29485	82226	90056
104	52711	38889	93074	60227	40011	85848	48767	52573
105	95592	94007	69971	91481	60779	53791	17297	59335
106	68417	35013	15529	72765	85089	57067	50211	47487
107	82739	57890	20807	47511	81676	55300	94383	14893
108	60940	72024	17868	24943	61790	90656	87964	18883
109	36009	19365	15412	39638	85453	46816	83485	41979
110	38448	48789	18338	24697	39364	42006	76688	08708
111	81486	69487	60513	09297	00412	71238	27649	39950
112	59636	88804	04634	71197	19352	73089	84898	45785
113	62568	70206	40325	03699	71080	22553	11486	11776
114	45149	32992	75730	66280	03819	56202	02938	70915
115	61041	77684	94322	24709	73698	14526	31893	32592
116	14459	26056	31424	80371	65103	62253	50490	61181
117	38167	98532	62183	70632	23417	26185	41448	75532
118	73190	32533	04470	29669	84407	90785	65956	86382
119	95857	07118	87664	92099	58806	66979	98624	84826
120	35476	55972	39421	65850	04266	35435	43742	11937
121	71487	09984	29077	14863	61683	47052	62224	51025
122	13873	81598	95052	90908	73592	75186	87136	95761
123	54580	81507	27102	56027	55892	33063	41842	81868
124	71035	09001	43367	49497	72719	96758	27611	91596
125	96746	12149	37823	71868	18442	35119	62103	39244
126	96927	19931	36089	74192	77567	88741	48409	41903
127	43909	99477	25330	64359	40085	16925	85117	36071
128	15689	14227	06565	14374	13352	49367	81982	87209
129	36759	58984	68288	22913	18638	54303	00795	08727
130	69051	64817	87174	09517	84534	06489	87201	97245
131	05007	16632	81194	14873	04197	85576	45195	96565
132	68732	55259	84292	08796	43165	93739	31685	97150
133	45740	41807	65561	33302	07051	93623	18132	09547
134	27816	78416	18329	21337	35213	37741	04312	68508
135	66925	55658	39100	78458	11206	19876	87151	31260
136	08421	44753	77377	28744	75592	08563	79140	92454
137	53645	66812	61421	47836	12609	15373	98481	14592
138	66831	68908	40772	21558	47781	33586	79177	06928
139	55588	99404	70708	41098	43563	56934	48394	51719
140	12975	13258	13048	45144	72321	81940	00360	02428
141	96767	35964	23822	96012	94591	65194	50842	53372
142	72829	50232	97892	63408	77919	44575	24870	04178
143	88565	42628	17797	49376	61762	16953	88604	12724
144	62964	88145	83083	69453	46109	59505	69680	00900
145	19687	12633	57857	95806	09931	02150	43163	58636
146	37609	59057	66967	83401	60705	02384	90597	93600
147	54973	86278	88737	74351	47500	84552	19909	67181
148	00694	05977	19664	65441	20903	62371	22725	53340
149	71546	05233	53946	68743	72460	27601	45403	88692
150	07511	88915	41267	16853	84569	79367	32337	03316